Long-Range Interacting Systems

Lecturers who contributed to this volume

Steven T. Bramwell
François Castella
Gilles Collette
Philippe W. Courteille
Daniel H.E. Dubin
Bérengère Dubrulle
Richard S. Ellis
Dominique F. Escande
Antoine Gerschenfeld
Pierre-Emmanuel Jabin
Michael Kastner
Michael K.-H. Kiessling
Jorge Kurchan
David Mukamel
T. Padmanabhan
Arkady Pikovsky
Gordon R.M. Robb
William C. Saslaw
Bruce Turkington
Abel Yang

École d'été de Physique des Houches

Session XC, 4–29 August 2008

École thématique du CNRS

Long-Range Interacting Systems

Edited by

T. Dauxois, S. Ruffo and L.F. Cugliandolo

OXFORD
UNIVERSITY PRESS

OXFORD
UNIVERSITY PRESS

Great Clarendon Street, Oxford OX2 6DP

Oxford University Press is a department of the University of Oxford.
It furthers the University's objective of excellence in research, scholarship,
and education by publishing worldwide in

Oxford New York

Auckland Cape Town Dar es Salaam Hong Kong Karachi
Kuala Lumpur Madrid Melbourne Mexico City Nairobi
New Delhi Shanghai Taipei Toronto

With offices in

Argentina Austria Brazil Chile Czech Republic France Greece
Guatemala Hungary Italy Japan Poland Portugal Singapore
South Korea Switzerland Thailand Turkey Ukraine Vietnam

Oxford is a registered trade mark of Oxford University Press
in the UK and in certain other countries

Published in the United States
by Oxford University Press Inc., New York

First published 2010

British Library Cataloguing in Publication Data
Data available

Library of Congress Cataloging in Publication Data
Data available

Printed in Great Britain
on acid-free paper by
CPI Antony Rowe, Chippenham, Wiltshire

ISBN 978–0–19–957462–9

1 3 5 7 9 10 8 6 4 2

École de Physique des Houches
Service inter-universitaire commun à l'Université Joseph Fourier de Grenoble et à l'Institut National Polytechnique de Grenoble

Subventionné par le Ministère de l'Éducation Nationale, de l'Enseignement Supérieur et de la Recherche, le Centre National de la Recherche Scientifique, le Commissariat à l'Énergie Atomique

Directeurs scientifiques de la session XC:
Thierry Dauxois, École Normale Supérieure de Lyon & CNRS, France; Stefano Ruffo, Università di Firenze, Italy

Previous sessions

I	1951	Quantum mechanics. Quantum field theory
II	1952	Quantum mechanics. Statistical mechanics. Nuclear physics
III	1953	Quantum mechanics. Solid state physics. Statistical mechanics. Elementary particle physics
IV	1954	Quantum mechanics. Collision theory. Nucleon-nucleon interaction. Quantum electrodynamics
V	1955	Quantum mechanics. Non equilibrium phenomena. Nuclear reactions. Interaction of a nucleus with atomic and molecular fields
VI	1956	Quantum perturbation theory. Low temperature physics. Quantum theory of solids. Ferromagnetism
VII	1957	Scattering theory. Recent developments in field theory. Nuclear and strong interactions. Experiments in high energy physics
VIII	1958	The many body problem
IX	1959	The theory of neutral and ionized gases
X	1960	Elementary particles and dispersion relations
XI	1961	Low temperature physics
XII	1962	Geophysics; the earth's environment
XIII	1963	Relativity groups and topology
XIV	1964	Quantum optics and electronics
XV	1965	High energy physics
XVI	1966	High energy astrophysics
XVII	1967	Many body physics
XVIII	1968	Nuclear physics
XIX	1969	Physical problems in biological systems
XX	1970	Statistical mechanics and quantum field theory
XXI	1971	Particle physics
XXII	1972	Plasma physics
XXIII	1972	Black holes
XXIV	1973	Fluids dynamics
XXV	1973	Molecular fluids
XXVI	1974	Atomic and molecular physics and the interstellar matter
XXVII	1975	Frontiers in laser spectroscopy
XXVIII	1975	Methods in field theory
XXIX	1976	Weak and electromagnetic interactions at high energy
XXX	1977	Nuclear physics with heavy ions and mesons
XXXI	1978	Ill condensed matter
XXXII	1979	Membranes and intercellular communication
XXXIII	1979	Physical cosmology

XXXIV	1980	Laser plasma interaction
XXXV	1980	Physics of defects
XXXVI	1981	Chaotic behaviour of deterministic systems
XXXVII	1981	Gauge theories in high energy physics
XXXVIII	1982	New trends in atomic physics
XXXIX	1982	Recent advances in field theory and statistical mechanics
XL	1983	Relativity, groups and topology
XLI	1983	Birth and infancy of stars
XLII	1984	Cellular and molecular aspects of developmental biology
XLIII	1984	Critical phenomena, random systems, gauge theories
XLIV	1985	Architecture of fundamental interactions at short distances
XLV	1985	Signal processing
XLVI	1986	Chance and matter
XLVII	1986	Astrophysical fluid dynamics
XLVIII	1988	Liquids at interfaces
XLIX	1988	Fields, strings and critical phenomena
L	1988	Oceanographic and geophysical tomography
LI	1989	Liquids, freezing and glass transition
LII	1989	Chaos and quantum physics
LIII	1990	Fundamental systems in quantum optics
LIV	1990	Supernovae
LV	1991	Particles in the nineties
LVI	1991	Strongly interacting fermions and high T_c superconductivity
LVII	1992	Gravitation and quantizations
LVIII	1992	Progress in picture processing
LIX	1993	Computational fluid dynamics
LX	1993	Cosmology and large scale structure
LXI	1994	Mesoscopic quantum physics
LXII	1994	Fluctuating geometries in statistical mechanics and quantum field theory
LXIII	1995	Quantum fluctuations
LXIV	1995	Quantum symmetries
LXV	1996	From cell to brain
LXVI	1996	Trends in nuclear physics, 100 years later
LXVII	1997	Modeling the earth's climate and its variability
LXVIII	1997	Probing the Standard Model of particle interactions
LXIX	1998	Topological aspects of low dimensional systems
LXX	1998	Infrared space astronomy, today and tomorrow
LXXI	1999	The primordial universe
LXXII	1999	Coherent atomic matter waves
LXXIII	2000	Atomic clusters and nanoparticles
LXXIV	2000	New trends in turbulence
LXXV	2001	Physics of bio-molecules and cells
LXXVI	2001	Unity from duality: Gravity, gauge theory and strings

LXXVII	2002	Slow relaxations and nonequilibrium dynamics in condensed matter
LXXVIII	2002	Accretion discs, jets and high energy phenomena in astrophysics
LXXIX	2003	Quantum entanglement and information processing
LXXX	2003	Methods and models in neurophysics
LXXXI	2004	Nanophysics: Coherence and transport
LXXXII	2004	Multiple aspects of DNA and RNA
LXXXIII	2005	Mathematical statistical physics
LXXXIV	2005	Particle physics beyond the Standard Model
LXXXV	2006	Complex systems
LXXXVI	2006	Particle physics and cosmology: the fabric of spacetime
LXXXVII	2007	String theory and the real world: from particle physics to astrophysics
LXXXVIII	2007	Dynamos
LXXXIX	2008	Exact methods in low-dimensional statistical physics and quantum computing
XC	2008	Long-range interacting systems

Publishers

- Session VIII: Dunod, Wiley, Methuen
- Sessions IX and X: Herman, Wiley
- Session XI: Gordon and Breach, Presses Universitaires
- Sessions XII–XXV: Gordon and Breach
- Sessions XXVI–LXVIII: North Holland
- Session LXIX–LXXVIII: EDP Sciences, Springer
- Session LXXIX–LXXXVIII: Elsevier
- Session LXXXIX– : Oxford University Press

Organizers

DAUXOIS Thierry, École Normale Supérieure de Lyon & CNRS, France
RUFFO Stefano, Università di Firenze, Italy
CUGLIANDOLO Leticia, Université Pierre et Marie Curie, Paris VI, France

Preface

The aim of the summer school was to present recent developments in the theoretical and experimental study of long-range interacting systems. Owing to the widespread presence of long-range interactions in many different physical systems, the approach was bound to be interdisciplinary. Lectures and seminars were therefore given by leading scientists in different fields of mathematics and physics: probability, transport theory, equilibrium and nonequilibrium statistical mechanics, condensed matter physics, astrophysics and cosmology, physics of plasmas, hydrodynamics, and cold atoms.

Systems with long-range interactions are characterized by a pair potential which decays at large distances as a power law, with an exponent smaller than the space dimension: examples are gravitational and Coulomb interactions. The thermodynamic and dynamical properties of such systems were poorly understood until a few years ago. Substantial progress has been made only recently, when it was realized that the lack of additivity induced by long-range interactions does not hinder the development of a fully consistent thermodynamic formalism. This has, however, important consequences: entropy is no more a convex function of macroscopic extensive parameters (energy, magnetization, etc.), and the set of accessible macroscopic states does not form a convex region in the space of thermodynamic parameters. This is at the origin of ensemble inequivalence, which in turn determines curious thermodynamic properties such as negative specific heat in the microcanonical ensemble, first discussed in the context of astrophysics. On the other hand, it has been recognized that systems with long-range interactions display universal nonequilibrium features. In particular, long-lived metastable states may develop, in which the system remains trapped for a long time before relaxing towards thermodynamic equilibrium.

The defining scientific focus of the school was the need, which emerged in several different scientific disciplines, to develop a consistent and comprehensive theoretical approach to long-range interaction that includes both equilibrium and nonequilibrium properties.

Historically, it was with the work of Emden and Chandrasekhar, and later Antonov, Henon, Lynden-Bell, and Thirring, in the context of astrophysics, that it was realized that for systems with long-range interactions the thermodynamic entropy might not have a global maximum, and therefore thermodynamic equilibrium itself could not exist. The appearance and meaning of negative temperature was first discussed in a seminal paper by Onsager on point vortices interacting via a long-range logarithmic potential in two dimensions. From the dynamical point of view, two-dimensional hydrodynamics displays a spontaneous organization of the flow into large-scale structures emerging out of random-vorticity initial distributions as a result of the long-range logarithmic interaction potential between vortices. Pure electron plasmas, free-electron lasers, and charged-particle beams exhibit the formation of multiple quasi-equilibria which persist over extremely long times. Condensed matter systems dominated by magnetic and electric dipolar interactions have a subtle thermodynamic limit and are characterized by a mean-field-type shape-dependent energy. One can also induce long-range interactions in trapped atom and ion systems and observe striking self-organization phenomena. Driven one-dimensional systems have been shown to display phase transitions even when the dynamics is local: this proves that the driving process builds up long-range interactions. For all these systems, the common ingredient, which is at the origin of the interesting macroscopic behavior, is the presence, at a microscopic level, of long-range interactions.

Some elements of the theory that might encompass long-range interactions (whose construction is still a challenge) have already emerged:

- A thermodynamic formalism which must consistently include inequivalence of statistical ensembles.
- Appropriate mathematical techniques to treat the thermodynamic limit of long-range interacting systems (large-deviation theory, for example).
- A transport theory based on long-range (mean-field) effects rather than on collisional processes.

Besides presenting the relevant physical properties of the various systems attacked in the different talks, all speakers, in one way or another, have dwelled upon these concepts. Their efforts towards the aim of concretely building up a theory of long-range interactions are summarized in their lecture notes. Although several theoretical aspects have been clarified recently, the field of long-range interactions still raises some controversial issues, of which the reader will find an echo in some of the manuscripts.

We warmly thank all contributors to this volume for the effort they made in preparing these chapters, which often cover much more material than the lectures themselves. We strongly hope that this set of lectures will provide, following the tradition of many earlier volumes of the Les Houches book series, a useful introduction to long-range interactions and serve as a classical reference for researchers and students in this very broad and rapidly evolving field. We are sure that the school has indeed provided a background for the future development of a new community, capable of addressing scientific problems where long-range interactions play a crucial role.

Fifty students attended the school (master and graduate students and postdoctoral fellows). One of the most rewarding experiences of this school was the lively atmosphere created by the participants through informal scientific discussions, music performances, mountain excursions, football competitions, ... We are convinced that, in addition to the deep knowledge of long-range interacting systems acquired during this session, the four weeks spent together will be decisive in creating a group of people with long-range interactions, more usually called friendship.

Leticia Cugliandolo, Stefano Ruffo, and Thierry Dauxois.

Acknowledgments

The 90th Les Houches summer school and the present volume were made possible thanks to the financial support of the following institutions, whose contributions are gratefully acknowledged:

- The "Université Européenne 2008" program of the French Ministère de l'Enseignement Supérieur et de la Recherche;
- La Formation Permanente du CNRS;
- L'Istituto Nazionale di Fisica Nucleare (INFN);
- L'École Doctorale de Physique et Astrophysique de Lyon;
- L'École Doctorale de Physique de Grenoble;
- Le GDR-CNRS Phénix;
- L'Institut des systèmes complexes de l'ENS Lyon (IXXI);
- C'Nano Rhône-Alpes, the Rhône-Alpes competence center in nanoscience;
- La Région Rhône-Alpes;
- The Université Joseph Fourier, the French Ministry of Research, and the Commissariat à l'Énergie Atomique, through their constant support of the Physics School.

Pictures illustrating this book were taken by Maxi Giuliani, Bruno Goncalves, and Thierry Dauxois.

The assistance of the permanent staff of the School, especially Murielle Gardette, Isabel Lelievre, and Brigitte Rousset, has been of invaluable help at every stage of the preparation and development of the school. We would like to warmly thank all of them on behalf of all the students and lecturers.

Murielle Gardette, Isabel Lelievre, and Brigitte Rousset.

1, Stefano Ruffo. 2, Rytis Paskauskas. 3, Jehan Boreux. 4, Camille Aron. 5, Antoine Venaille. 6, Antti Knowles. 7, Guglielmo Paoletti. 8, Luca Pretini. 9, Max Artomov. 10, Sergey Pogodin. 11, Fabio Staniscia. 12, Daniele Pinna. 13, Philippe Courteille. 14, Osamu Iguchi. 15, Togo Atsushi. 16, Abel Yang. 17, Laurent Chaput. 18, Hidetoshi Morita. 19, Antoine Gerschenfeld. 20, Dominique Escande. 21, Alessandro Vindigni. 22, Giacomo Gradenigo. 23, Bernard Barbara. 24, Lila Warszawski. 25, Stefan Rist. 26, Yevgen Kazakov. 27, Romain Mari. 28, Hanna Terletska. 29, Jixuan Hou. 30, Tineke Van Den Berg. 31, Filippo Simini. 32, Steven Bramwell. 33, Yasuhide Sota. 34, Eduardo Ezequiel Ferrero. 35, Viacheslav Shtyk. 36, Mario Alberto Annunziata. 37, Leticia Cugliandolo. 38, Marianne Heckmann. 39, Gilles Colette. 40, Thierry Dauxois. 41, Jean-Pierre Nguenang. 42, Juraj Szavits Nossan. 43, Lucas Nicolao. 44, Ori Hirschberg. 45, Michael Kiessling.

Contents

PART II HYDRODYNAMICS

PART III MATHEMATICAL ASPECTS

PART VI DIPOLAR INTERACTION IN CONDENSED MATTER

List of participants

Organizers

CUGLIANDOLO Leticia F.
Laboratoire de Physique Théorique et Hautes Energies, Université Pierre et Marie Curie, Paris VI, Tour 24, 5ème etage 4, Place Jussieu, 75252 Paris Cedex 05, France

DAUXOIS Thierry
Laboratoire de Physique, ENS-Lyon, 46 Allée d'Italie, F-69007 Lyon, France

RUFFO Stefano
Dipartimento di Energetica "S. Stecco" and CSDC, Università di Firenze and INFN, Via S. Marta, 3 I-50139, Firenze, Italy

Lecturers

BARBARA Bernard
Laboratoire Louis Néel, 25 Avenue des Martyrs, F-38042 Grenoble Cedex 9, France

BRAMWELL Steven
London Centre for Nanotechnology, University College London, 17–19 Gordon Street, London WC1H 0AH, United Kingdom

CASTELLA François
Institut de Recherche Mathématique de Rennes, Université de Rennes 1, Campus de Beaulieu, F-35042 Rennes Cedex, France

COURTEILLE Philippe
Physikalisches Institut, Eberhard-Karls-Universitat Tubingen, Auf der Morgenstelle 14, D-72076 Tubigen, Germany

DUBIN Dan
Department of Physics, University of California San Diego, La Jolla, CA 92093-0319, USA

DUBRULLE Bérengère
Groupe Instabilités et Turbulence, SPEC/DRECAM/DSM/CEA, F-91191 Gif sur Yvette Cedex, France

ELLIS Richard S.
Department of Mathematics and Statistics, Lederle Graduate Research Towers, Universty of Massachusetts, Amherst, MA 01003, USA

ESCANDE Dominique
PIIM, Case 321, Avenue Normandie Niemen, F-13397 Marseille Cedex 20, France

JABIN Pierre-Emmanuel
Université de Nice-Sophia Antipolis, Parc Valrose, F-06108 Nice Cedex 02, France
KASTNER Michael
Physikalisches Institut, Lehrstuhl für Theoretische Physik I, Universität Bayreuth, 95440 Bayreuth, Germany
KIESSLING Michael
Department of Mathematics, Hill Center, Busch Campus, Rutgers University, New Brunswick, NJ 08903, USA
KURCHAN Jorge
Laboratoire de Physique et Mécanique des Milieux Hétérogènes, Ecole Supérieure de Physique et de Chimie Industrielles, 10 rue Vauquelin, F-75231 Paris Cedex 05, France
MUKAMEL David
Department of Physics of Complex Systems, The Weizmann Institute of Science, Rehovot 76100, Israel
PADMANABHAN Thanu
IUCAA, Pune University Campus, Ganeshkhind, Pune 411 007, India
PIKOVSKY Arkady
Department of Physics and Astronomy, Potsdam University, Kare-Liebknecht-Str. 24/25, D-14476 Postdam-Golm, Germany
ROBB Gordon
Department of Physics, University of Strathclyde, Glasgow G4 0NG, United Kingdom
SASLAW Bill
Department of Astronomy, University of Virginia, Charlottesville, VA 22904, USA
TURKINGTON Bruce
Department of Mathematics and Statistics, Lederle Graduate Research Tower, University of Massachusetts, Amherst, MA 01003-9305, USA

Students and auditors

ANNUNZIATA Mario Alberto
Dipartimento di Fisica, Largo B. Pontecorvo 3, 56127 Pisa (PI), Italy
ARON Camille
Laboratoire de Physique Théorique et Hautes Energies, Tour 24-25, 5ème étage, Boite 126, 4 Place Jussieu, F-75252 Paris Cedex 05, France
ARTOMOV Max
Massachusetts Institute of Technology, 77 Massachusetts Avenue, Cambridge, MA 02139, USA
ATSUSHI Togo
Department of Materials Science, Yoshida-Honmachi, Sakyo, Kyoto 606-8317, Japan
BOREUX Jehan
Facultés Universitaires Notre Dame de La Paix (FUNDP), Rempart de la Vierge 8, 5000 Namur, Belgium

BOUCHET Freddy
INLN, CNRS, 1361 route des Lucioles Sophia Antipolis, 06560 Valbonne, France

CAMPA Alessandro
Health and Technology Department, Istituto Superiore di Sanità, Viale Regina, Elena 299, 00161 Roma, Italy

CHAPUT Laurent
Laboratoire de Physique et de spectroscopie électronique LPSE-UMR 7014, 4 rue des Frères Lumières, F-68093 Mulhouse Cedex, France

CHAVANIS Pierre-Henri
Laboratoire de Physique Théorique, IRSAMC, Université Paul Sabatier, 118 route de Narbonne, F-31062 Toulouse, France

COLLETTE Gilles
LSCE, Orme des Merisiers Batiment 701 CE Saclay, F-91191 Gif sur Yvette Cedex, France

DELLA FIORE Alberto
INFN, Sezione de Firenze, Via G. Sansone 1, I-50019 Sesto Fiorentino, Italy

DURING Gustavo
Laboratoire de Physique Statistique, Ecole Normale Supérieure (Paris), 24 rue Lhomond, F-75005 Paris Cedex 05, France

FERRERO Eduardo Ezequiel
Facultad de Matematica, Astronomia y Fisica, Universidad Nacional de Cordoba, Medina Allende y Haya de La Torre, Ciudad Universitaria, CP 5000 Cordoba, Argentina

GERSCHENFELD Antoine
Laboratoire de Physique Statistique, Ecole Normale Supérieure (Paris), 24 rue Lhomond, F-75005 Paris Cedex 05, France

GIULIANI Maximiliano
Department of Physics and Applied Mathematics, University of Navarra, Los Castanos Building, C/ Irunlarrea S/N, 31080 Pamplona, Spain

GONCALVES Bruno
School of Informatics and Center of Biocomplexity, 919 E. 10th St, Bloomington, IN 47406, USA

GRADENIGO Giacomo
University of Trento, Physics Department, Via Sommarive 14, I-38100 Povo (TN), Italy

HECKMANN Marianne
Institut für Festkörperphysik, Technische Universität Darmstadt, Hochschulstrasse 6, 64289 Darmstadt, Germany

HIRSCHBERG Ori
Department of Physics of Complex Systems, Weizmann Institute of Science, Rehovot 76100, Israel

HOU Jixuan
Laboratoire de Chimie, UMR 5182 CNRS, Ecole Normale Supérieure de Lyon, 46 Allée d'Italie, F-69364 Lyon Cedex 07, France

IGUCHI Osamu
Department of Physics, Ochanomizu University, 2-1-1 Ohtsuka, Bunkyo-ku, Tokyo 112-8610, Japan

KAZAKOV Yevgen
Kharkiv National University, Faculty of Physics and Technology, Plasma Physics Department, Svobody Square 4, 61077 Kharkiv, Ukraine

KNOWLES Antti
Paradiesstrasse 73, 4102 Binningen, Switzerland

MARI Romain
Laboratoire de Physique et Mécanique des Milieux Hétérogènes (UMR 7636), Ecole Supérieure de Physique et Chimie Industrielles, 10 rue Vauquelin, F-75231 Paris Cedex 5, France

MORITA Hidetoshi
Institut Non Linéaire de Nice, CNRS, 1361 route des Lucioles, F-06560 Valbonne, France

NGUENANG Jean-Pierre
Fundamental Physics Laboratory, Group of Nonlinear Physics and Complex Systems, Department of Physics, University of Douala, P.O. Box 24157, Douala, Cameroon

NICOLAO Lucas
Instituto de Fisica-UFRGS, Avenue Bento Goncalves, 9500, Caixa Postal 15051, 91501-970 Porto Alegre, RS, Brazil

OBEID Dina
Brown University, 182 Hope Street, Barus and Holley Bldg., Box 1843, Providence, RI 02912, USA

PAOLETTI Guglielmo
Dipartimento di Fisica, Università di Pisa and INFN, Largo Bruno Pontecorvo 3, 56127 Pisa, Italy

PASKAUSKAS Rytis
Sincrotone Trieste, S.S. 14 km 163.5, AREA Science Park, 34012 Basovizza TS, Italy

PINNA Daniele
Università di Bologna, Via Mario Fantin 15 Bologna, 40100, Italy

POGODIN Sergey
Department d'Enginyeria Quimica of the Universitat Rovira i Virgili, Avenue Paisos Catalans, 26, Tarragona 43007, Spain

PRETINI Luca
Facolta di Ingegneria, Università di Firenze, Dipartimento di Sistemi e Informatica, Via di Santa Marta 3, 50100 Firenze, Italy

RIBEIRO-TEIXEIRA Ana Carolina
Instituto de Fisica-UFRGS, Avenue Bento Goncalves, 9500, Caixa Postal 15051, 91501-970 Porto Alegre, RS, Brazil

RIST Stefan
Universitat Autonoma de Barcelona, Edifici CC, Facultat de Ciencies, Grup d'Optica, Departament Fisica, 08193 Bellaterra, Spain

SHTYK Viacheslav
National Academy of Sciences of Ukraine, Bogolyubov Institute for Theoretical Physics 14-b, Metrolohichna Street, Kyiv, 03680, Ukraine

SICARD Francois
LPNHE, 4 place Jussieu, Tour 43, RDC, F-75252 Paris Cedex 05, France

SIMINI Filippo
Università degli Studi di Padova, Dipartimento di Fisica G. Galilei, Via Marzolo 8, 35131 Padova, Italy

SOTA Yasuhide
Ochanomizu University, Department of Physics, Faculty of Science, 2-1-1 Ohtsuka, Bunkyo-ku 112-8610, Tokyo, Japan

STANISCIA Fabio
Sincrotrone Trieste, Strada Statale 14-km 163,5, AREA Science Park, 34012 Basovizza, Trieste, Italy

STEINER Jose Roberto
SQN 214, Bloco E, Apartment 607, CEP 7000-000, Brasilia DF, Brazil

SUGDEN Kate
SUPA, School of Physics, University of Edinburgh, United Kingdom.

SZAVITS NOSSAN Juraj
Institute of Physics, Bijenicka cesta 46, HR-10000 Zagreb, Croatia

TELES Tarcisio
Instituto de Fisica-UFRGS, Avenida Bento Goncalvez, 9500, 91501-970 Porto Alegre, RS, Brazil

TERLETSKA Hanna
The Florida State University, High Magnetic Field Laboratory, A-327 Condensed Matter Science, 1800 E. Paul Dirac Dr., Tallahassee, FL 32310-3706, USA

VAN DEN BERG Tineke
Centre de Physique Théorique, Université de Provence-CNRS, Luminy Case 907, F-13288 Marseille Cedex 9, France

VENAILLE Antoine
Coriolis LEGI, 21 Avenue des Martyrs, F-38000 Grenoble, France

VINDIGNI Alessandro
Laboratorium fur Festkoerperphysik, Wolfgang-Pauli-Str. 16, ETH Hoenggerberg, HPT C 5.1, 8093 Zurich, Switzerland

WARSZAWSKI Lila
School of Physics, The University of Melbourne, Parkville, Victoria 3010, Australia

YANG Abel
University of Virginia, 530 McCormick Road, Charlottesville, VA 22904, USA

Panorama from École des Houches: Aiguille Verte; the amphitheater during a lecture; long discussion at the blackboard after a lecture, with Gustavo During at the left; lively BabyFoot final; audience at the music concert; music band; the famous Les Houches 2008 soccer team; viewers of the womens' soccer final at the Beijing 2008 Olympic Games, USA versus Brazil. Dan Dubin (San Diego) is in the first row, while Roberto Steiner (Brasilia) is at the very back of the room: guess who won the game! (From left to right, top to bottom.)

Romain Mari & Camille Aron; Aviva Mukamel & Dina Obeid; Lucas Nicolao & Tineke Van Den Berg; Kate Sugden; Antoine Venaille; François Sicard; Hidetoshi Morita; Alessandro Vindigni; Lorenzo Ruffo; Juraj Szavits Nossan; Matias Kurchan; Stefan Rist; Guglielmo Paoletti; Antoine Gerschenfeld; Gilles Collette (from left to right, top to bottom).

Ana Carolina Ribeiro-Teixeira & Sergey Pogodin; Yasuhide Sota; Hanna Terletska; Antti Knowles; Camille Aron; Daniele Pinna; Bruno Goncalves; Mario Alberto Annunziata; Abel Yang; Filippo Simini; Eduardo Ezequiel Ferrero; Marianne Heckmann; Fabio Staniscia; Dauxois's family; Giacomo Gradenigo; Jixuan Hou; Jehan Boreux; Jean-Pierre Nguenang (from left to right, top to bottom).

Leticia Cugliandolo and her children; Aiguille de Bionnassay; Tarcisio Teles; Ori Hirschberg; Raphaële Andrault; Togo Atsushi; Osamu Iguchi; mountain view from the Les Houches School; Viacheslav Shtyk; Antonella Giaconi; Rytis Paskauskas; Ana Carolina Ribeiro-Teixeira; Yevgen Kazakov; Vasanthi Padmanabhan; Matilde Alessi; Maximiliano Giuliani; Max Artomov; Lucas Nicolao; Lila Warszawski (from left to right, top to bottom).

Abstracts of lectures

STEVEN T. BRAMWELL
London Centre for Nanotechnology, University College London, United Kingdom

Dipolar effects in condensed matter

The dipole–dipole interaction is ubiquitous in real condensed matter systems and leads to some interesting physics. This chapter starts with a basic introduction to the origin and properties of electric and magnetic dipoles, before specializing to the magnetic case. The equilibrium statistical mechanics of model dipolar systems is discussed in order to emphasize two defining characteristics of the interaction: first its intrinsic frustration and second its long-ranged nature. Two classes of dipolar system are then discussed in detail. The first is the dipolar ferromagnet, an old but interesting problem in condensed matter physics. The second is spin ice, a more topical problem in the physics of frustration, where many-body dipolar effects are seen to lead to highly degenerate states and emergent magnetic monopoles.

FRANÇOIS CASTELLA
IRMAR & IRISA, Université de Rennes 1, France

Solving ordinary differential equations when the coefficients have low regularity: A kinetic point of view

These notes gather together some known results on the solution of ordinary differential equations of the form $\dot{X}(t) = b(t, X(t))$, $X(0) = x$, in the case when the vector field $b = b(t, x)$ possesses *low regularity* in the variable x. This equation allows one to compute the flow induced by the vector field b, i.e. if $b(t, x)$ describes the velocity of a fluid particle sitting at x at time t, say, then $X(t)$ is called the flow induced by the vector field b: it represents the trajectory of a fluid particle that issues from x at time $t = 0$, as the particle is transported along the velocity field b.

While solving $\dot{X} = b(X)$ is a standard task when b is a smooth function, thanks to the well-known Cauchy–Lipschitz Theorem, solving this equation when b lacks regularity is an old problem, and an essentially optimal result was discovered by R. Di Perna and P. L. Lions in 1989. The goal of these notes is to present the main ideas of the Di Perna–Lions theory. We try to insist on the key ideas, and to point out those aspects or techniques that are of more general interest in the mathematical analysis of evolution equations.

The question of solving $\dot{X} = b(X)$ when b has low regularity in x is natural when, say, real-life fluids are considered, for which the velocity field $b(t, x)$ is given as the solution of, typically, the Euler equations, and therefore has limited regularity in x owing to strong, nonlinear energy exchanges between short scales and large scales. It is also a natural task in the modeling of various flows, such as polymeric flows, or even in the modeling of random flows, for which the velocity fields again lack smoothness.

PHILIPPE W. COURTEILLE
Physikalisches Institut, Eberhard-Karls-Universität Tübingen, Germany

Long-range interaction in cold-atom optics

Cold-atom optics provides ideal toy systems that allow one to study the essence of long-range interactions. Atomic gases are very pure, can completely be isolated from the environment, and can be brought to extremely low temperatures. Important quantities such as density and temperature and even the strength of short-range interactions can be tuned over wide ranges, allowing the impact of binary collisions on the system's dynamics to vanish. I shall discuss with several examples how long-range interactions can influence the stability of trapped atomic gases and trigger self-organization phenomena. I shall pay particular attention to the so-called collective atomic recoil laser, a self-synchronization effect observed in cold atomic clouds subject to a uniform interaction by mediation of the counterpropagating modes of a high-finesse ring cavity.

DANIEL H.E. DUBIN
University of California at San Diego, Physics Department, USA

Plasma collisional transport

These lectures discuss the collisional transport of particles, momentum, and energy across the magnetic field in a magnetized plasma. Collision operators and transport coefficients are derived from first principles that account for the long-range nature of the Coulomb interaction, which allows particles to transfer energy and momentum across large distances. These long-range interactions are particularly important in plasmas for which the Debye length is larger than the cyclotron radius, and several new results for transport in this regime are derived and compared with experiment. The lectures also discuss enhanced transport in nearly two-dimensional plasmas, and related effects associated with two-dimensional fluid dynamics.

BÉRENGÈRE DUBRULLE AND GILLES COLLETTE
Service de Physique de l'État Condensé, CEA Saclay, France

Statistical mechanics of turbulent von Kármán flows: Theory and experiments

We derive predictions concerning the equilibrium flow and the fluctuations in an axisymmetric Euler–Beltrami flow using tools borrowed from statistical mechanics. We link fluctuations to the mean flow and its response to perturbations. This provides an estimate of the fluctuation level as a function of the thermodynamic coefficients, namely the effective temperature, in a way reminiscent of the Fluctuation–Dissipation Theorem. These predictions are successfully confronted with observations drawn from stationary states of a turbulent von Kármán flow using particle image velocimetry and laser Doppler velocimetry measurements. This suggests that out-of-equilibrium steady states of a real turbulent flow may be described as equilibrium states of the Euler equation.

RICHARD S. ELLIS
Department of Mathematics and Statistics, University of Massachusetts Amherst, USA

The theory of large deviations and applications to statistical mechanics

This chapter starts by focusing on finite state spaces and examining Boltzmann's discovery of a statistical interpretation of relative entropy. We then prove a conditioned limit theorem involving relative entropy that elucidates a basic issue arising in many areas of application. What is the most likely way for an unlikely event to happen? After giving a general formulation of the large deviation principle, we discuss a number of basic results in the theory, including Cramér's Theorem, Sanov's Theorem, the Laplace principle, and the contraction principle. The theory of large deviations is then applied to study the phase-transition structure of three lattice spin models, the Curie–Weiss model, the Curie–Weiss–Potts model, and the mean-field Blume–Capel model. The chapter ends by applying large-deviation theory to treat equivalence and nonequivalence of ensembles for a general class of models.

DOMINIQUE F. ESCANDE
CNRS-Université de Provence, Marseille, France

Wave–particle interaction in plasmas: A qualitative approach

Plasmas are long-range interacting systems, are ubiquitous in the Universe, and are involved in many technologies. Many types of waves may be excited in such media by charged particles whose velocity is close to the phase velocity of the wave. This wave–particle interaction is the topic of this course, which focuses on its simplest instance involving electrostatic waves, the Langmuir waves whose typical frequency is the plasma frequency. This course describes this interaction at three levels of increasing complexity by relying on the mechanical intuition that can be gained from the harmonic oscillator and the nonlinear pendulum. The beam–plasma instability is described at length. When the beam is cold, the instability is related to the forcing of the electron harmonic oscillation underlying Langmuir waves, and its saturation is shown to occur by particle trapping. When the beam is warm, the instability occurs through the Landau effect, a transfer of momentum between a wave and particles close to being resonant with it, involving the average synchronization of particles with this wave. Hamiltonian chaos is introduced, and is shown to play an essential role in the saturation of the instability due to a warm beam, by inducing a diffusion of the particle velocities. Appropriate averages of the dynamics back up this diffusion picture, but chaotic dynamics is not at all stochastic, as shown by the existence of coherent structures embedded in Hamiltonian chaos.

PIERRE-EMMANUEL JABIN[1] AND ANTOINE GERSCHENFELD[2]
[1] Equipe Tosca, Inria, Sophia & Laboratoire Dieudonné, Université de Nice, France
[2] École Normale Supérieure, Paris

Mean-field limit for interacting particles

Mean-field limits are a very classical question in physics and mathematics. They are, moreover, crucial when very large numbers of interacting particles are considered but are still very ill-understood for some important physical interactions, Coulombian for instance. We explain here the classical results that have been obtained in the case of regular (Lipschitz or continuous) interaction kernels and present some new approaches to tackling the singular cases.

MICHAEL KASTNER
Physikalisches Institut, Universität Bayreuth, Germany

On the origin of phase transitions in long- and short-range interacting systems

In order to better understand the origin of a phase transition, we study the conditions under which the canonical free energy or the microcanonical entropy of a classical many-particle system can have nonanalytic points. In the first part of the lecture, analyticity properties are studied on a purely thermodynamic level. Different behavior is found for short- and long-range systems, and the differences are traced back to the differing concavity properties of the microcanonical entropy. In contrast to the long-range case, nonanalyticities of thermodynamic functions of a short-range interacting system appear to have their origin necessarily on a microscopic level of description. This microscopic mechanism for generating a nonanalyticity of the free energy or the entropy is discussed in the second part of the lecture. Nonanalyticities of the finite-system microcanonical entropy are shown to occur whenever there are stationary points of the energy landscape. In the thermodynamic limit of infinite particle number, most of these nonanalyticities disappear. Only those stationary points of the energy landscape whose curvature vanishes in the thermodynamic limit in a suitable way may induce a phase transition. This result is illustrated by means of exact results for simple classical spin models.

MICHAEL K.-H. KIESSLING
Department of Mathematics, Rutgers University, USA

A lecture on the relativistic Vlasov–Poisson equations

In this brief summary of my lecture I first introduce the equations of the "relativistic Vlasov–Poisson system," then I discuss their physical significance, and finally I summarize the main rigorous results of a recent joint paper by myself and A. S. Tahvildar-Zadeh which identify a sharp set of conditions for classical initial data to launch a global classical solution. Among the initial data which violate the sharp conditions, one finds some which lead to the formation of a singularity in finite time. The results are set in perspective with other results in the literature, and a list of intriguing open problems is given.

JORGE KURCHAN
PMMH-ESPCI, CNRS UMR 7636, 10 rue Vauquelin, 75005 Paris, France

Six out-of-equilibrium lectures

The purpose of these notes is to introduce a group of subjects in out-of-equilibrium statistical mechanics that have received considerable attention in the last fifteen years or so. They are mostly connected with time reversibility and its relation to entropy, are expressed in terms of large deviations, and involve at some level the notion of timescale separation.

DAVID MUKAMEL
Department of Physics, The Weizmann Institute of Science, Rehovot, Israel

Statistical mechanics of systems with long-range interactions

Thermodynamic and dynamical properties of systems with long-range pairwise interactions (LRI) which decay as $1/r^{d+\sigma}$ at large distances r in d dimensions are reviewed in these Notes. Two broad classes of such systems are identified. (a) Systems with a slow decay of the interactions, termed "strong" LRI, where the energy is super-extensive. These systems are characterized by unusual properties such as inequivalence of ensembles, negative specific heat, slow decay of correlations, and ergodicity breaking. (b) Systems with a faster decay of the interaction potential where the energy is additive, thus resulting in less dramatic effects. These interaction affect the thermodynamic behavior of systems near phase transitions, where long-range correlations are naturally present. Long-range correlations are often present in systems driven out of equilibrium when the dynamics involves conserved quantities. Steady-state properties of driven systems are considered within the framework outlined above.

T. PADMANABHAN
Inter-University Centre for Astronomy and Astrophysics, Pune, India

Statistical mechanics of gravitating systems: An overview

I review several issues related to the statistical description of gravitating systems in both static and expanding backgrounds. After briefly reviewing the results for the static background, I concentrate on gravitational clustering of collisionless particles in an expanding universe. In particular, I describe (a) how the nonlinear mode–mode coupling transfers power from one scale to another in Fourier space if the initial power spectrum is sharply peaked at a given scale, and (b) what are the asymptotic characteristics of gravitational clustering that are independent of the initial conditions. Numerical simulations and analytic work show that power transfer leads to a universal power spectrum at late times, somewhat reminiscent of the existence of the Kolmogorov spectrum in fluid turbulence.

ARKADY PIKOVSKY
Department of Physics, University of Potsdam, Germany

Synchronization of regular and chaotic oscillators

Many natural and human-made nonlinear oscillators exhibit the ability to adjust their rhythms as a result of weak interaction: two lasers, when coupled, start to generate with a common frequency; cardiac pacemaker cells fire simultaneously; violinists in an orchestra play in unison. Such coordination of rhythms is a manifestation of a fundamental nonlinear phenomenon—synchronization. Discovered in the seventeenth century by Christiaan Huygens, it has been observed in physics, chemistry, biology, and even social behavior; it has also found practical applications in engineering and medicine. The notion of synchronization has recently been extended to cover the adjustment of rhythms in chaotic systems, large ensembles of oscillating units, rotating objects, continuous media, etc. In spite of major progress in theoretical and experimental studies, synchronization remains a challenging problem of nonlinear science.

GORDON R.M. ROBB
SUPA, Department of Physics, University of Strathclyde, Glasgow, United Kingdom

Collective instabilities in light–matter interactions

Long-range interactions between particles can be induced by scattering of electromagnetic radiation. These interactions can give rise to collective instabilities where the particles scatter coherently owing to a process of spatial self-organization or bunching. Some examples of collective instabilities involving (i) relativistic electrons and (ii) cold atoms are described.

WILLIAM C. SASLAW[1,2] AND ABEL YANG[1]
[1] Department of Astronomy, University of Virginia, USA
[2] Institute of Astronomy, Cambridge, UK

Statistical mechanics of the cosmological many-body problem and its relation to galaxy clustering

The cosmological many-body problem is effectively an infinite system of gravitationally interacting masses in an expanding universe. Despite the interactions' long-range nature, an analytical theory of statistical mechanics describes the spatial and velocity distribution functions which arise in the quasi-equilibrium conditions that apply to many cosmologies. Consequences of this theory agree well with the observed distribution of galaxies. Further consequences such as thermodynamics provide insights into the physical properties of this system, including its robustness to mergers, and its transition from a grand canonical ensemble to a collection of microcanonical ensembles with negative specific heat.

BRUCE E. TURKINGTON
University of Massachusetts at Amherst, USA

Statistical mechanics of two-dimensional and quasi-geostrophic flows

One modern theory of coherent structures in two-dimensional, high-Reynolds-number flows is based on the equilibrium statistical mechanics of vorticity. This theory is able to predict the large-scale, long-lived flow structures that emerge and persist in freely decaying or weakly forced turbulent flows. In these lectures I first develop the older point-vortex formulation of the theory, and then I describe the more recent statistical equilibrium theory based on continuum vorticity dynamics. In both forms of the theory the long-range nature of vortex interactions leads to negative-temperature states, nonequivalent ensembles, and related stability issues. The lectures present a unified treatment of these features, relying primarily on the methodology of large-deviation theory. Finally, I develop the basic governing equations of geophysical fluid dynamics and apply the theory to model coherent structures in the oceans and atmosphere. In particular, the intense zonal shear flows and embedded oval vortices in the weather layer of Jupiter are realized as statistical equilibrium states.

Abstracts of seminars given by students and auditors

CAMILLE ARON
Université Pierre et Marie Curie, Paris VI, LPTHE, CNRS, Paris, France

Driven quantum coarsening
in collaboration with G. Biroli (IPhT, CEA) and L. F. Cugliandolo (LPTHE, Paris VI)

We study the driven dynamics of quantum coarsening. We analyze models of M-component rotors coupled to two electronic reservoirs at different chemical potentials that generate a current threading through the system. In the large-M limit we derive the dynamical phase diagram as a function of temperature, strength of quantum fluctuations, voltage, and coupling to the leads. We show that the slow relaxation in the ordering phase is universal. On large time and length scales the dynamics are analogous to stochastic classical dynamics, even for a quantum system driven out of equilibrium at zero temperature. We argue that our results apply to generic driven quantum coarsening.

MAX N. ARTYOMOV
Department of Chemistry, Massachusetts Institute of Technology, Cambridge, MA, USA

Molecular motors interacting with their own tracks
in collaboration with A. Yu. Morozov and A. B. Kolomeisky (Rice University)

The dynamics of molecular motors that move along linear lattices and interact with them via reversible destruction of specific lattice bonds is investigated theoretically by analyzing exactly solvable discrete-state "burnt-bridge" models. Molecular motors are viewed as diffusing particles that can asymmetrically break or rebuild periodically distributed weak links when passing over them. Our explicit calculations of dynamic properties show that coupling the transport of an unbiased molecular motor with the bridge-burning mechanism leads to a directed motion that lowers fluctuations and produces a dynamic transition in the limit of low concentration of weak links. Interaction between a backward-biased molecular motor and the bridge-burning mechanism yields a complex dynamic behavior. For reversible dissociation, the backward motion of the molecular motor is slowed down. There is a change in the direction of the molecular motor's motion for some range of parameters. The molecular motor also experiences nonmonotonic fluctuations due to the action of two opposing mechanisms: the reduced activity after the burned sites and the locking of large fluctuations. Large spatial fluctuations are observed when the two mechanisms are comparable. The properties of the molecular motor are different for the irreversible burning of bridges, where the velocity and fluctuations are suppressed for some concentration range, and a dynamic transition is also observed. The dynamics of the system is discussed in terms of the effective driving forces and transitions between different diffusional regimes.

EZEQUIEL E. FERRERO
Facultad de Matemática, Astronomía y Física, Universidad Nacional de Córdoba, Ciudad Universitaria, 5000 Córdoba, Argentina

Long-term ordering kinetics of the two-dimensional q-state Potts model
in collaboration with S. A. Cannas

We studied the nonequilibrium dynamics of the q-state Potts model in a square lattice after a quench from infinite temperature down to subcritical temperatures $T < T_c$. By means of a continuous-time Monte Carlo algorithm (nonconserved order parameter dynamics), we analyzed the long-term behavior of the energy and relaxation time for a wide range of quench temperatures and system sizes. For $q > 4$ we found the existence of different dynamical regimes, according to quench temperature range. First of all there is some characteristic temperature $T^* < T_c$, such that for $T^* < T < T_c$ simple coarsening (Lifshitz–Allen–Cahn domain growth behavior) dominates the relaxation, except very close to T_c, where evidence of nucleation relaxation mechanisms appears. For $T < T^*$ we found that, at long time scales, the LAC relaxation is interrupted when the system gets trapped in highly symmetric metastable states with finite probability, which induce activation in the domain growth, in agreement with early predictions of Lifshitz [*JETP*, **42**, 1354 (1962)]. We found two different types of such configurations: striped states and honeycomb-like structures. Finally, we found a temperature $T_g \ll T_c$, such that for $T < T_g$ the system always gets stuck at intermediate times in highly disordered metastable states with a finite lifetime, which have recently been identified as glassy states.

Ref.: E. E. Ferrero and S. A. Cannas, *Phys. Rev. E*, **76**, 031108 (2007).

MAXIMILIANO A. GIULIANI
Facultad de Ciencias, Universidad de Navarra, Pamplona, Spain

Contact line velocity in colloidal-crystal formation
in collaboration with W. González-Viñas

We present the results of a study of the velocity of the contact line during the formation of a colloidal crystal in a vertical deposition experiment. A cell, with a parallel-capacitor symmetry, was filled with a suspension of polystyrene particles and maintained at constant temperature and humidity. The characteristic velocities of the contact line were studied, and differences according to the initial concentration were observed. For low-concentration suspensions (0.1% w/w) one small velocity was detected, while for higher concentrations (0.3% and 0.5% w/w) an additional velocity appeared. Also, differences arise when the kinds of structures deposited are considered: low-concentration samples show mostly a sparse deposit, while at higher concentrations denser structures appear, with a higher proportion of multilayers as the concentration increases. The study of the temporal evolution and its comparison with the different structures obtained shows that multilayer and dense submonolayer structures are formed at fast rates, while low rates are associated with sparse submonolayer and compact monolayer structures. Current models do not predict these observations, but these models were developed for much larger temporal scales (days or weeks), and therefore it results necessary to modify these models or develop new ones to understand the process of formation at relatively short temporal scales (hours).

Bruno GONCALVES
Complex Systems Group, Indiana University, Bloomington, IN, USA

Ensemble inequivalence in k-regular graphs

We present a complete analytical solution of a system of Potts spins on a random k-regular graph in both the canonical and the microcanonical ensemble, using the large-deviation cavity method. The solution is shown to be composed of three different branches, resulting in a nonconcave entropy function. The analytical solution is confirmed with numerical Metropolis and Creutz simulations and our results clearly demonstrate the presence of a region with negative specific heat and, consequently, ensemble inequivalence between the canonical and microcanonical ensembles.

Giacomo GRADENIGO
Dipartimento di Fisica, Università di Trento, Italy

Fluctuating surface tensions in the mosaic picture of the glass transition
in collaboration with C. Cammarota, A. Cavagna, T. Grigera, and P. Verrocchio

We test a novel formulation of the mosaic scenario of the glass transition. This formulation is based on fluctuating surface tensions among metastable states. In the mosaic picture, metastable states are typically the local minima of the free energy in the temperature range where the complexity is nonzero ($T_K < T < T_{MC}$).

To test the behavior of the surface tension between states, we built configurations made of patches coming from different states at equal T: this was done for different patch sizes. We worked on a typical liquid former: a binary mixture of soft spheres.

At low temperatures ($T < T_{MC}$) we find a positive surface tension, with a domain-size-dependent distribution. We have some indications that the surface tension between finite domains is self-averaging with domain size. We find that the surface tension between states decreases with increasing temperature. This agrees with a transition from activated ($T < T_{MC}$, supercooled liquid) to nonactivated ($T > T_{MC}$, liquid) relaxational dynamics. The main approximation we rely on is to consider as states "inherent structures," namely configurations that are minima of the potential energy of our system, instead of free-energy minima.

MARIANNE HECKMANN
Institut für Festkörperphysik Darmstadt, University of Technology, Germany

Self-consistent field theory and strong stretching theory for block copolymer thin films

in collaboration with B. Drossel

We study the phase behavior of cylinder-forming diblock copolymers confined to thin films. We compute the free energy of possible morphologies in the intermediate segregation regime using self-consistent field theory and in the strong segregation regime using the framework of the strong segregation limit. We investigate the influence of the film thickness and the interaction of the walls with the polymer chains on the stability of different morphologies. We present phase diagrams showing the possible microphases for diblock copolymers with fixed volume fraction and fixed segregation parameter as a function of the film thickness and the affinity of the walls. Essentially, we find the same phase behavior with both methods and can show that only the simplest morphologies remain stable when the computation of the free energy is done with sufficient accuracy.

JI-XUAN HOU
Laboratoire de Physique, Ecole Normale Supérieure de Lyon, France

Scaled-particle theory for a hard-sphere fluid in a porous medium

in collaboration with W. Dong and M. Holovko

We first revisit the scaled-particle theory (SPT) for a bulk hard sphere (HS) and then generalize it to the case of an HS fluid confined in a random porous medium. The chemical potential of a scaled particle, whose size can be varied, is the work for inserting it into the system. The main idea of SPT is to use smooth extrapolation between two extreme regimes to get the work when the size of the scaled particle is equal to other hard spheres, and thus obtain the chemical potential of the fluid. One of the extreme regimes is when the size of the scaled particle is sufficiently small, and in this regime the work can be derived exactly by using statistical physics. In the other extreme regime, where the size of the scaled particle is sufficiently large, by using thermodynamic arguments, the work can be expressed by pressure–volume work and surface tension work. We have checked the accuracy of the theory by Monte Carlo simulations and our theory coincides with the simulation results very well.

YEVGEN O. KAZAKOV
Faculty of Physics and Technology, Kharkiv National University, Ukraine

Application of the phase integral method to the mode conversion problem
in collaboration with I. V. Pavlenko, I. O. Girka, and B. Weyssow

The phase integral method is a generalization of the commonly used WKB method to find an analytical solution of the wave (Schrödinger) equation. The matching procedure is done by tracing the evolution of WKB terms in the complex plane. The method is based on the Stokes phenomenon: the discontinuous change of the coefficient of a subdominant term upon crossing the Stokes line, $\psi = a_d\psi_d + a_s\psi_s \Rightarrow a_d\psi_d + (a_s + Ta_d)\psi_s$. The Stokes maps for different types of potential are shown. The phase integral method is applied to study the mode conversion process in a 1D nonuniform plasma in which two distinct modes can propagate simultaneously. In certain regions of the plasma these waves can interact efficiently, transferring energy from one mode to the other. The mode conversion coefficient C is found for the potential $Q(x) = k_A^2\,(1 - \Delta_1/(x - S_1) - \Delta_2/(x - S_2))$ with two singular points, which represent two coupling regions. It is shown that under certain conditions the mode conversion coefficient can be enhanced to reach 100%.

ANTTI KNOWLES
Institute for Theoretical Physics, ETH Zürich, Switzerland

An Egorov-type theorem for the mean-field limit of classical N-body dynamics
in collaboration with J. Fröhlich

Egorov's original semiclassical result, in the context of the quantum mechanics of a finite number of degrees of freedom, is that time evolution commutes with quantization, up to error terms that vanish as $\hbar$ tends to 0. It is formulated in the Heisenberg picture of quantum mechanics, thus allowing a great generality in the choice of states. I describe how the mean-field limit of classical N-body dynamics may be recast as an Egorov-type theorem. To this end, I start with the Vlasov equation and show that it is the Hamiltonian equation of motion of a classical Hamiltonian system. I quantize this system according to the usual rules of canonical quantization, and show that the resulting quantum theory is equivalent to classical N-body dynamics. The mean-field limit of classical N-body dynamics can now be understood as the semiclassical limit corresponding to this quantization, and the analogue of Egorov's theorem holds.

ROMAIN MARI
Laboratoire PMMH, Ecole Supérieure de Physique et Chimie Industrielles, Paris, France

Jamming versus glass transitions
in collaboration with J. Kurchan and F. Krzakala

Recent ideas based on the properties of assemblies of frictionless particles in mechanical equilibrium provide a perspective of amorphous systems different from that offered by the traditional approach originating in liquid theory. The relation, if any, between these two points of view, and the relevance of the former to the glass phase, has been difficult to ascertain. In this lecture we introduce a model for which both theories apply strictly: it exhibits on the one hand an ideal glass transition and on the other hand "jamming" features (fragility and soft modes) virtually identical to those of real systems. This allows us to disentangle the different contents and domains of applicability of the two physical phenomena.

Hidetoshi MORITA
Institut Non Linéaire de Nice, CNRS, France

Collective oscillations in long-range interacting systems
in collaboration with K. Kaneko (Tokyo) and F. Bouchet (Nice)

Metastable stationary states in long-range interacting systems were one of the main topics in the summer school. As an example of a metastable *nonstationary* state, we introduced a collective oscillation in particle Hamiltonian systems with a mean-field interaction, that is, a structure of macroscopic low-dimensional dissipative dynamics even in a microscopic high-dimensional conservative dynamics (Morita and Kaneko, *Phys. Rev. Lett.*, **96**, 050602 (2006)). We also introduced a similar macroscopically oscillating phenomenon observed in a two-dimensional fluid system, which is another example of a long-range interacting system. We discussed its relevance to the above collective oscillation, by noting the mathematical equivalence of two-dimensional fluid systems to particle Hamiltonian systems.

Lucas NICOLAO
Instituto de Física, Universidade Federal do Rio Grande do Sul, Porto Alegre, Brazil

Modulated phases in ultrathin magnetic films
in collaboration with D. A. Stariolo

We show results from simulations of the Langevin dynamics of a two-dimensional scalar model for ultrathin magnetic films, where the competition between local ferromagnetic and dipolar interactions leads to stripe patterns of magnetization out of the film plane. At low temperatures, the stripes form a two-dimensional array with both translational and orientational order. As the temperature increases, we observe the proliferation of different kinds of topological defects that ultimately drive the system to a disordered phase. In between disordered and straight stripes, we found evidence of two distinct low-temperature phases: a smectic phase, with quasi-long-range translational and orientational long-range order; and a nematic phase, with only orientational long-range order.

Ref.: L. Nicolao and D. A. Stariolo, *Phys. Rev. B*, **76**, 054453 (2007).

ANA C. RIBEIRO-TEIXEIRA
Instituto de Física, Universidade Federal do Rio Grande do Sul, Porto Alegre, Brazil.

Stripe-glass phase in a two-dimensional system without quenched disorder
in collaboration with D. A. Stariolo

Systems characterized by competing interactions are ubiquitous in nature. The competition of interactions on different length scales is known to produce modulated phases characterized by a large variety of mesoscopic patterns. The frustration induced by this kind of competition can stabilize a glassy phase in these systems. This is known to happen in three dimensions. Can this stripe-glass phase also be stabilized in two dimensions, e.g. a thin ferromagnetic film with dipolar interactions? Experimental evidence suggests this might be the case. To answer this question, we analyze the stripe-glass phase of a 2D system with competing interactions and no quenched disorder. Introducing replicas through averaging over a pinning field and within the self-consistent screening approximation, we analytically compute the configurational entropy as a function of temperature and the model parameters. The stripe-glass phase is connected to the appearance of a finite long-time replica correlation function in the system, below a crossover temperature, related to the mobility of defects in the sample.

STEFAN RIST
Departament de Fisica, Universitat Autònoma de Barcelona, Spain

Photon-mediated interaction between two distant atoms
in collaboration with G. Morigi, M. Hennrich, and J. Eschner

We study the photonic interaction between two distant atoms that are optically coupled by a lens focusing the radiation emitted in a certain solid angle into the other. Two regimes are identified, which are determined by the relation between the radiative instability of the atomic excited state with respect to the propagation time of a photonic excitation between the two atoms. We study the coherence properties of the emitted light and show that it carries information about the multiple-scattering processes the atomic spins undergo. The predictions are compared with the measured results reported in J. Eschner *et al.*, *Nature*, **413**, 495 (2001).

FRANÇOIS SICARD
LPNHE, Université Pierre et Marie Curie, Paris 6, France

1D gravity in infinite point distributions
in collaboration with A. Gabrielli and M. Joyce

The dynamics of infinite, asymptotically homogeneous distributions of purely self-gravitating particles in one spatial dimension provides a simple and interesting toy model for the analogous three-dimensional problem treated in cosmology. The well-known "Jeans swindle" is a prescription which, in three dimensions, makes the problem of Newtonian gravity well defined in the thermodynamical limit of infinite point distributions, i.e. $N, L \to +\infty, N/L > 0$. In practice it prescribes that the force on a particle should be calculated by summing symmetrically (e.g. in spheres) about the particle. Applied naively to the analogous problem in one dimension (i.e. a force independent of the separation of particles), the same prescription does not work, as the force on a point can never converge in any point distribution owing to the limiting nondecaying behavior of surface mass fluctuations. Following a discussion by Kiessling of the "Jeans swindle" in three dimensions, we show that the problem may be resolved by defining the force in an infinite point distribution as the limit of an exponentially screened pair interaction. We show explicitly that this prescription gives a well-defined (finite) force acting on particles in a class of perturbed infinite lattices, which are the point processes relevant to cosmological N-body simulations.

YASUHIDE SOTA
Department of Physics, Ochanomizu University, Japan &
Advanced Research Institute for Science and Engineering, Waseda University, Japan.

Self-organized relaxation in self-gravitating systems

in collaboration with O. Iguchi, T. Tashiro, and M. Morikawa

We propose a self-organized relaxation process driving a collisionless self-gravitating system to an equilibrium state satisfying the local virial (LV) relation, which is admitted only in special classes of stationary solutions in the Vlasov–Poisson system, such as Plummer's solution. If the system is not initially globally virialized, this causes a violent relaxation within a time interval as short as a few free-fall times, because of the effective potential oscillations. Here we numerically investigate the relaxation processes of N-body self-gravitating systems and demonstrate that:

- An N-body self-gravitating system converges to a critical state which is characterized by the LV relation through a violent relaxation process such as a cold collapse or two cluster mergers.
- Such self-adjusted phenomena are remarkably similar to the self-organized criticality in sandpiles, in which an avalanche of grains self-adjusts the slope of the sandpile to the critical value. The LV ratio $b = 1$ plays the same role as the critical slope of a sandpile, since a large amount of particles spread out at any place where the LV ratio exceeds the critical value, which adjusts the LV values to the critical value everywhere. A self-gravitating system, on the other hand, does not need energy injection from outside, but uses its own gravitational potential as a driving force and releases the energy toward the outside through escaping particles.

This seems to suggest that a state with the LV relation becomes an attractor for the violent relaxation process in collisionless self-gravitating systems under the condition that particles can escape infinitely.

JURAJ SZAVITS-NOSSAN
Institute of Physics, Zagreb, Croatia

Phase transitions in an asymmetric exclusion process with long-range hopping
in collaboration with K. Uzelac

We study an exclusion process in which particles may jump any distance $l \geq 1$ with a probability that decays as $l^{-(1+\sigma)}$. Besides the localization of a domain wall at a first-order phase transition, previous results [1] have shown a change in the continuous phase transition to the maximum-current phase. In particular, the exponent of the algebraic decay of the density profile differs from the short-range value 1/2 in the region $1 < \sigma < 2$, where its dependence on σ has been given by a conjecture based on numerical simulations. In the present work, we obtain the exact value of this exponent from a hydrodynamic equation for the density profile in the mean-field approximation [2]. For $\sigma > 2$, this equation is given by the viscous Burgers equation for the short-range case, but the usual diffusion term of this equation is replaced by a fractional one for $1 < \sigma < 2$. In the case of a translationally invariant system, the equation can be mapped onto the fractional Kardar–Parisi–Zhang equation [3], which predicts the value of the dynamical exponent $z = \min\{\sigma, 3/2\}$, in agreement with the results of our numerical simulations on a half-filled chain with periodic boundary conditions.

[1] J. Szavits-Nossan and K. Uzelac, *Phys. Rev. E*, **74**, 051104 (2006).
[2] J. Szavits-Nossan and K. Uzelac, *Phys. Rev. E*, **77**, 051116 (2008).
[3] E. Katzav, *Phys. Rev. E*, **68**, 031607 (2003).

TARCÍSIO N. TELES
Instituto de Física, Universidade Federal do Rio Grande do Sul, Porto Alegre, Brazil.

Collisionless relaxation in nonneutral plasmas
in collaboration with Yan Levin and Renato Pakter

In many applications where intense charged-particle beams are employed, it is desirable to have a knowledge of the final distribution of the particles in the beam. The type of interaction in these systems is long-range. So, in general, the system evolves to stationary states where the distribution function is not Maxwellian. In this work, a theory is presented which allows us to quantitatively predict the final stationary state achieved by a nonneutral plasma during a process of collisionless relaxation. As a specific application, the theory is used to study relaxation of charged-particle beams. It is shown that a fully matched beam relaxes to the Lynden-Bell distribution. However, when a mismatch is present and the beam oscillates, parametric resonances lead to a core–halo phase separation. The approach developed accounts for both the density and the velocity distributions in the final stationary state. [*Phys. Rev. Lett.*, **100**, 040604 (2008).]

ANTOINE VENAILLE
Laboratoire des Ecoulement Geophysiques et Industriels, Grenoble, France.

Ensemble inequivalence in two-dimensional flows

in collaboration with F. Bouchet (Nice)

Two-dimensional flows are long-range interacting systems, since the interaction potential is logarithmic (B. Turkington's lecture). Such systems are expected to have peculiar thermodynamical properties, such as negative specific heat (D. Mukamel's lecture).

We report the first explicit computation of ensemble inequivalence in such systems. We explain that the occurrence of ensemble inequivalence is related to peculiar phase transitions that lead to drastic changes in the flow structure (for instance a transition from a monopole to a dipole).

The method used to compute the phase transitions is very general (it is based on a search for the minimum of a quadratic functional conserved by the dynamics, with a linear constraint). It could be applied in other contexts, for instance in plasma physics (D. Dubin's lecture) and self-gravitating stars (T. Padmanabhan's lecture).

ALESSANDRO VINDIGNI
Swiss Federal Institute of Technology, ETH Zurich, Switzerland

Domain shrinking induced by localized fluctuations in ultrathin magnetic films

in collaboration with O. Portmann, N. Saratz, F. Cinti, P. Politi, and D. Pescia

Ultrathin Fe/Cu(001) films are magnetized out of plane and represent an experimental counterpart of the 2D dipolar frustrated Ising ferromagnet. The latter is a particular case of a more general model in which the ferromagnetic short-range (exchange) interaction competes with a long-range antiferromagnetic interaction decaying as $1/r^{d+\sigma}$. For $\sigma \leq 1$ the long-range interaction is able to avoid global phase separation, the uniformly magnetized state favored by the exchange interaction. In particular, the ground state consists of a monodimensional modulation of the order parameter resulting in a superlattice of positive and negative magnetic domains. As a result, the spins located at an interface between two oppositely magnetized domains are significantly more susceptible to thermal fluctuations than inside-domain spins. Thus the partial loss of translational invariance biases the balance between the ferromagnetic exchange and the antiferromagnetic long-range interaction, favoring the latter, which finally leads to domain shrinking with increasing temperature. We present mean-field calculations for an ideal stripe pattern with $d = 2$ and $\sigma = 1$ (dipolar interaction) which turn out to be in fairly quantitative agreement with experimental SEMPA images recorded on Fe/Cu(001) films. Preliminary Monte Carlo results supporting the generality of our arguments beyond the mean-field approach are also discussed.

LILA WARSZAWSKI
The University of Melbourne, Australia

Glitches in neutron stars

Pulsars are extremely reliable clocks, enabling astronomers to measure their period and spindown rate with great accuracy. Glitches are randomly timed increases in the spin period, accompanied by a change in the spindown rate. Glitches are thought to result from a rapid change in the coupling between the pulsar interior, containing a degenerate neutron superfluid, and its crust, mediated by pinned superfluid vortices. As the pulsar spins down, a lag builds up between the crust and the interior superfluid. When this lag exceeds some critical value, the vortices unpin, transferring angular momentum from the superfluid to the crust. We model this pinning and unpinning of vortices as a coherent noise process, in an effort to reproduce the observed power-law statistics of glitch sizes and Poissonian waiting times between successive glitches.

ABEL YANG
Department of Astronomy, University of Virginia, Charlottesville, VA, USA

Robustness of the galaxy distribution function

in collaboration with W. C. Saslaw, P. Chan, and B. Leong (National University, Singapore)

We model the evolution of the spatial counts-in-cells distribution of galaxies and show that the galaxy distribution function is robust to mergers and evolution. In particular, we show that the initial positions of merged galaxies follow a similar distribution to the galaxies themselves. We compute the evolution of the number of galaxies per unit volume with respect to redshift and compare our results with observations at high redshift. While the predicted average number of galaxies per cell $\overline{N}$ is much higher than the observed $\overline{N}$ for galaxies larger than $10^{10}m_{\odot}$, the predicted $\overline{N}$ for galaxies larger than $3 \times 10^{11}m_{\odot}$ agrees with observations to first-order. We also find that the predicted clustering measure b is smaller than the observed value of b at high redshifts. Our results also indicate that the average halo half-mass radius ϵ is sufficiently small that galaxies may be approximated as point masses for computations of the correlation potential energy. This would suggest that the value of b is largely independent of ϵ.

Part I

Statistical dynamics

1
Statistical mechanics of systems with long-range interactions

David Mukamel

Department of Physics, The Weizmann Institute of Science,
Rehovot, 76100, Israel
URL: http://www.weizmann.ac.il/home/fnmukaml/

1.1 Introduction

In these Notes we discuss some thermodynamic and dynamical properties of models of systems in which the two-body interaction potential between particles decays algebraically with their relative distance r at large distances. Typically the potential decays as $1/r^{d+\sigma}$ in d dimensions and it may be either isotropic or nonisotropic (as in the case of magnetic or electric dipolar interactions). In general, one can distinguish between two broad classes of long-range interactions (LRI): those with $-d \leq \sigma \leq 0$, which we term "strong" LRI, and those with $0 < \sigma \leq \sigma_c(d)$, which we term "weak" LRI. The parameter $\sigma_c(d)$ satisfies $\sigma_c(d) = 2$ for $d \geq 4$, and $\sigma_c(d) = d/2$ for $d < 4$, as will be discussed in more detail in the following sections. In systems with strong LRI the potential decays slowly with the distance, and this results in rather pronounced thermodynamic and dynamical effects. On the other hand, in systems with weak LRI the potential decays faster at large distances, resulting in distinct but less pronounced effects. Systems with $\sigma > \sigma_c$ behave thermodynamically like the more commonly studied systems with short-range interactions. For recent reviews on systems with long-range interactions see, for example, the proceedings of two workshops devoted to this subject (Dauxois *et al.* 2002; Campa *et al.* 2007*a*).

Long-range interactions are rather common in nature. Examples include self-gravitating systems ($\sigma = -2$) (Padmanabhan 1990; Chavanis 2002); dipolar ferroelectrics and ferromagnets, in which the interactions are anisotropic ($\sigma = 0$) (Landau and Lifshitz 1960), nonneutral plasmas ($\sigma = -2$) (Nicholson 1992); two-dimensional geophysical vortices which interact via a weak, logarithmically decaying potential ($\sigma = -2$) (Chavanis 2002); charged particles interacting via their mutual electromagnetic fields, such as in a free electron laser (Barré *et al.* 2004); and many others.

Let us first consider strong LRI. Such systems are nonadditive, and the energy of homogeneously distributed particles in a volume V scales superlinearly with the volume, as $V^{1-\sigma/d}$. The lack of additivity leads to many unusual properties, both thermal and dynamical, which are not present in systems with weak LRI or with short-range interactions. For example, as has first been pointed out by Antonov (1962) and later elaborated on by Lynden-Bell (Lynden-Bell and Wood 1968; Lynden-Bell 1999), Thirring and coworkers (Thirring 1970; Hertel and Thirring 1971; Posch and Thirring 2006), and others, the entropy S need not be a concave function of the energy E, yielding a negative specific heat within the microcanonical ensemble. Since the specific heat is always positive when calculated within the canonical ensemble, this indicates that the two ensembles need not be equivalent. Recent studies have suggested the inequivalence of ensembles is particularly manifested whenever a model exhibits a first-order transition within the canonical ensemble (Barré *et al.* 2001, 2002). Similar ensemble inequivalence between canonical and grand canonical ensembles has also been discussed (Misawa *et al.* 2006).

Typically, the entropy S, which is measured by the number of ways N particles with total energy E may be distributed in a volume V, scales linearly with the volume. This is irrespective of whether or not the interactions in the system are long-ranged. On the other hand, in systems with strong long-range interactions, the energy scales superlinearly with the volume. Thus, in the thermodynamic limit, the free energy $F = E - TS$ is dominated by the energy at any finite temperature T, suggesting that the entropy may be neglected altogether. This would result in trivial thermodynamics. However, in many real cases, when systems of finite size are considered, the temperature could be sufficiently high that the entropic term in the free energy, TS, becomes comparable to the energy E. In such cases the entropy may not be neglected and the thermodynamics is nontrivial. This is the case in some self-gravitating systems such as globular clusters (see, for example, Chavanis (2002)). In order to theoretically study this limit, it is convenient to rescale the energy by a factor $V^{\sigma/d}$ (or, alternatively, to rescale the temperature by a factor $V^{-\sigma/d}$), making the energy and the entropy contribution to the free energy of comparable magnitude. This is known as the Kac prescription (Kac *et al.* 1963). While systems described by this rescaled energy are extensive, they are nonadditive in the sense that the energy of two isolated subsystems is not equal to their total energy when they are combined together and are allowed to interact.

A special case is that of dipolar ferromagnets, where the interaction is anisotropic, and scales as $1/r^3$ ($\sigma = 0$). In this borderline case of strong long-range interactions, the energy depends on the shape of the sample. It is well known that for ellipsoidal magnets, the contribution of the long-distance part of the dipolar interaction leads to a mean-field-type term in the energy. This results in an effective Hamiltonian $H \to H - DM^2/V$, where M is the magnetization of the system and D is a shape-dependent coefficient known as the demagnetization factor. In this Hamiltonian, the long-range interaction between dipoles becomes independent of their distance, making it particularly convenient for theoretical studies (Campa *et al.* 2007*b*).

Studies of the relaxation processes in systems with strong long-range interactions in some models have shown that the relaxation of thermodynamically unstable states to the stable equilibrium state may be unusually slow, with a characteristic time which diverges with the number of particles, N, in the system (Antoni and Ruffo 1995; Latora *et al.* 1998, 1999; Yamaguchi 2003; Yamaguchi *et al.* 2004; Mukamel *et al.* 2005). This, too, is in contrast with relaxation processes in systems with short-range interactions, in which the relaxation time does not scale with N. As a result, long-lived quasi-stationary states (QSS) have been observed in some models, which do not relax to the equilibrium state in the thermodynamic limit.

Nonadditivity has been found to result, in many cases, in breaking of ergodicity. Here the phase space is divided into disjoint domains separated by finite gaps in macroscopic quantities, such as the total magnetization in magnetic systems (Mukamel *et al.* 2005; Fel'dman 1998; Borgonovi *et al.* 2004, 2006; Hahn and Kastner 2005, 2006; Bouchet *et al.* 2008). Within local dynamics, these systems are thus trapped in one of the domains.

Features which are characteristic of nonadditivity are not limited to systems with long-range interactions. In fact, finite systems with short-range interactions, in which the surface and bulk energies are comparable, are also nonadditive. Features such as negative specific heat in small systems (e.g. clusters of atoms) have been discussed in a number of studies (Lynden-Bell 1995, 1996; Gross 2000; Chomaz and Gulminelli 2002).

Let us turn now to systems with a weak long-range interaction, for which $0 < \sigma \leq \sigma_c(d)$. These systems are additive; namely, the energy of homogeneously distributed particles scales as V. As a result, the usual formulation of thermodynamics and statistical mechanics developed for systems with short-range interactions applies directly in this case. In particular, the specific heat is nonnegative as expected, and the various statistical-mechanical ensembles are equivalent. However, the long-range nature of the interaction becomes dominant at phase transitions, where long-range correlations are naturally built up in the system. For example, it is well known that one-dimensional (1D) systems with short-range interactions do not exhibit phase transitions and spontaneous symmetry breaking at finite temperatures. On the other hand, weak long-range interactions may result in phase transitions in one dimension at a finite temperature. A notable example of such a transition was introduced by Dyson in the 1960s (Dyson 1969*a*,*b*).

The critical behavior of systems near a continuous phase transition is commonly classified by universality classes. The critical exponents of each class do not depend on the details of the interactions but rather on some general features such as the spatial dimension of the system, its symmetry, and the range of forces. Weak long-range interactions modify the universality class of the system, leading to critical exponents which depend on the interaction parameter σ. It is also well known that the critical behavior of systems with short-range interactions become mean-field-like above a critical dimension d_c, where the effect of fluctuations may be neglected. For a generic critical point in a system with short-range interactions one has $d_c = 4$. Since long-range interactions tend to reduce fluctuations, weak LRI result in a smaller critical dimension, $d_c(\sigma)$, which is a function of the interaction parameter σ (see Fisher *et al.* 1972; Fisher 1974).

A distinct class of local interactions which are effectively weak and long-range in nature when applied to long polymers is that of excluded-volume interactions. This repulsive interaction, which is local in space, but which can take place between monomers which are far away along the polymer chain, can change the critical properties of the polymer, resulting in a distinct universality class. For example, while the average end-to-end distance of a random polymer of length L scales as $\sqrt{L}$, it scales with a different power law, L^ν, when a repulsive interaction between monomers is introduced. In general, the exponent $\nu(d)$ depends on the spatial dimension of the system and it becomes equal to the value corresponding to a random polymer, $1/2$, only at $d > 4$. Excluded-volume interactions have a profound effect on the thermodynamic behavior of polymer solutions, DNA denaturation, and many other systems of long polymers; see, for example, Des Cloizeax and Jannink (1990). We will not discuss excluded-volume effects in these Notes.

We now turn to systems out of thermal equilibrium. In many cases driven systems reach a nonequilibrium steady state in which detailed balance is not satisfied. Under rather broad conditions, such steady states in systems with conserving dynamics exhibit long-range correlations, even when the dynamics is local. In thermal equilibrium, the nature of the equilibrium state is independent of the dynamics involved. For example, an Ising model evolving under magnetization-conserving Kawasaki dynamics reaches the same equilibrium state as that reached by a nonconserving Glauber dynamics. The equilibrium state is uniquely determined by the Hamiltonian. This is not the case in nonequilibrium systems, and the resulting steady state is strongly affected by the details of the dynamics. Conserved quantities tend to introduce long-range correlations in the steady state, owing to the slow diffusive nature of their dynamics. At equilibrium, these long-range correlations are somehow canceled owing to detailed balance, which leads to the Gibbs equilibrium distribution. Under nonequilibrium conditions, such cancelation does not take place generically. Therefore, one expects properties of of equilibrium systems with long-range interactions to show up in steady states of nonequilibrium systems with conserving local interactions.

Long-range correlations have been studied in a large number of driven models. For example, as discussed above, one-dimensional equilibrium systems with short-range interactions do not exhibit phase transitions or spontaneous symmetry breaking at finite temperatures. In equilibrium, phase transitions in 1D systems take place only when long-range interactions are present. On the other hand, a number of models of driven one-dimensional systems have been shown to display phase transitions and spontaneous symmetry breaking even when their dynamics is local (see Mukamel 2000). This demonstrates the buildup of long-range correlations by the driving dynamical processes. An interesting model in this respect is what is known as the ABC model. This is a model of three species of particles, A, B, and C, which move on a ring with local dynamical rules, reaching a steady state in which the three species are spatially separated. The dynamical rules do not obey detailed balance, and hence the model is out of equilibrium. It has been demonstrated that the dynamics of this model result in effective long-range interactions, which can be explicitly calculated (Evans *et al.* 1998*a*,*b*). The model will be discussed in some detail in these Notes.

The Notes are organized as follows. In Section 1.2, we discuss properties of systems with strong long-range interactions. Some properties of weak long-range interactions, particularly the existence of long-range order in 1D and the upper critical dimension, are discussed in Section 1.3. Effective long-range interactions in driven models are demonstrated within the context of the ABC model in Section 1.4. Finally, a brief summary is given in Section 1.5.

1.2 Strong long-range interactions

1.2.1 General considerations

We start by presenting some general considerations concerning thermodynamic properties of systems with strong long-range interactions. In particular, we argue that in addition to a negative specific heat, or a nonconcave entropy curve, which could be realized in the microcanonical ensemble, this ensemble also yields a discontinuity in temperature whenever a first-order transition takes place.

Consider the nonconcave curve in Fig. 1.1. For a system with short-range interactions, this curve cannot represent the entropy $S(E)$. The reason is that, owing to additivity, the system represented by this curve is unstable in the energy interval $E_1 < E < E_2$. Entropy can be gained by phase-separating the system into two subsystems corresponding to E_1 and E_2, keeping the total energy fixed. The average energy and entropy densities in the coexistence region is given by the weighted average of the corresponding densities of the two coexisting systems. Thus the correct entropy curve in this region is given by the common tangent line, resulting in an overall concave curve. However, in systems with strong long-range interactions, the average energy

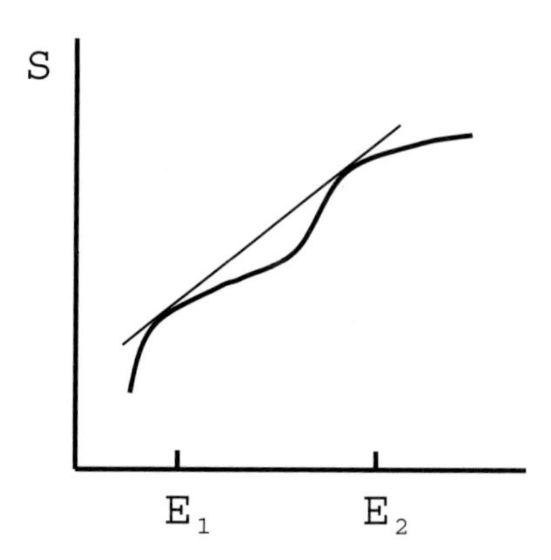

Fig. 1.1 A nonconcave entropy curve, which for additive systems is made concave by the common tangent line. In systems with long-range interactions, the nonconcave curve may represent the actual entropy of the system, yielding negative specific heat.

density of two coexisting subsystems is not given by the weighted average of the energy density of the two subsystems. Therefore, the nonconcave curve in Fig. 1.1 could, in principle, represent an entropy curve of a stable system, and phase separation need not take place. This results in a negative specific heat. Since within the canonical ensemble specific heat is nonnegative, the microcanonical and canonical ensembles are not equivalent. The above considerations suggest that the inequivalence of the two ensembles is particularly manifested whenever a coexistence of two phases is found within the canonical ensemble.

Another feature of systems with strong long-range interactions is that within the microcanonical ensemble, first-order phase transitions involve a discontinuity of temperature. To demonstrate this point consider, for example, a magnetic system which undergoes a phase transition from a paramagnetic to a magnetically ordered phase. Let M be the magnetization and $S(M, E)$ be the entropy of the system for a given magnetization and energy. A typical entropy vs. magnetization curve for a given energy close to a first-order transition is given in Fig. 1.2. It exhibits three local maxima, one at $M = 0$ and two other degenerate maxima at $\pm M_0$. At energies where the paramagnetic phase is stable, one has $S(0, E) > S(\pm M_0, E)$. In this phase the entropy is given by $S(0, E)$ and the temperature is obtained from $1/T = \partial S(0, E)/\partial E$. On the other hand, at energies where the magnetically ordered phase is stable, the entropy is given by $S(M_0, E)$ and the temperature is obtained from $1/T = \partial S(M_0, E)/\partial E$. At the first-order transition point, where $S(0, E) = S(\pm M_0, E)$, the two derivatives are generically not equal, resulting in a temperature discontinuity. A typical entropy vs. energy curve is given in Fig. 1.3. Note that these considerations do not apply to systems with short-range interactions. The reason is that in the case of short-range interactions, the entropy is a concave function of the two extensive parameters M and E, and the transition involves discontinuities of both parameters.

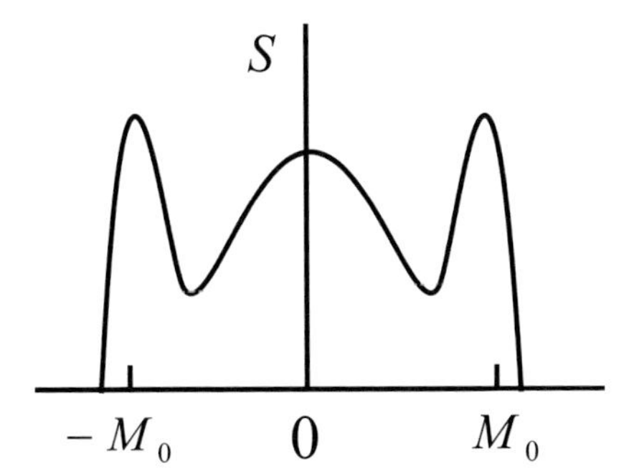

Fig. 1.2 A typical entropy vs. magnetization curve of a magnetic system with long-range interactions near a first-order transition at a given energy. As the energy varies, the heights of the peaks change and a first-order transition is obtained at the energy where the peaks are of equal height.

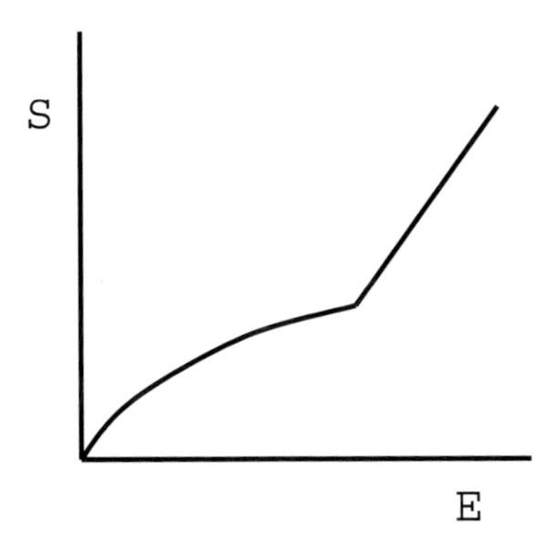

Fig. 1.3 A typical entropy vs. energy curve for a system with long-range interactions exhibiting a first-order transition. The slope discontinuity at the transition results in a temperature discontinuity.

Systems with long-range interactions are more likely to exhibit breaking of ergodicity owing to their nonadditive nature. This may be argued on rather general grounds. In systems with short-range interactions, the domain in the phase space of extensive thermodynamic variables, such as energy, magnetization, and volume, is convex. Let $\vec{X}$ be a vector whose components are the extensive thermodynamic variables over which the system is defined. Suppose that there exist microscopic configurations corresponding to two points $\vec{X}_1$ and $\vec{X}_2$ in this phase space. As a result of the additivity property of systems with short-range interactions, there exist microscopic configurations corresponding to any intermediate point between $\vec{X}_1$ and $\vec{X}_2$. Such microscopic configurations may be constructed by combining two appropriately weighted subsystems corresponding to $\vec{X}_1$ and $\vec{X}_2$, making use of the fact that for sufficiently large systems, surface terms do not contribute to bulk properties. Since systems with long-range interactions are nonadditive, such interpolation is not possible, and intermediate values of the extensive variables are not necessarily accessible. As a result, the domain in the space of extensive variables over which a system is defined need not be convex. When there exists a gap in phase space between two points corresponding to the same energy, local energy-conserving dynamics cannot take the system from one point to the other, and ergodicity is broken.

These and other features of canonical and microcanonical phase diagrams are explored in the following subsections by considering specific models.

1.2.2 Phase diagrams of models with strong long-range interactions

In order to obtain better insight into the thermodynamic behavior of systems with strong long-range interactions, it is instructive to analyze phase diagrams of representative models. A particularly convenient class of models is that where the long-range part of the interaction is of mean-field type. In such models $\sigma = -d$, and the interaction is independent of the distance between the particles. The insight obtained from studies of these models may, however, be relevant for other systems with $\sigma \leq 0$, since the main feature of these models, namely nonadditivity, is shared by all models with $\sigma \leq 0$.

In recent studies both the canonical and the microcanonical phase diagrams of some spin models with mean-field-type long-range interactions have been analyzed. Examples include discrete spin models such as the Blume–Emery–Griffiths model (Barré *et al.* 2001, 2002) and the Ising model with long- and short-range interactions (Mukamel *et al.* 2005), as well as continuous spin models of XY type (de Buyl *et al.* 2005; Campa *et al.* 2006). These models are simple enough that their thermodynamic properties can be evaluated in both ensembles. The common feature of these models is that their phase diagrams exhibit first- and second-order transition lines. It has been found that in all cases, the canonical and microcanonical phase diagrams differ from each other in the vicinity of the first-order transition line. A classification of possible types of inequivalent canonical and microcanonical phase diagrams in systems with long-range interactions is given in Bouchet and Barré (2005). In what follows we discuss in some detail the thermodynamics of one model, namely the Ising model with long- and short-range interactions (Mukamel *et al.* 2005).

Consider an Ising model defined on a ring with N sites. Let $S_i = \pm 1$ be the spin variable at site $i = 1, \ldots, N$. The Hamiltonian of the system is composed of two interaction terms and is given by

$$H = -\frac{K}{2}\sum_{i=1}^{N}(S_i S_{i+1} - 1) - \frac{J}{2N}\left(\sum_{i=1}^{N} S_i\right)^2 . \qquad (1.1)$$

The first term is a nearest-neighbor coupling, which can be either ferromagnetic ($K > 0$) or antiferromagnetic ($K < 0$). On the other hand, the second term is ferromagnetic,

$J > 0$, and it corresponds to a long-range, mean-field-type interaction. The reason for considering a ring geometry for the nearest-neighbor coupling is that this is more convenient for carrying out the microcanonical analysis. Similar features are expected to take place in higher dimensions as well.

The canonical phase diagram of this model was analyzed some time ago (Nagle 1970; Bonner and Nagle 1971; Kardar 1983). The ground state of the model is ferromagnetic for $K > -J/2$ and antiferromagnetic for $K < -J/2$. Since the system is one-dimensional, and since the long-range interaction term can only support ferromagnetic order, it is clear that for $K < -J/2$ the system is disordered at any finite temperature, and no phase transition takes place. However, for $K > -J/2$ one expects ferromagnetic order at low temperatures. Thus a phase transition takes place at some temperature to a paramagnetic, disordered phase (see Fig. 1.4). For large K the transition was found to be continuous, taking place at a temperature given by

$$\beta = e^{-\beta K} . \tag{1.2}$$

Here $\beta = 1/T$, $J = 1$ is assumed for simplicity, and $k_B = 1$ is taken for the Boltzmann constant. The transition becomes first-order for $K < K_{CTP}$, with a tricritical point located at an antiferromagnetic coupling $K_{CTP} = -\ln 3/2\sqrt{3} \simeq -0.317$. As usual, the first-order line has to be evaluated numerically. The first-order line intersects the $T = 0$ axis at $K = -1/2$. The (K, T) phase diagram is given in Fig. 1.4.

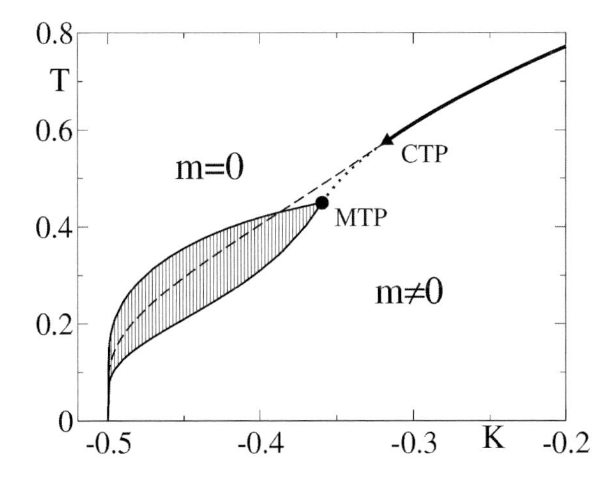

Fig. 1.4 The (K, T) phase diagrams of the model (1.1) within the canonical and microcanonical ensembles. In the canonical ensemble, the large-K transition is continuous (bold solid line) down to the tricritical point CTP, where it becomes first-order (dashed line). In the microcanonical ensemble, the continuous transition coincides with the canonical one at large K (bold line). It persists at lower K (dotted line) down to the tricritical point MTP, where it turns first-order, with a branching of the transition line (solid lines). The region between these two lines (shaded area) is not accessible.

Let us now analyze the phase diagram of the model within the microcanonical ensemble (Mukamel *et al.* 2005). To do this one has to calculate the entropy of the system for a given magnetization and energy. Let

$$U = -\frac{1}{2}\sum_i (S_i S_{i+1} - 1) \tag{1.3}$$

be the number of antiferromagnetic bonds in a given configuration characterized by N_+ up spins and N_- down spins, with $N_+ + N_- = N$. One would like to evaluate the number of microscopic configurations corresponding to (N_+, N_-, U). Such configurations are composed of $U/2$ segments of up spins which alternate with the same number of segments of down spins, where the total numbers of up and down spins are N_+ and N_-, respectively. The number of ways of dividing N_+ spins into $U/2$ groups is

$$\binom{N_+ - 1}{U/2 - 1}, \tag{1.4}$$

with a similar expression for the down spins. To leading order in N, the number of configurations corresponding to (N_+, N_-, U) is given by

$$\Omega(N_+, N_-, U) = \binom{N_+}{U/2}\binom{N_-}{U/2}. \tag{1.5}$$

Note that a multiplicative factor of order N has been neglected in this expression, since only exponential terms in N contribute to the entropy. This factor corresponds to the number of ways of placing the U ordered segments on the lattice. Expressing N_+ and N_- in terms of the number of spins N and the magnetization $M = N_+ - N_-$, and denoting $m = M/N$, $u = U/N$, and the energy per spin $\epsilon = E/N$, one finds that the entropy per spin, $s(\epsilon, m) = (1/N)\ln\Omega$, is given in the thermodynamic limit by

$$\begin{aligned} s(\epsilon, m) = {} & \frac{1}{2}(1+m)\ln(1+m) + \frac{1}{2}(1-m)\ln(1-m) \\ & - u\ln u - \frac{1}{2}(1+m-u)\ln(1+m-u) \\ & - \frac{1}{2}(1-m-u)\ln(1-m-u)\,, \end{aligned} \tag{1.6}$$

where u satisfies

$$\epsilon = -\frac{J}{2}m^2 + Ku\,. \tag{1.7}$$

By maximizing $s(\epsilon, m)$ with respect to m one obtains both the spontaneous magnetization $m_s(\epsilon)$ and the entropy $s(\epsilon) \equiv s(\epsilon, m_s(\epsilon))$ of the system for a given energy ϵ.

In order to analyze the microcanonical phase transition corresponding to this entropy we expand s in powers of m,

$$s = s_0 + Am^2 + Bm^4\,. \tag{1.8}$$

Here the zero-magnetization entropy is

$$s_0 = -\frac{\epsilon}{K}\ln\frac{\epsilon}{K} - \left(1-\frac{\epsilon}{K}\right)\ln\left(1-\frac{\epsilon}{K}\right), \tag{1.9}$$

the coefficient A is given by

$$A = \frac{1}{2}\left[\frac{1}{K}\ln\left(\frac{K-\epsilon}{\epsilon}\right) - \frac{\epsilon}{K-\epsilon}\right], \tag{1.10}$$

and B is another energy-dependent coefficient which can be easily evaluated. In the paramagnetic phase, both A and B are negative so that the $m = 0$ state maximizes the entropy. At the energy where A vanishes, a continuous transition to the magnetically ordered state takes place. Using the thermodynamic relation for the temperature

$$\frac{1}{T} = \frac{ds}{d\epsilon}, \tag{1.11}$$

the caloric curve in the paramagnetic phase is found to be

$$\frac{1}{T} = \frac{1}{K}\ln\frac{K-\epsilon}{\epsilon}. \tag{1.12}$$

This expression is also valid at the critical line, where $m = 0$. Therefore, the critical line in the (K,T) plane may be evaluated by taking $A = 0$ and using eqn (1.12) to express ϵ in terms of T. One finds that the expression for the critical line is the same as that obtained within the canonical ensemble, eqn (1.2).

The transition is continuous as long as B is negative, where the $m = 0$ state maximizes the entropy. The transition changes its character at a microcanonical tricritical point where $B = 0$. This takes place at $K_{MTP} \simeq -0.359$, which may be computed analytically using the expression for the coefficient B. The fact that $K_{MTP} < K_{CTP}$ means that while the microcanonical and canonical critical lines coincide up to K_{CTP}, the microcanonical line extends beyond this point into the region where, within the canonical ensemble, the model is magnetically ordered (see Fig. 1.4). In this region, the microcanonical specific heat is negative. For $K < K_{MTP}$ the microcanonical transition becomes first-order, and the transition line has to be evaluated numerically by maximizing the entropy. As discussed in the previous subsection, such a transition is characterized by a temperature discontinuity. The shaded region in the (K,T) phase diagram in Fig. 1.4 indicates an inaccessible domain resulting from the temperature discontinuity.

The main features of the phase diagram given in Fig. 1.4 are not peculiar to the Ising model defined by the Hamiltonian (1.1), but are expected to be valid for any system in which a continuous line changes its character and becomes first-order at a tricritical point. In particular, the lines of continuous transition are expected to be the same in both ensembles up to the canonical tricritical point. The microcanonical critical line extends beyond this point into the ordered region of the canonical phase diagram, yielding a negative specific heat. When the microcanonical tricritical point is reached, the transition becomes first-order, characterized by a discontinuity of the temperature. These features have been found in studies of other discrete spin models such as the spin-1 Blume–Emery–Griffiths model (Barré *et al.* 2001, 2002). They have also been found in continuous spin models such as the XY model with two- and four-spin mean-field-like ferromagnetic interaction terms (de Buyl *et al.* 2005), and in an XY model with long- and short-range, mean-field-type interactions (Campa *et al.* 2006).

1.2.3 Ergodicity breaking

Ergodicity breaking in models with strong long-range interactions has recently been explicitly demonstrated in a number of models such as a class of anisotropic XY models (Borgonovi *et al.* 2004, 2006), discrete spin Ising models (Mukamel *et al.* 2005), mean-field ϕ^4 models (Hahn and Kastner 2005, 2006), and isotropic XY models with four-spin interactions (Bouchet *et al.* 2008). Here we outline a demonstration of this feature for the Ising model with long- and short-range interactions defined in the previous section (Mukamel *et al.* 2005).

Let us consider the Hamiltonian (1.1) and take, for simplicity, a configuration of the spins with $N_+ > N_-$. The local energy U is, by definition, nonnegative. It also has an upper bound which, for the case $N_+ > N_-$, is $U \leq 2N_-$. This upper bound is achieved when the negative spins are isolated, each contributing two negative bonds to the energy. Thus $0 \leq u \leq 1 - m$. Combining this with eqn (1.7), one finds that for positive m the accessible states have to satisfy

$$\begin{aligned} &m \leq \sqrt{-2\epsilon}, \quad m \geq m_+, \quad m \leq m_-, \\ &\text{with } m_\pm = -K \pm \sqrt{K^2 - 2(\epsilon - K)}. \end{aligned} \tag{1.13}$$

Similar restrictions exist for negative m. These restrictions yield the accessible magnetization domain shown in Fig. 1.5 for $K = -0.4$.

The fact that the accessible magnetization domain is not convex results in non-ergodicity. At a given, sufficiently low energy, the accessible magnetization domain is composed of two intervals with large positive and large negative magnetization, respectively. Thus, starting from an initial condition which lies within one of these intervals, local dynamics, to be discussed in the next subsection, is unable to take the system to the other accessible interval, and ergodicity is broken. At intermediate energy values, another accessible magnetization interval emerges near the $m = 0$ state and three disjoint magnetization intervals are available. When the energy is increased, the three intervals join together and the model becomes ergodic.

1.2.4 Slow relaxation

In systems with short-range interactions, the relaxation from a thermodynamically unstable state is typically a fast process. For example, in a magnetic, Ising-like system, starting with a magnetically disordered state at a low temperature, where the stable state is the ordered one, the system will locally order in a short time. This leads to a domain structure in which the system is divided into up and down magnetic domains of some typical size. The domain-forming process is fast in the sense that its characteristic time does not scale with the system size. This domain structure is formed by fluctuations, when a locally ordered region reaches a critical size for which the loss in its surface free energy is compensated by the gain in its bulk free energy.

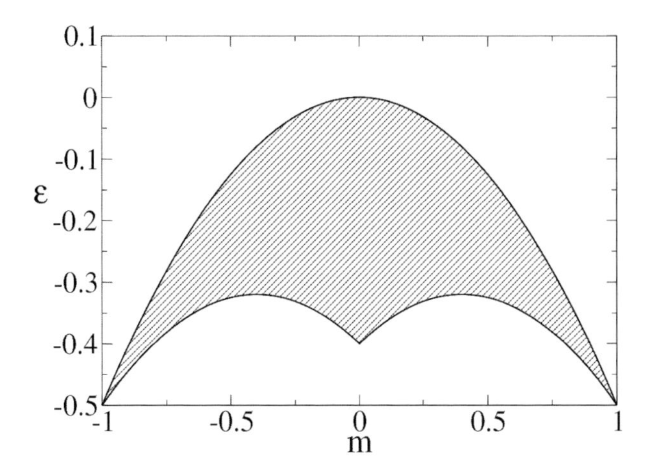

Fig. 1.5 Accessible region in the (m, ϵ) plane (shaded area) of the Hamiltonian (1.1) with $K = -0.4$. At low energies, the accessible domain is composed of two disjoint magnetization intervals, and at intermediate energies three such intervals exist, yielding ergodicity breaking. At higher energies the three intervals join together and ergodicity is restored.

This critical size is independent of the system size, leading to a finite relaxation time. Once the domain structure is formed, it exhibits a coarsening process in which the domains grow in size while their number is reduced. This process, which is typically slow, eventually leads to the ordered equilibrium state of the system.

This is very different from what happens in systems with strong long-range interactions. Here the initial relaxation from a thermodynamically unstable state need not be fast, and it could take place over a time scale which diverges with the system size. The reason is that in the case of strong long-range interactions one cannot define a critical size of an ordered domain, since the bulk and surface energies of a domain are of the same order. It is thus of great interest to study relaxation processes in systems with long-range interactions and to explore the types of behavior which might be encountered. In principle, the relaxation process may depend on the nature and symmetry of the order parameter; say, whether it is discrete, Ising-like, or one with a continuous symmetry such as in the XY model. It may also depend on the dynamical process, whether it is stochastic or deterministic. In this subsection we briefly review some recent results obtained in studies of the dynamics of some models with strong long-range interactions.

We start by considering the Ising model with long- and short-range interactions defined in Section 1.2.2. The relaxation processes in this model have recently been studied (Mukamel *et al.* 2005). Since Ising models do not have intrinsic dynamics, the common dynamics that one uses in studying them is Monte Carlo (MC) dynamics, which simulates the stochastic coupling of the model to a thermal bath. If one is interested in studying the dynamics of an isolated system, one has to resort to the microcanonical MC algorithm developed by Creutz (1983) some time ago. According to this algorithm, a demon with energy $E_d \geq 0$ is allowed to exchange energy with the system. One starts with a system with energy E and a demon with energy $E_d = 0$. The dynamics proceeds by selecting a spin at random and attempting to flip it. If, as a result of the flip, the energy of the system is reduced, the flip is carried out and the excess energy is transferred to the demon. On the other hand, if the energy of the system increases as a result of the attempted flip, the energy needed is taken from the demon and the move is accepted. If the demon does not have the necessary energy, the move is rejected. After a sufficiently long time and for a large system size N, the demon's energy will be distributed according to the Boltzmann distribution $\exp(-E_d/k_BT)$, where T is the temperature of the system with energy E. Thus, by measuring the energy distribution of the demon, one obtains the caloric curve of the system. Note that as long as the entropy of the system is an increasing function of its energy, the temperature is positive and the average energy of the demon is finite. The demon's energy is thus negligibly small compared with the energy of the system, which scales with its size. The energy of the system at any given time is $E - E_D$, and it exhibits fluctuations of finite width at energies just below E.

In applying microcanonical MC dynamics to models with long-range interactions, one should note that the Boltzmann expression for the energy distribution of the demon is valid only in the large-N limit. To next order in N one has

$$P(E_D) \sim \exp\left(-E_D/T - E_D^2/2C_VT^2\right), \tag{1.14}$$

where $C_V = O(N)$ is the system's specific heat. In systems with short-range interactions, the specific heat is nonnegative, and thus the next to leading term in the distribution function is a stabilizing factor which may be neglected for large N. On the other hand, in systems with long-range interactions, C_V may be negative in some regions of the phase diagram and, on the face of it, the next to leading term may destabilize the distribution function. However, the next to leading term is small, of order $O(1/N)$, and it is straightforward to argue that as long as the entropy is an increasing function of the energy, the next to leading term does not destabilize the distribution. The Boltzmann distribution for the energy of the demon is thus valid for large N.

Using the microcanonical MC algorithm, the dynamics of the model (1.1) has been studied in detail (Mukamel *et al.* 2005). Breaking of ergodicity in the region of the (K, ϵ) plane where it is expected to take place has been observed.

Microcanonical MC dynamics has also been applied to study the relaxation process of thermodynamically unstable states. It has been found that, starting with a zero-magnetization state at energies where this state is a local minimum of the entropy, the model relaxes to the equilibrium, magnetically ordered state on a time scale which diverges with the system size as $\ln N$. The divergence of the relaxation time is a direct result of the long-range interactions in the model.

The logarithmic divergence of the relaxation time may be understood by considering the Langevin equation which corresponds to the dynamical process. The equation for the magnetization m is

$$\frac{\partial m}{\partial t} = \frac{\partial s}{\partial m} + \xi(t)\,, \quad \langle \xi(t)\xi(t')\rangle = D\delta(t-t')\,, \tag{1.15}$$

where $\xi(t)$ is the usual white-noise term. The diffusion constant D scales as $D \sim 1/N$. This can be easily seen by considering the noninteracting case, in which the magnetization evolves by pure diffusion, where the diffusion constant is known to scale in this form. Since we are interested in the case of a thermodynamically unstable $m = 0$ state, which corresponds to a local minimum of the entropy, we may, for simplicity, consider an entropy function of the form

$$s(m) = am^2 - bm^4 \,, \tag{1.16}$$

with a and b nonnegative parameters. In order to analyze the relaxation process, we consider the corresponding Fokker–Planck equation for the probability distribution $P(m,t)$ of the magnetization at time t. It takes the form

$$\frac{\partial P(m,t)}{\partial t} = D\frac{\partial^2 P(m,t)}{\partial m^2} - \frac{\partial}{\partial m}\left(\frac{\partial s}{\partial m}P(m,t)\right), \tag{1.17}$$

This equation can be viewed as describing the motion of a particle whose coordinate, m, carries out an overdamped motion in a potential $-s(m)$ at temperature $T = D$. In order to probe the relaxation process from the $m = 0$ state, it is sufficient to consider the entropy (1.16) with $b = 0$. With the initial condition for the probability distribution $P(m,0) = \delta(m)$, the large-time asymptotic distribution is found to be (Risken 1996)

$$P(m,t) \sim \exp\left[-\frac{ae^{-at}m^2}{D}\right]. \tag{1.18}$$

This is a Gaussian distribution whose width grows with time. Thus, the relaxation time from the unstable state, τ_{us}, which corresponds to the width reaching a value of $O(1)$, satisfies

$$\tau_{us} \sim -\ln D \sim \ln N \,. \tag{1.19}$$

The logarithmic divergence with N of the relaxation time seems to be independent of the nature of the dynamics. Similar behavior has been found when the model (1.1) has been studied within a Metropolis-type canonical dynamics at fixed temperature (Mukamel *et al.* 2005).

The relaxation process from a metastable state (rather than an unstable state as discussed above) has been studied rather extensively in the past. Here the entropy has a local maximum at $m = 0$, while the global maximum is obtained at some $m \neq 0$. As one would naively expect, the relaxation time from the metastable $m = 0$ state, τ_{ms}, is found to grow exponentially with N (Mukamel *et al.* 2005):

$$\tau_{ms} \sim e^{N\Delta s} \,. \tag{1.20}$$

The entropy barrier corresponding to the nonmagnetic state, Δs, is the difference in entropy between that of the $m = 0$ state and the entropy at the local minimum separating it from the stable equilibrium state. Such exponentially long relaxation times are expected to occur independently of the nature of the order parameter or the type of dynamics (whether it is stochastic or deterministic). This has been found in the past in numerous studies of canonical Metropolis-type dynamics of the Ising model with mean-field interactions (Griffiths *et al.* 1966), in deterministic dynamics of the XY model (Antoni *et al.* 2004), and in models of gravitational systems (Chavanis and Rieutord 2003; Chavanis 2005).

A different, rather intriguing type of relaxation process has been found in studies of the Hamiltonian dynamics of the XY model with mean-field interactions (Antoni and Ruffo 1995; Latora *et al.* 1998, 1999; Yamaguchi 2003; Yamaguchi *et al.* 2004). This model has been termed the Hamiltonian mean-field (HMF) model. In this model, some nonequilibrium quasi-stationary states have been identified, whose relaxation time grows as a power of the system size N for some energy interval. These nonequilibrium quasi-stationary states (which become steady states in the thermodynamic limit) exhibit some interesting properties such as anomalous diffusion, which have been extensively studied (Latora *et al.* 1999; Yamaguchi 2003; Yamaguchi *et al.* 2004; Bouchet and Dauxois 2005). In other energy intervals the relaxation process has been found to be much faster, with a relaxation time which grows as $\ln N$ (Jain *et al.* 2007). In what follows, we briefly outline the main results obtained for the HMF model and for some generalizations of it.

The HMF model is defined on a lattice with each site occupied by an XY spin of unit length. The Hamiltonian takes the form

$$H = \sum_{i=1}^{N} \frac{p_i^2}{2} + \frac{1}{2N} \sum_{i,j=1}^{N} \left[1 - \cos(\theta_i - \theta_j)\right], \tag{1.21}$$

where θ_i and p_i are the phase and momentum, respectively, of the ith particle. In this model, the interaction is mean-field-like. The model exhibits a continuous transition at a critical energy $\epsilon_c = 3/4$ from a paramagnetic state at high energies to a ferromagnetic state at low energies. Within the Hamiltonian dynamics, the equations of motion of the dynamical variables are

$$\frac{d\theta_i}{dt} = p_i \,, \qquad \frac{dp_i}{dt} = -m_x \sin\theta_i + m_y \cos\theta_i \,, \tag{1.22}$$

where m_x and m_y are the components of the magnetization density

$$\vec{m} = \left(\frac{1}{N} \sum_{i=1}^{N} \cos\theta_i, \frac{1}{N} \sum_{i=1}^{N} \sin\theta_i \right). \tag{1.23}$$

The Hamiltonian dynamics obviously conserves both energy and momentum. The typical initial configuration for the nonmagnetic state is taken as the one where the phase variables are uniformly and independently distributed in the interval $\theta_i \in [-\pi, \pi]$. A particularly interesting case is that where the initial distribution of the momenta is uniform in an interval $[-p_0, p_0]$. This has been termed the "waterbag distribution." For such phase and momentum distributions, the initial energy density is given by $\epsilon = p_0^2/6 - 1/2$.

Extensive numerical studies of the relaxation of the nonmagnetic state with the waterbag initial distribution have been carried out. It has been found that in an energy interval just below ϵ_c this state is quasi-stationary, in the sense that the magnetization fluctuates around its initial value for some time τ_{qs} before it switches to the nonvanishing equilibrium value. This characteristic time has been found to scale as (Yamaguchi 2003; Yamaguchi *et al.* 2004)

$$\tau_{qs} \sim N^{\gamma}, \tag{1.24}$$

with $\gamma \simeq 1.7$.

A very useful insight into the dynamics of the HMF model is provided by analyzing the evolution of the probability distribution of the phase and momentum variables, $f(\theta, p, t)$, within the Vlasov equation approach (Yamaguchi *et al.* 2004). It has been found that in the energy interval $\epsilon^* < \epsilon < \epsilon_c$, with $\epsilon^* = 7/12$, the waterbag distribution is linearly stable. It is unstable for $\epsilon < \epsilon^*$. In the latter interval, the following growth law for the magnetization $m = \sqrt{m_x^2 + m_y^2}$ has been found (Jain *et al.* 2007):

$$m(t) \sim \frac{1}{\sqrt{N}} e^{\Omega t}, \tag{1.25}$$

where

$$\Omega = \sqrt{6(\epsilon^* - \epsilon)}. \tag{1.26}$$

The robustness of the quasi-stationary state to various perturbations has been explored in a number of studies. The anisotropic HMF model has recently been shown to exhibit similar relaxation processes to the HMF model itself (Jain *et al.* 2007). The anisotropic HMF model is defined by the Hamiltonian

$$H = \sum_{i=1}^{N} \frac{p_i^2}{2} + \frac{1}{2N} \sum_{i,j=1}^{N} [1 - \cos(\theta_i - \theta_j)] - \frac{D}{2N} \left[\sum_{i=1}^{N} \cos\theta_i \right]^2, \tag{1.27}$$

where the anisotropy term, with $D > 0$, represents a global coupling and favors order along the x-direction. The model exhibits a transition from a magnetically disordered to a magnetically ordered state along the x-direction at a critical energy $\epsilon_c = (3+D)/4$. An analysis of the Vlasov equation corresponding to this model shows that, as in the isotropic case, the waterbag initial condition is stable for $\epsilon^* < \epsilon < \epsilon_c$, where $\epsilon^* = (7 + D)/12$. In this energy interval, a quasi-stationary state has been observed numerically, with a power-law behavior (1.24) of the relaxation time. The exponent γ does not seem to change with the anisotropy parameter. Logarithmic growth in N of the relaxation time is found for $\epsilon < \epsilon^*$. A model with a local, on-site anisotropy term has also been analyzed along the same lines (Jain *et al.* 2007). This model is defined by the Hamiltonian

$$H = \frac{1}{2} \sum_{i=1}^{N} p_i^2 + \frac{1}{2N} \sum_{i,j=1}^{N} (1 - \cos(\theta_i - \theta_j)) + W \sum_{i=1}^{N} \cos^2\theta_i. \tag{1.28}$$

Here, too, both types of behavior have been found.

Other extensions of the HMF model include the addition of short-range, nearest-neighbor coupling to the Hamiltonian (Campa *et al.* 2006), and coupling of the HMF model to a thermal bath, making the dynamics stochastic (Baldovin and Orlandini

2006). In both cases, quasi-stationarity is observed with a power-law growth of the relaxation time (1.24), with an exponent γ which seems to vary with the interaction parameters of the models.

1.3 Weak long-range interactions

In this section we consider weak long-range interactions, where $0 < \sigma < \sigma_c(d)$. Systems with such interactions are additive and thus the special features found for strong long-range interactions do not occur. However, as discussed in the Introduction, owing to the long-range correlations which are built up in the vicinity of phase transitions, weak long-range interactions are expected to affect the thermodynamic properties of such systems close to the transition. Systems with weak long-range interactions have been extensively studied over the last four decades. Much is known about the collective behavior of these systems, the mechanism by which long-range order is induced in low dimensions, and their critical behavior at continuous phase transitions. In this section we discuss two features of these interactions: long-range order in 1D models, and the upper critical dimension above which the critical exponents of a second-order phase transition are given by the Landau, or mean-field, exponents.

1.3.1 Long-range order in a one-dimensional Ising model

Over 70 years ago Peierls (Peierls 1923, 1935) and Landau (Landau 1937; Landau and Lifshitz 1969) concluded that long-range order (or spontaneous symmetry breaking) does not take place in 1D systems with short-range interactions at finite temperatures. This may be easily argued by considering the 1D Ising model with nearest-neighbor interactions

$$H = -J \sum_{i=1}^{N-1} S_i S_{i+1} \,, \tag{1.29}$$

where $J > 0$ is a ferromagnetic coupling. The ground state of the model is ferromagnetic with all spins parallel, say in the up direction. Consider now an excitation where all spins in a segment of length n are flipped down. The energy cost of this excitation is $4J$, and is independent of the length n. On the other hand, the entropy of this excitation is $\ln N$, as the segment can be located at any point on the lattice. The free-energy cost of the excitation is thus

$$\Delta F = 4J - T \ln N \,, \tag{1.30}$$

which for sufficiently large N is negative at any given temperature $T > 0$. Thus at any finite temperature and in the thermodynamic limit, more excitations with arbitrarily large n are generated and the ferromagnetic long-range order of the $T = 0$ ground state is destroyed.

The nonexistence of phase transitions in 1D models with short-range interactions can also be demonstrated by considering the transfer matrix of the model. Since, owing to the short-range nature of the interactions the matrix is of finite order, and since all the matrix elements are positive, the Perron–Frobenius theorem guarantees that its largest eigenvalue is positive and nondegenerate (Bellman 1970; Ninio 1976). However, for a transition to take place, the largest eigenvalue has to become degenerate. Thus no transition takes place.

The argument presented above for the 1D case does not apply for the Ising model in higher dimensions. The reason is that the energy cost of a flipped droplet of linear size R scales as the surface area of the droplet, R^{d-1}. Generating large droplets is thus energetically costly, and long-range order can be maintained at sufficiently low temperatures. From the dynamical point of view, the increase of the energy cost with the droplet size means that there is a driving force on the droplet to shrink, and thus while droplets are spontaneously generated at finite temperatures, they tend to decrease in size with time. At low temperatures, where the rate at which droplets are generated is small, droplets do not have a chance to join together and flip the magnetization of the initial ground state before they shrink and disappear. Thus long-range order is preserved in the long-time limit. This is in contrast with the 1D case, where the energy cost of a droplet is independent of its size, and there is no driving force on a droplet to shrink.

In 1969, Dyson introduced an Ising model with a pairwise coupling which decreases algebraically with the distance between the spins (Dyson 1969*a*,*b*). He demonstrated that, depending on the power law of the coupling, the model may exhibit a phase transition and long-range order. The Hamiltonian of the Dyson model is

$$H = -\sum_{i,j} J(j) S_i S_{i+j} \,, \tag{1.31}$$

with

$$J(j) = \frac{J}{j^{1+\sigma}} \,, \tag{1.32}$$

where $J > 0$ is a constant. It has been shown that for $0 < \sigma < 1$, the model exhibits long-range order.

To argue for the existence of a phase transition in this model, we apply the argument given above for the case of short-range interactions to the Hamiltonian (1.31). To this end, we take the ferromagnetic ground state of the model and consider the excitation energy of a state in which, say, the leftmost n spins are flipped. The energy of this excitation is

$$E = 2J \sum_{k=1}^{n} \sum_{j=l-k+1}^{N-k} \frac{1}{j^{1+\sigma}} \,. \tag{1.33}$$

In order to estimate this energy, we replace the sums by integrals:

$$E \sim 2J \int_1^n dy \int_{n-y+1}^{N-y} dx \frac{1}{x^{1+\sigma}} \,. \tag{1.34}$$

The integrals may be readily evaluated to yield, in the limit $n \ll N$,

$$E \sim -\frac{2J}{\sigma} n N^{-\sigma} + \frac{2J}{\sigma(1-\sigma)} (n^{1-\sigma} - 1) \,. \tag{1.35}$$

For $0 < \sigma < 1$, the first term in eqn (1.35) vanishes in the thermodynamic limit and the excitation energy increases with the length of the droplet as $n^{1-\sigma}$. Therefore, large droplets tend to shrink in size. This is similar to the behavior of droplets in models with short-range interactions in dimensions higher than one. Thus, at low enough temperatures, the model is expected to exhibit spontaneous symmetry breaking. On the other hand, for $\sigma > 1$, the energy of a droplet is bounded, approaching $2J/\sigma(\sigma-1)$ in the large-n limit. As in the case of the short-range model, no spontaneous symmetry breaking is expected to take place here. It is interesting to note that for the case of strong long-range interactions, namely for $\sigma < 0$, the excitation energy increases with the system size as $N^{-\sigma}$ and the energy is superextensive.

The Dyson model has been a subject of extensive studies over the years. A particular point of interest is its behavior in the borderline case $\sigma = 1$, where logarithmic corrections to the power-law decay are significant. Also, as will be discussed in the following subsection, the critical exponents of the model are expected to be mean-field-like for $0 < \sigma < 1/2$ (see, for example, Aizenman and Fernandez (1988), Luijten and Blöte (1997), and Monroe (1998)).

1.3.2 Upper critical dimension

Perhaps the simplest approach for studying phase transitions in a given system is provided by the Landau theory (Landau and Lifshitz 1969). In this theory, one first identifies the order parameter of the transition, say the local magnetization m, in the case of a magnetic transition. One then uses the symmetry properties of the order parameter, expands the free energy in powers of the order parameter, and determines its equilibrium value by minimizing the free energy. In the case of a single-component, Ising-like magnetic transition, the free energy per unit volume f takes the form

$$f = \frac{1}{2}tm^2 + \frac{1}{4}um^4 + O(m^6)\,, \tag{1.36}$$

where t and $u > 0$ are phenomenological parameters. The fact that only even powers of m appear in this expansion is a result of the up–down symmetry of the magnetic order parameter. The equilibrium magnetization m is found by minimizing this free energy. This theory yields a phase transition at $t = 0$, with $m = 0$ for $t > 0$ and $m = \sqrt{-t/u}$ for $t < 0$. Thus the parameter t may be taken as temperature-dependent, with $t \propto (T - T_c)/T_c$ close to the critical temperature T_c. Below the transition, the order parameter grows as $m \propto (-t)^{\beta}$, where the order parameter critical exponent is $\beta = 1/2$. Similarly, one can obtain the other critical exponents associated with the transition. For example, the free energy per unit volume, f, of the model (1.36) is 0 for $t > 0$ and $-t^2/4u$ for $t < 0$. Thus the specific heat per unit volume, $C = -T\partial^2 f/\partial T^2$, exhibits a discontinuity at the transition. The critical exponent α associated with the specific-heat singularity, $C \propto t^{-\alpha}$, is thus $\alpha = 0$ within the Landau theory. Other critical exponents, corresponding to other thermodynamic quantities such as the magnetic susceptibility, correlation function, etc., can be easily calculated in a similar fashion. Within the Landau theory, the coefficients in the free energy (1.36) are taken as phenomenological parameters. For any given microscopic model, these coefficients may be calculated using the mean-field theory, in which fluctuations of the order parameter are neglected.

Theories which neglect fluctuations of the order parameter usually yield the exact free energy of the model only in the limit of infinite dimension. However, one can show that the mean-field theory, or the Landau theory, yields the correct critical exponents above a critical dimension, which for systems with short-range interactions is $d_c = 4$. This dimension is referred to as the upper critical dimension of the model. For $d < d_c$, the critical exponents become d-dependent, and the fluctuations of the order parameter need to be properly taken care of. This is usually done, for example, by applying renormalization group techniques. For reviews see, for example, Fisher (1974, 1998). Long-range interactions tend to suppress fluctuations. It is thus expected that long-range interactions should result in a smaller critical dimension, $d_c(\sigma)$, which could be a function of the interaction parameter σ. They can also modify the critical exponents below the critical dimension (see Fisher *et al.* 1972). The case of dipolar interactions is of particular interest. For these anisotropic interactions, the upper critical dimension remains $d_c = 4$; however, the critical exponents at dimensions below 4 are modified by the long-range nature of the interaction (see Aharony and Fisher 1974).

In this section, we consider the upper critical dimension of systems with weak long-range interactions. We first analyze the case of short-range interactions, and then extend the analysis to models with long-range interactions. To this end, one should extend the Landau theory to allow for fluctuations of the order parameter, and examine their behavior close to the transition. A convenient and fruitful starting point for this analysis is provided by constructing the coarse-grained effective Hamiltonian of the system. This Hamiltonian, referred to as the Landau–Ginzburg model, is expressed in terms of the long-wavelength degrees of freedom, and is obtained by averaging over the short-wavelength ones. For any given system, this model can be derived phenomenologically using the symmetry properties of the order parameter involved in the transition. For example, for systems with a single-component, Ising-like order parameter, say the magnetization, the effective Hamiltonian is expressed in terms of the local coarse-grained magnetization $m(\mathbf{r})$. For systems with short-range interactions, the Hamiltonian takes the form

$$\beta H = \int d^d r \left[\frac{1}{2} t m^2 + \frac{1}{4} u m^4 + \frac{1}{2} (\nabla m)^2 \right], \tag{1.37}$$

where t and $u > 0$ are phenomenological parameters, as in the Landau theory, and d is the spatial dimension. This Hamiltonian is obtained as an expansion in the small order parameter m, using the up–down symmetry of the microscopic interactions, noting that short-range local interactions result in a long-wavelength gradient term in the energy, of the form $(\nabla m)^2$. The partition sum Z corresponding to this Hamiltonian is obtained by carrying out a functional integral over all magnetization profiles $m(\mathbf{r})$,

$$Z = \int D[m(\vec{r})] e^{-\beta H}. \tag{1.38}$$

When spatial fluctuations of the order parameter are neglected, the model is reduced to the Landau theory discussed above.

Before demonstrating that the upper critical dimension of the model (1.37) is $d_c = 4$, we consider the model at $t > 0$ and evaluate the fluctuations of the order parameter around its average value $\langle m \rangle = 0$. To this end, we express the effective Hamiltonian in terms of the Fourier modes of the order parameter,

$$m(\mathbf{q}) = \int d^d r \, e^{i\mathbf{q}\cdot\mathbf{r}} m(\mathbf{r}) \,. \tag{1.39}$$

In terms of these modes, one has

$$m(\mathbf{r}) = \frac{1}{V} \sum_{\mathbf{q}} e^{-i\mathbf{q}\cdot\mathbf{r}} m(\mathbf{q}) \,, \tag{1.40}$$

and

$$\int d^d r \, m^2(\mathbf{r}) = \frac{1}{V} \sum_{\mathbf{q}} m(\mathbf{q}) m(-\mathbf{q}) \,. \tag{1.41}$$

Neglecting the fourth-order terms in the effective Hamiltonian (1.37), one is left with a Gaussian model which can be expressed in terms of the Fourier modes as

$$\beta H = \frac{1}{2V} \sum_{\mathbf{q}} (t + q^2) m(\mathbf{q}) m(-\mathbf{q}) \,. \tag{1.42}$$

In the limit $V \to \infty$, the sum can be expressed as an integral

$$\beta H = \frac{1}{2(2\pi)^d} \int d^d q \, (t + q^2) m(\mathbf{q}) m(-\mathbf{q}) \,. \tag{1.43}$$

To calculate the two-point order parameter correlation function, one first evaluates the average amplitudes of the Fourier modes using the Gaussian Hamiltonian (1.42),

$$\langle m(\mathbf{q}) m(\mathbf{q}') \rangle = \frac{V}{t + q^2} \delta_{\mathbf{q}, -\mathbf{q}'} \,. \tag{1.44}$$

The two-point correlation function is then given by

$$\langle m(\mathbf{r}) m(\mathbf{r}') \rangle = \frac{1}{V} \sum_{\mathbf{q}} \frac{1}{t + q^2} e^{-i\mathbf{q}(\mathbf{r}-\mathbf{r}')} \,, \tag{1.45}$$

which in the $V \to \infty$ limit becomes

$$\langle m(\mathbf{r}) m(\mathbf{r}') \rangle = \frac{1}{(2\pi)^d} \int d^d q \, \frac{e^{-i\mathbf{q}(\mathbf{r}-\mathbf{r}')}}{t + q^2} \,. \tag{1.46}$$

If $\mathbf{q}$ is scaled by $\sqrt{t}$, the integral (1.46) implies that the correlation function can be expressed in terms of a correlation length

$$\xi = t^{1/2} \,, \tag{1.47}$$

which diverges at the critical point with an exponent $\nu = 1/2$. It is easy to verify that at distances larger than ξ the correlation function decays exponentially with the distance as $e^{-|\mathbf{r}-\mathbf{r}'|/\xi}$ with a subleading power-law correction.

The Landau–Ginzburg effective Hamiltonian (1.37) may be used to calculate the local fluctuations of the order parameter on a coarse-grained scale of linear size ξ. From eqns (1.41) and (1.44) it follows that

$$\langle m^2(\mathbf{r})\rangle = \frac{1}{V}\sum_{\mathbf{q}} \frac{1}{t+q^2} \,. \tag{1.48}$$

Taking now the limit $V \to \infty$ and integrating over modes with a wavelength bigger than the correlation length ξ, namely $0 < q < \xi^{-1} = \sqrt{t}$, one finds

$$\langle m^2(\mathbf{r})\rangle = \int_{q<\sqrt{t}} \frac{d^d q}{(2\pi)^d}\frac{1}{t+q^2} \ \propto\ t^{d/2-1} \,. \tag{1.49}$$

In the Landau theory, the fluctuations of the order parameter are neglected. To check the validity of this assumption or, alternatively, to check at which dimensions this assumption is valid, we consider the fluctuations below the critical point ($t < 0$). Let

$$m_0 = \sqrt{-\frac{t}{u}} \tag{1.50}$$

be the order parameter at $t < 0$. To calculate the fluctuations of the order parameter, we consider small local deviations $m(\mathbf{r}) = m_0 + \delta m(\mathbf{r})$ around the average value. To second-order in $\delta m(\mathbf{r})$, the effective Landau–Ginzburg Hamiltonian is

$$\beta H = \frac{1}{2}(t + 3um_0^2)(\delta m(\mathbf{r}))^2 + \frac{1}{2}(\nabla \delta m(\mathbf{r}))^2 \,, \tag{1.51}$$

which, after using eqn (1.50), becomes

$$\beta H = |t|(\delta m(\mathbf{r}))^2 + \frac{1}{2}(\nabla \delta m(\mathbf{r}))^2 \,. \tag{1.52}$$

Thus the fluctuations of the order parameter around their average value below the transition are controlled by an effective Hamiltonian similar to that for the fluctuations of the order parameter above T_c. It therefore follows from eqn (1.49) that

$$\langle \delta m^2(\mathbf{r})\rangle \ \propto\ |t|^{d/2-1} \,. \tag{1.53}$$

For the Landau theory to be self-consistent near the transition, one requires that the fluctuations of the order parameter are negligibly small compared with the order parameter, namely

$$\langle \delta m^2(\mathbf{r})\rangle \ \ll\ m_0^2 \,. \tag{1.54}$$

This amounts to

$$|t|^{d/2-1} \ \ll\ |t| \,, \tag{1.55}$$

which is satisfied for small t as long as

$$d > d_c = 4 \,. \tag{1.56}$$

This analysis suggests that in systems with short-range interactions, for which the Landau–Ginzburg effective Hamiltonian (1.37) applies, the fluctuations of the order parameter are negligibly small, and the critical exponents corresponding to the transition are those given by the Landau theory. In dimensions less than 4, the inequality (1.55) can be satisfied only away from the critical point $|t| > |t_G|$, where t_G defines the Ginzburg temperature interval. This interval depends on the amplitudes of the power laws appearing in eqn (1.55), which can vary from one system to another. This is known as the Ginzburg criterion (Ginzburg 1960; Hohenberg 1968; Als-Nielsen and Birgeneau 1977). According to this criterion, the true critical exponents of a system in, say, dimension $d = 3$ can be observed only in a temperature interval $|t| < |t_G|$. Outside this interval, one should expect a crossover of the critical exponents to those of the Landau theory.

So far, we have discussed the upper critical dimension of a generic critical point of a system with short-range interactions. Let us now examine how this analysis is modified when one considers weak long-range interactions. For such systems the Landau–Ginzburg effective Hamiltonian takes the form

$$\beta H = \int d^d r \left[\frac{1}{2} t m^2 + \frac{1}{4} u m^4\right] + \int d^d r\, d^d r'\, m(\mathbf{r}) m(\mathbf{r}') \frac{1}{|\mathbf{r} - \mathbf{r}'|^{d+\sigma}}\,, \tag{1.57}$$

where the second integral yields the contribution of the long-range interaction to the energy. In terms of the Fourier components of the order parameter, this integral may be expressed as

$$\frac{1}{V} \sum_{\mathbf{q}} m(\mathbf{q}) m(-\mathbf{q}) \int d^d R \frac{e^{-i\mathbf{q}\cdot\mathbf{R}}}{R^{d+\sigma}}\,. \tag{1.58}$$

To leading orders in q, the Fourier transform of the long-range potential is of the form $(a + bq^\sigma + cq^2)/2$, where a, b, and c are constants. Note, though, that for integer values of σ a logarithmic correction to this form is present. For $\sigma = 2$, the form becomes $a + bq^2 \ln q$. These logarithmic corrections do not affect the considerations which will be presented below. Thus, to second-order in m, the Landau–Ginzburg effective Hamiltonian is

$$\beta H = \frac{1}{2V} \sum_{\mathbf{q}} (\bar{t} + bq^\sigma + cq^2) m(\mathbf{q}) m(-\mathbf{q})\,, \tag{1.59}$$

where $\bar{t} = t + a$.

It is straightforward to repeat the above analysis for the upper critical dimension in systems with short-range interactions and extend it to the case of weak long-range interactions. For $\sigma > 2$, the q^σ term in eqn (1.59) is dominated by the q^2 term and may thus be neglected. One is then back to the model corresponding to short-range interactions and the upper critical dimension is $d_c = 4$. On the other hand, for $0 < \sigma < 2$, the dominant term is q^σ. The correlation length in this case diverges as

$$\xi \propto |\bar{t}|^{-1/\sigma}\,. \tag{1.60}$$

Below the transition, the order parameter fluctuations around the mean-field value, $m_0 = (-\bar{t}/u)^{1/2}$, satisfy

$$\langle (\delta m(\mathbf{r}))^2 \rangle = \int_{q<1/\xi} \frac{d^d q}{(2\pi)^d} \frac{1}{2|\bar{t}| + q^\sigma} \propto |\bar{t}|^{d/\sigma - 1} . \tag{1.61}$$

Requiring that the fluctuations are much smaller than the order parameter (1.54),

$$|\bar{t}|^{d/\sigma-1} \ll |\bar{t}| , \tag{1.62}$$

one concludes that fluctuations may be neglected as long as

$$d > d_c = 2\sigma . \tag{1.63}$$

Alternatively, this implies that in $d < 4$ dimensions there exists a critical σ,

$$\sigma_c(d) = \frac{d}{2} , \tag{1.64}$$

such that for $0 < \sigma < \sigma_c(d)$ the critical exponents are mean-field-like. For $\sigma_c(d) < \sigma < 2$, the critical exponents are affected by the long-range nature of the interaction

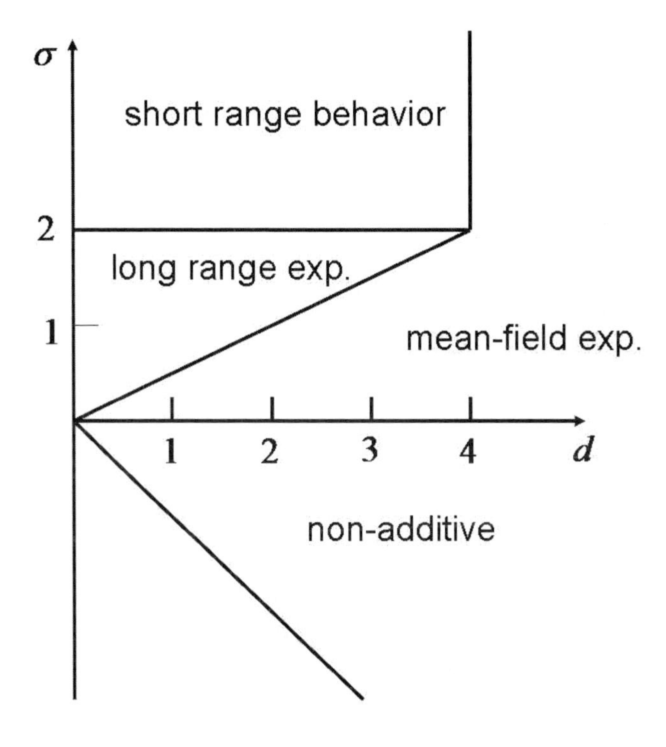

Fig. 1.6 The (d, σ) phase diagram, where the types of critical behavior in various regions in this plane are indicated. The system is nonadditive with strong long-range interactions for $d \leq \sigma \leq 0$. For $\sigma > 0$, the critical exponents can be either mean-field, short-range-like, or characteristic of long-range interactions depending on σ and d. See Sak (1973) for a detailed discussion of the transition line between long-range and short-range behavior. Note that in the case $\sigma > 2$, no phase transition takes place for $d \leq 1$.

and they become σ-dependent; see, for example, Fisher *et al.* (1972). For $\sigma > 2$, the critical exponents become those of short-range interactions. Applied to the Dyson model discussed in Section 1.3.1, these results suggest that in 1D the critical exponents of the model are mean-field-like for $0 < \sigma < 1/2$, and become of long-range type for $1/2 < \sigma < 1$. For $\sigma > 1$, the interaction is effectively short-range and no transition takes place.

The results of this analysis are presented in Fig. 1.6, where the types of critical behavior in various regions of the (d, σ) plane are indicated.

1.4 Long-range correlations in nonequilibrium driven systems

As discussed in the Introduction, steady states of driven systems are expected to exhibit long-range correlations when the dynamics involves one or more conserved variables. This takes place even though the dynamics is local, with transition rates which depend only on the local microscopic configuration of the dynamical variable. Such long-range correlations have been shown to lead to phase transitions and long-range order in a number of one-dimensional models. For reviews on steady-state properties of driven models see, for example, Schmittmann and Zia (1995), Schütz (2001), Derrida (2007), and Mukamel (2000). Features which are characteristic of strong long-range interactions, such as inequivalence of ensembles, have been reported in some cases (Grosskinsky and Schütz 2008). In this section we discuss a particular model, the *ABC* model, introduced by Evans and coworkers (Evans *et al.* 1998*a*,*b*), which exhibits spontaneous symmetry breaking and for which such correlations can be explicitly demonstrated. Moreover, for particular parameters defining this model, its dynamics obeys detailed balance. For this choice of the parameters the steady state becomes an equilibrium state, which can be expressed in terms of an effective Hamiltonian. This Hamiltonian has been explicitly expressed in terms of the dynamical variables of the model, and has been shown to display strong long-range interactions.

The model is defined on a 1D lattice of length N with periodic boundary conditions. Each site is occupied by either an A, B, or C particle. The evolution is governed by random sequential dynamics defined as follows: at each time step, two neighboring sites are chosen randomly and the particles on these sites are exchanged according to the following rates:

$$\begin{aligned} AB &\underset{1}{\overset{q}{\rightleftarrows}} BA\,, \\ BC &\underset{1}{\overset{q}{\rightleftarrows}} CB\,, \\ CA &\underset{1}{\overset{q}{\rightleftarrows}} AC\,. \end{aligned} \tag{1.65}$$

The rates are cyclic in A, B, and C and conserve the numbers of particles of each type N_A, N_B, and N_C, respectively.

For $q = 1$, the particles undergo symmetric diffusion and the system is disordered. This is expected since this is an equilibrium steady state. However, for $q \neq 1$, the particle exchange rates are biased. We will show that in this case the system evolves into a phase-separated state in the thermodynamic limit.

To be specific we take $q < 1$, although the analysis may be extended trivially to any $q \neq 1$. In this case the bias drives, say, an A particle to move to the left inside a B

domain, and to the right inside a C domain. Therefore, starting with an arbitrary initial configuration, the system reaches, after a relatively short transient time, a state of the type $\ldots AABBCCAAAB\ldots$ in which A, B, and C domains are located to the right of C, A, and B domains, respectively. Owing to the bias q, the domain walls $\ldots AB\ldots$, $\ldots BC\ldots$, and $\ldots CA\ldots$ are stable, and configurations of this type are long-lived. In fact, the domains in these configurations diffuse into each other and coarsen on a time scale of the order of q^{-l}, where l is a typical domain size in the system. This leads to a growth of the typical domain size as $(\ln t)/|\ln q|$. Eventually the system phase-separates into three domains of the different species of the form $A\ldots AB\ldots BC\ldots C$. A finite system does not stay in such a state indefinitely. For example, the A domain breaks up into smaller domains in a time of order $q^{-\min\{N_B,N_C\}}$. In the thermodynamic limit, however, when the density of each type of particle is nonvanishing, the time scale for the breakup of extensive domains diverges and we expect the system to phase-separate. Generically, the system supports particle currents in the steady state. This can be seen by considering, say, the A domain in the phase-separated state. The rates at which an A particle traverses a B (or C) domain to the right (or left) is of the order of q^{N_B} (or q^{N_C}). The net current is then of the order of $q^{N_B} - q^{N_C}$, vanishing exponentially with N. This simple argument suggests that for the special case $N_A = N_B = N_C$, the current is zero for any system size.

The special case of equal densities $N_A = N_B = N_C$ provides a very interesting insight into the mechanism leading to phase separation. We thus consider it in some detail. Examining the dynamics for these densities, we find that it obeys *detailed balance* with respect to some distribution function. Thus, in this case, the model is in fact in thermal equilibrium. It turns out, however, that although the dynamics of the model is *local*, the effective Hamiltonian corresponding to the steady-state distribution has *long-range interactions*, and may thus lead to phase separation. This particular mechanism is specific to equal densities. However, the dynamical argument for phase separation given above is more general, and is valid for unequal densities as well.

In order to specify the distribution function for equal densities, we define a local occupation variable $\{X_i\} = \{A_i, B_i, C_i\}$, where A_i, B_i, and C_i are equal to one if site i is occupied by particle A, B, or C, respectively, and zero otherwise. The probability of finding the system in a configuration $\{X_i\}$ is given by

$$W_N(\{X_i\}) = Z_N^{-1} q^{H(\{X_i\})}, \tag{1.66}$$

where H is the Hamiltonian

$$H(\{X_i\}) = \sum_{i=1}^{N-1} \sum_{k=i+1}^{N} (C_i B_k + A_i C_k + B_i A_k) - (N/3)^2, \tag{1.67}$$

and the partition sum is given by $Z_N = \sum q^{H(\{X_i\})}$. In this Hamiltonian, the site $i = 1$ can be arbitrarily chosen as one of the sites on the ring. It is easy to see that the Hamiltonian does not depend on this choice, and the Hamiltonian is translationally invariant as expected. The interactions in this Hamiltonian are strong long-range interactions, where the the strength of the interaction between two sites is independent of their distance. This corresponds to $\sigma = -1$ in the notation used in these Notes. The

Hamiltonian is thus superextensive, and the energy of macroscopic excitations scales as N^2.

In order to verify that the dynamics (1.65) obeys detailed balance with respect to the distribution function (1.66, 1.67), it is useful to note that the energy of a given configuration may be evaluated in an alternate way. Consider the fully phase-separated state

$$A \ldots AB \ldots BC \ldots C\,. \tag{1.68}$$

The energy of this configuration is $E = 0$, and this configuration and its translationally related configurations constitute the N-fold degenerate ground state of the system. We now note that nearest-neighbor (nn) exchanges $AB \to BA$, $BC \to CB$, and $CA \to AC$ cost one unit of energy each, while the reverse exchanges result in an energy gain of one unit. The energy of an arbitrary configuration may thus be evaluated by starting with the ground state and performing nn exchanges until the specified configuration is reached, keeping track of the energy changes at each step of the way. This procedure for obtaining the energy is self-consistent only when the densities of the three species are equal. To examine the self-consistency of this procedure consider, for example, the ground state (1.68), and move the leftmost particle A to the right by a series of nn exchanges until it reaches the right end of the system. Owing to translational invariance, the resulting configuration should have the same energy as the configuration (1.68), namely $E = 0$. On the other hand, the energy of the resulting configuration is $E = N_B - N_C$, since any exchange with a B particle yields a cost of one unit while an exchange with a C particle yields a gain of one unit of energy. Therefore, for self-consistency, the two densities N_B and N_C have to be equal and, similarly, they have to be equal to N_A.

The Hamiltonian (1.67) may be used to calculate steady-state averages corresponding to the dynamics (1.65). We start with an outline of the calculation of the free energy. Consider a ground state of the system (1.68). The low-lying excitations around this ground state are obtained by exchanging nn pairs of particles around each of the three domain walls. Let us first examine excitations which are localized around one of the walls, say AB. An excitation can be formed by one or more B particles moving into the A domain (or, equivalently, A particles moving into the B domain). A moving B particle may be considered as a walker. The energy of the system increases linearly with the distance traveled by the walker inside the A domain. An excitation of energy m at the AB boundary is formed by j walkers covering a total distance of m. Hence, the total number of states of energy m at the AB boundary is equal to the number of ways, $P(m)$, of partitioning an integer m into a sum of (positive) integers. This and related functions have been extensively studied in the mathematical literature over many years. Although no explicit general formula for $P(m)$ is available, its asymptotic form for large m is known (Andrews 1976):

$$P(m) \simeq \frac{1}{4m\sqrt{3}} \exp\left(\pi(2/3)^{1/2}\, m^{1/2}\right). \tag{1.69}$$

Also, a well-known result attributed to Euler yields the generating function

$$Y = \sum_{m=0}^{\infty} q^m P(m) = \frac{1}{(q)_\infty}\,, \tag{1.70}$$

where

$$(q)_\infty = \lim_{n\to\infty}(1-q)(1-q^2)\dots(1-q^n)\,. \tag{1.71}$$

This result may be extended to obtain the partition sum Z_N of the full model. In the limit of large N, the three domain walls basically do not interact. It has been shown that excitations around the different domain boundaries contribute additively to the energy spectrum (Evans *et al.* 1998*b*). As a result, in the thermodynamic limit the partition sum takes the form

$$Z_N = N/[(q)_\infty]^3\,, \tag{1.72}$$

where the multiplicative factor N results from the N-fold degeneracy of the ground state and the cubic power is related to the three independent excitation spectra associated with the three domain walls.

It is of interest to note that the partition sum is linear and not exponential in N, as is the case for systems with short-range interactions, leading to a nonextensive free energy. Since the energetic cost of macroscopic excitations is of order N^2 they are suppressed, and the equilibrium state is determined by the ground state and some local excitations around it.

Whether or not a system has long-range order in the steady state can be found by studying the decay of two-point density correlation functions. For example, the probability of finding an A particle at site i and a B particle at site j is

$$\langle A_i B_j\rangle = \frac{1}{Z_N}\sum_{\{X_k\}} A_i B_j\; q^{H(\{X_k\})}\,, \tag{1.73}$$

where the summation is over all configurations $\{X_k\}$ in which $N_A = N_B = N_C$. Owing to symmetry, many of the correlation functions will be the same, for example $\langle A_i A_j\rangle = \langle B_i B_j\rangle = \langle C_i C_j\rangle$. A sufficient condition for the existence of phase separation is

$$\lim_{r\to\infty}\;\lim_{N\to\infty}\left(\langle A_1 A_r\rangle - \langle A_1\rangle\langle A_r\rangle\right) > 0\,. \tag{1.74}$$

Since $\langle A_i\rangle = 1/3$, we wish to show that $\lim_{r\to\infty}\;\lim_{N\to\infty}\langle A_1 A_r\rangle > 1/9$. In fact it can be shown (Evans *et al.* 1998*b*) that for any given r and for sufficiently large N,

$$\langle A_1 A_r\rangle = 1/3 - O(r/N)\,. \tag{1.75}$$

This result demonstrates not only that there is phase separation, but also that each of the domains is pure. Namely, the probability of finding a particle a large distance inside a domain of particles of another type is vanishingly small in the thermodynamic limit.

The ABC model exhibits phase separation and long-range order as long as $q \neq 1$. The parameter q is in fact a temperature variable in the case of equal densities, with $\beta = -\ln q$ as can be seen from eqn (1.66). Thus the model leads to a phase-separated state at any finite temperature. A very interesting limit is that of $q \to 1$, which amounts to taking the infinite-temperature limit. To probe this limit, Clincy and coworkers (Clincy *et al.* 2003) studied the case $q = e^{-\beta/N}$. This amounts to either scaling the temperature by N or, alternatively, scaling the Hamiltonian (1.67) by $1/N$, as is done in the Kac prescription. It has been shown that in this case the model exhibits a phase transition at inverse temperature $\beta_c = 2\pi\sqrt{3}$, where the system is homogeneous at high temperatures and phase-separated at low temperatures.

The analysis of the ABC model presented in this section indicates that, rather generally, one should expect features which are characteristic of long-range interactions to show up in steady states of driven systems. In the ABC model, for equal densities, where the effective Hamiltonian governing the steady state can be explicitly written, the interactions are found to be of strong long-range (in fact mean-field) nature. By continuity, this is expected to hold even for the nonequal-densities case, where no effective interaction can be written. It is of interest to explore in more detail steady-state properties of driven systems within the framework of systems with long-range interactions outlined above.

1.5 Summary

In these Notes, some properties of systems with long-range interactions have been discussed. Two broad classes can be identified in systems with pairwise interactions which decay as $1/r^{d+\sigma}$ at large distances r: those with $-d \le \sigma \le 0$, which we term "strong" long-range interactions, and those with $0 < \sigma \le \sigma_c(d)$, which are termed "weak" long-range interactions. Here $\sigma_c(d) = 2$ for $d \ge 4$ and $\sigma_c(d) = d/2$ for $d < 4$. Some thermodynamic and dynamical features which are characteristic of these two classes have been discussed and demonstrated in representative models.

Systems with strong LRI are nonadditive and, as a result, they do not share many of the common features of systems with short-range interactions. In particular, the various ensembles need not be equivalent, the microcanonical specific heat can be negative, and a temperature discontinuity can occur as the energy of the system is varied. These systems also display distinct dynamical behavior, with slow relaxation processes where the characteristic relaxation time diverges with the system size. In addition, they are found to exhibit breaking of ergodicity, which is induced by the long-range nature of the interaction. Some general considerations arguing that such phenomena should take place have been presented. Simple models (the Ising model with long- and short-range interactions and the XY model) have been analyzed, where these features are explicitly demonstrated. While the models are mean-field-like with $\sigma = -d$, the characteristic features they display are expected to hold in the broader class of systems with $-d \le \sigma \le 0$.

Systems with weak LRI are additive, and therefore the general statistical-mechanical framework of systems with short-range interactions applies here as well. However, near phase transitions, where long-range correlations are naturally built up, weak long-range interactions are effective in modifying the system's thermodynamic properties. For example, unlike short-range interactions, weak LRI can induce long-range order in one-dimensional systems. They also affect the upper critical dimension of a system, above which the universality class of the transition is that of the mean-field approximation.

Systems driven out of equilibrium tend to exhibit long-range correlations when their dynamics involves conserved variables. Thus such driven systems are expected to display some of the features of equilibrium systems with long-range interactions.

A simple model of three species of particles, the ABC model, has been discussed in these Notes, where its long-range correlations can be explicitly expressed in terms of long-range interactions. The interactions are found to be of strong long-range nature. Thus, exploring features characteristic of long-range interactions in driven systems could lead to useful insight and better understanding of their collective behavior.

Acknowledgments

I thank Ori Hirschberg and A. C. D. van Enter for comments on these Notes. Support from the Israel Science Foundation (ISF) and the Minerva Foundation, with funding from the Federal German Ministry for Education and Research, is gratefully acknowledged.

References

Aharony A. and Fisher M. E. (1974). *Phys. Rev. B*, **8**, 3323.

Aizenman M. and Fernandez A. (1988). *Lett. Math. Phys.*, **16**, 39.

Als-Nielsen J. and Birgeneau R. J. (1977). *Am. J. Phys.*, **45**, 554.

Andrews G. E. (1976). *The Theory of Partitions*, Encyclopedia of Mathematics and its Applications **2**, p. 1, Addison-Wesley.

Antoni M. and Ruffo S. (1995). *Phys. Rev. E*, **52**, 2361.

Antoni M., Ruffo S., and Torcini A. (2004). *Europhys. Lett.*, **66**, 645.

Antonov V. A. (1962). *Vest. Leningrad Univ.*, **7**, 135. Translation in *IAU Symposium/Symp.-Int. Astron. Union*, **113**, 525 (1995).

Baldovin F. and Orlandini E. (2006). *Phys. Rev. Lett.*, **96**, 240602.

Barré J., Mukamel D., and Ruffo S. (2001). *Phys. Rev. Lett.*, **87**, 030601.

Barré J., Mukamel D., and Ruffo S. (2002). "Ensemble inequivalence in mean-field models of magnetism," in *Dynamics and Thermodynamics of Systems with Long-Range Interactions*, edited by Dauxois T., Ruffo S., Arimondo E., and Wilkens M., Lecture Notes in Physics **602**, pp. 45–67, Springer-Verlag.

Barré J., Dauxois T., De Ninno G., Fanelli D., and Ruffo S. (2004). *Phys. Rev. E*, **69**, 045501(R).

Bellman R. (1970). *Introduction to Matrix Analysis*, McGraw-Hill.

Bonner J. C. and Nagle J. F. (1971). *J. Appl. Phys.*, **42**, 1280.

Borgonovi F., Celardo G. L., Maianti M., and Pedersoli E. (2004). *J. Stat. Phys.*, **116**, 1435.

Borgonovi F., Celardo G. L., Musesti A., Trasarti-Battistoni R., and Vachal P. (2006). *Phys. Rev. E*, **73**, 026116.

Bouchet F. and Barré J. (2005). *J. Stat. Phys.*, **118**, 1073.

Bouchet F. and Dauxois T. (2005). *Phys. Rev. E*, **72**, 045103.

Bouchet F., Dauxois T., Mukamel D., and Ruffo S. (2008). *Phys. Rev. E*, **77**, 011125.

Campa A., Giansanti A., Mukamel D., and Ruffo S. (2006). *Physica A*, **365**, 120.

Campa A., Giansanti A., Morigi G., and Sylos Labini F. (Eds.) (2007*a*). *Dynamics and Thermodynamics of Systems with Long Range Interactions: Theory and Experiments*, AIP Conf. Proc. **970**.

Campa A., Khomeriki R., Mukamel D., and Ruffo S. (2007*b*). *Phys. Rev. B*, **76**, 064415.

Chavanis P. H. (2002). “Statistical mechanics of two-dimensional vortices and three-dimensional stellar systems,” in *Dynamics and Thermodynamics of Systems with Long-Range Interactions*, edited by Dauxois T., Ruffo S., Arimondo E., and Wilkens M., Lecture Notes in Physics **602**, pp. 208–289, Springer-Verlag.
Chavanis P. H. (2005). *Astron. Astrophys.*, **432**, 117.
Chavanis P. H. and Rieutord M. (2003). *Astron. Astrophys.*, **412**, 1.
Chomaz P. and Gulminelli F. (2002). “Phase transitions in finite systems,” in *Dynamics and Thermodynamics of Systems with Long-Range Interactions*, edited by Dauxois T., Ruffo S., Arimondo E., and Wilkens M., Lecture Notes in Physics **602**, pp. 68–129, Springer-Verlag.
Clincy M., Derrida B., and Evans M. R. (2003). *Phys. Rev. E*, **67**, 066115.
Creutz M. (1983). *Phys. Rev. Lett.*, **50**, 1411.
Dauxois T., Ruffo S., Arimondo E., and Wilkens M. (Eds.) (2002). *Dynamics and Thermodynamics of Systems with Long-Range Interactions*, Lecture Notes in Physics **602**, Springer-Verlag.
de Buyl P., Mukamel D., and Ruffo S. (2005). *AIP Conf. Proc.*, **800**, 533.
Derrida B. (2007). *J. Stat. Mech.*, P07023.
des Cloizeaux J. and Jannink G. (1990). *Polymers in Solution: Their Modelling and Structure*, Clarendon Press.
Dyson F. J. (1969*a*). *Commun. Math. Phys.*, **12**, 91.
Dyson F. J. (1969*b*). *Commun. Math. Phys.*, **12**, 212.
Evans M. R., Kafri Y., Koduvely H. M., and Mukamel D. (1998*a*). *Phys. Rev. Lett.*, **80**, 425.
Evans M. R., Kafri Y., Koduvely H. M., and Mukamel D. (1998*b*). *Phys. Rev. E*, **58**, 2764.
Fel'dman E. B. (1998). *J. Chem. Phys.*, **108**, 4709.
Fisher M. E. (1974). *Rev. Mod. Phys.*, **46**, 597.
Fisher M. E. (1998). *Rev. Mod. Phys.*, **70**, 653.
Fisher M. E., Ma S. K., and Nickel B. G. (1972). *Phys. Rev. Lett.*, **29**, 917.
Ginzburg V. L. (1960). *Sov. Phys. Solid State*, **2**, 1824.
Griffiths R. B., Weng C. Y., and Langer J. S. (1966). *Phys. Rev.*, **149**, 1.
Gross D. H. E. (2000). *Microcanonical Thermodynamics: Phase Transitions in “Small” Systems*, World Scientific.
Grosskinsky S. and Schütz G. M. (2008). *J. Stat. Phys.*, **132**, 77.
Hahn I. and Kastner M. (2005). *Phys. Rev. E*, **72**, 056134.
Hahn I. and Kastner M. (2006). *Eur. Phys. J. B*, **50**, 311.
Hertel P. and Thirring W. (1971). *Ann. Phys.*, **63**, 520.
Hohenberg P. C. (1968). “Critical phenomena and their bearing on the superconducting transition,” in *Proc. Conf. Fluctuations in Superconductors*, edited by Goree W. S. and Chilton F., p. 305, Stanford Research Institute.
Jain K., Bouchet F., and Mukamel D. (2007). *J. Stat. Mech.*, P11008.
Kac M., Uhlenbeck G. E., and Hemmer P. C. (1963). *J. Math. Phys.*, **4**, 216.
Kardar M. (1983). *Phys. Rev. B*, **28**, 244.
Landau L. D. (1937). *Phys. Z. Sowjet.*, **11**, 26.

Landau L. D. and Lifshitz E. M. (1960). *Course of Theoretical Physics. Vol. 8, Electrodynamics of Continuous Media*, Pergamon.

Landau L. D. and Lifshitz E. M. (1969). *Course of Theoretical Physics. Vol. 5, Statistical Physics*, second edition, Pergamon.

Latora V., Rapisarda A., and Ruffo S. (1998). *Phys. Rev. Lett.*, **80**, 692.

Latora V., Rapisarda A., and Ruffo S. (1999). *Phys. Rev. Lett.*, **83**, 2104.

Luijten E. and Blöte H. W. J. (1997). *Phys. Rev. B*, **56**, 8945.

Lynden-Bell D. (1999). *Physica A*, **263**, 293.

Lynden-Bell D. and Wood R. (1968). *Mon. Not. R. Astron. Soc.*, **138**, 495.

Lynden-Bell R. M. (1995). *Mol. Phys.*, **86**, 1353.

Lynden-Bell R. M. (1996). "Negative specific heat in clusters of atoms," in *Gravitational Dynamics*, edited by Lahav O., Terlevich E., and Terlevich R. J., pp. 35–42, Cambridge University Press.

Misawa T., Yamaji Y., and Imada M. (2006). *J. Phys. Soc. Jpn.*, **75**, 064705.

Monroe J. L. (1998). *J. Math. A: Math. Gen.*, **31**, 9809.

Mukamel D. (2000). *Soft and Fragile Matter: Nonequilibrium Dynamics, Metastability and Flow*, edited by Cates M. E. and Evans M. R., Institute of Physics Publishing.

Mukamel D., Ruffo S., and Schreiber N. (2005). *Phys. Rev. Lett.*, **95**, 240604.

Nagle J. F. (1970). *Phys. Rev. A*, **2**, 2124.

Nicholson D. R. (1992). *Introduction to Plasma Physics*, Krieger.

Ninio F. (1976). *J. Phys. A: Math. Gen.*, **9**, 1281.

Padmanabhan T. (1990). *Phys. Rep.*, **188**, 285.

Peierls R. E. (1923). *Helv. Phys. Acta*, **7**, 81.

Peierls R. E. (1935). *Ann. Inst. Poincaré*, **5**, 177.

Posch H. and Thirring W. (2006). *Phys. Rev. E*, **75**, 051103.

Risken H. (1996). *The Fokker–Planck Equation*, Springer-Verlag, p. 109.

Sak J. (1973). *Phys. Rev. B*, **8**, 281.

Schmittmann B. and Zia R. K. P. (1995). *Phase Transitions and Critical Phenomena*, Vol. 17, Academic Press.

Schütz G. M. (2001). *Phase Transitions and Critical Phenomena*, Vol. 19, Academic Press.

Thirring W. (1970). *Z. Phys.*, **235**, 339.

Yamaguchi Y. Y. (2003). *Phys. Rev. E*, **68**, 066210.

Yamaguchi Y. Y., Barré J., Bouchet F., Dauxois T., and Ruffo S. (2004). *Physica A*, **337**, 36.

2
Six out-of-equilibrium lectures

Jorge Kurchan

PMMH-ESPCI, CNRS UMR 7636, 10 rue Vauquelin, 75005 Paris, France

The purpose of these notes is to introduce a group of subjects in out-of-equilibrium statistical mechanics that have received considerable attention in the last fifteen years or so. They are mostly connected with time reversibility and its relation to entropy, are expressed in terms of large deviations, and involve at some level the notion of timescale separation.

We shall consider systems that contain, in their dynamic rules, an element of noise. There are good reasons for this. On the practical side, a driven system needs to dissipate heat if it is to reach a stationary regime rather than heating up indefinitely. Stochastic systems, where energy is provided by a bath through time-dependent, random forces, are well-studied physical models of heat reservoirs. A second, equally important reason is that, very often, dynamical systems in the presence of noise are much easier to study than purely deterministic ones, because then very subtle ergodicity considerations become trivial. If the aim of ergodic theory is to understand how randomness arises from deterministic constituents, once stochasticity is added "by hand" the question is artificially bypassed. One may then concentrate on the issues that are specific to nonequilibrium systems with many degrees of freedom, just as one postpones ergodicity questions in the day-to-day practice of equilibrium statistical mechanics. One last consideration is that even purely deterministic systems are sometimes more clearly understood as a small-noise limit. This stochastic stability approach is very natural and appealing, not only from the physical, but also from the mathematical point of view (see e.g. Cowieson and Young 2005).

Out-of-equilibrium statistical mechanics is a domain shared between theoretical physicists, mathematical physicists, and probabilists, a fact reflected by a severely fragmented literature.[1] Workers in each of these fields have in mind a different network of relations between subjects and techniques. Two ideas that look similar to a physicist may seem very distant to a probabilist, and vice-versa. The physicist's point of view—the one I adopt—stresses the relations between dynamics and statistical mechanics in space–time, and between stochastic evolution and quantum mechanics, and is on the alert for hidden symmetries and for scaling.

The first lecture introduces stochastic dynamics in a formalism that uses as much as possible the analogy with quantum mechanics, on the assumption that the reader is already familiar with Schrödinger's equation. Next, we discuss the consequences of time reversibility (in particular detailed balance), and how this is intimately related to thermodynamic equilibrium. Crucial for these notes is the fact that the term responsible for the breaking of a time-reversal symmetry in an out-of-equilibrium system is directly related to entropy production. In lecture three, we discuss timescale separation and metastability. In particular, we present through an example a general formalism for metastability (Gaveau and Schulman 1996*a*,*b*, 1998; Gaveau *et al.* 1999; Bovier *et al.* 2000, 2002, 2004) based on the spectral decomposition of the dynamical operator.

[1] I have followed, as general references, the following: the books by Risken (1984), Gardiner (1983), and van Kampen (1981) for the general stochastic context; Parisi's book (Parisi 1988) in several places, in particular the relation between stochastic and quantum mechanics; and Zinn-Justin's book (Zinn-Justin 1996) for technical background on path integrals and a more field-theoretic point of view. I have also found very illuminating the lecture notes of H. Hilhorst (2009) (in French) and of J. Cardy (2009), where the reader may bridge the main gap of these lectures: renormalization. The review by Hänggi *et al.* (1990) has a very comprehensive view of activation processes.

It is quite elementary and intuitive, and is unjustifiably little known. We also describe very briefly the hydrodynamic limit, mainly to present an example where fluctuations become weak through coarse-graining, rather than through low temperatures. This allows us to carry over the "low-noise" results, which we introduce for simplicity in the case of low temperatures, for smooth, large-scale fluctuations: this is the *macroscopic fluctuation theory* (Bertini *et al.* 2001, 2002). In lecture five, we present two forms of large deviations: (i) low-noise: we stress as much as possible the complete analogy between the "Freidlin–Wentzell" (Freidlin and Wentzell 1984; Kitahara 1975; Graham and Tél 1984; Fogedby *et al.* 2004) theory and the standard WKB semiclassical approach in quantum mechanics; and (ii) deviations of long-time averages: in the context of glasses, this is referred to as *space–time thermodynamics* (Jack *et al.* 2006), because the long-time large-deviation functions are in complete analogy with $(d+1)$-dimensional thermodynamics. We shall show that the phase transitions encountered in systems with slow dynamics within this formalism are closely related to the spectral manifestations of metastability discussed in lecture three. With all these elements in hand, and playing with the time-reversal symmetry and its breaking, we obtain the Fluctuation Theorem and Jarzynski's equality, the subject of the last lecture.

The main aim of this short course is to stimulate curiosity. If it feels incomplete, I will judge it successful.

2.1 Trajectories, distributions, and path integrals

In this section we introduce equations of motion containing a deterministic part, and a stochastic thermal bath. Next, we make the passage from a description in terms of trajectories to one in terms of distributions. The evolution of "probability clouds" in space is formally very close to (imaginary-time) quantum mechanics. We do our best to exploit this analogy as much as possible because it opens the way for the application of all the methods in quantum mechanics and field theory.

2.1.1 From trajectories to distributions

Trajectories. Let us start by considering a system satisfying Hamilton's equations. In addition, we shall allow for the possibility of external forces that do not derive from a potential acting on the system; throughout this work we shall denote them $\mathbf{f}(\mathbf{q})$. Nonconservative forces do work, and tend to heat the system up. If we wish that the system will eventually become stationary, we need a thermal bath to absorb energy, both theoretically and in practice. The simplest thermostat is the Langevin bath, consisting of a friction term proportional to the velocity and white noise. The equations of motion read:

$$\begin{cases} \dot{q}_i = \dfrac{\partial \mathcal{H}}{\partial p_i} = \ldots \text{ unless otherwise stated here } \ldots = \dfrac{p_i}{m} = v_i\,, \\ \dot{p}_i = -\dfrac{\partial \mathcal{H}}{\partial q_i} - \underbrace{f_i(\mathbf{q})}_{\text{Forcing}} + \underbrace{\eta_i(t) - \gamma p_i}_{\text{Thermal bath}}\,. \end{cases} \tag{2.1}$$

We shall use $v_i = p_i/m$ throughout. If the thermal bath is itself in equilibrium, the noise intensity and friction coefficient are related by

$$\langle \eta_i(t)\eta_j(t')\rangle = 2\gamma T \delta_{ij}\delta(t-t'), \tag{2.2}$$

which, as we shall see, allows the system to equilibrate at temperature T in the absence of forcing. The parameter γ measures the intensity of coupling to the bath. In an unforced system, it does not affect the stationary distribution—the Gibbs–Boltzmann distribution $\sim e^{-\beta H}$—but it does control the dynamics: compare, for example, the cases of a system of particles interacting with a potential V in equilibrium with a medium at temperature T, when the medium is made of air or of honey (small and large γ, respectively).

Equation (2.1) can be justified in a number of ways. In Section 2.2.2, we shall see how this can be done.

Let us now consider the overdamped case of large γ. We can formally (and somewhat dangerously) neglect the acceleration term as follows:

$$\begin{gathered} m\ddot{q}_i + \gamma\dot{q}_i + \frac{\partial V}{\partial q_i} + f_i = \eta_i(t) \\ \Downarrow \\ \gamma\frac{dq_i}{dt} + \frac{\partial V}{\partial q_i} + f_i = \eta_i(t) \\ \Downarrow \\ \frac{dq_i}{d\tau} + \frac{\partial V}{\partial q_i} + f_i = \eta_i(\tau). \end{gathered} \tag{2.3}$$

In the last step we have rescaled time as $\gamma\tau = t$, which yields $\langle \eta_i(\tau)\eta_j(\tau')\rangle = 2T\delta_{ij}\delta(\tau-\tau')$. A few things to note in the passage to the overdamped equation (2.3) are:

- We now have half the dynamic variables.
- The velocity $\dot{q}_i$ is now discontinuous in time, as is the noise itself. We have to be careful what we mean by eqn (2.3), if for example we are going to programme it on a computer. We adopt the *Îto convention*, which means, for example, in one dimension,

$$q(t+\delta t) - q(t) + \delta t\,(V' + f)(q(t)) = (\delta t)^{1/2}\eta(t). \tag{2.4}$$

  The simplicity comes from the fact that the force is evaluated at the old time t, and does not anticipate the result at the new time $(t+\delta t)$ (van Kampen 1981; Gardiner 1983; Risken 1984; Hilhorst 2009).
- As we shall see, *these ambiguities*[2] *disappear when we consider the evolution of distribution functions instead of the trajectories.*
- Because the velocity is discontinuous, so is the quantity $\sum_i f_i\dot{q}_i$: neglecting inertia makes power become a subtle business. Indeed, some derivations, in particular involving work, become more transparent if inertia is kept, which makes velocities a smooth function of time.

Distributions. Let us now change our point of view, and consider the system not from the point of view of individual trajectories with particular noise realizations, but

[2]Which are the analogue of factor-ordering ambiguities in quantum mechanics.

as a "probability cloud" evolving in space. The passage from the former to the latter description can be done in several ways, and it is instructive to see their relations.

Consider the Langevin equation (2.3). We wish to obtain the equation of motion of the probability $P(\mathbf{q},t)$. Let us first do this separately for a process in the absence of forces, and for a process of advection in the absence of noise. We obtain, respectively,

$$\dot{q}_i = \eta_i \qquad \longrightarrow \qquad \frac{dP}{dt} = T\nabla^2 P\,,$$
$$\dot{q}_i = -\left(\frac{\partial V}{\partial q_i} + f_i\right) \qquad \longrightarrow \qquad \frac{dP}{dt} = \sum_i \frac{\partial}{\partial q_i}\left[\frac{\partial V}{\partial q_i} + f_i\right] P\,.$$

The first equation is just diffusion. The second uses the familiar fact that if a distribution is carried by a flow $\dot{q}_i = g_i(\mathbf{q})$, its evolution is given by the advective derivative $\dot{P} = \sum_i \partial(g_i P)/\partial q_i$.

A useful remark. A trick that is often implicitly used in computer simulations is that whenever two processes act simultaneously and their effect in a small time interval is small, one gets the same result by alternating short intervals with each acting alone. If the separate evolutions are $dP/dt = -H_1 P$ and $dP/dt = -H_2 P$, we can, by the same argument, recompose this as $dP/dt = (H_1 + H_2)P$. Applied to the previous situation, this means that the evolution of the probability for the full Langevin equation (2.3) is

$$\dot{q}_i = -\left(\frac{\partial V}{\partial q_i} + f_i\right) + \eta_i \qquad \longrightarrow \qquad \frac{dP}{dt} = \sum_i \frac{\partial}{\partial q_i}\left[T\frac{\partial}{\partial q_i} + \frac{\partial V}{\partial q_i} + f_i\right] P\,,$$

$$\dot{P}(\mathbf{q},t) = -H_{FP} P(\mathbf{q},t)\,, \tag{2.5}$$

where we have defined the generator

$$H_{FP} = -\sum_i \frac{\partial}{\partial q_i}\left[T\frac{\partial}{\partial q_i} + \frac{\partial V}{\partial q_i} + f_i\right]. \tag{2.6}$$

Writing eqn (2.5) as a continuity equation, we identify the current

$$\dot{P} = -\mathrm{div}\,\mathbf{J} \quad \text{with the definition} \quad J_i(\mathbf{q}) = \left[T\frac{\partial}{\partial q_i} + \frac{\partial V}{\partial q_i} + f_i\right] P\,. \tag{2.7}$$

Exactly the same procedure can be used to derive the evolution of the probability for the case with inertia (2.1). In order to split the system into a diffusion and an advection term, we need to work in phase space. The result is the Kramers equation,

$$\dot{P}(\mathbf{q},\mathbf{p},t) = -H_K P(\mathbf{q},\mathbf{p},t)\,, \tag{2.8}$$

with

$$H_K = \underbrace{\frac{\partial \mathcal{H}}{\partial p_i}\frac{\partial}{\partial q_i} - \frac{\partial \mathcal{H}}{\partial q_i}\frac{\partial}{\partial p_i}}_{H_{\mathrm{Liouville}}:\ \{\ ,\ \}} \quad \underbrace{-\gamma\frac{\partial}{\partial p_i}\left(T\frac{\partial}{\partial p_i} + \frac{p_i}{m}\right)}_{H_b:\ \text{bath}} \quad \underbrace{-\frac{\partial}{\partial p_i} f_i(\mathbf{q})}_{H_f:\ \text{forcing}} \tag{2.9}$$

(summation convention), where we recognize the Poisson bracket associated with Hamilton's equations, plus a bath, and (finally) a forcing term.

Again, writing the Kramers equation as a continuity equation

$$\frac{\partial P(\mathbf{q},\mathbf{p},t)}{\partial t} = -H_K P(\mathbf{q},\mathbf{p},t) = -\mathrm{div}\,\mathbf{J} = -\sum_i \left(\frac{\partial J_{q_i}}{\partial q_i} + \frac{\partial J_{p_i}}{\partial p_i} \right), \tag{2.10}$$

we identify a current

$$J_{q_i} = \frac{\partial \mathcal{H}}{\partial p_i} P(\mathbf{q},\mathbf{p},t)\,, \qquad J_{p_i} = -\left(\gamma T \frac{\partial}{\partial p_i} + \gamma \frac{\partial \mathcal{H}}{\partial p_i} + \frac{\partial \mathcal{H}}{\partial q_i} + f_i \right) P(\mathbf{q},\mathbf{p},t)\,. \tag{2.11}$$

A surprise is that, even in an unforced, $\mathbf{f} = 0$ equilibrium stationary state, the phase-space current is nonzero. This suggests that we should define an alternative quantity, the *reduced phase-space current*, as (Tailleur *et al.* 2004)

$$J_{q_i}^{\mathrm{red}} \equiv J_{q_i} + T \frac{\partial P(\mathbf{q},\mathbf{p})}{\partial p_i}\,, \qquad J_{p_i}^{\mathrm{red}} = J_{p_i} - T \frac{\partial P(\mathbf{q},\mathbf{p})}{\partial q_i}\,. \tag{2.12}$$

The currents (2.11) and (2.12) differ by a term without divergence, and hence their fluxes over closed surfaces coincide. The interesting property of eqn (2.12) is that it is zero for the canonical distribution, as one can easily check. Furthermore, in a case with metastable states it is small everywhere, and it is concentrated along reaction paths.

Gaussian thermostat

In some situations one wishes to study a system with a thermostat that preserves the energy. This can be done with a deterministic (noiseless) "Gaussian" (Hoover 1986) thermostat, extensively used in the context of entropy production and the Gallavotti–Cohen Theorem. As we shall see later, it is in some cases convenient to have in addition a small amount of energy-preserving noise. We thus consider

$$\begin{cases} \dot{q}_i = \dfrac{p_i}{m}\,, \\ \dot{p}_i = -\dfrac{\partial \mathcal{H}}{\partial q_i} - g_{ij}(\eta_j - f_j(\mathbf{q})) = -\dfrac{\partial \mathcal{H}}{\partial q_i} + \underbrace{g_{ij}\eta_j}_{\substack{\text{conservative}\\ \text{noise}}} - \underbrace{f_i(\mathbf{q})}_{\text{forcing}} + \underbrace{\gamma(t) p_i}_{\text{thermostat}}\,. \end{cases} \tag{2.13}$$

Here:

- η_j are white, independent noises of variance ϵ, unrelated to temperature, since the energy is fixed.
- $g_{ij} = \delta_{ij} - p_i p_j / \mathbf{p}^2$ is the projector onto the space tangential to the energy surface.
- Multiplying the first equation of eqn (2.13) by $\partial V(\mathbf{q})/\partial q_i$, the second by p_i/m, and adding, one concludes that energy is conserved provided $\gamma(t) = \mathbf{f} \cdot \mathbf{p}/\mathbf{p}^2$.

The product $g_{ij}(\mathbf{p})\eta_j$ is rather ill-defined because both g_{ij} and η_j are discontinuous functions of time. The ambiguity is removed by discretizing time (Risken 1984) or by specifying the evolution of probability, as we now do.

Repeating the steps leading to the Fokker–Planck and Kramers equations, we find that the probability evolves through

$$\dot{P}(\mathbf{q}, \mathbf{p}) = -H_G P(\mathbf{q}, \mathbf{p}) , \tag{2.14}$$

where H is the operator

$$H_G = \frac{p_i}{m}\frac{\partial}{\partial q_i} - \frac{\partial V(\mathbf{q})}{\partial q_i}\frac{\partial}{\partial p_i} + \frac{\partial}{\partial p_i}[\gamma p_i] - \frac{\partial}{\partial p_i} f_i - \epsilon \frac{\partial}{\partial p_j} g_{ij} g_{il} \frac{\partial}{\partial p_l} , \tag{2.15}$$

where the summation convention is assumed. The precise factor ordering in the last term is important, and specifies the meaning of eqn (2.13). In the absence of driving, $\mathbf{f} = 0$, it is easy to check that H annihilates any function that depends on the phase-space coordinates only through the energy $\mathcal{H} = \mathbf{p}^2/2m + V$. Hence, the noise respects the microcanonical measure in that case.

Other spaces. Doi–Peliti variables. The Hilbert spaces associated with probability distributions of the Fokker–Planck and Kramers equations are different, as the former consists of functions in an N-dimensional space $P(\mathbf{q})$ and the latter of functions in a $2N$-dimensional space $P(\mathbf{q}, \mathbf{p})$. In fact, other spaces appear naturally (Doi 1976; Peliti 1985; Hilhorst 2009) when the dynamic variables are not continuous. This does not bring in any new conceptual feature, but it allows us to write other stochastic problems in a familiar "quantum" notation. Let us give two examples.

Bosons. We consider particles on a lattice, with no exclusion. Denoting by n_i the number of particles in site i, the complete set of configurations is spanned by the space

$$|n\rangle = |n_1, \ldots, n_N\rangle = \otimes_{i=1}^{N} |n_i\rangle , \tag{2.16}$$

so that a probability distribution can be written as

$$P = \sum_{n_1,\ldots,n_N} c_{n_1,\ldots,n_N} |n_1, \ldots, n_N\rangle . \tag{2.17}$$

We can write any evolution operator in this space by introducing the generators

$$\begin{aligned} a_i|n_i\rangle &= n_i|n_i - 1\rangle , \\ a_i^\dagger|n_i\rangle &= |n_i + 1\rangle , \\ a_i|0\rangle &= 0 . \end{aligned} \tag{2.18}$$

Note that $a_i^\dagger$ and a_i are not mutually Hermitian conjugates. For example, for simple diffusion on a one-dimensional lattice we have $\dot{P} = -HP$, with

$$H = -\sum_i (a_{j+1}^\dagger + a_{j-1}^\dagger - 2a_j^\dagger) a_j . \tag{2.19}$$

Spins: The simple symmetric exclusion process. Boson variables do not lend themselves easily to processes where occupation of sites is limited, because particles exclude one another. In those cases, fermions and spins, which have a finite Hilbert space, appear naturally. To be more definite, consider the "simple symmetric exclusion process" (SSEP) on a one-dimensional lattice. This corresponds to a stochastic process on the lattice $\{1, \ldots, N\}$ where particles jump to neighboring sites, but with the limitation that each site can accommodate at most one particle. Configurations $n \in \{0,1\}^N$ are then identified with ket states

$$|n\rangle = |n_1, \ldots, n_N\rangle = \otimes_{i=1}^{N} |n_i\rangle\,, \tag{2.20}$$

which specify the occupation number of each site, namely $n_i \in \{0,1\}$. The bulk evolution is given by the transition rates

$$\begin{aligned} w(n^{i+1,i}, n) &= -\langle n^{i,i+1}|H_B|n\rangle = (1-n_i)n_{i+1}\,, \\ w(n^{i,i+1}, n) &= -\langle n^{i,i+1}|H_B|n\rangle = n_i(1-n_{i+1})\,, \end{aligned} \tag{2.21}$$

where $n^{i,j}$ is the configuration which is obtained from the configuration n by removing a particle from i and adding it to j.

We introduce the operators S, which act as

$$\begin{aligned} S_i^+|n_i\rangle &= (1-n_i)|n_i+1\rangle\,, \\ S_i^-|n_i\rangle &= n_i|n_i-1\rangle\,, \\ S_i^0|n_i\rangle &= \left(n_i - \frac{1}{2}\right)|n_i\rangle\,, \end{aligned} \tag{2.22}$$

and satisfy the SU(2) algebra

$$\begin{aligned} [S_i^0, S_i^\pm] &= \pm S_i^\pm\,, \\ [S_i^-, S_i^+] &= -2S_i^0\,. \end{aligned} \tag{2.23}$$

Note again that, in this representation, $S^\pm$ are not mutually Hermitian conjugates.

In terms of these, the evolution of the SSEP is generated by a spin-one-half ferromagnet (Schütz and Sandow 1994)

$$-H_B = \sum_{i=1}^{N-1} \left(S_i^+ S_{i+1}^- + S_i^- S_{i+1}^+ + 2S_i^0 S_{i+1}^0 - \frac{1}{2} \right). \tag{2.24}$$

2.1.2 Hilbert space

We shall throughout these lectures use the bracket notation, following Kadanoff and Swift (1986). Doing this, we uncover the similarities and differences between stochastic and quantum dynamics, and it allows us to import many techniques developed in quantum many-body and field theory. Here we do everything for the Fokker–Planck case; the generalization to other dynamics is straightforward. We define, as usual, the $\mathbf{q}$ representation:

$$P(\mathbf{q}) = \langle \mathbf{q}|\psi\rangle \,. \tag{2.25}$$

The evolution equation becomes

$$\frac{d}{dt}|\psi\rangle = -H_{FP}|\psi\rangle \,, \tag{2.26}$$

whose solution is

$$|\psi(t)\rangle = e^{-tH_{FP}}|\psi_0\rangle \quad \rightarrow \quad P(\mathbf{q},t) = \langle \mathbf{q}|e^{-tH_{FP}}|\psi_0\rangle \,. \tag{2.27}$$

The transition probability is given by the matrix element

$$P(\mathbf{q},t\,;\mathbf{q}_0,t_0) = \langle \mathbf{q}|e^{-(t-t_0)H_{FP}}|\mathbf{q}_0\rangle \,. \tag{2.28}$$

As in quantum mechanics, it is useful to consider the spectrum of H_{FP}. Because H_{FP} is not self-adjoint, we have to distinguish right and left eigenvectors:

$$H_{FP}|\psi_a^R\rangle = \lambda_a|\psi_a^R\rangle \,, \qquad \langle\psi_a^L|H_{FP} = \lambda_a\langle\psi_a^L| \,. \tag{2.29}$$

The resolution of the identity is

$$\langle\psi_a^L|\psi_b^R\rangle = \delta_{ab} \,, \qquad \mathbf{I} = \sum_a |\psi_a^R\rangle\langle\psi_a^L| \,. \tag{2.30}$$

Defining the (in general, unnormalizable) "flat" state $\langle -|$ as a constant over all configurations,

$$\langle -|\mathbf{q}\rangle = 1 \,, \tag{2.31}$$

and using the fact that conservation of probability implies that at all times

$$\begin{aligned} 1 = &\sum_{\text{all configurations}} \langle \mathbf{q}|\psi(t)\rangle = \langle -|\psi(t)\rangle \\ \rightarrow \quad &\frac{d}{dt}\langle -|\psi(t)\rangle = -\langle -|H_{FP}|\psi(t)\rangle = 0 \quad \forall\psi(t) \,, \end{aligned} \tag{2.32}$$

and the fact that a stationary state satisfies

$$\frac{d}{dt}|\text{stat}\rangle = -H_{FP}|\text{stat}\rangle = 0 \,, \tag{2.33}$$

we have that the "flat" and stationary states are the left and right zero-eigenvalue eigenvectors

$$\langle -|H_{FP} = 0 \,, \qquad H_{FP}|\text{stat}\rangle = 0 \,. \tag{2.34}$$

Finally, writing

$$|\psi_0\rangle = \sum_a c_a \, |\psi_a^R\rangle \,, \tag{2.35}$$

we have

$$P(\mathbf{q},t) = \langle \mathbf{q}|\psi(t)\rangle = \sum_a c_a \, \langle \mathbf{q}|\psi_a^R\rangle e^{-t\lambda_a} \,. \tag{2.36}$$

Equation (2.36) already shows us that λ_a cannot be negative, and that the long-time properties of the probabilities are encoded in the eigenvectors with small eigenvalues.

In particular, if all eigenvalues have real parts larger than zero, there is no stationary state: this happens typically if the system is unbounded.

In the Kramers case, all of what we have said applies, provided one considers functions

$$P(\mathbf{q},\mathbf{p}) = \langle \mathbf{q},\mathbf{p}|\psi\rangle\,. \tag{2.37}$$

Perron–Frobenius theorem

P is a probability distribution, it has to be positive everywhere at all times. If one λ_a has a negative real part, it dominates the sum (2.36) for large times, its coefficient going to infinity. Because $\langle -|\psi_a\rangle = 0$, $\langle \mathbf{q}|\psi_a\rangle$ necessarily takes positive and negative values, and this will make $P(t)$ at large times not everywhere positive, contrary to the assumption.

Correlations and responses. Correlation functions are averages over many realizations of a process, with different realizations of the random noise each time. They can be expressed in this notation as

$$C_{AB}(t,t') = \langle A(t)B(t')\rangle_{\text{realization}} = \langle -|A\, e^{-(t-t')H}\; B\, e^{-t'H}|\text{init}\rangle\,. \tag{2.38}$$

Here H may be the Fokker–Planck, Kramers, or in general any Doi–Peliti operator. Similarly, for the response functions,

$$R_{AB}(t,t') = \frac{\delta\langle A(t)\rangle}{\delta h_B(t')} = \langle -|A\, e^{-(t-t')H}\, \frac{dH}{dh_B}\, e^{-t'H}|\text{init}\rangle\,, \tag{2.39}$$

where the average $\langle A\rangle$ is over noise realizations. Eigenvalues with small real parts have something to say about the decay of long-time correlations:

$$\begin{aligned} C(t,t') = \langle A(t)A(t')\rangle = \langle -|Ae^{-(t-t')H}A|\text{init}\rangle &= \sum_a \langle -|A|\psi_a^R\rangle\langle\psi_a^L|A|\text{init}\rangle e^{-\lambda_a t} \\ &\sim \langle -|A|\text{stat}\rangle\langle -|A|\text{init}\rangle + \langle -|A|\psi_1^R\rangle\langle\psi_1^L|A|\text{init}\rangle e^{-\lambda_1 t}\,, \end{aligned} \tag{2.40}$$

which decays to the asymptotic value as an exponential of $\operatorname{Re}\lambda_1$, the first eigenvector with nonzero real part. The existence of a gap between the lowest and the first eigenvalue leads to exponential decays of correlations.

Conserved quantities. Consider a probability distribution that is concentrated on an energy shell, for example in phase space, $E(\mathbf{q},\mathbf{p}) = E_0$. This can be expressed as an eigenvalue equation

$$E(\mathbf{q},\mathbf{p})P(\mathbf{q},\mathbf{p},t) = E_0 P(\mathbf{q},\mathbf{p},t)\,. \tag{2.41}$$

Consider further an evolution that conserves energy, generated by some H. By assumption, there are no transitions between shells of different energies, so the matrix of H is of block form, each block corresponding to a value of E_0. This in turn implies that E commutes with the operator H:

$$[E,H] = 0\,. \tag{2.42}$$

Clearly, what we have said applies to any conserved quantity.

The fact that the stochastic process can be now broken into subspaces of fixed energy entails the existence for each E_0 of a different stationary state $|\text{stat}_{E_0}\rangle$, and the corresponding flat measure $\langle -_{E_0}|$.

Analogy with quantum mechanics. Making the identification

$$\hat{q}_i^{\text{op}} \longrightarrow T\frac{\partial}{\partial q_i}\,, \tag{2.43}$$

we may write

$$\begin{aligned}\tilde{H} \equiv TH_{FP} &= -\underbrace{\left(T\frac{\partial}{\partial q_i}\right)}_{\downarrow}\left[\underbrace{\left(T\frac{\partial}{\partial q_i}\right)}_{\downarrow}+\frac{\partial V}{\partial q_i}+f_i\right]\\ &= \quad -\hat{q}_i^{\text{op}} \quad \left[\hat{q}_i^{\text{op}}+\frac{\partial V}{\partial q_i}+f_i\right],\end{aligned}$$

and the evolution is

$$\dot{P} = \underbrace{-\frac{1}{T}\tilde{H}}_{H_{FP}}P\,. \tag{2.44}$$

The situation is analogous to quantum mechanics, with T playing the role of $\hbar$ and the $\hat{q}_i^{\text{op}}$ the role of "momentum operators." The quantum-like Hamiltonian comes from a "classical" Hamiltonian

$$\tilde{H} = -\hat{q}_i\left[\hat{q}_i+\frac{\partial V}{\partial q_i}+f_i\right] \tag{2.45}$$

in a phase space with twice the number of variables. This "classical" problem with a double number of degrees of freedom will play a key role in the next sections.

2.1.3 Functional approach

Let us first consider a general and extremely powerful trick to convert an algebraic problem of finding the roots of a system of equations $Q_a(\mathbf{x}) = 0$ into a statistical problem of calculating a partition function. We compute the sum over roots of a function A:

$$\langle A(\mathbf{x})\rangle \equiv \sum A(\mathbf{x})|_{\text{roots of } Q_a}\,. \tag{2.46}$$

This can be written

$$\begin{aligned}\langle A(\mathbf{x})\rangle &= \int d\mathbf{x}\; A(\mathbf{x})\;\delta(Q_a(\mathbf{x}))\;\det\left|\frac{\partial Q_c}{\partial x_b}(\mathbf{x})\right|\\ &= \int d\mathbf{x}\int_{-i\infty}^{i\infty} d\hat{\mathbf{x}}\; A(\mathbf{x})\;\det\left|\frac{\partial Q_c}{\partial x_b}(\mathbf{x})\right|\; e^{\sum_a \hat{x}_a Q_a(\mathbf{x})}\,.\end{aligned} \tag{2.47}$$

The determinant is there to ensure that we count each root with a unit weight. We have thus obtained an expression that has the form of a partition function.

The Martin–Siggia–Rose/De Dominicis–Janssen (Martin *et al.* 1973; De Dominicis 1976; Janssen 1976; De Dominicis and Peliti 1978) approach consists of applying this technique to differential, rather than ordinary equations, in this case the Langevin equation. To do this properly, one must discretize time and write a delta function, imposing eqn (2.4) at each time. One obtains a sum over paths:

$$P(q,t) = \overbrace{\int D[q]}^{\substack{\text{paths with}\\ \text{appropriate b.c.}}} \quad \underbrace{D[\eta]}_{\text{noise}} \; \overbrace{\delta\left[\dot{q} + \frac{\partial V}{\partial q} + \eta\right]}^{\text{functional delta}} \; \underbrace{e^{-\int dt\, \eta^2/4T}}_{\text{path probability}} . \tag{2.48}$$

Although for the moment we stay in one dimension, and omit the forcing term, the generalization to several dimensions and $\mathbf{f} \neq 0$ is straightforward. The determinant is unity in the Îto convention (2.4), because the matrix of second derivatives (containing the derivative of the equation of motion at time t_i with respect to $q(t_j)$) is upper triangular with ones in the diagonal.

Introducing the integral representation of the delta with a "hat" variable (one per time), we obtain

$$P(q,t) = \int D[q]\; D[\eta] \int_{-i\infty}^{+i\infty} \overbrace{D[\hat{q}]}^{\text{"response" field}} \quad e^{\int dt\, \left[\hat{q}_i(\dot{q}+\partial V/\partial q+\eta)\; -\eta^2/4T\right]} . \tag{2.49}$$

The boundaries are free for $\hat{q}$. Integrating the Gaussian noise away, we obtain

$$P(q,t) = \int D[q]\; D[\hat{q}] \;\exp \int dt \left[\hat{q}\dot{q} - \underbrace{\left\{-\hat{q}\left(T\hat{q} + \frac{\partial V}{\partial q}\right)\right\}}_{\substack{\downarrow\\ H}}\right]$$

$$= \int D[q,\hat{q}]\; e^{\pm \int dt\, [\hat{q}\dot{q} - H]} . \tag{2.50}$$

Making the change of variables $\hat{q} \to \hat{q}/T$, the evolution equation becomes, for many degrees of freedom,

$$P(q,t) = \int D[\mathbf{q},\hat{\mathbf{q}}] \;\exp \frac{1}{T} \int dt \left[\hat{q}_i\dot{q}_i - \underbrace{\left\{-\hat{q}_i\left(\hat{q}_i + \frac{\partial V}{\partial q_i}\right)\right\}}_{\substack{\downarrow\\ \tilde{H}}}\right] \tag{2.51}$$

(summation convention). The exponent is multiplied by $1/T$, which thus plays the role of $1/\hbar$. From what we know of the path integral representation of quantum mechanics, we recognize the action associated with the evolution equation (2.44). The analogy with (imaginary-time) quantum mechanics can be carried further *in the absence of forcing*, by making the transformation

$$H_h = e^{\beta V/2}\, H_{FP}\, e^{-\beta V/2} = \frac{2}{T}\sum_i \left[-\frac{T^2}{2}\frac{\partial^2}{\partial q_i^2} + \frac{1}{8}\left(\frac{\partial V}{\partial q_i}\right)^2 - \frac{T}{4}\frac{\partial^2 V}{\partial q_i^2} \right]. \tag{2.52}$$

We now have an H_h which is truly of the Schrödinger form, albeit with a new potential $V_{\text{eff}} = \frac{1}{8}\left(\partial V/\partial q_i\right)^2 - \frac{T}{4}\partial^2 V/\partial q_i^2$.

All in all, the situation is as in Fig. 2.1: we can go from the Langevin equation to the Fokker–Planck equation for the distribution directly, and then construct a path integral representation of the evolution using the analogy between the Fokker–Planck and Schrödinger equations. Alternatively, we can go straight from the Langevin equation to the path integral, through the Martin–Siggia–Rose/De Dominicis–Janssen construction. The latter procedure is much more flexible, and is thus extensively used in the physics literature.

Lagrangian (Onsager–Machlup) form. As in all phase-space problems where the momenta appear quadratically (in this case the "hat" variables $\hat{q}_i$), we have the possibility of integrating them away, thus going to the Lagrangian representation (Onsager and Machlup 1953; Machlup and Onsager 1953). Alternatively, we can go straight to this representation by integrating eqn (2.48) over the noise *before* introducing the "hat" variables. The result is

$$P(q,t) = \int D[\mathbf{q}]\, \exp\left[-\frac{1}{4T}\int dt\, \left(\dot{q}_i + \frac{\partial V}{\partial q_i} + f_i \right)^2 \right]. \tag{2.53}$$

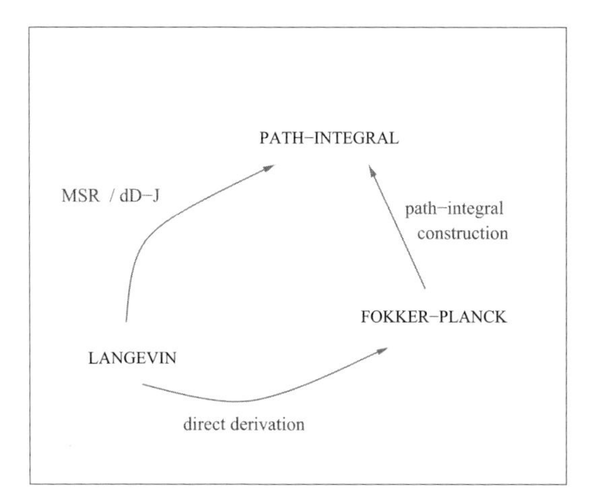

Fig. 2.1 With the Martin–Siggia–Rose/De Dominicis–Janssen construction, one can go straight from the stochastic equation to the path integral. Otherwise, given the equation of motion for the probability distribution, one can express it as a path integral as in quantum mechanics textbooks.

Note that we have inherited the Îto convention from eqn (2.4), which means a definite prescription on how to discretize the time integral. This sum takes the form of a partition function at temperature $\sim T$. This means we can in principle compute it using well-established Monte Carlo methods: *a nonequilibrium problem for the configurations becomes an "equilibrium" problem in trajectory space.*

In the absence of forcing, $\mathbf{f} = 0$, we can expand the square in the exponent, and, recognizing that the double product is a total derivative, we have

$$P(q,t) = \int D[\mathbf{q}] \, \exp\left\{ -\frac{1}{4T} \int dt \left[\dot{q}_i^2 + \left(\frac{\partial V}{\partial q_i}\right)^2 - 2T\left(\frac{\partial^2 V}{\partial q_i^2}\right)\right]\right\} e^{-(1/2T)[V(\mathbf{q})-V(\mathbf{q}_0)]} . \tag{2.54}$$

The second derivative in the exponent comes from the correct discretization (see box). We recognize the action of a polymer—the monomer index being the time—at temperature T in a potential $\propto |\nabla V|^2$. This analogy can be exploited fruitfully (Chandler and Wolynes 1981).

Conventions

Suppose we had started from eqn (2.52) and constructed the Lagrangian path integral as in quantum mechanics textbooks. We would have obtained eqn (2.54), the exponentials of the potential at the ends of the trajectory being just the change of basis leading to eqn (2.52). The extra term $-\frac{1}{2}\int dt \; \partial^2 V/\partial q_i^2$ in the exponent appears naturally. In our case, it appears as part of the integral $\int dt \; \dot{q}_i \, (\partial V/\partial q_i)$, once we specify what this means in terms of a discrete sum.

The question of discretization conventions must be treated with care, as they may lead to errors in the results. However, one should not exaggerate the physical importance of the whole problem, as it disappears as soon as the system is regularized by including inertia.

Kramers equation. All the steps above can be performed for a system with inertia. A somewhat confusing fact is that our extended space consists of $4N$ coordinates, including the $2N$ original coordinates *and momenta* $(\mathbf{q}, \mathbf{p})$, playing the role of coordinates in the extended space, and the $2N$ "hat" variables $(\hat{\mathbf{q}}, \hat{\mathbf{p}})$, playing the role of momenta in the extended space, associated now with the operators

$$\hat{q}_i^{\mathrm{op}} \to T\frac{\partial}{\partial q_i}, \qquad \hat{p}_i^{\mathrm{op}} \to T\frac{\partial}{\partial p_i} . \tag{2.55}$$

2.2 Time reversal and equilibrium

As we have seen in the previous section, stochastic dynamics leads to an evolution that is quite close to a generic quantum problem, often with a non-Hermitian Hamiltonian. Such generality is in a way bad news, since one can hardly expect to find a method to solve a problem that is so general. A special and important class is that of systems that evolve in contact with an equilibrium thermal bath, and are under the action of conservative forces. There is then a symmetry that can be interpreted as time reversal, which has important consequences. We shall also see that if this symmetry is broken by forces that do work, the symmetry-breaking term can be interpreted as an entropy production rate: this will be the basis of the nonequilibrium theorems of Section 2.6.

2.2.1 Detailed balance

The detailed-balance property is a relation between the probabilities of going from a configuration a to a configuration b and vice versa:

$$e^{-\beta V(a)} P_{a\to b} = e^{-\beta V(b)} P_{b\to a} \,. \tag{2.56}$$

The name "detailed" comes from the fact that if we ask only for the Gibbs–Boltzmann distribution to be stationary, we need only that eqn (2.56) holds when added over configurations, and not term by term. We can telescope eqn (2.56) to obtain the following for a chain of configurations:

$$P_{a_1\to a_2} P_{a_2\to a_3} \dots P_{a_{m-1}\to a_m} = e^{-\beta[V(a_m)-V(a_1)]} P_{a_m\to a_{(m-1)}} \dots P_{a_3\to a_2} P_{a_2\to a_1} \,, \tag{2.57}$$

which means that

$$\textbf{Probability [path]} = \; e^{-\beta[V(\text{final})-V(\text{initial})]} \; \textbf{Probability [reversed path]} \,.$$

And, in particular, for all closed circuits,

$$\textbf{Probability [circuit]} = \textbf{Probability [reversed circuit]} \,.$$

In other words, we have the Onsager–Machlup reversibility:

- The probability of any path going from a to b is equal to the probability of the time-reversed path, times a constant that depends only on the endpoints a, b.
- Hence, if for some reason there is essentially one type of path that takes the system from a to b, then there is also essentially one path that takes it from b to a, and this is its time-reversed path.

These properties are directly observable experimentally (see Andrieux *et al.* 2007).

Fokker–Planck. Detailed balance holds for the Fokker–Planck evolution without forcing, $f_i = 0$. Let us see the implication this has:

$$\begin{aligned} \langle \mathbf{q}' | e^{-tH_{FP}} | \mathbf{q} \rangle \, e^{-\beta V(\mathbf{q})} &= \langle \mathbf{q} | e^{-tH_{FP}} | \mathbf{q}' \rangle \, e^{-\beta V(\mathbf{q}')} \,, \\ \langle \mathbf{q}' | e^{-tH_{FP}} e^{-\beta V} | \mathbf{q} \rangle &= \langle \mathbf{q} | e^{-tH_{FP}} e^{-\beta V} | \mathbf{q}' \rangle = \langle \mathbf{q}' | e^{-\beta V} e^{-tH_{FP}^{\dagger}} | \mathbf{q} \rangle \,. \end{aligned} \tag{2.58}$$

Since this is true $\forall (\mathbf{q}, \mathbf{q}')$, we have

$$e^{\beta V} e^{-tH_{FP}} e^{-\beta V} = e^{-tH_{FP}^{\dagger}} \quad \forall t \qquad \rightarrow \qquad e^{\beta V} H_{FP} e^{-\beta V} = H_{FP}^{\dagger} \,, \tag{2.59}$$

which in turn means that

$$H_h = e^{\beta V/2} H_{FP} e^{-\beta V/2} \tag{2.60}$$

is Hermitian, a fact that we have already checked (see eqn (2.52)). Equation (2.59) gives also a direct relation between right and left eigenvectors:

$$|\psi_\alpha^L\rangle = e^{\beta V} |\psi_\alpha^R\rangle \,. \tag{2.61}$$

We now understand why a Hermitian form cannot be obtained when there is forcing: detailed balance is then lost.

Kramers. Let us now see how this generalizes to a process with inertia, having an energy $\mathcal{H} = \mathbf{p}^2/2m + V(\mathbf{q})$. In this case, something like detailed balance holds, but on the condition that we reverse the velocities:

$$e^{-\beta \mathcal{H}(a)} P_{a \to b} = e^{-\beta \mathcal{H}(\bar{b})} P_{\bar{b} \to \bar{a}} \,, \tag{2.62}$$

where $\bar{a}, \bar{b}$ are the configurations a, b with the velocities reversed. Also, in the Kramers case we can telescope eqn (2.62) to obtain relations for trajectories and for closed circuits, as we did above for the Fokker–Planck evolution.

In operator notation, eqn (2.62) reads

$$\langle \mathbf{q}', \mathbf{p}' | e^{-tH_K} | \mathbf{q}, \mathbf{p} \rangle \, e^{-\beta \mathcal{H}(\mathbf{q},\mathbf{p})} = \langle \mathbf{q}, -\mathbf{p} | e^{-tH_K} | \mathbf{q}', -\mathbf{p}' \rangle \, e^{-\beta \mathcal{H}(\mathbf{q}')} \,, \tag{2.63}$$

which leads to

$$\Pi e^{\beta \mathcal{H}} \, H_K \, e^{-\beta \mathcal{H}} \Pi^{-1} = H_K^\dagger \,, \tag{2.64}$$

where we have introduced the operator that reverses velocities, $\Pi \mathbf{p} \Pi^{-1} = -\mathbf{p}$, and similarly with derivatives. H_K cannot, in general, be taken to a Hermitian form. Applying to eqn (2.64) steps analogous to those leading from eqn (2.59) to eqn (2.60), one finds that Hermiticity is broken because $\Pi^{1/2}$ is not real.

Another form of time reversal: The adjoint

A driven overdamped system admits a form of time reversal (Bertini *et al.* 2001, 2002; Chetrite and Gawedzki 2008) that is not, however, a symmetry. Consider the Fokker–Planck operator with nonconservative forces, and assume we know its stationary distribution:

$$H|\text{stat}\rangle = -\frac{\partial}{\partial q_i}\left(T\frac{\partial}{\partial q_i} + f_i\right)|\text{stat}\rangle = 0 \,. \tag{2.65}$$

Put $\langle \mathbf{q} | \text{stat} \rangle \equiv \phi(\mathbf{q})$ and compute

$$\phi^{-1} H \phi = -\left[T\frac{\partial}{\partial q_i} + \left(2T\frac{\partial \ln \phi}{\partial q_i} + f_i\right)\right]\frac{\partial}{\partial q_i} = H_{\text{adj}}^\dagger \,, \tag{2.66}$$

where we have defined the adjoint

$$H_{\text{adj}} = -\frac{\partial}{\partial q_i}\left[T\frac{\partial}{\partial q_i} - \left(2T\frac{\partial \ln \phi}{\partial q_i} + f_i\right)\right]. \tag{2.67}$$

This describes another diffusion problem at temperature T in a new force field,

$$f_i^{\text{rev}} = -\left(2T\frac{\partial \ln \phi}{\partial q_i} + f_i\right), \tag{2.68}$$

which only coincides with the original one when $\mathbf{f}$ is derived from a potential. This formula is the basis of the Hatano–Sasa formula (Hatano and Sasa 2001).

A similar but stronger form of time reversibility arises if we accept that some variables change signs, as velocities do. We refer the reader to Graham (1980). Note that in all these cases, we need to know the stationary distribution a priori, so the formulae are moderately useful in practice.

2.2.2 Equilibrium theorems: Reciprocity and fluctuation–dissipation

Detailed balance is a form of time-reversal symmetry in the trajectories. It cannot come as a surprise that in equilibrium it implies time-reversal symmetry in the correlation functions. Let us examine this for the Kramers equation. Denoting by $|GB\rangle$ the Gibbs–Boltzmann distribution, we obtain

$$\begin{aligned} C_{AB}(t-t') &= \langle -|A\, e^{-(t-t')H_K}\; B|GB\rangle = \langle GB|B\, e^{-(t-t')H_K^\dagger}\; A|-\rangle \\ &= \langle -|e^{-\beta\mathcal{H}}B\, e^{-(t-t')H_K^\dagger}A\, e^{\beta\mathcal{H}}|GB\rangle \\ &= \langle -|\underbrace{\Pi e^{-\beta\mathcal{H}}Be^{\beta\mathcal{H}}\Pi^{-1}}_{\bar{B}}\; e^{-(t-t')H_K}\; \underbrace{\Pi e^{-\beta\mathcal{H}}Ae^{\beta\mathcal{H}}\Pi^{-1}}_{\bar{A}}|GB\rangle \\ &= C_{\bar{B}\bar{A}}(t-t'), \end{aligned} \tag{2.69}$$

where we have used eqn (2.64) and we have defined $\bar{A}(\mathbf{q},\mathbf{p}) = A(\mathbf{q},-\mathbf{p})$. The same derivation can be done for the Fokker–Planck case for observables dependent on coordinates only.

Another important equilibrium property is the Fluctuation–Dissipation Theorem. This relates the response of the expectation of an observable A produced by a kick given by a field conjugate to an observable B to the corresponding two-time correlation. Putting $\mathcal{H}_{h_B} = \mathcal{H} - h_B B$, we compute

$$R_{AB}(t-t') = \frac{\delta\langle A(t)\rangle}{\delta h_B(t')} = \langle -|A\, e^{-(t-t')H}\; \left.\frac{dH}{dh_B}\right|_{h_B=o} |GB\rangle\,. \tag{2.70}$$

Because the equilibrium distribution in the presence of a field $|GB\,(h_B)\rangle$ satisfies for all h_B the condition that $H_{h_B}|GB\,(h_B)\rangle = 0$, we have

$$\begin{aligned} \left.\left(\frac{dH}{dh_B}\right)|GB\rangle\right|_{h_B=0} &= -H\left.\left(\frac{d}{dh_B}|GB\rangle\right)\right|_{h_B=0} = \beta H\,(B - \langle B\rangle)\,|GB\rangle_{h_B=0} \\ &= \beta\, HB\,|GB\rangle_{h_B=0} \end{aligned} \tag{2.71}$$

and, substituting in eqn (2.70),

$$
\begin{aligned}
R_{AB}(t-t') &= \beta\langle -|A\, e^{-(t-t')H}\; H\; B\;|GB\rangle \\
&= \beta\frac{\partial}{\partial t'}C_{AB}(t-t')\,.
\end{aligned} \tag{2.72}
$$

The Fluctuation–Dissipation Theorem of the first and second kind. A derivation of the Langevin equation. At the outset we started with a bath that contained friction proportional to γ and noise whose variance was γT. This very precise relation between noise and friction is often referred to as the "Fluctuation–Dissipation Theorem of the first kind," because it relates the dissipation and fluctuations *of the bath*, rather than of the system. If the bath satisfies this relation, a system in contact with it will eventually equilibrate (this might take a long time) and will then satisfy the Fluctuation–Dissipation Theorem (2.72)—of the "second kind"—for its observables.

Next, assume there is a large number $\alpha = 1, \ldots, M$ of independent copies of such systems, with coordinates $\mathbf{q}^\alpha, \mathbf{p}^\alpha$, all in equilibrium with a bath. We intend to use them in turn as baths for a further system of coordinates $\mathbf{q}', \mathbf{p}'$ and energy $\mathcal{H}'(\mathbf{q}', \mathbf{p}')$. To do this we couple them, for example through a term

$$
\mathcal{H} = \underbrace{\sum_\alpha \mathcal{H}_\alpha}_{\text{bath}} + \underbrace{\mathcal{H}'(\mathbf{q}', \mathbf{p}')}_{\text{system}} \underbrace{- M^{-1/2}\sum_\alpha \mathbf{q}^\alpha \cdot \mathbf{q}'}_{\text{coupling}}\,. \tag{2.73}
$$

We may ask what is the condition for the coupling term to constitute a legitimate thermal bath for the primed system. The equations of motion of the primed variables are

$$
\ddot{q}' = -\frac{\partial \mathcal{H}'}{\partial q_i'} - \mathbf{h(t)}\,, \tag{2.74}
$$

where the field $\mathbf{h}$ is given by $\mathbf{h} = M^{-1/2}\sum_\alpha q_\alpha$. The large-$M$ limit allows us to treat each $M^{-1/2}\mathbf{q}^\alpha \cdot \mathbf{q}'$ as a small perturbation to the system $\mathcal{H}_\alpha$, and to invoke the central limit theorem to say that $M^{-1/2}\mathbf{q}^\alpha$ is a Gaussian. Assuming the expectation values of $\langle q_\alpha \rangle = 0$ in the absence of coupling, the field $\mathbf{h}$ has two contributions for large M: (i) a random Gaussian noise $\eta(t)$ with correlation $C_{\alpha\alpha}(t,t') = \langle \eta(t)\eta(t')\rangle = \langle q_\alpha(t) q_\alpha(t')\rangle$, and (ii) a drift due to the back effect of the q', which acts as a field on the q_α. Again, because M is large, the average response of the ensemble α is

$$
M^{-1/2}\langle q^\alpha \rangle = \int_0^t dt'\; R_{\alpha\alpha}(t,t') q'(t')\,. \tag{2.75}
$$

The equation of motion of the primed variable becomes

$$
\ddot{q}' = -\frac{\partial \mathcal{H}'}{\partial q'} + \eta(t) + \int_0^t dt'\; R_{\alpha\alpha}(t,t') q'(t')\,. \tag{2.76}
$$

This is in fact the generalized Langevin equation, describing the primed system in contact with a thermal bath with colored noise, and friction with memory.[3] The condition that the bath is a good equilibrium one is precisely

[3] The white-noise limit (2.1) is obtained when $C_{\alpha\alpha}(t-t') \propto \delta(t-t')$.

$$TR_{\alpha\alpha}(t,t') = \frac{\partial}{\partial t'} C_{\alpha\alpha}(t,t') \,. \tag{2.77}$$

The primed system will, under the action of this dynamics, equilibrate to the Gibbs–Boltzmann distribution.[4]

- The Fluctuation–Dissipation Theorem of the second kind for the q_α has become a Fluctuation–Dissipation Theorem of the first kind when they are considered as a bath for the primed variable q'.
- The friction is the back reaction of the bath to the perturbation exerted by the system, while the noise is given by the incoherent addition of motions of the bath's constituents.
- The fluctuation–dissipation relation regulates the balance between these two effects in such a way as to guarantee that equilibrium is "passed on" to a coupled system.

2.2.3 Time-reversal violations and entropy production

In the previous section we have seen that there is a symmetry related to time reversal in systems that are in contact with an equilibrium thermal bath and have only conservative forces. Once forcing is allowed, this symmetry is broken: in an interesting manner. As we shall now see, the symmetry-breaking term turns out to be proportional to the power injected by the nonconservative forces, divided by a temperature: it can be interpreted as an entropy production.[5]

For simplicity, we shall do this calculation in two cases with inertia, in order to avoid unnecessary complications brought about by the fact mentioned above that the power in an overdamped case is a discontinuous function of time.

Kramers equation. Let us attempt to obtain a relation like eqn (2.64) in the presence of forcing. Referring to eqn (2.9), $H_K = H_{\text{Liouville}} + H_b + H_f$, we compute $\Pi e^{\beta E} H_K e^{-\beta E} \Pi^{-1}$ in detail. First, we make the similarity transformation

$$\begin{aligned}
e^{\beta \mathcal{H}} H_{\text{Liouville}} e^{-\beta \mathcal{H}} &= H_{\text{Liouville}} \\
e^{\beta \mathcal{H}} H_b e^{-\beta \mathcal{H}} &= \gamma \left(T \frac{\partial}{\partial p_i} - \frac{p_i}{m} \right) \frac{\partial}{\partial p_i} \\
e^{\beta \mathcal{H}} H_f e^{-\beta \mathcal{H}} &= -f_i \left(\frac{\partial}{\partial p_i} - \beta \frac{p_i}{m} \right) .
\end{aligned} \tag{2.78}$$

Next, Hermitian conjugation changes signs of derivatives, and reverses the order of factors. Finally, velocity reversal transforms ($p_i \to -p_i$) and $\partial/\partial p_i \to -\partial/\partial p_i$. All in all, we get

$$\left[\Pi e^{\beta \mathbf{H}} H_K e^{-\beta \mathbf{H}} \Pi^{-1} \right] = H_K^\dagger - \underbrace{\beta \sum_i f_i v_i}_{\text{Power}/T} \tag{2.79}$$

[4] With an energy that includes a contribution $\langle [q^\alpha]^2 \rangle q'^2$ coming from the interaction term (the factor $\langle [q^\alpha]^2 \rangle$ is the equilibrium expectation for a single isolated $\mathcal{H}_\alpha$). This term can compensated by an opposite one in $\mathcal{H}'$.

[5] Subtle questions about when one can call this entropy are extensively addressed in the literature (Maes 2003); here, we shall not get into these matters.

(here again, $v_i = p_i/m$ is the velocity). As announced, we find that the relation (2.64) has now an extra term corresponding to power divided by temperature, an entropy production rate.

Gaussian thermostat. We may repeat the calculation for the Gaussian thermostat (2.15). Only velocity reversal is necessary, and we obtain

$$\begin{aligned} \left[\Pi H_G \Pi^{-1}\right]^\dagger &= H_G + [\gamma p_i]\frac{\partial}{\partial p_i} - \frac{\partial}{\partial p_i}[\gamma p_i] \\ &= H_G - (N-1)\frac{\mathbf{f}\cdot\mathbf{p}}{\mathbf{p}^2} \end{aligned} \tag{2.80}$$

(summation convention), where $2N$ is the dimension of phase space. Again, this has the interpretation of a power divided by a kinetic temperature ($\sim \mathbf{p}^2$).

2.2.4 What does a stationary out-of-equilibrium distribution look like?

A driven system, if coupled to a thermostat, will reach a stationary distribution. The problem with out-of-equilibrium statistical mechanics is that there is no simple, general expression for this distribution. If the thermostat keeps the energy constant, as does the Gaussian one defined above, even if the probability is then restricted to an energy shell, it does not cover it in a uniform, microcanonical way. In fact, if the thermostat is deterministic, the distribution will be a fractal. If, on the other hand, the thermostat involves noise, the fractal will be blurred, but the distribution is still nonuniform on the energy shell. Can we get some intuition about this? The purpose of this section is to see, in a relatively simple example (Bonetto and Gallavotti 1997; Kurchan 2007), what happens when a system is forced.

We shall consider a Lorentz gas, or, equivalently, a particle in a billiard as in Fig. 2.2, under the effect of a constant field, with periodic boundary conditions. We assume that there is a Gaussian thermostat, which fixes the modulus of the velocity to be constant; we take this as unity. The trajectories between bounces are given by

$$\begin{aligned} \dot{p}_x &= E - \frac{E p_x^2}{p^2}\,, \\ \dot{p}_y &= -\frac{E p_x p_y}{p^2}\,, \\ p\dot{\theta} &= -E\sin\theta = -\frac{d}{d\theta}[-E\cos\theta]\,, \end{aligned} \tag{2.81}$$

where we have defined the angle $(p_x, p_y) = p(\sin\theta, \cos\theta)$ (Fig. 2.3). If there were no obstacles, there would be two stationary situations: when the velocity is parallel to the field (stable) and when it is antiparallel (unstable). Such trajectories constitute the *attractor* and the *repellor*, respectively.

Consider now the effect of obstacles. When the field is off, in the presence of obstacles the system is known (Sinai 1963*a*,*b*, 1970) to be ergodic. Phase-space points on the energy shell—the Cartesian product of the allowed configurations times the velocity sphere—are visited uniformly. At the other extreme, if the field is very strong,

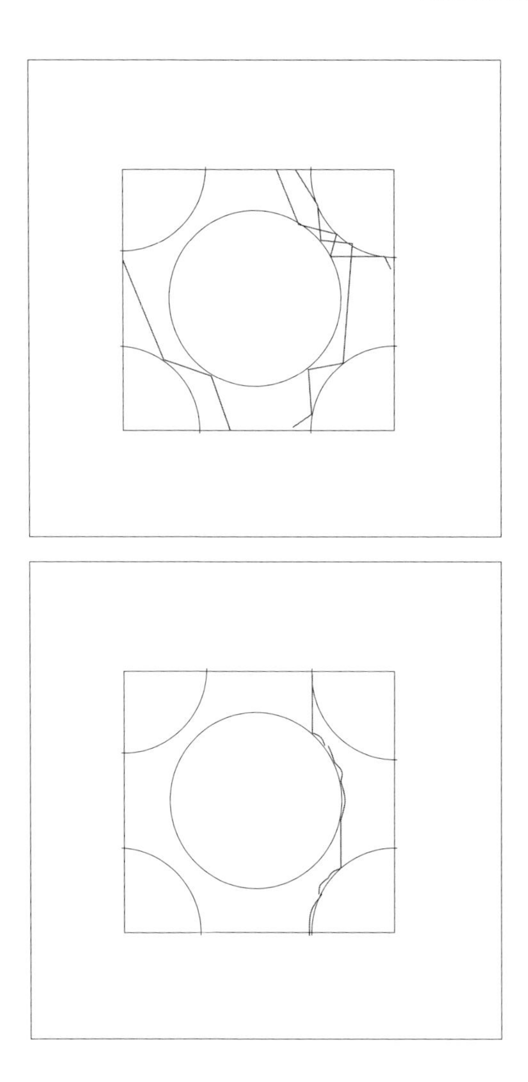

Fig. 2.2 Left: a trajectory at zero field. Right: a trajectory under a very strong field, pointing in the downward direction: the particle follows, through short bounces, the surface of the obstacle, and escapes when the field becomes tangent to it.

the trajectories continue to bounce close to the surface of the obstacles until they escape along a tangent direction parallel to the field, only to hit a new obstacle—see the bottom panel of Fig. 2.2. If we consider a stationary situation, these "trickling-down" trajectories involve *only a very restricted part of configuration space.* Adding energy-conserving noise to eqn (2.81) does not dramatically change the situation. In conclusion, the stronger the forcing field E, the more focused the *attractor* reached at long times is on a subset of the energy shell. As we shall see later, the Gallavotti–Cohen fluctuation relation is, in a certain sense, a measure of this focusing.

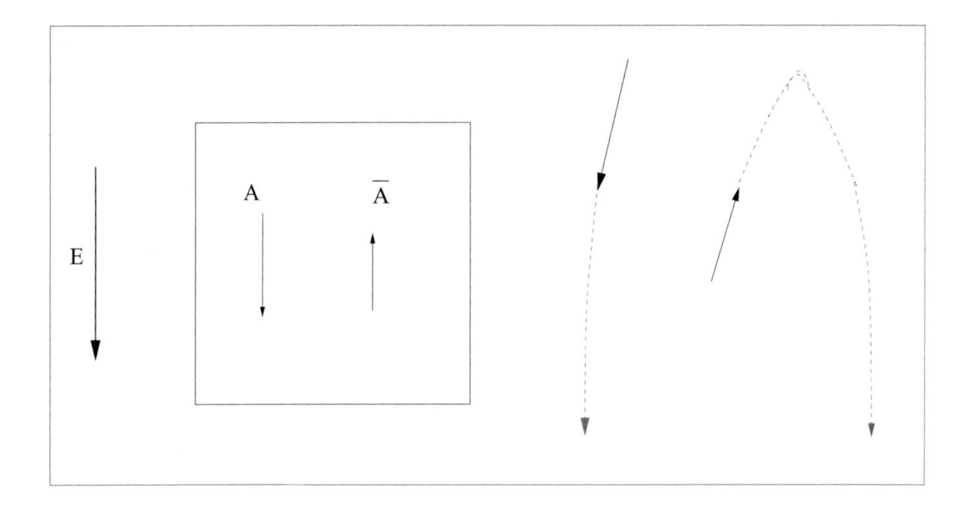

Fig. 2.3 Free trajectory under a field. The attractor A and repellor $\bar{A}$ are parallel and antiparallel, respectively, to the field.

2.2.5 Spectra of Fokker–Planck, Kramers, and Liouville operators

As we have seen above, the detailed-balance property implies that the Fokker–Planck operator can be taken to a Hermitian form H_h via the similarity transformation (2.60). This implies that its eigenvalues are real. If, on the other hand, detailed balance is violated, then there is no such transformation, and some eigenvalues come in complex conjugate pairs. In any case, the Perron–Frobenius theorem mentioned above implies that the real parts of eigenvalues are nonnegative.

At the other extreme, the Liouville operator corresponding to pure Hamiltonian dynamics,

$$H_{\text{Liouville}} = \frac{\partial \mathcal{H}}{\partial p_i}\frac{\partial}{\partial q_i} - \frac{\partial \mathcal{H}}{\partial q_i}\frac{\partial}{\partial p_i}, \tag{2.82}$$

(summation convention) would seem to be, at least superficially, *anti*-Hermitian, since it has only first derivatives. Its spectrum would be pure imaginary, except that we have to be careful when we define the space of wavefunctions on which it acts. Clearly, the Kramers operator, which in a sense interpolates between the two operators, is neither Hermitian nor anti-Hermitian, and has pairs of complex eigenvalues even in the conservative case, as we have mentioned already.

Consider a strongly chaotic, "mixing" Hamiltonian system. If we start from an initial configuration distributed in a small probability cloud, the probability distribution mixes completely in phase space—like a drop of ink in a stirred liquid, hence the name. Another way of obtaining the same result is to start from a single configuration, but to subject the dynamics to a small amount of noise. The fast mixing of the probability implies a concomitant fast decay of correlation functions to their stationary value. As discussed above (see eqn (2.40)), this would imply that there is a gap in the real part of the spectrum. Hence, if the deterministic system is sufficiently chaotic, we expect that, in the presence of a small amount of noise, its spectrum will have a gap in the real part of the lowest eigenvalues. When the noise is strictly zero the situation is more subtle, as there are many unstable periodic orbits, on which the correlations are of course periodic in time.

A very interesting question is, then, what happens when we start from a Hamiltonian system with stochastic noise, and gradually decrease the noise's intensity. This can be done with the Kramers equation without forcing, or, better, with the energy-conserving thermostat (2.15), letting $\epsilon \to 0$. The remarkable result, consistent with the discussion above, is that if the system is sufficiently chaotic, as we let $\epsilon \to 0$ some eigenvalues do not become imaginary: they retain a positive real part. These eigenvalues, and the corresponding eigenvectors, are the Ruelle–Pollicott resonances (Ruelle 1986; Pollicott 1986): the ones that have smaller real part are responsible for the long-time relaxation to equilibrium. A simple example of this phenomenon is discussed in the box below.

An instructive example is the particle in a harmonic potential with noise. The diagonalization of the corresponding Kramers operator is simple, and can be found in Risken's book (Risken 1984, Chapter 10). The spectrum is as follows. The eigenvalues are labeled by $n_1, n_2 = 0, 1, 2, \ldots$.

• **Stable case,** $V = \frac{1}{2}\omega^2 q^2$
Eigenvalues $\lambda_{n_1,n_2} = \frac{1}{2}\gamma(n_1 + n_2) + (i/2)\sqrt{4\omega^2 - \gamma^2}\,(n_1 - n_2)$. The spectrum becomes imaginary in the $\gamma \to 0$ limit of zero coupling to the bath, when the system becomes an undamped harmonic oscillator. This is compatible with the existence of many stable, periodic orbits.

• **Unstable case,** $V = -\frac{1}{2}\omega^2 q^2$
Eigenvalues $\lambda_{n_1,n_2} = \frac{1}{2}\sqrt{\gamma^2 + 4\omega^2}\,(n_1+n_2+1)+\frac{1}{2}\gamma(n_1-n_2-1)$. All eigenvalues are real and larger than zero. The latter is to be expected, given that there is no stationary state. The fact that even in the limit $\gamma \to 0$ the spectrum stays real may come as a surprise: it underlines the fact that although the Liouville operator seems superficially anti-Hermitian (and would thus lead us to expect pure imaginary eigenvalues), in fact the Hilbert space in which it acts makes it not so. Again, we find that the spectrum retains a real part as $\gamma \to 0$ when it is has *unstable* orbits ($q(t) = 0$, $p(t) = 0$, in this case) that are destroyed by any amount of noise (see also Gaspard *et al.* 1995; Gaspard 2003).

2.3 Separation of timescales

Many systems have processes happening on very different timescales. Often, what is interesting is what takes long to happen, while the rapid fluctuations are relatively featureless. Consider the following examples:

- *Metastability.* Chemical reactions often have metastable states. Consider a mixture of oxygen and hydrogen. It takes a very short time for this gas to become "equilibrated" into a mixture $O_2 + 2H_2$, which stays in a stationary state until the reaction $O_2 + 2H_2 \to 2H_2O$ starts somewhere—an extremely unlikely event—and then rapidly propagates throughout. The true equilibrium state, water vapor, is then reached. Similarly, diamond eventually decays into graphite, but this process is fortunately slow.

- *Hydrodynamic limit.* Systems with soft modes have a slow evolution along these modes, while the "hard" ones relax much faster. The typical example is a liquid, whose macroscopic motion is visible and slow, while the density fluctuations at the molecular scale evolve fast. As we shall see below, in some cases one can give a "hydrodynamic" description, with the fast fluctuations acting as a noise whose intensity goes to zero with the coarse-graining scale.
- *Coarsening and glasses.* Suppose one quenches a ferromagnet to a low, but nonzero temperature. Domains of positive and negative magnetization start growing. Inside each domain, the system resembles a pure ferromagnetic state with fast magnetization fluctuations. The domain walls, however, evolve slowly, the slower the larger the domains. A more subtle case is that of glasses, which have a fast evolution ("cage motion") and slow, collective rearrangements ("aging").

In all these cases, our ambition is to concentrate as much as possible on what is slow and interesting. In some cases, when we know a priori which are the slow coordinates, we may attempt to eliminate the fast fluctuations by allowing them to thermalize, at fixed values of the slow coordinates, which we then treat adiabatically. One thus obtains a "free-energy landscape" for the slow variables (see e.g. E *et al.* 2002).

What happens when we do not know exactly what is fast and what is slow? In the next sections we introduce a general approach to metastability, first in detail in a simple warming-up context, and then mention briefly how it works in general. In the last section, we describe the hydrodynamic limit of a transport problem, and how it takes us to a low-noise (quasi-deterministic) situation.

2.3.1 Metastability

The simple case of weak noise. Consider an overdamped Langevin system. We wish to analyze the spectrum of its Fokker–Planck evolution operator in the weak-noise limit. Because we know that T plays a role analogous to $\hbar$, we shall use what we know from semiclassical quantum mechanics. Let us first transform H_{FP} to its Hermitian basis (2.60):

$$H_h = e^{\beta V/2}\, H_{FP}\, e^{-\beta V/2} = \frac{2}{T}\sum_i \left[-\frac{T^2}{2}\frac{\partial^2}{\partial q_i^2} + \underbrace{\frac{1}{8}\left(\frac{\partial V}{\partial q_i}\right)^2 - \frac{T}{4}\frac{\partial^2 V}{\partial q_i^2}}_{V_{\text{eff}}} \right]. \tag{2.83}$$

Consider first a one-dimensional harmonic potential,

$$V = \frac{1}{2}aq^2, \qquad H_{FP} = -\frac{d}{dq}\left[T\frac{d}{dq} + aq\right]. \tag{2.84}$$

We have

$$H_h = e^{\beta V/2}\, H_{FP}\, e^{-\beta V/2} = \frac{2}{T}\left[-\frac{T^2}{2}\frac{d^2}{dq^2} + \frac{1}{2}\left(\frac{a}{2}\right)^2 q^2 - \frac{T}{4}a\right]. \tag{2.85}$$

Apart from the global factor $2/T$, we recognize the Hamiltonian of a quantum oscillator with $\omega = |a|/2$, $\hbar = T$, and $\hbar\omega = Ta/2$. The eigenvalues are then $\hbar\omega(n + \frac{1}{2})$, that is,

$$\lambda = \frac{2}{T}\left[\left(n + \frac{1}{2}\right)\frac{T|a|}{2} - \frac{Ta}{4}\right] = \begin{cases} 0,\ |a|,\ 2|a|, \dots & \text{if } a > 0\,, \\ |a|,\ 2|a|, \dots & \text{if } a < 0\,. \end{cases} \tag{2.86}$$

As expected, the lowest eigenvalue is zero in the stable case and positive in the unstable case. The gap between eigenvalues is proportional to the curvature of the potential.

We can extend this result to the stationary point of any potential, using the fact that at low temperature only the neighborhood of the saddle points contributes. Expanding V_{eff} around a minimum, which we assume is at $q = 0$, and putting $q/\sqrt{T} \to x$, we have

$$\begin{aligned} V_{\text{eff}} &= V''_{\text{eff}}(0)\frac{q^2}{2} + V'''_{\text{eff}}(0)\frac{q^3}{6} + \dots \\ &= V''_{\text{eff}}(0)\frac{x^2}{2} + \underbrace{V'''_{\text{eff}}(0)\sqrt{T}\frac{x^3}{6}}_{\text{subdominant}} + \dots \end{aligned} \tag{2.87}$$

so that

$$H_h = 2\left[-\frac{1}{2}\frac{d^2}{dx^2} + \frac{1}{2}\left(\frac{|V''(0)|}{2}\right)^2 x^2 - \frac{1}{4}V''(0) + \text{subdominant} \dots\right]. \tag{2.88}$$

The eigenvalues are given by eqn (2.86) with $V''(0)$ playing the role of a.

The generalization to many dimensions is straightforward. Since one has to expand the potential only to second-order around a saddle, one can then go to the basis where the second-derivative matrix is diagonal, and treat each mode as an independent oscillator. Every saddle point yields to this order an independent spectrum in the Fokker–Planck operator, but only local minima have zero eigenvalues. There is then exactly one zero eigenvalue per minimum. If the calculation is done exactly, one finds that the degeneracy is lifted by a small amount inversely proportional to the passage times between states, in this case exponentially small in the inverse temperature. If there are p local minima, there are then p "almost zero" eigenvalues $\lambda_1, \dots, \lambda_p$, with the associated escape times $\lambda_1^{-1}, \dots, \lambda_p^{-1}$ bounded by the smallest escape time $t_{\text{pass}} = \min\{\lambda_1^{-1}, \dots, \lambda_p^{-1}\}$.

Let us now turn to the eigenvectors corresponding to the "almost zero" eigenvalues. The construction we have done above for eigenvalues can be completed to obtain (approximate) right and left eigenvectors. Close to a minimum, we expect the right eigenvector to be correspond to a Gibbs distribution peaked around it. Similarly, we expect the left eigenvector to be essentially a constant. This is in fact the case, but the result is stronger, as it holds throughout the basin of attraction of each minimum. The situation is depicted in Fig. 2.4. For small temperatures, it is as if an infinitely high, thin wall encloses each basin, within which the eigenvectors and eigenvalues are those of the isolated region. After some time, the finite-temperature corrections which split the degeneracy of "almost zero" eigenvalues also mix the approximate eigenvectors localized in each basin.

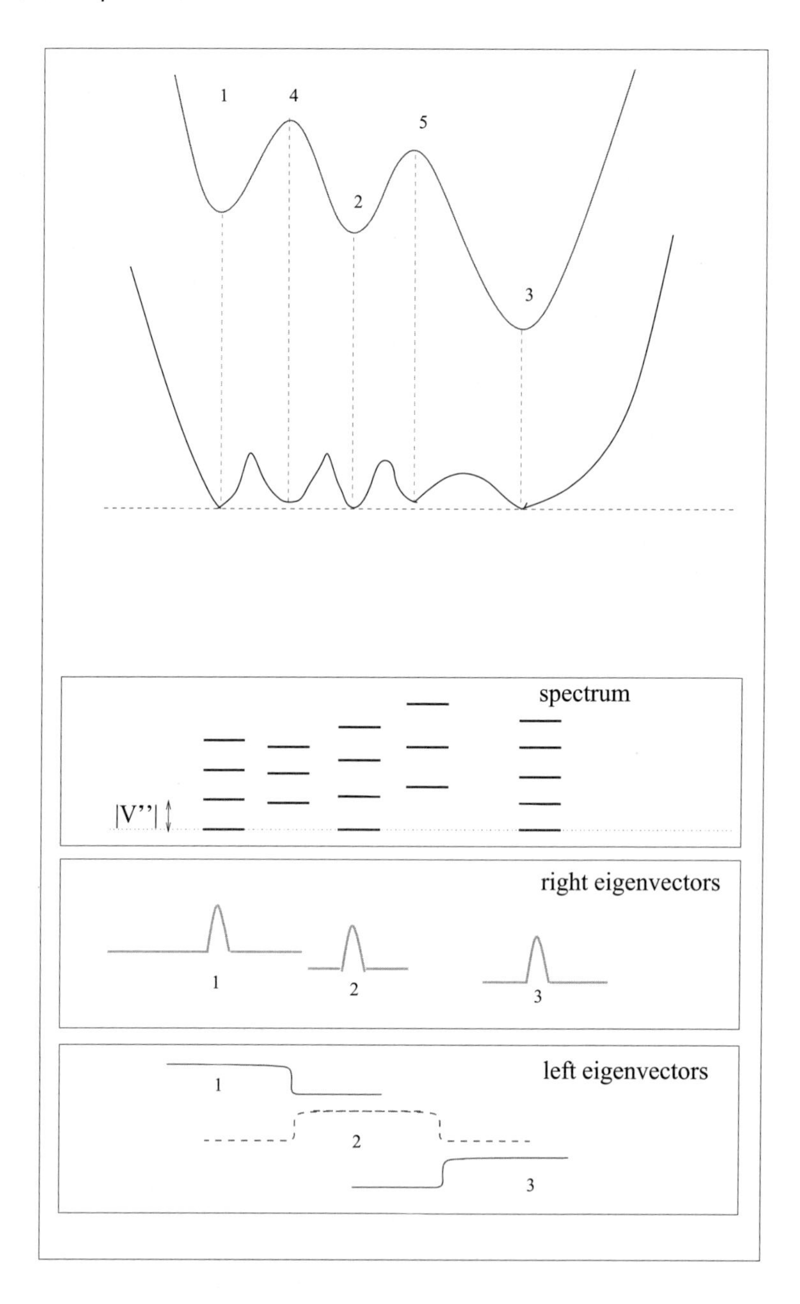

Fig. 2.4 Low-temperature spectrum for a multivalleyed potential V (top), and the associated effective potential V_{eff} (bottom). Only the minima of V contribute, with near-zero eigenvalues. The near-zero (in this case three-dimensional) subspace is spanned by right eigenfunctions that are essentially the equilibrium distribution in each basin, and left eigenfunctions (the *committors*) that are essentially constant within each basin. The places where committors take the value one-half are the *transition states*.

General approach to metastability. Spectra, states, and committor functions. It turns out that the low-temperature situation is just an instance of a very general scenario. Indeed, one can turn things around and use the existence of a gap in the spectrum to give a general and useful definition of metastability (Gaveau and Schulman 1996*a*,*b*, 1998; Gaveau *et al.* 1999; Bovier *et al.* 2000, 2002, 2004) (see Biroli and Kurchan (2001) for an application to metastability in glasses). Consider an evolution operator having the lowest p eigenvalues $\lambda_1, \dots, \lambda_p$, whose real part is separated by a gap from the others $\lambda_{p+1}, \lambda_{p+2}, \dots$, (see Fig. 2.5). There are two characteristic times, $t_{\text{pass}} = \min\{\lambda_1^{-1}, \dots, \lambda_p^{-1}\}$ and $t^* = \max\{\lambda_{p+1}^{-1}, \lambda_{p+2}^{-1}, \dots\}$: they will be interpreted as the minimal time needed to escape a metastable state and to equilibrate within a state, respectively. In the previous section, t^* was of order one and t_{pass} was exponentially large in T. At times of order $t \gg t^*$, the dynamics projects completely to the space below the gap, but if $t \ll t_{\text{pass}}$ there is still no time for discerning between different eigenvectors below the gap. Clearly, the operator $\exp[-tH]$ for $t^* \ll t \ll t_{\text{pass}}$ is essentially a projector onto the space "below the gap" (up to terms of order $\exp[-t\lambda_a]$, with $a > p$).

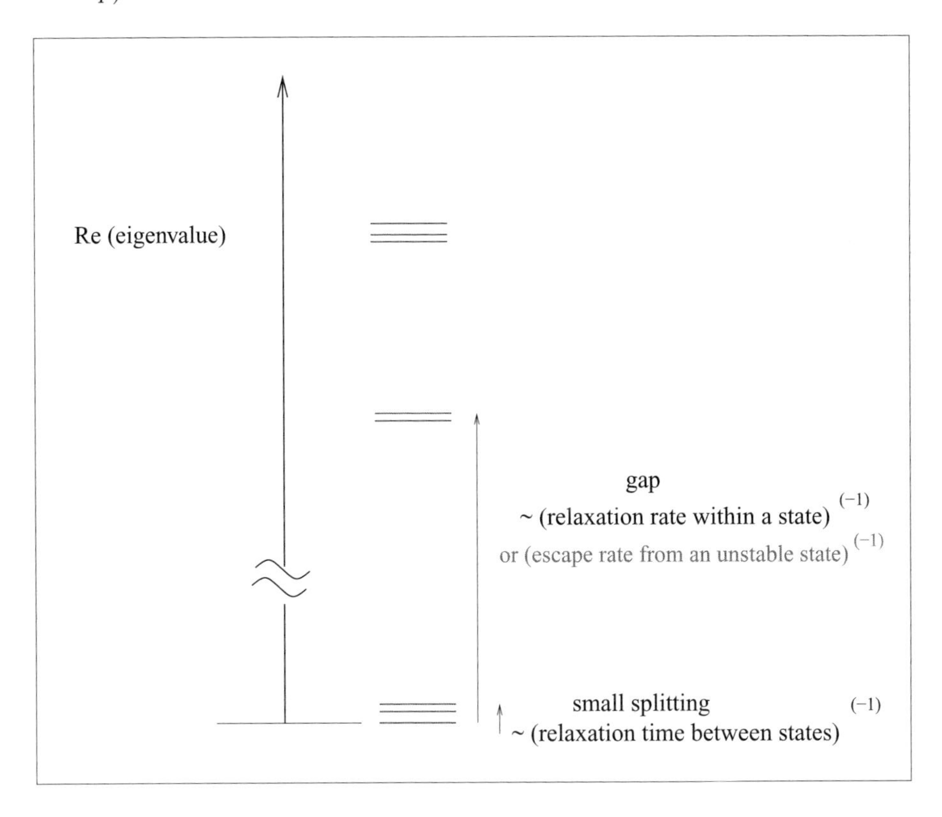

Fig. 2.5 The general situation in a system with metastability, whatever the origin of timescale separation. The states $|P_a\rangle$ that can be interpreted as metastable probability distributions are linear combinations of the right eigenvectors "below the gap."

Within the same accuracy, it turns out that one can then find a basis of p right eigenvectors $|P_a\rangle$ which are:

- positive: $P_a(\mathbf{q}) = \langle \mathbf{q}|P_a\rangle \geq 0$;
- almost stationary: $H|P_a\rangle \sim 0 \quad \forall\, a = 1, \ldots, p$;
- normalized and not zero in nonoverlapping regions of space.

As a consequence, the vectors $|P_a\rangle$ have all the good properties of metastable states: they are positive normalized distributions, are nonzero only in different regions of the configuration space, and are stationary on timescales less than t_{pass}. The last property is related to the fact that one can also find a basis of p almost-stationary ($\langle Q_a|H \sim 0$) left eigenvectors $\langle Q_a|$. They satisfy the approximate orthogonality and normalization conditions

$$\langle Q_a|P_b\rangle \sim \delta_{ab}\,. \tag{2.89}$$

One can also write approximately

$$e^{-tH} \sim \sum_a |P_a\rangle\langle Q_a|\,. \tag{2.90}$$

Note that neither $\langle Q_a|$ nor $|P_b\rangle$ are exact eigenvectors "below the gap," but they are linear combinations of them.

The $Q_a(\mathbf{q})$ are the *committor functions* (Bolhuis *et al.* 2002; Du *et al.* 1998; Dellago *et al.* 2002) of the states, giving the probability that, starting in a certain point $\mathbf{q}$, the dynamics (again, in times $t^* \ll t \ll t_{\text{pass}}$) goes to the state a. This can be easily seen as follows. The probability of ending in a point $\mathbf{q}'$ starting from a point $\mathbf{q}$ is, at times of order $t^* \ll t \ll t_{\text{pass}}$,

$$\text{Probability} \sim \langle \mathbf{q}'|e^{-tH}|\mathbf{q}\rangle \sim \sum_i \langle \mathbf{q}'|P_a\rangle\langle Q_a|\mathbf{q}\rangle\,. \tag{2.91}$$

If the point $\mathbf{q}'$ is well within the state "a," then $P_a(\mathbf{q}')$ is large and the other P_b ($b \neq a$) are small. There is only one nonzero term in this sum, and we conclude that the probability to fall in the state "a" is proportional to $Q_a(\mathbf{q})$. Each $Q_a(\mathbf{q}) = \langle Q_a|\mathbf{q}\rangle$ is essentially one within the basin of attraction of the state a, and almost zero everywhere else. The places where the $Q_a(\mathbf{q}) \sim \frac{1}{2}$ are called the *transition states.*

Given any observable A, we can calculate its average within the state "a" as

$$\langle A\rangle_a = \langle Q_a|A|P_a\rangle\,. \tag{2.92}$$

Again, the situation described above can be summarized by saying that for times $t^* \ll t \ll t_{\text{pass}}$ everything happens as if there is an infinite wall enclosing each basin of attraction. In the proof of this, as in the simple example of the previous subsection, the definition is unavoidably linked to the timescales (t^*, t_{pass}): if one considers really infinite times, before any other limit, then the distinction between states vanishes. In real life, this separation of timescales might be controlled by a parameter, or it might be just a more or less valid approximation. As an example of the former, consider an Ising ferromagnet of size L: the longest relaxation time within a state is the time for a domain to grow to the size of the system, $t^* \sim L^2$. The longest overall time is the time needed to flip the global magnetization, which requires jumping the highest energy barrier, $t_{\text{pass}} \sim e^{cL^{d-1}}$. For large L, there is an ample regime $t^* \ll t \ll t_{\text{pass}}$ where there are two well-defined states. Many examples for which in fact this construction is most interesting have no parameter that controls the separation of t_{pass} and t^*: the time for relaxation within a metastable state and the time for escaping it are different but not infinitely so.

2.3.2 Transition currents

Given a metastable state $|P\rangle$ constructed as above, we can find the probability current "leaking" from it directly. For example, in a Fokker–Planck case, putting $P(\mathbf{q}) = \langle \mathbf{q}|P\rangle$,

$$J_i(\mathbf{q}) \propto \left(T\frac{\partial}{\partial q_i} + \frac{\partial V}{\partial q_i} + f_i \right) P(\mathbf{q}) \,. \tag{2.93}$$

Because every metastable state P is a linear combination of the exact eigenvectors $\psi^R_\alpha(\mathbf{q}) = \langle \psi^R_\alpha|\mathbf{q}\rangle$ with eigenvalues below the gap,

$$P(\mathbf{q}) = \sum_{\alpha=1}^{p} c_\alpha \psi^R_\alpha(\mathbf{q}) \,, \tag{2.94}$$

the associated escape currents are (see Fig. 2.6) linear combinations of currents obtained by acting on each of them,

$$J^\alpha_i \equiv \left(T\frac{\partial}{\partial q_i} + \frac{\partial V}{\partial q_i} + f_i \right) \psi^R_\alpha(\mathbf{q}) \,. \tag{2.95}$$

If the stationary state has no current, then there are only $p-1$ independent currents, rather than p of them.

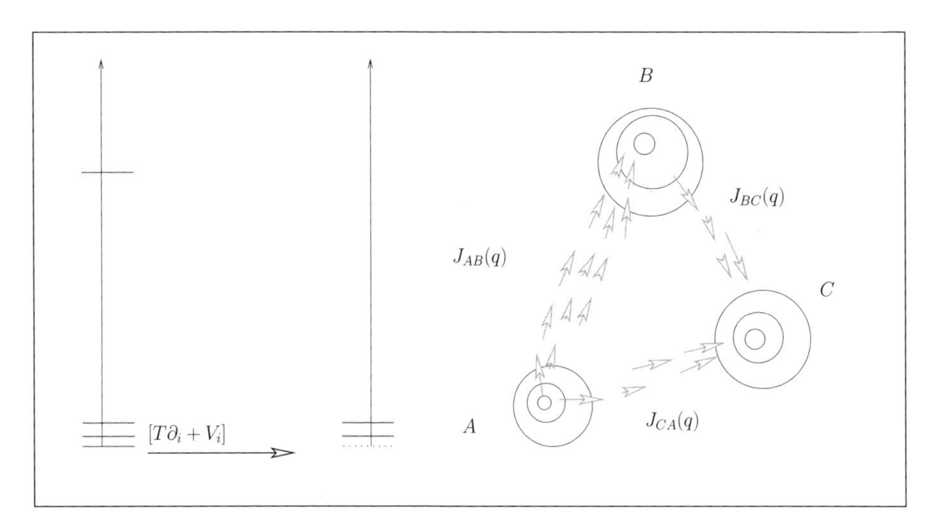

Fig. 2.6 Spectrum and escape currents. By acting with $[T\partial/\partial q_i + \partial V/\partial q_i]$ on the eigenvectors "below the gap," one obtains a basis for all the interstate currents.

In many physical situations, we have some idea of the reaction path followed by the current, but we do not know the intensity of such a current or the reaction rate. Let us derive a formula for the rate in terms of an unnormalized current distribution $\mathbf{J}(q)$. The only assumption we make is that the reaction time is slower then any other relaxation (Tailleur *et al.* 2004). Suppose one has a current $\mathbf{J}$ escaping a metastable state in an overdamped Langevin problem $P(\mathbf{q})$ with only conservative forces:

$$J_i(\mathbf{q}) \propto \left(T\frac{\partial}{\partial q_i} + \frac{\partial V}{\partial q_i}\right) P(\mathbf{q}) \quad \text{with} \quad P(\mathbf{q}) = \sum_{\alpha=1}^{p} c_\alpha \psi_\alpha^R(\mathbf{q}) \tag{2.96}$$

$$\text{and} \quad \int d^N\mathbf{q}\, \psi_\alpha^L(\mathbf{q})\psi_\delta^R(\mathbf{q}) = \delta_{\alpha\delta} = \int d^N\mathbf{q}\, e^{\beta V}\, \psi_\alpha^R(\mathbf{q})\psi_\delta^R(\mathbf{q}) . \tag{2.97}$$

The last equality in eqn (2.97) is deduced from the relation between right and left eigenvectors (2.61). We first compute

$$\begin{aligned}
\int d^Nq\, e^{\beta V}\, \mathbf{J}^2 &= \int d^Nq \left\{ \left(T\frac{\partial}{\partial q_i} + \frac{\partial V}{\partial q_i}\right) P \right\} e^{\beta V} \left\{ \left(T\frac{\partial}{\partial q_i} + \frac{\partial V}{\partial q_i}\right) P \right\} \\
&= T \int d^Nq \left\{ \left(T\frac{\partial}{\partial q_i} + \frac{\partial V}{\partial q_i}\right) P \right\} \frac{\partial}{\partial q_i} \left(e^{\beta V} P\right) \\
&= T \int d^Nq\, P\, e^{\beta V}\, (H_{FP} P) = \sum_{\alpha=1} \lambda_\alpha c_\alpha^2 ,
\end{aligned} \tag{2.98}$$

and similarly

$$\int d^Nq\, e^{\beta V}\, (\text{div}\, \mathbf{J})^2 = \int d^Nq\, (H_{FP}P)\, e^{\beta V}\, (H_{FP}P) = \sum_{\alpha=1} \lambda_\alpha^2 c_\alpha^2 , \tag{2.99}$$

where we have used the eigenvalue equation and the normalization (2.97). Let us now assume for simplicity that there is only one metastable state, so that there are $p = 2$ eigenvalues "below the gap." At large times $t^* \ll t \ll t_{\text{pass}} \sim \lambda_m^{-1}$, only the first nonzero eigenvalue λ_m contributes to the sums, and we get

$$t_{\text{activ}} = \lambda_m^{-1} = \frac{\sum \lambda_\alpha c_\alpha^2}{\sum \lambda_\delta^2 c_\delta^2} = \frac{\int d^Nx\, e^{\beta V}\, \mathbf{J}^2}{T \int d^Nq\, e^{\beta V}\, (\text{div}\, \mathbf{J})^2} . \tag{2.100}$$

Note that the normalization of the current is irrelevant. *This formula is valid on the assumption of separation of timescales, irrespective of its cause.*

For the Kramers equation, a similar expression can be obtained in the same way, in terms of the *reduced current* (2.12). A tedious but straightforward calculation yields (Tailleur *et al.* 2004)

$$t_{\text{activ}} = \text{Re}\, \lambda_{\max}^{-1} = \frac{\gamma \int d^N\mathbf{q}\, d^N\mathbf{p}\, e^{\beta E} \left[\sum_i J_{q_i}^{\text{red}}(\mathbf{q}, \mathbf{p}) J_{q_i}^{\text{red}}(\mathbf{q}, -\mathbf{p})\right]}{T \int d^N\mathbf{q}\, d^N\mathbf{p}\, e^{\beta E}\, \text{div}_{\mathbf{q}}, \mathbf{p}[J^{\text{red}}(\mathbf{q}, \mathbf{p})] \text{div}_{\mathbf{q}}, \mathbf{p}[J^{\text{red}}(\mathbf{q}, -\mathbf{p})]} . \tag{2.101}$$

Note that the sum in the numerator runs over coordinates and not over momenta. These formulae are useful because they do not depend on a global normalization of the current.

Arrhenius formula. The Arrhenius expression for activated passage in the low-temperatures case can be easily derived from the formula (2.100), using the fact that the numerator is dominated the neighborhood of the barrier top, and the numerator by the neighborhood of the bottom of the starting well. Let us show how this works for a one-dimensional double well as in Fig. 2.7. The current starts in the metastable state, is then approximately constant, and falls in the stable state.

The divergence of the current $J(q)$ of a state $\psi^R(q)$ satisfies

$$\text{div} J = H_{FP}\psi(q) = \lambda_m \psi(q)\,, \tag{2.102}$$

where λ_m is the first nonzero eigenvalue. Now, we have seen that if $\psi(q)$ is a metastable equilibrium distribution in the departure state, it is proportional there to the Gibbs–Boltzmann distribution *in that basin*, so that

$$\text{div} J \propto \psi(q) \propto e^{-\beta V} \tag{2.103}$$

up to a normalization. The current in a point q within the metastable basin is then

$$J(q) = \int_{-\infty}^{q} dq'\, \text{div} J(q') = \int_{-\infty}^{q} dq'\, e^{-\beta V} \sim e^{-\beta V_{\min}} \int_{-\infty}^{q} dq'\, e^{-\beta V''_{\min} q'^2/2}\,, \tag{2.104}$$

where we have used a Gaussian approximation. Around the barrier, we get

$$J_{\text{barrier}} \sim \left(\frac{2\pi}{\beta V''_{\min}}\right)^{1/2} e^{-\beta V_{\min}}\,. \tag{2.105}$$

The denominator of eqn (2.100) is dominated by the neighborhood of the minimum:

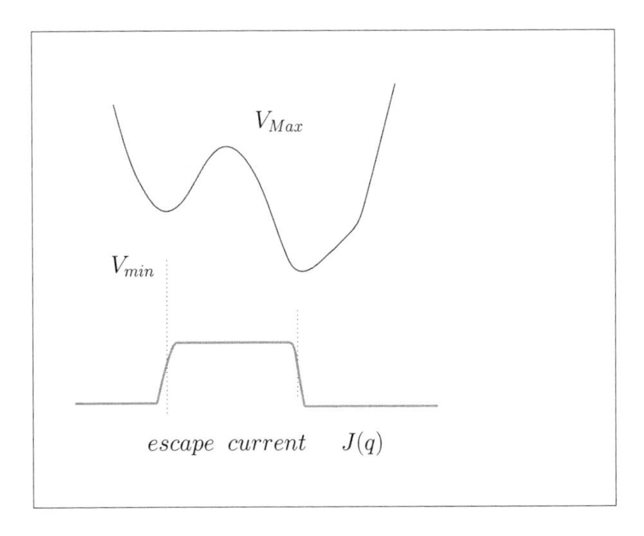

Fig. 2.7 Escape from a metastable state of height $V_{\min}$ through a barrier of height V_{Max}. Bottom: the current distribution.

$$\int_{\text{basin}} dq\, e^{\beta V}\, (\text{div}\, J)^2 \sim \int dq\, e^{-\beta V} \sim e^{-\beta V_{\min}} \left(\frac{2\pi}{\beta V''_{\min}} \right)^{1/2} . \tag{2.106}$$

The numerator in eqn (2.100) becomes, using the fact that J is essentially constant around the maximum,

$$\int_{\text{barrier}} dq\, e^{\beta V}\, J^2 \sim J^2_{\text{barrier}} \int_{\text{barrier}} dq\, e^{\beta V} \sim e^{\beta V_{\text{Max}}} \left(\frac{2\pi}{\beta V''_{\text{Max}}} \right)^{1/2} J^2_{\text{barrier}} \, . \tag{2.107}$$

Putting numerator and denominator together, we get

$$t_{\text{activ}} = \lambda_{\max}^{-1} = \frac{\int dq\, e^{\beta V} J^2}{T \int dq\, e^{\beta V} (\text{div}\, J)^2} \sim 2\pi\, \frac{e^{\beta (V_{\text{Max}} - V_{\min})}}{\sqrt{|V''_{\text{Max}} V''_{\min}|}} , \tag{2.108}$$

which is the Arrhenius formula (Hänggi *et al.* 1990), with the correct prefactor. Note how simple the argument is, once we have eqn (2.100).

2.3.3 Hydrodynamic limit and macroscopic fluctuations

The simple symmetric exclusion process of eqn (2.24) and Fig. 2.8 has a separation of timescales, this time brought about by a local conservation law (of particle number) rather than by barriers. Consider an L-site long, isolated chain, with average occupation one-half. Suppose now that we make a vacancy of 20 contiguous unoccupied sites. For this to happen spontaneously is an extremely rare event ($\sim 2^{-20}$), and left on its own, the vacancy will be covered rapidly, in a time of order one. At the other extreme, consider a slowly varying average density profile, say a sinusoidal oscillation of length L. Such a fluctuation will take a time of order L^2 to die out.

This separation of timescales manifests itself in the stochastic evolution operator, which has eigenvalues of order 1 for the steepest and L^{-2} for the smoothest spatial fluctuations.

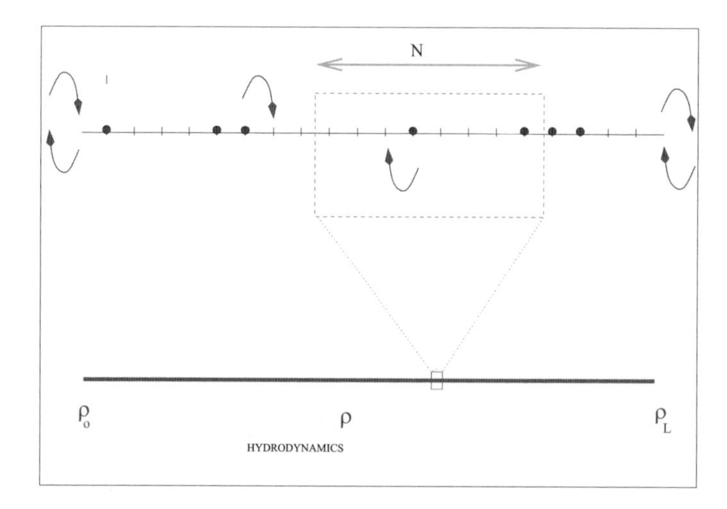

Fig. 2.8 The simple symmetric exclusion process, and its coarse-grained hydrodynamic limit.

We may now choose to study only the smooth fluctuations, corresponding to the lowest eigenvalues (see Spohn 1983; Bertini *et al.* 2001, 2002). To do this we introduce a parameterization of space[6] $x_k = k/L$ and rescale the time as $t \to L^2 t$. In the rescaled time, steep fluctuations disappear immediately. At the macroscopic level, the density profiles we consider are smooth functions, and discrete gradients can be replaced by continuous ones:

$$\rho_{k+1} - \rho_k \to \frac{\nabla \rho(x_k)}{L}, \quad \hat{\rho}_{k+1} - \hat{\rho}_k \to \frac{\nabla \hat{\rho}(x_k)}{L}, \quad \frac{1}{L}\sum_{k=1}^{L-1} \to \int_0^1 dx\,. \tag{2.109}$$

It can be shown that in terms of these variables, the evolution of the smooth density fluctuations is given by the stochastic equation (Spohn 1983)

$$\dot{\rho} = \frac{1}{2}\nabla^2 \rho + \nabla[\sqrt{\rho(1-\rho)}\eta]\,, \qquad \rho(0) = \rho_0\,, \qquad \rho(1) = \rho_1\,, \tag{2.110}$$

where η is a white noise of variance $1/(2L)$. This is the formula for the fluctuating hydrodynamics of the exclusion process (Spohn 1983).

We can now construct the action with the Martin–Siggia–Rose (Martin *et al.* 1973; De Dominicis 1976; Janssen 1976; De Dominicis and Peliti 1978) formalism. Performing the usual steps as in Section 2.1.3, introducing conjugate fields $\hat{\rho}(x,t)$, we get

$$P(\rho^{\mathbf{f}}, t_f; \rho^{\mathbf{i}}, 0) = \int D[\rho]\, D[\hat{\rho}]\; e^{-2L\{\int_0^{t_f}\int_0^1 dx\, dt\, (\hat{\rho}\dot{\rho} - \tilde{H})\}}\,. \tag{2.111}$$

This time, L^{-1} plays the role of the temperature, or of $\hbar$, and we have the "classical" Hamiltonian density

$$\tilde{H} = \frac{1}{2}\int dx\, \left[\rho(1-\rho)(\nabla\hat{\rho})^2 - \nabla\hat{\rho}\nabla\rho\right]. \tag{2.112}$$

The paths are constrained to be $\rho_i(x)$ and $\rho_f(x)$ at the initial and final times, respectively. The values of $\hat{\rho}$ are unconstrained, which is in agreement with the fact that this is a Hamiltonian problem with two sets of boundary conditions.

The message of this section is that *we may transfer all the "semiclassical" low-noise (small-T) techniques to the coarse-grained limit* to obtain a "macroscopic fluctuation theory" (Bertini *et al.* 2001, 2002; Jordan *et al.* 2004).

2.4 Large deviations

In equilibrium statistical mechanics, we are given the probability of being in any particular configuration. For a dynamical system, we may wish to ask similar questions concerning *histories*, rather than configurations: what is the probability that the system visits a sequence of configurations at given times, or that during a time interval its average energy has a given value, and so on. Often these events are *rare*; their probability is small. In spite of this, they may be important: for example, what is interesting

[6]Note that x here is an index labeling the field $\rho(x)$. Comparing this with the examples in the first sections, we should use the correspondence $(q_i, i) \to (\rho(x), x)$.

in chemistry are reactions that are slow compared with thermal vibrations. In this section we shall study two types of large deviations: those that are rare because they are induced by (weak) noise, and those that are rare because they are sustained for a long time. Technically, this provides us with two small parameters: a noise intensity and an inverse time-span.

2.4.1 Climbing to unusual heights

We now study the probability of finding the system in unusual configurations. The formalism is essentially the WKB theory for semiclassical quantum mechanics; mathematicians know it as the Freidlin–Wentzell (Freidlin and Wentzell 1984) formalism. For simplicity, the discussion in this section will be for the overdamped case, but one can do the same for any other stochastic equation.

In the small-noise limit, whatever its origin (low temperature or hydrodynamic limit), the equations of motion are essentially deterministic. If we ask for the probability $P(\mathbf{q}_0, t_0 \to \mathbf{q}, t)$ of the system meeting an appointment at time t in $\mathbf{q}$, given that it started in $\mathbf{q}_0$ at time t_0, we may get two sorts of answers: (i) essentially *one* if a deterministic path passes precisely through the given points at the given times; (ii) exponentially small, $\sim e^{-(1/T)\mathcal{F}(\mathbf{q},t)}$, otherwise: only thanks to the noise can the system get out of its deterministic schedule in order meet the appointment (this includes being in the right place at the wrong time). In order to calculate probabilities, we go back to the path-integral expression (2.51),

$$P(\mathbf{q}_0, t_0 \to \mathbf{q}, t) = \int D[\mathbf{q}, \hat{\mathbf{q}}] \; e^{(1/T)\int dt \; [\sum_i \hat{q}_i \dot{q}_i - \tilde{H}]} \, . \tag{2.113}$$

As we mentioned in Section 2.1, this is an imaginary-time path integral with the "classical" Hamiltonian

$$\tilde{H} = -\sum \hat{q}_i \left(\hat{q}_i + \frac{\partial V}{\partial q_i} + f_i \right) . \tag{2.114}$$

The path integral is dominated by the extremal trajectories, which satisfy Hamilton's equations,

$$\begin{cases} \dot{q}_i = \dfrac{\partial H}{\partial \hat{q}_i} & = -\left(\hat{q}_i + \dfrac{\partial V}{\partial q_i} + f_i \right) - \hat{q}_i \, , \\ \dot{\hat{q}}_i = -\dfrac{\partial H}{\partial q_i} & = \displaystyle\sum_j \left(\frac{\partial^2 V}{\partial q_j \partial q_i} \right) \hat{q}_j + \frac{\partial f_j}{\partial q_i} \hat{q}_j \, . \end{cases} \tag{2.115}$$

The probability now takes the large-deviation form

$$\ln \{ P(\mathbf{q}_0, t_0 \to \mathbf{q}, t) \} = -\frac{1}{T} \mathcal{F}_t = -\frac{1}{T} \text{Action} \, , \tag{2.116}$$

which defines the large-deviation function $\mathcal{F}_t(\mathbf{q})$. The action is the integral

$$\text{Action} = -\int dt \left[\sum_i \hat{q}_i \dot{q}_i - \tilde{H}(\mathbf{q}, \hat{\mathbf{q}}) \right] , \tag{2.117}$$

where $(\mathbf{q}, \hat{\mathbf{q}})$ is the solution of eqn (2.115).

Noiseless solution. A family of solutions of eqn (2.115) can be easily found:

$$\hat{q}_i = 0\,, \qquad \dot{q}_i = -\frac{\partial V}{\partial q_i} - f_i\,. \tag{2.118}$$

The dynamics of the original (hatless) variables is just the noiseless equation of motion (2.3). The action is zero, as can be easily checked. This means that the large-deviation function will also be zero—in fact, its smallest value.

Other solutions. Other solutions of eqn (2.115) can be found with $\hat{q}_i \neq 0$. They do not correspond to motion in the original force field (2.3), signaling the fact that noise is playing an important role. The action is positive, a fact that can be best appreciated in the Lagrangian formalism (2.53). The large-deviation function is now positive, and the probability is exponentially suppressed in $1/T$, again an indication that noise is playing a role.

Large times. In most cases, we are interested in the long-time limit $\mathcal{F}_t$ at large times:

$$e^{-(1/T)F(\mathbf{q})} = \lim_{t\to\infty} P(\mathbf{q}_0, t_0 \to \mathbf{q}, t) \tag{2.119}$$

or, in other words, in the probability of finding a stationary system in a configuration $\mathbf{q}$. The corresponding large-time deviation function (which we shall denote simply as $\mathcal{F}$) becomes independent of the initial condition. How can this be?

Consider first a quadratic potential $V = \frac{1}{2}q^2$ in one dimension, as in Fig. 2.9. The Hamiltonian is $\tilde{H} = -\hat{q}(\hat{q} + q)$. The "classical" trajectories are given by

$$\begin{cases} \dot{q} = -2\hat{q} - q\,, \\ \dot{\hat{q}} = \hat{q}\,. \end{cases} \tag{2.120}$$

The solution with $\hat{q}(t) = 0$ corresponds to the relaxation to the minimum. If we now start at time t_0 at, say, $q = 1$ and ask that at time $t > t_0$ we should be at $q = -1$, we have to take one of the trajectories with $\hat{q} > 0$; the one that arrives at $q = -1$ at the right time. If we now consider very large times t, the solution will be one that passes close to the (hyperbolic) point $(q = 0\,,\, \hat{q} = 0)$, in the vicinity of which it will spend a long time. In the limit $t \to \infty$, the trajectory is the succession of a gradient descent to the origin and an "uphill" trajectory emerging from the origin. Similarly, the trajectory that starts at $q = 1$ and ends at large times at $q = +2$ is composed of a gradient descent towards the *left*, followed by an uphill motion to the *right*—the limit of a $\hat{q} < 0$ "bounce."

In conclusion, in order to calculate the stationary large-deviation function we have to consider the "downhill" trajectory (the *anti-instanton*) with $\hat{q} = 0$ from the initial to the stationary point, followed by an "uphill" (*instanton*) trajectory from the stationary point to the final point. The large-deviation function is the sum of the downhill and the uphill actions. As mentioned above, the former is zero: when calculating the probability of reaching a configuration at large times, we may consider that we started from a stationary point—a physically intuitive result since, in a short time at the beginning, the system goes from the initial to a stationary configuration.

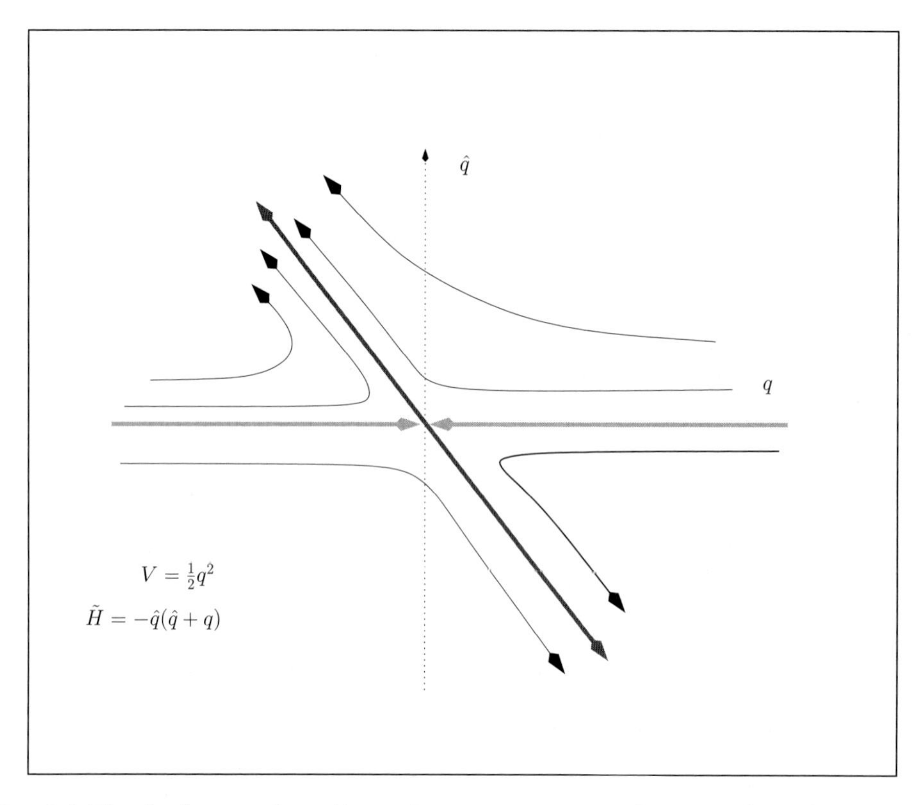

Fig. 2.9 The $(q, \hat{q})$ space for a Langevin process in a one-dimensional quadratic potential. The incoming straight lines indicate the noiseless "downhill" trajectories, the outgoing straight lines the "uphill" trajectories. Curved trajectories missing the origin are relevant for finite-time large deviations.

We need the "uphill," instanton trajectories to calculate, via their action, the probability of a rare configuration. *For generic dynamics, this is hard problem to solve analytically, and even numerically.*

Detailed balance and Onsager–Machlup symmetry. As we saw in Section 2.2, if there is detailed balance, and given that one trajectory dominates for the downhill process, we should expect the time-reversed trajectory to dominate for the uphill process. Let us check this explicitly for a multidimensional system with a potential V and no forcing, $f_i = 0$. We propose, as a partial solution of eqn (2.115),

$$\dot{q}_i = +\frac{\partial V}{\partial q_i} . \tag{2.121}$$

Inserting this into the first equation of eqn (2.115), this implies $\hat{q}_i = -\partial V/\partial q_i$, which, when replaced in the second equation of eqn (2.115), gives

$$\dot{\hat{q}}_i = \sum_j \frac{\partial^2 V}{\partial q_j \partial q_i} \hat{q}_j \,. \tag{2.122}$$

Replacing $\hat{q}_i = -\partial V / \partial q_i$ in this equation, and using

$$\frac{d}{dt}\left(-\frac{\partial V}{\partial q_i}\right) = -\sum_j \frac{\partial^2 V}{\partial q_j \partial q_i} \dot{q}_j \,, \tag{2.123}$$

we obtain an identity. We conclude that the time-reversed dependence[7] (2.121) is indeed a solution. We may compute the action of the uphill trajectory, which turns out to depend exclusively on the initial and final potentials:

$$\begin{aligned} \text{action} &= \int dt \sum_i \underbrace{\hat{q}_i}_{-\frac{\partial V}{\partial q_i}} \dot{q}_i + \int dt \sum_i \hat{q}_i \underbrace{\left(\hat{q}_i + \frac{\partial V}{\partial q_i}\right)}_{=0} \\ &= \int dt \, \frac{dV}{dt} = V(\mathbf{q}) - V(\mathbf{q}_{\min}) \,, \end{aligned} \tag{2.124}$$

in accordance with the general situation for detailed balance discussed in Section 2.2.1. Up to a multiplicative constant, this implies via eqn (2.116) that the stationary probability is the Gibbs–Boltzmann weight, as expected.

On the other hand, repeating the calculation in the case in which there is a generic force term f_i, we find that the second equation of eqn (2.115) is satisfied by the time-reversed trajectory if $\sum_j (\partial f_i / \partial q_j - \partial f_j / \partial q_i) f_j = 0$, which is in general not true. Hence, we reach the important conclusion that *in the presence of forces that do not derive from a potential, relaxations into and excursions out of the stationary state are not the time-reversed forms of one another.* Furthermore, we have to calculate the action on the basis of the explicit solution, and this is no longer miraculously given exclusively in terms of the initial and final configurations.

The Arrhenius law again. In the preceding paragraphs we have assumed that in order to reach the final configuration, only one downhill and one uphill trajectory suffice. This is clearly the case when there is only one stable state. In the presence of metastability, the trajectory may be a sequence of downhill and uphill segments.

For example, in the case of a double-well potential, the passage probability (the Arrhenius law) can be obtained as the probability of falling onto a stable point, then climbing up to an unstable point, and then descending to the next saddle, as in Fig. 2.10. The probability of descents being of order unity, we are left only with the uphill path. Again, if the system derives from a potential, the probability of climbing depends only on the difference in height between the valley and the saddle, and we recover the Arrhenius law calculated in previous sections (2.108). In the presence of nonconservative forces, then the path joining the stable and unstable saddles has to be computed, and from its action we get the probability.

[7] Note that time reversal applies to the variables q, and not the $\hat{q}$.

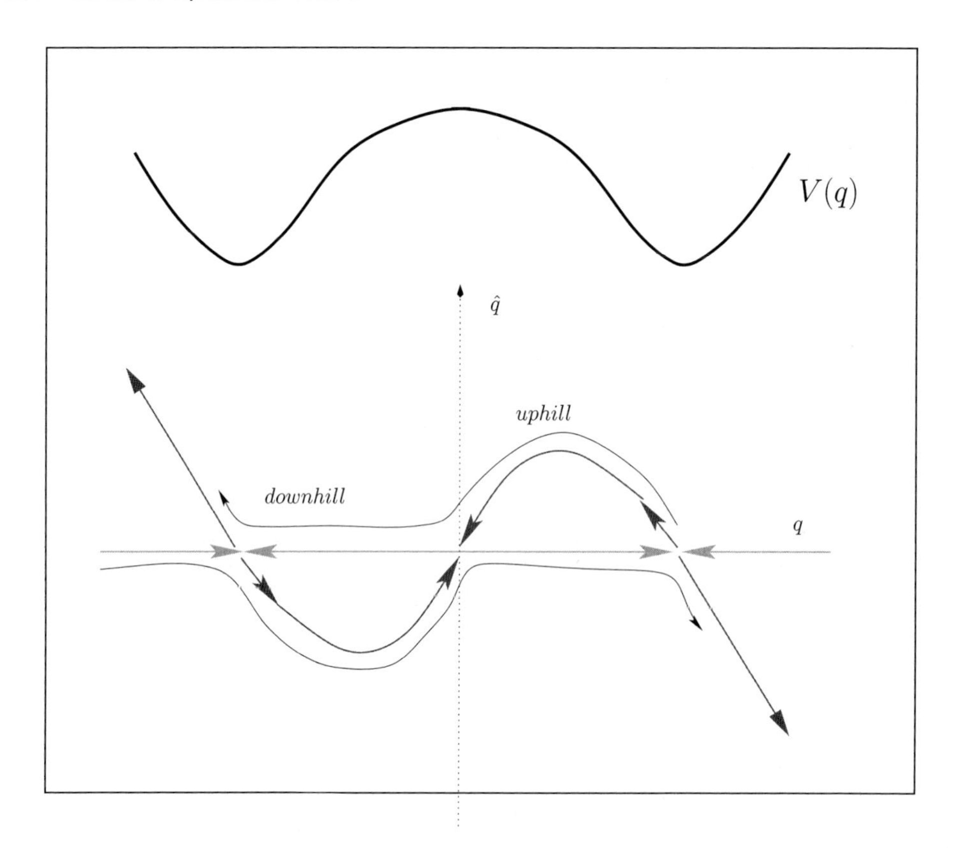

Fig. 2.10 Trajectories in $(q, \hat{q})$ space for a double-well potential.

In many-dimensional energy landscapes, one may wonder if the most probable path still goes through saddles with only one unstable direction. A moment's thought shows this to be the case, as discussed by Murrel and Laidler (1968).

Low noise in phase space: Kramers and thermostatted. Let us mention briefly how one proceeds in the case of phase-space dynamics with noise. One obtains, by following the same steps as in the previous paragraph, "classical" equations in an extended space $(q_i, p_i, \hat{q}_i, \hat{p}_i)$. There are solutions of these equations that have $\hat{q}_i = \hat{p}_i = 0$: they correspond to the original noiseless equations in the original space and have zero action. The rest of the solutions have nonzero $(\hat{q}_i, \hat{p}_i)$ and positive action.

In some cases, one may go to a "Lagrangian"[8] description involving only (q_i, p_i), but then the equations obtained contain second time-derivatives, a relic of the noise in the original phase space.

[8]Note that $(\mathbf{q}, \mathbf{p})$ are the "coordinates" and $(\hat{\mathbf{q}}, \hat{\mathbf{p}})$ the "momenta" in the $4N$-dimensional phase space.

Periodic orbits, complexities, and traces

Classical mechanics

What follows is very sketchy; its only purpose is to stimulate the curiosity of the reader, who will find an excellent reference on the subject (Cvitanovic *et al.* 2008). As we have seen in the previous sections, the spectrum of an evolution operator contains all the information on the ergodic properties of the system. One way to study the spectrum of any operator H is to compute the trace $\mathrm{Tr}\, e^{-tH}$, and then obtain the resolvent, which has poles in the eigenvalues λ of H, through

$$\sum_i \frac{1}{\lambda - \lambda_a} = \mathrm{Tr}[\lambda - H]^{-1} = \int_0^\infty dt\, \mathrm{Tr}\, e^{-t(\lambda - H)}\,. \tag{2.125}$$

The trace on the right-hand side is a sum over paths, just as seen in the previous sections; the only difference is that we have to consider *closed* trajectories in which the initial and final configurations coincide. For an evolution taking place in phase space $(\mathbf{q}, \mathbf{p})$, it is an integral of probabilities of return after time t $\mathrm{Tr}\, e^{-tH} \sim \int d\mathbf{q}\, d\mathbf{p}\, P(\mathbf{q}, \mathbf{p}, t\,;\, \mathbf{q}, \mathbf{p}, t = 0)$.

As mentioned in Section 2.2, one can study the chaoticity properties of Hamiltonian dynamics by studying the spectrum of its evolution operator in the presence of noise, and then letting the noise level go to zero: one thus uncovers Ruelle–Pollicott resonances. This poses the problem that noise will make energy nonconserved, and generate a slow diffusion in energy. To avoid this, we may use the energy-conserving noise of the Gaussian thermostat described in Section 2.1, leading to H_G, given by

$$H_G = \sum_i \left[\frac{p_i}{m}\frac{\partial}{\partial q_i} - \frac{\partial V(\mathbf{q})}{\partial q_i}\frac{\partial}{\partial p_i}\right] - \epsilon \sum_{jl}\left[\frac{\partial}{\partial p_j} g_{jl} \frac{\partial}{\partial p_l}\right] \tag{2.126}$$

(compare eqn (2.15) with $\mathbf{f} = 0$).

A trace in the path integral then becomes a sum of periodic orbits on the energy shell. In the small-noise limit those that dominate are the ones having zero action, and these are just the periodic orbits of the original Hamilton's equations. One thus expresses $\mathrm{Tr}\, e^{-tH_G}$ as a sum over orbits of period t.

The final product is that eqn (2.125) becomes a relation between the resolvent, containing the information on eigenvalues, and a sum of periodic orbits of all periods. The interested reader will find this done properly in Cvitanovic *et al.* (2008).

Complexity in statistical mechanics

A different application of the same idea (Biroli and Kurchan 2001) is to count the number of metastable states. With the assumptions of Section 2.3.1, picking a time t intermediate between the time needed to equilibrate within a state, t^*, and the time needed to escape it, t_{pass}, the number of states is given by

$\mathcal{N}_{\text{states}} = \operatorname{Tr} e^{-tH}$ with $t^* \ll t \ll t_{\text{pass}}$. This becomes a sum over all periodic trajectories of period t.

Sampling transition paths in practice

A problem of great interest in physics and chemistry is that of computing in practice the escape route (and rate) from a metastable state, for example the decay of a metastable molecule. In many cases we have a separation of timescales between the molecular vibrations and the time needed for the actual decay to occur. In order to compute the probability $P(a, t; b, t_0)$ of starting at t_0 in a configuration a belonging to the metastable state and reaching a configuration b belonging to the stable state at time t, we may sum over paths going from a to b, with their appropriate weights given by the action.

Depending on the dynamics, different approaches may be more practical. In the case of a system that is strongly coupled to a thermal bath and follows a Langevin equation, one can follow the Lagrangian "polymer" analogy (2.53), and use any Monte Carlo simulation method for a system in equilibrium at temperature T to sample the paths.

For a system that is closer to being deterministic, a well-developed technique (TPS, "transition path sampling") (Bolhuis *et al.* 2002) uses an algorithm that samples trajectories by modifying them slightly at the barrier. This kind of change allows to obtain a new path that still has good chances of being a transition (going from state a to state b), at least if the system is not too chaotic.

There is a large body of literature on the subject (see e.g. Bolhuis *et al.* 2002; E *et al.* 2002; Micheletti *et al.* 2004), since the potential applications in chemistry, biochemistry, and physics are huge.

2.4.2 Unusual time averages

In this section we study a different type of large deviation. Instead of asking for the probability of the system reaching an unusual place, we ask for the probability that it sustains an unusual time average for an observable during a long interval,

$$\frac{1}{t} \int dt' \, A(t') = \overline{A} \,. \tag{2.127}$$

The most celebrated example is the average power, which can be interpreted in some cases as entropy production:

$$\sigma_t = -\frac{1}{t} \int_0^t dt' \, \frac{\mathbf{f} \cdot \mathbf{v}}{T} \,, \tag{2.128}$$

where $\mathbf{v} = \dot{\mathbf{q}}$. Another example is the average potential energy,

$$\overline{V} = \frac{1}{t} \int_0^t dt' \, V(t') \,. \tag{2.129}$$

Expressing the probability in terms of trajectories,

$$P(A) = \sum_{\text{Trajec.}} (\text{Prob. Trajectory})\, \delta\left(t\bar{A} - \int_0^t dt'\; A(t')\right), \tag{2.130}$$

and writing the delta function as an exponential,

$$\delta\left(t\bar{A} - \int_0^t dt'\; A(t')\right) = \int_{-i\infty}^{+i\infty} d\mu\; e^{\mu\left(t\bar{A} - \int_0^t dt'\; A(t')\right)}, \tag{2.131}$$

we get

$$\begin{aligned} P(A) &= \int_{-i\infty}^{+i\infty} d\mu\; e^{\mu t\bar{A}} \underbrace{\int D[\mathbf{q}]\; (\text{Prob. Trajectory})\, e^{-\mu \int_0^t dt'\; A(t')}}_{\downarrow} \\ &= \int_{-i\infty}^{+i\infty} d\mu\; e^{\mu t\bar{A}}, \qquad\qquad \times\, e^{-tG(\mu)} \end{aligned}$$

which defines the large-deviation function $G(\mu)$. For example, in the Fokker–Planck case, the probability reads

$$e^{-tG(\mu)} = \int D[\mathbf{q}, \hat{\mathbf{q}}]\; e^{(1/T)\int dt'\; \left[\sum_i \hat{q}_i \dot{q}_i - \tilde{H}_{FP} - \mu \int_0^t dt'\; A(t')\right]}. \tag{2.132}$$

What we have done is nothing but the analogue of a passage from a microcanonical calculation of "entropy" $= \ln \bar{A}$ to a canonical calculation of "free energy" $G(\mu)/\mu$ at an "inverse temperature" μ. The "space" in our problem is in fact the time, and the "extensive quantities" are those that are proportional to time: we extracted a time in the definition of $G(\mu)$ in order to make it "intensive." For large t, in analogy with the thermodynamic limit, assuming that $G(\mu)$ has a good limit, we may evaluate the integral over μ by saddle point, to obtain

$$\ln P(A) \sim t[\mu^* A - G(\mu^*)], \tag{2.133}$$

with

$$A(\mu^*) = \left.\frac{dG}{d\mu}\right|_{\mu^*}. \tag{2.134}$$

This is the Legendre transform that takes us from the canonical to the microcanonical calculation. Note that $A(\mu^*)$ plays the role of $E(\beta)$ in ordinary thermodynamic systems. Now, by simple comparison, eqn (2.132) can be brought back to operator language:

$$e^{-tG(\mu)} = \langle \text{final}| e^{-t[H_{FP} + \mu A(\mathbf{q})]} |\text{init}\rangle. \tag{2.135}$$

A similar expression holds for the Kramers equation for H_K. In the language of the analogy with one-dimensional models, this is just the transfer matrix formalism (Huang

1987), along the time dimension. If we now introduce the right and left eigenvectors of $H + \mu A$,

$$[H + \mu A(\mathbf{q})] \, |\psi_i^R(\mu)\rangle = \lambda_i(\mu)|\psi_i^R(\mu)\rangle \,, \qquad \langle|\psi_i^L(\mu)| \, [H + \mu A(\mathbf{q})] = \lambda_i(\mu)\langle|\psi_i^L(\mu)| \,,$$

we obtain

$$e^{-tG(\mu)} = \sum_a \langle \text{final}|\psi_a^R(\mu)\rangle\langle\psi_a^L(\mu)|\text{init}\rangle \; e^{-t\lambda_a(\mu)} \tag{2.136}$$

and, as $t \to \infty$,

$$G(\mu) = \lambda_{\min}(\mu) \,, \tag{2.137}$$

where $\lambda_{\min}(\mu)$ is the eigenvalue with the smallest real part. Again, this is in complete analogy with the expression of free-energy densities in terms of the lowest eigenvalue of the transfer matrix.

Note the fundamental difference between these large-deviation functions and those of Section 2.4.1: in that case, by "large" deviations we meant that they were exponentially small in the temperature (or the coarse-graining size), while here we mean that they are sustained for long times, and the only large parameter is precisely the time t.

2.4.3 Simulating large deviations (and quantum mechanics)

Large deviations can be measured by evolving the system with its real dynamics, and then making a histogram of the deviations obtained. However, eqn (2.135) suggests that we should try to simulate directly a system that evolves through $H_{FP} + \mu A$ (Giardina *et al.* 2006; Lecomte and Tailleur 2007; Aldous and Vazirani 1994; Grassberger *et al.* 1998):

$$\dot{P} = -[H_{FP} + \mu A]P \quad \rightarrow \quad P(\mathbf{q}, t) = e^{-t[H_{FP} + \mu A]} \,. \tag{2.138}$$

Clearly, as $t \to \infty$ the distribution P tends to the eigenvalue with the lowest real part.

We are dealing with a dynamics without probability conservation. In fact, we can reproduce it by using a large number of noninteracting walkers, each performing the original (Langevin) dynamics with independent noises, occasionally giving birth to another walker starting in the same place, and occasionally dying. A negative or positive value of $\mu A(\mathbf{q})$ gives a probability $|\mu A(\mathbf{q})| \, dt$ of making a clone or of dying, respectively, in a time interval dt. At each time, the global number of clones $M(t)$ changes, in such a way that for long times, $M(t)/M(t=0) \sim e^{-\lambda_{\min}(\mu)t}$. In practice, one can normalize the total number periodically by cloning or decimating all walkers with a random factor. The factor needed to keep the population constant is, again, the exponential of the lowest eigenvalue.

Notice that the imaginary-time Schrödinger equation is precisely of the form (2.138), with no drift in the Langevin process and where μA is the quantum potential. Indeed, the method described above was developed for precisely this case, and is called "diffusion Monte Carlo" (DMC) (Grimm and Storer 1971; Anderson 1975).

2.5 Metastability and dynamical phase transitions

In several places above we have pointed out that the stochastic dynamics can be seen as a kind of "thermodynamics in space–time." Trajectories contribute with a weight given by the (Onsager–Machlup) action, much as the energy determines the Gibbs–Boltzmann weight in thermodynamics. Large-deviation theory just consists of biasing the measure with an extra weight added to the dynamic action and computing a new sum, which then loses the meaning of a transition probability and becomes a large-deviation function.

Systems with nontrivial dynamical properties sometimes show very little in their static (time-independent) structure. The typical example is that of glasses, which are virtually indistinguishable from liquids from the point of view of organization of the molecules, until one looks at their dynamics, which is dramatically slower. This situation has motivated some researchers (Jack *et al.* 2006) to look to space–time thermodynamics—the statistical properties of trajectories—for the missing structure. One considers the large-deviation theory of systems that are dynamically nontrivial, and indeed it turns out that one often finds (Garrahan *et al.* 2007) that there is a rich structure of "dynamic" phase transitions in the large-deviation functions.

In this section, we shall see that space–time transitions are closely related to the approach to metastability of Section 2.3, into which they provide useful insights.

2.5.1 A simple example

Let us start with a simple example of a particle in a potential V performing over-damped Langevin motion at a low temperature T (see Fig. 2.11). We shall consider the large-deviation function of the energy $G(\mu)$, associated with the probability of observing an average energy $\bar{V}$ (2.129) over long time intervals. As we saw in the previous section, $G(\mu)$ is obtained from the lowest eigenvalue of $H_{FP} - \mu V$:

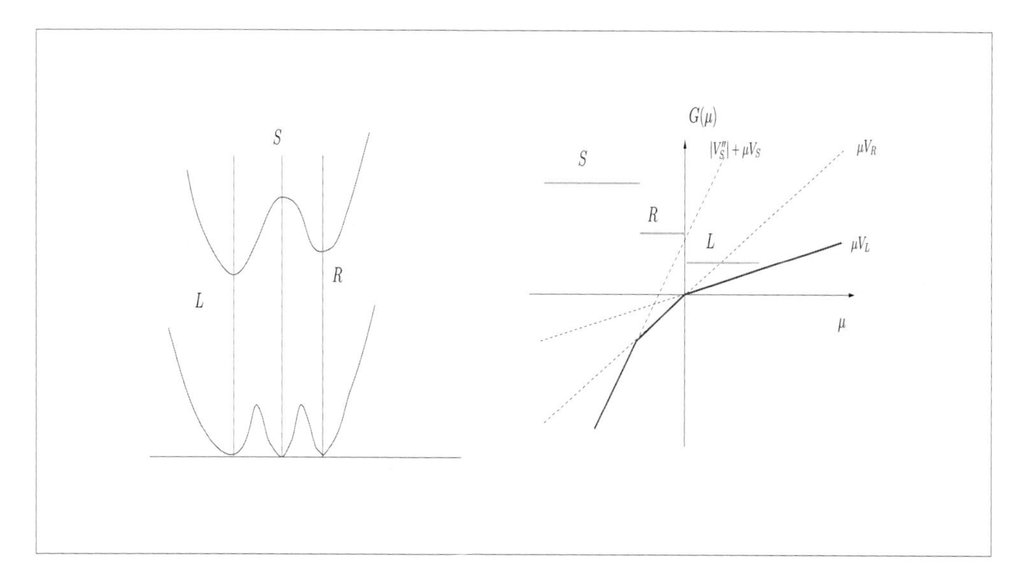

Fig. 2.11 Left: the potential V and the related effective potential V_{eff}. Right: energy (horizontal lines) and the associated large-deviation function $G(\mu)$. The system has two first-order transitions, at $\mu = 0$ and at $\mu = -(V_s - V_r)/|V_s''|$.

$$H_{FP} + \mu V = \frac{2}{T}\left[-T\frac{d^2}{dx^2} + \frac{V'^2}{8} - \frac{T}{4}V'' + \frac{\mu}{2}TV\right] \tag{2.139}$$

(see Benzi *et al.* (1985) for a similar application). The lowest eigenvalue at low temperatures is given by the same arguments as in Section 2.3.1, except that now we have to add the extra term proportional to μ in eqn (2.139). At small T, each minimum in the effective potential

$$V_{\text{eff}}(\mu) = \frac{V'^2}{8} - \frac{T}{4}V'' + \frac{\mu}{2}TV \tag{2.140}$$

contributes separately, just as in Section 2.3.1. To leading order in T, the contribution of saddle points of $V(q^s)$ at q^s is $\lambda \sim \mu V(q^s)$ if it is a minimum, and $\lambda \sim [|V''(q^s)| + \mu V(q^s)]$ if it is a maximum. The lowest amongst all eigenvalues dominates:

$$G(\mu) = \lambda_{\text{min}} = \min \begin{cases} \lambda_L = \mu V_L\,, \\ \lambda_R = \mu V_R\,, \\ \lambda_S = |V_S''| + \mu V_S\,. \end{cases} \tag{2.141}$$

The values of $\bar{V}$ are given by the Legendre transform $V(\mu^*) = dG/d\mu$, and read

$$V(\mu^*) = \begin{cases} V_L\,, & \mu^* > 0\,, \\ V_R\,, & \mu^* < 0 \text{ and } \mu^* > -|V_s''|/(V_s - V_r)\,, \\ V_S\,, & \mu^* < -|V_s''|/(V_s - V_r)\,. \end{cases} \tag{2.142}$$

There are two first-order phase-transitions (see Fig. 2.11).

Let us pause and analyze this physically. The scenario is typical first-order, with three homogeneous "phases" in time, corresponding to the three values of eqn (2.142). When we condition a long trajectory to have a time-averaged energy V_0, this is realized by the system by a "phase coexistence" of periods t_L, t_S, t_R spent in each of the three stationary points, such that $t = t_L + t_S + t_R$ and $V_0 t = V_L t_L + V_S t_S + V_R t_R$. This is strictly analogous to the ice/water coexistence when the total energy is fixed. *An important lesson is that if the system is conditioned to have a value of an observable intermediate between those values which it takes in two metastable phases, it prefers to achieve this by spending some time in each state, rather than all the time in an intermediate situation.*

2.5.2 Spectral properties and phase transitions

One can in fact show that the situation we have seen above is very general. In particular, it is quite normal that we should have a first-order transition at $\mu = 0$. To see this, let us use the formalism of Section 2.3. Consider the operator H. If there are p independent long-lived states, with eigenvalues $\lambda_a < t_{\text{pass}}^{-1} \sim 0$, and a time t^* to thermalize inside a state, we can construct a basis of p right and left eigenvectors P_a, Q_a, with $\langle Q_a|P_b\rangle \sim \delta_{ab}$ having essentially a zero eigenvalue.

Let us now calculate the eigenvalues of $H + \mu A$, for small μ, but still where $\mu \gg t_{\text{pass}}^{-1}$. We may use (non-Hermitian) first-order perturbation theory, to get

$$\lambda_a \sim \lambda_a(\mu = 0) + \mu \langle Q_a | A | P_a \rangle \sim \mu \langle A \rangle_{\text{in state } a} \,. \tag{2.143}$$

In other words, the quasi-degenerate eigenvalues split proportionally to the expectation value of the observable A in each state, and the phase that dominates is

$$G_\mu = \min_{\text{states } a} \{ \langle A \rangle_a \} \,. \tag{2.144}$$

Remarkably, the distribution function has in fact pinned down a "pure state" P_a, Q_a, using the observable A. When the sign of μ is reversed, the "minimum" in eqn (2.144) is transformed into a "maximum" and the selected state is changed. There is hence always a first-order phase transition. By playing with a different observable, we may make a different transition that selects any state. Hence, we conclude that the dynamic-phase-transition approach is in fact equivalent to the metastability approach of Section 2.3, but it gives us new tools and a practical perspective.

2.6 Fluctuation Theorems and Jarzynski's equality

The nonequilibrium work relations known as the Fluctuation Theorem (Evans *et al.* 1993; Evans and Searles 1994; Cohen and Gallavotti 1999; Gallavotti and Cohen 1995) and Jarzynski's equality (Jarzynski 1997*a*,*b*) are very general results valid for strongly out-of-equilibrium systems. They concern large deviations of work. As such they are closely related to—and enrich our perspective of—the Second Law of Thermodynamics. The two subjects are quite similar, and in fact may in some cases be encompassed by a single, more general result: Crooks's equality (Crooks 1998), which we shall not review here. These results are very recent—surprisingly so, given their technical simplicity—and have received enormous attention in the last fifteen years. They are both based on the relation (Section 2.2.3) between time-reversal symmetry breaking on one side, and work and entropy on the other.

2.6.1 The Fluctuation Theorem(s)

We shall consider here only Langevin processes with inertia, and the Kramers equation. This makes the discussion simpler, because of the fact mentioned in previous sections that power, being a product of a force times a velocity, is only a continuous function of time when there is inertia.

We have seen in Section 2.2.3 that the "time-reversal" symmetry becomes, in the presence of forcing,

$$\left[\Pi e^{\beta H} H_K e^{-\beta H} \Pi^{-1} \right]^\dagger = H_K^\dagger + \frac{d(t\sigma_t)}{dt} \,. \tag{2.145}$$

The violation of the symmetry is proportional to a quantity that has the form of an entropy production (see eqns (2.128) and (2.79)):

$$\sigma_t = \frac{\text{power}}{T} = -\frac{1}{tT} \int_0^t dt' \, \mathbf{f} \cdot \mathbf{v} \,. \tag{2.146}$$

Equation (2.145) is the basis for the results that we shall discuss. In fact, there are several variants of time reversal, and each gives different identities (Chetrite and Gawedzki 2008).

General: The implications of an explicitly broken symmetry in statistical mechanics

The Fluctuation Theorem makes use of the explicit breaking of a discrete symmetry, the detailed-balance relation. In fact, whenever we have a system composed of a part that is symmetric under a transformation, plus an *anti*symmetric perturbation, we can derive a relation for large deviations of the perturbations. Consider for example the statistical mechanics of a system with variables $s_1, \ldots, s_N$ and energy E_0, having the discrete symmetry $E_0(\mathbf{s}) = E_0(-\mathbf{s})$. This symmetry implies the vanishing of all odd correlation spin functions. Now, let us perturb the energy with a field $E(\mathbf{s}) = E_0(\mathbf{s}) - (h/2)M(\mathbf{s})$, conjugate to a term $M(\mathbf{s}) = \sum_i s_i$ with $M(-\mathbf{s}) = -M(\mathbf{s})$, so that now

$$E(-\mathbf{s}) = E(\mathbf{s}) + hM(\mathbf{s}) \,. \tag{2.147}$$

Can we conclude something in the presence of $h \neq 0$, when the symmetry is explicitly broken? Indeed we can: consider the distribution of the symmetry-breaking term

$$P\left[M(\mathbf{s}) = -M\right] = \int d\mathbf{s}\, \delta\left[M(\mathbf{s}) + M\right] e^{-\beta(E_0 - (h/2)M)} \,. \tag{2.148}$$

Changing variables $\mathbf{s} \to -\mathbf{s}$, and using the symmetry of E_0,

$$\begin{aligned} P\left[M(\mathbf{s}) = -M\right] &= \int d\mathbf{s}\, \delta\left[-M(\mathbf{s}) + M\right] e^{-\beta(E_0 + h/M)} \\ &= e^{-\beta h M} P\left[M(\mathbf{s}) = M\right] , \end{aligned} \tag{2.149}$$

or

$$\frac{P\left[M(\mathbf{s}) = M\right]}{P\left[M(\mathbf{s}) = -M\right]} = e^{\beta h M} \,. \tag{2.150}$$

One wonders if this elementary property had ever been used in other fields of physics before the Fluctuation Theorem. On the other hand, a derivation of the Fluctuation Theorem that makes close contact with this thermodynamic property has been given by Narayan and Dhar (2004).

The Fluctuation Theorem is a statement about the distribution $P(\sigma_t)$ of the average quantity (2.146) when the experimental protocol is repeated many times. It reads

$$\frac{P(\sigma_t)}{P(-\sigma_t)} \sim e^{t\sigma_t} \,. \tag{2.151}$$

A relation like this was first proposed by Evans *et al.* (1993).

The Second Law of Thermodynamics states that the work done on a system *over long times* must be positive. Equation (2.151) is then a statement about the "violations" of the Second Law when the average work has the opposite sign.[9] The factor t in the exponent to a certain extent quantifies the suppression of the probability of such processes when the time is large, thus giving a better perspective on the Second Law.

Equation (2.151) can be reexpressed by multiplying it by $e^{-t\mu\sigma_t}$ and integrating over σ_t, as a property of the large-deviation function $G(\mu) = (1/t) \int d\sigma_t \; P(\sigma_t) e^{-\mu\sigma_t t}$ (see eqn (2.132)):

$$G(\mu) = G(1 - \mu) . \tag{2.152}$$

The result in eqns (2.151) and (2.152) is extremely general (Evans *et al.* 1993; Gallavotti and Cohen 1995; Kurchan 1998; Lebowitz and Spohn 1999; Maes 1999), independent of the model's parameters, and valid for several types of dynamics. Two different settings have to be distinguished:

- *Transient.* Each measurement of σ_t is made starting from a thermalized system at temperature T, a configuration chosen with the Gibbs–Boltzmann distribution. At time $t = 0$, nonconservative forces are switched on, and σ_t is proportional to the work they do during a time t, which need not be long. The system may be isolated or connected to a thermostat, which then has to be at the same temperature T.
- *Stationary.* Here, the sampling is of a system that, by assumption, has achieved stationarity in the presence of forcing. For this to be possible, it needs a thermostat to absorb heat. The system is not in equilibrium, and the fluctuation relation is valid only in the limit of long sampling periods $t \to \infty$.

Another distinction we can make is whether the dynamics is *stochastic* (e.g. Langevin) or *deterministic* (the Gaussian thermostat (2.13) without energy-conserving noise). The only hard case is the *stationary and deterministic* one, treated by the Gallavotti–Cohen (Gallavotti and Cohen 1995) Theorem. Not only is it technically more subtle, but it relies on a real physical condition that the system has to meet, as we shall see.

Transient, with or without bath. Equation (2.79) can be rewritten, for all μ (for the moment an arbitrary number), as

$$\left[H_K - \mu\frac{\mathbf{f}\cdot\mathbf{v}}{T}\right]^\dagger = \left(\Pi e^{\beta\mathcal{H}}\right)\left[H_K - (1-\mu)\frac{\mathbf{f}\cdot\mathbf{v}}{T}\right]\left(e^{-\beta\mathcal{H}}\Pi^{-1}\right). \tag{2.153}$$

Using the expression (2.135) for the large-deviation function, starting from the Gibbs–Boltzmann distribution $|GB\rangle$, we compute

$$\begin{aligned} e^{-tG(\mu)} &= \langle -|e^{-t[H_K-(\mu/T)\mathbf{f}\cdot\mathbf{v}]}|GB\rangle = \langle GB|e^{-t[H_K-(\mu/T)\mathbf{f}\cdot\mathbf{v}]^\dagger}|-\rangle \\ &= \langle -|e^{-\beta\mathcal{H}}e^{-t[H_K-(\mu/T)\mathbf{f}\cdot\mathbf{v}]^\dagger}e^{\beta\mathcal{H}}|GB\rangle \\ &= \langle -|\Pi e^{-t[H_K+((1-\mu)/T)\mathbf{f}\cdot\mathbf{p}]}\Pi^{-1}|GB\rangle \\ &= \langle -|e^{-t[H_K-((1-\mu)/T)\mathbf{f}\cdot\mathbf{p}]}|GB\rangle = e^{-tG(1-\mu)} , \end{aligned} \tag{2.154}$$

[9] The quotation marks are just to remind us that since the Second Law applies to the limit of long times $tN \to \infty$ for a single-instance experiment, these are not true violations.

i.e. eqn (2.152). Note that we did not have to use any assumptions either about times or about the dynamics.

Stationary with bath. To compute the large-deviation function for long times, we proceed as in the previous section, and introduce eigenvectors and eigenvalues as in eqn (2.136):

$$\left[H_K - \frac{\mu}{T}\mathbf{f}\cdot\mathbf{v}\right]|\psi_i^R\rangle = \lambda_i|\psi_i^R\rangle\,, \qquad \langle\psi_i^L|\left[H_K - \frac{\mu}{T}\mathbf{f}\cdot\mathbf{v}\right] = \lambda_i\langle\psi_i^L|\,. \tag{2.155}$$

We get

$$e^{-tG(\mu)} = \langle -|e^{-t[H_K-(\mu/T)\mathbf{f}\cdot\mathbf{v}]}|\text{init}\rangle = \sum_a \langle -|\psi_a^R\rangle\langle\psi_a^L|\text{init}\rangle\, e^{-t\lambda_a(\mu)}\,.$$

If the spectrum has a gap, the eigenvalue with the lowest real part dominates, and we have that $G(\mu) = \lambda_{\min}(\mu)$ as $t \to \infty$.

Now, because $[H_K - (\mu/T)\mathbf{f}\cdot\mathbf{v}]$ and $[H_K - ((1-\mu)/T)\mathbf{f}\cdot\mathbf{v}]^\dagger$ are related by a similarity transformation (2.153), their spectra are the same. Hence, we have the following result, to the extent that to leading order in the time, $G(\mu)$ depends only on eigenvalues and not on eigenvectors:

$$G(\mu) = G(1-\mu)\,. \tag{2.156}$$

In this case, eqn (2.156) is valid only at large times. Where can this fail? The problem in the large-time evaluation of eqn (2.156) arises if $\langle\psi^L(\lambda_{\min})|\text{init}\rangle \sim 0$. At zero noise intensity this may well happen, because eigenvectors may in that case be completely localized.

Gallavotti–Cohen Theorem. We shall not derive here the Gallavotti–Cohen Theorem, but just say a few words. The Gallavotti–Cohen Theorem is the stationary Fluctuation Theorem for a system in contact with a *deterministic* Gaussian thermostat, like the one we introduced in eqn (2.13) but with exactly zero noise. It turns out that unlike the stochastic and the transient case, which are essentially always valid, the Gallavotti–Cohen result breaks down in some systems, and in most cases when the forcing is very strong. This is not a defect of the proof, but a reflection of a physical fact: *the fluctuation relation in the deterministic case holds only if the system has certain "ergodic" properties.*

To have a perspective on this, one can consider approaching the deterministic case as a limit of the noisy thermostatted case (2.13). One can reproduce the derivation above for this case (Kurchan 2007), and easily conclude that for every level of noise ϵ the theorem is valid, in the limit of large time windows $t \gg t_{\min}(\epsilon)$, for some time $t_{\min}(\epsilon)$. The trouble comes from the fact that as $\epsilon \to 0$ it may happen that $t_{\min}(\epsilon)$ diverges: in other words, the time $t_{\min}(\epsilon)$ needed for the correct sampling of fluctuations may become infinite in the deterministic limit. The result of Gallavotti and Cohen proves that this is not the case for a class of very chaotic systems. For this class, their theorem implies that $\lim_{\epsilon\to 0} t_{\min}(\epsilon) < \infty$.

The timescale $t_{\min}(\epsilon)$ has a clear physical meaning, which we shall just hint at here. A driven deterministic system has an *attractor* on the energy surface (see Section

2.2.4). Winding the dynamics back in time defines a *repellor*, which is stationary but unstable. In chaotic Hamiltonian (undriven) systems, the attractor and repellor are intertwined; they occupy the same region in phase space. As the drive is turned on, the attractor distribution focuses on a region of the energy shell, as described in Section 2.2.4. So does the repellor, in a region that may be nonoverlapping.

In the presence of weak noise, even if the attractor and repellor are in principle separate, the system may occasionally switch from attractor to repellor and back: this is exactly analogous to the "coexistence" we found in lecture 5. The typical time to do this (Kurchan 2007; see also Bonetto and Gallavotti 1997) is precisely $t_{\rm min}(\epsilon)$ for an otherwise ergodic system. The condition for $t_{\rm min}(\epsilon)$ to remain finite as the noise goes to zero is then that the attractor and repellor overlap sufficiently in phase space that no noise is needed to jump from one to the other. This is an extra condition the system must satisfy in order that the Gallavotti–Cohen Fluctuation Theorem holds. *In a word, the applicability of the Gallavotti–Cohen Fluctuation Theorem is, for a chaotic deterministic system, a symptom that the attractor and repellor have not divorced under the effect of forcing.*

2.6.2 Jarzynski's equality

Jarzynski's equality is a remarkable generalization of the second principle. Consider a system with an energy dependent upon a parameter (volume, magnetic field,...), which we denote by α. We start from a Gibbs–Boltzmann equilibrium corresponding to the parameter $\alpha_{\rm initial}$ and then evolve while changing $\alpha(t)$ at arbitrary speed up to a time t, with $\alpha_{\rm final} = \alpha(t)$. The equality is then

$$e^{-\beta[F(\alpha_{\rm final})-F(\alpha_{\rm initial})]} = \langle e^{-\beta\ {\rm work}} \rangle_{\alpha_{\rm initial} \to \alpha_{\rm final}} . \tag{2.157}$$

The average is over trajectories starting from equilibrium at $t = 0$, but otherwise arbitrary. *Note the surprising appearance of $F(\alpha_{\rm final})$, an equilibrium quantity, despite the fact that the system is not in equilibrium at time t.* We shall prove this result for a Langevin system with inertia. It is valid independent of the friction coefficient γ, and indeed even if $\gamma = 0$ and the system is isolated.

From eqn (2.157) we can go to the second principle, taking logarithms on both sides and using Jensen's inequality, $\langle e^A \rangle \geq e^{\langle A \rangle}$:

$$F(\alpha_{\rm final}) - F(\alpha_{\rm initial}) \leq {\rm work}\,. \tag{2.158}$$

The time-dependent energy is

$$\mathcal{H}_\alpha = \mathcal{H}(\mathbf{q}, \mathbf{p}, \alpha) = \sum_i \frac{p_i^2}{2m} + V(\mathbf{q}, \alpha)\,. \tag{2.159}$$

If the parameter α depends on time, it does work:

$$\text{force} \times \text{velocity} = \text{energy change} - \text{work}\,,$$

$$\int dt \sum_i \dot{q}_i \frac{\partial V}{\partial q_i} = \int dt \left(\frac{d\mathcal{H}_\alpha}{dt} - \frac{\partial \mathcal{H}_\alpha}{\partial \alpha} \dot{\alpha} \right) = \mathcal{H}\Big|_{\rm initial}^{\rm final} - \int dt \left(\frac{\partial \mathcal{H}_\alpha}{\partial t} \right).$$

We assume we start from the equilibrium configuration $|GB(\alpha_1)\rangle$ corresponding to a given value α_1. The total evolution over a time t can be written by breaking the time into short intervals as in Fig. 2.12:

$$U(t) = e^{-\delta t H_{\alpha M}} \dots e^{-\delta t H_{\alpha 2}} e^{-\delta t H_{\alpha 1}} . \tag{2.160}$$

Because $\langle -|H_\alpha = \langle -|$, we have

$$\begin{aligned} 1 &= \langle -|e^{-\delta t H_{\alpha M}} \dots e^{-\delta t H_{\alpha 2}} e^{-\delta t H_{\alpha 1}} |GB(\alpha_1)\rangle \\ &= \langle GB\ (\alpha_1)| e^{-\delta t H^\dagger_{\alpha 1}} e^{-\delta t H^\dagger_{\alpha 2}} \dots e^{-\delta t H^\dagger_{\alpha M}} |-\rangle \\ &= \frac{Z_{\alpha M}}{Z_{\alpha_1}} \langle -|e^{-\beta \mathcal{H}_{\alpha_1}} e^{-\delta t H^\dagger_{\alpha 1}} e^{-\delta t H^\dagger_{\alpha 2}} \dots e^{-\delta t H^\dagger_{\alpha M}} e^{\beta \mathcal{H}_{\alpha M}} |GB\ (\alpha_M)\rangle . \end{aligned} \tag{2.161}$$

As in all these theorems, we wish to introduce time reversal. We proceed as follows. We insert factors between every two exponentials,

$$e^{-\beta \mathcal{H}_{\alpha_1}} e^{-\delta t H^\dagger_{\alpha 1}} \underset{\left(e^{\beta \mathcal{H}_{\alpha_1}} e^{-\beta \mathcal{H}_{\alpha_1}}\right)}{\uparrow} \overset{\left(e^{\beta \mathcal{H}_{\alpha_2}} e^{-\beta \mathcal{H}_{\alpha_2}}\right)}{\downarrow} e^{-\delta t H^\dagger_{\alpha 2}} \dots e^{-\delta t H^\dagger_{\alpha M}} e^{\beta \mathcal{H}_{\alpha M}} , \tag{2.162}$$

and use eqn (2.64) in each factor:

$$e^{-\beta \mathcal{H}_{\alpha_r}} e^{-\delta t H^\dagger_{\alpha_r}} e^{\beta \mathcal{H}_{\alpha_r}} = \Pi e^{-\delta t H_{\alpha_r}} \Pi^{-1} . \tag{2.163}$$

Putting everything into eqn (2.161), using the fact that $\Pi^2 = 1$ and that $e^{-\delta t H_{\alpha M}}$ $|GB\ (\alpha_M)\rangle = |GB\ (\alpha_M)\rangle$, we get

$$1 = \frac{Z_{\alpha M}}{Z_{\alpha_1}} \langle -|e^{-\delta t H_{\alpha 1}} e^{-\beta(\mathcal{H}_{\alpha_1} - \mathcal{H}_{\alpha_2})} e^{-\delta t H_{\alpha 2}} e^{-\beta(\mathcal{H}_{\alpha_2} - \mathcal{H}_{\alpha_3})} \dots e^{-\delta t H_{\alpha M}} |GB\ (\alpha_M)\rangle . \tag{2.164}$$

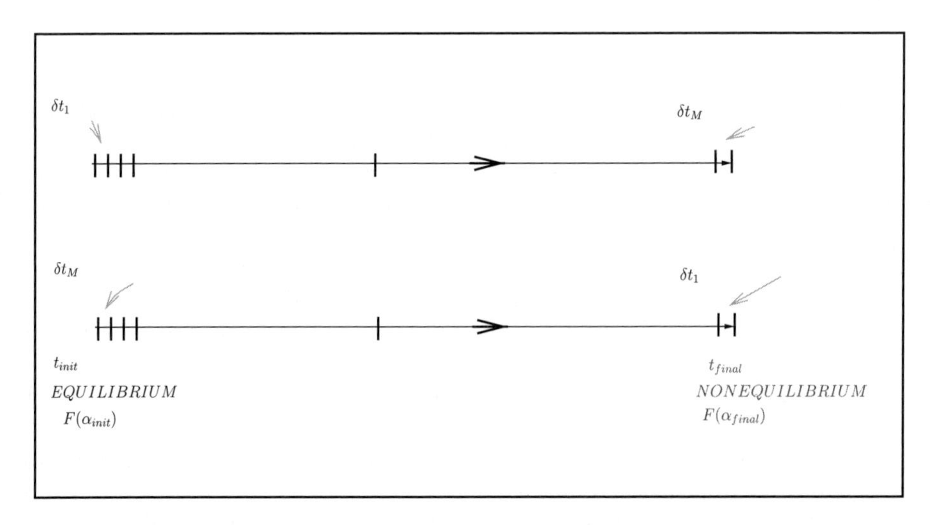

Fig. 2.12 Breaking the time interval.

Then because $\mathcal{H}_{\alpha_r} - \mathcal{H}_{\alpha_{(r+1)}} \propto \delta\alpha$, and using, as usual, the fact that $e^{\delta t A} e^{\delta t B} \sim e^{\delta t(A+B)+O[(\delta(t))^2]}$ for small δt, we obtain

$$1 = \frac{Z_{\alpha_M}}{Z_{\alpha_1}} \langle -| e^{-\delta t H_{\alpha_1} - \beta(\mathcal{H}_{\alpha_1} - \mathcal{H}_{\alpha_2})} e^{-\delta t H_{\alpha_2} - \beta(\mathcal{H}_{\alpha_2} - \mathcal{H}_{\alpha_3})} \times \ldots e^{-\delta t H_{\alpha_M} - \beta(\mathcal{H}_{\alpha_{M-1}} - \mathcal{H}_{\alpha_M})} | GB\,(\alpha_M) \rangle$$
$$= \frac{Z_{\alpha_M}}{Z_{\alpha_1}} \langle -| e^{-\delta t H_{\alpha_1} - \beta(\partial \mathcal{H}_{\alpha_1}/\partial\alpha)\delta\alpha} e^{-\delta t H_{\alpha_2} - \beta(\partial \mathcal{H}_{\alpha_2}/\partial\alpha)\delta\alpha} \times \ldots e^{-\delta t H_{\alpha_M} - \beta(\partial \mathcal{H}_{\alpha_{M-1}}/\partial\alpha)\delta\alpha} | GB\,(\alpha_M) \rangle , \qquad (2.165)$$

which, by simple comparison, means that

$$1 = e^{\beta(F_{\alpha_M} - F_{\alpha_1})} \langle e^{-\beta \int dt\, \partial\mathcal{H}/\partial t} \rangle , \qquad (2.166)$$

where the average is over trajectories starting from initial points chosen with the equilibrium distribution at (α_M) and ending anywhere where evolution and noise take them.

We finally get

$$e^{-\beta(F_{\text{final}} - F_{\text{initial}})} = \langle e^{-\beta \int dt\, \partial\mathcal{H}/\partial t} \rangle_{\alpha_{\text{initial}} \to \text{any}} \qquad (2.167)$$

The average is over trajectories starting from equilibrium at α_{initial} and ending anywhere that the dynamics takes them. The interpretation of work $= \int dt\, \partial\mathcal{H}/\partial t$ has generated controversy (Peliti 2008); it is the work done by the system on the external sources (e.g. pistons) of the "fields" α. Here again, subtly different equalities can be obtained depending on the precise expression used for work (see Jarzynski 2007).

2.6.3 A paradox

The Jarzynski equality has been criticized (Sung 2005; Crooks and Jarzynski 2007) on the basis that it it is supposed to fail for a process of free expansion. This paradox was completely resolved by Crooks and Jarzynski. The resolution is in itself instructive, because it highlights the role of rare fluctuations. The argument goes as follows: if we open the tap of a bottle of gas, letting the gas freely expand into an empty room, there is no work done, and yet there is a change in the free energies before and after: Jarzynski's equality must be violated.

Let us first distinguish two ways of making a free expansion: with a sliding wall (Fig. 2.13) or with a rapidly receding piston (Fig. 2.14). In the case of the sliding wall we see immediately where the catch is: we are supposed to start with an equilibrium initial condition, but this requires that the right-hand compartment in Fig. 2.13 be also full. If that is the case, when we slide the wall open there is neither work done nor a free-energy change.

This seems like a cheat, because we could have a solid piston receding infinitely fast as in Fig. 2.14, and in that case no need of an equilibrated right-hand compartment. Following Crooks and Jarzynski, suppose then that the piston is pulled backwards at a huge velocity v. If $v \gg (kT)^{1/2}$, it is highly unlikely that any gas particle will hit the

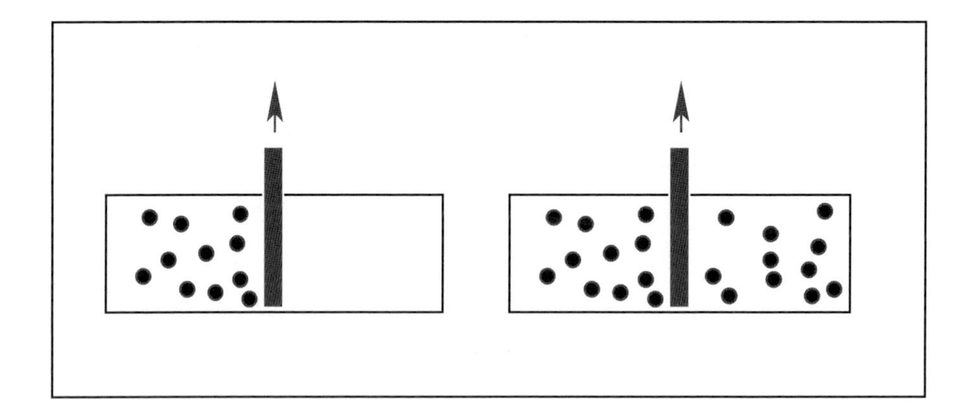

Fig. 2.13 Sliding wall. Left and right: nonequilibrium and equilibrium initial conditions.

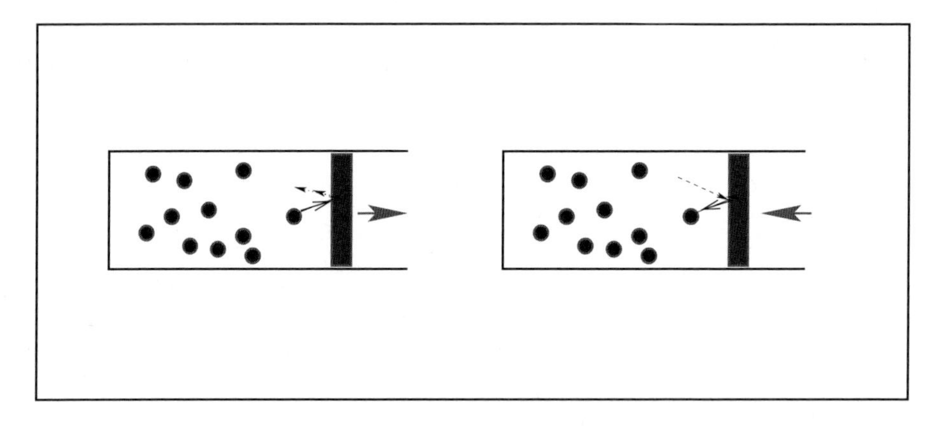

Fig. 2.14 Pulling back a piston fast. The process and its time-reversed version.

piston as it recedes, so in almost any run of the experiment no work is done. There are, however, very rare realizations of the experiment in which an unusually energetic particle catches up with the piston, bounces, and loses most of its velocity in the process. This can be best seen if one considers the time-reversed process (Fig. 2.14) in which a rapidly incoming piston hits a slow-moving particle. It turns out (Sung 2005; Crooks and Jarzynski 2007) that the rare realizations in which this strong "particle cooling" happens suffice to account for the exponential in the free-energy difference.

This example beautifully illustrates the role of rare fluctuations in Jarzynski's equality, and the extent to which what dominates the equation can be an extremely atypical process that violates our intuition, which is designed to apply to probable events.

2.6.4 Experimental work

A serious account of both the Fluctuation Theorem and Jarzynski's equality should discuss the by now quite considerable experimental work. We shall not do this here, but just refer the interested reader to the References (Ritort 2008).

Let us make, however, a few remarks from a theoretician's perspective. The Fluctuation Theorem is, as its name suggests, a theorem. As such, it need not be tested experimentally. The question is, of course, to what extent a specific physical system satisfies the hypotheses of the theorem. Clearly, no real system has a Gaussian thermostat. Even a Langevin thermostat is an idealization—a question that has been brought up in the context of Jarzynski's equality (see below). The usual way out is to say that the thermostat is "far away" and then its nature becomes irrelevant. This would normally sound extremely reasonable, but bear in mind that, as the previous section shows, intuition may be misleading when applied to large deviations. At any rate, if the nature of the thermostat is indeed not important, then we might as well suppose it is a stochastic one, so that the Fluctuation Theorem holds without any assumption related to ergodicity.

Consider, for example, the Lyon experiment (Ciliberto and Laroche 1998), where a liquid is enclosed between two horizontal plates, the lower at a higher temperature than the upper one, i.e. $T_d > T_u$. Heat is transmitted by (Rayleigh–Bénard) convection. The Fluctuation Theorem establishes a relation between having a heat flux $\mathcal{J}$ in one sense and in the other, $P(\mathcal{J})/P(-\mathcal{J}) \sim e^{-(\beta_u - \beta_d)\mathcal{J}t}$ (Eckmann *et al.* 1999; Rey-Bellet and Thomas 2002). Surely, one argues, the top and bottom plates can be considered to be in contact with a good Langevinian bath, and the nature of Rayleigh–Bénard convection cannot depend on that. If this assumption is correct, then the validity of the Fluctuation Theorem is not in doubt, as no ergodicity properties are required of the system when there is a Langevin bath. But, in fact, the experiment does not find a fluctuation relation with the true top and bottom temperatures (and indeed, with such temperatures the violations of the Second Law would be unobservable). Why is that? It would seem—but this needs further elucidation—that what the experiment is in fact testing is a pre-asymptotic result, because the time window is not long enough. One can of course ask if at this pre-asymptotic level there is, for some other reason, *another* fluctuation relation, with a higher effective temperature, but this requires a different theory (Bonetto and Gallavotti 1997).

Let us now turn very briefly to the Jarzynski equality. In this case, experimental work (Ritort 2008) has been done not so much to test the equality, as in the case of the Fluctuation Theorem, but to *use* it to evaluate free-energy differences with fast, out-of-equilibrium measurements. For example, two RNA strands may be unzipped by pulling with optical tweezers, and the work that this costs is measured (Ritort 2006). The experiment is repeated many times, and from the work distribution the equilibrium free-energy difference is measured. If we move too fast, it is the rarer and rarer runs that dominate, so the experiment has to be repeated many times. Here again, the example of the piston in the previous section is very illuminating: if the piston recedes very fast, only in very rare repetitions of the experiment will a fast particle catch up with it and be stopped: but this event dominates the average in Jarzynski's equality!

Let us say a final word concerning Jarzynski's equality in the presence of a (water) bath. It has been argued (Cohen and Mauzerall 2004, 2005; Jarzynski 2004) that the water cannot be considered as a Langevinian—or in fact as any other equilibrium—

bath while the system is evolving fast. This is indeed so, but the problem is solved (Jarzynski 1997*a,b*) by "moving the bath away," i.e. by including the surrounding water as part of the system.

Acknowledgment

I wish to thank J. Tailleur for reading of the manuscript and for suggestions.

References

Aldous D. and Vazirani U. (1994). "Go with the winners algorithms," in *Proc. 35th IEEE Symp. on Foundations of Computer Science.*
Anderson J. (1975). *J. Chem. Phys.*, **63**, 1499.
Andrieux D., Gaspard P., Ciliberto S., *et al.* (2007). *Phys. Rev. Lett.*, **98**, 150601.
Benzi R., Paladin G., *et al.* (1985). *J. Phys. Math. Gen.*, **18**, 2157.
Bertini L., De Sole A., Gabrielli D., Jona-Lasinio G., and Landim C. (2001). *Phys. Rev. Lett.*, **87**, 40601.
Bertini L., De Sole A., Gabrielli D., Jona-Lasinio G., and Landim C. (2002). *J. Stat. Phys.*, **107**, 635.
Biroli G. and Kurchan J. (2001). *Phys. Rev. E*, **64**, 016101.
Bolhuis, P. G., Chandler D., Dellago C., and Geissler P. (2002). *Ann. Rev. Phys. Chem.*, **59**, 291.
Bonetto F. and Gallavotti G. (1997). *Commun. Math. Phys.*, **189**, 263.
Bonetto F., Gallavotti G., and Garrido P. L. (1997). *Physica D*, **105**, 226.
Bovier A., Eckhoff M., Gayrard V., and Klein, M. (2000). *J. Phys. A., Math. Gen.*, **33**, L447.
Bovier A., Eckhoff M., Gayrard V., and Klein, M. (2002). *Commun. Math. Phys.*, **228**, 219.
Bovier A., Eckhoff M., Gayrard V., and Klein, M. (2004). *J. Eur. Math. Soc.*, **6**, 399.
Cardy J. (1999). *Field Theory and Non-Equilibrium Statistical Mechanics*, Lectures presented as part of the Troisieme Cycle de la Suisse Romande, Spring, www-thphys.physics.ox.ac.uk/people/JohnCardy/.
Chandler D. and Wolynes P. (1981). *J. Chem. Phys.*, **74**, 4078.
Chetrite R. and Gawedzki K. (2008). *Commun. Math. Phys.*, **282**, 469.
Ciliberto S. and Laroche C. (1998). *J. de Physique IV*, **8**, 215.
Cohen E. G. D. and Gallavotti G. (1999). *J. Stat. Phys.*, **96**, 1343.
Cohen E. G. D. and Mauzerall D. (2004). *J. Stat. Mech.*, P07006.
Cohen E. G. D. and Mauzerall D. (2005). *Mol. Phys.*, **103**, 2923.
Cowieson W. and Young L. S. (2005). *Ergodic Theor. Dyn. Syst.*, **25**, 1115.
Crooks G. E. (1998). *J. Stat. Phys.*, **90**, 1481.
Crooks G. E. and Jarzynski C. (2007). *Phys. Rev. E*, **75**, 021116.
Cvitanovic P. *et al.* (2008). *Chaos, Classical and Quantum*, http://chaosbook.org/.
De Dominicis C. (1976). *J. Phys. C*, **1**, 247.
De Dominicis C. and Peliti L. (1978). *Phys. Rev. B*, **18**, 353.
Dellago C., Bolhuis P. G., and Geissler P. L. (2002). *Adv. Chem. Phys.*, **123**, 1.
Doi M. (1976). *J. Phys. A*, **9**, 1479.

Du R., Pande V. S., Grosberg A. Y., Tanaka T., and Shaknovich E. (1998). *J. Chem. Phys.*, **108**, 334.
E W., Ren W., and Vanden-Eijnden E. (2002). *Phys. Rev. B*, **66**, 052301.
Eckmann J. P., Pillet C. A., and Rey-Bellet L. J. (1999). *Stat. Phys.*, **95**, 305.
Evans D. J. and Searles D. J. (1994). *Phys. Rev. E*, **50**, 1645.
Evans D. J., Cohen E. G. D., and Morriss G. P. (1993). *Phys. Rev. Lett.*, **71**, 2401.
Fogedby H. C., Hertz J., and Svane A. (2004). *Phys. Rev. E*, **70**, 031105.
Freidlin M. I. and Wentzell A. D. (1984). *Random Perturbations of Dynamical Systems*, Springer.
Gallavotti G. and Cohen E. G. D. (1995). *Phys. Rev. Lett.*, **74**, 2694.
Gardiner C. W. (1983). *Handbook of Stochastic Methods for Physics, Chemistry and the Natural Sciences*, Springer-Verlag.
Garrahan J. P., Jack R. L., Lecomte V., Pitard E., van Duijvendijk K., and van Wijland F. (2007). *Phys. Rev. Lett.*, **98**, 195702.
Gaspard P. (2003). "Dynamical systems theory of irreversibility," in *Proceedings of the NATO Advanced Study Institute, International Summer School on Chaotic Dynamics and Transport in Classical and Quantum Systems, Cargèse, Corsica*, edited by Collet P., Courbage M., and Metens, S.
Gaspard P., Nicolis G., Provata A., and Tasaki S. (1995). *Phys. Rev. E*, **51**, 74.
Gaveau B. and Schulman L. S. (1996*a*). *J. Math. Phys.*, **37**, 3897.
Gaveau B. and Schulman L. S. (1996*b*). *Phys. Lett. A*, **229**, 347.
Gaveau B. and Schulman L. S. (1998). *J. Math. Phys.*, **39**, 1517.
Gaveau B., Lesne A., and Schulman L. S. (1999). *Phys. Lett. A*, **258**, 222.
Giardina C., Kurchan J., and Peliti L. (2006). *Phys. Rev. Lett.*, **96**, 120603.
Graham R. (1980). *Z. Phys. B*, **40**, 149.
Graham R. and Tél. T. (1984). *J. Stat. Phys.*, **35**, 729.
Grassberger P., Frauenkron H., and Nadler W. (1998). "A Monte Carlo strategy for simulating polymers and other things," in *Monte Carlo Approach to Biopolymers and Protein Folding*, edited by Grassberger P. *et al.*, World Scientific.
Grimm R. C. and Storer R. G. (1971). *J. Comput. Phys.*, **7**, 134.
Hänggi P., Talkner P., and Borkovec M. (1990). *Rev. Mod. Phys.*, **62**, 251.
Hatano T. and Sasa S. (2001). *Phys. Rev. Lett.*, **86**, 3463.
Hilhorst H. (2009). http://www.th.u-psud.fr/page_perso/Hilhorst/.
Hoover W. G. (1986). *Molecular Dynamics*, Lecture Notes in Physics **258**, Springer.
Huang K. (1987). *Statistical Mechanics*, second edition, Wiley.
Jack R. L., Garrahan J. P., and Chandler D. (2006). *J. Chem. Phys.*, **125**, 184509.
Janssen H. K. (1976). *Z. Phys.*, **24**, 113.
Jarzynski C. (1997*a*). *Phys. Rev. Lett.*, **78**, 2690.
Jarzynski C. (1997*b*). *Phys. Rev. E*, **56**, 5018.
Jarzynski C. (2004). *J. Stat. Mech.: Theor. Exp.*, P09005.
Jarzynski C. (2007). *C. R. Phys.*, **8**, 495, arXiv:cond-mat/0612305.
Jordan A. N., Sukhorukov E. V., and Pilgram S. (2004). *J. Math. Phys.*, **45**, 4386.
Kadanoff L. P. and Swift J. (1986). *Phys. Rev.*, **165**, 310.
Kitahara (1975). *Adv. Chem. Phys.*, **29**, 85.
Kurchan J. (1998). *J. Phys. A, Math. Gen.*, **31**, 3719.

Kurchan J. (2007). *J. Stat. Phys.*, **128**, 1307.
Lebowitz J. L. and Spohn H. (1999). *J. Stat. Phys.*, **95**, 333.
Lecomte V. and Tailleur J. (2007). *J. Stat. Mech.*, P03004.
Machlup S. and Onsager L. (1953). *Phys. Rev.*, **91**, 1512.
Maes C. (1999). *J. Stat. Phys.*, **95**, 367.
Maes C. (2003). *On the Origin and the Use of Fluctuation Relations for the Entropy*, Poincaré Seminar, books.google.com.
Martin P. C., Siggia E. D., and Rose H. H. (1973). *Phys. Rev. A*, **8**, 423.
Micheletti C., Laio A., and Parrinello M. (2004). *Phys. Rev. Lett.*, **92**, 170601.
Murrel J. N. and Laidler K. (1968). *Trans. Faraday Soc.*, **64**, 371.
Narayan O. and Dhar A. (2004). *J. Phys. A, Math. Gen.*, **37**, 63.
Onsager L. and Machlup S. (1953). *Phys. Rev.*, **91**, 1505.
Parisi G. (1988). *Statistical Field Theory*, Frontiers in Physics **66**, Addison-Wesley.
Peliti L. (1985). *J. de Physique*, **46**, 1469.
Peliti L. (2008). *Phys. Rev. Lett.*, **101**, 098903.
Pollicott M. (1986). *Invent. Math.*, **85**, 147.
Rey-Bellet L. and Thomas L. E. (2002). *Ann. Henri Poincaré*, **3**, 483.
Risken H. (1984). *The Fokker–Planck equation*, Springer-Verlag.
Ritort F. (2006). *J. Phys. C Condens. Matt.*, **18**, R531.
Ritort R. (2008). *Advances in Chemical Physics*, Vol. 137, Wiley, arXiv:0705.0455.
Ruelle D. (1986). *Phys. Rev. Lett.*, **56**, 405.
Schütz G. and Sandow S. (1994). *Phys. Rev. E*, **49**, 2726.
Sinai Ya. G. (1963*a*). *Dokl. Akad. Nauk SSSR*, **151**, 1261.
Sinai Ya. G. (1963*b*). *Sov. Mat. Dokl.*, **4**, 1818.
Sinai Ya. G. (1970). *Russ. Math. Surv.*, **25**, 137.
Spohn H. (1983). *J. Phys. A, Math. Gen.*, **16**, 4275.
Sung J. (2005). cond-mat/0510119.
Tailleur J., Tanase-Nicola S., and Kurchan J. (2004). *J. Stat. Phys.*, **122**, 557.
Tanase-Nicola S. and Kurchan J. (2004). *J. Stat. Phys.*, **116**, 1201.
van Kampen N. G. (1981). *Stochastic Processes in Physics and Chemistry*, North-Holland.
Zinn-Justin J. (1996). *Quantum Field Theory and Critical Phenomena*, Oxford University Press.

3

Synchronization of regular and chaotic oscillators

Arkady Pikovsky

Department of Physics, University of Potsdam,
Karl-Liebknecht-Str 24/25, D-14478 Potsdam-Golm, Germany
URL: www.stat.physik.uni-potsdam.de/~pikovsky

3.1 Historical remarks and overview

3.1.1 First observations

The history of synchronization goes back to the seventeenth century when the famous Dutch scientist Christiaan Huygens reported on his observation of synchronization of two pendulum clocks. He briefly described his findings in his book *Horologium Oscillatorium: The Pendulum Clock, or Geometrical Demonstrations Concerning the Motion of Pendula as Applied to Clocks* (Huygens 1673):

> It is quite worth noting that when we suspended two clocks so constructed from two hooks imbedded in the same wooden beam, the motions of each pendulum in opposite swings were so much in agreement that they never receded the least bit from each other and the sound of each was always heard simultaneously. Further, if this agreement was disturbed by some interference, it reestablished itself in a short time. For a long time I was amazed at this unexpected result, but after a careful examination finally found that the cause of this is due to the motion of the beam, even though this is hardly perceptible.

According to a letter of Huygens to his father, the observation of synchronization was made while Huygens was sick and stayed in bed for a couple of days watching two clocks hanging on a wall. Besides an exact description, he also gave a qualitative explanation of this effect of *mutual synchronization*; he correctly understood that the conformity of the rhythms of the two clocks had been caused by an imperceptible motion of the beam. In modern terminology, this would mean that the clocks were synchronized in antiphase owing to *coupling* through the beam.

In the middle of the nineteenth century, in his book *The Theory of Sound*, Lord Rayleigh (Rayleigh 1945) described a remarkable phenomenon of synchronization in acoustical systems:

> When two organ-pipes of the same pitch stand side by side, complications ensue which not unfrequently give trouble in practice. In extreme cases the pipes may almost reduce one another to silence. Even when the mutual influence is more moderate, it may still go so far as to cause the pipes to speak in absolute unison, in spite of inevitable small differences.

Thus, Rayleigh observed not only mutual synchronization when two distinct but similar pipes began to sound in unison, but also the related effect of oscillation death, when the coupling resulted in the suppression of oscillations of interacting systems. It is worth noting that quite recently the historical observations of Huygens and Rayleigh have been reproduced in experiments (Bennett *et al.* 2002; Abel *et al.* 2006).

Although it was probably the oldest scientifically studied nonlinear effect, synchronization was understood only in the 1920s when Edward Appleton and Balthasar Van der Pol systematically—theoretically and experimentally—studied the synchronization of triode generators. This new stage in the investigation of synchronization was related to the development of electrical and radio physics (now these fields belong to engineering). In 1920, W. H. Eccles and J. H. Vincent applied for a British Patent confirming their discovery of the synchronization property of a triode generator. In their experiments, Eccles and Vincent coupled two generators which had slightly different frequencies and demonstrated that the coupling forced the systems to vibrate with a common frequency.

A few years later, Appleton and van der Pol replicated and extended the experiments of Eccles and Vincent and took the first step in the theoretical study of this effect (Appleton 1922; van der Pol 1927). Considering the simplest case, they showed that the frequency of a generator can be entrained, or synchronized, by a weak external signal of a slightly different frequency. These studies were of great practical importance because triode generators became the basic elements of radio communication systems. The synchronization phenomenon was used to stabilize the frequency of a powerful generator with the help of one which was weak but very precise.

To conclude the historical introduction, we cite the Dutch physician Engelbert Kaempfer (Kaempfer 1727), who, after his voyage to Siam in 1680, wrote (citation taken from Buck and Buck (1968)):

The glowworms ... represent another shew, which settle on some Trees, like a fiery cloud, with this surprising circumstance, that a whole swarm of these insects, having taken possession of one Tree, and spread themselves over its branches, sometimes hide their Light all at once, and a moment after make it appear again with the utmost regularity and exactness.

This very early observation reports on synchronization in a large population of oscillating systems. The same physical mechanism that causes the insects to keep in sync is responsible for the emergence of synchronous clapping in a large audience and for the onset of rhythms in neuronal populations.

3.1.2 Brief overview

We understand by "synchronization" an adjustment of the rhythms of oscillating objects due to a weak interaction. It is important to emphasize that synchronization is an essentially nonlinear effect. In contrast to many classical physical problems, where consideration of nonlinearity gives a correction to a linear theory, here taking account of nonlinearity is crucial: the phenomenon occurs only in *self-sustained* systems, which operate nonlinearly far from equilibrium.

In our presentation below we first describe the oscillating objects, capable of being synchronized, as self-sustained oscillators (Section 3.2.1). Their main feature is that the dynamics can be reduced to one variable, the phase. Its motion obeys universal equations, and allows one to describe in a unified manner the effects of synchronization by a periodic external force (Section 3.2.2) and the effects due to a weak mutual coupling of two systems (Section 3.2.3). After this, we focus on the properties of large systems. We describe two classes of them: oscillator lattices with a local coupling (Section 3.3), and globally coupled ensembles (Section 3.4).

Modern concepts also cover such objects as rotators and chaotic systems (Section 3.5); in the latter case one distinguishes between different forms of synchronization: complete, phase, master–slave, etc. In this lecture we first demonstrate how the notion of the phase can be extended to chaotic systems, and describe in Section 3.5.1 the effect of phase synchronization of chaos. Then we present a statistical theory of complete synchronization (Section 3.5.2). The latter can be considered as a particular symmetric state of a chaotic system. Remarkably, the same approach can be straightforwardly extended to describe *synchronization by common noise* (Section 3.5.4). An interesting link to a modern statistical mechanics of nonequilibrium systems is provided by a consideration of complete synchronization of space–time chaos in Section 3.5.5.

Of course, in these lectures only basic ideas can be presented; for details and further references, see Pikovsky *et al.* (2001), Mosekilde *et al.* (2002), Strogatz (2003), and Acebron *et al.* (2005).

3.2 Elementary synchronization

3.2.1 Self-sustained oscillators

Self-sustained oscillators are models of natural oscillating objects, and these models are essentially nonlinear. Mathematically, such an oscillator is described by an autonomous (i.e. without explicit time dependence) nonlinear dynamical system. It differs both from linear oscillators (which, if damping is present, can oscillate only because of an external forcing) and from nonlinear energy-conserving systems, whose dynamics essentially depends on the initial state. Some examples of self-sustained oscillatory systems are the electronic circuits used for the generation of radio-frequency power, lasers, the Belousov–Zhabotinsky and other oscillatory chemical reactions, the pacemakers (sino-atrial nodes) of human hearts, the artificial pacemakers that are used in the case of cardiac pathologies, and many other natural and artificial systems. An outstanding common feature of such systems is their ability to be synchronized.

The dynamics of oscillators is typically described in the state space $\vec{x}$:

$$\frac{d\vec{x}}{dt} = \vec{f}(\vec{x}) \,. \tag{3.1}$$

A stable limit cycle is an isolated closed trajectory describing periodic oscillations (see Fig. 3.1). In a vicinity of the limit cycle, one can introduce local coordinates: the phase θ, which serves as a coordinate along the cycle, and transverse coordinates A_i, which we will refer to as "amplitudes."

One can always choose the phase to be proportional to a fraction of the period, i.e. in such a way that it grows uniformly in time,

$$\frac{d\theta}{dt} = \omega_0 \,, \tag{3.2}$$

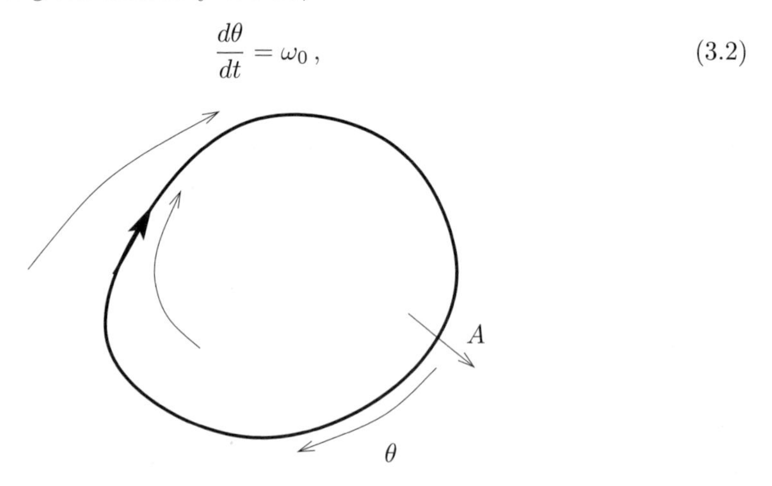

Fig. 3.1 A limit cycle in a projection on a plane. The phase and the amplitude variables are depicted schematically.

where ω_0 is the natural frequency of oscillations. The phase is neutrally stable: its perturbations neither grow nor decay. In terms of nonlinear dynamics, neutral stability means that the phase is a variable that corresponds to a zero Lyapunov exponent of the dynamical system. In contrast to this, the amplitude variables A_i relax to their values on the limit cycle, which corresponds to negative Lyapunov exponents.

This neutrality of the phase means that even an infinitely small perturbation (e.g. an external periodic forcing or a coupling to another system) can cause large deviations of the phase—in contrast to the amplitude, which is only slightly perturbed, owing to the transverse stability of the cycle. The main consequence of this fact is that the phase can be very easily adjusted by an external action, and as a result the oscillator can be synchronized!

A simple solvable example is the so-called Landau–Stuart oscillator, which in terms of the complex variable a is defined by the equation

$$\frac{da}{dt} = i\omega_0 a + (\mu - |a|^2)a \; . \tag{3.3}$$

The phase and the amplitude here are simply the argument and the modulus: $a = Ae^{i\theta}$. One can easily see that on the limit cycle, $A = \sqrt{\mu}$, and the phase rotates uniformly with a velocity ω_0.

3.2.2 Entrainment by external force

We begin our discussion of synchronization phenomena by considering the simplest case, the entrainment of a self-sustained oscillator by an external periodic force. We recall that the phase of an oscillator is neutrally stable and can be adjusted by a small action, whereas the amplitude is stable. This property allows a description of the effect of a small forcing/coupling within the framework of the phase approximation. This is in fact the first-order perturbation theory in the amplitude of the forcing ε: in this order only the phase is affected (because it is neutral), while the amplitude remains unchanged (because it is stable). [For more details on the reduction, see Kuramoto (1984) and Pikovsky *et al.* (2001).] Considering the simplest case of a limit cycle oscillator, driven by a periodic force with frequency ω and amplitude ε, we can write the equation for the perturbed phase dynamics in the form

$$\frac{d\theta}{dt} = \omega_0 + \varepsilon G(\theta, \omega t) \, , \tag{3.4}$$

where the coupling function G depends on the form of the limit cycle and of the forcing. As the states with phases θ and $\theta + 2\pi$ are physically equivalent, the function G is 2π-periodic in both its arguments, and therefore can be represented as a double Fourier series

$$G(\theta, \omega t) = \sum_{k,l} G_{k,l} e^{ik\theta + il\omega t} \; .$$

If the frequency of the external force is close to the natural frequency of the oscillator, $\omega \approx \omega_0$, then the series contains fast-oscillating and slowly varying terms; the latter are those with $l = -k$, and can be written as

$$g(\theta - \omega t) = \sum_k G_{k,-k} e^{ik(\theta - \omega t)} .$$

Only these slow terms contribute to the long-time evolution of the phase. Keeping them, which is equivalent to the averaging of the original equation (3.4) over the oscillation period, we get

$$\frac{d\theta}{dt} = \omega_0 + \varepsilon g(\theta - \omega t) . \tag{3.5}$$

Introducing the difference between the phases of the oscillation and of the forcing $\varphi = \theta - \omega t$, we obtain the following basic equation for the phase dynamics:

$$\frac{d\varphi}{dt} = -(\omega - \omega_0) + \varepsilon g(\varphi) . \tag{3.6}$$

The function g is 2π-periodic, and in the simplest case, where $g(\cdot) = \sin(\cdot)$, eqn (3.6) is called the Adler equation:

$$\frac{d\varphi}{dt} = -\delta\omega + \varepsilon \sin \varphi . \tag{3.7}$$

This equation has two essential parameters: the frequency mismatch $\delta\omega = \omega - \omega_0$ and the strength of the forcing ε. Its analysis is rather straightforward. One can easily see that, in the plane of the parameters of the external forcing (ω, ε), there exists a region $\varepsilon g_{\min} < \omega - \omega_0 < \varepsilon g_{\max}$, where eqn (3.6) has a stable stationary solution (called the Arnold tongue; see Fig. 3.2). This solution with $\varphi = \text{const}$ corresponds to the conditions of phase locking (the phase θ just follows the phase of the force, i.e. $\theta = \omega t + \text{constant}$) and frequency entrainment (the observed frequency of the oscillator $\Omega = \langle \dot{\theta} \rangle$ exactly coincides with the forcing frequency ω; the brackets $\langle\rangle$ denote time averaging).

One can view the Adler equation (3.6) as the equation of an overdamped particle in an inclined periodic potential:

$$\frac{d\varphi}{dt} = -\frac{dU}{d\varphi} , \qquad U = -\delta\omega\varphi + \varepsilon \cos \varphi . \tag{3.8}$$

Then the synchronous case $|\delta\omega| < \varepsilon$ corresponds to a potential with minima, while in the asynchronous case $|\delta\omega| > \varepsilon$, the potential has no minima and a "particle" slides

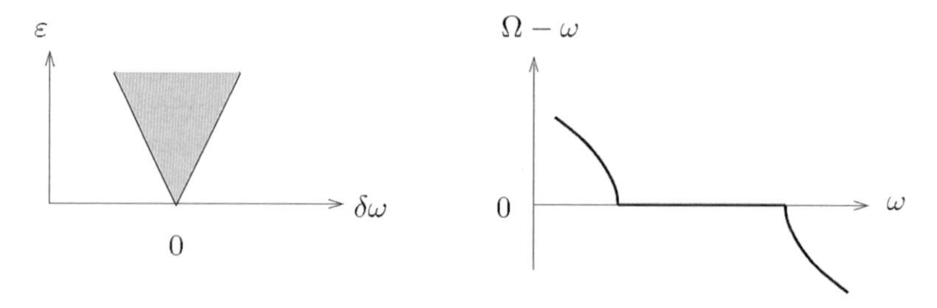

Fig. 3.2 Left panel: Arnold tongue on the plane of the parameters "frequency mismatch" and "forcing amplitude." Right panel: dependence of the observed frequency on the forcing frequency, for a particular amplitude of forcing.

down. This picture allows one to describe qualitatively what can be expected in the presence of noise. In the sliding, asynchronous regime, the effect of noise is weak. But in the synchronous state a small amount of noise results in occasional jumps over the potential barrier to the next minimum. As a result, the perfect synchrony is destroyed: relatively long epochs when the phase of the oscillator θ follows the phase of the external force ωt intermingle with phase slips when the phase of the oscillator shifts by $\pm 2\pi$ with the respect to the external phase. In contrast to the noise-free case, the frequency deviates slightly from the external frequency, but nevertheless the synchronization effect is pronounced.

Above, we have analyzed the Adler equation, which appears after averaging. In the full phase equation (3.4), synchronization is observed for high-order resonances $n\omega \approx m\omega_0$ as well. In this case the dynamics of the generalized phase difference $\psi = m\theta - n\omega t$, where n, m are integers, is described by an equation similar to eqn (3.6), namely

$$\frac{d\psi}{dt} = -(n\omega - m\omega_0) + \varepsilon\tilde{g}(\psi) ,$$

with some effective coupling function $\tilde{g}$. The synchronous regime then means perfect entrainment of the oscillator frequency at a rational multiple of the forcing frequency, $\Omega = (n/m)\omega$, as well as phase locking, where $m\theta = n\omega t + \text{constant}$. The overall picture can be shown in the (ω, ε) plane: there exist a family of triangular-shaped synchronization regions touching the ω-axis at the rational multiples of the natural frequency $(m/n)\omega_0$; these regions are usually called Arnold tongues (Fig. 3.3(a)). This picture is preserved for a moderate forcing, although now the shape of the tongues generally differs from exactly triangular. For a fixed amplitude of the forcing ε and a varied driving frequency ω, one observes different phase-locking intervals where the motion is periodic, whereas in between them it is quasiperiodic. The curve of Ω vs. ω thus consists of horizontal plateaus at all possible rational frequency ratios; this fractal curve is called the devil's staircase (Fig. 3.3(b)). A famous example of such a curve is the voltage–current plot for a Josephson junction in an AC electromagnetic field; in this context, the synchronization plateaus are called Shapiro steps. Note that a junction can be considered as a rotator (where rotations are maintained by a DC current); this example demonstrates that the synchronization properties of rotators are very close to those of oscillators.

Finally, we note that the phase difference in the synchronous state is not necessarily constant, but may oscillate around a constant value. Indeed, the solution $m\theta - n\omega t =$ constant was obtained from eqn (3.4) by means of averaging, i.e. by neglecting the fast-oscillating terms. If we take these terms into account, then we have to reformulate the condition of phase locking as $|m\theta - n\omega t| <$ constant. Thus, in the synchronous regime the phase difference is bounded; otherwise, it grows infinitely.

3.2.3 Phase dynamics of two coupled oscillators

The synchronization of two coupled self-sustained oscillators can be described in a similar way. A weak interaction affects only the phases of two oscillators θ_1 and θ_2, and eqn (3.4) generalizes to

(a)

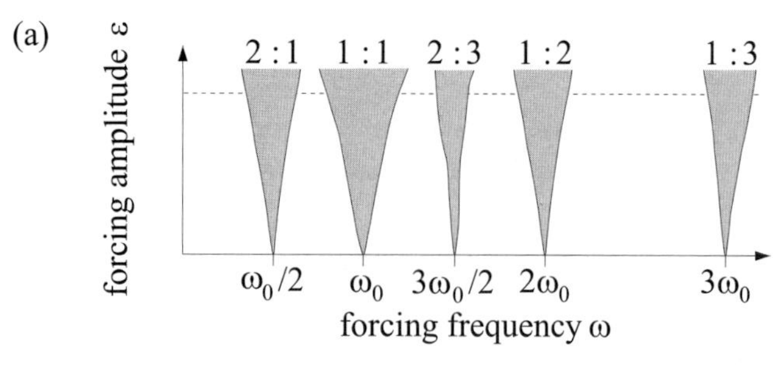

(b)

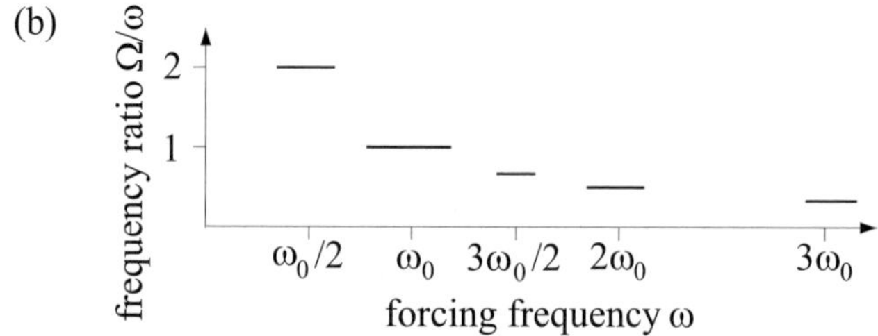

Fig. 3.3 A family of synchronization regions, or Arnold tongues (shown schematically). The numbers on top of each tongue indicate the order of locking, e.g. "2 : 3" means that the relation $2\omega = 3\Omega$ is fulfilled. (b) The Ω/ω vs. ω plot for a fixed amplitude of the force (shown by the dashed line in (a)) has a characteristic shape, known as the *devil's staircase.* (In this schematic illustration, the variation of the frequency ratio between the main plateaus of the staircase is not shown.)

$$\frac{d\theta_1}{dt} = \omega_1 + \varepsilon G_1(\theta_1, \theta_2)\,, \qquad \frac{d\theta_2}{dt} = \omega_2 + \varepsilon G_2(\theta_2, \theta_1)\,. \tag{3.9}$$

Similarly to the analysis of eqn (3.4) above, for $\omega_1 \approx \omega_2$ we average the r.h.s. of eqn (3.9), leaving only slow terms in the functions $G_{1,2}$, i.e. the terms that depend on the phase differences only, to obtain

$$\frac{d\theta_1}{dt} = \omega_1 + \varepsilon g_1(\theta_1 - \theta_2)\,, \qquad \frac{d\theta_2}{dt} = \omega_2 + \varepsilon g_2(\theta_2 - \theta_1)\,. \tag{3.10}$$

For the phase difference $\varphi = \theta_2 - \theta_1$, one obtains

$$\frac{d\varphi}{dt} = \delta\omega + \varepsilon g(\varphi)\,, \tag{3.11}$$

where $\delta\omega = \omega_2 - \omega_1$ is the frequency mismatch, and

$$g(\varphi) = g_2(\varphi) - g_1(-\varphi)\,.$$

Equation (3.11) is the same Adler equation (3.6) as that for forced oscillations. Synchronization now means that two nonidentical oscillators start to oscillate with the same frequency (or, more generally, with rationally related frequencies). To be more specific, let us assume a symmetric coupling containing only one Fourier component:

$$g_1(\psi) = g_2(\psi) = \frac{1}{2}\sin(\alpha - \psi)\,.$$

Then the interaction between the two oscillators is described by

$$\frac{d\theta_1}{dt} = \omega_1 + \frac{\varepsilon}{2}\sin(\theta_2 - \theta_1 + \alpha)\,, \qquad \frac{d\theta_2}{dt} = \omega_2 + \frac{\varepsilon}{2}\sin(\theta_1 - \theta_2 + \alpha)\,, \tag{3.12}$$

and the Adler equation (3.11) takes the form

$$\frac{d\varphi}{dt} = \delta\omega - \varepsilon\cos\alpha\sin\varphi\,. \tag{3.13}$$

We see that the parameter α effectively renormalizes the strength of the coupling ε (which we assume for definiteness to be positive). If $\cos\alpha > 0$, the coupling is attractive, and for small mismatches $|\delta\omega| < |\varepsilon\cos\alpha|$ the phase difference $\bar{\varphi} = \arcsin(\delta\omega/(\varepsilon\cos\alpha))$, lying in the interval $-\pi/2 < \varphi < \pi/2$, is stable; if $\cos\alpha < 0$, the coupling is repulsive and the stable phase difference $\bar{\varphi} = \pi - \arcsin(\delta\omega/(\varepsilon\cos\alpha))$ lies in the interval $\pi/2 < \varphi < 3\pi/2$. One refers to these two types of synchrony as "in phase" and "in antiphase," respectively. If $\alpha = \pm\pi/2$, the coupling term in eqn (3.11) disappears: now the interaction is "neutral"; it leads to a common shift of the frequencies of both oscillators, but adjusts neither the phases nor the frequencies. In the synchronous case, the common frequency of the oscillators Ω can be obtained by substituting the stable phase differences in eqn (3.10), which yields

$$\Omega = \frac{\omega_1 + \omega_2 + \varepsilon\sin\alpha\cos\bar{\varphi}}{2}\,.$$

In the case of nearly neutral coupling, this frequency may be shifted away from the interval (ω_1, ω_2); otherwise, it lies within this interval.

It is worth mentioning that locking of the phases and frequencies implies no restrictions on the amplitudes. In fact, the synchronized oscillators may have very different amplitudes and waveforms (e.g. the oscillations may be of relaxation type (pulse-like) or quasi-harmonic); in this case, of course, the coupling functions g_1 and g_2 are different. We conclude the discussion of mutual synchronization of two coupled systems by mentioning that, similarly to the case of a periodic forcing, synchronization of order $n : m$ is also possible.

3.3 Synchronization in oscillator lattices

In many natural situations, more than two oscillating objects interact. If two oscillators can adjust their rhythms, we can expect that a large number of systems could do the same. One example has already been mentioned in Section 3.1: a large population of flashing fireflies constitutes what we can call an ensemble of mutually coupled oscillators, and can flash in synchrony. A firefly communicates via light pulses with all other insects in the population. In this case one speaks of *global* (all-to-all) coupling. There are other situations, when oscillators are ordered into chains or lattices, where each element interacts only with several neighbors. Such structures are common for man-made systems; examples are laser arrays and series of Josephson junctions; but they may also be encountered in nature. In this section we discuss synchronization effects in large spatially ordered ensembles of oscillators, and then proceed in the next section with ensembles of globally coupled elements.

The simplest example of a regular spatial structure is a chain, where each element interacts with its nearest neighbors. Generally, both the spatial ordering and the interaction may be more complicated; for example, the oscillators could interact with several neighbors. Describing the interaction in the phase approximation, we generalize eqn (3.10) to

$$\frac{d\theta_k}{dt} = \omega_k + \varepsilon g(\theta_{k+1} - \theta_k) + \varepsilon g(\theta_{k-1} - \theta_k) , \qquad k = 1, \ldots, N , \tag{3.14}$$

where k is the position of the oscillator in the lattice. To analyze eqn (3.14), it is convenient to consider the phase differences $\psi_k = \theta_{k+1} - \theta_k$ and to write

$$\frac{d\psi_k}{dt} = \delta_k + \varepsilon[g(\psi_{k+1}) + g(-\psi_k) - g(-\psi_{k-1}) - g(\psi_k)] , \tag{3.15}$$

where $\delta_k = \omega_{k+1} - \omega_k$ is the difference of the natural frequencies between two neighboring oscillators. Similarly to the case of two interacting oscillators, we take $g(\psi) = \sin(\alpha - \psi)$, and obtain for the lattice

$$\frac{d\psi_k}{dt} = \delta_k + \varepsilon \cos\alpha[\sin\psi_{k+1} + \sin\psi_{k-1} - 2\sin\psi_k] + \varepsilon \sin\alpha[\cos\psi_{k-1} - \cos\psi_{k+1}] . \tag{3.16}$$

We see that the interaction is composed of two terms, one dissipative, $\propto \cos\alpha$, and one conservative, $\propto \sin\alpha$. While the general case, where both couplings are present, has been less studied, the situations of a purely dissipative ($\sin\alpha = 0$) and a purely conservative ($\cos\alpha = 0$) interaction have been discussed in the literature.

3.3.1 Dissipative coupling: Formation of clusters

Suppose the oscillators have slightly different frequencies that are distributed somehow over the ensemble. What kind of collective behavior can be expected in such a population? Certainly, if the interaction is very weak, there will be no synchronization, so that all the systems will oscillate with their own frequencies. We can also imagine that sufficiently strong coupling can synchronize the whole ensemble, provided the natural frequencies are not too different. For an intermediate coupling or for a broader distribution of the natural frequencies of the elements, we can expect some partially synchronous states. Indeed, it may happen that several oscillators synchronize and oscillate with a common frequency, whereas their neighbors have their own, different, frequencies. There may appear several such groups, or *clusters*, of synchronized elements.

To describe the formation of clusters, we take the model (3.16) with $\cos\alpha = 1$:

$$\frac{d\psi_k}{dt} = \delta_k + \varepsilon[\sin\psi_{k+1} + \sin\psi_{k-1} - 2\sin\psi_k] . \tag{3.17}$$

First we look for a stationary solution $\dot{\psi}_k = 0$. Then the variables ψ_k can be found as

$$\psi_k = \sin^{-1}(u_k^0) \,, \tag{3.18}$$

where u_k^0 is the solution of the linear tridiagonal equation

$$u_{k-1} + u_{k+1} - 2u_k = -\varepsilon^{-1}\delta_k \,, \qquad k = 1, \ldots, N-1 \,. \tag{3.19}$$

If $|u_k^0| < 1$ for all k, then a solution exists. In fact, there are 2^{N-1} solutions (3.18) because of the two roots of $\sin^{-1}$, but only one of them is stable (Ermentrout and Kopell 1984). This regime, which exists for large enough ε in any finite lattice, corresponds to one synchronization cluster: because all phase differences ψ_k are constants, the frequencies of all oscillators coincide.

Let us consider a particular case where the frequency differences δ_k are random. In this case, in the cluster solution the variables ψ_k are random as well, and one can expect that the correlations between them will decay. Then the phase profile $\theta_k = \sum_{n<k} \psi_n$ is "rough." This is not surprising, because the equation for the phase dynamics (3.14) in the case of random frequencies ω_k is very close to the Kardar–Parisi–Zhang equation for a roughening interface (Barabási and Stanley 1995) in the presence of static (quenched) disorder.

For a smaller coupling, the one-cluster state (3.18), (3.19) disappears at some critical parameter value, where the first $|u_k^0|$ reaches one. This means that at some "weak link" in the lattice the coupling cannot support the full synchrony and the lattice breaks into two clusters, each having its own frequency. When the coupling strength ε decreases further, more and more clusters appear. Moreover, the dynamics may become complex: nonlinear interaction of different clusters may lead to chaotic states. In the limit $\varepsilon \to 0$, all oscillators become independent and the regime is quasiperiodic, with N independent frequencies. We illustrate this in Fig. 3.4, where cluster formation in a lattice of 32 oscillators with purely dissipative coupling is shown.

3.3.2 Conservative coupling: Compactons in oscillator lattices

Here we describe some quite interesting dynamical regimes that can be observed in homogeneous oscillator lattices with purely conservative coupling, i.e. as described

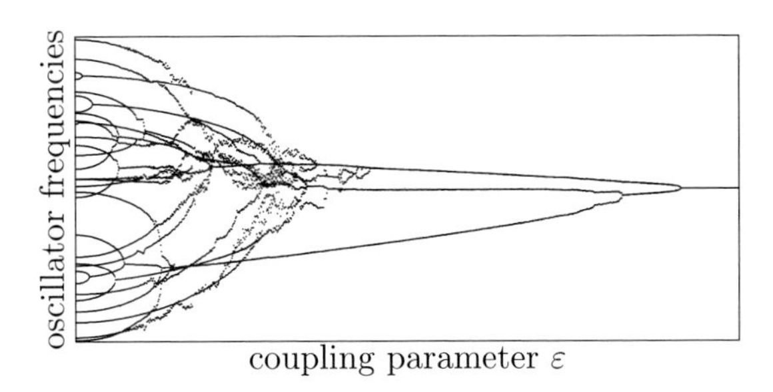

Fig. 3.4 Formation of clusters in a chain of 32 dissipatively coupled oscillators with a random distribution of natural frequencies. For a small coupling, one observes many independent oscillations which form clusters; for large couplings, only a few clusters are present. For a very large coupling, all oscillators are synchronized. For an intermediate coupling, the clusters interact strongly and behave, presumably, in a chaotic manner so that their frequencies are determined statistically.

by eqn (3.16) with $\sin\alpha = -1$, $\delta_k = 0$ (Rosenau and Pikovsky 2005; Pikovsky and Rosenau 2006). By rescaling the time with the value of the coupling ε, we obtain an equation containing no parameters,

$$\frac{d\psi_k}{dt} = \cos\psi_{k+1} - \cos\psi_{k-1} . \tag{3.20}$$

Equation (3.20) looks like a nonlinear wave equation, but with a remarkable property: for small perturbations of the homogeneous state $\psi_k^0 = 0$, there are no linear waves—all perturbations are essentially nonlinear.

To understand what happens in the lattice, we approximate the lattice model with a continuous one, i.e. with a partial differential equation. Introducing $\psi(x,t)$ and approximating the difference operator by use of the first and the third derivatives, we obtain a so-called quasi-continuous approximation,

$$\frac{\partial\psi}{\partial t} = \left(2\frac{\partial}{\partial x} + \frac{1}{3}\frac{\partial^3}{\partial x^3}\right)\cos\psi . \tag{3.21}$$

The approximation is "quasi-continuous" because we do not have here any small parameter: the waves we shall obtain will be not long in comparison with the lattice spacing. Thus the procedure we follow is not justified mathematically; it can only be supported by a comparison with numerics.

To proceed further let us consider waves of small amplitude, where we can use $\cos\psi \approx 1 - \psi^2/2$, to obtain

$$\frac{\partial\psi}{\partial t} + \frac{\partial\psi^2}{\partial x} + \frac{1}{6}\frac{\partial^3\psi^2}{\partial x^3} = 0 . \tag{3.22}$$

This equation resembles very much the famous Korteweg–de Vries equation, but the dispersion term containing the third derivative in space is nonlinear. Rosenau and Hyman (1993) first realized that this is the essential ingredient that ensures strongly localized solutions. To find traveling-wave solutions, we make the usual ansatz $\psi(x,t) = \psi(x - Vt) = \psi(s)$ and, after integrations, get

$$\psi^2\left(-V\psi + \frac{3}{4}\psi^2 + \frac{1}{2}\left(\frac{d\psi}{ds}\right)^2\right) = 0 . \tag{3.23}$$

This equation has a trivial solution $\psi = 0$ and a nontrivial one, satisfying

$$-V\psi + \frac{3}{4}\psi^2 + \frac{1}{2}\left(\frac{d\psi}{ds}\right)^2 = 0 ,$$

which is nothing else but an equation for energy conservation in a parabolic potential, a solution that has one zero point and can be explicitly written as $\psi(s) = (4V/3)\cos^2\sqrt{3/8}s$. This solution touches zero and, therefore, two solutions coincide in the same point: in the usual context of ordinary differential equations, such a nonuniqueness of a solution is a problem. Here, however, the nonuniqueness of the solution is directly related to the fact that the term with the largest derivative in eqn (3.22) is nonlinear and becomes singular at $\psi = 0$. Therefore one can combine the two solutions into one solitary wave

$$\psi(s) = \begin{cases} \frac{4V}{3}\cos^2\sqrt{\frac{3}{8}}s & \text{if} \quad s < \sqrt{\frac{2}{3}}\pi \, , \\ 0 & \text{if} \quad s > \sqrt{\frac{2}{3}}\pi \, , \end{cases} \tag{3.24}$$

which has a compact support and therefore is called a *compacton*. The width of this wave does not depend on the amplitude, and the velocity is proportional to the amplitude. We have obtained a compacton analytically for small amplitudes. The analysis of the full equation (3.21) for larger amplitudes is quite similar to that performed above; the only difference is that the potential is nonparabolic and the form of the compacton must be obtained via an integration (see the details in Rosenau and Pikovsky 2005; Pikovsky and Rosenau 2006; Ahnert and Pikovsky 2008).

To justify the existence of a compacton, one has to analyze the genuine lattice problem. Substituting the corresponding traveling-wave ansatz $\psi_k(t) = \psi(t - \lambda^{-1}k)$ into eqn (3.20), we arrive at the integral equation

$$\psi(t) = \int_{t-\lambda^{-1}}^{t+\lambda^{-1}} (1 - \cos\psi(s))\, ds \, . \tag{3.25}$$

One can solve this equation numerically to obtain localized solutions with rapidly decaying tails. The decay rate can be easily estimated by considering, in the limit $t \to \infty$, the ansatz $\psi \sim e^{-f(t)}$. Approximating the integral in eqn (3.25) using Watson's lemma by the value at $t - \lambda^{-1}$, we get for f the relation $f(t) \approx 2f(t - \lambda^{-1})$, which has an exponential solution $f = C\exp(t\lambda \ln 2)$. The resulting expression for the tails of the solitary wave is

$$\psi \sim \exp[-C\exp(t\lambda \ln 2)] \, , \tag{3.26}$$

which means a *superexponential* decay of the tails. Thus, although the correct solitary traveling wave in the lattice (3.20) is not strictly compact, its tails decay so strongly that one can consider the compacton solution of the continuous model (3.21) as a good approximation. We illustrate some compact solutions in Fig. 3.5.

In the context of the original problem of interacting oscillators (3.14), the compacton solution can be interpreted as follows. The states with $\psi = 0$ on both sides of the compacton correspond to two homogeneous states in the lattice, with phases θ_l and θ_r. The moving compacton corresponds to a moving kink between these two homogeneous states.

We conclude the discussion of compactons with one simple example of a realistic physical system where such solitary waves appear. This is a chain of elastic balls, known as the toy called Newton's cradle (Fig. 3.6). The elastic force between the balls is, according to Hertz's law, proportional to the displacement to the power 3/2. Thus, there is no linear restoring force in the lattice, which is described by the equations

$$\ddot{q}_k = (q_{k+1} - q_k)^{3/2} - (q_k - q_{k-1})^{3/2} \, . \tag{3.27}$$

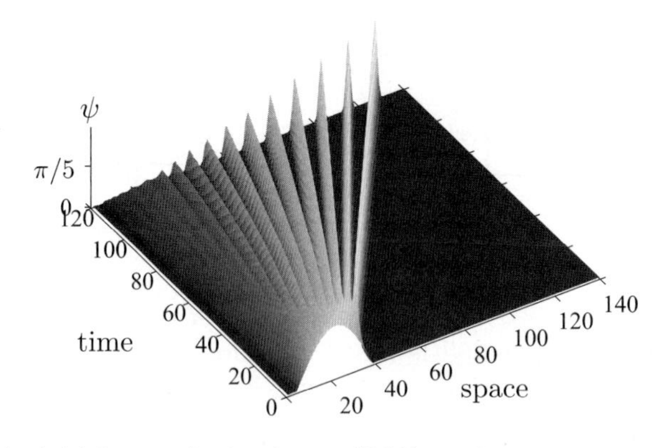

Fig. 3.5 An initial perturbation in eqn (3.20) produces numerous compactons.

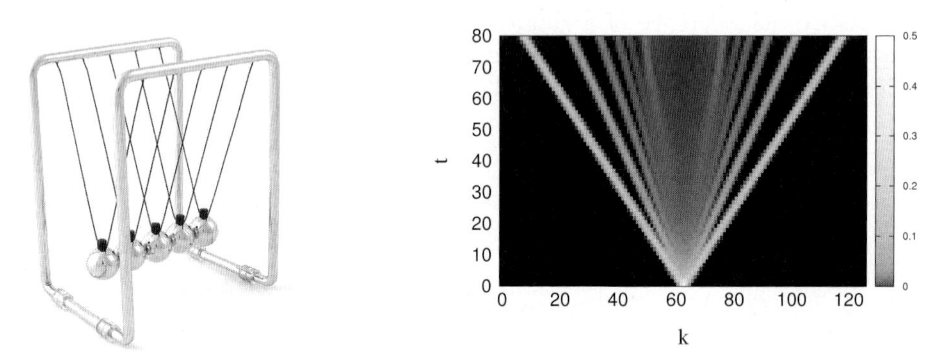

Fig. 3.6 Left panel: a lattice of elastic balls is the simplest strongly nonlinear lattice. A localized initial perturbation produces numerous compactons in a lattice with nonlinearity index $n = 3$ (right panel). Here the local energy is shown by a grayscale.

If, instead of the displacements q_k, we introduce their differences $Q_k = q_{k+1} - q_k$, the lattice dynamics is described by

$$\ddot{Q}_k = |Q_{k+1}|^n - 2\,|Q_k|^n + |Q_{k-1}|^n \ , \tag{3.28}$$

where we have introduced a general nonlinearity index n, because a solution can be found for arbitrary $n > 1$. Equation (3.28) is quite similar to eqn (3.20); the only difference is that the time derivative and the spatial difference appear squared. In the quasi-continuous approximation, one writes a partial differential equation

$$[Q(x,t)]_{tt} = [Q^n(x,t)]_{xx} + \frac{1}{12}\,[Q^n(x,t)]_{xxxx} \ , \tag{3.29}$$

which has purely compact traveling-wave solutions

$$Q(x-\lambda t) = \begin{cases} |\lambda|^m \left(\frac{n+1}{2n}\right)^{1/(1-n)} \cos^m\left(\sqrt{3}\,\frac{n-1}{n}s\right) & \text{if} \quad |s| < \frac{\pi}{2\sqrt{3}(n-1)/n}\,, \\ 0 & \text{else}\,, \end{cases} \tag{3.30}$$

with $m = 2/(n-1)$. Also, an analysis of the true lattice waves can be performed similarly to the case of oscillator lattices above. A traveling-wave solution $Q(t-\lambda^{-1}k)$ satisfies an integral equation

$$Q(t) = \int_{t-\lambda^{-1}}^{t+\lambda^{-1}} (\lambda^{-1} - |t-\xi|)Q^n(\xi)\,d\xi\,, \tag{3.31}$$

which is similar to eqn (3.25). Here also, solutions can be found numerically for different nonlinearity indices n. Estimation of the tails yields a superexponential decay of the field. For details of the analysis of strongly oscillating lattices (3.27), see Ahnert and Pikovsky (2009) and Nesterenko (2001).

3.4 Globally coupled oscillators

3.4.1 Basic ideas

Now we study synchronization phenomena in large ensembles of oscillators, where each element interacts with all others. This is usually described as *global*, or all-to-all, coupling. As a representative example, we have already mentioned synchronous flashing in a population of fireflies. A very similar phenomenon, self-organization in a large applauding audience, can be experienced in a theater. Indeed, if the audience is large enough, then one can often hear a rather fast (several oscillatory periods) transition from noise to rhythmic, nearly periodic, applause. This happens when the majority of the public applaud in unison, or synchronously (an experimental study of synchronous clapping is reported in Néda *et al.* (2000)).

The all-to-all interaction is also described as a *mean-field* coupling. Each firefly is influenced by the light field that is created by the whole population. Similarly, each applauding person hears the sound that is produced by all other people in the hall. Thus, we can say that all elements are exposed to a common force which results from the summation of the outputs of all elements. Clearly, this force can entrain many oscillators if their frequencies are close. The problem is that this force (the mean field) is not predetermined, but arises from interaction within the ensemble. This force determines whether the systems synchronize, but it itself depends on their oscillation—it is a typical example of self-organization (Haken 1993) or of a nonequilibrium phase transition. To explain qualitatively the appearance of this force, one should consider this problem self-consistently.

First, assume for the moment that the mean field is zero. Then all the elements in the population oscillate independently, and their contributions to the mean field nearly cancel each other. Even if the frequencies of these oscillations are identical, but their phases are independent, the average of the outputs of all elements of the ensemble is small if compared with the amplitude of a single oscillator. (According to the law of large numbers, it tends to zero when the number of interacting oscillators tends to infinity; the fluctuations of the mean field are of the order $N^{-1/2}$.) Thus, the asynchronous, zero-mean-field state obeys the self-consistency condition.

Next, to demonstrate that synchronization in the population is also possible, we suppose that the mean field is nonvanishing. Then, naturally, it entrains at least some part of the population, the outputs of these entrained elements add up coherently, and the mean field is indeed nonzero, as assumed. Which of these two states—synchronous or asynchronous—is realized, or, in other words, which one is stable, depends on the strength of the interaction between each pair and on how different the elements are. The interplay between these two factors, the coupling strength and the distribution of the natural frequencies, also determines how many oscillators are synchronized, and, hence, how strong the mean field is.

We present here a particular approach that allows one to proceed rather far in the analytical treatment of the problem. The results are not universally applicable, but give nevertheless a flavor of what can happen in more general situations.

3.4.2 Ensemble of identical forced phase oscillators

We start with a rather simple problem: an ensemble of identical phase oscillators ($\omega_k = \omega$, $\forall k$), all subject to the same forcing. We just take the equation for a single oscillator (3.5), take the simplest form of the coupling function $\sin(\cdot)$, and write

$$\frac{d\theta_k}{dt} = \omega + \operatorname{Im}(Ze^{-i\theta_k}) \,, \qquad k = 1, \ldots, N \,. \tag{3.32}$$

The complex external field $Z(t)$ describes the forcing, which is generally time-dependent. In a seminal paper by Watanabe and Strogatz (1994), a remarkable ansatz for this ensemble has been found: one makes a transformation to new time-dependent real variables $\rho(t)$, $\Psi(t)$, and $\Phi(t)$ and to N constants ψ_k, according to

$$e^{i(\theta_k - \Phi)} = \frac{\rho + e^{i(\psi_k - \Psi)}}{\rho e^{i(\psi_k - \Psi)} + 1} \,, \tag{3.33}$$

and, after tedious calculations, it is possible to prove that eqn (3.32) is exactly satisfied if the new variables obey the following set of equations:

$$\dot{\rho} = \frac{1-\rho^2}{2} \operatorname{Re}(Ze^{-i\Phi}) \,, \tag{3.34}$$

$$\dot{\Phi} = \omega + \frac{1+\rho^2}{2\rho} \operatorname{Im}(Ze^{-i\Phi}) \,, \tag{3.35}$$

$$\dot{\Psi} = \frac{1-\rho^2}{2\rho} \operatorname{Im}(Ze^{-i\Phi}) \,. \tag{3.36}$$

This remarkable result means that the dynamics of the ensemble is partially integrable: it can be parameterized by only three variables plus constants of motion, independently of how large the ensemble is (there are of course some minor restrictions; see Watanabe and Strogatz (1994) for details). The variables ρ, Φ, and Ψ have the following physical meaning. Oscillators form a "bunch," the width of which is inversely proportional to

ρ: if $\rho = 0$ then the oscillators are distributed nearly uniformly on the unit circle, while for $\rho = 1$ they form a cluster (their phases are equal). The variable Φ determines the position of the bunch, while Ψ describes the motion of individual oscillators with respect to the bunch. We shall denote the state with $\rho = 0$ as desynchronized, the state with $\rho = 1$ as synchronized, and the states with $0 < \rho < 1$ as partially synchronized.

It is convenient to introduce new variables $\zeta = \Phi - \Psi$ and $z = \rho e^{i\Phi}$. Then eqns (3.34)–(3.36) can be rewritten as

$$\dot{z} = i\omega z + \frac{1}{2}Z - \frac{1}{2}Z^* z^2 , \tag{3.37}$$

$$\dot{\zeta} = \omega + \mathrm{Im}(Zz^*) . \tag{3.38}$$

Here the variable z is in fact the complex order parameter, which determines the collective synchrony in the ensemble, although in a nontrivial way. Typically one introduces a complex mean field of a population θ_k of oscillators as

$$\varsigma = \frac{1}{N}\sum_k e^{i\theta_k} . \tag{3.39}$$

Substituting here the representation (3.33), we get

$$\varsigma = z\left[1 + (1 - |z|^{-2})\sum_{l=1}^{\infty}(-z^* e^{i\zeta})^l C_l\right] , \tag{3.40}$$

where

$$C_l = \frac{1}{N}\sum_{k=1}^{N} e^{il\psi_k} . \tag{3.41}$$

The situation is rather complicated for a general distribution of constants ψ_k; thus we restrict ourselves in our further considerations to the simplest case where all C_l vanish, which corresponds to a uniform distribution of constants ψ_k in the thermodynamic limit $N \to \infty$. In this case eqn (3.40) reduces to $\varsigma = z$, i.e. the variable z directly characterizes the mean field of the population.

3.4.3 Identical oscillators with linear and nonlinear global coupling

We now introduce the mean-field coupling between the oscillators in the population. The simplest setup is where the complex force $Z(t)$ acting on the oscillators in the ensemble is proportional to the complex mean field of the population:

$$Z = \varepsilon e^{-i\alpha} z = \varepsilon e^{-i\alpha}\frac{1}{N}\sum_k e^{i\theta_k} , \tag{3.42}$$

where $\varepsilon > 0$ and α are parameters of the coupling. Then eqn (3.32) can be rewritten as

$$\frac{d\theta_k}{dt} = \omega + \varepsilon\frac{1}{N}\sum_l \sin(\theta_l - \theta_k - \alpha) , \qquad k = 1, \ldots, N . \tag{3.43}$$

We see that this model, first introduced by Kuramoto (1984), is a direct generalization of eqn (3.12), describing the interaction between two oscillators.

Substituting eqn (3.42) in eqn (3.34), we obtain a closed equation for the amplitude of the order parameter:

$$\frac{d\rho}{dt} = \varepsilon \frac{\rho(1-\rho^2)}{2} \cos\alpha \ . \tag{3.44}$$

If the coupling between oscillators is attractive ($\cos\alpha > 0$), then the synchronous state $\rho = 1$ is stable: all oscillators synchronize and form a cluster. If the interaction is repulsive ($\cos\alpha < 0$), then the asynchronous state $\rho = 0$ is the only attractor in the system.

We now generalize the model (3.43) by introducing a *nonlinear* coupling. In eqn (3.42) it is assumed that the force Z acting on each oscillator depends linearly on the mean field in the ensemble, with a factor $\varepsilon e^{-i\alpha}$. A more general setup appears when the relation between Z and z is nonlinear (Rosenblum and Pikovsky 2007; Pikovsky and Rosenblum 2009). We discuss here the simplest nontrivial case, where the phase factor α depends on the amplitude of the mean field so that $\alpha = \alpha_0 + \varepsilon^2|z|^2$. Physically, this means that the nature of the coupling—whether it is attractive or repulsive—depends on the strength of the force. With this minimal modification, all the analysis above remains valid, with the following modification of the equation for the amplitude of the mean field (3.43):

$$\frac{d\rho}{dt} = \varepsilon \frac{\rho(1-\rho^2)}{2} \cos(\alpha_0 + \varepsilon^2\rho^2) \ . \tag{3.45}$$

Now an interesting case occurs if $\cos\alpha_0 > 0$ and $\cos(\alpha_0 + \varepsilon^2) < 0$. Then both the asynchronous state with $\rho = 0$ and the fully synchronous state with $\rho = 1$ are unstable: the former because the coupling for small ρ is attractive, and the latter because the coupling for large ρ is repulsive. As a result, a partially synchronous state at which $\cos(\alpha_0 + \varepsilon^2\bar{\rho}^2) = 0$ is established. In this state, the oscillators are not synchronized but nevertheless are coherent, forming a nonzero mean field. According to eqn (3.36), the oscillators move with a finite velocity with respect to the bunch, and thus the frequency of an individual oscillator differs from that of the mean field. For more details on this regime of partial synchronization, see Rosenblum and Pikovsky (2007) and Pikovsky and Rosenblum (2009).

3.4.4 Kuramoto model: Continuous distribution of frequencies

We now proceed to the less trivial case of *heterogeneous* populations of oscillators. It is convenient for us just to extend the framework above by assuming that we have many subpopulations of identical oscillators. We will consider the simplest case, where the oscillators differ in their natural frequencies only, and they all are forced with the same field Z; therefore we can rewrite eqn (3.37) in terms of a frequency-dependent complex order parameter $z(\omega)$:

$$\frac{dz(\omega)}{dt} = i\omega z(\omega) + \frac{1}{2}Z - \frac{1}{2}Z^* z(\omega)^2 \ . \tag{3.46}$$

We have to complement this with a relation for the forcing field Z. Generalizing eqn (3.42) to the heterogeneous case, we write

$$Z = \varepsilon \int d\omega \; n(\omega) z(\omega) \ , \tag{3.47}$$

where we assume a continuous distribution of natural frequencies with density $n(\omega)$ (which we take to be symmetric with respect to the middle frequency $\bar{\omega}$), and for simplicity set $\alpha = 0$. Then, substituting

$$Z = \varepsilon R e^{i\delta t} , \qquad z = \rho(\omega) e^{i\delta t + i\psi} ,$$

(here R is real) in eqns (3.46) and (3.47), we obtain a set of equations

$$R = \varepsilon R \int_{-\pi/2}^{\pi/2} d\psi \, \cos^2\psi \, n(\delta + R\sin\psi) , \tag{3.48}$$

$$0 = \varepsilon R \int_{-\pi/2}^{\pi/2} d\psi \, \cos\psi \sin\psi \, n(\delta + R\sin\psi) . \tag{3.49}$$

We notice that with the choice $\delta = \bar{\omega}$ eqn (3.49) is satisfied exactly owing to symmetry, and then eqn (3.48) gives the self-consistency condition

$$R = \varepsilon R \int_{-\pi/2}^{\pi/2} d\psi \, \cos^2\psi \, n(\bar{\omega} + R\sin\psi) , \tag{3.50}$$

which has one trivial solution $R = 0$ for $\varepsilon < \varepsilon_c = 2(\pi n(\bar{\omega}))^{-1}$ and, additionally, a nontrivial solution for $\varepsilon > \varepsilon_c$. This is the Kuramoto synchronization transition (Kuramoto 1984).

Remarkably, in some particular cases even the dynamics of the mean field Z can be found in an analytic form. Ott and Antonsen (2008) have recently noticed that in some cases the integration in eqn (3.47) can be performed using analytic properties of the functions $n(\omega)$ and $z(\omega)$. They argued that $z(\omega)$ has no poles in the upper half-plane of complex ω, and, for a Lorentzian distribution of frequencies $n(\omega) = \pi^{-1}\gamma(\gamma^2 + (\omega - \bar{\omega})^2)^{-1}$, they calculated the integral via the residue at the point $\omega = \bar{\omega} + i\gamma$. This gives a relation $Z = \varepsilon z(\bar{\omega} + i\gamma)$. Taking the corresponding equation in eqn (3.46), Ott and Antonsen obtained a closed equation for the evolution of the complex mean field (cf. eqn (3.3)),

$$\frac{dZ}{dt} = i\bar{\omega} - \gamma Z + \varepsilon \frac{Z}{2}(1 - \varepsilon^{-2}|Z|^2) .$$

In this equation, the Kuramoto transition is just a Hopf bifurcation from the zero steady state to a nontrivial limit cycle with frequency $\bar{\omega}$ and amplitude

$$|Z| = \sqrt{\varepsilon(\varepsilon - 2\gamma)} .$$

For a further discussion of the Kuramoto model, see Acebron *et al.* (2005) and references therein.

3.5 Chaotic systems

An exciting development in recent decades has been an extension of studies of synchronization to chaotic systems. Here one distinguishes different types of synchrony: phase synchronization is mostly similar to classical synchronization, while complete synchronization means full identity and is possible in a symmetric case only. Below, we describe these effects. Furthermore, we discuss synchronization induced by common external noise.

3.5.1 Phase synchronization

Nowadays it is well known that self-sustained oscillators can generate rather complex, *chaotic* signals. The mathematical image of such chaotic oscillations is not a limit cycle, but a strange attractor. Recent studies have revealed that such systems, when they are coupled, are also capable of undergoing synchronization. Certainly, in this case we have to specify this notion more precisely, because it is not obvious how to characterize the rhythm of a chaotic oscillator. It is helpful that sometimes chaotic waveforms are rather simple, so that a signal is "almost periodic"; we can consider it as consisting of similar cycles with varying amplitude and period (which can be roughly defined as the time interval between adjacent maxima). Taking a large time interval τ, we can count the number of cycles within this interval N_τ, compute the *mean frequency*

$$\omega_0 = \lim_{\tau \to \infty} 2\pi \frac{N_\tau}{\tau} , \tag{3.51}$$

and use it for characterization of the chaotic oscillatory process.

With the help of the mean frequencies, we can describe the collective behavior of interacting chaotic systems in the same way as we did for periodic oscillators. If the coupling is large enough, the mean frequencies of two oscillators become equal, and one can obtain a synchronization region, exactly as in the case of periodic systems. It is important that the coincidence of mean frequencies does not imply that the signals coincide as well. It turns out that a weak coupling does not affect the chaotic nature of the two oscillators; the amplitudes remain irregular and uncorrelated, whereas the frequencies are adjusted in a fashion that allows us to speak of a phase shift between the signals. This regime is denoted as *phase synchronization* of the chaotic systems.

The phase synchronization of chaotic systems is mostly close to the classical locking phenomena. It is based on the observation that many chaotic self-sustained oscillators admit determination of the instantaneous phase and the corresponding mean frequency. Below, we illustrate this with a simple mathematical model, called the Rössler system:

$$\begin{aligned} \dot{x} &= -y - z , \\ \dot{y} &= x + 0.15y , \\ \dot{z} &= 0.4 + z(x - 8.5) . \end{aligned} \tag{3.52}$$

For the given parameter values, this system demonstrates chaotic dynamics, as shown in Fig. 3.7(a).

In the two-dimensional projection on the (x, y) plane, the strange attractor looks like a smeared limit cycle; this allows one to introduce the phase θ as a variable that gains 2π with each rotation of the phase space trajectory. The evolution of this phase is rather similar to that of the phase of the periodic oscillator (2.14); the only difference is that it does not grow uniformly. Indeed, owing to the chaoticity, one can hardly expect that the periods of rotation along different loops of the chaotic trajectory will be exactly equal: in general, they depend on the amplitude, and the latter is chaotic. For this reason, one can characterize the phase dynamics of a chaotic oscillator as a composition of a uniform growth with an average frequency ω_0 and of a random

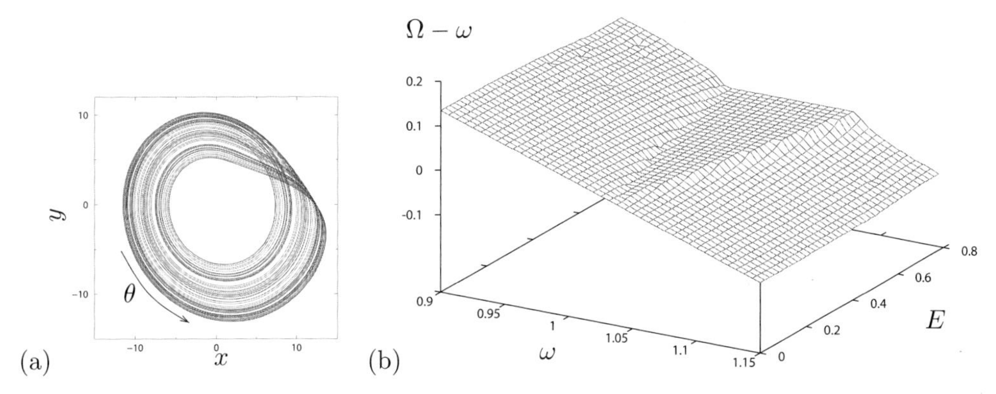

Fig. 3.7 (a) The projection of the state space of the Rössler attractor on the (x, y) plane. Trajectories rotate around a center, allowing one to introduce a phase variable θ. (b) Frequency entrainment by an external force according to eqn (3.53). Here the measured frequency Ω is shown as a function of the amplitude and frequency of the forcing. In a triangular region, the oscillator frequency Ω coincides with ω—this is the phase synchronization of chaos.

walk. (This feature makes the synchronization properties of chaotic systems close to the properties of noisy periodic oscillators.) The average frequency corresponds to the peak in the spectrum of the chaotic signal. The intensity of the random walk corresponds to the width of this peak, or, in other words, characterizes the suitability of the oscillator to serve as a clock.

Having introduced the phase and frequency for chaotic oscillators, we can characterize their synchronization. Now it becomes rather obvious that the effects of phase locking and frequency entrainment, known for periodic self-sustained oscillators, can be observed for chaotic systems as well. The simplest case is that of phase locking by an external periodic signal, for example for the Rössler model (3.52), obtained by adding the external force to the equations:

$$\begin{aligned} \dot{x} &= -y - z + E\cos(\omega t) , \\ \dot{y} &= x + 0.15y , \\ \dot{z} &= 0.4 + z(x - 8.5) . \end{aligned} \tag{3.53}$$

When the chaotic oscillator is driven by a signal with a frequency ω close to ω_0, the forcing affects the evolution of the phase, and the observed (mean) frequency Ω becomes adjusted to the external frequency as is shown in Fig. 3.7(b). This region is a complete analog of the synchronization regions (Arnold tongues) for periodic oscillators.

It is important to emphasize that the chaos itself is not suppressed by the external force. What happens is not a disappearance of chaos, but an adjustment of the mean oscillation frequency. Chaos may be destroyed by a strong force, but a small forcing affects only the phase, entraining the frequency of its rotation.

The mutual phase synchronization of chaotic oscillators is also quite similar to the classical case. To demonstrate this, one can couple two chaotic oscillators (3.52). The quantities to be observed are the phases of the oscillators: for each oscillator, one has to extract these phases from portraits like those shown in Fig. 3.7(a). Then a calculation of the phase difference and the observed frequencies Ω_1, Ω_2 characterizes the entrainment. For large enough coupling and for small mismatch of the natural frequencies, one observes that the frequencies become equal, $\Omega_1 = \Omega_2$, as in the case of periodic oscillators described in Section 3.2.3.

Furthermore, a synchronization transition in a population of chaotic oscillators can be observed as well. The mechanism here is the same as described in Section 3.4 above: owing to interaction, some oscillators become entrained and start to oscillate with the same frequency, the fields of these oscillators sum coherently, and the resulting mean field maintains synchrony. Considering chaotic oscillators, we additionally have to take into account that only the phases of the oscillators are adjusted, whereas the individual oscillations remain chaotic. Thus, the mean field arises because of contributions from mutually entrained oscillators with nearly equal phases but with chaotic, different amplitudes. Summation of these contributions leads to a periodic field with some average amplitude—chaos is "washed out" owing to the averaging over the ensemble. As a result, the synchronization transition in an ensemble of coupled chaotic oscillators manifests itself as the appearance of a periodic macroscopic mean field, while each individual oscillator remains chaotic (Pikovsky *et al.* 1996). Such a synchronization transition has been observed in experiments with an ensemble of 64 electrochemical chaotic oscillators (Kiss *et al.* 2002).

3.5.2 Complete synchronization: Basic ideas

Strong mutual coupling of chaotic oscillators leads to their *complete synchronization* when two or more chaotic systems have exactly the same states, and these identical states vary irregularly in time. In contrast to phase synchronization, this can be observed in any chaotic system, not necessarily an autonomous one, and, in particular, in periodically driven oscillators and in discrete-time systems (maps). In fact, this phenomenon is not close to the classical synchronization of periodic oscillations, as here we do not have an adjustment of rhythms. Instead, complete synchronization means a suppression of differences in coupled *identical* systems. Therefore, this effect cannot be described as entrainment or locking; it is closer to the onset of symmetry. Maybe another word instead of "synchronization" would serve better for underlining this difference; we will follow the accepted terminology of today, using the adjective "complete" to avoid ambiguity.

The main precondition for complete synchronization is that the interacting systems are identical, i.e. they are described by exactly the same equations of motion. This identity implies that if the initial states of these systems are equal, then during the evolution they remain equal at all times. However, in practice this coincidence of states will be realized only if such a regime is stable, i.e. if it is restored after a small violation. This imposes a condition on the strength of the coupling between the systems.

To be more concrete in our discussion, let us consider a coupled system of the type

$$\frac{d\vec{x}}{dt} = \vec{F}(\vec{x}) + \varepsilon(\vec{y} - \vec{x}) , \qquad \frac{d\vec{y}}{dt} = \vec{F}(\vec{y}) + \varepsilon(\vec{x} - \vec{y}) . \tag{3.54}$$

Here $\vec{x}$ and $\vec{y}$ are two identical systems, described by the same equations with the same $\vec{F}$, and we will assume that the solutions are chaotic. The coupling parameter ε describes a so-called diffusive coupling, which tends to equalize the states of the two systems.

While the coupling tends to equalize the states of the two systems, another mechanism prevents this. This mechanism is the sensitive dependence on initial conditions that is inherent in chaos. Suppose that $\varepsilon = 0$, and then we have two uncoupled identical systems; they can be regarded as two realizations of one system with different initial conditions. Because chaotic motions depend sensitively on the initial conditions (this phenomenon is often called the "butterfly effect"), the values $\vec{y}(t)$ and $\vec{x}(t)$ will differ significantly after some time, even if $\vec{y}(0) \approx \vec{x}(0)$.

Summarizing, we see two counterplaying tendencies in the diffusive interaction of two identical chaotic systems: intrinsic chaotic instability tends to make the states of the systems different, while coupling tends to equalize them. As a result, there exists a critical value of the coupling ε_c, such that for stronger coupling a completely synchronized state $\vec{y}(t) = \vec{x}(t)$ sets in. In this regime the coupling term in eqn (3.54) vanishes, and, hence, each of the systems varies chaotically in time as if they were uncoupled. Let us look at a small perturbation of this state, by writing the evolution equation for the difference $\vec{v} = \vec{x} - \vec{y}$ in an approximation linear in $\vec{v}$:

$$\frac{d\vec{v}}{dt} = \frac{\partial \vec{F}}{\partial \vec{x}}\vec{v} - 2\varepsilon\vec{v} . \tag{3.55}$$

Substituting $\vec{v} = \vec{v}'e^{-2\varepsilon t}$ here, we can reduce eqn (3.55) to

$$\frac{d\vec{v}'}{dt} = \frac{\partial \vec{F}}{\partial \vec{x}}\vec{v}' . \tag{3.56}$$

But this is exactly the equation for a perturbation of a chaotic solution in a single, uncoupled system. Thus, such a perturbation grows with the largest Lyapunov exponent; $|\vec{v}'| \propto e^{\lambda t}$. For the difference of the two systems, we thus obtain

$$|\vec{v}| \propto e^{(\lambda - 2\varepsilon)t} . \tag{3.57}$$

We see that the critical coupling in this simple example is just $\varepsilon_c = \lambda/2$. Thus, complete synchronization is a threshold phenomenon: it occurs only when the coupling exceeds some critical level, proportional to the largest Lyapunov exponent of the individual system. Below the threshold, the states of the two chaotic systems are different but close to each other.

We can illustrate these theoretical ideas by the results of Roy and Thornburg (1994), who observed synchronization of the chaotic intensity fluctuations of two Nd:YAG lasers with modulated pump beams. The coupling was implemented by overlapping the intracavity laser fields, and varied during the experiment. For strong coupling, the intensities became identical, although they continued to vary in time chaotically (Fig. 3.8).

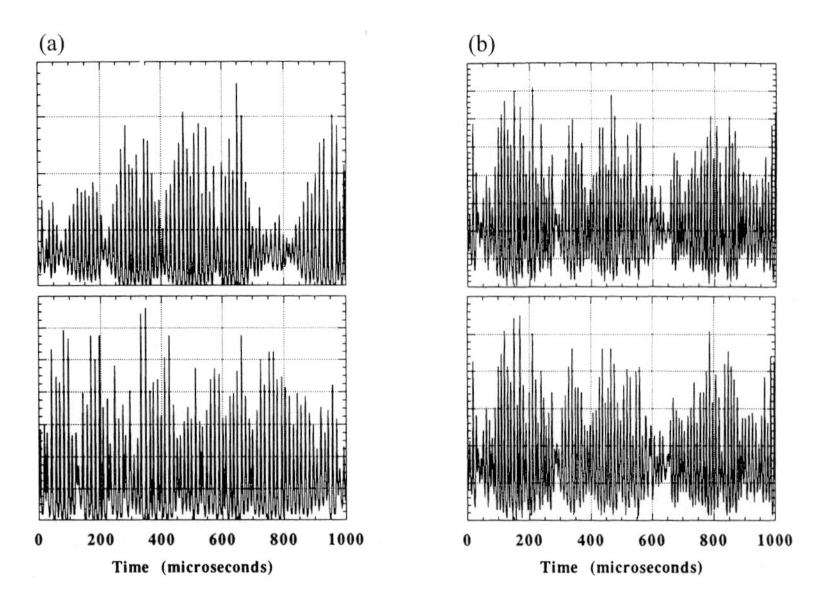

Fig. 3.8 Complete synchronization in coupled lasers. (a) Under-threshold coupling: the intensities (arbitrary units) of the uncoupled lasers fluctuate chaotically, and the behavior in time is different, although both lasers experience the same pump modulation. (b) Under strong coupling, beyond the synchronization threshold, both lasers remain chaotic, but now the oscillations are nearly identical, i.e. complete synchronization sets in. After Roy and Thornburg (1994).

3.5.3 Statistical analysis of the simplest model

In this section we present a statistical theory of the complete synchronization transition. Our main goal is to show the thermodynamic formalism—the theory of large deviations—"in action" (Pikovsky and Grassberger 1991). As a particular example, we consider a simple chaotic system, called the "skewed tent map" of an interval $0 \le x \le 1$ onto itself, given by the following piecewise linear transformation:

$$x_{n+1} = (x_n) = \begin{cases} \dfrac{x_n}{a} & 0 \le x \le a \;, \\ \dfrac{1 - x_n}{1 - a} & a \le x_n \le 1 \;. \end{cases} \tag{3.58}$$

Here $a < 1$ is a parameter defining the "skewness" of the map; the case $a = 1/2$ corresponds to a symmetric tent map. It is quite straightforward to follow a small perturbation of the initial state x_0 in the linear approximation, where we get

$$\delta x_n = f'(x_{n-1})\, \delta x_{n-1} = f'(x_{n-1}) f'(x_{n-2}) \dots f'(x_0)\, \delta x_0 \;, \tag{3.59}$$

and because all derivatives are larger than one, i.e. $|f'(x_k)| > 1$, the perturbation grows. To quantify this, one takes the mean growth rate of the logarithm of the perturbation and defines the Lyapunov exponent as

$$\lambda = \lim_{n\to\infty} \frac{1}{n} \ln \left| \frac{\delta x_n}{\delta x_0} \right| = \langle \ln |f'(x)| \rangle \ . \tag{3.60}$$

In the case of the skewed tent map, the invariant distribution of x is uniform, and thus

$$\lambda = -a \ln a - (1-a)\ln(1-a) \ .$$

We now introduce the coupling according to

$$\begin{pmatrix} x_{n+1} \\ y_{n+1} \end{pmatrix} = \begin{pmatrix} 1-\varepsilon & \varepsilon \\ \varepsilon & 1-\varepsilon \end{pmatrix} \begin{pmatrix} x_n \\ y_n \end{pmatrix} = \begin{pmatrix} f(x_n) + \varepsilon(f(y_n) - f(x_n)) \\ f(y_n) + \varepsilon(f(x_n) - f(y_n)) \end{pmatrix} \ . \tag{3.61}$$

We are interested in the synchronized state, where $x_n \approx y_n$. Therefore we introduce new variables

$$u = \frac{x+y}{2} \ , \qquad v = \frac{x-y}{2} \ , \tag{3.62}$$

and, assuming v to be small, obtain to first order

$$u_{n+1} = \frac{f(u_n + v_n) + f(u_n - v_n)}{2} \approx f(u_n) \ , \tag{3.63}$$

$$v_{n+1} = (1-2\varepsilon)\frac{f(u_n + v_n) - f(u_n - v_n)}{2} \approx (1-2\varepsilon) f'(u_n) v_n \ . \tag{3.64}$$

Equation (3.64) is quite similar to eqn (3.59), and thus it is convenient to introduce $z_n = \ln |v_n|$:

$$z_{n+1} = z_n + \ln|1-2\varepsilon| + \ln|f'(u_n)| \ . \tag{3.65}$$

We see that z_n performs a random walk, which is biased to the positive or negative direction depending on the sign of the so-called transverse Lyapunov exponent $\lambda_\perp$, where

$$\lambda_\perp = \ln|1-2\varepsilon| + \langle \ln|f'(u_n)| \rangle = \ln|1-2\varepsilon| + \lambda \ . \tag{3.66}$$

Thus the synchronization threshold is determined by the condition $\lambda_\perp = 0$, which yields

$$\varepsilon_c = \frac{1-e^{-\lambda}}{2} \ . \tag{3.67}$$

To have a more detailed description of the dynamics near the synchronization threshold, we introduce a local transverse Lyapunov exponent according to

$$\Lambda_n = \ln|1-2\varepsilon| + \ln|f'(u_n)| - \lambda_\perp = \ln|f'(u_n)| - \langle \ln|f'(u_n)| \rangle \ .$$

This is nothing else but the fluctuating part of the steps of the random walker z_n. According to this definition, after T time steps the walker's displacement is

$$z_T - z_0 = T\lambda_\perp + \sum_{j=0}^{T-1} \Lambda_j = T\lambda_\perp + TX_T \ , \tag{3.68}$$

where

$$X_T = \frac{1}{T}\sum_{j=0}^{T-1}\Lambda_j \ .$$

To calculate the statistical characteristics of X_T, we introduce the generating function $c(t)$ of the random variable Λ according to

$$e^{c(t)} = \langle e^{t\Lambda}\rangle \ .$$

Then the generating function for X is

$$\langle e^{tX}\rangle = \left\langle \exp\left[t\frac{1}{T}\sum_{j=0}^{T-1}\Lambda_j\right]\right\rangle = \prod_{j=0}^{T-1}\left\langle e^{(t/T)\lambda_j}\right\rangle = e^{Tc(t/T)} \ , \tag{3.69}$$

where we have used the fact that the Λ_j are independent, which is valid for the skewed tent map. We look for a distribution of X in the form $P(X) \approx e^{-TI(X)}$. Calculation of the generating function of X from this ansatz leads to

$$\langle e^{tX}\rangle \approx \int dX\ e^{-TI(X)+tX} = \int dX\ e^{-T(I(X)-(t/T)X)} \ .$$

Denoting $\tau = t/T$ and using eqn (3.69), we obtain

$$e^{Tc(\tau)} \approx \int dX\ e^{-T(I(X)-\tau X)} \ .$$

Evaluation of the integral for large T leads to the following relation between $c(\tau)$ and $I(X)$:

$$c(\tau) = -\min_X (I(X) - \tau X) \ , \tag{3.70}$$

which is nothing else but the Legendre transform. For a smooth concave $I(X)$, it can be written as

$$c(\tau) + I(X^*) = \tau X^* \ , \qquad I'(X^*) = \tau \ .$$

The relations above allow us to find the distribution $W(z)$ of the random walker z. The following equation for W follows from eqn (3.68):

$$W(z) = \int dX\ w(z - T\lambda_\perp - TX)P(X) \ . \tag{3.71}$$

Applying the ansatz

$$W(z) \approx e^{\kappa z} \ , \tag{3.72}$$

we obtain

$$1 = \int dX\ e^{-T(I(X)+\lambda_\perp\kappa + X\kappa} \ .$$

Evaluation of the integral for large T leads to the relations

$$I(X^*) - (X^* + \lambda_\perp)I'(X^*) = 0 \ , \qquad \kappa = -I'(X^*) \ , \tag{3.73}$$

allowing us to find the exponent κ.

For the skewed tent map, the scaling function $I(X)$ follows from the binomial distribution; in a parametric form, it reads

$$I(h) = -h \ln \frac{a}{h} - (1-h) \ln \frac{1-a}{1-h} , \tag{3.74}$$

$$X(h) = -h \ln a - (1-h) \ln(1-a) + a \ln a + (1-a) \ln(1-a) . \tag{3.75}$$

Inserting this in eqn (3.73) yields

$$\kappa(h) = \frac{\ln(1-h) - \ln h}{\ln(1-a) - \ln a} - 1 , \tag{3.76}$$

$$\lambda_{\perp}(h) = -X(h) - \frac{I(h)}{\kappa} . \tag{3.77}$$

The value of the exponent κ is positive for positive $\lambda_{\perp}$, and negative for negative values of the transverse Lyapunov exponent. It is restricted to the interval $\ln a + \lambda < \lambda_{\perp} < \ln(1-a) + \lambda$ (where we assume $a < 1 - a$). The physical meaning of this restriction is the following: in this interval, the steps of the random walker can change sign, and thus large deviations in both the positive and negative directions are possible. Outside of this domain, the walker moves persistently in one direction, and no solution of the type (3.72) is possible.

Returning back to the variable $|v| = |x - y| = e^z$, we see that the solutions constructed are power-law distributions of the difference between the two interacting chaotic systems:

$$\text{Prob}(|x-y|) \sim |x-y|^{\kappa - 1} .$$

This solution is nonnormalizable: it diverges at small $|x - y|$ if $\kappa < 0$, i.e. for negative transverse Lyapunov exponents, when the synchronous state is stable; and it diverges at large $|x - y|$ if $\kappa > 0$, i.e. for positive transverse Lyapunov exponents, when the synchronous state is unstable. In both cases, one can introduce cutoffs to get a normalizable stationary distribution. The cutoff at small $|x - y|$ may be due to a small nonidentity of the coupled systems, and the cutoff at large $|x - y|$ appears naturally when one goes beyond the linear approximation for the dynamics of the difference variable v.

The theory above shows that the thermodynamic formalism, or the theory of large deviations, can be quite naturally applied to a statistical description of chaos. Moreover, one can make a link between the statistical description and the geometrical properties of the corresponding strange attractors (see Pikovsky and Grassberger (1991) and Pikovsky *et al.* (2001) for details).

3.5.4 Complete synchronization by common noise

Here we describe an interesting phenomenon of synchronization by common noise, which is rather close to the effect of complete synchronization presented in the previous section. Let us consider two dynamical systems which are identical and are driven by the same irregular forcing ξ_n (below, we interpret this as noise; it may also be a chaotic force):

$$x_{n+1} = f(x_n, \xi_n) , \tag{3.78}$$

$$y_{n+1} = f(y_n, \xi_n) . \tag{3.79}$$

Comparing this with eqn (3.61), we see that this system also possesses a symmetric solution $x_n = y_n$, and thus here a symmetric, completely synchronized state is possible even without coupling. Therefore, we can proceed similarly and write equations for the variables of eqn (3.62):

$$u_{n+1} = f(u_n, \xi_n) , \qquad v_{n+1} = f'(u_n, \xi_n) v_n , \tag{3.80}$$

where we have used a linear approximation in v. This system is quite similar to eqns (3.63) and (3.64), and the growth or decay of the difference field v depends on the sign of

$$\lambda = \langle \ln |f'(u_n, \xi_n)| \rangle , \tag{3.81}$$

which is nothing else but the usual Lyapunov exponent of the noise-driven system (because there is no coupling, the usual Lyapunov exponent is the same as the transverse one). Depending on the sign of λ, one observes synchronization for negative λ and desynchronization if $\lambda > 0$. It is noteworthy that by observing just one system one cannot decide whether it has a positive or a negative Lyapunov exponent: because of the noise, in all cases the dynamics is irregular. The distinction between the two cases becomes visible only if two (or more) systems are observed.

Preparing two or several identical systems and applying the same noise to them seems to be a rather artificial setup. Remarkably, however, this has been done in neuroscience with a small modification: there, one performs several experiments by acting with one and the same noisy forcing (which, for this experiment, is prerecorded for all runs) on one system (a neuron or a group of neurons). In this experiment, the two variables x and y in eqns (3.78) and (3.79) correspond to just two runs of the driving experiment. Synchrony then means that in both runs, the response of the neurons to the noise is the same. Such synchronization by common noise is called in a neurophysiological context "reliability" (Mainen and Sejnowski 1995).

The consideration of two systems as in eqns (3.78) and (3.79) can be straightforwardly generalized to an ensemble of them. All the systems in the ensemble are subject to the same noise, but they have different initial conditions.

An analytical calculation of the Lyapunov exponent for a noisy system can be quite a complicated task, but numerically it is performed in the same way as for chaotic dynamics—by following the norm of the solution of the linearized equation (3.80). Noise can generally affect the Lyapunov exponent in both directions: it can increase or decrease it. In particular, it may happen that a noise-free chaotic dynamical system in the presence of noise will be "stabilized" in the sense that the Lyapunov exponent will become negative. Also, the opposite can happen: a regular dynamical system with a negative or zero Lyapunov exponent may become chaotic (in the sense of a positive Lyapunov exponent) owing to a noisy forcing. However, for a small forcing we expect some continuous dependence on the noise intensity. This means that a chaotic dynamical system will have a positive Lyapunov exponent for a small amount of noise—here, small common noise does not synchronize. In contrast, a stable steady state having a negative Lyapunov exponent will maintain this property for a small amount of noise, which will synchronize.

A nontrivial case is that of periodic oscillations: the largest Lyapunov exponent on the limit cycle is exactly zero, and what happens for small noise is a nontrivial question. As have been argued and demonstrated in Pikovsky (1984*a*,*b*) and Goldobin and Pikovsky (2005*a*,*b*), small noise acting on a limit cycle always leads to a negative Lyapunov exponent, i.e. to a synchronization. Physically, this follows from the property already discussed, that weakly perturbed limit cycle oscillations can be described with the phases only. The equation for the phase is a one-dimensional nonautonomous one, and in such an equation chaos is impossible: the dimension of the state space is too small to allow instability of trajectories. Thus the Lyapunov exponent for any driven phase equation is nonpositive.

For a noisy force that is not very small, the restriction above is no longer valid and the Lyapunov exponent may change sign. We illustrate this with a simple model of a noise-driven neuron, described by the FitzHigh–Nagumo model (Goldobin and Pikovsky 2006):

$$\frac{dv}{dt} = \varepsilon^{-1}[(3 - v^2)v - w] + \xi(t) , \qquad \langle \xi(t)\xi(t + t') \rangle = 2\sigma^2 \delta(t') , \tag{3.82}$$

$$\frac{dw}{dt} = v - v_0 . \tag{3.83}$$

This model describes, for $v_0 < -1$, an excitable system which, under the influence of noise, generates spikes. Numerical evaluation of the Lyapunov exponent gives a negative exponent for a very small noise level, a positive exponent for a moderate noise level, and again a negative exponent for strong noise. Thus, identical neurons driven by common noise will be synchronized if the noise is weak or strong, and will be desynchronized for a moderate noise level. This is illustrated in Fig. 3.9. The spiking of an individual neuron is depicted by a small vertical bar, and one can easily see synchrony as perfect vertical lines and asynchrony when these lines are broken. Similar pictures have been observed in experiments with real neurons (Mainen and Sejnowski 1995).

3.5.5 Complete synchronization of space–time chaos

The general discussion of complete synchronization in Section 3.5.2 above is not restricted to low-dimensional chaotic dynamics; it applies to extended systems that demonstrate space–time chaos as well. We illustrate this with the following numerical example. We take two partial differential equations yielding space–time chaos, namely the Kuramoto–Sivashinsky equations for two fields $q(x, t)$ and $p(x, t)$, and couple them:

$$\frac{\partial q}{\partial t} + \frac{\partial^2 q}{\partial x^2} + \frac{\partial^4 q}{\partial x^4} + q\frac{\partial q}{\partial x} = \varepsilon(p - q) , \tag{3.84}$$

$$\frac{\partial p}{\partial t} + \frac{\partial^2 p}{\partial x^2} + \frac{\partial^4 p}{\partial x^4} + p\frac{\partial p}{\partial x} = \varepsilon(q - p) . \tag{3.85}$$

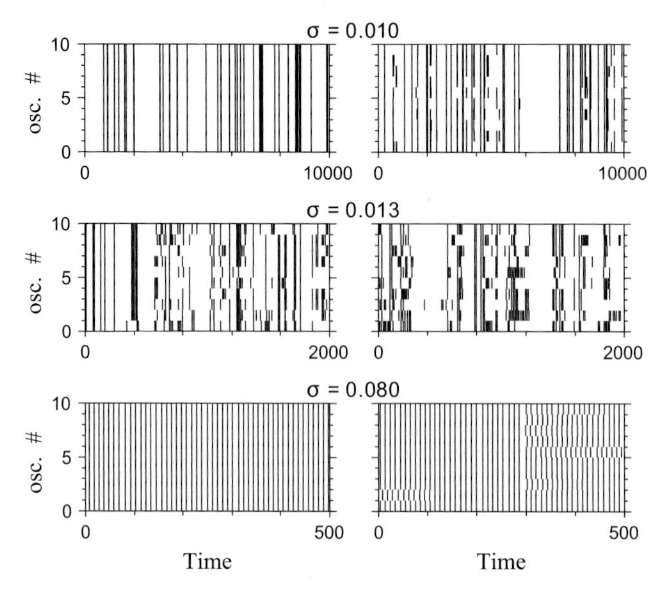

Fig. 3.9 Demonstration of synchronization and desynchronization of an ensemble of ten identical neurons (3.82). In each panel, small vertical bars represent a spike of one neuron. Left column: purely identical systems driven by common noise. Right column: the noise is not completely identical for all neurons, but has small neuron-to-neuron deviations. As a result, the ideal picture in the left column is slightly smeared in the right column. The rows show three different levels of noise, for which the largest Lyapunov exponent is negative (top and bottom rows) or positive (middle row). Ideally, vertical lines in the top and bottom rows indicate synchronization of neurons; scattered bars in the middle row indicate desynchronization (Goldobin and Pikovsky 2006).

The effect of the coupling is demonstrated in Fig. 3.10. The two fields are uncoupled until time $t = 200$, and the coupling $\varepsilon = 0.1$, which is larger than critical value, is switched on at $t = 200$. One can see that the space–time dynamics in the two subsystems becomes identical quite fast, while remaining chaotic.

The statistical theory of the synchronization transition is based on the same ideas as in Section 3.5.3 above. In analogy with eqn (3.64), we write for the difference field $v(x,t) = |p(x,t) - q(x,t)|$ the following model equation:

$$\frac{\partial v(x,t)}{\partial t} = \xi(x,t)w(x,t) + \frac{\partial^2 v(x,t)}{\partial x^2} \, . \tag{3.86}$$

Here $\xi(x,t)$ is the local transverse perturbation growth rate (corresponding to the term $|(1-2\varepsilon)f'(u_n)|$ in eqn (3.64)). Additionally, we account for spatial coupling by including the diffusion of the perturbation field. Similarly to the transformation to eqn (3.65), it is convenient to introduce $h(x,t) = \ln v(x,t)$ to get

$$\frac{\partial h}{\partial t} = \xi(t,x) + \frac{\partial^2 h}{\partial x^2} + \left(\frac{\partial h}{\partial x}\right)^2 \, . \tag{3.87}$$

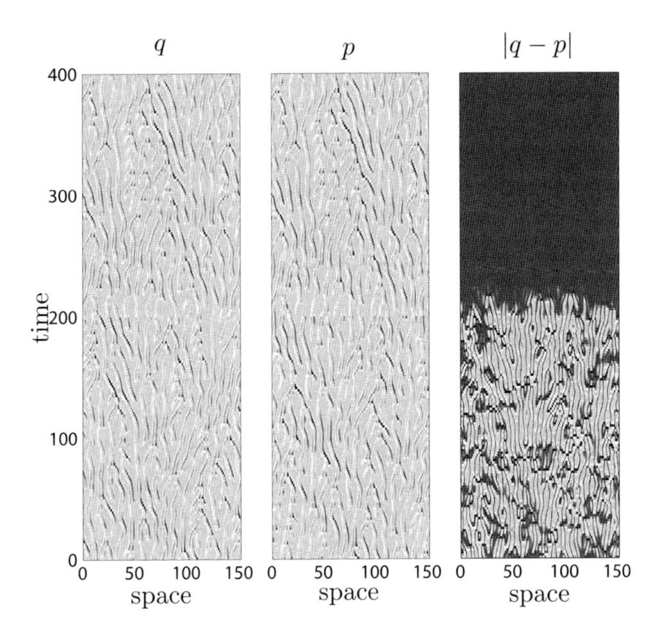

Fig. 3.10 Demonstration of a complete synchronization transition for space–time chaos. After the coupling is switched on at $t = 200$, the two fields very soon become identical, while both continue to demonstrate irregularity both in space and in time.

This is nothing else but the famous Kardar–Parisi–Zhang equation, widely discussed in the context of roughening interfaces (Kardar *et al.* 1986; Barabási and Stanley 1995). Qualitatively, this means that the perturbation field $v(x,t)$ is the exponential of a rough interface, and thus it has sharp peaks at the places where the interface profile $h(x,t)$ has maxima, and is extremely (exponentially) small at other places. This means that near the synchronization threshold, when desynchronization is weak, the deviations from synchrony take the form of localized bursts. Quantitatively, one can use the well-developed theory of the Kardar–Parisi–Zhang equation to calculate the scaling properties of the critical coupling and of the statistical properties of the perturbation field in the thermodynamic limit of large system size (see Ahlers and Pikovsky 2002).

3.6 Conclusion

In spite of its long history, the theory of synchronization remains a rapidly developing branch of nonlinear science. From the viewpoint of theoretical physics, synchronization transitions are strongly nonequilibrium phase transitions, which in many cases can hardly be compared to equilibrium phenomena. Nevertheless, many important questions, such as those of finite-size effects, are inspired by corresponding developments in the realm of equilibrium statistical physics.

Recent theoretical developments have been strongly influenced by interdisciplinary studies, especially of widely growing applications to biological and medical problems. It turns out that synchronization is very frequently encountered in living systems (Glass

2001; Pikovsky *et al.* 2001). In particular, it is believed that the mechanism of the Kuramoto transition plays an important role in the dynamics of neural ensembles and is responsible for the emergence of such severe pathologies as epilepsy and Parkinson's disease. Finally, we should mention that ideas from synchronization theory are being used in the analysis of multivariate experimental data. The goal of such analysis is to detect weak interactions between oscillatory systems, for example to reveal coordination between respiratory and cardiac rhythms in humans or to localize the source of pathological brain activity in Parkinson's disease.

References

Abel, M., Bergweiler, S., and Gerhard-Multhaupt, R. (2006). Synchronization of organ pipes: Experimental observations and modeling. *J. Acoust. Soc. Am.*, **119**(4), 2467–2475.

Acebron, J. A., Bonilla, L. L., Vicente, C. J. P., Ritort, F., and Spigler, R. (2005). The Kuramoto model: A simple paradigm for synchronization phenomena. *Rev. Mod. Phys.*, **77**(1), 137–175.

Ahlers, V. and Pikovsky, A. S. (2002). Critical properties of the synchronization transition in space–time chaos. *Phys. Rev. Lett.*, **88**(25), 254101.

Ahnert, K. and Pikovsky, A. (2008). Traveling waves and compactons in phase oscillator lattices. *CHAOS*, **18**(3), 037118.

Ahnert, K. and Pikovsky, A. (2009). Compactons and chaos in strongly nonlinear lattices. *Phys. Rev. E*, **79**, 026209.

Appleton, E. V. (1922). The automatic synchronization of triode oscillator. *Proc. Cambridge Phil. Soc. (Math. Phys. Sci.)*, **21**, 231–248.

Barabási, A.-L. and Stanley, H. E. (1995). *Fractal Concepts in Surface Growth.* Cambridge University Press, Cambridge.

Bennett, M., Schatz, M., Rockwood, H., and Wiesenfeld, K. (2002). Huygens' clocks. *Proc. R. Soc. (A)*, **458**(2019), 563–579.

Buck, J. and Buck, E. (1968). Mechanism of rhythmic synchronous flashing of fireflies. *Science*, **159**, 1319–1327.

Ermentrout, G. B. and Kopell, N. (1984). Frequency plateaus in a chain of weakly coupled oscillators, I. *SIAM J. Math. Anal.*, **15**(2), 215–237.

Glass, L. (2001). Synchronization and rhythmic processes in physiology. *Nature*, **410**, 277–284.

Goldobin, D. S. and Pikovsky, A. (2005*a*). Synchronization and desynchronization of self-sustained oscillators by common noise. *Phys. Rev. E*, **71**, 045201(R).

Goldobin, D. S. and Pikovsky, A. (2005*b*). Synchronization of self-sustained oscillators by common white noise. *Physica A*, **351**, 126.

Goldobin, D. S. and Pikovsky, A. (2006). Antireliability of noise-driven neurons. *Phys. Rev. E*, **73**, 061906.

Haken, H. (1993). *Advanced Synergetics: Instability Hierarchies of Self-Organizing Systems.* Springer, Berlin.

Huygens, C. (1673). *Horologium Oscillatorium.* Apud F. Muguet, Parisiis, France. English translation: *The Pendulum Clock*, Iowa State University Press, Ames, 1986.

Kaempfer, E. (1727). *The History of Japan (With a Description of the Kingdom of Siam)*. Sloane, London. Posthumous translation; reprint by McLehose, Glasgow, 1906.

Kardar, M., Parisi, G., and Zhang, Y.-C. (1986). Dynamic scaling of growing interfaces. *Phys. Rev. Lett.*, **56**, 889–892.

Kiss, I. Z., Zhai, Y., and Hudson, J. L. (2002). Emerging coherence in a population of chemical oscillators. *Science*, **296**, 1676–1678.

Kuramoto, Y. (1984). *Chemical Oscillations, Waves and Turbulence.* Springer, Berlin.

Mainen, Z. F. and Sejnowski, T. J. (1995). Reliability of spike timing in neocortical neurons. *Science*, **268**, 1503.

Mosekilde, E., Maistrenko, Yu., and Postnov, D. (2002). *Chaotic Synchronization: Applications To Living Systems.* World Scientific, Singapore.

Néda, Z., Ravasz, E., Brechet, Y., Vicsek, T., and Barabási, A.-L. (2000). Tumultuous applause can transform itself into waves of synchronized clapping. *Nature*, **403**(6772), 849–850.

Nesterenko, V. F. (2001). *Dynamics of Heterogeneous Materials.* Springer, New York.

Ott, E. and Antonsen, T. M. (2008). Low dimensional behavior of large systems of globally coupled oscillators. *CHAOS*, **18**(3), 037113.

Pikovsky, A. S. (1984*a*). Synchronization and stochastization of nonlinear oscillations by external noise. In *Nonlinear and Turbulent Processes in Physics* (ed. R. Z. Sagdeev), Harwood Academic, Singapore, pp. 1601–1604.

Pikovsky, A. S. (1984*b*). Synchronization and stochastization of the ensemble of autogenerators by external noise. *Radiophys. Quantum Electron.*, **27**(5), 576–581.

Pikovsky, A. S. and Grassberger, P. (1991). Symmetry breaking bifurcation for coupled chaotic attractors. *J. Phys. A: Math., Gen.*, **24**(19), 4587–4597.

Pikovsky, A. and Rosenau, P. (2006). Phase compactons. *Physica D*, **218**, 56–69.

Pikovsky, A. and Rosenblum, M. (2009). Self-organized partially synchronous dynamics in populations of nonlinearly coupled oscillators. *Physica D*, **238**(1), 27–37.

Pikovsky, A., Rosenblum, M., and Kurths, J. (1996). Synchronization in a population of globally coupled chaotic oscillators. *Europhys. Lett.*, **34**(3), 165–170.

Pikovsky, A., Rosenblum, M., and Kurths, J. (2001). *Synchronization: A Universal Concept in Nonlinear Sciences.* Cambridge University Press, Cambridge.

Rayleigh, J. (1945). *The Theory of Sound.* Dover, New York.

Rosenau, P. and Hyman, J. M. (1993). Compactons: Solitons with finite wavelength. *Phys. Rev. Lett.*, **70**(5), 564–567.

Rosenau, P. and Pikovsky, A. (2005). Phase compactons in chains of dispersively coupled oscillators. *Phys. Rev. Lett.*, **94**, 174102.

Rosenblum, M. and Pikovsky, A. (2007). Self-organized quasiperiodicity in oscillator ensembles with global nonlinear coupling. *Phys. Rev. Lett.*, **98**, 064101.

Roy, R. and Thornburg, K. S. (1994). Experimental synchronization of chaotic lasers. *Phys. Rev. Lett.*, **72**, 2009–2012.

Strogatz, S. H. (2003). *Sync: The Emerging Science of Spontaneous Order.* Hyperion, New York.

van der Pol, B. (1927). Forced oscillators in a circuit with non-linear resistance. (Reception with reactive triode). *Philos. Mag.*, **3**, 64–80.

Watanabe, S. and Strogatz, S. H. (1994). Constants of motion for superconducting Josephson arrays. *Physica D*, **74**, 197–253.

Part II

Hydrodynamics

4
Statistical mechanics of two-dimensional and quasi-geostrophic flows

Bruce E. TURKINGTON

University of Massachusetts at Amherst, USA

4.1 Introduction

4.1.1 Overview

Among the systems with long-range interactions that have been investigated in recent years, two-dimensional fluid dynamics and its relatives in geophysical fluid dynamics provide some of the most interesting examples. The special properties of the transport of vorticity by such flows results in the self-organization of coherent structures, which can be shear layers, isolated vortices, or more complex flows. These structures are observed in freely decaying 2D turbulence, both numerically and experimentally, and they also organize in other more complicated fluid motions with weak forcing and dissipation. An often-quoted physical example is the weather layer on the giant planet Jupiter: its striking banded structure marks the east–west reversals in its persistent zonal winds, and within these zonal shear flows roll the long-lived oval vortices, such as the Great Red Spot. These strong zonal winds and intense vortices are ultimately energized by thermal convection on (relatively) small scales, but their large-scale dynamics is dictated by two-dimensional potential vorticity transport.

In these lectures we describe how statistical mechanics can be adapted to model the long-lived, large-scale structures in effectively two-dimensional turbulent fluid flows. We confine ourselves to equilibrium statistical theories based on vorticity formulations of the governing fluid dynamics. The development of comparable nonequilibrium theories is a very worthwhile goal, but a complete and systematic mathematical theory does not yet exist. Accordingly, we do not treat nonequilibrium models here. Moreover, we do not attempt to extend the theory to fully 3D flows, because the phenomenology of three-dimensional turbulence is fundamentally different.

For historical and pedagogical reasons, we begin by developing the simplest and oldest statistical theory in this field, which is based on the classical point-vortex idealization. A microstate in this theory is an assembly of N point vortices that behave somewhat like particles interacting via a logarithmic potential. At sufficiently high energy, the appropriately scaled continuum limit ($N \to \infty$) predicts the formation of macrovortices, which are stable equilibrium states having negative absolute temperature. Most of the essential ideas used in the modern statistical-theory-based continuum vorticity dynamics may be illustrated in the simpler context of the point-vortex dynamics. We therefore develop the entire theory first in this context. In particular, we invoke the theory of large deviations to derive the continuum limits, characterize the equivalence or nonequivalence of Gibbs ensembles, and refine stability theory to include states in the nonequivalent regimes. We also give an algorithm to compute equilibrium states.

But the point-vortex discretization is a rather drastic approximation, and so we proceed to develop the modern statistical theories of a continuum vorticity field. These are exact mean-field theories which have qualitative features similar to the point-vortex theory, but they incorporate the full set of dynamical invariants. They are able to predict the organized structures that eventually form in two-dimensional flows after a long turbulent evolution. We discuss some subtle points in the formulation of these models, stemming from the treatment of arbitrarily small scales.

We then turn our attention to geophysical applications, which largely motivate our theoretical development. The large-scale, low-frequency motions of an ocean or atmosphere resemble 2D fluid dynamics thanks to the constraints imposed by a small vertical/horizontal aspect ratio, a rapidly rotating frame of reference, and a density stratification. We focus on quasi-geostrophic dynamics, the most important asymptotic regime for modeling circulation in an ocean or atmosphere, which is entirely described by the self-induced advection of potential vorticity. To make these notes relatively self-contained, we include in summary form the derivations of the standard governing equations of geophysical fluid dynamics.

The greater complexity of the geophysical problems results in a richer set of predicted behaviors. Briefly, we illustrate how the typical solutions depend on various physical parameters. As a concrete application we construct a model of the weather layer of Jupiter, and we obtain the prominent coherent structures (jets and vortices) as statistical-equilibrium macrostates. Interestingly, the canonical and microcanonical ensembles for the model are not equivalent in the range of parameters for which the Great Red Spot emerges in the zonal jet structure. We explain how a refined stability analysis of steady coherent structures in quasi-geostrophic turbulence is related to the nonequivalence issue. These applications complete our presentation, and in so doing they make a compelling case for the predictive capacity of the statistical-equilibrium theory of coherent structures.

4.1.2 Fluid dynamics background

We are interested in high-Reynolds-number flows of incompressible fluids. These flows are "turbulent" in the sense that they excite many nonlinearly interacting scales of motion, and consequently they require a statistical description. The basic governing equations are the Navier–Stokes equations (Batchelor 1967; Majda and Bertozzi 2001)

$$\nabla \cdot v = 0\,,$$
$$\frac{\partial v}{\partial t} + v \cdot \nabla v + \frac{1}{\rho}\nabla p = f + \nu\,\triangle v\,. \tag{4.1}$$

In this primitive-variable formulation, the unknowns are the velocity field $v = (v_1, v_2, v_3)$ and the pressure p, both of which are functions of a spatial variable $x \in D \subset R^3$ and the time t. The fluid density ρ is a assumed to be constant. The coefficient of the diffusive term $\nu\,\triangle v$ is the kinematic viscosity, $\nu = \mu/\rho$, where μ is the shear viscosity of the fluid; $\triangle = \nabla^2$ denotes the Laplacian operator.

The first equation in eqn (4.1) expresses the conservation of mass. Indeed, if B^t denotes any moving volume within the fluid domain D, which is materially transported by the flow v, then eqn (4.1) means that

$$\frac{d}{dt}\int_{B^t} \rho\, dx = \int_{B^t} \rho\nabla \cdot v\, dx\,.$$

Similarly, the second (vector) equation in eqn (4.1) expresses the conservation of momentum. Integrated over a moving volume B^t, it becomes

$$\frac{d}{dt}\int_{B^t} \rho v\, dx = \int_{B^t} \rho\left(\frac{\partial v}{\partial t} + v \cdot \nabla v\right)\, dx = \int_{\partial B^t}\left[-pn + \mu\frac{\partial v}{\partial n}\right]\, ds + \int_{B^t} \rho f\, dx\,.$$

From this form, it is evident that the body forces f are those acting per unit mass on the fluid (such as gravity or Coriolis forces), while the pressure and viscous stresses act at surfaces within the fluid.

These equations are supplemented by boundary conditions on the solid bounding walls, which for a viscous fluid are the no-slip conditions $v = 0$ on a fixed boundary ∂D. With these conditions, the Navier–Stokes equations (4.1) are a complete set of governing equations for (v, p). The incompressibility requirement, $\nabla \cdot v = 0$, is an idealization that amounts to setting the speed of acoustic (sound) waves to infinity. Consequently, $\partial p/\partial t$ does not appear in the governing equations, and the pressure p instantaneously adjusts to the fluid motion v.

Viscous flows can be driven by applied pressure gradients, moving walls, or body forces. These sources of energy are counteracted by the dissipation due to viscosity, which transfers the macroscopic kinetic energy of the flow to heat. Among the very wide range of applications of these equations, we are concerned with flows such as those that occur in planetary atmospheres and the Earth's oceans, and which have large spatial scales and long time scales. In these regimes of motion, there is a huge separation of scales between those of the relevant fluid motion and that of the action due to the dissipation through molecular viscosity. In technical terms, we are interested in flows having a very large Reynolds number, which is defined to be the dimensionless ratio $\mathrm{Re} = VL/\nu$, where L and V are characteristic length and velocity scales, respectively. Re quantifies the typical magnitude of the advective term $\partial v/\partial t + v \cdot \nabla v$ relative to the dissipative term $\nu \,\Delta v$ in eqn (4.1). In geophysical flows, the Reynolds number (based on molecular diffusion, not eddy diffusion) is typically enormous (ranging from 10^8 to 10^{12}). Thus, it is the nonlinear advection of fluid properties on time scales of order $T = L/V$ that dominates our attention.

By introducing the appropriately scaled, nondimensionalized variables

$$x' = \frac{x}{L}, \; v' = \frac{v}{V}, \; t' = \frac{t}{T}, \; p' = \frac{p}{\rho V^2}, \; f' = \frac{Vf}{T},$$

the Navier–Stokes equations take the nondimensionalized form

$$\nabla \cdot v = 0\,,$$
$$\frac{\partial v}{\partial t} + v \cdot \nabla v + \nabla p = f + \frac{1}{\mathrm{Re}} \Delta v\,, \tag{4.2}$$

in which we drop the primes after rescaling the variables.

The phenomenology of high-Reynolds-number flows ($\mathrm{Re} \gg 1$) is an extensive and difficult subject, especially for fully 3D flows (Chorin 1994; Majda and Bertozzi 2001; Marchioro and Pulvirenti 1994). Let it suffice for us to indicate a few general properties. We consider the total kinetic energy

$$E = \frac{1}{2} \int_D v^2 \, dx\,, \tag{4.3}$$

and we find by a straightforward calculation that

$$\frac{dE}{dt} = \int_D v \cdot f \, dx - \frac{1}{\mathrm{Re}} \int_D |\nabla v|^2 \, dx\,. \tag{4.4}$$

The dissipative term in this expression is closely related to the vorticity field in the flow, namely $\omega = \nabla \times v$. For this reason, we introduce the so-called enstrophy, which is the integral

$$A = \frac{1}{2} \int_D \omega^2 \, dx = \frac{1}{2} \int_D |\nabla v|^2 \, dx \,. \tag{4.5}$$

The second expression for A uses the fact that v is divergence-free. Combining eqn (4.4) with eqn (4.5), we find that

$$\frac{dE}{dt} = \int_D v \cdot f \, dx - \frac{1}{\mathrm{Re}} A \,, \tag{4.6}$$

which shows that energy is dissipated at a rate proportional to the enstrophy. In turn, the behavior of A depends upon the advection and dissipation of vorticity in the fluid. The equation governing this behavior is obtained by taking the curl of the momentum equation. A standard calculation leads to the important identity

$$\frac{\partial \omega}{\partial t} + v \cdot \nabla \omega = \omega \cdot \nabla v + \frac{1}{\mathrm{Re}} \triangle \omega + \nabla \times f \,. \tag{4.7}$$

This equation says that as vorticity is transported with a fluid parcel it evolves under the influence of stretching and bending ($\omega \cdot \nabla v$), dissipation ($(1/\mathrm{Re})\triangle\omega$), and rotational driving ($\nabla \times f$).

In high-Reynolds-number flows, the predominant mechanism is the stretching and bending of vortex lines, leading to the intensification of vorticity. The complicated, 3D, nonlocal, self-induced process is not well understood, but certainly it is known to generate highly fluctuating vorticity fields and to cause the rapid growth of the enstrophy A. Phenomenologically, energy is transferred to small scales by the stretching and bending process, where it is dissipated by viscosity. The vortex structures involved in this transfer process are all inherently unstable, and it is through their instabilities that a flux of energy to small scales is accomplished. In the Kolmogorov picture of forced 3D turbulence, the statistical properties of this process are expected to be stationary and the flux of energy in wavenumber space is supposed to be approximately local in the wavenumber k. Then simple dimensional-analysis arguments produce a universal spectrum of energy scaling like $k^{-5/3}$. Thus, 3D turbulence is a strongly nonequilibrium process mediated by short-lived vortex structures on a range of intermediate scales, the so-called inertial range. Whether the concepts of equilibrium statistical mechanics can be used to capture any of the universal features of 3D turbulence is unclear, with most evidence and opinion suggesting that they cannot (Chorin 1994).

4.1.3 Motivation for equilibrium statistical theory in 2D

Unlike the general situation in 3D, flows that are constrained to be two-dimensional have energy fluxes toward large scales and therefore they tend to organize into long-lived, coherent macrostates. The prediction and description of these macrostates provides a reasonable goal for equilibrium statistical theory. This 2D phenomenology underlies all of our development of a statistical mechanics of scalar vorticity and its geophysical cousin, potential vorticity.

We now restrict our attention to a purely two-dimensional fluid flow in a domain $D \subset R^2$. The key point is that now the vorticity is essentially a scalar; indeed, if $v = (v_1, v_2, 0)$ and depends on $x = (x_1, x_2)$ and the time t, then $\omega = (0, 0, \zeta)$, where

$$\zeta = \frac{\partial v_2}{\partial x_1} - \frac{\partial v_1}{\partial x_2}. \tag{4.8}$$

So, when discussing 2D fluid dynamics, we write simply $v = (v_1, v_2)$ and $\zeta = \text{ curl } v$. The equation governing the evolution of vorticity then reduces to the scalar equation

$$\frac{\partial \zeta}{\partial t} + v \cdot \nabla \zeta = \frac{1}{\text{Re}} \triangle \zeta + \text{ curl } f\,. \tag{4.9}$$

Crucially, the stretching and bending term in eqn (4.7) has disappeared thanks to two-dimensionality. (A weak form of stretching will reappear in the potential-vorticity transport equation later.) One consequence is that the enstrophy (4.5) now decays in the absence of rotational driving; if curl $f = 0$, then

$$\frac{dA}{dt} = -\frac{1}{\text{Re}} \int_D |\nabla \zeta|^2 \, dx \leq 0\,.$$

The energy is therefore nearly constant, since

$$\frac{dE}{dt} = -\frac{1}{\text{Re}} A(t) \leq -\frac{1}{\text{Re}} A(0) \ll 1 \qquad \text{for} \quad \text{Re} \gg 1\,.$$

Whereas in 3D turbulence, energy is transferred downscale and then dissipated, in 2D flow, energy is effectively trapped in the large scales of motion, where the effects of dissipation are often negligible.

On the other hand, there is a flux of enstrophy toward small scales in 2D turbulence, as is observed in numerical and physical experiments. In physical space, the scalar vorticity field ζ develops high gradients, especially near points of high strain. But its oscillations are bounded between maximum and minimum values that are non-increasing during a free evolution in view of the advection equation (4.9). Thus, the self-induced advection of vorticity in 2D turbulence tends to produce a vorticity field that fluctuates on a range of small scales, but is constrained by the conservation of the total energy. When expressed in terms of ζ, the energy functional is

$$H = \frac{1}{2} \int_D \zeta \psi \, dx = \frac{1}{2} \int_D \int_D \zeta(x)\, g(x, x')\, \zeta(x')\, dx\, dx'\,, \tag{4.10}$$

using the Green's function $g(x, x')$ for the Laplacian operator $-\triangle$ on D with homogeneous Dirichlet boundary conditions on ∂D. The streamfunction ψ is related to the vorticity ζ by

$$-\triangle \psi = \zeta \;\text{ in }\; D, \qquad \text{with} \qquad \psi = 0 \;\text{ on }\; \partial D\,, \tag{4.11}$$

or, equivalently,

$$\psi(x) = \int_D g(x, x')\, \zeta(x')\, dx'\,. \tag{4.12}$$

The velocity field is derived from the streamfunction by $v_1 = \partial \psi / \partial x_2\,,\; v_2 = -\partial \psi / \partial x_1$; for this relation, we write simply $v = \text{Curl}\, \psi$.

These general and phenomenological considerations lead us to a conceptual picture of the end state of 2D turbulence. We imagine the limiting regime in which $\mathrm{Re} \to \infty$ and $t \to \infty$, so that the 2D flow has sufficient time to develop vorticity fluctuations on fine scales and to organize into coherent vortex structures on coarse scales. Then the vorticity field ζ resembles a random set of "spins" $\zeta(x)$ at "sites" x, which we may imagine to constitute a fine-grained lattice over the fluid domain D. The fluctuating spins are spatially uncorrelated, owing to the effects of chaotic advection, but are conditioned by the conservation of the total energy H and the total "spin" C, that is, the total circulation

$$C = \int_{\partial D} v \cdot dx = \int_D \zeta \, dx \,. \tag{4.13}$$

Thus, without yet making these concepts precise, we glimpse the outline of a statistical-equilibrium theory of coherent macrostates—namely, a theory characterizing the coarse-grained structure that emerges from fine-grained vorticity fluctuations owing to the constraints imposed by the conservation of total energy and circulation. These two dynamical invariants of inviscid 2D incompressible flow constrain the coherent macrostate because they are not sensitive to spatial coarse-graining, C being linear in ζ and H being dependent on the *long-range interactions* of ζ through the Green's function $g(x, x')$, whose principal part is the logarithmic potential $-(1/2\pi)\log|x - x'|$. The statistics of the fine-grained fluctuations of vorticity are presumably controlled by the other dynamical invariants of the inviscid version of eqn (4.9). These are the generalized enstrophy integrals

$$A = \int_D a(\zeta) \, dx \,, \tag{4.14}$$

which are conserved for any suitably regular real function a; the classical enstrophy (4.5) corresponds to $a(z) = \frac{1}{2}z^2$. Together these exhaust the inviscid invariants, apart from those associated with special symmetries (linear impulse from translational symmetry, or angular impulse from rotational symmetry).

This heuristic discussion motivates the remainder of these lectures. In them we will derive and then analyze Gibbsian equilibrium ensembles for vortex systems. We begin with the point-vortex theory, in which the inviscid dynamics for a continuum vorticity field ζ is replaced by a singular discretization into point vortices. This dynamics is manifestly Hamiltonian, and so it leads directly to a statistical theory of equilibrium. But, as is suggested by our opening remarks, we shall later formulate lattice models that more closely represent the true dynamics of continuum vorticity dynamics, from which we construct more faithful equilibrium statistical theories of coherent macrostates.

4.2 Statistical theory of point-vortex systems

4.2.1 Point-vortex dynamics

The governing equations for inviscid, incompressible flow in a two-dimensional domain D can be written in the form

$$\frac{\partial \zeta}{\partial t} + [\zeta, \psi] = 0, \qquad \psi = G\zeta \,, \tag{4.15}$$

which emphasizes that they are equivalent to a single, nonlinear transport equation for the scalar vorticity ζ. In eqn (4.15), $G : \zeta \mapsto \psi$ denotes the Green's operator defined by eqn (4.11), which corresponds to the Green's function

$$g(x, x') = -\frac{1}{2\pi} \log |x - x'| + h(x, x'), \tag{4.16}$$

where $h(x, x')$ is harmonic in both of its arguments x and x' and is determined so that g satisfies the boundary condition $g(x, x') = 0$ for all $x \in \partial D$ and every $x' \in D$. In other words, g satisfies the family of boundary value problems

$$-\triangle_x g(x, x') = \delta(x - x') \quad \text{for} \quad x \in D, \qquad \text{and} \qquad g(x, x') = 0 \quad \text{for} \quad x \in \partial D$$

for every $x' \in D$. And, $[\zeta, \psi]$ denotes the Poisson bracket

$$[\zeta, \psi] = \frac{\partial \zeta}{\partial x_1} \frac{\partial \psi}{\partial x_2} - \frac{\partial \zeta}{\partial x_2} \frac{\partial \psi}{\partial x_1}.$$

The 2D Euler equation (4.15) is then equivalent to the inviscid, unforced version of eqn (4.9), in which the velocity field v is derived from the streamfunction ψ according to

$$v_1 = \frac{\partial \psi}{\partial x_2}, \qquad v_2 = -\frac{\partial \psi}{\partial x_1}.$$

Thus, the entire content of 2D fluid dynamics in this context is contained in the self-induced advection of vorticity: at each instant of time t, ζ determines a velocity field $v = \text{Curl}\, G\zeta$ and, in turn, ζ is transported by this velocity as if it were a material tracer. In particular, if $B^t \subset D$ is any moving area transported by the flow, then the circulation in B^t is constant:

$$\frac{d}{dt} \int_{B^t} \zeta(x, t)\, dx = 0 \qquad \text{for all} \quad t.$$

The most transparent way to visualize this self-induced transport is to adopt the classical idealization of the vorticity field into a finite number N of point vortices. Namely, we suppose that

$$\zeta(x, t) = \sum_{i=1}^{N} \gamma_i \delta(x - z_i(t)), \tag{4.17}$$

where $\gamma_i \in R$ is the circulation (or vortex strength) of the point vortex located at position $z_i \in D$. The property of the continuum dynamics that circulation is conserved along the flow implies that each γ_i is constant in time. The concentration of this circulation at the moving point z_i means that the corresponding streamfunction is

$$\psi(x, t) = \sum_{i=1}^{N} \gamma_i g(x, z_i(t)). \tag{4.18}$$

But then the induced velocity due to the vortex at z_i diverges like $1/|x - z_i|$ as x approaches the vortex center z_i. Moreover, the kinetic energy of each vortex is infinite (logarithmically divergent near the vortex). The classical theory of Helmholtz, Kirchhoff, and Routh regularizes the dynamics by noting that the singular logarithmic term in the Green's function induces a purely circular flow around each vortex center, and hence does not itself induce any motion of the center (Marchioro and Pulvirenti 1994). Consequently, this singular term is dropped from the equations of motion for the vortex centers z_i, $i = 1, \dots N$. Thus, we arrive at the equations of motion for the N centers:

$$\frac{dz_i}{dt} = \sum_{j \neq i} \gamma_j \, \mathrm{Curl}_x g(x, z_j)|_{x=z_i} + \gamma_i \, \mathrm{Curl}_x h(x, z_i)|_{x=z_i} \, . \tag{4.19}$$

This equation says that vortex i moves under the influence of all the other vortices j, with $j \neq i$, and under the influence of the boundary via h.

This dynamics can be justified rigorously from the continuum equations of motion by showing that it results from taking the limit of initial conditions that approach (in the appropriate weak sense) a set of delta functions (Marchioro and Pulvirenti 1994; Turkington 1987). Such a result is valid for any time interval during which the point vortices do not collide. However, collisions are possible for certain initial configurations of point vortices. And, under continuum fluid dynamics, near-collisions result in vortex mergers, in which two vortices coalesce into one vortex and eject a fraction of their total circulation into filamentation. This is an important process that is not properly modeled in the classical point-vortex dynamics outlined here.

4.2.2 Statistical mechanics of N identical point vortices

Except for very simple, or symmetrical, configurations of a few ($N \leq 3$) vortices, the point-vortex dynamics is typically chaotic. Our interest therefore attaches to understanding the statistical mechanics of point-vortex systems with large N, not to following the details of a particular evolving configuration. To this end, let us consider the special case of N identical vortices, each having circulation $\gamma = \Gamma/N$, for some total circulation $\Gamma > 0$. To establish the connection with classical statistical mechanics, let us write the governing dynamics in a Hamiltonian form—namely,

$$\gamma \frac{dz_i}{dt} = \mathrm{Curl}_{z_i} H_N(z_1, \dots, z_N) \qquad \text{for} \quad i = 1, \dots, N \, , \tag{4.20}$$

where the Hamiltonian is

$$H_N(z_1, \dots, z_N) = \frac{\gamma^2}{2} \sum_{\substack{i,j=1 \\ i \neq j}}^{N} g(z_i, z_j) + \frac{\gamma^2}{2} \sum_{i=1}^{N} h(z_i, z_i) \, . \tag{4.21}$$

In fact, if we define generalized positions and momenta by $z_i = \gamma^{-1/2}(q_i, p_i)$, then this dynamics is a canonical Hamiltonian system. A peculiarity of 2D point-vortex dynamics is that the two spatial coordinates of each vortex center provide the conjugate canonical variables in its Hamiltonian structure.

The phase space for eqn (4.20) is D^N, whose typical point is a configuration $z = (z_1, \dots, z_N)$. The phase volume element $dz = \prod_{i=1}^{N} dz_i$ on D^N is invariant under the phase flow for this (canonical) Hamiltonian system. Moreover, H and Γ are invariants of the dynamics, and unless special spatial symmetries are present, they are the only dynamical invariants. For simplicity, we proceed by assuming that D does not have a translational or rotational symmetry; extending the theory to include the associated invariants of linear and angular impulse is straightforward. Finally, let us also assume that the domain D is normalized so that its area is unity, that is, $|D| = 1$. Then the phase volume measure

$$P_N(dz) = \prod_{i=1}^{N} dz_i|_D \tag{4.22}$$

is a probability measure.

Given this classical structure, we introduce the two fundamental Gibbs distributions (or ensembles). The canonical distribution with inverse temperature $\beta \in R$ is the probability measure

$$P_{N,\beta}(dz) = \exp\left(-N\beta H_N(z) - N\phi_N(\beta)\right) P_N(dz)\,, \tag{4.23}$$

which is normalized by the function

$$\phi_N(\beta) = \frac{1}{N}\log\int_{D^N} \exp\left(-N\beta H_N(z)\right) P_N(dz)\,.$$

The microcanonical distribution with energy E is the conditional probability measure

$$P_N^E(dz) = P_N(dz|\, H_N = E) = W_N(E)^{-1}\delta(H_N(z) - E)\,. \tag{4.24}$$

As a delta distribution concentrated on the energy surface, it is normalized by

$$W_N(E) = \int_{H_N=E} \frac{d^{N-1}S}{|\nabla_z H_N|}\,.$$

In the canonical ensemble (4.23), we have scaled the inverse temperature by N. This scaling is included so that in the continuum limit when $N \to \infty$ the mean energy in the canonical distribution remains finite. This is the natural thermodynamic limit in the analysis of point-vortex systems, and it allows the comparison of the canonical and microcanonical statistical-equilibrium models in a fixed, finite domain with finite total energy and circulation.

4.2.3 Large-deviation analysis of continuum limit

We now turn to the analysis of the continuum limit ($N \to \infty$) of the two statistical models we have introduced, one governed by the canonical ensemble and the other by the microcanonical ensemble. Unlike the case in many other branches of statistical mechanics, we cannot presuppose the equivalence of these two models—indeed, the long-range nature of the vortex interaction leads to nonequivalence in cases of real physical interest (Dauxois *et al.* 2002).

We base our analysis on the modern theory of large deviations, which can be viewed as a probabilistic theory of entropy (Ellis 1985, 1995). Large-deviation principles and their applications over a range of physical problems are discussed in the lectures by R. S. Ellis in this volume. We will draw on this theory directly to analyze the canonical ensemble (4.23) and the microcanonical ensemble (4.24) in the large-N limit.

The first step is to establish an entropy concept which connects the microscopic states of the system to an appropriate macroscopic description. This key concept is the same for both ensembles, as it is determined by the product measure P_N defined in eqn (4.22). With respect to P_N, $z_1, \ldots, z_N$ are independent, identically distributed random vectors taking values in D, and their common distribution is uniform over D. We form an empirical measure which represents the density of the identical point vortices, namely,

$$L_N(x) = \frac{1}{N} \sum_{i=1}^{N} \delta(x - z_i) \,. \tag{4.25}$$

Intuitively, this empirical measure is just a histogram of the random locations z_i; precisely, for any Borel subset B of D, $L_N(B) = (\text{number of } z_i \in B)/N$.

By the law of large numbers, we expect that as N approaches infinity, L_N should become very nearly uniform over D and that deviations from this expected behavior should be very rare. The large-deviation principle that forms the basis of our analysis is the Sanov Theorem (Ellis 1985, 1995), which constitutes the asymptotically precise expression of this idea. We refer to Ellis's lectures for a mathematically rigorous statement; here we give a simplified expression of the essential content of the theorem. Let ρ be any probability density on D; then the following asymptotic statement characterizes the probability with respect to P_N that L_N is close to the density ρ:

$$P_N \{ L_N \approx \rho \} \sim \exp(-N I(\rho)) \qquad \text{as} \quad N \to \infty \,, \tag{4.26}$$

where

$$I(\rho) = \int_D \rho(x) \log \rho(x) \, dx \,. \tag{4.27}$$

I is the so-called rate functional for this large-deviation principle. We refer to it as the negentropy functional, as $-I(\rho)$ is the (Shannon) entropy of the density ρ.

The idea behind this theorem is a familiar calculation of probabilities that has been used in statistical physics for a century. Partition the domain into disjoint cells $B_1, \ldots B_M$, for some M with $1 \ll M \ll N$, and let N_k denote the number of z_i in each B_k. If $L_N \approx \rho$, then $N_k \approx N\rho(y_k)|B_k|$, where y_k and $|B_k|$ denote the centroid and area, respectively, of B_k. Combinatorics gives the desired probability

$$P_N \{ L_N \approx \rho \} = \frac{N!}{N_1! \cdots N_M!} |B_1|^{N_1} \cdots |B_M|^{N_M} \,,$$

whose exponential asymptotics follows from Stirling's formula, taking $N \to \infty$. The integral I emerges in the limit $M \to \infty$ as the partition is refined.

The next step is to represent the Hamiltonian H_N in terms of L_N, in the sense that the energy of a microscopic configuration $z = (z_1, \ldots, z_N)$ is approximated by the energy of the "cloud" of vortices, each carrying circulation Γ/N,

$$\zeta_N(x) = \Gamma L_N(x) = \frac{\Gamma}{N} \sum_{i=1}^{N} \delta(x - z_i) \,.$$

To accomplish this, it is necessary to regularize the vortex interaction by replacing the Green's function $g(x, x')$ by a smoothing $g^{(\epsilon)}(x, x')$ that is nonsingular as $|x - x'| \to 0$ and approaches $g(x, x')$ as $\epsilon \to 0$. This complication arises from the point-vortex idealization, which produces a divergent velocity field at each vortex location and infinite self-energy in each vortex; this technical difficulty is not met with in the continuum theories derived later. Let $H_N^{(\epsilon)}$ denote the regularization of the Hamiltonian (4.21), and let $H^{(\epsilon)}$ denote the corresponding regularization of the energy functional

$$H(\zeta) = \frac{1}{2} \int_D \int_D g(x, x')\, \zeta(x)\, \zeta(x')\, dx\, dx' \,. \tag{4.28}$$

Then

$$H_N^{(\epsilon)}(z) = H^{(\epsilon)}(\zeta_N) - \frac{\Gamma^2}{2N^2} \sum_{i=1}^{N} g^{(\epsilon)}(z_i, z_i) = H^{(\epsilon)}(\zeta_N) + O\left(\frac{1}{N} |\log \epsilon|\right), \tag{4.29}$$

assuming that the regularization is parameterized by a finite vortex core of scale ϵ. Thus, provided that we choose the regularization $\epsilon(N) \to 0$ slowly enough as $N \to \infty$ that the error term in eqn (4.29) goes to zero, we have

$$H_N(z) = H^{(\epsilon(N))}(\zeta_N) + o(1) \qquad \text{as} \quad N \to \infty \,.$$

This approximation in terms of ζ_N, combined with the large-deviation principle for L_N with respect to P_N, allows us to evaluate the continuum limit in either the canonical or the microcanonical case.

In the canonical ensemble, we appeal to the Laplace asymptotics for the large-deviation principle (4.26) to calculate

$$\begin{aligned} \phi_N(\beta) &= \frac{1}{N} \log \int_{D^N} \exp(-N\beta H_N(z))\, P_N(dz) \\ &\approx \frac{1}{N} \log \int_{D^N} \exp(-N\beta H^{(\epsilon(N))}(\Gamma L_N))\, P_N(dz) \\ &\to \sup_{\rho} \left[-\beta H(\Gamma\rho) - I(\rho)\right] \qquad \text{as } N \to \infty \,, \end{aligned} \tag{4.30}$$

where the supremum is over all densities ρ in D with $\int_D \rho\, dx = 1$. In addition, with respect to $P_{N,\beta}$, the empirical measure L_N satisfies the large-deviation principle

$$P_{N,\beta}\{\, L_N \approx \rho \,\} \sim \exp(-N R_\beta(\rho)) \qquad \text{as} \quad N \to \infty \,, \tag{4.31}$$

with the rate functional

$$R_\beta(\rho) = I(\rho) + \beta H(\Gamma\rho) \;-\; \min_{\tilde\rho} \left[I(\tilde\rho) + \beta H(\Gamma\tilde\rho)\right].$$

Thus, we arrive at the variational principle that characterizes the most probable macroscopic states of the point-vortex system with respect to the canonical ensemble. Namely, these are the vortex densities ρ on D that satisfy $R_\beta(\rho) = 0$, and they solve

$$\text{minimize} \quad \int_D \rho(x) \log \rho(x)\, dx + \frac{\beta\Gamma^2}{2} \int_D \int_D g(x,x')\rho(x)\rho(x')\, dx\, dx'$$
$$\text{subject to} \quad \int_D \rho\, dx = 1 \quad \text{and} \quad \rho \geq 0 \text{ on } D. \tag{4.32}$$

The corresponding microcanonical ensemble may be analyzed in a similar way. In particular, we determine the asymptotics of the conditional distributions P_N^E using the large-deviation principle (4.26) as follows: for any vortex density ρ having the prescribed energy $E = H(\Gamma\rho)$,

$$\begin{aligned} P_N^E\{L_N \approx \rho\} &= P_N\{L_N \approx \rho \,|\, H_N(z) = E\} \\ &= P_N\left\{L_N \approx \rho \,|\, H^{(\epsilon(N))}(\Gamma L_N) = E + o(1)\right\} \\ &= \frac{P_N\{L_N \approx \rho\}}{P_N\left\{H^{(\epsilon(N))}(\Gamma L_N) = E + o(1)\right\}}. \end{aligned}$$

Hence the large-deviation principle for the microcanonical ensemble follows directly from eqn (4.26):

$$P_N^E\{L_N \approx \rho\} \sim \exp(-N[I(\rho) + s(E)]) \quad \text{as} \quad N \to \infty, \tag{4.33}$$

where $s(E)$ is the microcanonical (specific) entropy

$$s(E) = \lim_{N\to\infty} \frac{1}{N} \log P_N\{H(\Gamma L_N) = E\} = \sup_\rho\{-I(\rho) \mid H(\Gamma\rho) = E\}. \tag{4.34}$$

Thus, we obtain the variational principle characterizing the most probable macrostates for the microcanonical ensemble:

$$\text{minimize} \quad \int_D \rho(x) \log \rho(x)\, dx \quad \text{subject to}$$
$$\frac{\Gamma^2}{2} \int_D \int_D g(x,x')\rho(x)\rho(x')\, dx\, dx' = E, \quad \int_D \rho\, dx = 1, \quad \text{and} \quad \rho \geq 0 \text{ on } D. \tag{4.35}$$

4.2.4 Properties of the equilibrium states

In this and the subsequent two sections we examine the solutions to the canonical and microcanonical variational principles derived above, summarizing their general properties and then giving a numerical method to compute their solutions. For the sake of simplicity, we now normalize the total circulation to be $\Gamma = 1$; this normalization can always be achieved by rescaling the vorticity. Then the vortex density ρ coincides with the continuum vorticity ζ, and for this reason we will express the variational principles in terms of ζ throughout our discussion.

The Euler–Lagrange equation for either of the governing variational problems (4.32) or (4.35) takes the same form, namely

$$\zeta = \exp(-\beta\psi - \mu) \quad \text{with} \quad \psi = G\zeta, \tag{4.36}$$

where μ is the Lagrange multiplier associated with the circulation constraint. In the canonical model, β is prescribed, while in the microcanoncial model β is the Lagrange multiplier for the energy constraint. In either model, there is a one-parameter family of solutions ζ, depending on β or E, respectively. The mean-field equation is a nonlinear elliptic partial differential equation for ψ in which μ is determined along with ψ to satisfy $\int_D \exp(-\beta\psi - \mu)\, dx = 1$. The dependence on β gives this boundary value problem the character of a nonlinear eigenvalue problem.

The qualitative properties of the mean field ζ, which is the predicted coherent structure, depend strongly on the sign of β. Unlike the case for most models in statistical mechanics, the inverse temperature β can be positive or negative, and a negative temperature can be stable and physically realizable. For $\beta = 0$, the mean field is uniform, which, under our normalizations, means that $\zeta = 1$ throughout the domain D. When $\beta > 0$, the mean vorticity increases toward the boundary of D, and for large positive β the solution is a strong shear layer near the boundary with a nearly irrotational interior. On the other hand, when $\beta < 0$, the mean vorticity organizes in the interior, and for large negative β the solution is a strong vortex surrounded by nearly irrotational flow.

In the canonical model, these solutions are minimizers of the functional $I(\zeta) + \beta H(\zeta)$ subject to the constraints $\int_D \zeta\, dx = 1$ and $\zeta(x) \geq 0$ for all $x \in D$. When $\beta > 0$, this functional is strictly convex in ζ and the constraint is linear in ζ. Hence the solution exists and is unique for each positive β. The physical interpretation of these solutions is somewhat delicate, however, because they correspond to shear layers supported near the boundary walls. Presumably viscous effects should be included in some sublayer of these flows in a realistic model with a no-slip boundary condition.

Most interest, therefore, is attached to the solutions with $\beta < 0$, since these solutions capture the self-organization of vortex structures observed in high-Reynolds-number turbulence. In this case, though, the functional $I(\zeta) + \beta H(\zeta)$ is generally not convex and so solutions are not guaranteed to be unique, and solution branches may bifurcate.

In the microcanonical model, the same type of behavior is also exhibited. In this case, $\beta = \partial s/\partial E$, where $s(E)$ is the microcanonical entropy (4.34). This entropy is maximal for $\zeta = 1$ in D, and this uniform mean vorticity defines a critical energy E_0. For $E < E_0$, $\beta > 0$, while for $E > E_0$, $\beta < 0$. Thus, the transition from low to high energy in the microcanonical model implies that the inverse temperature passes from positive to negative values. This is the famous observation of Onsager, who argued from the microcanonical ensemble but without the benefit of the mean-field equation (Onsager 1949). For finite N, he considered the phase volume $V(E) = \text{vol}\{z \in D^N : H_N(z) \leq E\}$, and noted that $V(E)$ increases from 0 to $1 = |D|^N$ as E increases from 0 to $+\infty$. Consequently, $V(E)$ must have an inflection point at some $E = E_0$, where $d^2V/dE^2 = 0$. But, since by definition $s(E) = \log(dV/dE)$, this means that $\beta(E_0) = 0$. Moreover, for $E > E_0$, $V(E)$ would typically be concave (or else it would become unbounded), and so $\beta(E) < 0$. In this way, he argued that negative-temperature macrostates would organize in the point-vortex system and that they would represent coalescence into coherent macrovortices within the flow interior. Onsager's early intuitive arguments have been confirmed by mathematical analysis of solution branches and by numerical computations (Kraichnan and Montgomery 1980; Montgomery and Joyce 1974; Lungren and Pointin 1977;

Smith and O'Neil 1990; Eyink and Spohn 1993; Kiessling 1993; Caglioti *et al.* 1992, 1995; Kiessling and Lebowitz 1997).

4.2.5 Ensemble equivalence and stability of solutions

Perhaps the most remarkable property of this statistical-equilibrium theory in the negative-temperature regime is that the microcanonical and canonical variational principles do not necessarily produce the same solutions. In fact, the microcanonical equilibrium macrostates always include the canonical equilibrium macrostates, but some microcanonical macrostates may not be realized as canonical macrostates. We now turn to this question and to the related question of the nonlinear stability of the equilibrium macrostates, especially those microcanonical equilibria that are not canonical equilibria.

We can easily demonstrate that any canonical equilibrium ζ_β is necessarily a microcanonical equilibrium with the corresponding energy value $E_\beta = H(\zeta_\beta)$. If ζ is any mean vorticity that is admissible in the microcanonical variational principle (4.35) with energy E_β, then $I(\zeta) + \beta H(\zeta) \geq I(\zeta_\beta) + \beta H(\zeta_\beta)$, since ζ_β is a minimizer of $I + \beta H$. But $H(\zeta) = H(\zeta_\beta)$, and so $I(\zeta_\beta) \leq I(\zeta)$. Hence the canonical equilibrium ζ_β is a microcanonical equilibrium.

On the other hand, examples show that the converse is false in general. In order for equivalence of ensembles to hold, it is necessary and sufficient that the microcanonical entropy $s(E)$ be concave. To see that this condition is sufficient, we note that the concave function $s(E)$ has a supporting tangent line at any E:

$$s(E') \leq s(E) + \beta(E' - E) \qquad \text{for all } \; E' ,$$

where $\beta = ds/dE$ at E. Given this property, we argue as follows. If ζ^E is a microcanonical equilibrium with energy E, then for any ζ with any energy $E' = H(\zeta)$,

$$\begin{aligned} I(\zeta^E) + \beta H(\zeta^E) &= -s(E) + \beta E \\ &\leq -s(E') + \beta E' \;\leq\; I(\zeta) + \beta H(\zeta) \,. \end{aligned}$$

This establishes that ζ^E minimizes the functional $I + \beta H$ and hence is a canonical equilibrium with inverse temperature β.

When this global concavity condition on the microcanonical entropy is not satisfied, then there will be microcanonical equilibria that are not realized as canonical equilibria for any value of β. In particular, for any energy value E lying in the region where the concave hull of s does not coincide with s itself, there are no values of β such that $I + \beta H$ is minimized by the microcanonical equilibrium corresponding to E. We refer to Ellis *et al.* (2000) and Ellis *et al.* (2002*b*) for a proof of the necessity of the concavity condition. In those papers, the borderline cases when $s(E)$ is not strictly concave are also considered, and the theorems are formulated by means of sets of equilibria in view of the fact that solutions may not be unique within the separate ensembles themselves.

In the case when the canonical ensemble is not equivalent to the microcanonical ensemble, it is, however, possible to modify the canonical ensemble and thereby to obtain one that is equivalent to the microcanonical ensemble. The idea behind this construction is the observation that, even if there is no supporting tangent line to $s(E)$ at E, there will be a supporting parabola in the sense that

$$s(E') \leq s(E) + \beta(E' - E) + \frac{c}{2}(E' - E)^2 \quad \text{for all } E'$$

for some sufficiently large positive constant c. This condition requires merely that $s(E)$ be a smooth function, which is generically the case. Rearranging this supporting-parabola condition gives

$$s(E') - \beta E' - \frac{c}{2}(E' - E)^2 \leq s(E) - \beta E \quad \text{for all } E',$$

and hence the above argument is easily modified to show that the microcanoncial equilibrium ζ^E minimizes the augmented functional

$$I(\zeta) + \beta H(\zeta) + \frac{c}{2}[H(\zeta) - E]^2 .$$

In turn, this augmented functional is the rate functional for a so-called Gaussian ensemble, in which the term βH in the canonical Gibbs distribution is replaced by $\beta H + c[H - E]^2/2$ or, more simply, by $bH + cH^2/2$ for appropriate $b \in R$ and $c > 0$. Then the above argument shows that this Gaussian ensemble realizes all the microcanonical equilibria when c is fixed large enough and b is varied. Precise statements of these results, as well as generalizations to conditions weaker than the smoothness of $s(E)$, are given in Costeniuc *et al.* (2005) and Costeniuc *et al.* (2006).

We now address the related question of the stability of the steady mean flows induced by any equilibrium macrostate ζ. Interestingly, these hydrodynamic-stability theorems are intimately related to the large-deviation analysis and its rate functionals, and they parallel the criteria for equivalence/nonequivalence of ensembles (Ellis *et al.* 2002*a*,*b*).

If ζ_β is an equilibrium macrostate for the canonical ensemble, then it is a critical point of the functional

$$J_\beta(\zeta) = I(\zeta) + \beta H(\zeta) ,$$

meaning that the first variation $\delta J_\beta = 0$ at $\zeta = \zeta_\beta$. Moreover, ζ_β is a unconstrained minimizer, and so the second variation $\delta^2 J_\beta \geq 0$ at $\zeta = \zeta_\beta$. Now let us assume that this second variation is strictly positive definite, which is a technical nondegeneracy requirement that may be verified by a linear analysis around the equilibrium. Then J_β is a Lyapunov functional for ζ_β, since it is composed of conserved quantities for the continuum dynamics (4.15) and has a strict minimum at ζ_β. This conclusion relies on the conservation of the generalized enstrophy integral (4.14) for $a(\zeta) = \zeta \log \zeta$, which is identical to the negentropy $I(\zeta)$ in the equilibrium theory for point-vortex dynamics. The estimates that realize the Lyapunov stability argument are contained in the inequalities

$$k \int_D |\zeta - \zeta_\beta|^2 \, dx \leq J_\beta(\zeta) \leq K \int_D |\zeta - \zeta_\beta|^2 \, dx \quad \text{for some } 0 < k < K < \infty .$$

For $\beta > 0$, the functional J_β is the sum of two strictly convex functionals, and hence the equilibrium solution ζ_β is necessarily nondegenerate. Thus, as a steady solution of the two-dimensional ideal flow equations, ζ_β is nonlinearly (or Lyapunov) stable. This is the content of the first Arnold stability theorem, which was originally formulated for a deterministic steady-state ζ rather than the mean field for a statistical-equilibrium model (Arnold 1965, 1978; Holm *et al.* 1985). For a range of inverse temperature, say $\beta_- < \beta < 0$, convexity will continue to hold and solutions to the unconstrained minimization problem that defines canonical equilibria will continue to exist. These solutions will be nonlinearly stable, provided only that the technical conditions of nondegeneracy of the second variation of J_β are verified. The nonlinear stability of ζ_β for this range of β is the content of the second Arnold stability theorem (Arnold 1965, 1978; Holm *et al.* 1985).

Thus, we see that the famous Arnold stability theorems, which are nonlinear improvements of the classical linear stability theorems of Rayleigh, are satisfied by the canonical equilibria, and that the required Lyapunov functional is precisely the rate function in the large-deviation principle for the canonical ensemble. In short, the self-organization of the many-point-vortex system produces a coherent macrostate that is both steady and stable. But the statistical-equilibrium theory is different from traditional stability theory, in which one constructs families of steady flows (such as shear flows) and then tests their stability to infinitesimal disturbances. Rather, the canonical statistical theory selects only those steady states that are most probable with respect to microscopic fluctuations and stable with respect to macroscopic perturbations. Moreover, the statistical theory is not restricted to base flows with special symmetries.

When we turn to the microcanonical statistical-equilibrium theory, we discover a surprising and important extension of the Arnold stability theorems (Ellis *et al.* 2002*a*,*b*). If the microcanonical equilibria are realized as canonical equilibria, then their stability follows from the Lyapunov argument based on the function J_β, where $\beta = ds/dE$. Thus, nonlinear stability holds through the regime of equivalence of ensembles. But many interesting coherent states lie outside this regime; they are microcanonically constrained equilibria that are not realized as canonical equilibria. In this situation, neither the first nor the second Arnold theorem applies. Nonetheless, all of these microcanonical equilibria are indeed nonlinearly stable (provided that they are nondegenerate extrema). This conclusion follows from using the augmented functional

$$J^E(\zeta) = I(\zeta) + \beta H(\zeta) + \frac{c}{2}[H(\zeta) - E]^2$$

as the Lyapunov functional for a microcanonical equilibrium macrostate ζ^E. For $c > 0$ fixed large enough, J^E satisfies $\delta J^E = 0$ and $\delta^2 J^E > 0$ at $\zeta = \zeta^E$ with respect to variations $\delta\zeta$ that are unconstrained. By contrast, $I + \beta H$ has a positive second variation at ζ^E only for those variations $\delta\zeta$ tangential to the energy constraint $H = E$, and not necessarily for variations transverse to the energy constraint. But stability is not lost in this situation, because both I and H are independently conserved by ideal fluid dynamics. Rather than attempt to carry out an energy-constrained Lyapunov argument, we have introduced the quadratic term into J^E, which penalizes departures from the energy surface $H = E$. In this way, we obtain a functional having all the properties required in the usual Lyapunov analysis, and therefore we establish nonlinear stability of the microcanonically constrained equilibria. (In the constrained-optimization literature, the construction J^E is called the augmented Lagrangian. In our situation, the Lagrangian functional is the combination $I + \beta H$ with Lagrange multiplier β.)

This procedure supplies a significant extension of the known stability criteria for steady flows. In essence, it relies on the simple idea that the independent conservation of energy H and generalized enstrophy I contains more information about the growth of perturbations than does the conservation of a linear combination $I+\beta H$. Taking this extra information into account in the Lyapunov argument therefore leads to refined stability theorems. Moreover, this refinement is needed whenever the microcanonical ensemble is not equivalent to the canonical ensemble, and this occurs in many cases of physical interest, not merely in some exceptional situations.

From a conceptual point of view, we are led to conclude that the microcanonical formulation of the statistical-equilibrium theory is the more natural and useful one. First, some steady, stable states arising as microcanonical equilbria are not realized as canonical equilibria. Second, the microcanonical equilibrium states are parameterized by their energy values, which can be directly evaluated, unlike the inverse temperature. Third, the energy invariant is robust, because it involves the long-range interactions between vortices and is insensitive to the small-scale fluctuations of the many-vortex system. Moreover, for a small, finite viscosity the rate of dissipation of energy is much slower than the dissipation of negentropy, or any other convex enstrophy, and in the infinite-Reynolds-number limit energy is conserved even though enstrophy is dissipated. This limiting behavior is consistent with the statistical-equilibrium theory based on ideal fluid dynamics in which I is minimized subject to a strict constraint $H = E$.

4.2.6 Algorithm to compute equilibria

Here we include a numerical method for computing the microcanonical equilibria for the statistical point-vortex theory. The structure of this algorithm is quite general, and similar algorithms may also be used to compute the equilibrium states of the statistical theories based on continuum vorticity dynamics developed later in this chapter. We present it here in the specific form that applies to the point-vortex statistical-equilibrium theory.

We recall that the problem at hand is the following:

$$\begin{aligned} &\text{minimize} \quad I(\zeta) = \int_D \zeta \log \zeta \, dx \qquad \text{over} \quad \zeta \geq 0 \\ &\text{subject to} \; H(\zeta) = \frac{1}{2} \int_D |\nabla \psi|^2 \, dx = E, \quad C(\zeta) = \int_D \zeta \, dx = 1 \,, \end{aligned} \tag{4.37}$$

in which $\psi = G\zeta$ solves the linear elliptic boundary value problem $-\Delta\psi = \zeta$ in D, with $\psi = 0$ on ∂D.

This optimization problem has two constraints: a quadratically nonlinear constraint on the energy, and a linear constraint on the circulation. Associated with these constraints are the Lagrange multipliers β and μ, respectively. Any minimizer for eqn (4.37) satisfies the nonlinear elliptic eigenvalue problem (4.36).

One might therefore be inclined to try a naive iterative method of solution as follows. From an approximate solution triple (ψ^n, β^n, μ^n), imagine solving the Poisson equation $-\triangle\psi^{n+1} = \exp(-\beta\psi^n - \mu)$ in D along with its boundary condition $\psi^{n+1} = 0$ on ∂D, for a range of parameters β and μ, presumably near to β^n and μ^n, respectively. Update the pair (β^{n+1}, μ^{n+1}) by enforcing the constraints $H(\zeta^{n+1}) = E$ and $C(\zeta^{n+1}) = 1$; this involves solving a pair of real equations in the two parameters. Iterate to convergence.

We now define an iterative algorithm that has the general format of this naive method, but which has provable convergence properties that are derived from the optimization structure of the subproblem solved at each iteration. In short, we construct our algorithm to respect the mathematical structure of the constrained optimization problem that we are solving (Whitaker and Turkington 1994; Turkington and Whitaker 1996; DiBattista *et al.* 1998).

The key idea is to linearize the energy constraint around the current iterate, and to update the solution by minimizing I subject to the two linear constraints. Namely, given ζ^n, or equivalently ψ^n, we take ζ^{n+1} to be the unique solution of

$$\text{minimize} \quad I(\zeta) \quad \text{over} \quad \zeta \geq 0 \quad \text{subject to}$$
$$H(\zeta^n) + \int_D \psi^n(\zeta - \zeta^n)\, dx = E, \qquad C(\zeta) = \int_D \zeta\, dx = 1\,.$$

We note that $\delta H/\delta\zeta = \psi$. The multipliers for these linear constraints are β^{n+1} and μ^{n+1}, and they determine the solution $\zeta^{n+1} = \exp(-\beta^{n+1}\psi^n - \mu^{n+1})$. In turn, these two multipliers are determined by imposing the linear constraints, which form a pair of equations in (β, μ):

$$\int_D \psi^n \exp(-\beta\psi^n - \mu)\, dx = E + H(\zeta^n)\,, \qquad \int_D \exp(-\beta\psi^n - \mu)\, dx = 1\,.$$

This 2×2 nonlinear system has a gradient form, meaning that it is equivalent to solving the unconstrained convex minimization problem

$$\text{minimize} \quad \int_D \exp(-\beta\psi^n - \mu)\, dx + [E + H(\zeta^n)]\beta + \mu \qquad \text{over} \quad (\beta, \mu) \in R^2\,,$$

which determines (β^{n+1}, μ^{n+1}) uniquely given ψ^n. In the language of optimization theory, the minimization problem in the multipliers is dual to the primal problem that determines ζ^{n+1}. The dual problem can be solved by implementing a (quasi-)Newton-type method in the two-vector (β, μ). The computational expense is mainly associated with the evaluation of the spatial integrals that define this vector. Some care must be exercised in choosing step sizes in the Newton search because of the presence of the exponential function.

We thus have a well-defined iterative step, $\zeta^n \mapsto \zeta^{n+1}$, in which the associated Lagrange multipliers (β^{n+1}, μ^{n+1}) are uniquely and consistently updated. Provided that this sequence of iterations converges, it certainly produces a solution of the statistical-equilibrium problem (4.37).

This natural iterative algorithm possesses some desirable properties, which we now briefly outline. We will limit ourselves to the most interesting regime: microcanonical macrostates with $\beta < 0$. When $\beta > 0$, these macrostates can be realized as canonical macrostates which can be found by unconstrained minimization of $I + \beta H$, since equivalence of ensembles holds throughout this regime. In fact, the above algorithm converges in much of the positive-temperature regime for the microcanoncial model anyway, although the argument to follow proves its convergence properties only for the negative-temperature regime.

We claim that, for any initial guess ζ^0 satisfying $H(\zeta^0) \geq E$ and $C(\zeta^0) = 1$, the following two inequalities hold for all $n \geq 1$:

$$(A) \quad H(\zeta^n) \geq E \,, \qquad\qquad (B) \quad I(\zeta^{n+1}) \leq I(\zeta^n) \,. \tag{4.38}$$

We immediately see that (A) holds because the Hamiltonian H is convex in ζ, and so at iteration n the linearized constraint implies that

$$H(\zeta^{n+1}) \;\geq\; H(\zeta^n) + \int_D \psi^n (\zeta^{n+1} - \zeta^n)\, dx \;=\; E \,.$$

The convexity of I as a functional of ζ then implies (B) by virtue of the following calculation:

$$\begin{aligned} I(\zeta^n) &\geq I(\zeta^{n+1}) + \int_D \left(\log \zeta^{n+1} + 1\right)(\zeta^n - \zeta^{n+1})\, dx \\ &= I(\zeta^{n+1}) - \beta^{n+1} \int_D \psi^n (\zeta^n - \zeta^{n+1})\, dx \\ &= I(\zeta^{n+1}) - \beta^{n+1} \left[H(\zeta^n) - E \right] \\ &\geq I(\zeta^{n+1}) \,, \end{aligned}$$

using the assumption that $\beta^{n+1} \leq 0$. Thus, we find that the iterative algorithm monotonically decreases the objective functional I, while it maintains a one-sided energy constraint.

A further analysis in which second-order terms in the above expansion of I are retained shows that in fact

$$I(\zeta^n) - I(\zeta^{n+1}) \;\geq\; \frac{1}{2M} \int_D (\zeta^n - \zeta^{n+1})^2 \, dx \,,$$

where M is an upper bound on $\max_D \zeta^{n+1}$. The monotonic decrease of negentropy ensures that the iterative sequence ζ^n cannot become unbounded. Thus, we see that $\zeta^n - \zeta^{n+1}$ approaches zero in the $L^2(D)$ norm. Consequently, the linearized energy constraint implies that $H(\zeta^n)$ approaches E (from above) as $n \to \infty$. Finally, any limit point in $L^2(D)$ of the iterative sequence ζ^n produces a solution of the statistical-equilibrium problem.

In view of these global properties, this iterative algorithm offers a particularly attractive approach to solving the maximum-entropy problem posed by the micro-canonical model, which is a nontrivial optimization problem with a nonlinear objective functional and a nonlinear constraint. In implemented computations, this method converges with a linear rate of convergence over a wide range of model parameter choices from initial guesses that need not be close to the converged solutions. Moreover, the properties (4.38) hold even though solutions in the negative-temperature regime may not be unique, and the algorithm can converge to different solutions (local constrained minima) from different initial guesses. Finally, we mention that the iterative step is computationally efficient in that it involves (i) a Poisson solve to update ψ, which can be accomplished with any fast solver, and (ii) a solution of the convex dual problem in two real variables, amenable to standard quasi- or damped-Newton methods.

4.3 Statistical theory of 2D ideal flows

Even though the point-vortex model offers a convenient conceptual framework for developing a statistical mechanics of two-dimensional flow, it suffers from a number of deficiencies. The principal complaints that one can raise about the point-vortex theory are:

1. Each point vortex induces a divergent flow at its center and hence has infinite self-energy. Also, all generalized enstrophies are infinite.
2. The close interactions of concentrated, but finite vortex cores are not faithfully modeled by point vortices, and vortex mergers are not permitted.
3. Different point-vortex discretizations of the same continuum vorticity field lead to different predicted equilibrium states, and there is no universal way to assign the circulations to the many individual point vortices in the model.
4. The theory based on point vortices applies only to purely two-dimensional fluid dynamics, and does not extend conveniently to other more complicated models such as those arising in geophysical fluid dynamics.

These shortcomings of the point-vortex model are not all insurmountable within the conceptual framework of a finite collection of moving vortex centers. For instance, finite-core vortices can be introduced, and empirical rules for merging them can be imposed. But we will not pursue this line of development, because there is a theoretically attractive alternative now available. Namely, we will describe a statistical mechanics based on the actual continuum vorticity dynamics, following ideas first put forward by Miller (Miller 1990; Miller *et al.* 1992) and Robert and Sommeria (Robert 1991; Robert and Sommeria 1991). This theory, along with its subsequent refinements and extensions, forms a satisfactory theory of the equilibrium properties of coherent structures in 2D turbulence and its relatives in geophysical fluid dynamics. Our primary goal in these lectures is to present an accessible summary of this modern theory and its applications. Our lengthy development of the older point-vortex theory was intended as a prelude to the that exposition and development, to which we now turn.

4.3.1 Hamiltonian structure of 2D vorticity dynamics

We now adopt incompressible, inviscid fluid flow on a domain $D \subset R^2$ as our underlying microscopic dynamics. This dynamics has a formal Hamiltonian structure that contains the point-vortex dynamics as a singular limit (Holm *et al.* 1985). We sketch the outline of this structure next, recognizing that we require only certain key features for our subsequent development.

Let $Z = Z(x,t)$ denote the (scalar) vorticity $Z = \text{ curl } V$ of an ideal flow with velocity $V = V(x,t)$ at $x \in D$ and time $t \geq 0$, and let Ψ be the associated stream-function, so that $V = \text{ Curl } \Psi$ and $Z = -\triangle\Psi$ in D, with $\Psi = 0$ on ∂D. (We generally use capital letters to denote the microscopic flow quantities, which eventually will be treated as random variables in the statistical theory.)

The governing equation of 2D ideal flow is the transport equation (4.15), expressed now in the unknown Z:

$$\frac{\partial Z}{\partial t} + [\,Z, \Psi\,] = 0 \qquad \text{in} \quad D \times [0, \infty)\,. \tag{4.39}$$

Here, $[\cdot,\cdot]$ is the bracket on R^2 introduced after eqn (4.15). The space of all (admissible) vorticity fields Z forms a phase space for 2D ideal vorticity dynamics, and the governing equation (4.39) is derivable formally from a Hamiltonian as follows. Consider the total (kinetic) energy as a functional of Z, namely

$$H(Z) = \frac{1}{2}\int_D\int_D Z(x)\, g(x,x')\, Z(x')\, dx\, dx'\,, \tag{4.40}$$

and define the Poisson bracket

$$\{F, H\} \;=\; \int_D Z(x)\, \left[\frac{\delta F}{\delta Z}(x)\,,\,\frac{\delta H}{\delta Z}(x)\right]\, dx\,.$$

This expression involves the functional derivatives of F and H, which we assume are sufficiently smooth functionals on the phase space of vorticity fields. (In these notes, we will not enter into the technical details of the functional analysis that is needed to justify these constructions mathematically.) In terms of these formal notions, the governing dynamics is equivalent to the family of identities

$$\frac{d}{dt}F(Z) = \{F(Z), H(Z)\} \qquad \text{for all test functionals } \; F\,.$$

To verify this statement, we notice that

$$\frac{\delta H}{\delta Z}(x) = \int_D g(x,x')\, Z(x')\, dx' \;= \Psi(x)$$

and we make the calculation

$$\begin{aligned}\frac{d}{dt}F(Z(\cdot,t)) &= \int_D \frac{\delta F}{\delta Z}\,\frac{\partial Z}{\partial t}\, dx \\ &= -\int_D \frac{\delta F}{\delta Z}\,[\,Z, \Psi\,]\, dx \;=\; \int_D Z(x)\,\left[\frac{\delta F}{\delta Z}(x)\,,\,\Psi\right]\, dx\,,\end{aligned}$$

using integration by parts in the last equality. These identities indicate that the continuum vorticity transport equation is an infinite-dimensional Hamiltonian structure. This formal structure is not, however, in a canonical form when the natural phase point Z is used.

The special case in which the test functional is the linear functional $F(Z) = \int_D \phi(x)Z(x)\,dx$ for any smooth function ϕ on D results in the so-called weak form of the equations of motion. Namely,

$$\frac{d}{dt}\int_D \phi Z\,dx - \int_D Z\,[\phi, \Psi]\,dx = 0 \qquad \text{for all test functions } \phi .$$

As is well known, this weak form is the basis of any spectral truncation of the continuum vorticity dynamics. Let ϕ_k, $k = 1, 2, \ldots$, be an orthonormal basis for $L^2(D)$ consisting of eigenfunctions of the Laplacian: $-\triangle\phi_k = \lambda_k\phi_k$ in D with $\phi_k = 0$ on ∂D. Then the Fourier components of Z, namely $\hat{Z}_k = \int_D \phi_k Z\,dx$, $k = 1, 2, \ldots$, evolve according to

$$\frac{d}{dt}\hat{Z}_k = \sum_{p,q=1}^{\infty}\left[\int_D \phi_p[\phi_k, \phi_q]\,dx\right]\cdot \hat{Z}_p\,\lambda_q^{-1}\hat{Z}_q\,.$$

When we truncate the Fourier series after N terms, so that k, p, q in this identity run from 1 to N, and $Z = \sum_1^N \hat{Z}_k\phi_k$, then we obtain an N-dimensional dynamics having the following exactly conserved quantities:

$$H_N = \frac{1}{2}\sum_1^N \lambda_k^{-1}\hat{Z}_k^2 \quad \text{(energy)}, \qquad A_N = \frac{1}{2}\sum_1^N \hat{Z}_k^2 \quad \text{(enstrophy)},$$

$$d\hat{Z} = \prod_1^N d\hat{Z}_k \qquad \text{(phase volume element)}.$$

Thus, the N-mode truncated system defines a microscopic dynamics to which the methods of equilibrium statistical mechanics may be applied, using its two conserved quantities, energy and enstrophy, and its invariant phase volume. Moreover, a Gibbs distribution in which both energy and enstrophy are treated canonically is a multivariate Gaussian in $\hat{Z}$, making it easy to analyze. Thus, we arrive at a direct and straightforward construction of a statistical-equilibrium theory, in which the vorticity field is a random superposition of finitely many Fourier modes (Kraichnan 1975; Kraichnan and Montgomery 1980).

One may ask: what is the continuum limit of this theory and what fluid dynamical significance do the predicted macrostates have? Alas, in the context of 2D flow in a domain, this canonical model does not yield a very significant result, and this shortcoming may be attributed to a lack of equivalence of ensembles. Briefly, here is the argument. The Gibbs distribution in question is

$$P_{N,\beta,\alpha}(d\hat{Z}) = \exp\Big(-N\beta H_N(\hat{Z}) - N\alpha A_N(\hat{Z}) - f(\beta,\alpha,N)\Big)\,d\hat{Z}\,,$$

where the canonical parameters β and α conjugate to the energy and enstrophy invariants are scaled by N, so that the mean energy and mean enstrophy have finite limits as $N \to \infty$. With respect to this Gaussian distribution in $\hat{Z} \in R^N$, these mean values are computed to be

$$\langle H_N\rangle = \frac{1}{2N}\sum_1^N \frac{1}{\beta + \lambda_k\alpha}\,, \qquad \langle A_N\rangle = \frac{1}{2N}\sum_1^N \frac{1}{\beta\lambda_k^{-1} + \alpha}\,.$$

From these expressions, we see that the ensemble exists for $\alpha > 0$ and $\beta > -\lambda_1\alpha$. But since the eigenvalues λ_k increase to infinity as $k \to \infty$, we find that in the continuum limit, $\langle H \rangle = 0$ and $\langle A \rangle = (2\alpha)^{-1}$. Thus, only zero mean energy is achievable—a particularly strong case of nonequivalence of ensembles—and the predicted mean state is $\zeta = \langle Z \rangle = 0$. In other words, the main result derived from this simple model is that, for large N, enstrophy equipartitions among the Fourier modes in independent Gaussian fluctuations containing negligible energy. This conclusion implies that, in terms of the microscopic vorticity $Z(x) = \sum_1^N \hat{Z}_k \phi_k(x)$, the fluctuations are asymptotically uncorrelated in space:

$$\langle Z(x)Z(x') \rangle = \sum_1^N \langle \hat{Z}_k^2 \rangle \, \phi_k(x)\phi_k(x') \to 0 \qquad \text{for} \quad x \neq x' \, .$$

If, alternatively, the energy invariant is treated microcanonically, then a more informative statistical equilibrium ensues. Since the necessary analysis is subsumed in our subsequent development (in which generalized enstrophies are included), here we merely state the result. In the corresponding continuum limit as $N \to \infty$, the most probable macrostate ζ is found to be the vorticity field that minimizes $\alpha A(\zeta)$ subject to $H(\zeta) = E$. This macrostate is proportional to the first eigenfunction of the Laplacian on D, as is evident from the mean-field equation

$$0 = \alpha \frac{\delta A}{\delta \zeta} + \beta \frac{\delta H}{\delta \zeta} = \alpha\zeta + \beta\psi \, , \qquad \text{where} \;\; \psi = G\zeta \, ,$$

in which β is the Lagrange multiplier for the energy constraint. Thus, in the microcanonical model a nonzero coherent macrostate is obtained with $\beta = -\lambda_1\alpha$, and this macrostate contains the specified energy E. The remainder of the enstrophy, not associated with this macrostate, resides in fluctuations which, as in the canonical model, are asymptotically equipartitioned over the modes.

This statistical-equilibrium model therefore suggests that an initial vorticity field with energy E_0 and enstrophy A_0 will relax after a turbulent evolution to a macrostate with final energy $E = E_0$ and a final enstrophy A that is as small as possible. In the course of this evolution, the excess enstrophy $A_0 - A$ that is lost on the large scales eventually goes into small-scale, spatially uncorrelated fluctuations of the vorticity field. Such a phenomenology is consistent, at least qualitatively, with the observed behavior of 2D turbulence, in which energy tends to flux toward large scales while enstrophy tends to flux toward small scales. In fact, this statistical-equilibrium theory realizes the extreme case of the 2D cascade picture—the energy is trapped in the largest scales of motion (the domain scale), and the portion of the enstrophy not contained in the energetic macrostate is transferred to (arbitrarily) small scales. In a high-Reynolds-number flow, these fluctuations are dissipated by viscosity. (We caution that the cascade picture properly applies to a forced flow that is in a statistically stationary state, not to one that approximates a statistical-equilibrium state after a long, unforced evolution. And the statistical-equilibrium theory does not produce energy or enstrophy spectra that agree with the phenomenological theory of forced turbulence in 2D.)

Attractive though these simple energy–enstrophy models are, they fail to respect the full set of invariants of continuum vorticity dynamics. Even the circulation $C(Z) = \int_D Z(x)\,dx$, a linear invariant, is not necessarily conserved under truncation. The circulation may be considered the simplest of the generalized enstrophy invariants $A(Z) = \int_D a(Z)\,dx$, for any regular real function $a(z)$. All these integrals are conserved exactly by the continuum dynamics, because the vorticity is advected by the incompressible velocity field that it induces, but none of them are respected by the spectrally truncated dynamics for finite N, unless $a(z) = \text{const} \cdot z^2$. For this reason, we now turn to a modern formulation of statistical equilibrium that incorporates all these continuum invariants in a unified and elegant manner.

4.3.2 Miller–Robert–Sommeria theory

The central new element in a statistical-equilibrium theory of vorticity dynamics that includes all generalized enstrophy invariants $A(Z)$ is a proper macroscopic description of the scalar vorticity field Z. In essence, what is needed is the appropriate coarse-graining of the vorticity. The theory proposed and developed by Miller and, independently, by Robert and Sommeria distinguishes the microscopic vorticity $Z = Z(x)$, which fluctuates rapidly from point to point in space, from a macroscopic state that varies slowly in space. For this macrostate, the theory utilizes the local, single-point distribution of the microscopic vorticity. Specifically, let $\rho(x, z)$ be the probability density function at a spatial point $x \in D$ over values of Z near x. Here, and throughout our subsequent discussion, z denotes a real variable that runs over the range of the microscopic vorticity (which is invariant in time). At any moment of time, consider a sequence of small neighborhoods centered at a point x, say disks $B_r(x)$ of radius r around x, and calculate the histogram of the fluctuating vorticity Z over these neighborhoods; that is, for macroscopically small $r > 0$,

$$\frac{|\{\, x' \in B_r(x) \;:\; u \leq Z(x') \leq v \,\}|}{|B_r(x)|} \approx \int_u^v \rho(x, z)\, dz\,.$$

Thus, $\rho(x, z)\,dz$ represents the volume fraction near x over which Z takes values in the interval $[z, z + dz]$. Of course, this is not a rigorous mathematical statement, because there is the implicit assumption that Z fluctuates on a range of scales smaller than r, yet r is taken small enough that ρ is almost independent of r. Formally, one might define the macrostate ρ as the following long-time or ensemble average:

$$\rho(x, z) = \langle \delta(z - Z(x)) \rangle\,,$$

which effectively expresses the same concept but without explicitly invoking a finite coarse-graining scale r. This formal statement means that for all continuous test functions $\chi(x, z)$ on the product $D \times R$ of the domain and range of the microscopic vorticity, there holds

$$\int_D \int_R \chi(x, z)\, \rho(x, z)\, dx\, dz = \int_D \langle \chi(x, Z(x)) \rangle\, dx\,.$$

We expect that the averaging over time or over the ensemble of the fluctuating microstate Z results in a macrostate $\rho(x, z)$ that varies smoothly with x. For the sake of clarity, we write ρ as a density with respect to dz, but the local probability distribution over z may or may not be absolutely continuous with respect to the uniform measure dz.

To make this key concept more precise, we next introduce a discretization of the microstate from the domain D to a fine lattice over D, which we denote by Λ_N, using N to index the number of lattice sites (or microcells). For conceptual purposes, we may imagine that $Z(s,t)$, for $s \in \Lambda_N$, is a highly resolved numerical solution of the vorticity transport equation on a fine grid Λ_N of N spatial grid points. As is well known from high-Reynolds-number simulations of this type, in which one allows for the effects of a small physical viscosity or of numerical hyperviscosities introduced to stabilize the numerical scheme, the microscopic vorticity develops finite-order fluctuations on a range of small scales, and enstrophy (a bulk measure of these fluctuations) is fluxed toward the smallest resolved scale on the grid. Nonetheless, there also a flux of energy toward large scales and, consequently, coherent structures are formed in the coarse-grained vorticity field. The purpose of the macrostate ρ is to capture these coarse-grained structures in a way that is appropriate for statistical-equilibrium theory.

To this end, we also introduce a coarse lattice over D, which we denote by Δ_M, having M sites (or macrocells), where $1 \ll M \ll N$. We suppose that this coarse lattice is subdivided consistently with respect to the fine lattice, so that each macrocell in Δ_M contains exactly N/M microcells of the fine lattice. For instance, this setup can be easily obtained for a rectangular domain D and regular rectangular fine and coarse lattice subdivisions. The theory does not strictly rely upon such uniformity, which we suppose here merely for the sake of definiteness in our exposition.

Now, with respect to (random) microstates $Z(s)$, $s \in \Lambda_N$, we define the coarse-graining process

$$W_{N,M}(x, dz) = \frac{M}{N} \sum_{s \in B_i} \delta(z - Z(s))\, dz \qquad \text{for} \quad x \in B_i \quad (i = 1, \ldots, M), \tag{4.41}$$

where B_i denotes any macrocell in the coarse lattice Δ_M. Thus, $W_{N,M}$ is the empirical measure of the microscopic vorticity over each macrocell of the coarse lattice. In the appropriately scaled continuum limit of a statistical-equilibrium lattice model, we expect that this coarse-graining $W_{N,M}$ should approach the continuum macrostate $\rho(x,z)$ in the sense that

$$W_{N,M}(x, dz)\, dx \to \rho(x,z)\, dx\, dz \qquad \text{as} \quad N, M, N/M \to \infty\,,$$

where the convergence is in the weak sense of measures on $D \times R$. Thus, the goal of the Miller–Robert–Sommeria theory is to derive variational principles which are similar to the maximum-entropy principles described in the previous section for the point-vortex model, and which characterize the most probable macrostates ρ in statistical equilibrium. Such equilibrium macrostates will quantify the large-scale coherent structures that emerge and persist in two-dimensional turbulence.

4.3.3 Special case of a vortex patch

Before presenting the full theory, let us first consider the simplest case of the Miller–Robert–Sommeria theory, when the microscopic vorticity takes only two values, say

$Z = 0$ and $Z = \lambda > 0$. The 2D ideal equations of motion are known to have unique solutions even in this mildly singular case in which the vorticity is discontinuous across an interface (Marchioro and Pulvirenti 1994; Majda and Bertozzi 2001). The evolution of such "vortex patches" has been extensively studied numerically. The typical behavior is that a patch with a smooth interface separating a rotational interior where $Z = \lambda$ from an irrotational exterior where $Z = 0$ will develop instabilities, which will lead to strong distortions of the patch interface. After a sufficiently long evolution, the interface will be highly filamented on a range of small scales, to an extent that it becomes impractical to maintain a microscopic description of it. Rather, an initially sharp interface evolves into a blurred interface, which on a coarse-grained scale looks like a continuous transition from the interior vorticity value of λ to the exterior value of 0. In other words, on a macroscopic scale there is a mixing of the microscopic levels 0 and λ until a final equilibrium macrostate is formed.

Formally, the macrostate for such a vortex patch is given by

$$\rho(x, z) = \rho_0(x)\delta(z) + \rho_\lambda(x)\delta(z - \lambda),$$

where $\rho_0(x)$ and $\rho_\lambda(x)$ are the probabilities that $Z = 0$ and $Z = \lambda$, respectively, in a small neigborhood of x. The macroscopic mean vorticity is then $\zeta = \lambda\rho_\lambda$. Consequently, the macrostate ρ is entirely determined by ζ, since $\rho_0 = 1 - \zeta/\lambda$ and $\rho_\lambda = \zeta/\lambda$. Heuristically, if the interior of the patch, where $Z = \lambda$, is colored black, and the exterior of the patch, where $Z = 0$, is colored white, then ζ is a gray level representing mixing on the macroscale.

The mixing entropy of a macrostate ζ is straightforwardly computed by standard combinatorics. Namely, if $n = N/M$ is the number of microcells in a macrocell at x with $\zeta = \zeta(x)$, then the number of realizations of a state with $n_\lambda = n\zeta/\lambda$ black microcells and $n_0 = n(1 - \zeta/\lambda)$ white microcells is equal to $n!/n_0!n_\lambda!$. Asymptotically for large n, we have

$$\log \frac{n!}{n_0!n_\lambda!} = -n\left[\left(1 - \frac{\zeta}{\lambda}\right)\log\left(1 - \frac{\zeta}{\lambda}\right) + \frac{\zeta}{\lambda}\log\frac{\zeta}{\lambda}\right] + \cdots$$

Thus, we define the entropy $S(\zeta) = -I(\zeta)$, where

$$I(\zeta) = \int_D \left[\left(1 - \frac{\zeta(x)}{\lambda}\right)\log\left(1 - \frac{\zeta(x)}{\lambda}\right) + \frac{\zeta(x)}{\lambda}\log\frac{\zeta(x)}{\lambda}\right] dx\,. \tag{4.42}$$

This negentropy is integrated over the space variable, meaning that the negentropy of each macrocell contributes independently to the total negentropy. The most probable macrostate ζ over D is accordingly the one that minimizes $I(\zeta)$ over all admissible macrostates, since then it has overwhelmingly the most microscopic realizations.

This minimization is conditioned by the conservation of total energy and circulation. Thus, we seek to express these dynamical constraints in terms of the macrostate ζ. First, we note that the circulation $C(Z) = C(\zeta)$ is unchanged under coarse-graining, being linear in Z. This is not true of any other of the generalized enstrophy invariants, because they are nonlinear in Z. But, for a vortex patch, the circulation determines all the generalized enstrophy invariants, since the conservation of the patch area is the only dynamical invariant inferred from the incompressible rearrangement of vorticity. Consequently, this macroscopic description combined with the circulation constraint on ζ fully encodes the vorticity invariants.

Next, we consider the energy constraint. As in the point-vortex case, we note that it is a large-scale invariant that is not sensitive to small-scale vorticity fluctuations. More precisely, we use the fact that the Green's function is compatible with local averaging in the sense that

$$\lim_{r,r'\to 0} \left\{ \frac{1}{|B_r(x)| \cdot |B_{r'}(x')|} \int_{B_r(x)} \int_{B_{r'}(x')} g(y,y')\, dy\, dy' \; - \; g(x,x') \right\} = 0$$

in $L^2(D \times D)$. Consequently, the energy functional $H(Z)$ in eqn (4.40) is continuous with respect to the local averaging that defines coarse-graining. Thus, $H(Z) \approx H(\zeta)$, and this approximation becomes exact in the continuum limit.

With these ingredients in hand, we can now articulate a variational principle that characterizes the most probable macrostate of the Miller–Robert–Sommeria model. Namely, the mean vorticity of the macroscopic state in the microcanonical model for a vortex patch is the solution to the constrained optimization problem

$$\text{minimize} \quad I(\zeta) \qquad \text{subject to} \;\; H(\zeta) = E, \;\; C(\zeta) = \Gamma. \tag{4.43}$$

Just as in our discussion of the point-vortex theory, it is possible to derive this maximum-entropy principle from a rigorous large-deviation analysis; but we shall omit the details at this point of our development (Robert 1991; Michel and Robert 1994; Boucher *et al.* 2000). Suffice it to say that, in such a large-deviation principle, the rate function is the entropy deficit, $I(\tilde{\zeta}) - I(\zeta)$, between any macrostate $\tilde{\zeta}$ satisfying the constraints and the equilibrium macrostate ζ.

The mean-field equation satisfied by the equilibrium macrostate ζ that solves eqn (4.43) is immediately seen to be

$$0 = -\frac{1}{\lambda} \log\left(1 - \frac{\zeta(x)}{\lambda}\right) + \frac{1}{\lambda} \log \frac{\zeta(x)}{\lambda} + \beta\psi(x) + \mu\,,$$

where β and μ are the Lagrange multipliers for the energy and circulation constraints, respectively. Rearranging this equation and using $\zeta = -\Delta\psi$, we obtain the nonlinear elliptic eigenvalue problem for the streamfunction ψ of the most probable state:

$$-\Delta\psi = \lambda \left\{ \frac{1}{2} + \frac{1}{2} \tanh(-\lambda[\beta\psi + \mu]) \right\} \quad \text{in } D, \qquad \psi = 0 \quad \text{on } \partial D\,. \tag{4.44}$$

The predictive content of the statistical-equilibrium theory is contained in this boundary value problem, in which the eigenvalue-type parameters β and μ are found along with the solution ψ by imposing the constraints $H(\zeta) = E$ and $C(\zeta) = \Gamma$.

In this way, the statistical theory predicts that, given an initial vortex patch Z^0 with energy $E = H(Z^0)$ and circulation $\Gamma = C(Z^0)$, the turbulent evolution will produce a final state that is represented on the macroscale by ζ and has the same energy and circulation. Unlike the initial microstate Z^0, which has a discontinuous jump in vorticity between 0 and λ across the patch boundary, the final macrostate ζ is a continuous function lying between the extreme vorticity levels of 0 and λ. The inverse temperature β effectively determines the sharpness of the transitional layer where the patch boundary is mixed on small scales. The "chemical potential," μ, adjusts to determine the location of the transition layer.

We thus obtain a natural modification of the point-vortex theory, now with a finite vorticity intensity λ. In fact, the Onsager–Montgomery–Joyce theory is realized as the limiting case of this Miller–Robert–Sommeria theory as the parameter $\lambda \to \infty$. In this so-called "dilute limit," a fragmentary vortex patch with fixed total circulation and large vorticity intensity looks like an assembly of many concentrated vortices. We omit the mathematical derivation which shows that the Miller–Robert–Sommeria theory becomes the point-vortex theory in the dilute limit.

For finite λ and given Γ, there is a maximum realizable energy $E_{\max}(\lambda, \Gamma)$ in the microcanonical vortex patch model. As the prescribed energy E approaches $E_{\max}$, the equilibrium macrostate $\zeta^{E,\Gamma}$ approaches a steady vortex patch with circulation Γ and vorticity λ, and $\beta \to -\infty$. This is a zero-temperature limit, where the statistical-equilibrium state becomes deterministic. These limiting states therefore play the role of the single concentrated point vortex in the Onsager theory, which is realized as the energy $E \to +\infty$ and the appropriately scaled inverse temperature $\beta \to \beta_\infty > -\infty$ (Caglioti *et al.* 1992, 1995).

A virtue of the Miller–Robert–Sommeria theory is that it allows quantitative verification against direct numerical simulations of high-Reynolds-number decaying 2D turbulence. Approximations to vortex patches are feasible initial data for highly resolved numerical computations, such as pseudo-spectral methods on doubly periodic domains. (Double periodicity is convenient numerically, and it eliminates any walls and hence any boundary layers.) Necessarily, $\Gamma = 0$ in this geometry, and so the model is modified to apply to microstates taking the values λ and $-\lambda$. The equilibrium macrostates are parameterized by λ and E; for given λ, the low-energy states are parallel shear layers, while the high-energy states are opposite-signed, diagonal vortex pairs. These predictions have been shown to be in good qualitative and quantitative agreement with direct numerical simulations of long-time behavior when, for instance, the initial vorticity field is chosen to be a "checkerboard," in which $Z^0 = \lambda$ and $Z^0 = -\lambda$ alternately in a uniformly subdivided square domain (Segre and Kida 1998).

One can also examine the predictive capacity of the theory to detect qualitative transitions in equilibrium behavior. For instance, the roll-up of a parallel shear layer can be modeled using the vortex patch formulation of the statistical-equilibrium theory (Turkington and Whitaker 1996). In a unit-area rectangular domain with slip conditions on parallel bounding walls at $x_2 = \pm 0.5$ and periodic conditions in the x_1-direction with period 1, consider a patch of area 0.1 with $Z^0 = \lambda = 10$, and $Z^0 = 0$ elsewhere. Then $\Gamma = 1$ and E depends upon the shape of the Z^0 vortex patch. If an x_1-translational symmetry is imposed (allowing only states independent of x_1), then, necessarily, shear layers are obtained as the most probable macrostates. But if this symmetry condition is relaxed, then the macrostate is able to roll up into a coherent vortex. The x_1-independent solution and the organized macrovortex have exactly the same invariants (E, Γ, and λ) and, accordingly, the existence of the rolled-up state can be interpreted as a prediction of the end state of a shear-layer instability—the equilibrium shear profile loses its stability to streamwise disturbances and eventually forms into a coherent macrovortex. The finite-β maximum-entropy state $\zeta^{E,\Gamma}$ not only predicts the large-scale structure of the final macrovortex but also quantifies the degree of filamentation that is present in the microscopic vorticity field, which results from the instability and subsequent vorticity entrainment throughout the roll-up process.

With respect to this model, we note that there is another conserved quantity in such a domain, owing to x_1-translation invariance. This linear impulse invariant is related to the x_1-component of the linear momentum of the flow, given by $L_1 = \int_D x_2 \zeta \, dx$. The inclusion of this invariant into the constraints of the maximum-entropy principle produces an x_1-translational velocity (a Lagrange multiplier). If, as in Turkington and Whitaker (1996), the predicted macrostates have an even symmetry under $x_2 \mapsto -x_2$, this translational velocity is zero. Otherwise, translating macrostates can be studied by imposing a nonzero L_1.

The statistical-equilibrium theory can also be implemented to model another fundamental and generic process in 2D turbulence—namely, the vortex merger. In the inverse cascade process, coherent vortices that emerge within the turbulent field interact somewhat like point vortices until a pair of vortices approach closely enough that their mutual interaction causes them to merge into a single vortex. Typically, the circulation of the resulting vortex is less than the combined circulations of the two vortices that merge, with the remainder of the circulation going into small-scale filamentary vorticity in the surrounding flow field. It is particularly interesting to ask whether the statistical-equilibrium theory can capture this merger phenomenon. In this context, we note that it is less natural to do so within the point-vortex theory because (1) the point-vortex theory incorrectly represents close interactions at the microscopic scale, and (2) the point-vortex statistical theory allows the merged vortex to have a mean vorticity greater than the maximum vorticity of the pair before the merger, in contradiction to continuum vorticity dynamics.

The problem of the merger phenomenon may be posed on the entire plane or on a circular domain with sufficiently large radius. On such rotationally symmetric domains, the angular impulse $M = \int_D |x|^2 \zeta \, dx$ is a conserved quantity. Clearly, this invariant plays an important role in the merger process, since vorticity that is initially separated from the center of the merger (taken to be the origin $x = 0$) coalesces toward the center. The existence of this invariant immediately implies that some fraction of the circulation of the merging pair must move to relatively large distances from the center. Thus, the formulation of the statistical-equilibrium model of a vortex merger contains an energy, a circulation, and an angular-impulse constraint.

This application of the Miller–Robert–Sommeria vortex patch theory has been investigated in Whitaker and Turkington (1994). The main result is that the statistical-equilibrium theory merges vortex patches without regard to their separation. For two identical circular patches with a given separation (between centers), the theory predicts that the equilibrium macrostate is a rotationally symmetric macrovortex surrounded by a "fringe" of filamentary vorticity that contains a small fraction of the circulation but a substantial portion of the angular impulse. Thus, the theory is unable to detect a critical aspect ratio (between the vortex separation and the vortex radii) above which a merger does not occur. This lack of agreement between the equilibrium statistical theory and physical and numerical experiments on well-separated vortex patches is probably due to the lack of complete ergodicity in the evolving flows. The statistical-equilibrium theory in its pure form expects that the unresolved microscopic dynamics does explore all of the available phase space. Presumably, the migration of filaments from the unstable vortex boundaries is suppressed by dynamical effects in the actual evolution, making ergodicity incomplete. If the statistical predictions are computed on a circular domain with a boundary that is close enough to the vortex pair, then the outward migration of filamentary vorticity is confined and the merger is indeed suppressed. The origin of the lack of ergodicity is a hidden constraint imposed by a separating streamline in the corotating frame, across which the vorticity may not cross. An effort to impose isolating spatial constraints of this kind in the equilibrium theory has been made (Chavanis and Sommeria 1998).

4.3.4 Peak vorticity theory

The Miller–Robert–Sommeria theory is based on a mixing entropy I that results from counting microcells in which the microscopic vorticity Z takes distinct values. This entropy connects a definite fine-grained description to an associated coarse-grained description, as in any statistical-mechanical theory. But, at least in the context of ideal flow, there is no definite fine scale at which the values of an evolving microscopic vorticity $Z(\cdot, t)$ can be discerned. Rather, the self-induced rearrangement of vorticity can produce fluctuations on a range of fine scales, making it impossible to distinguish the sharp vorticity values on any particular fine scale. Moreover, there is no discrete dynamics on a fine grid that conserves all the invariants of continuum ideal flow. Consequently, the derivation of the Miller–Robert–Sommeria theory is incomplete, because it requires a discretization to define its entropy principle, but it imposes constraints that hold only for the continuum dynamics.

Considerations of this kind have led to an alternative formulation of the theory, which we now sketch in the special case of vortex patches (Turkington 1999). An evolving continuum microstate Z that takes the values 0 and λ appears, when observed on any finite microscale (by averaging on that scale), to take values in the interval $0 \leq Z \leq \lambda$, not just at the upper and lower limits of this interval. Moreover, an analysis of spectrally truncated dynamics transposed onto any spatial grid Λ_N (as in a pseudo-spectral numerical method) shows that the finite-dimensional phase volume $\prod_{s\in\Lambda_N} dZ(s)$ is invariant. For these reasons, it is natural to consider a macroscopic description of the vorticity field to be a probability density $\rho(x, z)$ on the values of Z near the spatial point $x \in D$, with respect to the uniform measure dz on the range $[0, \lambda]$. Given that the invariance of the uniform measure on the range of the vorticity is the only a priori information in the macroscopic description, the negentropy expression appropriate to this formulation is

$$I(\rho) = \int_D \int_0^\lambda \rho(x, z) \log \rho(x, z)\, dx\, dz\,. \tag{4.45}$$

The local mean vorticity associated with the macrostate ρ is then

$$\zeta(x) = \int_0^\lambda z\rho(x,z)\,dz\,.$$

The energy and circulation constraints are imposed on the mean vorticity through the functionals $H(\zeta)$ and $C(\zeta)$ as before. Accordingly, the equilibrium macrostate for the microcanonical version of the statistical theory with this alternative formulation is defined to be the maximizer ρ of the entropy $-I(\rho)$ in eqn (4.45) subject to $H(\zeta) = E$ and $C(\zeta) = \Gamma$.

This formulation contrasts with the Miller–Robert–Sommeria formulation, for which ρ is a combination of point masses concentrated at $z = 0$ and $z = \lambda$. In essence, the original model imposes a definite fine scale on which microstates behave as discretized patches, while the alternative model allows mixing on all fine scales and observes the continuum dynamics through a sequence of "windows" corresponding to those fine scales. The alternative theory allows for the loss of some information—the blurring of the vorticity field on fine scales—which is not acknowledged in the original Miller–Robert–Sommeria theory. Since only the vorticity extrema are retained, we refer to the alternative theory as the peak vorticity model.

The mean-field equation for this formulation is computed as before. The result is a different equilibrium vorticity–streamfunction relationship, namely,

$$\zeta = \frac{df}{d\sigma}(-\beta\psi - \mu) \qquad \text{with} \quad f(\sigma) = \log\int_0^\lambda e^{\sigma z}\,dz \quad (\sigma \in R)\,.$$

Calculating this relationship explicitly gives the nonlinear elliptic eigenvalue problem for the most probable state:

$$-\Delta\psi = \lambda\left\{\frac{1}{2} + \frac{1}{2}L(-\lambda[\beta\psi + \mu])\right\} \quad \text{in } D, \qquad \psi = 0 \quad \text{on } \partial D\,, \tag{4.46}$$

where $L(u) = \coth u - 1/u$ is the Langevin function. Qualitatively, $L(u)$ is similar to $\tanh u$, which appears in the original model, but it approaches its asymptotes of ± 1 as u approaches $\pm\infty$ with an algebraic decay rather than an exponential decay. This difference in the profile functions that determine the mean-field equation in the statistical models is the main manifestation of the different ways that the fine-scale fluctuations are handled.

A complete discussion of the alternative theory is presented in Turkington (1999), where it is formulated in general, not merely for vortex patches. The key idea is that all convex generalized enstrophy functionals, $A(Z) = \int_D a(Z)\,dx$ with convex $a(z)$, are decreased by averaging of the vorticity on *any* fine scale. Thus, the exact conservation of all generalized enstrophies by an ideal continuum vorticity dynamics is replaced by upper bounds on the family of all convex enstrophies. The macrostate $\rho(x,z)$ is a probability density in z with respect to the uniform measure dz over the range of the vorticity, and the negentropy has the form (4.45) with the interval $[0,\lambda]$ replaced by the range of the vorticity. As in each of the models discussed here, the energy and circulation constraints are imposed on the mean vorticity field ζ derived from ρ, but the equilibrium macrostate is the minimizer of the negentropy subject to equality constraints on the energy and circulation and *inequality* constraints on all convex generalized enstrophies.

Such a formulation acknowledges that some portion of any convex generalized enstrophy may be lost to subgrid-scale fluctuations on any finite grid, as is demonstrated in all numerical simulations of 2D turbulence. An advantage of the alternative theory is that, under typical circumstances, many of the convex-enstrophy constraints will be nonbinding in equilibrium, meaning that they are satisfied as strict inequalities by the equilibrium state. Then they play no role in the resulting mean-field equation, since their Lagrange multipliers necessarily vanish. For instance, for a vorticity field that takes on peak values $\pm\lambda$ and also the value 0—say, two patches of opposite-signed vorticity surrounded by irrotational flow—the alternative theory shows that the total absolute circulation, $A = \int_D |Z|\, dx$, is the only generalized enstrophy that can be active in equilibrium (Turkington 1999).

This result is especially interesting in the limit as $\lambda \to \infty$, with finite energy E and absolute circulation A. In this dilute limit, we arrive at an alternative to the Onsager–Montgomery–Joyce theory for an equal number of positive and negative point vortices moving in an irrotational flow. The point-vortex model with a unit circulation distribution of positive vortices, $\int_D \zeta_+\, dx = 1$, and of negative vortices, $\int_D \zeta_-\, dx = 1$, uses the negentropy expression

$$\int_D \zeta_+ \log \zeta_+\, dx + \int_D \zeta_- \log \zeta_-\, dx\,,$$

which is minimized over ζ_+ and ζ_- subject to the energy constraint $H(\zeta_+ - \zeta_-) = E$. The result is the mean-field equation

$$-\Delta\psi = 2e^{-\mu} \sinh(-\beta\psi)\,.$$

This prediction has been tested against direct simulation in a doubly periodic geometry, and good qualitative agreement has been established: the predicted end state is a diagonal pair of opposite-signed macrovortices, and the vorticity–streamfunction relation shows a hyperbolic-sine shape (Montgomery *et al.* 1992). The limiting case of the peak vorticity theory for the same situation yields the maximum-entropy principle

$$\text{maximize } I(\rho) \qquad \text{subject to} \qquad H(\zeta) = E, \qquad \int_D \int_{-\infty}^{+\infty} |z| \rho(x,z)\, dx\, dz = 2\,,$$

where I is as in eqn (4.45) except that the integration over the vorticity range now extends from $-\infty$ to $+\infty$. The result of this alternative model of dilute intense vorticity in an irrotational flow leads to the mean-field equation

$$-\Delta\psi = \frac{-2\beta\psi}{\alpha^2 - \beta^2\psi^2}\,,$$

where α now enters as the Lagrange multiplier for the absolute-circulation constraint. Qualitatively, the solutions to the alternative theory are similar to the more well-known point-vortex theory. But the vorticity–streamfunction relationship has a sharper peak at the center of the dipolar pair of macrovortices. An unpublished quantitative comparison of this statistical-equilibrium solution with the result of a long-time direct numerical simulation with an initial condition representative of two opposite-signed patches of intense vorticity ($\lambda \to \infty$), each of unit absolute circulation, shows a remarkably good fit between theory and numerical experiment, both at the extreme values of the vorticity and at its intermediate values.

This dilute limit of the peak vorticity theory provides a completely self-consistent alternative to the point-vortex theory. Both seek to model the organization of small areas of intense vorticity surrounded by irrotational background flow into coherent macrostates. Essentially, the alternative theory (the dilute limit of the peak vorticity model) resembles a mixed version of the point-vortex model in which a collection of vortices with a random distribution of vortex strengths is used, rather than a collection of vortices all with the same vortex strength (Berdichevsky 1995). But the vorticity–streamfunction relationship predicted by the peak vorticity model is more intrinsic and universal than any relationship derived from a particular version of the point-vortex theory.

4.3.5 General formulation with a prior distribution on vorticity

Next we turn to yet another variant of the Miller–Robert–Sommeria theory, in which the exact conservation of all generalized enstrophies is replaced by the specification of a probability distribution on fine-scale vorticity fluctuations. Our preceding development has suggested that in many practical applications it is not appropriate to retain all the information that is contained in the family of generalized enstrophy invariants. On the one hand, these invariants are very sensitive to any nonideal effects, such as viscous dissipation. On the other hand, they are often not relevant to an application of the theory which seeks to understand how branches of equilibrium solutions behave when model parameters are varied; then the details of the initial condition (often not even known) carried by the enstrophy invariants are not pertinent. In fact, the statistical-equilibrium theory of coherent structures has found its greatest success in applications for which the fine-scale fluctuations of vorticity can be crudely parameterized by a statistical distribution, and the coarse-grained behavior can be tracked by the values of a few invariants (Majda and Wang 2006). For these reasons, we now reformulate the statistical theory in a version that is most useful for real applications (Ellis *et al.* 2002*a*,*b*).

It has long been known that the ideal invariants of 2D fluid dynamics fall into two categories: the "rugged" invariants and the "fragile" invariants (Kraichnan and Montgomery 1980). An invariant is rugged if it is not destroyed by averaging on fine scales, while it is fragile if it is. Our foregoing analysis has already stressed this distinction, and we have shown that the energy is a rugged invariant by virtue of the long-range interactions between vortices. The circulation is also rugged, being linear. But all generalized enstrophies A corresponding to a nonlinear real function $a(z)$ are sensitive to averaging on the small scale, simply because $\langle a(Z)\rangle \neq a(\langle Z\rangle)$, where the angle brackets denote some kind of averaging over fine scales. The same distinction is made in phenomenological theories of 2D turbulence, in which some ideal invariants condition the energy flux toward large scales while others condition the enstrophy flux toward small scales.

We can combine all of these insights from turbulence modeling, statistical mechanics, and mathematical analysis into a natural and useful formulation of the statistical-equilibrium theory. Specifically, our approach is based on the following:

1. The rugged invariants (energy, circulation, and linear or angular impulse) are treated microcanonically; they condition the large scales of motion.
2. The fragile invariants (all nonlinear generalized enstrophies) are treated canonically; they condition the fine-scale fluctuations of vorticity, which is modeled as a bath of random vorticity.
3. The appropriately scaled statistical ensemble defined by these rugged and fragile invariants is analyzed by the Cramér Theorem of large-deviation theory; the negentropy functional that connects the fine-grained fluctuations to the coherent coarse-grained structures is the large-deviation rate functional.

As in our introductory discussion of the Miller–Robert–Sommeria theory, we consider a lattice Λ_N of N sites s, and let $Z = Z(s)$ denote the random microscopic vorticity field on this lattice. We introduce the "prior distribution" $\pi(dz)$, a probability measure on the range of the vorticity (which we take to be the real line R), and we let

$$P_N(dZ) = \prod_{s\in\Lambda_N} \pi(dZ(s))\,. \tag{4.47}$$

The idea is that $P_N(dZ)$ encodes the statistical behavior of the fine-scale vorticity fluctuations in a way that is consistent with the conservation of all generalized enstrophies. To be more specific, let $a_1(z), \ldots, a_m(z)$ be any finite number of independent generalized-enstrophy integrands, and consider an arbitrary linear combination $a(z) = \alpha_1 a_1(z) + \cdots + \alpha_m a_m(z)$, which is merely another generalized-enstrophy integrand. The lattice enstrophy for $a(z)$, namely

$$A_N(Z) = \frac{1}{N}\sum_{s\in\Lambda_N} a(Z(s))$$

determines the appropriately scaled canonical distribution

$$\exp[-NA_N(Z)]\prod_{s\in\Lambda_N} dZ(s) \;=\; \prod_{s\in\Lambda_N}\exp[-a(Z(s))]\,dZ(s)\,.$$

From this product expression we see immediately that, if we identify $\pi(dz)$ with $\exp(-a(z))\,dz$, then we have realized the canonical ensemble as eqn (4.47). Moreover, the single-point distribution $\pi(z)$ can be arbitrary, because the corresponding generalized-enstrophy integrand $a(z)$ can be arbitrary. Thus, we have a formal justification for simply specifying a prior distribution $\pi(dz)$ on lattice-scale fluctuations of the microscopic vorticity, which determines a spatially homogeneous canonical ensemble $P_N(dZ)$. The real parameters $\alpha_1, \ldots, \alpha_m$ that define the density of the prior distribution are canonical thermal parameters for the bath of fine-scale vorticity fluctuations. In short, the prior distribution parameterizes the "enstrophy bath" defined by the canonical ensemble (4.47).

We then proceed just as in the other formulations of these statistical theories. Namely, we construct the Gibbs distribution

$$P_N^{E,\Gamma}(dZ) = P_N(dZ \mid H_N(Z) = E,\, C_N(Z) = \Gamma)\,, \tag{4.48}$$

where H_N and C_N are lattice versions of the energy and circulation functionals. This distribution is the governing statistical ensemble for the model—it is microcanonical in the rugged invariants (energy and circulation) and canonical in the fragile invariants (generalized enstrophies).

To analyze the continuum limit of eqn (4.48), we introduce a coarse-graining process as in eqn (4.41); recall that we subdivide the domain D into M macrocells B_i with $1 \ll M \ll N$. Now, however, we may work with the simpler process

$$Y_{N,M}(x) = \frac{M}{N}\sum_{s\in B_i} Z(s) \;=\; \int_{-\infty}^{+\infty} zW_{N,M}(x,dz) \qquad \text{for} \quad x \in B_i\,, \tag{4.49}$$

which is precisely the local average of Z over each macrocell B_i $(i = 1,\ldots,M)$. Since the process $Y_{N,M}$ is the expectation with respect to z of $W_{N,M}$, it satisfies a large-deviation principle that is the contraction of the large-deviation principle for the empirical measures $W_{N,M}$. The relevant large-deviation principle for $Y_{N,M}$ is therefore built from the Cramér Theorem, which applies to the sample means of independent, identically distributed random variables. This theorem is stated and discussed in the lecture notes by Ellis. The rate functional is defined by the Legendre transform of the cumulant-generating function for the prior distribution $\pi(dz)$. Namely, let

$$f(\sigma) = \log \int_{-\infty}^{+\infty} e^{\sigma z}\, \pi(dz)\,, \qquad f^*(\zeta) = \sup_{\sigma\in R}[\,\zeta\sigma - f(\sigma)\,]\,. \tag{4.50}$$

(Abusing notation, here we use ζ as a real variable on the range of the mean vorticity.) The rate functional for the doubly indexed process $Y_{N,M}$ with respect to the product measure P_N in the continuum limit as $N, M, N/M \to \infty$ is the negentropy integral

$$I(\zeta) = \int_D f^*(\zeta(x))\, dx\,, \tag{4.51}$$

which is a functional of any admissible mean vorticity field ζ on D.

The large-deviation principle with respect to the governing Gibbs distribution (4.48) can be stated as follows: for any $\zeta \in L^2(D)$ with $H(\zeta) = E$ and $C(\zeta) = \Gamma$, there holds

$$\frac{1}{N}\log P_N^{E,\Gamma}\{Y_{N,M} \approx \zeta\} \;\to\; \min_{\tilde\zeta\in L^2(D)} \{I(\tilde\zeta) : H(\tilde\zeta) = E, C(\tilde\zeta) = \Gamma\} \;-\; I(\zeta) \tag{4.52}$$

in the continuum limit.

Consequently, we arrive at the maximum-entropy principle that determines the most probable mean vorticity field ζ that emerges and persists within a random field of vorticity fluctuations with prior distribution π, and has a prescribed energy E and circulation Γ. The equilibrium macrostate $\zeta^{E,\Gamma}$ that minimizes the negentropy I for those values of the energy and circulation satisfies the mean-field equation

$$\zeta = \frac{df}{d\sigma}(-\beta\psi - \mu) = \frac{\int z e^{-(\beta\psi+\mu)z}\,\pi(dz)}{\int e^{-(\beta\psi+\mu)z}\,\pi(dz)}. \tag{4.53}$$

Thus, the model establishes an explicit connection between the statistical properties of the fine-scale fluctuations and the nonlinearity of the mean-field equation for the coarse-scale coherent structure: $\pi(dz) \mapsto f'(\sigma)$. For any given prior distribution π the vorticity–streamfunction relationship is a monotonic function, since f and f^* are convex functions. The Lagrange multipliers β and μ are determined along with the solution ζ or, equivalently, ψ.

With a given prior distribution, the equilibrium theory produces a microcanonical thermodynamical formalism. The entropy is

$$S(E,\Gamma) = -I(\zeta^{E,\Gamma}) = \lim_{N\to\infty} \frac{1}{N} \log P_N\{\, H_N = E, C_N = \Gamma \,\},$$

and hence $\beta = \partial S/\partial E$ and $\mu = \partial S/\partial \Gamma$. As in the point-vortex theory, $S(E,\Gamma)$ is not necessarily concave in its arguments. Rather, there are regimes in parameter space where the associated canonical ensemble, either in energy or in circulation or both, is not equivalent to the microcanonical ensemble. Moreover, the properties discussed above concerning the relations between the ensembles carry over to this setting. In particular, the microcanonical ensemble always has a richer set of equilibrium states, in the sense that there are equilibria omitted by the canonical ensemble in the regime of nonequivalence (Ellis *et al.* 2000, 2002*b*; Costeniuc *et al.* 2005).

The nonlinear hydrodynamic stability of the microcanonical equilibrium macrostates, which are distinguished steady solutions of the 2D ideal flow equations, can be established, using the conserved functionals I, H, C to build Lyapunov functionals. A full development of these results is given in Ellis *et al.* (2002*a*) and Ellis *et al.* (2002*b*). As in the case of the point-vortex theory, these stability theorems refine the Arnold stability criteria in the nonequivalent regimes. In fact, the arbitrary choice of the prior distribution π produces the most general vorticity–streamfunction relation in equilibrium, and thus we obtain the widest possible extension of the Arnold stability theorems. The equilibrium macrostates are overwhelmingly most probable with respect to fine-scale fluctuations—this is the physical content of the large-deviation principle—and they are nonlinearly stable with respect to finite perturbations—this is the content of the refined stability theorems. Finally, the general vorticity–streamfunction relation is derived from the statistical properties of the microscopic vorticity fluctuations, not merely postulated or modeled from macroscopic considerations. The importance of this attribute of the statistical-equilibrium theory with a prior distribution is illustrated in a physical application in the next section.

4.4 Applications to geophysical fluid dynamics

A primary motivation for developing a statistical-equilibrium theory for vorticity advection is to model the coherent structures that are observed in geophysical fluid dynamics. In atmosphere or ocean dynamics, the small ratios between the vertical and horizontal scales, and between the relative vorticity and the rotation of the planet, combine to enforce near two-dimensionality on the flow field. For this reason, the statistical theory described in the preceding sections applies naturally to geophysical flows in that asymptotic regime which is most relevant for real applications. In the next subsection we summarize these governing equations and sketch their derivation from basic physical principles. We then proceed to implement the statistical-equilibrium theory for this underlying dynamics, first in an energy–enstrophy model and then in a model with a non-Gaussian prior distribution. Our lectures culminate in a model of the midlatitude bands of the weather layer on Jupiter.

4.4.1 Derivation of quasi-geostrophic dynamics

The simplest model that is relevant to the circulation of an ocean or atmosphere on a rotating planet is the rotating-shallow-water (RSW) model (Pedlosky 1987; Vallis 2006). We therefore base our discussion on the RSW model, for a single layer of homogeneous fluid. The so-called "primitive equations" of geophysical fluid dynamics, which are the basis of dynamic meteorology and physical oceanography, can be conceptualized as being analogous to the RSW equations but for a continuous stratification of density (temperature) in the vertical direction. We confine our discussion to the RSW prototype. [We use the standard notation of geophysical fluid dynamics in the following section, and consequently some symbols now have a different meaning than in the preceding sections.]

We consider a rotating planet, which is a sphere of radius r_0 with angular velocity $\mathbf{\Omega}$. Let θ and ϕ be the latitude and longitude, respectively. We are interested in a midlatitude band of the planetary atmosphere or ocean centered at θ_0 and ϕ_0. Accordingly, we introduce rectangular coordinates about the central point, using $x = r_0 \cos\theta\,(\phi - \phi_0)$, $y = r_0(\theta - \theta_0)$, $z = r - r_0$; x increases eastward, y increases northward, and z increases upward. Of particular importance is the component of $2\mathbf{\Omega}$ normal to the planetary surface, which is the Coriolis parameter

$$f = 2\Omega \sin\theta \approx f_0 + \beta_0 y\,,$$

where $f_0 = 2\Omega \sin\theta_0$ and $\beta_0 = 2\Omega r_0^{-1} \cos\theta_0$. In this so-called β-plane approximation, we work in the local rectangular coordinates x, y and retain the variation of the Coriolis parameter to the first-order.

In the vertical direction, we assume that the fluid layer is bounded below by a fixed surface at $z = -H(x, y)$ and above by a free surface at $z = \eta(x, y, t)$. Thus, the fluid depth at any point is $\eta + H$, and this is occupied by fluid of constant density ρ. The fluid motion is described by a velocity field (u, v, w) and a pressure field p.

The equations governing an inviscid, incompressible fluid, expressed in the frame that rotates with the planet, are

$$\frac{\partial u}{\partial x} + \frac{\partial v}{\partial y} + \frac{\partial w}{\partial z} = 0\,, \tag{4.54}$$

$$\frac{Du}{Dt} - fv = -\frac{1}{\rho}\frac{\partial p}{\partial x}\,, \qquad \frac{Dv}{Dt} + fu = -\frac{1}{\rho}\frac{\partial p}{\partial y}\,, \tag{4.55}$$

$$\frac{Dw}{Dt} = -\frac{1}{\rho}\frac{\partial p}{\partial z} - g\,, \tag{4.56}$$

where g denotes the gravitational acceleration, which acts in the local vertical direction. Here we use the shorthand $D/Dt = \partial/\partial t + u\partial/\partial x + v\partial/\partial y + w\partial/\partial z$ for the material derivative. These equations describe the conservation of mass (4.54), horizontal momentum (4.55), and vertical momentum (4.56). In addition, they are supplemented by kinematic equations at the top and bottom surfaces

$$\frac{D}{Dt}(z-\eta) = 0 \quad \text{at} \quad z = \eta\,, \qquad \frac{D}{Dt}(z+H) = 0 \quad \text{at} \quad z = -H\,. \tag{4.57}$$

Finally, the governing system is completed by a free-boundary condition

$$p = p_0 \quad \text{at} \quad z = \eta\,, \tag{4.58}$$

where p_0 is a constant ambient pressure above the free surface.

The RSW equations are an approximation to these fundamental equations valid for small aspect ratio $\delta = D/L$, where D and L are characteristic length scales in the vertical and horizontal directions, respectively. Typically, $\delta \sim 10^{-2}$ for the oceans or the atmosphere. If V and W denote characteristic velocity scales in the horizontal and vertical directions, respectively, then it follows from eqn (4.54) that $W \sim \delta V$. Using this and comparing the pressure scale implied by eqn (4.55), it follows that the vertical acceleration can be neglected in eqn (4.56). Namely, $\partial p/\partial z = -\rho g + O(\delta^2)$, which is the hydrostatic approximation that essentially defines the RSW equations. Thus, we replace eqn (4.56) by the approximation

$$p(x,y,z,t) = p_0 + \rho g[\,\eta(x,y,t) - z\,]\,. \tag{4.59}$$

Consequently, the horizontal pressure gradient is independent of z, and so it is consistent to consider the horizontal velocity field as independent of z. The horizontal momentum equations therefore become

$$\left[\frac{\partial}{\partial t} + u\frac{\partial}{\partial x} + v\frac{\partial}{\partial y}\right] u - fv = -g\frac{\partial \eta}{\partial x}\,, \quad \left[\frac{\partial}{\partial t} + u\frac{\partial}{\partial x} + v\frac{\partial}{\partial y}\right] v + fu = -g\frac{\partial \eta}{\partial y}\,. \tag{4.60}$$

Thus, the RSW dynamics can be visualized as consisting of moving columns of fluid whose vertical expansion or contraction is dictated by the convergence or divergence of the horizontal velocity (u,v). Eliminating w in terms of (u,v) and inserting it into the kinematic conditions (4.57) yields the governing equation for the free-surface height η:

$$\left[\frac{\partial}{\partial t} + u\frac{\partial}{\partial x} + v\frac{\partial}{\partial y}\right](\eta + H) + (\eta + H)\left(\frac{\partial u}{\partial x} + \frac{\partial v}{\partial y}\right) = 0\,. \tag{4.61}$$

Equations (4.60) and (4.61) constitute the RSW equations, which close in terms of (u,v) and η.

There is a remarkable exact conserved quantity for the RSW equations. Namely, the potential vorticity (PV)

$$Q = \frac{\zeta + f}{\eta + H}, \tag{4.62}$$

where $\zeta = \partial v/\partial x - \partial u/\partial y$ is the relative vorticity. The RSW equations imply that the PV satisfies

$$\left[\frac{\partial}{\partial t} + u\frac{\partial}{\partial x} + v\frac{\partial}{\partial y}\right] Q = 0 .$$

The pointwise horizontal advection of Q is entirely analogous to the case of purely two-dimensional flow ($\eta = 0, f = 0, H = 1$), which conserves ζ.

We are interested in a physical regime in which the horizontal pressure gradients nearly balance the Coriolis force—that is, a regime that is near to geostrophic balance. To quantify this asymptotic regime, we introduce the dimensionless groups

$$\epsilon = \frac{V}{f_0 L} \ll 1 , \qquad \lambda = \frac{\sqrt{gD}}{f_0 L} \approx 1 , \tag{4.63}$$

namely, the Rossby number ϵ and the relative deformation scale $\lambda = L_d/L$, where $L_d = \sqrt{gD}/f_0$ is the Rossby deformation radius. In addition, we scale the Coriolis gradient β_0 and the bottom depth H according to

$$\frac{L\beta_0}{f_0} = \epsilon\beta , \qquad H(x, y) = D[1 - \epsilon b(x, y)] .$$

Under these scalings, which are generally satisfied in geophysical applications, the nondimensionalized β and the fractional bottom topography b are taken to be independent of the small parameter ϵ (Pedlosky 1987; Vallis 2006).

We derive the quasi-geostrophic equations in the limit of small Rossby number ($\epsilon \to 0$) by first writing the RSW equations (4.60) and (4.61) in terms of the nondimensionalized variables

$$(x, y) = L(\hat{x}, \hat{y}), \quad t = \frac{L}{V}\hat{t}, \quad (u, v) = V(\hat{u}, \hat{v}), \quad \eta = \frac{f_0 V L}{g}\hat{\eta} .$$

Next we expand the unknowns in these nondimensionalized governing equations in power series in ϵ. That is, after removing the hats from the unknowns and variables, we invoke the expansions $u = u_0(x, y, t) + \epsilon u_1(x, y, t) + O(\epsilon^2)$ and so forth for v and η.

This is a singular perturbation expansion, in which the material derivatives of u, v, and η are multiplied by the small parameter ϵ. As a result, the lowest-order terms yield conditions on u_0, v_0, η_0 that do not entirely determine those fields. Namely, we obtain exact geostrophic balance

$$(u_0, v_0) = \left(-\frac{\partial \eta_0}{\partial y}, \frac{\partial \eta_0}{\partial x}\right) . \tag{4.64}$$

To determine the geostrophic streamfunction η_0, it is necessary to consider the next order in the expansion, which we omit here. All that we require is that these equations are solvable, since the solvability condition then determines η_0 and hence closes the approximate dynamics to leading order. The solvability condition reduces elegantly to the single scalar equation

$$\left[\frac{\partial}{\partial t} + u_0\frac{\partial}{\partial x} + v_0\frac{\partial}{\partial y}\right]\left(\zeta_0 - \lambda^{-2}\eta_0 + \beta y + b(x,y)\right) = 0\,. \tag{4.65}$$

Noting that $\zeta_0 = \triangle\eta_0$, we see that eqns (4.64) and (4.65) together form the transport equation

$$\left[\frac{\partial}{\partial t} + u_0\frac{\partial}{\partial x} + v_0\frac{\partial}{\partial y}\right] q_0 = 0\,, \qquad \text{with } q_0 = (\triangle - \lambda^{-2})\eta_0 + \beta y + b(x,y)\,.$$

This means that the so-called quasi-geostrophic potential vorticity q_0 is advected by the streamfunction η_0, which is itself determined instantaneously by q_0 via an inhomogeneous Helmholtz equation with appropriate boundary conditions.

In what follows, we will use this quasi-geostrophic (QG) dynamics as our underlying microscopic dynamics for a statistical-mechanical theory. In keeping with usual practice and in analogy with our presentation of 2D vorticity dynamics, we will write the QG dynamics in the form

$$\frac{\partial q}{\partial t} + [\psi, q] = 0\,, \qquad q = (\triangle - \lambda^{-2})\psi + \beta y + b(x,y)\,, \tag{4.66}$$

using ψ to denote the geostrophic streamfunction (identical to η_0). [We note that this streamfunction has a different sign from the one used in the 2D theory above.] The QG PV q is directly related to the RSW PV through the asymptotic formula $Q = (f_0/D)[1 + \epsilon q + O(\epsilon^2)]$. Obviously, the 2D vorticity transport equation (4.15) is the special case of eqn (4.66) when $\lambda = \infty, \beta = 0, b = 0$.

The QG dynamics have a Hamiltonian structure similar to that of 2D dynamics, with energy and generalized-enstrophy integrals

$$H(q) = \frac{1}{2}\int \left[|\nabla\psi|^2 + \lambda^{-2}\psi^2\right]\, dx\, dy\,, \tag{4.67}$$

$$A(q) = \int a(q)\, dx\, dy\,. \tag{4.68}$$

For the purposes of understanding the self-interaction of the PV field, it is useful to reexpress the Hamiltonian directly in terms of q as

$$H(q) = \frac{1}{2}\int [q - \beta y - b](\triangle - \lambda^{-2})^{-1}[q - \beta y - b]\, dx\, dy\,,$$

from which we see that the Green's function for the operator $\triangle - \lambda^{-2}$ defines the interaction. This introduces a finite range λ, which is like a screening length. In geophysical terms, λ is the length scale at which potential and kinetic energy are comparable, or the scale at which buoyancy and Coriolis forces balance.

4.4.2 Effects of β, topography, and deformation scale

The statistical-equilibrium theory discussed in Section 3.5 for the 2D case easily extends to the QG case. Given a prior distribution $\pi(dq)$ on fine-grained potential-vorticity fluctuations, we construct the cumulant-generating function

$$f(\sigma) = \log \int_{-\infty}^{+\infty} e^{\sigma q}\, \pi(dq) \qquad (\sigma \in R)$$

and its Legendre–Fenchel transform $f^*(s) = \sup_\sigma [s\sigma - f(\sigma)]$. The coherent macrostates that form in QG turbulence are characterized as the solutions to the following constrained minimization problem:

$$\text{minimize} \quad I(\bar{q}) = \int_D f^*(\bar{q})\, dx\, dy \qquad \text{subject to} \quad H(\bar{q}) = E, \;\; C(\bar{q}) = \Gamma \tag{4.69}$$

over all local mean fields $\bar{q}$. The virtue of this formulation is that it immediately produces the mean-field equation that relates the mean PV to the geostrophic streamfunction ψ; namely,

$$\bar{q} = f'(\theta\psi - \mu) \quad \text{with} \quad (\triangle - \lambda^{-2})\, \psi + \beta y + b(x, y) = \bar{q}\,, \tag{4.70}$$

in which θ and μ are the "inverse temperature" and "chemical potential" associated with the energy and circulation constraints; that is, the Lagrange multipliers at the minimizer $\bar{q}$. The governing constrained optimization problem (4.69) can be solved numerically by using a version of the algorithm described in Section 2.6 (DiBattista *et al.* 1998).

The solutions $\bar{q}$ and their corresponding mean flow patterns ψ depend on three different sets of parameters: (1) the total energy and circulation values specified in the microcanonical variational principle (4.69); (2) the prior distribution on PV fluctuations $\pi(dq)$; and (3) the nondimensional geophysical parameters λ, β, and b. This third set of parameters distinguishes the QG theory from the 2D theory, and by including these effects and varying these parameters the statistical-equilibrium theory produces a much richer set of predictions (Majda and Wang 2006).

To understand this behavior, it is useful to outline the predictions of energy–enstrophy theory in the QG case (Carnevale and Frederiksen 1987; Salmon *et al.* 1976; Bretherton and Haidvogel 1976). The choice of a Gaussian prior distribution

$$\pi(dq) = \sqrt{\frac{\alpha}{2\pi}}\, e^{-\alpha q^2/2}$$

with a specified $\alpha > 0$ simplifies the statistical-equilibrium theory. The negentropy I in eqn (4.69) then becomes the quadratic enstrophy and the mean-field equation (4.70) is linear; thus,

$$\bar{q} = \frac{\theta\psi - \mu}{\alpha}\,, \qquad \text{var}\; q = \frac{1}{\alpha}\,.$$

We may set $\alpha = 1$ without loss, by rescaling θ and μ, and so the theory is governed by the linear partial differential equation

$$\triangle\psi - \lambda^{-2}\psi + \beta y + b(x,y) = \theta\psi - \mu\,.$$

Solutions can therefore be found by Fourier series expansion. Consider the case when $\mu = 0$, which amounts to relaxing the circulation constraint as in traditional energy–enstrophy theory. The total topography (including the β-term as an effective bottom inclination) has an expansion $\tilde{b} = b(x,y) + \beta y = \sum_m B_m\phi_m(x,y)$, using the eigenfunctions defined by $-\triangle\phi_m = \kappa_m^2\phi_m$ in D with appropriate boundary conditions on ∂D. The solution $\psi(x,y) = \sum_m \Psi_m\phi_m(x,y)$ is then given by

$$\Psi_m = \frac{1}{\kappa_m^2 + \lambda^{-2} + \theta}B_m \qquad (m = 1, 2, \ldots)\,.$$

Assuming that $B_1 \neq 0$, we find a family of solutions for $\theta > -(\kappa_1^2 + \lambda^{-2})$. In terms of the prescribed energy value, θ in this range is uniquely determined by the equation

$$\frac{1}{2}\sum_{m=1}^{\infty}\frac{\kappa_m^2}{[\kappa_m^2 + \lambda^{-2} + \theta]^2}B_m^2 = E\,.$$

As θ approaches $-(\kappa_1^2 + \lambda^{-2})$, the corresponding energy E tends to infinity.

Thus, we see that the total topography $\tilde{b}$ determines the mean-flow structure at each value of the energy E. Moreover, this macroscopic structure is correlated with the topography, especially when the deformation scale λ is small; indeed, the Fourier coefficients Ψ_m and B_m for low modes m are then approximately proportional. In particular, this means that maxima of $\tilde{b}$ are correlated with anticyclonic flows, while minima are correlated with cyclonic flows. This behavior is a well-known prediction of the energy–enstrophy theory with topography that is confirmed by observation and simulations in oceanography (Holloway 1986).

Similarly, the β effect is reflected in the equilibrium macrostates, in the sense that with increasingly strong β the flows tend to become increasingly zonal. For instance, for fixed values of the other parameters and constraints, a family of equilibria will transition from a coherent vortex embedded in a zonal shear to a purely zonal shear as β is increased. This phenomenon is displayed in DiBattista *et al.* (1998), where its relation to the Rhines scale $\sqrt{V/\beta}$, a length scale limiting the formation of vortices in QG turbulence, is discussed (Rhines 1979).

Finally, we mention that nonequivalence of ensembles occurs for QG flows, and in fact the nonequivalent regimes of the statistical-equilibrium theory expand as λ decreases. In zonal flow geometry (as is discussed in the following section), there is a strong nonequivalence associated with the circulation constraint, even when the prior distribution is Gaussian. For a non-Gaussian prior, there is also a somewhat weaker nonequivalence in the energy constraint when the inverse temperature θ is negative. A prototypical QG example is given in Ellis *et al.* (2002*b*) and others in DiBattista *et al.* (1998).

4.4.3 Model of Jupiter's weather layer

We conclude our development of the statistical mechanics of vorticity by applying it to model the persistent zonal shear flows and embedded oval vortices in the atmosphere of the planet Jupiter. Since Jupiter is a giant, rapidly rotating planet, we are able to use the quasi-geostrophic approximation with some confidence (the Rossby number ϵ is smaller than 0.1). The observed flows, which have been so vividly pictured by the Voyager and Galileo satellites, reside in the "weather layer." Roughly speaking, this active layer is analogous to the Earth's troposphere: it is bounded above by a strongly stratified layer and contains convection. But the Jovian weather layer lies above a deep lower layer of fluid that is in motion. A key point in the formulation of a dynamical model is the influence of the largely unknown lower layer on the observed upper layer. Two issues arise:

1. The upper layer feels the influence of the lower-layer mean flow as an effective bottom topography.
2. The upper-layer flow is energized by small-scale convection from below, which conditions the statistical properties of its fluctuating eddy field.

With these physical issues in mind, let us now formulate the $1+1/2$-layer model, which is the simplest geophysical-fluid-dynamics model that captures the essential features of the weather layer (Dowling 1995*a*; Marcus 1993). This is a QG model that is derived from the RSW equations in exactly the same way that we derived the QG dynamics for a single layer over a fixed topography in Section 4.1. The difference is that now we consider two layers: an active upper layer of density ρ_1 and average depth D_1, and a passive lower layer of density $\rho_2 > \rho_1$ and average depth $D_2 \gg D_1$. The free surface is now the interface between the layers, across which the pressure is continuous. We take the vertical coordinate z to be such that $z = 0$ is the rigid top surface, $z = -D_1$ is the undisturbed interface, and $z = -D_1 - D_2$ is the rigid bottom surface. The free-surface interface is then described by $z = -D_1 + \eta(x, y, t)$, with η being the displacement.

Even though D_2 is large compared with D_1, we nevertheless assume that both are small compared with the horizontal length scale of motion, and consequently the RSW equations hold in each layer. In terms of the horizontal gradient $\nabla = (\partial/\partial x, \partial/\partial y)$, a straightforward analysis of the hydrostatic balance in the upper and lower layers yields

$$\nabla p_1 = \nabla P\,, \qquad \nabla p_2 = \nabla P + g(\rho_2 - \rho_1)\nabla\eta\,,$$

where $P = P(x, y, t)$ is the pressure perturbation at the top of the weather layer, $z = 0$. We introduce the "reduced gravity" $g' = g(\rho_2 - \rho_1)/\rho_1$, and we make the Boussinesq approximation, in which we consider $\rho_1 \approx \rho_2$ except in the buoyancy terms. We may bring the governing equations for the two layers into a form analogous to the single-layer RSW equations by introducing the new field variables

$$\eta_1 = \frac{P}{g'\rho_1}\,, \qquad \eta_2 = \eta_1 + \eta\,.$$

This substitution allows us to eliminate the horizontal pressure gradients ∇p_1 and ∇p_2 in favor of η_1 and η_2, thereby obtaining a pair of RSW equations in η_1 and η_2, which may be considered as effective height perturbations in the upper and lower layers, respectively. In these two-layer RSW equations, $D_1 - \eta_2$ plays the role of the effective total depth in the upper layer, as does $D_2 - \eta_1$ in the lower layer. For lack of space, we do not display these equations.

We now carry out the quasi-geostrophic scaling and singular-perturbation argument for the two-layer system. The Rossby number is as before, but now there are two deformation scales:

$$\epsilon = \frac{V}{f_0 L}, \qquad \lambda_i^2 = \frac{g' D_i}{f_0 L^2} \quad (i = 1, 2).$$

A rather lengthy calculation results in the two-layer QG system for this reduced-gravity system. That is, we obtain geostrophic streamfunctions ψ_i, $i = 1, 2$, for the horizontal flow in each layer, and corresponding potential vorticities

$$q_1 = \Delta\psi_1 + \lambda_1^{-2}(\psi_2 - \psi_1) + \beta y, \qquad q_2 = \Delta\psi_1 + \lambda_2^{-2}(\psi_1 - \psi_2) + \beta y.$$

The QG dynamics simply expresses advection of PV in each layer:

$$\frac{\partial q_1}{\partial t} + [\psi_1, q_1] = 0, \qquad \frac{\partial q_2}{\partial t} + [\psi_2, q_2] = 0.$$

For our Jupiter model, we are interested in a $1 + 1/2$-layer model, which results from taking the limit $D_2/D_1 \to \infty$ in the two-layer model. Then, $\lambda_2 \to \infty$, and so in this limit $q_2 = \Delta\psi_2 + \beta y$. Consequently, the PV in the lower layer is independent of ψ_1, and so the deep lower-layer flow decouples from the upper-layer dynamics. We may therefore impose a lower-layer flow. In oceanic applications of these models, the lower layer is often taken to be static. But on Jupiter there is evidence of strong zonal winds below the weather layer (Dowling and Ingersoll 1988). Hence we assume that the lower-layer flow is steady and zonal, meaning that $\psi_2 = \psi_2(y)$ and $q_2 = d^2\psi_2/dy^2 + \beta y$. The lower layer is passive in the sense that its flow is unaffected by the dynamics of the upper layer. The influence of the lower layer is therefore isomorphic to an effective bottom topography $b(y) = \lambda_1^{-2}\psi_2(y)$ resulting from the pressure perturbations of the lower-layer zonal shear flow. Thus, we finally arrive at a governing equation for the upper layer that is identical to eqn (4.66). Our statistical-equilibrium model of the coherent states in the weather layer is therefore governed by the variational principle (4.69).

With the statistical-equilibrium theory in hand, it is not difficult to find physically relevant parameter regimes for a model in which coherent vortices embedded in zonal shears are the most probable states. For instance, the flow displayed in Fig. 1 of Ellis *et al.* (2002*a*) was chosen to illustrate a solution at particular values of E and Γ, which contains an anticyclonic oval vortex that is embedded in a zonal shear flow consisting of an anticyclonic band and a cyclonic band. In this example, the total circulation Γ is cyclonic, and the prior distribution is a gamma distribution that is nearly Gaussian but has a small anticyclonic skewness (normalized third-order moment). Also, there is an effective bottom topography that is sinusoidal, representing a lower-layer flow having an anticyclonic band and a cyclonic band. Consequently, the flow in the upper layer tends to follow the lower-layer pattern, especially since $\lambda = 0.2$ is small. The flow configuration and the physical parameters were set to represent a zonal domain on Jupiter that contains an anticyclonic zone north of a cyclonic belt.

Branches of solutions were examined in Ellis *et al.* (2002*a*) by varying E and Γ. For smaller E the flow is purely zonal, while for larger E the flow transitions to a coherent anticyclonic vortex rolling in the anticyclonic band of the shear. Thus, the statistical-equilibrium theory produces a flow field that is qualitatively similar to the flow in a band containing the Great Red Spot (GRS) on Jupiter. This state is a microcanonical equilibrium, and hence the refined stability theorems guarantee its nonlinear stability. In fact, the transition from purely zonal shear to an embedded oval vortex occurs at values of E and Γ that are well within the regime of nonequivalence of ensembles for this configuration. These physically important Jovian flows are not realized as canonical equilibria, and their stability cannot be deduced from the standard Arnold stability theorems. This fact explains the confusing issue sometimes raised in the literature that the permanent jet structure of Jupiter's weather layers seems to lie beyond the range of validity of the classical stability criteria (sufficient conditions) for shear flows on a β-plane.

While it is gratifying to exhibit flow fields that emulate those observed on the Jovian planets, it is desirable to use the statistical-equilibrium theory in a truly predictive manner. For instance, one can ask:

1. Why does a great oval vortex not emerge and persist in any of the anticyclonic zonal shears at other latitudes, say in the northern hemisphere of Jupiter?
2. Why are the most persistent and coherent vortices anticyclones rather than cyclones (which are only observed as weakly organized "brown barges")?

The answers to these questions bring us back to the two issues highlighted at the beginning of this section—namely, the combined influence of the effective bottom topography and the prior distribution of fine-grained potential-vorticity fluctuations. By resolving these two issues, we obtain a definite and predictive application of the statistical-equilibrium theory.

We follow the development given in Turkington *et al.* (2001) (see also Majda and Wang 2006). Namely, we consider two specific midlatitude bands on Jupiter, for which the latitude lies in the range from Θ_- to Θ_+. Each band is selected to contain two belts (cyclonic regions) and two zones (anticyclonic regions). The southern-hemisphere band lies between 36.6° S and 13.7° S; the northern-hemisphere band lies between 23.1° N to 42.5° N. (On both these domains, we use northern-hemisphere sign conventions to make them isomorphic problems.)

The characteristic length scale L is defined to be half the channel width, $L = |\Theta_+ - \Theta_-| r_0/2$, where $r_0 = 7 \times 10^7$ m represents the radius of Jupiter. The characteristic velocity scale V is defined as the r.m.s. velocity of the centered Limaye profile, $\tilde{u}$, which is the permanent zonally averaged wind field in the upper layer (Limaye 1986). We pose a nondimensionalized problem on the domain $-1 < y < +1$, with periodicity in x (with period 4). Jupiter's angular speed of rotation $\Omega = 1.76 \times 10^{-4}\,\mathrm{s}^{-1}$ determines f_0 and β_0 and hence the dimensionless parameter β and the Rossby number, which is $\epsilon \approx 0.014$ for the southern-hemisphere domain and $\epsilon \approx 0.021$ for the northern-hemisphere domain. The dimensionless deformation scale is given by $\lambda = c/f_0 L$, where $c = 454\,\mathrm{s}^{-1}$ is the estimated gravity-wave speed on Jupiter (Dowling 1995*b*). This estimate gives roughly $\lambda \approx 0.2$ in both the northern and the southern bands.

Following Dowling (1995*b*), we infer the streamfunction ψ_2 for the lower layer from the observed zonally averaged Limaye profile in the upper layer, $\tilde{u}$. Specifically, we define ψ_2 so that the streamfunction $\tilde{\psi}$ of the observed zonal shear $\tilde{u}$ coincides with the minimum-enstrophy state with energy and circulation derived from $\tilde{u}$. Using the QG mean-field equation, this means that ψ_2 is derived from the identity

$$\frac{d^2\tilde{\psi}}{dy^2} - \lambda^{-2}\tilde{\psi} + \beta y + \lambda^{-2}\psi_2 = \theta\tilde{\psi} - \mu\,,$$

with multipliers θ and μ determined by the energy and circulation constraints, respectively. This choice puts the QG equilibrium state at the critical value with respect to Arnold's stability criterion, in agreement with Dowling's analysis of observations. We stress that, since the Limaye profile of east–west jets has a different structure in the northern and the southern domains (for reasons that we do not know), the effective bottom topography is different for these two domains.

It remains to justify a choice of the prior distribution $\pi(dq)$ for Jupiter's weather layer. Fortunately there also exists evidence that the mean PV–streamfunction relation in the band containing the GRS is roughly linear (Dowling 1995*b*). This suggests that in the statistical-equilibrium theory, the prior distribution should be nearly Gaussian. But, in addition, an analysis of satellite observations shows that the fine-grained turbulence in the weather layer is not symmetric, but, rather, the most intense eddies are anticyclones, even in belts where the overall circulation is cyclonic (Ingersoll *et al.* 2000). Thus, we are led to impose an anticyclonically skewed, nearly Gaussian, distribution on the fine-grained PV fluctuations. That is, we take a prior $\pi_s(dq)$ satisfying

$$\int q\pi_s(dq) = 0\,, \qquad \int q^2\pi_s(dq) = 1\,, \qquad \int q^3\pi_s(dq) = 2s < 0$$

for small negative s. An appropriately scaled and centered gamma distribution is used in Turkington *et al.* (2001). Then the mean-field equation takes the form

$$\bar{q} = \frac{1}{s}\left[\frac{1}{1 - s(\theta\psi - \mu)} - 1\right] = (\theta\psi - \mu) + s(\theta\psi - \mu)^2 + \cdots\,,$$

with the nonlinearity scaled by the skewness.

This choice of prior distribution has a physical basis. On a rapidly rotating planet, horizontal divergence produces anticyclonic vorticity. Such divergence results from upward flow that expands into the weather layer. The modern explanation of the energy source for the wind field in the weather layer is that it is driven by thermal convection from below—in fact, thunderstorms and lightning have been observed (Ingersoll *et al.* 2000). Since the flows generated by these convective plumes are small in horizontal extent compared with the scale of the bands, they are most naturally included in the statistical description of the small-scale eddies. Thus, the prior distribution $\pi_s(dq)$ models the effect of thermal forcing on the weather layer's turbulent flow field. The skewness parameter s is thus a surrogate for the magnitude of this forcing and is the only adjustable parameter in the model.

Finally, we are ready to conduct a numerical experiment. In either the northern or the southern domain, we first set the parameters and the constraint values in the maximum-entropy principle so that for $s = 0$ (corresponding to quadratic enstrophy) the observed Limaye profile (zonally averaged jet structure) is the microcanonical equilibrium state. This is necessary because, otherwise, there would not be a consistent procedure by which to determine the unobserved lower-layer flow. We then compute branches of solutions with $s < 0$ by continuation, using an algorithm that is an adaptation of the one described in Section 2.6. In other words, we gradually increase the thermal forcing that drives the QG turbulence, which in turn organizes into coherent states at large scales. A remarkable result ensues. In the southern domain, an anticyclonic oval vortex vortex emerges centered at a latitude that corresponds to 23° S, which is the latitude of the Great Red Spot. In the northern domain, the predicted flow remains a zonal shear for all (admissible) values of s. The streamline patterns for these flows are given in the figures in Turkington *et al.* (2001). Thus, the statistical-equilibrium model with an inferred effective bottom topography and a physically motivated prior distribution does predict the formation of a GRS-type vortex in the southern band and, equally significantly, does not predict any vortex in the northern band.

This result is a particularly sharp application of the general theory. It shows that the location of the GRS is dependent on the jet structure of the lower-layer flow. It also demonstrates that a prior distribution on fine-grained PV is able to capture the key features of small-scale turbulence energized by thermal convection. Most importantly, it exemplifies the use of the statistical-equilibrium theory as a link between the modeled statistical properties on the fine scales and the predicted coherent structures on the coarse scales.

Moreover, one can attempt to fashion a prior distribution that even more closely matches the small-scale dynamics than the gamma distribution chosen in Turkington *et al.* (2001). For instance, Bouchet and Sommeria (2002) argue for a PV distribution in the Miller–Robert–Sommeria theory that has bounded support, with bounds related to the maximum vorticity differences generated by the thermal convection process. Crucially, their prior distribution also has an anticyclonic skewness. When the theory is implemented with such a prior distribution, it is able to capture some of the features of the internal structure of the GRS vortex, especially the strong circular jet that forms the boundary of the vortex and surrounds its relatively quiescent core. And, the role of the deformation radius is revealed in this structure, in the sense that the width of the boundary jet is approximately λ.

Thus, our lectures conclude with a very satisfying outcome—a striking physical application of an elegant and complete mathematical theory.

References

Arnold, V. I. (1965). Conditions for nonlinear stability of stationary plane curvilinear flows of an ideal fluid. *Dokl. Akad. Nauk. SSSR*, **162**, 975–978.

Arnold, V. I. (1978). *Mathematical Methods of Classical Mechanics.* Springer-Verlag.

Batchelor, G. K. (1967). *An Introduction to Fluid Dynamics.* Cambridge University Press.

Berdichevsky, V. L. (1995). Statistical mechanics of point vortices. *Phys. Rev. E*, **51**, 4432–4452.

Boucher, C., Ellis, R. S., and Turkington, B. (2000). Derivation of maximum entropy principles in two-dimensional turbulence via large deviations. *J. Stat. Phys.*, **98**,

1235–1278.

Bouchet, F. and Sommeria, J. (2002). Emergence of intense jets and Jupiter's Great Red Spot as maximum-entropy structures. *J. Fluid Mech.*, **464**, 165–207.

Bretherton, F. P. and Haidvogel, D. B. (1976). Two-dimensional turbulence over topography. *J. Fluid Mech.*, **78**, 129–154.

Caglioti, E., Lions, P. L., Marchioro, C., and Pulvirenti, M. (1992). A special class of stationary flows for two dimensional Euler equations: A statistical mechanics description. *Commun. Math. Phys.*, **143**, 501–525.

Caglioti, E., Lions, P. L., Marchioro, C., and Pulvirenti, M. (1995). A special class of stationary flows for two dimensional Euler equations: A statistical mechanics description. Part ii. *Commun. Math. Phys.*, **174**, 229–260.

Carnevale, G. F. and Frederiksen, J. S. (1987). Nonlinear stability and statistical mechanics of flow over topography. *J. Fluid Mech.*, **175**, 157–181.

Chavanis, P.-H. and Sommeria, J. (1998). Classification of robust isolated vortices in two-dimensional hydrodynamics. *J. Fluid Mech.*, **356**, 259–296.

Chorin, A. J. (1994). *Vorticity and Turbulence.* Springer-Verlag.

Costeniuc, M., Ellis, R. S., Touchette, H., and Turkington, B. (2005). The generalized canonical ensemble and its universal equivalence with the microcanonical ensemble. *J. Stat. Phys.*, **119**, 1283–1329.

Costeniuc, M., Ellis, R. S., Touchette, H., and Turkington, B. (2006). Generalized canonical ensembles and ensemble equivalence. *Phys. Rev. E*, **73**, 026105.

Dauxois, T., Ruffo, S., Arimondo, E., and Wilkens, M. (eds.) (2002). *Dynamics and Thermodynamics of Systems with Long Range Interactions*, Lecture Notes in Physics, **602**. Springer-Verlag.

DiBattista, M., Majda, A., and Turkington, B. (1998). Prototype geophysical vortex structures via large-scale statistical theory. *Geophys. Astrophys. Fluid Dyn.*, **89**, 235–283.

Dowling, T. (1995*a*). Dynamics of Jovian atmospheres. *Ann. Rev. Fluid Mech.*, **27**, 293–334.

Dowling, T. (1995*b*). Estimate of Jupiter's deep zonal-wind profile from Shoemaker–Levy 9 data and Arnold's second stability criterion. *Icarus*, **117**, 439–442.

Dowling, T. and Ingersoll, A. P. (1988). Potential vorticity and layer thickness variations in the flow around Jupiter's Great Spot and White Oval BC. *J. Atmos. Sci.*, **46**, 1380–1396.

Ellis, R. S. (1985). *Entropy, Large Deviations, and Statistical Mechanics.* Springer-Verlag. Reprinted in Classics of Mathematics series, 2006.

Ellis, R. S. (1995). An overview of the theory of large deviations and applications to statistical mechanics. Scand. Actuarial J., **1**, 97–142.

Ellis, R. S., Haven, K., and Turkington, B. (2000). Large deviation principles and complete equivalence and nonequivalence results for pure and mixed ensembles. *J. Stat. Phys.*, **101**, 999–1064.

Ellis, R. S., Haven, K., and Turkington, B. (2002*a*). Nonequivalent statistical equilibrium ensembles and refined stability theorems for most probable flows. *Nonlinearity*, **15**, 239–255.

Ellis, R. S., Turkington, B., and Haven, K. (2002*b*). Analysis of statistical equilibrium models of geostrophic turbulence. *J. Appl. Math. Stoch. Anal.*, **15**, 341–361.

Eyink, G. L. and Spohn, H. (1993). Negative-temperature states and large-scale, long-lived vortices in two-dimensional turbulence. *J. Stat. Phys.*, **70**, 833–886.

Holloway, G. (1986). Eddies, waves, circulation and mixing: Statistical geofluid mechanics. *Ann. Rev. Fluid Mech.*, **18**, 91–147.

Holm, D., Marsden, J. E., Ratiu, T., and Weinstein, A. (1985). Nonlinear stability of fluid and plasma equilibria. *Phys. Rep.*, **123**, 1–116.

Ingersoll, A. P., Gierasch, P. J., Banfield, C., Vasavada, A. R., and the Galileo Imaging Team (2000). Moist convection as an energy source for the large-scale motions in Jupiter's atmosphere. *Nature*, **403**, 630–632.

Kiessling, M. K.-H. (1993). Statistical mechanics of classical particles with logarithmic interactions. *Commun. Pure Appl. Math.*, **46**, 27–56.

Kiessling, M. K.-H. and Lebowitz, J. L. (1997). The micro-canonical point vortex ensemble: Beyond equivalence. *Lett. Math. Phys.*, **42**, 43–56.

Kraichnan, R. H. (1975). Statistical dynamics of two-dimensional flows. *J. Fluid Mech.*, **67**, 155–175.

Kraichnan, R. H. and Montgomery, D. (1980). Two-dimensional turbulence. *Rep. Prog. Phys.*, **43**, 547–619.

Limaye, S. S. (1986). Jupiter: New estimates of the mean zonal flow at the cloud level. *Icarus*, **65**, 335–352.

Lungren, T. S. and Pointin, Y. B. (1977). Statistical mechanics of two-dimensional vortices. *J. Stat. Phys.*, **17**, 323–355.

Majda, A. J. and Bertozzi, A. (2001). *Vorticity and Incompressible Flow.* Cambridge University Press.

Majda, A. J. and Wang, X. (2006). *Nonlinear Dynamics and Statistical Theories for Basic Geophysical Flows.* Cambridge University Press.

Marchioro, C. and Pulvirenti, M. (1994). *Mathematical Theory of Incompressible Nonviscous Fluids.* Springer-Verlag.

Marcus, P. S. (1993). Jupiter's Great Red Spot and other vortices. *Ann. Rev. Astrophys.*, **31**, 523–573.

Michel, J. and Robert, R. (1994). Large deviations for Young measures and statistical mechanics of infinite-dimensional dynamical systems with conservation law. *Commun. Math. Phys.*, **159**, 195–215.

Miller, J. (1990). Statistical mechanics of Euler equations in two dimensions. *Phys. Rev. Lett.*, **65**, 2137–2140.

Miller, J., Weichman, P., and Cross, M. C. (1992). Statistical mechanics, Euler's equations, and Jupiter's red spot. *Phys. Rev. A*, **45**, 2328–2359.

Montgomery, D. and Joyce, G. (1974). Statistical mechanics of negative temperature states. *Phys. Fluids*, **17**, 1139–1145.

Montgomery, D., Matthaeus, W. H., Martinez, D., and Oughton, S. (1992). Relaxation in two dimensions and the sinh-Poisson equation. *Phys. Fluids A*, **4**, 3–6.

Onsager, L. (1949). Statistical hydrodynamics. *Nuovo Cimento*, **6**, 279–287. Supplemento, no. 2 (Convegno Internazionale di Meccanica Statistica).

Pedlosky, J. (1987). *Geophysical Fluid Dynamics*, second edition. Springer-Verlag.

Rhines, P. (1979). Geostrophic turbulence. *Ann. Rev. Fluid Mech.*, **11**, 404–441.

Robert, R. (1991). A maximum-entropy principle for two-dimensional perfect fluid dynamics. *J. Stat. Phys.*, **65**, 531–553.

Robert, R. and Sommeria, J. (1991). Statistical equilibrium states for two-dimensional flows. *J. Fluid Mech.*, **229**, 291–310.

Salmon, R., Holloway, G., and Hendershott, M. C. (1976). The equilibrium statistical mechanics of simple quasi-geostrophic models. *J. Fluid Mech.*, **75**, 691–701.

Segre, E. and Kida, S. (1998). Late states of incompressible 2D decaying vorticity fields. *Fluid Dyn. Res.*, **23**, 89–112.

Smith, R. A. and O'Neil, T. M. (1990). Nonaxisymmetric thermal equilibria of a cylindrically bounded guiding center plasma or discrete vortex system. *Phys. Fluids B*, **2**, 2961–2975.

Turkington, B. (1987). On the evolution of a concentrated vortex in an ideal fluid. *Arch. Rat. Mech. Anal.*, **97**, 75–87.

Turkington, B. (1999). Statistical equilibrium measures and coherent states in two-dimensional turbulence. *Commun. Pure Appl. Math.*, **52**, 781–809.

Turkington, B. and Whitaker, N. (1996). Statistical equilibrium computations of coherent structures in turbulent shear layers. *SIAM J. Sci. Comput.*, **17**, 1414–1433.

Turkington, B., Majda, A., Haven, K., and DiBattista, M. (2001). Statistical equilibrium predictions of jets and spots on Jupiter. *Proc. Natl. Acad. Sci. USA*, **98**, 12346–12350.

Vallis, G. K. (2006). *Atmospheric and Oceanic Fluid Dynamics: Fundamentals and Large-Scale Circulation.* Cambridge University Press.

Whitaker, N. and Turkington, B. (1994). Maximum entropy states for rotating vortex patches. *Phys. Fluids*, **6**, 3963–3973.

5

Statistical mechanics of turbulent von Kármán flows: Theory and experiments

Bérengère DUBRULLE and Gilles COLLETTE

Service de Physique de l'État Condensé, DSM, CEA Saclay, CNRS URA 2464, 91191 Gif-sur-Yvette, France

5.1 Introduction

In a turbulent flow, the number of degrees of freedom scales like $Re^{9/4}$ and can reach 10^{22} for atmospheric-like flows. This is beyond the present capacity of computers. For example, numerical simulations of a von Kármán turbulent flow at $Re = 10^6$, a standard laboratory flow used for turbulence studies (see below), would require resolutions of the order of 10^{15} grid points and integration times of the order 6×10^5 years of cpu with current computers. This conclusion justifies the introduction of turbulence models, to reduce the number of degrees of freedom and make turbulence amenable to numerical simulation or theoretical understanding. This goal cannot be reached unless the different components of turbulence and their interactions are identified.

Turbulence being intrinsically a stochastic process, it can be decomposed into two components: the mean flow, and fluctuations around it. A good turbulence model should therefore be able to predict both the structure of the mean flow and its influence on and through fluctuations, within a reduced number of degrees of freedom. This kind of information is typically provided by statistical mechanics. Can we adapt statistical methods to deal with the turbulence problem?

For a long time, the answer has been negative, despite early attempts by Onsager (1949). In the last decade, the answer has been positive for (rather artificial) 2D or quasi-2D situations (Chavanis 2003), thanks to the theoretical breakthroughs of Robert and Sommeria (1991) and Miller (1990). More recently, Leprovost *et al.* have shown that the 2D formalism could actually be extended to a typical 2D 1/2 situation, of an axisymmetric flow (Leprovost *et al.* 2006). Using a thermodynamic approach, they predict the shape of stationary (mean-flow) solutions, a prediction that has been tested and confirmed experimentally for a turbulent von Kármán flow by Monchaux *et al.* (2006). In the present lecture course, we show how to extend these computations one step further by considering also fluctuations. We use mean-field theory to link these fluctuations to the mean flow and its response to perturbations. This provides an estimate of the fluctuation level as a function of the thermodynamic coefficients, namely the effective temperature, in a way reminiscent of the Fluctuation–Dissipation Theorem. These predictions are tested for a turbulent von Kármán flow using particle image velocimetry and laser Doppler velocimetry measurements.

5.2 Hypotheses and theoretical framework

5.2.1 Turbulence, Navier–Stokes equations, and classical statistical mechanics

A turbulent flow is described by the Navier–Stokes equations

$$\partial_t \mathbf{u} + \mathbf{u} \cdot \nabla \mathbf{u} = -\frac{1}{\rho_0} \nabla P + \nu \nabla^2 \mathbf{u} + f\,, \tag{5.1}$$

where, $\mathbf{u}$ is the velocity, P is the pressure, ν is the viscosity, and f is a forcing. In the absence of forcing, the velocity decays to zero owing to the dissipation, so that turbulence occurs in an intrinsically out-of-equilibrium system. In the sequel, we focus on the simplest situation, where forcing and dissipation equilibrate on average, so that stationary solutions can develop. The goal of the present paper is to describe these stationary solutions and the fluctuations around these solutions using tools borrowed from classical statistical mechanics. Specifically, we are going to follow what we call the "Onsager program": select a Hamiltonian system, perform equilibrium or near-equilibrium statistical mechanics, and compute its equilibria. We compare these equilibria with actual experimental turbulent flow, and show that they resemble each other. In other words, it seems that stationary states of an out-of equilibrium system can be described by equilibrium states of a well-chosen Hamiltonian system.

5.2.2 Stationary Navier–Stokes solutions vs. solutions of Euler equations

Since forcing and dissipation equilibrate on average for stationary solutions of the Navier–Stokes equations, it seems natural to consider this limiting case first, in our quest for a framework suitable for classical statistical mechanics. In such a limit, we get the Euler equations

$$\partial_t \mathbf{u} + \mathbf{u} \cdot \nabla \mathbf{u} = -\frac{1}{\rho_0} \nabla P \,. \tag{5.2}$$

This is indeed a Hamiltonian system as long as one considers regular solutions, such as those based on finite Galerkin expansions. In 2D, the consideration of Euler solutions to describe Navier–Stokes stationary solutions is well accepted, based on the observation that the vorticity cannot blow up and that the limit $\nu \to 0$ is usually well behaved under a reasonable regularity hypothesis. In 3D, this hypothesis is still controversial, ever since Onsager. One major problem is that one cannot exclude the possibility that vorticity blows up in 3D, which would make the limit $\nu \to 0$ singular. A signature of this is the famous $4/5$ law of homogeneous turbulence, which links energy dissipation to the third moment of the velocity increments, independently of any viscosity. For this reason, Onsager suggested that one should consider weak solutions of the Euler equations to describe stationary states of the Navier–Stokes equations, thereby allowing a finite amount of energy dissipation even in the absence of viscosity. This suggestion was illustrated recently in an elegant way by Duchon and Robert (2000). However, weak solutions are not directly amenable to the methods of classical statistical physics, and nobody has yet succeeded in following to the end Onsager's suggestion. In the present case, we overcome this difficulty by considering only regular solutions of the Euler equations. The upside is that we deal with a Hamiltonian system. The downside is that we may have lost any connection with actual turbulence (Robert 2004).

5.2.3 The Euler systems and conservation laws

The Euler equations for regular solutions are characterized by a number of conservation laws, which depend on the geometry and the dimension of the system. In 2D, for example, conservation laws exist for the kinetic energy $E = \int u^2 \, d\mathbf{x}$; the enstrophy $\Omega = \int \omega^2 \, d\mathbf{x}$, where $\omega = \nabla \times u$ is the vorticity; and, more generally, any function of the vorticity. In 3D, the generic conserved quantities are the kinetic energy E and the helicity $H = \int u \cdot \omega \, d\mathbf{x}$. Additional conservation laws are possible in the presence of additional symmetries, such as axisymmetry (see Leprovost *et al.* 2004). However, in the sequel, we focus on systems in which the only conserved quantities are the energy E and the helicity H. This property holds, for example, for the Beltrami flow, a very simple solution of the Euler equations, such that the velocity and the vorticity are aligned everywhere in the flow. By extension, we therefore call our Hamiltonian system with conservation of energy and helicity the "Beltrami system."

5.2.4 Euler equation in the axisymmetric case

Equations. In the axisymmetric case, the Euler equations take the form

$$\frac{1}{r}\partial_r(ru_r) + \partial_z u_z = 0\,, \tag{5.3}$$

$$\partial_t u_r + u_r\partial_r u_r + u_z\partial_z u_r - \frac{u_\theta^2}{r} = -\frac{1}{\rho}\partial_r p\,, \tag{5.4}$$

$$\partial_t u_\theta + u_r\partial_r u_\theta + u_z\partial_z u_\theta + \frac{u_r u_\theta}{r} = 0\,, \tag{5.5}$$

$$\partial_t u_z + u_r\partial_r u_z + u_z\partial_z u_z = -\frac{1}{\rho}\partial_z p\,, \tag{5.6}$$

where (u_r, u_θ, u_z) are the velocity components in a cylindrical reference frame (r, θ, z). It was shown in Leprovost *et al.* (2006) that these axisymmetric, incompressible Euler equations can be rewritten in a simplified form in terms of σ, ξ, and ψ, where $\sigma = ru_\theta$ is the angular momentum, ξ is related to the azimuthal component of the vorticity by $\xi = \omega_\theta/r$, and ψ is the streamfunction associated with the poloidal component of the velocity, so that $\mathbf{u} = u_\theta\hat{\mathbf{e}}_\theta + \nabla\times((\psi/r)\hat{\mathbf{e}}_\theta)$. Note that $u_r = -\partial_z\psi/r$ and $u_z = \partial_r\psi/r$. The axisymmetric Euler equations can then be recast as

$$\partial_t\sigma + \{\psi, \sigma\} = 0\,, \tag{5.7}$$

$$\partial_t\xi + \{\psi, \xi\} = \partial_z\left(\frac{\sigma^2}{4y^2}\right), \tag{5.8}$$

$$\Delta_*\psi \equiv \frac{1}{2y}\partial_{zz}\psi + \partial_{yy}\psi = -\xi\,, \tag{5.9}$$

where $y = r^2/2$, $\{\psi, \phi\} = \partial_y\psi\partial_z\phi - \partial_z\psi\partial_y\phi$ are the Poisson brackets, and Δ_* is a pseudo-Laplacian.

Stationary states. A few general remarks are in order regarding this special case.

(i) One sees from eqn (5.7) that the angular momentum is conserved along the fluid lines, and can only be mixed through the Euler dynamics. This is the analog of vorticity mixing in 2D, and justifies the introduction of a mixing entropy (see below).

(ii) The stationary solutions of the Euler axisymmetric equations (5.7)–(5.9) have been established in Leprovost *et al.* (2006). They satisfy

$$\sigma = f(\psi)\,, \tag{5.10}$$

$$-\Delta_*\psi = \xi = \frac{f(\psi)f'(\psi)}{2y} + g(\psi)\,, \tag{5.11}$$

where f and g are arbitrary functions. We see therefore that the axisymmetric Euler equations admit an infinite number of stationary solutions, specified by the two functions f and g. To select which solutions are appropriate to our system, we use statistical mechanics, which will provide specific forms for f and g as we shall see below.

Conservation laws. Axisymmetric inviscid flows satisfy an infinite number of conserved quantities, which can be found in Leprovost *et al.* (2006). In the present course, we consider only two of them, namely the kinetic energy E and the helicity H. In the axisymmetric case, these two quantities can be written simply as

$$E = \frac{1}{2}\int (u_r^2 + u_\theta^2 + u_z^2) r\, dr\, dz = \frac{1}{2}\int \xi\psi\, dy\, dz + \frac{1}{4}\int \frac{\sigma^2}{y} dy\, dz\,, \tag{5.12}$$

$$H = \int (u_r\omega_r + u_\theta\omega_\theta + u_z\omega_z) r\, dr\, dz = 2\int \xi\sigma\, dy\, dz\,. \tag{5.13}$$

5.3 Statistical mechanics of Euler–Beltrami system

5.3.1 Basic setup

In the sequel, we consider the following Euler–Beltrami system, with a velocity field $u = \overline{u} + u'$, where $\overline{u}$ is the averaged velocity field. We assume, furthermore, that $\overline{u}$ is axisymmetric and that the total system evolves so as to conserve the averaged energy $E = \int \overline{u^2}\, d\mathbf{x}$ and the averaged helicity $H = \int \overline{u \cdot \omega}\, d\mathbf{x}$. We can therefore introduce the probability of fluctuations $\rho(u', \omega', x)$, i.e. the probability to observe a fluctuation of velocity u' and vorticity ω' at point x, and write the conserved quantities as

$$E = \frac{1}{2}\int \overline{\xi\psi}\, dy\, dz + \frac{1}{4}\int \frac{\overline{\sigma}^2}{y} dy\, dz + \frac{1}{4\pi}\int \rho(u', \omega', x) u'^2\, d\theta\, dy\, dz\, du'\, d\omega'\,, \tag{5.14}$$

$$H = \int u \cdot \omega = \int \overline{\xi}\overline{\sigma}\, dy\, dz + \frac{1}{2\pi}\int \rho(u', \omega', x) u' \cdot \omega'\, d\theta\, dy\, dz\, du'\, d\omega'\,. \tag{5.15}$$

In the sequel, we consider a simple model, to compute the average flow and the fluctuations so as to satisfy the energy and helicity constraints. Following Onsager's program, we introduce into our model a suitable entropy, and maximize it under the energy and helicity conservation constraints, so as to obtain the Gibbs states. From these Gibbs states, we derive relations for the mean flow and for the fluctuations.

5.3.2 Mean-field theory

When we are studying first- or second-order fluctuations in statistical mechanics, a convenient launch point is provided by a mean-field theory. This method is traditionally very efficient in systems of high dimensionality, or with long-range forces, a condition met with in fluid mechanics. In our system, we have at our disposal a privileged direction, the azimuthal direction, which is the direction of symmetry. In a real turbulent system, we have been able to observe that the fluctuations in the azimuthal direction are much larger than in the other directions. It is therefore natural, as a first elementary step, to consider a model in which only azimuthal fluctuations are considered, and to suppose that the fluctuations in the other directions are simply frozen. This is the core of the mean-field theory that we describe now. Less simple models, taking into account fluctuations in other directions, are described in Naso *et al.* (2009).

Let us therefore assume that the fluctuations are mainly in the azimuthal direction, so that $u'_r = u'_z = 0$. In that case, u' is made only from fluctuations of u'_θ or, equivalently, fluctuations of σ. The corresponding probability distribution function $\rho_\xi(\mathbf{x}, \eta)$ is the probability that $\sigma = \eta$ at position $\mathbf{x} = (y, z)$. As for the velocity field in the r, z-direction, this is purely axisymmetric and can be described by $\xi = \overline{\xi}$ and $\psi = \overline{\psi}$. To proceed further, we need to introduce an entropy. If the fluctuations of σ were purely axisymmetric, it would be natural to introduce the mixing entropy

$$S[\rho] = -\int \rho \ln \rho \, dy \, dz \, d\eta \, . \tag{5.16}$$

Indeed, since σ is conserved along the field lines in the axisymmetric case, this entropy will increase under the dynamics until the flow has achieved a stationary state. In our case, the fluctuations of σ are not necessarily axisymmetric, but we can observe that $S[\rho]$ defined as in eqn (5.16) is the neginformation (the opposite of the information). Maximizing this neginformation under given constraints is just the simplest procedure we can adopt to compute the fluctuations, according to information theory and its application to statistical mechanics by Jaynes (1957). We therefore adopt this procedure.

Using our above hypothesis, it can be checked that the conserved quantities can be written as

$$\overline{E} = \int \frac{\overline{\xi}\,\overline{\psi}}{2} dy \, dz + \int \frac{\eta^2}{4y} \rho \, dy \, dz \, d\eta \, , \tag{5.17}$$

$$\overline{H} = 2 \int \overline{\xi} \eta \rho \, dy \, dz \, d\eta \, . \tag{5.18}$$

One sees that these two quantities depend only on ρ and $\overline{\xi}$ (since $\overline{\psi}$ is directly slaved to $\overline{\xi}$). Therefore, the most probable distribution at metaequilibrium is obtained by maximizing over ρ and $\overline{\xi}$ the mixing entropy $S[\rho]$ at fixed $\overline{E}$, $\overline{H}$ and with the local normalization $\int \rho \, d\eta = 1$. Introducing Lagrange multipliers, one can write the variational principle as

$$\delta S - \beta_\xi \, \delta \overline{E} - \mu_\xi \, \delta \overline{H} = 0 \, . \tag{5.19}$$

Here, we have introduced two Lagrange parameters β and μ, and labeled them with a ξ to recall that this is a procedure performed assuming that the field in the r, z-direction is not fluctuating or, equivalently, that ξ is smooth.

The variations on $\overline{\xi}$ imply

$$\beta_\xi \overline{\psi} + \mu_\xi \overline{\sigma} = 0 \, , \tag{5.20}$$

while the variations on ρ yield the Gibbs state

$$\rho_\xi(\mathbf{x}, \eta) = \frac{1}{Z_\xi} e^{-\beta_\xi \eta^2/4y - \mu_\xi \overline{\xi} \eta} \, , \tag{5.21}$$

where the "partition function" is determined via the normalization condition

$$Z_\xi(\mathbf{x}) = \int e^{-\beta_\xi \eta^2/4y - \mu_\xi \overline{\xi} \eta} d\eta \, . \tag{5.22}$$

Note that $\rho_\xi(\mathbf{x}, \eta)$, the distribution of the fluctuations of σ, is Gaussian, with

$$\overline{\sigma} = -\frac{2y\mu_\xi\overline{\xi}}{\beta_\xi}, \tag{5.23}$$

$$\sigma_2 \equiv \overline{\sigma^2} - \overline{\sigma}^2 = \frac{2y}{\beta_\xi}. \tag{5.24}$$

5.3.3 Comments

Considering the equilibrium solution given by eqns (5.20) and (5.23), one sees that they are of the form of eqns (5.10) and (5.11) with f linear and where $f(x) = \lambda x$, i.e. $\lambda = -\mu_\xi/\beta_\xi$ and $g = 0$. This means that the equilibrium state is a stationary state of the Euler equation. Furthermore, with such a simple form, one can check that the vorticity and velocity are aligned everywhere, i.e. $\omega = \lambda u$. Such a simple state is called a Beltrami state. We can also provide an interesting interpretation of our fluctuation relation (5.24), which can easily be recast into

$$\overline{u_\theta^2} - \overline{u_\theta}^2 = \frac{1}{\beta_\xi}, \tag{5.25}$$

predicting uniformity of azimuthal velocity fluctuations. Equation (5.25) shows that the azimuthal velocity fluctuations define an effective statistical temperature $1/\beta_\xi$. The equations may be regarded as formally analogous to fluctuation–dissipation relations (FDRs) since they link fluctuations and temperature. These predictions enable the measurements of effective temperatures of turbulence through the fluctuations of u_θ in a Beltrami flow. Because variances are positive, β_ξ is always positive, unlike in the 2D situation, where the temperature can be negative In contrast, μ_ξ can take positive or negative values, depending on the sign of the helicity.

The analogy between our predictions and FDRs can actually be pushed forward. Indeed, another possible way to derive eqn (5.24) is to introduce, as in classical statistical mechanics, a partition function Z_ξ describing the Beltrami equilibrium state in the mean-field approximation:

$$\overline{\sigma^2} - \overline{\sigma}^2 = \frac{1}{\mu_\xi^2}\frac{\delta^2 \log Z_\xi}{\delta\overline{\xi}^2} = -\frac{1}{\mu_\xi}\frac{\delta\overline{\sigma}}{\delta\overline{\xi}}. \tag{5.26}$$

Formally, the mathematical objects $\delta\overline{\sigma}/\delta\overline{\xi}$ can be seen as response functions. From this point of view, eqn (5.26) again reflects a formal analogy with FDRs since another classical way to write it down is to link the fluctuations of a field to its response to a perturbation.

5.4 Experimental study of a turbulent von Kármán flow

We now turn to the confrontation of these theoretical results with real turbulence. Namely, we would like to check whether the stationary solutions of a real turbulent system have anything to do with the equilibrium states of our idealized Beltrami–Euler system. For this, we shall consider both average states and fluctuations.

5.4.1 Experimental setup

Our experimental setup consisted of a Plexiglas cylinder (radius $R_c = 100$ mm) filled with water. The fluid was mechanically stirred by a pair of impellers driven by two independent motors in exact counterrotation. The resulting flow belonged to the von Kármán class of flows, with a mean flow divided into two toric cells separated by an azimuthal shear layer. We defined the Reynolds number as $Re = UL/\nu = 2\pi F R_c^2/\nu$, where F is the impeller frequency and ν the fluid viscosity. By rotating the impellers at speeds from 2 to 8 Hz and using glycerol at different dilution ratios, we could achieve Reynolds numbers from 100 to 500 000. Our two models of impellers, TM60 and TM73, were flat discs of respective radii $0.925R_c$ and $0.75R_c$, fitted with radial blades of height $0.2R_c$ and respective curvatures $0.50R_c$ and $0.925R_c$. The inner faces of the discs were $1.8R_c$ apart. Different forcings were associated with the convex or concave face of the blades moving forward, denoted in the sequel by the senses (+) and (−). From now on, all physical quantities and equations will be considered in their nondimensionalized form using F and R_c.

5.4.2 Technical details

The velocity measurements presented here were performed either with a DANTEC laser Doppler velocimetry (LDV) system or with a stereoscopic particle image velocimetry (SPIV) system. Before describing the results it is useful to give a few technical details, so that we can realize the amount of the work needed to fulfill the innocent-looking job of comparing theory with experiments.

SPIV. The SPIV system (Monchaux 2007) was supplied by DANTEC Dynamics. The cylinder was mounted inside a water-filled square Plexiglas container in order to reduce optical distortion. Two digital cameras were aimed at a meridian plane of the flow through two perpendicular faces of the square container, giving a 2D, three-component velocity field map. Correlation calculations were done on 32×32 pixels^2 windows with 50% overlap. As a result, each velocity was averaged over a 4.16×4.16 mm^2 window over the 1.5 mm laser sheet thickness. The spatial resolution was 2.08 mm. A basic measurement was a set of 5000 acquisitions at a rate of 4 images per second. From this set, we computed the time averages and fluctuations of the three velocity components.

LDV. The LDV data provided only the axial and azimuthal velocity components on a 170 point grid covering half a meridian plane, through a time series of about 200 000 randomly sampled values at each grid point. From this time series, it was straightforward to get time-averaged axial and azimuthal velocities. The remaining radial component of the mean velocity field could then be obtained using incompressibility and axisymmetry: this procedure was later validated through direct measurements of the radial velocity with a particle image velocimetry (PIV) system on the same flow. Figure 5.1 provides examples of the mean and instantaneous flows computed using the PIV and LDV systems.

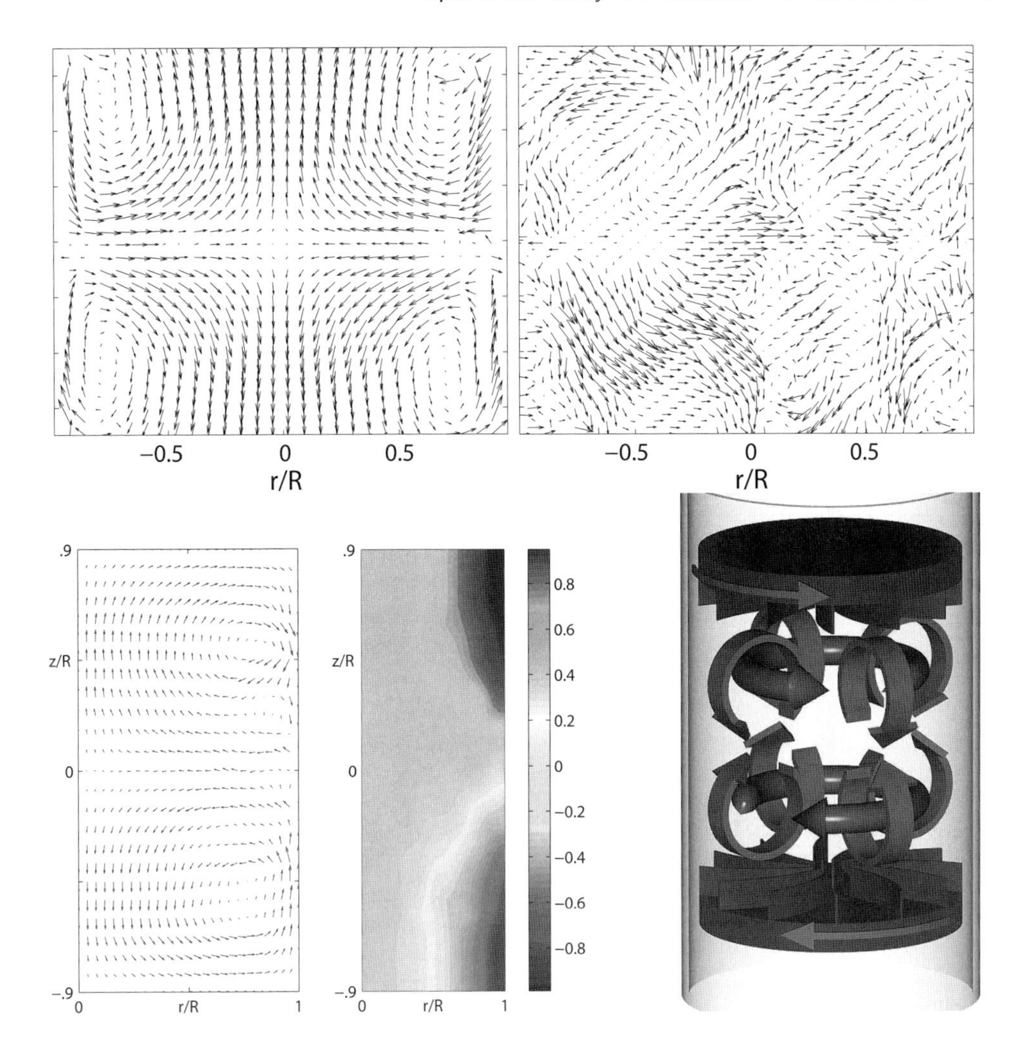

Fig. 5.1 Top: PIV measurements. The left part shows the poloidal velocity for the mean flow, and the right part shows the poloidal velocity for the instantaneous flow. Bottom: LDV measurements. The left part shows the velocity field for $Re = 2.5\times10^5$, using colored contours for the azimuthal component and an arrow representation for the poloidal part, and the right part displays a sketch of the two toric recirculating cells.

5.4.3 Study of mean flows

In the first stage, we compared experimental average flows with theoretical equilibrium flows. For this, we plotted σ as a function of ψ. If the average coincides with an equilibrium state of the axisymmetric Euler equation, this should define a function f in eqn (5.10). This function should, furthermore, be linear if the turbulent average state is an equilibrium solution of the axisymmetric Euler–Beltrami system. Once f

has been defined, one can then complete the check by searching for the function g defining the Euler equilibrium state. For this, we plotted $\xi - ff'/r^2$ as a function of ψ (see Section 5.3.3). If a function is defined, this is g in eqn (5.11). In the case of Euler–Beltrami equilibrium, $g = 0$. After these considerations, let us first apply naively our procedure over an arbitrary velocity field. The result is provided in Fig. 5.2. One sees that when the whole flow is considered, there is significant scatter and it is not easy to get a function f from this mess. This is not too surprising in a way: near boundaries or near the impeller, the forcing and/or the dissipation are at their largest, and maybe it is not such a good idea to try to compare data obtained in those areas with a model where forcing and dissipation have been removed. So this motivates a second test, performed over the 50% portion of the flow obtained by removing regions close to the impellers and along the surface of the vessel. One sees that the restricted data gather into a cubic-shaped function fitted by a two-parameter cubic: $f(\psi) = p_1\psi + p_3\psi^3$. This fit can then be used to obtain g as shown also in Fig. 5.2, through $\xi - ff'/r^2$ as a

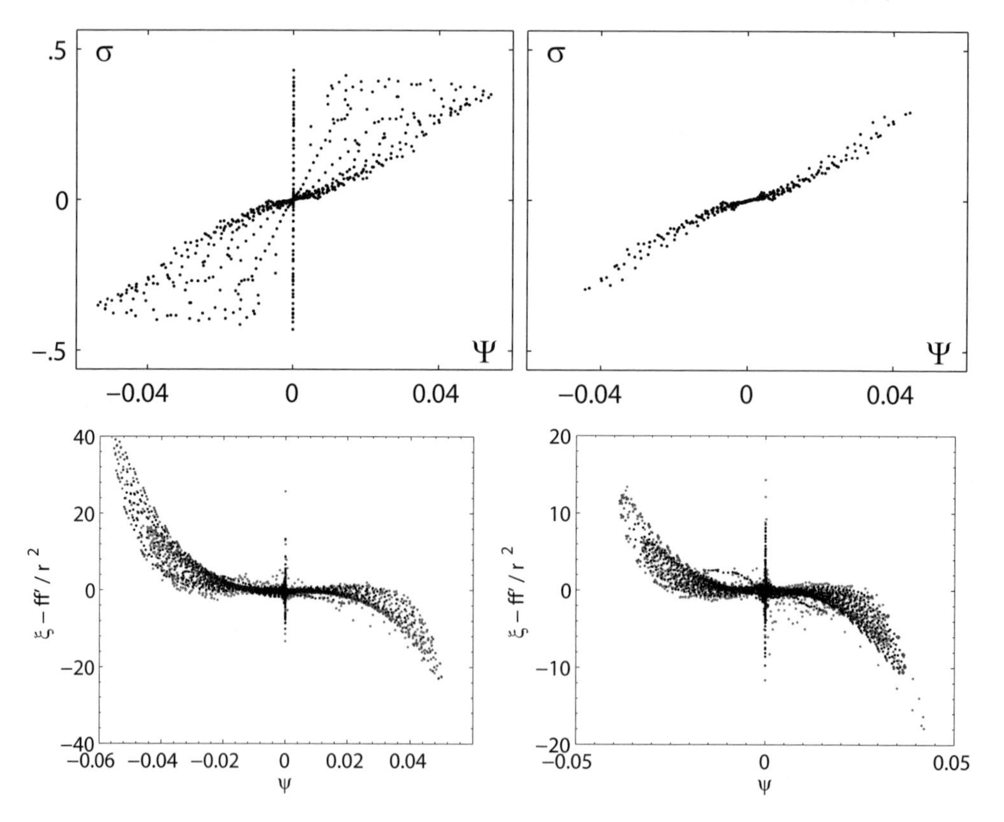

Fig. 5.2 Top: σ vs. ψ for experimental time-averaged flow in direction (+) at $Re = 2100$. Left, for the whole flow. Right, for $r \leq 0.81, 0.56 \leq z \leq 0.56$, corresponding to 50% of the flow volume. The remaining points are clearly gathering along a cubic-shaped function f. Bottom: $\xi - ff'/r^2$ versus ψ for time-averaged field direction (−) at left and (+) at right. Light color, for the whole flow; black, for $r \leq 0.81, 0.56 \leq z \leq 0.56$.

function of ψ. In the sequel, we restrict our comparison between theory and experiment to the 50% portion of the flow away from the boundaries.

With these tools at hand, we can now explore what is going on in our system by considering different Reynolds numbers and different forcing conditions.

We present results computed for the two directions of rotation, at different Reynolds numbers. Figure 5.3 presents the f fits obtained in each case. We have included in each graph a straight line tangential to the experimental curves for $\psi = 0$ as a reference for the Beltrami flow. In Fig. 5.4, we present the g fits obtained under the same conditions. Cubic-shaped functions are required for the time-averaged g. The plots of g consist of wide, noisy bands surrounding the fits in both cases. The noise for g is larger than for f, since it is a result of a two-step procedure including a spatial derivative.

As the Reynolds number increases, the cubic curves for f collapse onto the Beltrami line. At $Re = 4 \times 10^5$, it is not even possible to distinguish the f fit from a straight

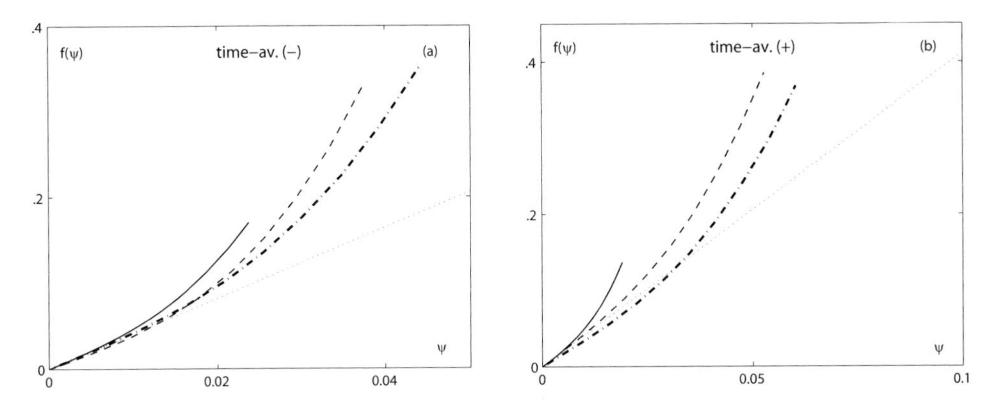

Fig. 5.3 f fit. (a), (b) Time-averaged field for directions ($-$) and ($+$), respectively. Legend for all figures: $Re = 150$ (thick —), $Re = 2100$ ($--$), $Re = 2.5 \times 10^5$ ($\cdot - \cdot$). The thin dotted straight line is for the Beltrami flow. As f is odd, we display it for positive values of ψ only.

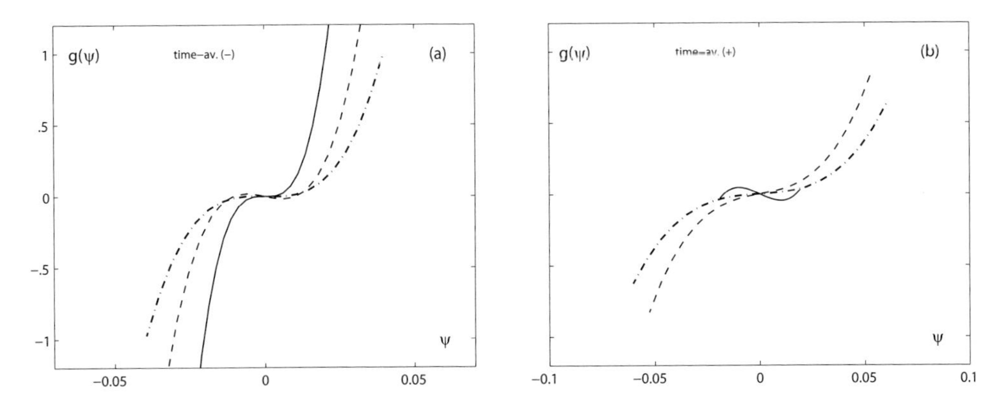

Fig. 5.4 g fit. (a), (b) Time-averaged field for directions ($-$) and ($+$), respectively. Same legend as in Fig. 5.3.

line, and this is true for the two directions of rotation (not shown). In the case of g, all the fits show a *plateau*, and as the Reynolds number increases, there is a wider range of ψ—or, equivalently, the volume of the flow considered—where g is very close to zero, i.e. very close to the Beltrami g function. This is better seen by looking directly at ξr^2 as a function of σ, which should be linear in the case of a Beltrami flow. This is shown in Fig. 5.5, for $Re = 4 \times 10^5$.

A second observation is the dependence of f and g on the forcing. Comparison of the right and left parts of Fig. 5.4, obtained for two different forcings, shows that the slope of f depends slightly on the forcing ($p_1 = 5$ for $(-)$ direction, and $p_1 = 4$ for $(+)$ direction). For any forcing, g remains close to zero on the measurement scale.

The conclusion of this first series of measurements is that (i) the stationary flow is an equilibrium solution of the axisymmetric Euler equation, and (ii) for large enough Reynolds numbers, the stationary flow is even an equilibrium solution of the axisymmetric Euler–Beltrami system. What about fluctuations?

Test of fluctuation relations. The theory that we have presented holds only in the Euler–Beltrami case. We therefore now consider only large enough Reynolds numbers, and test the fluctuation relations. Figure 5.6 presents an analysis of the fluctuation relations for the angular velocity u_θ. Over the whole flow, the velocity fluctuations are roughly constant. The relative scatter, which mostly tracks the z-dependance, increases with r. Focusing again on the bulk flow, we observe a reduced scatter and can measure $1/\beta_\xi$ with a simple average. We conclude that, for these von Kármán flows, the variance of the azimuthal velocity is constant in the bulk flow. This result is compatible with the mean-field analysis set out in the first part of this course for a Beltrami flow. The standard error estimates are typically 10%, ranging from 5 to 20%.

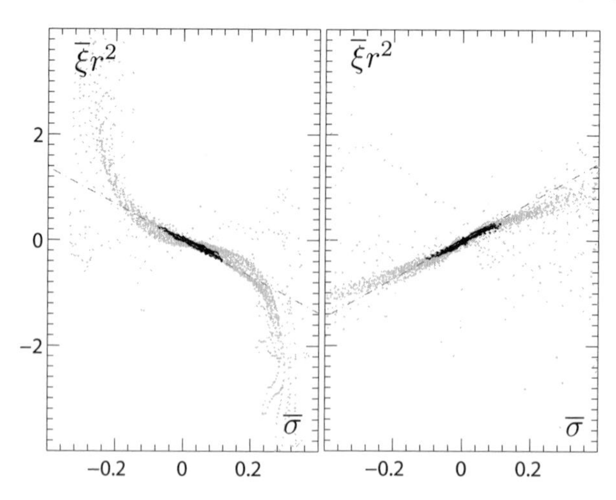

Fig. 5.5 $\overline{\xi} r^2$ versus $\overline{\sigma}$ for two experimental von Kármán flows with TM60 impellers at $Re = 4 \times 10^5$, sense $(+)$ at left, $(-)$ at right. The black dots correspond to data for the bulk flow ($|z| \leq 0.5$, $|r| \leq 0.5$) and define mostly linear functions. The slopes of the dot–dashed lines are given by the first-order coefficient of an odd cubic fit of the data.

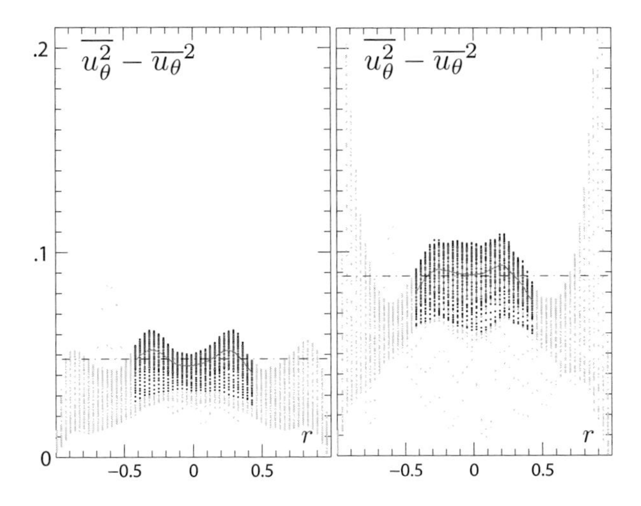

Fig. 5.6 Evolution of azimuthal velocity fluctuations given by eqn (5.25) with the radial coordinate r for two experimental von Kármán flows with TM60 impellers at $Re = 4 \times 10^5$, sense (+) at left, (−) at right. The black dots correspond to data for the bulk flow ($|z| \leq 0.5$, $|r| \leq 0.5$). The corresponding mean values and standard deviations at each z are plotted by black lines and error bars. Horizontal dot–dashed lines show the averages over the bulk flow, i.e. measured values for $1/\beta_\xi$.

5.5 Conclusion

We have derived predictions concerning the equilibrium flow and the fluctuations in an axisymmetric Euler–Beltrami flow using tools borrowed from statistical mechanics. We have then confronted these predictions with observations drawn from stationary states of a turbulent von Kármán flow, and have shown that they coincide. This result is a priori unexpected since the theory, which follows from the Euler equation, does not explicitly take into account the forcing and the dissipation, which *implicitly* determine the form of the steady state. This is additional confirmation that out-of-equilibrium steady states of a real turbulent flow may be described as equilibrium states of the Euler equation as suggested by Monchaux *et al.* (2006). This also suggests that in the present case, the possible singularities of the Navier–Stokes equations do not play a major role. This conclusion, which is well accepted in 2D turbulence, is quite intriguing in the present case. We note that the von Kármán flow differs from the traditional ideal, homogeneous, isotropic 3D turbulence by the presence of a well-defined mean flow. We conjecture that this strong mean flow, which possibly fully enforces the dynamics of the small or fluctuating scales through nonlocal processes akin to RDT (Dubrulle *et al.* 2004), is responsible for "regularization" of solutions of the Navier–Stokes equations, and makes the problem amenable to studies through tools borrowed from traditional equilibrium statistical physics. This is quite a cause for hope in the field of "turbulence modeling and applications," since most industrial and natural turbulence does include such a mean flow, which may equally play a regularizing role.

References

Chavanis P. H. (2003). *Phys. Rev. E*, **68**, 036108.
Dubrulle B., Laval J.-P., Nazarenko S., and Zaboronski O. (2004). *J. Fluid Mech.*, **520**, 1–21.
Duchon J. and Robert R. (2000). *Nonlinearity*, **13**(1), 249–255.
Fielding S. and Sollich P. (2002). *Phys. Rev. Lett.*, **88**, 050603.
Jaynes E. T. (1957). *Phys. Rev.*, **106**, 620.
Leprovost N., Dubrulle B., and Chavanis P. H. (2005). *Phys. Rev. Lett.*, **71**, 036311.
Leprovost N., Dubrulle B., and Chavanis P. H. (2006). *Phys. Rev. E*, **73**, 046308.
Onsager L. (1949). *Nuovo Cimento*, Suppl. **6**, 279.
Miller J. (1990). *Phys. Rev. Lett.*, **65**, 2137.
Monchaux R. (2007). PhD Thesis, University of Paris.
Monchaux R., Ravelet F., Dubrulle B., Chiffaudel A., and Daviaud F. (2006). *Phys. Rev. Lett.*, **96**, 124502.
Monchaux R., Chavanis P.-H., Chiffaudel A., Cortet P.-P., Daviaud F., Diribarne P., and Dubrulle B. (2008). *Phys. Rev. Lett.*, **101**, 174502.
Naso A., Monchaux R., Chavanis P.-H., and Dubrulle B. (2009). In preparation.
Ravelet F., Marié L., Chiffaudel A., and Daviaud F. (2004). *Phys. Rev. Lett.*, **93**, 164501.
Robert R. (2004). *Images des Mathématiques*, CNRS, pp. 91–99.
Robert R. and Sommeria J. (1991). *J. Fluid Mech.*, **229**, 291.

Part III

Mathematical aspects

6

The theory of large deviations and applications to statistical mechanics

Richard S. ELLIS

Department of Mathematics and Statistics,
University of Massachusetts Amherst, USA

6.1 Introduction

The theory of large deviations studies the exponential decay of probabilities in certain random systems. It has been applied to a wide range of problems in which detailed information on rare events is required. One is often interested not only in the probability of rare events but also in the characteristic behavior of the system as the rare event occurs. For example, in applications to queueing theory and communication systems, the rare event could represent an overload or breakdown of the system. In this case, large deviation methodology can lead to an efficient redesign of the system so that the overload or breakdown does not occur. In applications to statistical mechanics, the theory of large deviations gives precise, exponential-order estimates that are perfectly suited for asymptotic analysis.

This paper will present a number of topics in the theory of large deviations and several applications to statistical mechanics, all united by the concept of relative entropy. This concept entered human culture through the first large deviation calculation in science, carried out by Ludwig Boltzmann. Stated in a modern terminology, his discovery was that the relative entropy expresses the asymptotic behavior of certain multinomial probabilities. This statistical interpretation of entropy has the following crucial physical implication (Ellis 2006, Section 1.1). Entropy is a bridge between a microscopic level, on which physical systems are defined in terms of the complicated interactions among the individual constituent particles, and a macroscopic level, on which the laws describing the behavior of the system are formulated.

Boltzmann and, later, Gibbs asked a fundamental question. How can one use probability theory to study equilibrium properties of physical systems such as an ideal gas, a ferromagnet, or a fluid? These properties include such phenomena as phase transitions, for example, the liquid–gas transition and spontaneous magnetization in a ferromagnet. Another example arises in the study of freely evolving, inviscid fluids, for which one wants to describe coherent states. These are steady, stable mean flows composed of one or more vortices that persist amidst the turbulent fluctuations of the vorticity field. The answer to this fundamental question, which led to the development of classical equilibrium statistical mechanics, is that one studies equilibrium properties via probability measures on configuration space known today as the microcanonical ensemble and the canonical ensemble. For background in statistical mechanics, I recommend Ellis (2006), Lanford (1973), and Wightman (1979), which cover a number of topics relevant to these notes.

Boltzmann's calculation of the asymptotic behavior of multinomial probabilities in terms of relative entropy was carried out in 1877 as a key component of his paper that gave a probabilistic interpretation of the Second Law of Thermodynamics (Boltzmann 1877). This momentous calculation represents a revolutionary moment in human culture, during which both statistical mechanics and the theory of large deviations were born. Boltzmann based his work on the hypothesis that atoms exist. Although this hypothesis is universally accepted today, one might be surprised to learn that it was highly controversial during Boltzmann's time (Lindley 2001, pp. vii–x).

Boltzmann's work is put in historical context by W. R. Everdell in his book *The First Moderns*, which traces the development of the modern consciousness in nineteenth- and twentieth-century thought (Everdell 1997). Chapter 3 focuses on the mathematicians of Germany in the 1870's—namely Cantor, Dedekind, and Frege—who "would become the first creative thinkers in any field to look at the world in a fully twentieth-century manner" (p. 31). Boltzmann is then presented as the man whose investigations in stochastics and statistics made possible the work of the two other great founders of twentieth-century theoretical physics, Planck and Einstein. As Everdell writes, "he was at the center of the change" (p. 48).

Like many areas of mathematics, the theory of large deviations has both a left hand and a right hand; the left hand provides heuristic insight while the right hand provides rigorous proofs. Although the theory is applicable in many diverse settings, the right-hand technicalities can be formidable. Recognizing this, I would like to supplement the rigorous, right-hand formulation of the theory with a number of basic results presented in a left-hand format useful to the applied researcher. For a review of the theory emphasizing applications to statistical mechanics, Touchette (2009) is recommended. The Web document Ellis (2008) is an expanded version of this paper containing additional technical results and applications.

Here is an overview of this paper. In Section 6.2, a basic probabilistic model for random variables having a finite state space is introduced. In Section 6.3, we explain Boltzmann's discovery of the asymptotic behavior of multinomial probabilities in terms of relative entropy. Section 6.4 proves a conditioned limit theorem involving relative entropy that elucidates a basic issue arising in many areas of application. What is the most likely way for an unlikely event to happen? In Section 6.5, we introduce the general concepts of a large deviation principle and a Laplace principle together with related results. In the last two sections, the theory of large deviations is applied to several problems in statistical mechanics. In Section 6.6, the theory is used to study equilibrium properties of a basic model of ferromagnetism known as the Curie–Weiss model, which is a mean-field approximation to the much more complicated Ising model. We then use the insights gained in treating the Curie–Weiss model to derive the phase-transition structure of two other basic models, the Curie–Weiss–Potts model and the mean-field Blume–Capel model. Our work in the preceding section leads in Section 6.7 to the formulation of a general procedure for applying the theory of large deviations to the analysis of an extensive class of statistical-mechanical models, an analysis that allows us to address the fundamental problem of equivalence and nonequivalence of ensembles.

6.2 A basic probabilistic model

In this section, we introduce a basic probabilistic model for random variables having a finite state space. In later sections, a number of questions in the theory of large deviations will be investigated in the context of this model. Let $\alpha \geq 2$ be an integer, $y_1 < y_2 < \ldots < y_\alpha$ a set of α real numbers, and $\rho_1, \rho_2, \ldots, \rho_\alpha$ a set of α positive real numbers summing to 1. We think of $\Lambda = \{y_1, y_2, \ldots, y_\alpha\}$ as the set of possible outcomes of a random experiment in which each individual outcome y_k has the probability ρ_k of occurring. The vector $\rho = (\rho_1, \rho_2, \ldots, \rho_\alpha)$ is an element of the set of probability vectors

$$\mathcal{P}_\alpha = \left\{ \gamma = (\gamma_1, \gamma_2, \ldots, \gamma_\alpha) \in \mathbb{R}^\alpha : \gamma_k \geq 0, \sum_{k=1}^{\alpha} \gamma_k = 1 \right\}.$$

Any vector $\gamma \in \mathcal{P}_\alpha$ also defines a probability measure on the set of subsets of Λ via the formula $\gamma = \sum_{k=1}^{\alpha} \gamma_k \, \delta_{y_k}$, where for $y \in \Lambda$, $\delta_{y_k}\{y\} = 1$ if $y = y_k$ and equals 0 otherwise. For $B \subset \Lambda$, we define $\gamma\{B\} = \sum_{k=1}^{\alpha} \gamma_k \, \delta_{y_k}\{B\} = \sum_{y_k \in B} \gamma_k$.

For each positive integer n, the configuration space for n independent repetitions of the experiment is $\Omega_n = \Lambda^n$, a typical element of which is denoted by $\omega = (\omega_1, \omega_2, \ldots, \omega_n)$. For each $\omega \in \Omega_n$, we define

$$P_n\{\omega\} = \prod_{j=1}^{n} \rho\{\omega_j\} = \prod_{j=1}^{n} \rho_{k_j} \quad \text{if } \omega_j = y_{k_j}\,.$$

We then extend this to a probability measure on the set of subsets of Ω_n by defining

$$P_n\{B\} = \sum_{\omega \in B} P_n\{\omega\} \text{ for } B \subset \Omega_n\,.$$

P_n is called the product measure with one-dimensional marginals ρ and is written ρ^n.

An important special case occurs when each ρ_k equals $1/\alpha$. Then for each $\omega \in \Omega_n$, $P_n\{\omega\} = 1/\alpha^n$, and for any subset B of Ω_n, $P_n\{B\} = \text{card}(B)/\alpha^n$, where $\text{card}(B)$ denotes the cardinality of B, i.e. the number of elements in B.

We return to the general case. With respect to P_n, the coordinate functions $X_j(\omega) = \omega_j, j = 1, 2, \ldots, n$, are independent identically distributed (i.i.d.) random variables with common distribution ρ; that is, for any subsets $B_1, B_2, \ldots, B_n$ of Λ,

$$\begin{aligned} &P_n\{\omega \in \Omega_n : X_j(\omega) \in B_j \text{ for } j = 1, 2, \ldots, n\} \\ &= \prod_{j=1}^{n} P_n\{\omega \in \Omega_n : X_j(\omega) \in B_j\} = \prod_{j=1}^{n} \rho\{B_j\}\,. \end{aligned}$$

Example 6.1. Random phenomena that can be studied via this basic model include standard examples such as coin tossing and die tossing and also include a discrete ideal gas.

(a) *Coin tossing.* In this case, $\Lambda = \{1, 2\}$ and $\rho_1 = \rho_2 = 1/2$.

(b) *Die tossing.* In this case, $\Lambda = \{1, 2, \ldots, 6\}$ and each $\rho_k = 1/6$.

(c) *Discrete ideal gas.* Consider a discrete ideal gas consisting of n identical, non-interacting particles, each having α equally likely energy levels $y_1, y_2, \ldots, y_\alpha$; in this case each ρ_k equals $1/\alpha$. The coordinate functions X_j represent the random energy levels of the molecules of the gas. The statistical independence of these random variables reflects the fact that the molecules of the gas do not interact. ■

It is worthwhile to reiterate the basic probabilistic model in the context of the general framework in probability theory, which involves five quantities $(\Omega, \mathcal{F}, P, X_j, \Lambda)$. Ω is a configuration space, $\mathcal{F}$ is a σ-algebra of subsets of Ω (i.e. a class of subsets of Ω containing Ω and closed under complements and countable unions), P is a probability measure on $\mathcal{F}$ (i.e. a countably additive set function satisfying $P\{\Omega\} = 1$), and X_j is a sequence of random variables (i.e. measurable functions) taking values in a state space Λ. The triplet $(\Omega, \mathcal{F}, P)$ is called a probability space. In the present section, we make the following choices for $n \in I\!N$: (a) $\Omega = \Omega_n = \Lambda^n$, where $\Lambda = \{y_1, y_2, \ldots, y_\alpha\}$; (b) $\mathcal{F}$ is the set of all subsets of Λ^n; and (c) P is the product measure $P_n = \rho^n$, where $\rho = \sum_{k=1}^{\alpha} \rho_k \delta_{y_k}$, each $\rho_k > 0$, and $\sum_{k=1}^{\alpha} \rho_k = 1$; $X_j(\omega) = \omega_j$ for $\omega \in \Omega_n$.

In the next section, we examine Boltzmann's discovery of a statistical interpretation of entropy.

6.3 Boltzmann's discovery and relative entropy

In its original form, Boltzmann's discovery concerns the asymptotic behavior of certain multinomial coefficients. For the purpose of applications in this paper, it is advantageous to formulate it in terms of a probabilistic quantity known as the empirical vector. We use the notation of the preceding section. Thus, let $\alpha \geq 2$ be an integer; let $y_1 < y_2 < \ldots < y_\alpha$ be a set of α real numbers; let $\rho_1, \rho_2, \ldots, \rho_\alpha$ be a set of α positive real numbers summing to 1; let Λ be the set $\{y_1, y_2, \ldots, y_\alpha\}$; and let P_n be the product measure on $\Omega_n = \Lambda^n$ with one-dimensional marginals $\rho = \sum_{k=1}^{\alpha} \rho_k \delta_{y_k}$. For $\omega = (\omega_1, \omega_2, \ldots, \omega_n) \in \Omega_n$, we let $\{X_j, j = 1, \ldots, n\}$ be the coordinate functions defined by $X_j(\omega) = \omega_j$. The X_j form a sequence of i.i.d. random variables with common distribution ρ.

We now turn to the object under study in the present section. For $\omega \in \Omega_n$ and $y \in \Lambda$, define

$$L_n(y) = L_n(\omega, y) = \frac{1}{n} \sum_{j=1}^{n} \delta_{X_j(\omega)}\{y\}.$$

Thus $L_n(\omega, y)$ counts the relative frequency with which y appears in the configuration ω; in symbols, $L_n(\omega, y) = n^{-1} \cdot \text{card}\{j \in \{1, \ldots, n\} : \omega_j = y\}$. We then define the empirical vector

$$\begin{aligned} L_n = L_n(\omega) &= (L_n(\omega, y_1), \ldots, L_n(\omega, y_\alpha)) \\ &= \frac{1}{n} \sum_{j=1}^{n} \left(\delta_{X_j(\omega)}\{y_1\}, \ldots, \delta_{X_j(\omega)}\{y_\alpha\}\right). \end{aligned}$$

L_n equals the sample mean of the i.i.d. random vectors $(\delta_{X_j(\omega)}\{y_1\}, \ldots, \delta_{X_j(\omega)}\{y_\alpha\})$. It takes values in the set of probability vectors

$$\mathcal{P}_\alpha = \left\{ \gamma = (\gamma_1, \gamma_2, \ldots, \gamma_\alpha) \in \mathbb{R}^\alpha : \gamma_k \geq 0, \sum_{k=1}^{\alpha} \gamma_k = 1 \right\}.$$

The limiting behavior of L_n is straightforward to determine. Let $\|\cdot\|$ denote the Euclidean norm on $\mathbb{R}^\alpha$. For any $\gamma \in \mathcal{P}_\alpha$ and $\varepsilon > 0$, we define the open ball

$$B(\gamma, \varepsilon) = \{\nu \in \mathcal{P}_\alpha : \|\gamma - \nu\| < \varepsilon\}.$$

Since the X_j have the common distribution ρ, for each $y_k \in \Lambda$

$$E^{P_n}\{L_n(y_k)\} = E^{P_n}\left\{\frac{1}{n}\sum_{j=1}^{n}\delta_{X_j}\{y_k\}\right\} = \frac{1}{n}\sum_{j=1}^{n}P_n\{X_j = y_k\} = \rho_k\,,$$

where E^{P_n} denotes the expectation with respect to P_n. Hence, by the weak law of large numbers for the sample means of i.i.d. random variables, for any $\varepsilon > 0$

$$\lim_{n\to\infty} P_n\{L_n \in B(\rho,\varepsilon)\} = 1\,. \tag{6.1}$$

It follows that for any $\gamma \in \mathcal{P}_\alpha$ not equal to ρ and for any $\varepsilon > 0$ satisfying $0 < \varepsilon < \|\rho - \gamma\|$,

$$\lim_{n\to\infty} P_n\{L_n \in B(\gamma,\varepsilon)\} = 0\,. \tag{6.2}$$

As we will see, Boltzmann's discovery implies that these probabilities converge to 0 exponentially fast in n. The exponential decay rate is given in terms of the relative entropy, which we now define.

Definition 6.1. (Relative entropy.) *Let $\rho = (\rho_1,\ldots,\rho_\alpha)$ denote the probability vector in $\mathcal{P}_\alpha$ in terms of which the basic probabilistic model is defined. The relative entropy of $\gamma \in \mathcal{P}_\alpha$ with respect to ρ is defined by*

$$I_\rho(\gamma) = \sum_{k=1}^{\alpha}\gamma_k \log\frac{\gamma_k}{\rho_k}\,.$$

Several properties of the relative entropy are given in the next lemma. The proof is typical of proofs of analogous results involving relative entropy (e.g. Proposition 6.4.3) in that we use a global, convexity-based inequality rather than calculus to determine where I_ρ attains its infimum over $\mathcal{P}_\alpha$. In the present case the global inequality is that for $x \geq 0$, $x\log x \geq x - 1$ with equality if and only if $x = 1$.

Lemma 6.3.1. *For $\gamma \in \mathcal{P}_\alpha$, $I_\rho(\gamma)$ measures the discrepancy between γ and ρ in the sense that $I_\rho(\gamma) \geq 0$ and $I_\rho(\gamma) = 0$ if and only if $\gamma = \rho$. Thus $I_\rho(\gamma)$ attains its infimum of 0 over $\mathcal{P}_\alpha$ at the unique measure $\gamma = \rho$. In addition, I_ρ is strictly convex on $\mathcal{P}_\alpha$; that is, for $0 < \lambda < 1$ and any $\mu \neq \nu$ in $\mathcal{P}_\alpha$,*

$$I_\rho(\lambda\mu + (1-\lambda)\nu) < \lambda I_\rho(\mu) + (1-\lambda)I(\nu)\,.$$

Proof. For $x \geq 0$, the graph of the strictly convex function $x\log x$ has the tangent line $y = x - 1$ at $x = 1$. Hence $x\log x \geq x - 1$, with equality if and only if $x = 1$. It follows that for any $\gamma \in \mathcal{P}_\alpha$,

$$\frac{\gamma_k}{\rho_k}\log\frac{\gamma_k}{\rho_k} \geq \frac{\gamma_k}{\rho_k} - 1\,, \tag{6.3}$$

with equality if and only if $\gamma_k = \rho_k$. Multiplying this inequality by ρ_k and summing over k yields

$$I_\rho(\gamma) = \sum_{k=1}^{\alpha} \gamma_k \log \frac{\gamma_k}{\rho_k} \geq \sum_{k=1}^{\alpha} (\gamma_k - \rho_k) = 0\,.$$

We now prove that $I_\rho(\gamma) = 0$ if and only if $\gamma = \rho$. If $\gamma = \rho$, then the definition of the relative entropy shows that $I_\rho(\gamma) = 0$. Now assume that $I_\rho(\gamma) = 0$. Then

$$\begin{aligned} 0 &= \sum_{k=1}^{\alpha} \gamma_k \log \frac{\gamma_k}{\rho_k} \\ &= \sum_{k=1}^{\alpha} \left(\gamma_k \log \frac{\gamma_k}{\rho_k} - (\gamma_k - \rho_k) \right) \\ &= \sum_{k=1}^{\alpha} \rho_k \left(\frac{\gamma_k}{\rho_k} \log \frac{\gamma_k}{\rho_k} - \left(\frac{\gamma_k}{\rho_k} - 1 \right) \right). \end{aligned}$$

We now use the facts that $\rho_k > 0$ and that for $x \geq 0$, $x \log x \geq x - 1$ with equality if and only if $x = 1$. It follows that for each k, $\gamma_k = \rho_k$ and thus that $\gamma = \rho$. This completes the proof that $I_\rho(\gamma) \geq 0$ and $I_\rho(\gamma) = 0$ if and only if $\gamma = \rho$, which is the first assertion in the proposition.

Since

$$I_\rho(\gamma) = \sum_{k=1}^{\alpha} \rho_k \frac{\gamma_k}{\rho_k} \log \frac{\gamma_k}{\rho_k}\,,$$

the strict convexity of I_ρ is a consequence of the strict convexity of $x \log x$ for $x \geq 0$. ■

We are now ready to give the first formulation of Boltzmann's discovery, which we state using a heuristic notation and which we label, in recognition of its formal status, as a pseudo-theorem. However, the formal calculations used to motivate the pseudo-theorem can easily be turned into a rigorous proof of an asymptotic theorem. That theorem is stated in Theorem 6.3.3. From Boltzmann's momentous discovery, both the theory of large deviations and the Gibbsian formulation of equilibrium statistical mechanics grew. The notation $P_n\{L_n \in d\gamma\}$ represents the probability that L_n is close to γ.

Pseudo-theorem 6.3.2. (Boltzmann's discovery—formulation 1.) *For any* $\gamma \in \mathcal{P}_\alpha$,

$$P_n\{L_n \in d\gamma\} \approx \exp[-n I_\rho(\gamma)] \quad \text{as } n \to \infty\,.$$

Heuristic proof. Since $\gamma \in \mathcal{P}_\alpha$, $\sum_{k=1}^{\alpha} \gamma_k = 1$. By elementary combinatorics,

$$\begin{aligned} P_n\{L_n \in d\gamma\} &= P_n\left\{ \omega \in \Omega_n : L_n(\omega) \sim \frac{1}{n}(n\gamma_1, n\gamma_2, \ldots, n\gamma_\alpha) \right\} \\ &\approx P_n\{\text{card}\{\omega_j = y_1\} \sim n\gamma_1, \ldots, \text{card}\{\omega_j = y_\alpha\} \sim n\gamma_\alpha\} \\ &\approx \frac{n!}{(n\gamma_1)!(n\gamma_2)! \cdots (n\gamma_\alpha)!}\, \rho_1^{n\gamma_1}\, \rho_2^{n\gamma_2} \cdots \rho_\alpha^{n\gamma_\alpha}\,. \end{aligned}$$

Stirling's formula in the weak form $\log(n!) = n\log n - n + \mathrm{O}(\log n)$ yields

$$\begin{aligned}
&\frac{1}{n}\log P_n\{L_n \in d\gamma\} \\
&\approx \frac{1}{n}\log\left(\frac{n!}{(n\gamma_1)!(n\gamma_2)!\cdots(n\gamma_\alpha)!}\right) + \sum_{k=1}^{\alpha}\gamma_k\log\rho_k \\
&= \frac{1}{n}\log\left(\frac{n^n e^{-n}}{(n\gamma_1)^{n\gamma_1}e^{-n\gamma_1}\cdots(n\gamma_\alpha)^{n\gamma_\alpha}e^{-n\gamma_\alpha}}\right) + \sum_{k=1}^{\alpha}\gamma_k\log\rho_k + \mathrm{O}\left(\frac{\log n}{n}\right) \\
&= \frac{1}{n}\log\left(\frac{1}{\gamma_1^{n\gamma_1}\cdots\gamma_\alpha^{n\gamma_\alpha}}\right) + \sum_{k=1}^{\alpha}\gamma_k\log\rho_k + \mathrm{O}\left(\frac{\log n}{n}\right) \\
&= -\sum_{k=1}^{\alpha}\gamma_k\log\gamma_k + \sum_{k=1}^{\alpha}\gamma_k\log\rho_k + \mathrm{O}\left(\frac{\log n}{n}\right) \\
&= -\sum_{k=1}^{\alpha}\gamma_k\log\frac{\gamma_k}{\rho_k} + \mathrm{O}\left(\frac{\log n}{n}\right) \;=\; -I_\rho(\gamma) + \mathrm{O}\left(\frac{\log n}{n}\right).
\end{aligned}$$

The term $\mathrm{O}(\log n/n)$ converges to 0 as $n \to \infty$. Hence, multiplying both sides of the last equation by n and exponentiating yields the result. ■

Pseudo-theorem 6.3.2 has the following interesting consequence. Let γ be any vector in $\mathcal{P}_\alpha$ which differs from ρ. Since $I_\rho(\gamma) > 0$ (Lemma 6.3.1), it follows that

$$P_n\{L_n \in d\gamma\} \approx \exp[-nI_\rho(\gamma)] \to 0 \;\text{ as } n \to \infty\,,$$

a limit which, if rigorous, would imply eqn (6.2).

Let A be a Borel subset of $\mathcal{P}_\alpha$, i.e. A is a member of the Borel σ-algebra of $\mathcal{P}_\alpha$, which is the smallest σ-algebra containing the open sets. The class of Borel subsets includes all open subsets of $\mathcal{P}_\alpha$ and all closed subsets of $\mathcal{P}_\alpha$. If ρ is not contained in the closure of A, then by the weak law of large numbers

$$\lim_{n\to\infty} P_n\{L_n \in A\} = 0\,,$$

and, by analogy with the heuristic asymptotic result given in Pseudo-theorem 6.3.2, we expect that these probabilities converge to 0 exponentially fast with n. This is in fact the case. In order to express the exponential decay rate of such probabilities in terms of the relative entropy, we introduce the notation $I_\rho(A) = \inf_{\gamma\in A} I_\rho(\gamma)$. The range of $L_n(\omega)$ for $\omega \in \Omega_n$ is the set of probability vectors having the form k/n, where $k \in \mathbb{R}^\alpha$ has nonnegative integer coordinates summing to n; hence the cardinality of the range does not exceed n^α. Since

$$P_n\{L_n \in A\} = \sum_{\gamma\in A} P_n\{L_n \in d\gamma\} \approx \sum_{\gamma\in A}\exp[-nI_\rho(\gamma)]$$

and

$$\exp[-nI_\rho(A)] \le \sum_{\gamma\in A} \exp[-nI_\rho(\gamma)] \le n^\alpha \exp[-nI_\rho(A)],$$

one expects that to exponential order,

$$P_n\{L_n \in A\} \approx \exp[-nI_\rho(A)] \text{ as } n \to \infty. \tag{6.4}$$

As formulated in Corollary 6.3.4, this asymptotic result is indeed valid. It is a consequence of the following rigorous reformulation of Boltzmann's discovery, known as Sanov's Theorem, which expresses the large deviation principle for the empirical vectors L_n. That concept is defined in general in Definition 6.2, and a general form of Sanov's Theorem is stated in Theorem 6.5.6. The special case of Sanov's Theorem stated next is proved in (Ellis 2006, Theorem VIII.2).

Theorem 6.3.3. (Boltzmann's discovery—formulation 2.) *The sequence of empirical vectors L_n satisfies the large deviation principle on $\mathcal{P}_\alpha$ with rate function I_ρ in the following sense.*

(a) Large deviation upper bound. *For any closed subset F of $\mathcal{P}_\alpha$,*

$$\limsup_{n\to\infty} \frac{1}{n} \log P_n\{L_n \in F\} \le -I_\rho(F).$$

(b) Large deviation lower bound. *For any open subset G of $\mathcal{P}_\alpha$,*

$$\liminf_{n\to\infty} \frac{1}{n} \log P_n\{L_n \in G\} \ge -I_\rho(G).$$

Comments on the proof. For $\gamma \in \mathcal{P}_\alpha$ and $\varepsilon > 0$, $B(\gamma, \varepsilon)$ denotes the open ball with center γ and radius ε, and $\overline{B}(\gamma, \varepsilon)$ denotes the corresponding closed ball. Since $\mathcal{P}_\alpha$ is a compact subset of $\mathbb{R}^\alpha$, any closed subset F of $\mathcal{P}_\alpha$ is automatically compact. By a standard covering argument, it is not hard to show that the large deviation upper bound holds for any closed set F provided that one obtains the large deviation upper bound for any closed ball $\overline{B}(\gamma, \varepsilon)$:

$$\limsup_{n\to\infty} \frac{1}{n} \log P_n\{L_n \in \overline{B}(\gamma, \varepsilon)\} \le -I_\rho(\overline{B}(\gamma, \varepsilon)).$$

Likewise, the large deviation lower bound holds for any open set G provided one obtains the large deviation lower bound for any open ball $B(\gamma, \varepsilon)$:

$$\liminf_{n\to\infty} \frac{1}{n} \log P_n\{L_n \in B(\gamma, \varepsilon)\} \ge -I_\rho(B(\gamma, \varepsilon)).$$

The bounds in the last two equations can be proved via combinatorics and Stirling's formula as in the heuristic proof of Pseudo-theorem 6.3.2; one can easily adapt the calculations given in (Ellis 2006, Section I.4). The details are omitted in the present work. ■

Given A a Borel subset of $\mathcal{P}_\alpha$, we denote by A° the interior of A relative to $\mathcal{P}_\alpha$ and by $\overline{A}$ the closure of A. For a class of Borel subsets, the next corollary gives a rigorous version of the asymptotic formula (6.4). This class consists of sets A such that $\overline{A^\circ}$ equals $\overline{A}$. Any open ball $B(\gamma, \varepsilon)$ or closed ball $\overline{B}(\gamma, \varepsilon)$ satisfies this condition. For any Borel subset A, the continuity of I_ρ on $\mathcal{P}_\alpha$ guarantees that $I_\rho(A^\circ) = I_\rho(\overline{A^\circ})$. Hence if $\overline{A^\circ} = \overline{A}$, then $I_\rho(A^\circ) = I_\rho(\overline{A})$, which is the hypothesis of Theorem 6.5.2. The next corollary is a consequence of that theorem and the large deviation principle given in Theorem 6.3.3.

Corollary 6.3.4. *Let A be any Borel subset of $\mathcal{P}_\alpha$ satisfying $\overline{A^\circ} = \overline{A}$. Then*

$$\lim_{n \to \infty} \frac{1}{n} \log P_n\{L_n \in A\} = -I_\rho(A).$$

The next corollary of Theorem 6.3.3 allows one to conclude that a large class of probabilities involving L_n converge to 0. The general version of this corollary given in Theorem 6.5.3 is extremely useful in applications. For example, we will use it in Section 6.6 to analyze several lattice spin models and in Section 6.7 to motivate the definitions of the sets of equilibrium macrostates for the canonical ensemble and the microcanonical ensemble for a general class of systems (Theorems 6.7.1(c) and 6.7.2(c)).

Corollary 6.3.5. *Let A be any Borel subset of $\mathcal{P}_\alpha$ such that $\overline{A}$ does not contain ρ. Then $I_\rho(\overline{A}) > 0$, and for some $C < \infty$*

$$P_n\{L_n \in A\} \leq C \exp[-nI_\rho(\overline{A})/2] \to 0 \quad \textit{as } n \to \infty.$$

Proof. Since $I_\rho(\gamma) > I_\rho(\rho) = 0$ for any $\gamma \neq \rho$, the positivity of $I_\rho(\overline{A})$ follows from the continuity of I_ρ on $\mathcal{P}_\alpha$. The second assertion is an immediate consequence of the large deviation upper bound applied to $\overline{A}$ and the positivity of $I_\rho(\overline{A})$. ■

Take any $\varepsilon > 0$. Applying Corollary 6.3.5 to the complement of the open ball $B(\rho, \varepsilon)$ yields $P_n\{L_n \notin B(\rho, \varepsilon)\} \to 0$ or, equivalently,

$$\lim_{n \to \infty} P_n\{L_n \in B(\rho, \varepsilon)\} = 1.$$

Although this rederives the weak law of large numbers for L_n expressed in eqn (6.1), this second derivation relates the order-1 limit for L_n to the point $\rho \in \mathcal{P}_\alpha$ at which the rate function I_ρ attains its infimum. In this context, we call ρ the equilibrium value of L_n with respect to the measures P_n.

In the next section, we will present a limit theorem for L_n whose proof is based on the precise, exponential-order estimates given by the large deviation principle in Theorem 6.3.3.

6.4 The most likely way for an unlikely event to happen

In this section, we prove a conditioned limit theorem that elucidates a basic issue arising in many areas of application. What is the most likely way for an unlikely event to happen? For example, in applications to queueing theory and communication systems, the unlikely event could represent an overload or breakdown of the system. If one knows the most likely way that the overload could occur, then one could efficiently redesign the system so that the overload does not happen.

We use the notation of the preceding section. Thus, let $\alpha \geq 2$ be an integer; let $y_1 < y_2 < \ldots < y_\alpha$ be a set of α real numbers; let $\rho_1, \rho_2, \ldots, \rho_\alpha$ be a set of α positive real numbers summing to 1; let Λ be the set $\{y_1, y_2, \ldots, y_\alpha\}$; and let P_n be the product measure on $\Omega_n = \Lambda^n$ with one-dimensional marginals $\rho = \sum_{k=1}^{\alpha} \rho_k \delta_{y_k}$. For $\omega = (\omega_1, \omega_2, \ldots, \omega_n) \in \Omega_n$, we let $\{X_j, j = 1, \ldots, n\}$ be the coordinate functions defined by $X_j(\omega) = \omega_j$. The X_j form a sequence of i.i.d. random variables with common distribution ρ. For $\omega \in \Omega_n$ and $y \in \Lambda$, we also define

$$L_n(y) = L_n(\omega, y) = \frac{1}{n} \sum_{j=1}^{n} \delta_{X_j(\omega)}\{y\}$$

and the empirical vector

$$L_n = L_n(\omega) = (L_n(\omega, y_1), \ldots, L_n(\omega, y_\alpha)) = \frac{1}{n} \sum_{j=1}^{n} \left(\delta_{X_j(\omega)}\{y_1\}, \ldots, \delta_{X_j(\omega)}\{y_\alpha\}\right).$$

L_n takes values in the set of probability vectors

$$\mathcal{P}_\alpha = \left\{ \gamma = (\gamma_1, \gamma_2, \ldots, \gamma_\alpha) \in \mathbb{R}^\alpha : \gamma_k \geq 0, \sum_{k=1}^{\alpha} \gamma_k = 1 \right\}.$$

The main result in this section is the conditioned limit theorem for L_n given in Theorem 6.4.1. This theorem has the bonus of giving insight into a basic construction in statistical mechanics. As we show in Section 5 of Ellis (2008), it motivates the form of the canonical ensemble for the discrete ideal gas and, by extension, for any statistical-mechanical system characterized by conservation of energy. These unexpected theorems are the first indication of the power of Boltzmann's discovery, which gives precise exponential-order estimates for probabilities of the form $P_n\{L_n \in A\}$.

The conditioned limit theorem that we will consider has the following form. Suppose that one is given a particular set A for which $P_n\{L_n \in A\} > 0$ for all sufficiently large n. One wants to determine a set B belonging to a certain class (e.g. open balls) such that the conditioned limit

$$\lim_{n\to\infty} P_n\{L_n \in B \,|\, L_n \in A\} = \lim_{n\to\infty} P_n\{L_n \in B \cap A\} \cdot \frac{1}{P_n\{L_n \in A\}} = 1$$

is valid. Since, to exponential order,

$$P_n\{L_n \in B \cap A\} \cdot \frac{1}{P_n\{L_n \in A\}} \approx \exp[-n(I_\rho(B \cap A) - I_\rho(A))],$$

one should obtain the conditioned limit if B satisfies $I_\rho(B \cap A) = I_\rho(A)$. If one can determine the point in A where the infimum of I_ρ is attained, then one picks B to contain this point. As we explain in Proposition 6.4.3, such a minimizing point can be determined for an important class of sets A.

In order to formulate the conditioned limit theorem, we define

$$S_n = \sum_{j=1}^{n} X_j \quad \text{and} \quad \bar{y} = \sum_{k=1}^{\alpha} y_k \rho_k = E^{P_n}\{X_1\}\,.$$

By the weak law of large numbers, for any $\varepsilon > 0$

$$\lim_{n\to\infty} P_n\{|S_n/n - \bar{y}| \geq \varepsilon\} = 0\,.$$

Given a small positive number a, we choose z to satisfy $y_1 < z - a < z < \bar{y}$. The conditioned limit theorem involves positive numbers $\{\rho_k^*, k = 1, \ldots, \alpha\}$ summing to 1 and satisfying

$$\rho_k^* = \lim_{n\to\infty} P_n\{X_1 = k \,|\, S_n/n \in [z-a, z]\}\,. \tag{6.5}$$

By the law of large numbers, the event on which we are conditioning—namely, that $S_n/n \in [z-a, z]$ for all n—is a rare event converging to 0 as $n \to \infty$. A similar result would hold if we assumed that $S_n/n \in [z, z+a]$, where $\bar{y} < z < z + a < y_\alpha$.

The limit (6.5) will be seen to follow from the following more easily answered question: conditioned on the event $\{S_n/n \in [z-a, z]\}$, determine the most likely configuration $\rho^* = (\rho_1^*, \ldots, \rho_\alpha^*)$ of L_n in the limit $n \to \infty$. In other words, we want $\rho^* \in \mathcal{P}_\alpha$ such that for any $\varepsilon > 0$,

$$\lim_{n\to\infty} P_n\{L_n \in B(\rho^*, \varepsilon) \,|\, S_n/n \in [z-a, z]\} = 1\,.$$

In Theorem 6.4.1 we give the form of ρ^*, which depends on z through a parameter β. In order to indicate this dependence, we write $\rho^{(\beta)}$ in place of ρ^*. The form of $\rho^{(\beta)}$ is independent of a. In the proof of the analogous result given in Ellis (2008), we write $-\beta$ instead of β in the definition of $\rho^{(\beta)}$ in order to be consistent with conventions in statistical mechanics.

Theorem 6.4.1. *Let $z \in (y_1, \bar{y})$ be given, and choose $a > 0$ such that $z - a > y_1$. The following conclusions hold.*

(a) *There exists a unique $\rho^{(\beta)} \in \mathcal{P}_\alpha$ such that for every $\varepsilon > 0$,*

$$\lim_{n\to\infty} P_n\{L_n \in B(\rho^{(\beta)}, \varepsilon) \,|\, S_n/n \in [z-a, z]\} = 1\,. \tag{6.6}$$

The quantity $\rho^{(\beta)} = (\rho_1^{(\beta)}, \ldots, \rho_\alpha^{(\beta)})$ has the form

$$\rho_k^{(\beta)} = \frac{1}{\sum_{j=1}^{\alpha} \exp[\beta y_j]\, \rho_j} \cdot \exp[\beta y_k]\, \rho_k\,,$$

where $\beta = \beta(z) < 0$ is the unique value of β satisfying $\sum_{k=1}^{\alpha} y_k \rho_k^{(\beta)} = z$.

(b) *For any continuous function f mapping $\mathcal{P}_\alpha$ into $\mathbb{R}$,*

$$\lim_{n\to\infty} E^{P_n}\{f(L_n) \,|\, S_n/n \in [z-a, z]\} = f(\rho^{(\beta)})\,.$$

(c) *For each* $k \in \{1, \dots, \alpha\}$,

$$\lim_{n\to\infty} P_n\{X_1 = y_k \mid S_n/n \in [z-a, z]\} = \rho_k^{(\beta)}.$$

In order to prove the theorem, for $t \in \mathbb{R}$ we introduce

$$c(t) = \log E^{P_n}\{e^{tX_1}\} = \log\left(\sum_{k=1}^{\alpha} \exp[ty_k]\,\rho_k\right). \tag{6.7}$$

The function c, which equals the logarithm of the moment-generating function of X_1, is also known as the cumulant-generating function of X_1. In order to show that $\rho^{(\beta)}$ is well defined, we need the following lemma.

Lemma 6.4.2. *The cumulant-generating function* c *has the following properties.*

(a) $c''(t) > 0$ *for all* t, *i.e.* c *is strictly convex on* $\mathbb{R}$.

(b) $c'(0) = \sum_{k=1}^{\alpha} y_k \rho_k = \bar{y}$.

(c) $c'(t) \to y_1$ *as* $t \to -\infty$, *and* $c'(t) \to y_\alpha$ *as* $t \to \infty$.

(d) *The range of* $c'(t)$ *for* $t \in \mathbb{R}$ *is the open interval* (y_1, y_α), *which is the interior of the smallest interval containing the support* $\{y_1, y_2, \dots, y_\alpha\}$.

Proof.

(a) We define

$$\langle y \rangle_t = \frac{1}{\sum_{j=1}^{\alpha} \exp[ty_j]\,\rho_j} \cdot \sum_{k=1}^{\alpha} y_k \exp[ty_k]\,\rho_k$$

and

$$\langle (y - \langle y \rangle_t)^2 \rangle_t = \frac{1}{\sum_{j=1}^{\alpha} \exp[ty_j]\,\rho_j} \cdot \sum_{k=1}^{\alpha} (y_k - \langle y \rangle_t)^2 \exp[ty_k]\,\rho_k,$$

and calculate

$$c'(t) = \frac{1}{\sum_{j=1}^{\alpha} \exp[ty_j]\,\rho_j} \cdot \sum_{k=1}^{\alpha} y_k \exp[ty_k]\,\rho_k = \langle y \rangle_t$$

and

$$\begin{aligned} c''(t) &= \frac{1}{\sum_{j=1}^{\alpha} \exp[ty_j]\,\rho_j} \cdot \sum_{k=1}^{\alpha} y_k^2 \exp[ty_k]\,\rho_k - \langle y \rangle_t^2 \\ &= \langle (y - \langle y \rangle_t)^2 \rangle_t > 0. \end{aligned}$$

The last line gives part (a). This calculation shows that $c'(t)$ equals the mean of the probability measure $\exp[ty_k]\rho_k / \sum_{j=1}^{\alpha} \exp[ty_j]\,\rho_j$, and $c''(t)$ equals the variance of this probability measure.

(b) This follows from the formula for $c'(t)$ in part (a).

(c) Since $y_1 < y_j$ for all $j = 2, \dots, \alpha$,

$$\lim_{t \to -\infty} c'(t) = \lim_{t \to -\infty} \frac{1}{\sum_{j=1}^{\alpha} \exp[t(y_j - y_1)]\, \rho_j} \cdot \sum_{k=1}^{\alpha} y_k \exp[t(y_k - y_1]\, \rho_k = y_1 \, .$$

One similarly proves that $\lim_{t \to \infty} c'(t) = y_\alpha$.

(d) According to part (a), $c'(t)$ is a strictly increasing function of t. Hence the limits in part (c) show that the range of $c'(t)$ for $t \in \mathbb{R}$ equals the open interval (y_1, y_α). This completes the proof of the lemma. ■

We now prove Theorem 6.4.1.

Proof of Theorem 6.4.1. We first prove that $\rho^{(\beta)}$ is well defined. This follows immediately from Lemma 6.4.2, which shows that there exists a unique $\beta = \beta(z)$ satisfying

$$\begin{aligned} c'(\beta) &= \frac{1}{\sum_{j=1}^{\alpha} \exp[\beta y_j]\, \rho_j} \cdot \sum_{k=1}^{\alpha} y_k \exp[\beta y_k]\, \rho_k \\ &= \sum_{k=1}^{\alpha} y_k \rho_k^{(\beta)} = z \, , \end{aligned} \tag{6.8}$$

as claimed. Since $y_1 < z < \bar{y}$, $\beta = \beta(z)$ is negative.

Assuming the limit in part (a), we first prove the limits in parts (b) and (c). We then prove the limit in part (a).

(b) This limit follows from part (a) and the continuity of f. In order to see this, given $\delta > 0$, choose $\varepsilon > 0$ so that whenever $\gamma \in \mathcal{P}_\alpha$ lies in the open ball $B(\rho^{(\beta)}, \varepsilon)$, we have $|f(\gamma) - f(\rho^{(\beta)})| < \delta$. Then

$$\begin{aligned} &\left| E^{P_n}\{f(L_n) \,|\, S_n/n \in [z-a, z]\} - f(\rho^{(\beta)}) \right| \\ &\quad \le E^{P_n}\{|f(L_n) - f(\rho^{(\beta)})| \,|\, S_n/n \in [z-a, z]\} \\ &\quad = E^{P_n}\{|f(L_n) - f(\rho^{(\beta)})|\, 1_{B(\rho^{(\beta)}, \varepsilon)}(L_n) \,|\, S_n/n \in [z-a, z]\} \\ &\qquad + E^{P_n}\{|f(L_n) - f(\rho^{(\beta)})|\, 1_{[B(\rho^{(\beta)}, \varepsilon)]^c}(L_n) \,|\, S_n/n \in [z-a, z]\} \\ &\quad \le \delta + 2\|f\|_\infty P\{L_n \in [B(\rho^{(\beta)}, \varepsilon)]^c \,|\, S_n/n \in [z-a, z]\} \, . \end{aligned}$$

By part (a), the probability in the last line of the above equation converges to 0 as $n \to \infty$. Since $\delta > 0$ is arbitrary, part (b) is proved.

(c) By symmetry,

$$\begin{aligned} &P_n\{X_1 = y_k \,|\, S_n/n \in [z-a, z]\} \\ &\quad = E^{P_n}\{\delta_{X_1}\{y_k\} \,|\, S_n/n \in [z-a, z]\} \\ &\quad = E^{P_n}\left\{ \frac{1}{n} \sum_{j=1}^{n} \delta_{X_j}\{y_k\} \;\middle|\; S_n/n \in [z-a, z] \right\} \\ &\quad = E^{P_n}\{L_n(y_k) \,|\, S_n/n \in [z-a, z]\} \, . \end{aligned}$$

Now define f_k to be the continuous function on $\mathcal{P}_\alpha$ that maps γ to γ_k. Part (b) yields the desired limit:

$$\begin{aligned}
&\lim_{n\to\infty} P_n\{X_1 = y_k \mid S_n/n \in [z-a, z]\} \\
&= \lim_{n\to\infty} E^{P_n}\{L_n(y_k) \mid S_n/n \in [z-a, z]\} \\
&= \lim_{n\to\infty} E^{P_n}\{f_k(L_n) \mid S_n/n \in [z-a, z]\} \\
&= f_k(\rho^{(\beta)}) = \rho_k^{(\beta)} .
\end{aligned}$$

(a) In order to prove the limit

$$\lim_{n\to\infty} P_n\{L_n \in B(\rho^{(\beta)}, \varepsilon) \mid S_n/n \in [z-a, z]\} = 1 ,$$

we rewrite the event $\{S_n/n \in [z-a, z]\}$ in terms of L_n. Define the closed convex set

$$\Gamma(z) = \left\{ \gamma \in \mathcal{P}_\alpha : \sum_{k=1}^{\alpha} y_k \gamma_k \in [z-a, z] \right\} ,$$

which contains $\rho^{(\beta)}$. Since for each $\omega \in \Omega_n$

$$\frac{1}{n} S_n(\omega) = \sum_{k=1}^{\alpha} y_k \, L_n(\omega, y_k) ,$$

it follows that $\{\omega \in \Omega_n : S_n(\omega)/n \in [z-a, z]\} = \{\omega \in \Omega_n : L_n(\omega) \in \Gamma(z)\}$ and thus that

$$P_n\{L_n \in B(\rho^{(\beta)}, \varepsilon) \mid S_n/n \in [z-a, z]\} = P_n\{L_n \in B(\rho^{(\beta)}, \varepsilon) \mid L_n \in \Gamma(z)\} .$$

We first motivate the desired limit, using the formal notation

$$P_n\{L_n \in A\} \approx \exp[-n I_\rho(A)] \quad \text{as } n \to \infty , \tag{6.9}$$

which was introduced in eqn (6.4). Sanov's Theorem (Theorem 6.5.6) is the rigorous statement of the exponential behavior of the distribution of L_n; a special case that applies to the current setup is given in Theorem 6.3.3. For large n, we have by eqn (6.9)

$$\begin{aligned}
&P_n\{L_n \in B(\rho^{(\beta)}, \varepsilon) \mid S_n/n \in [z-a, z]\} \\
&= P_n\{L_n \in B(\rho^{(\beta)}, \varepsilon) \mid L_n \in \Gamma(z)\} \\
&= P_n\{L_n \in B(\rho^{(\beta)}, \varepsilon) \cap \Gamma(z)\} \cdot \frac{1}{P_n\{L_n \in \Gamma(z)\}} \\
&\approx \exp[-n(I_\rho(B(\rho^{(\beta)}, \varepsilon) \cap \Gamma(z)) - I_\rho(\Gamma(z)))] .
\end{aligned}$$

The last expression, and thus the probability in the first line of the equation, are of order 1 provided

$$I_\rho(B(\rho^{(\beta)}, \varepsilon) \cap \Gamma(z)) = I_\rho(\Gamma(z)) . \tag{6.10}$$

Proposition 6.4.3 shows that I_ρ attains its infimum over $\Gamma(z)$ at the unique point $\rho^{(\beta)}$. This gives eqn (6.10) and motivates the fact that for large n,

$$P_n\{L_n \in B(\rho^{(\beta)}, \varepsilon) \mid S_n/n \in [z-a, z]\} \approx 1\,.$$

We now convert these formal calculations into a proof of the limit

$$\lim_{n\to\infty} P_n\{L_n \in B(\rho^{(\beta)}, \varepsilon) \mid S_n/n \in [z-a, z]\} = 1\,.$$

We prove this by showing that

$$\begin{aligned}&\lim_{n\to\infty} P_n\{L_n \in [B(\rho^{(\beta)}, \varepsilon)]^c \mid S_n/n \in [z-a, z]\} \\ &\quad = \lim_{n\to\infty} P_n\{L_n \in [B(\rho^{(\beta)}, \varepsilon)]^c \mid L_n \in \Gamma(z)\} = 0\,.\end{aligned} \tag{6.11}$$

The key is to use Corollary 6.3.4, which states that if A is any Borel subset of $\mathcal{P}_\alpha$ satisfying $\overline{A^\circ} = \overline{A}$, then

$$\lim_{n\to\infty} \frac{1}{n} \log P_n\{L_n \in A\} = -I_\rho(A)\,.$$

Since both sets $[B(\rho^{(\beta)}, \varepsilon)]^c \cap \Gamma(z)$ and $\Gamma(z)$ satisfy the hypothesis of Corollary 6.3.4, it follows that

$$\begin{aligned}&\lim_{n\to\infty} \frac{1}{n} \log P_n\{L_n \in [B(\rho^{(\beta)}, \varepsilon)]^c \mid L_n \in \Gamma(z)\} \\ &= \lim_{n\to\infty} \frac{1}{n} P_n\{L_n \in [B(\rho^{(\beta)}, \varepsilon)]^c \cap \Gamma(z)\} - \lim_{n\to\infty} \frac{1}{n} \log P_n\{L_n \in \Gamma(z)\} \\ &= -I_\rho([B(\rho^{(\beta)}, \varepsilon)]^c \cap \Gamma(z)) + I_\rho(\Gamma(z))\,.\end{aligned}$$

According to Proposition 6.4.3, I_ρ attains its infimum over $\Gamma(z)$ at the unique point $\rho^{(\beta)}$. It follows that $I_\rho([B(\rho^{(\beta)}, \varepsilon)]^c \cap \Gamma(z)) - I_\rho(\Gamma(z)) > 0$. In combination with the last equation, this yields the desired limit (6.11), showing in fact that the convergence to 0 is exponentially fast.

The proof of Theorem 6.4.1 will be complete after we prove the next proposition, which is carried out without calculus by using properties of the relative entropy. In order to motivate the proposition, we consider the calculus problem of determining critical points of $I_\rho(\gamma)$ subject to the constraints that $\sum_{k=1}^{\alpha} \gamma_k = 1$ and $\sum_{k=1}^{\alpha} y_k \gamma_k = z$. Let λ and $-\beta$ be Lagrange multipliers corresponding to these two constraints. Then, for each k,

$$\begin{aligned}0 &= \frac{\partial(I_\rho(\gamma) + \lambda\,(\sum_{k=1}^{\alpha} \gamma_k - 1) - \beta\,(\sum_{k=1}^{\alpha} y_k\gamma_k) - z)}{\partial \gamma_k} \\ &= \log \gamma_k + 1 - \log \rho_k + \lambda - \beta y_k\,.\end{aligned}$$

It follows that $\gamma_k = \rho_k^{(\beta)}$, where $\beta = \beta(z)$ is as specified in the proposition.

Proposition 6.4.3. *Let $z \in (y_1, \bar{y})$ be given, and choose $a > 0$ such that $z - a > y_1$. Then I_ρ attains its infimum over*

$$\Gamma(z) = \{\gamma \in \mathcal{P}_\alpha : \sum_{k=1}^{\alpha} y_k \gamma_k \in [z - a, z]\}$$

at the unique point $\rho^{(\beta)} = (\rho_1^{(\beta)}, \ldots, \rho_\alpha^{(\beta)})$ defined in part (a) of Theorem 6.4.1.

Proof. We recall that for each $k \in \{1, \ldots, \alpha\}$,

$$\frac{\rho_k^{(\beta)}}{\rho_k} = \frac{1}{\sum_{j=1}^{\alpha} \exp[\beta y_j]\, \rho_j} \cdot \exp[\beta y_k] = \frac{1}{\exp[c(\beta)]} \cdot \exp[\beta y_k]\,,$$

where c is the cumulant-generating function defined in eqn (6.7). Hence, for any $\gamma \in \Gamma(z)$,

$$\begin{aligned} I_\rho(\gamma) = \sum_{k=1}^{\alpha} \gamma_k \log \frac{\gamma_k}{\rho_k} &= \sum_{k=1}^{\alpha} \gamma_k \log \frac{\gamma_k}{\rho_k^{(\beta)}} + \sum_{k=1}^{\alpha} \gamma_k \log \frac{\rho_k^{(\beta)}}{\rho_k} \\ &= I_{\rho^{(\beta)}}(\gamma) + \beta \sum_{k=1}^{\alpha} y_k \gamma_k - c(\beta)\,. \end{aligned}$$

Since $I_{\rho^{(\beta)}}(\rho^{(\beta)}) = 0$ and $\sum_{k=1}^{\alpha} y_k \rho_k^{(\beta)} = z$, it follows that

$$I_\rho(\rho^{(\beta)}) = I_{\rho^{(\beta)}}(\rho^{(\beta)}) + \beta \sum_{k=1}^{\alpha} y_k \rho_k^{(\beta)} - c(\beta) = \beta z - c(\beta)\,. \tag{6.12}$$

Now consider any $\gamma \in \Gamma(z)$, $\gamma \neq \rho^{(\beta)}$. Since $\beta < 0$, $\sum_{k=1}^{\alpha} y_k \gamma_k \leq z$, and $I_{\rho^{(\beta)}}(\gamma) \geq 0$ with equality if and only if $\gamma = \rho^{(\beta)}$ (Lemma 6.3.1), we obtain

$$\begin{aligned} I_\rho(\gamma) &= I_{\rho^{(\beta)}}(\gamma) + \beta \sum_{k=1}^{\alpha} y_k \gamma_k - c(\beta) \\ &> \beta \sum_{k=1}^{\alpha} y_k \gamma_k - c(\beta) \geq \beta z - c(\beta) = I_\rho(\rho^{(\beta)})\,. \end{aligned}$$

We conclude that for any $\gamma \in \Gamma(z)$, $I_\rho(\gamma) \geq I_\rho(\rho^{(\beta)})$ with equality if and only if $\gamma = \rho^{(\beta)}$. Thus I_ρ attains its infimum over $\Gamma(z)$ at the unique point $\rho^{(\beta)}$. The proof of the proposition and thus the proof of Theorem 6.4.1 are complete. ■

In the next section, we formulate the general concepts of a large deviation principle and a Laplace principle. Subsequent sections will apply the theory of large deviations to study interacting systems in statistical mechanics.

6.5 Generalities: Large deviation principle and Laplace principle

In Theorem 6.3.3, we formulate Sanov's Theorem, which is the large deviation principle for the empirical vectors L_n on the space $\mathcal{P}_\alpha$ of probability vectors in $\mathbb{R}^\alpha$. In Subsection 6.6.2, we apply Sanov's Theorem to analyze the phase-transition structure of a model of ferromagnetism known as the Curie–Weiss–Potts model. Applications of the theory of large deviations to other models in statistical mechanics require large deviation principles in different settings. As we will see in Subsection 6.6.1, analyzing the phase-transition structure of the model of ferromagnetism known as the Curie–Weiss model involves Cramér's Theorem. This theorem states the large deviation principle for the sample means of i.i.d. random variables, which in the case of the Curie–Weiss model take values in the closed interval $[-1, 1]$. Analyzing the Ising model in dimensions $d \geq 2$ is much more complicated. It involves a large deviation principle on the space of translation-invariant probability measures on $\{-1, 1\}^{\mathbb{Z}^d}$ (Ellis 1995, Section 11).

In order to define the general concept of a large deviation principle, we need some notation. First, for each $n \in \mathbb{N}$ let $(\Omega_n, \mathcal{F}_n, P_n)$ be a probability space. Thus Ω_n is a set of points, $\mathcal{F}_n$ is a σ-algebra of subsets of Ω_n, and P_n is a probability measure on $\mathcal{F}_n$. An example is given by the basic model in Section 6.2, where $\Omega_n = \Lambda^n = \{y_1, y_2, \ldots, y_\alpha\}^n$, $\mathcal{F}_n$ is the set of all subsets of Ω_n, and P_n is the product measure with one-dimensional marginals ρ.

Second, let $\mathcal{X}$ be a complete, separable metric space or, as it is often called, a Polish space. Thus there exists a function m, called a metric, mapping $\mathcal{X} \times \mathcal{X}$ into $[0, \infty)$ and having the properties that for all x, y, and z in $\mathcal{X}$, $m(x, y) = m(y, x)$ (symmetry), $m(x, y) = 0 \Leftrightarrow x = y$ (identity), and $m(x, z) \leq m(x, y) + m(y, z)$ (triangle inequality); furthermore, with respect to m, any Cauchy sequence in $\mathcal{X}$ converges to an element of $\mathcal{X}$ (completeness) and $\mathcal{X}$ has a countable dense subset (separability). Elementary examples are $\mathcal{X} = \mathbb{R}^d$ for $d \in \mathbb{N}$; $\mathcal{X} = \mathcal{P}_\alpha$, the set of probability vectors in $\mathbb{R}^\alpha$; and, in the notation of the basic probabilistic model in Section 6.2, $\mathcal{X}$ equal to the closed bounded interval $[y_1, y_\alpha]$. In all three cases the metric is the Euclidean distance.

Third, for each $n \in \mathbb{N}$ let Y_n be a random variable mapping Ω_n into $\mathcal{X}$. For example, in the notation of the basic probability model in Section 6.2, with $\mathcal{X} = \mathcal{P}_\alpha$, let $Y_n = L_n$ or, with $\mathcal{X} = [y_1, y_\alpha]$, let $Y_n = \sum_{j=1}^n X_j/n$, where $X_j(\omega) = \omega_j$ for $\omega \in \Omega_n = \Lambda^n$.

Before continuing, we need several standard definitions from topology. A subset G of $\mathcal{X}$ is said to be open if for any x in G there exists $\varepsilon > 0$ such that the open ball $B(x, \varepsilon) = \{y \in \mathcal{X} : m(y, x) < \varepsilon\}$ is a subset of G. A subset F of $\mathcal{X}$ is said to be closed if the complement, F^c, is open or, equivalently, if for any sequence x_n in F converging to some $x \in \mathcal{X}$, we have $x \in F$. Finally, a subset K of $\mathcal{X}$ is said to be compact if for any sequence x_n in K, there exists a subsequence of x_n converging to a point in K. Equivalently, K is compact if, whenever K is a subset of the union of a collection $\mathcal{C}$ of open sets, then K is a subset of the union of finitely many open sets in $\mathcal{C}$. If $\mathcal{X} = \mathbb{R}^d$, then K is compact if and only if K is closed and bounded.

A class of Polish spaces arising naturally in applications is obtained by taking a Polish space $\mathcal{Y}$ and considering the space $\mathcal{P}(\mathcal{Y})$ of probability measures on $(\mathcal{Y}, \mathcal{B}(\mathcal{Y}))$. $\mathcal{B}(\mathcal{Y})$ denotes the Borel σ-algebra of $\mathcal{Y}$, which is the σ-algebra generated by the open subsets of $\mathcal{Y}$. We say that a sequence $\{\Pi_n, n \in \mathbb{N}\}$ in $\mathcal{P}(\mathcal{Y})$ converges weakly to

$\Pi \in \mathcal{P}(\mathcal{Y})$, and write $\Pi_n \Rightarrow \Pi$, if $\int_{\mathcal{Y}} f d\Pi_n \to \int_{\mathcal{Y}} f d\Pi$ for all bounded, continuous functions f mapping $\mathcal{Y}$ into $\mathbb{R}$. A fundamental fact is that there exists a metric m on $\mathcal{P}(\mathcal{Y})$ such that $\Pi_n \Rightarrow \Pi$ if and only if $m(\Pi, \Pi_n) \to 0$ and $\mathcal{P}(\mathcal{Y})$ is a Polish space with respect to m (Ethier and Kurtz 1986, Section 3.1).

Let I be a function mapping the complete, separable metric space $\mathcal{X}$ into $[0, \infty]$. I is called a rate function if I has compact level sets, i.e. for all $M < \infty$, $\{x \in \mathcal{X} : I(x) \leq M\}$ is compact. This technical regularity condition implies that I has closed level sets or, equivalently, that I is lower semicontinuous, i.e. if $x_n \to x$, then $\liminf_{n\to\infty} I(x_n) \geq I(x)$. In particular, if $\mathcal{X}$ is compact, then the lower semicontinuity of I implies that I has compact level sets. When $\mathcal{X} = \mathcal{P}_\alpha$, an example of a rate function is the relative entropy I_ρ with respect to ρ; when $\mathcal{X} = [y_1, y_\alpha]$, any continuous function I mapping $[y_1, y_\alpha]$ into $[0, \infty)$ is a rate function.

We next define the concept of a large deviation principle. If Y_n satisfies the large deviation principle with rate function I, then we summarize this by the formal notation

$$P_n\{Y_n \in dx\} \asymp \exp[-nI(x)]\, dx\,.$$

For any subset A of $\mathcal{X}$, we define $I(A) = \inf_{x \in A} I(x)$.

Definition 6.2. (Large deviation principle.) *Let $\{(\Omega_n, \mathcal{F}_n, P_n), n \in \mathbb{N}\}$ be a sequence of probability spaces, $\mathcal{X}$ a complete, separable metric space, $\{Y_n, n \in \mathbb{N}\}$ a sequence of random variables such that Y_n maps Ω_n into $\mathcal{X}$, and I a rate function on $\mathcal{X}$. Then Y_n satisfies the large deviation principle on $\mathcal{X}$ with rate function I if the following two limits hold.*

Large deviation upper bound. *For any closed subset F of $\mathcal{X}$,*

$$\limsup_{n\to\infty} \frac{1}{n} \log P_n\{Y_n \in F\} \leq -I(F)\,.$$

Large deviation lower bound. *For any open subset G of $\mathcal{X}$,*

$$\liminf_{n\to\infty} \frac{1}{n} \log P_n\{Y_n \in G\} \geq -I(G)\,.$$

We shall explore several consequences of this definition. It is reassuring that a large deviation principle has a unique rate function. The following result is proved in Dupuis and Ellis (1997, Theorem 1.3.1).

Theorem 6.5.1. *If Y_n satisfies the large deviation principle on $\mathcal{X}$ with rate function I and with rate function J, then $I(x) = J(x)$ for all $x \in \mathcal{X}$.*

The next theorem gives a condition that guarantees the existence of large deviation limits.

Theorem 6.5.2. *Assume that Y_n satisfies the large deviation principle on $\mathcal{X}$ with rate function I. Let A be a Borel subset of $\mathcal{X}$ having closure $\overline{A}$ and interior A° and satisfying $I(\overline{A}) = I(A^\circ)$. Then*

$$\lim_{n\to\infty} \frac{1}{n} \log P\{Y_n \in A\} = -I(A)\,.$$

Proof. We evaluate the large deviation upper bound for $F = \overline{A}$ and the large deviation lower bound for $G = A^\circ$. Since $\overline{A} \supset A \supset A^\circ$, it follows that $I(\overline{A}) \leq I(A) \leq I(A^\circ)$ and

$$
\begin{aligned}
I(\overline{A}) &\geq \limsup_{n\to\infty} \frac{1}{n} \log P\{Y_n \in \overline{A}\} \geq \limsup_{n\to\infty} \frac{1}{n} \log P\{Y_n \in A\} \\
&\geq \liminf_{n\to\infty} \frac{1}{n} \log P\{Y_n \in A\} \geq \liminf_{n\to\infty} \frac{1}{n} \log P\{Y_n \in A^\circ\} \geq I(A^\circ) .
\end{aligned}
$$

By hypothesis, the two extreme terms are equal to each other and to $I(A)$, and so the theorem follows. ■

The next theorem states useful facts concerning the infimum of a rate function over the entire space and the use of the large deviation principle to show the convergence of a class of probabilities to 0. Part (b) generalizes Corollary 6.3.5.

Theorem 6.5.3. *Suppose that Y_n satisfies the large deviation principle on $\mathcal{X}$ with rate function I. The following conclusions hold.*

(a) *The infimum of I over $\mathcal{X}$ equals 0, and the set of $x \in \mathcal{X}$ for which $I(x) = 0$ is nonempty and compact.*

(b) *Define $\mathcal{E}$ to be the nonempty, compact set of $x \in \mathcal{X}$ for which $I(x) = 0$, and let A be a Borel subset of $\mathcal{X}$ such that $\overline{A} \cap \mathcal{E} = \emptyset$. Then $I(\overline{A}) > 0$, and for some $C < \infty$,*

$$
P_n\{Y_n \in A\} \leq C \exp[-nI(\overline{A})/2] \to 0 \quad \text{as } n \to \infty .
$$

Proof.

(a) We evaluate the large deviation upper bound for $F = \mathcal{X}$ and the large deviation lower bound for $G = \mathcal{X}$. Since $P\{Y_n \in \mathcal{X}\} = 1$, we obtain $I(\mathcal{X}) = 0$. We now prove that I attains its infimum of 0 by considering any infimizing sequence x_n that satisfies $I(x_n) \leq 1$ and $I(x_n) \to 0$ as $n \to \infty$. Since I has compact level sets, there exists a subsequence $x_{n'}$ and a point $x \in \mathcal{X}$ such that $x_{n'} \to x$. Hence, by the lower semicontinuity of I,

$$
0 = \lim I(x_{n'}) \geq I(x) \geq 0 .
$$

It follows that I attains its infimum of 0 at x and thus that the set of $x \in \mathcal{X}$ for which $I(x) = 0$ is nonempty and compact. This gives part (a).

(b) If $I(\overline{A}) > 0$, then the desired upper bound follows immediately from the large deviation upper bound. We prove that $I(\overline{A}) > 0$ by contradiction. If $I(\overline{A}) = 0$, then there exists a sequence x_n such that $\lim_{n\to\infty} I(x_n) = 0$. Since I has compact level sets and $\overline{A}$ is closed, there exists a subsequence $x_{n'}$ converging to an element $x \in \overline{A}$. Since I is lower semicontinuous, it follows that $I(x) = 0$ and thus that $x \in \mathcal{E}$. This contradicts the assumption that $\overline{A} \cap \mathcal{E} = \emptyset$. The proof of the proposition is complete. ■

We next state Cramér's Theorem, which is the large deviation principle for the sample means of i.i.d. random variables taking values in $\mathbb{R}^d$. The rate function is defined by a variational formula that in general cannot be evaluated explicitly. We denote by $\langle \cdot, \cdot \rangle$ the inner product on $\mathbb{R}^d$. The theorem is derived from the Gärtner–Ellis Theorem in Ellis (2006, Section VII.5); a direct proof is given in Ellis (2008, Section 7).

Theorem 6.5.4. (Cramér's Theorem.) *Let $\{X_j, j \in I\!N\}$ be a sequence of i.i.d. random vectors taking values in $\mathbb{R}^d$ and satisfying $E\{\exp\langle t, X_1\rangle\} < \infty$ for all $t \in \mathbb{R}^d$. We define the sample means $S_n/n = \sum_{j=1}^n X_j/n$ and the cumulant-generating function $c(t) = \log E\{\exp\langle t, X_1\rangle\}$. The following conclusions hold.*

(a) *The sequence of sample means S_n/n satisfies the large deviation principle on $\mathbb{R}^d$ with rate function $I(x) = \sup_{t \in \mathbb{R}^d}\{\langle t, x\rangle - c(t)\}$.*
(b) *I is a convex, lower semicontinuous function on $\mathbb{R}^d$, and it attains its infimum of 0 at the unique point $x_0 = E\{X_1\}$.*

A nice application of Cramér's Theorem is to derive the special case of Sanov's Theorem given in Theorem 6.3.3 (Ellis 2006, Section VIII.2). The latter states the large deviation principle for the empirical vectors of i.i.d. random variables having a finite state space. For application in Subsection 6.6.1, we next state a special case of Cramér's Theorem, for which the rate function can be given explicitly.

Corollary 6.5.5. *In the basic probability model of Section 6.2, let $\Lambda = \{-1, 1\}$ and $\rho = (\frac{1}{2}, \frac{1}{2})$, which corresponds to the probability measure $\rho = \frac{1}{2}\delta_{-1} + \frac{1}{2}\delta_1$ on Λ. For $\omega \in \Omega_n$, define $S_n(\omega) = \sum_{j=1}^n \omega_j$. Then the sequence of sample means S_n/n satisfies the large deviation principle on the closed interval $[-1, 1]$ with rate function*

$$I(x) = \tfrac{1}{2}(1-x)\log(1-x) + \tfrac{1}{2}(1+x)\log(1+x)\,. \tag{6.13}$$

Proof. In this case, $c(t) = \log(\frac{1}{2}[e^t + e^{-t}]$. The function $c(t)$ satisfies $c''(t) > 0$ for all t, and the range of c' equals $(-1, 1)$. Hence, for any $x \in (-1, 1)$, the supremum in the definition of I is attained at the unique $t = t(x)$ satisfying $c'(t(x)) = x$. One can easily verify that $t(x) = \frac{1}{2}\log[(1+x)/(1-x)]$ and that $I(x) = t(x) \cdot x - c(t(x))$ is given by eqn (6.13). When $x = 1$ or $x = -1$, the supremum in the definition of $I(x)$ is not attained but equals $\log 2$, which coincides with the value of the right side of eqn (6.13). ■

Corollary 6.5.5 is easy to motivate using the formal notation of Pseudo-theorem 6.3.2. For any $x \in [-1, 1]$, $S_n(\omega)/n \sim x$ if and only if approximately $(n/2)(1-x)$ of the ω_j's equal -1 and approximately $(n/2)(1+x)$ of the ω_j's equal 1. For any probability vector $\gamma = (\gamma_1, \gamma_2)$,

$$I_\rho(\gamma) = \gamma_1 \log(2\gamma_1) + \gamma_2 \log(2\gamma_2)\,.$$

Hence

$$\begin{aligned} P_n\{S_n/n \sim x\} &\approx P_n\{L_n(-1) = \tfrac{1}{2}(1-x), L_n(1) = \tfrac{1}{2}(1+x)\} \\ &\approx \exp[-nI_\rho(\tfrac{1}{2}(1-x), \tfrac{1}{2}(1+x))] = \exp[-nI(x)]\,. \end{aligned}$$

We next state a general version of Sanov's Theorem, which gives the large deviation principle for the sequence of empirical measures of i.i.d. random variables. Let $(\Omega, \mathcal{F}, P)$ be a probability space, $\mathcal{Y}$ a complete, separable metric space, ρ a probability measure on $\mathcal{Y}$, and $\{X_j, j \in I\!N\}$ a sequence of i.i.d. random variables mapping Ω into $\mathcal{Y}$ and having the common distribution ρ. For $n \in I\!N$, $\omega \in \Omega$, and A any Borel subset of $\mathcal{Y}$, we define the empirical measure

$$L_n(A) = L_n(\omega, A) = \frac{1}{n} \sum_{j=1}^{n} \delta_{X_j(\omega)}\{A\},$$

where for $y \in \mathcal{Y}$, $\delta_y\{A\}$ equals 1 if $y \in A$ and 0 if $y \notin A$. For each ω, $L_n(\omega, \cdot)$ is a probability measure on $\mathcal{Y}$. Hence the sequence L_n takes values in the complete, separable metric space $\mathcal{P}(\mathcal{Y})$. For $\gamma \in \mathcal{P}(\mathcal{Y})$, we write $\gamma \ll \rho$ if γ is absolutely continuous with respect to ρ, i.e. if $\rho\{A\} = 0$ for a Borel subset A of $\mathcal{Y}$, then $\gamma\{A\} = 0$. If $\gamma \ll \rho$, then $d\gamma/d\rho$ denotes the Radon–Nikodym derivative of γ with respect to ρ.

Theorem 6.5.6. (Sanov's Theorem.) *The sequence L_n satisfies the large deviation principle on $\mathcal{P}(\mathcal{Y})$ with a rate function given by the relative entropy with respect to ρ. For $\gamma \in \mathcal{P}(\mathcal{Y})$, this quantity is defined by*

$$I_\rho(\gamma) = \begin{cases} \int_{\mathcal{Y}} \left(\log \frac{d\gamma}{d\rho}\right) d\gamma & \text{if } \gamma \ll \rho, \\ \infty & \text{otherwise}. \end{cases}$$

This theorem is proved, for example, in Dembo and Zeitouni (1998, Section 6.2) and in Dupuis and Ellis (1997, Chapter 2). If the support of ρ is a finite set $\Lambda \subset \mathbb{R}$, then Theorem 6.5.6 reduces to Theorem 6.3.3.

The concept of a Laplace principle will be useful in the analysis of statistical-mechanical models. As we will see in Section 6.7, where a general class of statistical-mechanical models are studied, the Laplace principle gives a variational formula for the canonical free energy (Theorem 6.7.1(a)).

Definition 6.3. (Laplace principle.) *Let $\{(\Omega_n, \mathcal{F}_n, P_n), n \in I\!N\}$ be a sequence of probability spaces, $\mathcal{X}$ a complete, separable metric space, $\{Y_n, n \in I\!N\}$ a sequence of random variables such that Y_n maps Ω_n into $\mathcal{X}$, and I a rate function on $\mathcal{X}$. Then Y_n satisfies the Laplace principle on $\mathcal{X}$ with rate function I if for all bounded, continuous functions f mapping $\mathcal{X}$ into $\mathbb{R}$,*

$$\lim_{n\to\infty} \frac{1}{n} \log E^{P_n}\{\exp[nf(Y_n)]\} = \sup_{x\in\mathcal{X}}\{f(x) - I(x)\}.$$

Suppose that Y_n satisfies the large deviation principle on $\mathcal{X}$ with rate function I. Then substituting $P_n\{Y_n \in dx\} \asymp \exp[-nI(x)]\, dx$ gives

$$
\begin{aligned}
\lim_{n\to\infty}\frac{1}{n}\log E^{P_n}\{\exp[nf(Y_n)]\} &= \lim_{n\to\infty}\frac{1}{n}\log\int_{\Omega_n}\exp[nf(Y_n)]\,dP_n \\
&= \lim_{n\to\infty}\frac{1}{n}\log\int_{\mathcal{X}}\exp[nf(x)]\,P_n\{Y_n\in dx\} \\
&\approx \lim_{n\to\infty}\frac{1}{n}\log\int_{\mathcal{X}}\exp[nf(x)]\,\exp[-nI(x)]\,dx \\
&= \lim_{n\to\infty}\frac{1}{n}\log\int_{\mathcal{X}}\exp[n(f(x)-I(x)]\,dx\,.
\end{aligned}
$$

Since the asymptotic behavior of the last integral is determined by the largest value of the integrand, the last line of the above equation suggests the following limit:

$$
\lim_{n\to\infty}\frac{1}{n}\log E^{P_n}\{\exp[nf(Y_n)]\} = \sup_{x\in\mathcal{X}}\{f(x)-I(x)\}\,.
$$

Hence it is plausible that Y_n satisfies the Laplace principle with rate function I, a fact first proved by Varadhan (1966). In fact, we have the following stronger result, which shows that the large deviation principle and the Laplace principle are equivalent.

Theorem 6.5.7. *Y_n satisfies the large deviation principle on $\mathcal{X}$ with rate function I if and only if Y_n satisfies the Laplace principle on $\mathcal{X}$ with rate function I.*

We have just motivated the suggestion that the large deviation principle with rate function I implies the Laplace principle with the same rate function. In order to motivate the converse, let A be an arbitrary Borel subset of $\mathcal{X}$ and consider the function

$$
\varphi_A = \begin{cases} 0 & \text{if } x\in A\,, \\ -\infty & \text{if } x\in A^c\,. \end{cases}
$$

Clearly φ_A is not a bounded, continuous function on $\mathcal{X}$. If it were, then evaluating the Laplace expectation $E\{\exp[nf(Y_n)]\}$ for $f=\varphi_A$ would yield the large deviation limit

$$
\begin{aligned}
\lim_{n\to\infty}\frac{1}{n}\log P_n\{Y_n\in A\} &= \lim_{n\to\infty}\frac{1}{n}\log E^{P_n}\{\exp[n\varphi_A(Y_n)]\} \\
&= \sup_{x\in A}\{\varphi_A(x)-I(x)\} \\
&= -\inf_{x\in A} I(x) = -I(A)\,.
\end{aligned}
$$

The proof that the Laplace principle implies the large deviation principle involves approximating φ_A by suitable bounded, continuous functions. Theorem 6.5.7 is proved in Dupuis and Ellis (1997, Theorems 1.2.1 and 1.2.3).

We end this section by presenting two ways to obtain large deviation principles from existing large deviation principles. In the first theorem, we show that a large deviation principle is preserved under continuous mappings.

Theorem 6.5.8. (Contraction principle.) *Assume that Y_n satisfies the large deviation principle on $\mathcal{X}$ with rate function I and that ψ is a continuous function mapping*

$\mathcal{X}$ into a complete, separable metric space $\mathcal{Y}$. Then $\psi(Y_n)$ satisfies the large deviation principle on $\mathcal{Y}$ with rate function

$$J(y) = \inf\{I(x) : x \in \mathcal{X}, \psi(x) = y\} = \inf\{I(x) : x \in \psi^{-1}(y)\}.$$

Proof. Since I maps $\mathcal{X}$ into $[0, \infty]$, J maps $\mathcal{Y}$ into $[0, \infty]$. The fact that J has compact level sets in $\mathcal{Y}$ follows from the definition of J (Ellis 2008, Theorem 6.12). The large deviation upper bound is proved next. If F is a closed subset of $\mathcal{Y}$, then since ψ is continuous, $\psi^{-1}(F)$ is a closed subset of $\mathcal{X}$. Hence, by the large deviation upper bound for Y_n,

$$\begin{aligned}
\limsup_{n\to\infty} \frac{1}{n} \log P_n\{\psi(Y_n) \in F\} &= \limsup_{n\to\infty} \frac{1}{n} \log P_n\{Y_n \in \psi^{-1}(F)\} \\
&\leq - \inf_{x\in\psi^{-1}(F)} I(x) \\
&= - \inf_{y\in F} \{\inf\{I(x) : x \in \mathcal{X}, \psi(x) = y\} \\
&= - \inf_{y\in F} J(y) = -J(F).
\end{aligned}$$

The large deviation lower bound is proved similarly. The proof of the theorem is complete. ■

In the next theorem, we show that a large deviation principle is preserved if the probability measures P_n are multiplied by suitable exponential factors and then normalized. This result will be applied in Sections 6.6 and 6.7 when we prove the large deviation principle for statistical-mechanical models with respect to the canonical ensemble (Theorems 6.6.1, 6.6.3, 6.6.5, and 6.7.1).

Theorem 6.5.9. *Assume that with respect to the probability measures P_n, Y_n satisfies the large deviation principle on $\mathcal{X}$ with rate function I. Let ψ be a bounded, continuous function mapping $\mathcal{X}$ into $\mathbb{R}$. For $A \in \mathcal{F}_n$, we define new probability measures*

$$P_{n,\psi}\{A\} = \frac{1}{\int_{\mathcal{X}} \exp[-n\psi(Y_n)]\, dP_n} \cdot \int_A \exp[-n\psi(Y_n)]\, dP_n.$$

Then, with respect to $P_{n,\psi}$, Y_n satisfies the large deviation principle on $\mathcal{X}$ with rate function

$$I_\psi(x) = I(x) + \psi(x) - \inf_{y\in\mathcal{X}}\{I(y) + \psi(y)\}.$$

Proof. We omit the straightforward proof showing that since I has compact level sets and ψ is bounded and continuous, I_ψ has compact level sets. Since I_ψ maps $\mathcal{X}$ into $[0, \infty]$, it follows that I_ψ is a rate function. We complete the proof by showing that with respect to $P_{n,\psi}$, Y_n satisfies the equivalent Laplace principle with rate function I_ψ (Theorem 6.5.7). Let f be any bounded, continuous function mapping $\mathcal{X}$ into $\mathbb{R}$.

Since $f + \psi$ is bounded and continuous and since, with respect to P_n, Y_n satisfies the Laplace principle with rate function I, it follows that

$$\begin{aligned}
&\lim_{n\to\infty} \frac{1}{n} \log \int_{\Omega_n} \exp[nf(Y_n)]\, dP_{n,\psi} \\
&= \lim_{n\to\infty} \frac{1}{n} \log \int_{\Omega_n} \exp[n(f(Y_n) - \psi(Y_n))]\, dP_n \\
&\qquad - \lim_{n\to\infty} \frac{1}{n} \log \int_{\Omega_n} \exp[-n\psi(Y_n)]\, dP_n \\
&= \sup_{x\in\mathcal{X}} \{f(x) - \psi(x) - I(x)\} - \sup_{y\in\mathcal{X}} \{-\psi(y) - I(y)\} \\
&= \sup_{x\in\mathcal{X}} \{f(x) - I_\psi(x)\}\,.
\end{aligned}$$

Thus, with respect to $P_{n,\psi}$, Y_n satisfies the Laplace principle with rate function I_ψ, as claimed. This completes the proof. ■

This completes our discussion of the large deviation principle, the Laplace principle, and related general results. In the next section, we begin our study of statistical-mechanical models by considering the Curie–Weiss spin model and other mean-field models.

6.6 The Curie–Weiss model and other mean-field models

Mean-field models in statistical mechanics lend themselves naturally to a large deviation analysis. We illustrate this by first studying the Curie–Weiss model of ferromagnetism, one of the simplest examples of an interacting system in statistical mechanics. After treating this model, we also outline a large deviation analysis of two other mean-field models, the Curie–Weiss–Potts model and the mean-field Blume–Capel model. As we will see in the next section, using the theory of large deviations to analyze these models suggests how one can apply the theory to analyze much more complicated models.

6.6.1 Curie–Weiss model

The Curie–Weiss model is a spin model defined on the complete graph on n vertices $1, 2, \ldots, n$. It is a mean-field approximation to the Ising model and related, short-range, ferromagnetic models (Ellis 2006, Section V.9). In the Curie–Weiss model the spin at site $j \in \{1, 2, \ldots, n\}$ is denoted by ω_j, a quantity taking values in $\Lambda = \{-1, 1\}$; the value -1 represents spin-down and the value 1 spin-up. The configuration space for the model is the set $\Omega_n = \Lambda^n$, containing all configurations or microstates $\omega = (\omega_1, \omega_2, \ldots, \omega_n)$ with each $\omega_j \in \Lambda$.

Let $\rho = \frac{1}{2}(\delta_{-1} + \delta_1)$ and let P_n denote the product measure on Ω_n with one-dimensional marginals ρ. Thus $P_n\{\omega\} = 1/2^n$ for each $\omega \in \Omega_n$. For $\omega \in \Omega_n$, we define the spin per site $S_n(\omega)/n = \sum_{j=1}^{n} \omega_j/n$. The Hamiltonian, or energy function, is defined by

$$H_n(\omega) = -\frac{1}{2n} \sum_{i,j=1}^{n} \omega_i \omega_j = -\frac{n}{2}\left(\frac{S_n(\omega)}{n}\right)^2, \tag{6.14}$$

and the probability of ω corresponding to an inverse temperature $\beta > 0$ is defined by the canonical ensemble

$$\begin{aligned} P_{n,\beta}\{\omega\} &= \frac{1}{Z_n(\beta)} \cdot \exp[-\beta H_n(\omega)]\, P_n\{\omega\} \\ &= \frac{1}{Z_n(\beta)} \cdot \exp\left[\frac{n\beta}{2}\left(\frac{S_n(\omega)}{n}\right)^2\right] P_n\{\omega\}, \end{aligned} \tag{6.15}$$

where $Z_n(\beta)$ is the partition function

$$Z_n(\beta) = \int_{\Omega_n} \exp[-\beta H_n(\omega)]\, P_n(d\omega) = \int_{\Omega_n} \exp\left[\frac{n\beta}{2}\left(\frac{S_n(\omega)}{n}\right)^2\right] P_n(d\omega).$$

$P_{n,\beta}$ models a ferromagnet in the sense that the maximum of $P_{n,\beta}\{\omega\}$ over $\omega \in \Omega_n$ occurs at the two microstates having all coordinates ω_i equal to -1 or all coordinates equal to 1. Furthermore, as $\beta \to \infty$ all the mass of $P_{n,\beta}$ concentrates on these two microstates.

A distinguishing feature of the Curie–Weiss model is its phase transition. Namely, the alignment effects incorporated in the canonical ensemble $P_{n,\beta}$ persist in the limit $n \to \infty$. This is most easily seen by evaluating the $n \to \infty$ limit of the distributions $P_{n,\beta}\{S_n/n \in dx\}$. For $\beta \leq 1$ the alignment effects are relatively weak, and as we will see, S_n/n satisfies a law of large numbers, concentrating on the value 0. However, for $\beta > 1$ the alignment effects are so strong that the law of large numbers breaks down, and the limiting $P_{n,\beta}$ distribution of S_n/n concentrates on two points $\pm m(\beta)$ for some $m(\beta) \in (0, 1)$ (see eqn (6.17)). The analysis of the Curie–Weiss model to be presented below can be easily modified to handle an external magnetic field h. The resulting probabilistic description of the phase transition yields the predictions of mean-field theory (Ellis 2006, Section V.9; Parisi 1988, Section 3.2).

We calculate the $n \to \infty$ limit of $P_{n,\beta}\{S_n/n \in dx\}$ by establishing a large deviation principle for the spin per site S_n/n with respect to $P_{n,\beta}$. For each n, S_n/n takes values in $[-1, 1]$. According to the special case of Cramér's Theorem given in Corollary 6.5.5, with respect to the product measures P_n, S_n/n satisfies the large deviation principle on $[-1, 1]$ with rate function

$$I(x) = \tfrac{1}{2}(1-x)\log(1-x) + \tfrac{1}{2}(1+x)\log(1+x). \tag{6.16}$$

Because of the form of $P_{n,\beta}$ given in eqn (6.15), the large deviation principle for S_n/n with respect to $P_{n,\beta}$ is an immediate consequence of Theorem 6.5.9 with $\mathcal{X} = [-1, 1]$ and $\psi(x) = -\frac{1}{2}\beta x^2$ for $x \in [-1, 1]$. The proof of that theorem uses the exponential form of the measures to derive the equivalent Laplace principle. We record the large deviation principle in the next theorem.

Theorem 6.6.1. *With respect to the canonical ensemble $P_{n,\beta}$ defined in eqn (6.15), the spin per site S_n/n satisfies the large deviation principle on $[-1,1]$ with rate function*

$$I_\beta(x) = I(x) - \tfrac{1}{2}\beta x^2 - \inf_{y\in[-1,1]}\{I(y) - \tfrac{1}{2}\beta y^2\},$$

where $I(x)$ is defined in eqn (6.16).

The limiting behavior of the distributions $P_{n,\beta}\{S_n/n \in dx\}$ is now determined by examining where the nonnegative rate function I_β attains its infimum of 0 or, equivalently, where $I(x) - \frac{1}{2}\beta x^2$ attains its infimum $\inf_{y\in[-1,1]}\{I(y) - \frac{1}{2}\beta y^2\}$ (Ellis 2006, Section IV.4). Global minimum points x^* satisfy

$$I_\beta'(x^*) = 0 \quad \text{or} \quad I'(x^*) = \beta x^*.$$

The second equation is equivalent to the mean-field equation $x^* = (I')^{-1}(\beta x^*) = \tanh(\beta x^*)$ (Ellis 2006, Section V.9; Parisi 1988, Section 3.2). Figure 6.1 motivates the next theorem, which is a consequence of the following easily verified properties of I: (a) $I''(0) = 1$; (b) I' is convex on $[0,1]$ and $\lim_{x\to 1} I'(x) = \infty$; (c) I' is concave on $[-1,0]$ and $\lim_{x\to -1} I'(x) = -\infty$.

Theorem 6.6.2. *For each $\beta > 0$, we define*

$$\begin{aligned}\mathcal{E}_\beta &= \{x \in [-1,1] : I_\beta(x) = 0\}\\ &= \left\{x \in [-1,1] : I(x) - \tfrac{1}{2}\beta x^2 \text{ is minimized}\right\}.\end{aligned}$$

The following conclusions hold.

(a) *For $0 < \beta \le 1$, $\mathcal{E}_\beta = \{0\}$.*
(b) *For $\beta > 1$, there exists $m(\beta) > 0$ such that $\mathcal{E}_\beta = \{\pm m(\beta)\}$. The function $m(\beta)$ is monotonically increasing on $(1,\infty)$ and satisfies $m(\beta) \to 0$ as $\beta \to 1^+$, and $m(\beta) \to 1$ as $\beta \to \infty$.*

According to part (b) of Theorem 6.5.3, if A is any closed subset of $[-1,1]$ such that $A \cap \mathcal{E}_\beta = \emptyset$, then $I(A) > 0$ and, for some $C < \infty$,

$$P_{n,\beta}\{S_n/n \in A\} \le C \exp[-nI(A)/2] \to 0 \text{ as } n \to \infty.$$

In combination with Theorem 6.6.2, we are led to the following weak limits:

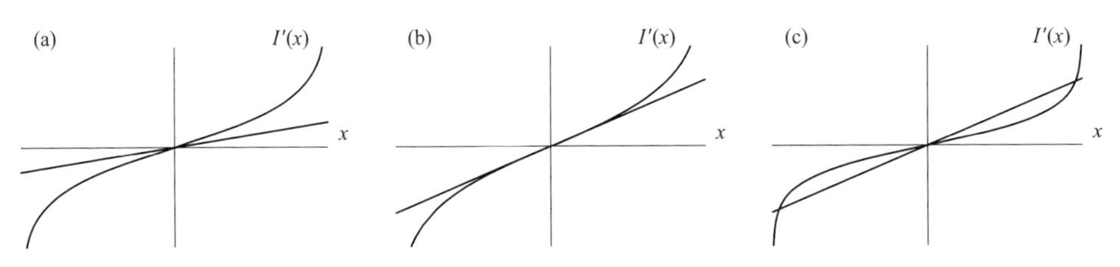

Fig. 6.1 Solutions of $I'(x^*) = \beta x^*$: (a) $\beta < 1$, (b) $\beta = 1$, (c) $\beta > 1$.

$$P_{n,\beta}\{S_n/n \in dx\} \Longrightarrow \begin{cases} \delta_0 & \text{if } 0 < \beta \le 1\,, \\ \frac{1}{2}\delta_{m(\beta)} + \frac{1}{2}\delta_{-m(\beta)} & \text{if } \beta > 1\,. \end{cases} \tag{6.17}$$

We call $m(\beta)$ the spontaneous magnetization for the Curie–Weiss model and $\beta_c = 1$ the critical inverse temperature (Ellis 2006, Section IV.4). It is worth remarking that it is much easier to derive the weak limits in eqn (6.17) from the large deviation principle for S_n/n with respect to $P_{n,\beta}$ rather than to prove the weak limits directly.

The limit (6.17) justifies calling $\mathcal{E}_\beta$ the set of canonical equilibrium macrostates for the spin per site S_n/n in the Curie–Weiss model. Because $m(\beta) \to 0$ as $\beta \to 1^+$ and 0 is the unique equilibrium macrostate for $0 < \beta \le 1$, the phase transition at β_c is said to be continuous or second-order.

The phase transition in the Curie–Weiss model and in related models arises as a result of two competing microscopic effects. The first effect tends to randomize the system. It is caused by thermal excitations and is measured by entropy. The second effect tends to order the system. It is caused by attractive forces of interaction and is measured by the energy. At sufficiently low temperatures and thus for sufficiently large values of β, the energy effect predominates and a phase transition becomes possible. This phenomenology is reflected in the form of $\mathcal{E}_\beta$ given in Theorem 6.6.2. For $0 < \beta \le 1$, I_β has a unique global minimum point coinciding with the unique global minimum point of I at 0. For $\beta > 1$, I_β has two global minimum points at $\pm m(\beta)$, a structure that is consistent with the facts that $-\frac{1}{2}\beta x^2$ has two global minimum points on $[-1, 1]$ at 1 and -1 and that $m(\beta) \to 1$ as $\beta \to \infty$. For x near 0, $I(x) \sim \frac{1}{2}x^2 + \frac{1}{12}x^4$ and thus $I(x) - \frac{1}{2}\beta x^2 \sim \frac{1}{2}(1-\beta)x^2 + \frac{1}{12}x^4$. The set of global minimum points of the latter function bifurcates continuously from $\{0\}$ for $\beta \le 1$ to $\{\pm\sqrt{3(\beta-1)}\}$ for $\beta > 1$. This behavior is consistent with the continuous phase transition described in Theorem 6.6.2 and suggests that as $\beta \to 1^+$, $m(\beta) \sim \sqrt{3(\beta-1)} \to 0$.

Before we leave the Curie–Weiss model, there are several additional points that should be emphasized. The first is to focus on what makes possible the large deviation analysis of the phase transition in the model. In eqn (6.14), we write the Hamiltonian as a quadratic function of the spin per site S_n/n, which by the version of Cramér's Theorem given in Corollary 6.5.5 satisfies the large deviation principle on $[-1, 1]$ with respect to the product measures P_n. The equivalent Laplace principle allows us to convert this large deviation principle into a large deviation principle with respect to the canonical ensemble $P_{n,\beta}$. The form of the rate function I_β allows us to complete the analysis. In this context, the sequence S_n/n is called the sequence of macroscopic variables for the Curie–Weiss model. In the next section, we will generalize these steps to formulate a large deviation approach to a wide class of models in statistical mechanics.

Our large deviation analysis of the phase transition in the Curie–Weiss model has the attractive feature that it directly motivates us to attach physical importance to $\mathcal{E}_\beta$. This set is the support of the $n \to \infty$ limit of the distributions $P_{n,\beta}\{S_n/n \in dx\}$. An analogous fact is true for a large class of statistical-mechanical models (Ellis *et al.* 2000, Theorem 2.5).

As shown in eqn (6.17), the large deviation analysis of the Curie–Weiss model yields the limiting behavior of the $P_{n,\beta}$ distributions of S_n/n. For $0 < \beta \le 1$, this limit corresponds to the classical weak law of large numbers for the sample means of i.i.d. random variables and suggests examining the analogues of other classical limit results such as the central limit theorem. We end this section by summarizing these limit results for the Curie–Weiss model, referring the reader to (Ellis 2006, Section V.9) for proofs. If $\theta \in (0,1)$ and f is a nonnegative integrable function on $\mathbb{R}$, then the notation $P_{n,\beta}\{S_n/n^\theta \in dx\} \Longrightarrow f(x)\,dx$ means that the distributions of S_n/n^θ converge weakly to the probability measure on $\mathbb{R}$ having a density proportional to f with respect to the Lebesgue measure.

In the Curie–Weiss model for $0 < \beta < 1$, the alignment effects among the spins are relatively weak, and the analogue of the central limit theorem holds (Ellis 2006, Theorem V.9.4):

$$P_{n,\beta}\left\{\frac{S_n}{n^{1/2}} \in dx\right\} \Longrightarrow \exp\left[\frac{-\frac{1}{2}x^2}{\sigma^2(\beta)}\right] dx\,,$$

where $\sigma^2(\beta) = 1/(1-\beta)$. However, when $\beta = \beta_c = 1$, the limiting variance $\sigma^2(\beta)$ diverges, and the central-limit scaling $n^{1/2}$ must be replaced by $n^{3/4}$, which reflects the onset of long-range order at β_c. In this case we have (Ellis 2006, Theorem V.9.5)

$$P_{n,\beta_c}\left\{\frac{S_n}{n^{3/4}} \in dx\right\} \Longrightarrow \exp\left[-\frac{1}{12}x^4\right] dx\,.$$

Finally, for $\beta > \beta_c$, $(S_n - n\tilde{z})/n^{1/2}$ satisfies a central-limit-type theorem when S_n/n is conditioned to lie in a sufficiently small neighborhood of $\tilde{z} = m(\beta)$ or $\tilde{z} = -m(\beta)$; see Theorem 2.4 in Ellis *et al.* (1980) with $k = 1$.

The results discussed in this subsection have been extensively generalized to a number of models, including the Curie–Weiss–Potts model (Costeniuc *et al.* 2005*a*; Ellis and Wang 1990), the mean-field Blume–Capel model (Costeniuc *et al.* 2007; Ellis *et al.* 2005), and the Ising and related models (Dobrushin and Shlosman 1994; Föllmer and Orey 1987; Olla 1988), which exhibit much more complicated behavior and are much harder to analyze. For the Ising and related models, refined large deviations at the surface level have been studied; see Dembo and Zeitouni (1998, p. 339) for references.

This completes our discussion of the Curie–Weiss model. In order to reinforce our understanding of the large deviation analysis of that model, in the next two subsections we present the large deviation analysis of the Curie–Weiss–Potts model and the Blume–Capel model. This, in turn, will yield the phase-transition structure of the two models as in Theorem 6.6.2.

6.6.2 Curie–Weiss–Potts model

Let $q \geq 3$ be a fixed integer, and define $\Lambda = \{y_1, y_2, \ldots, y_q\}$, where the y_k are any q distinct vectors in $\mathbb{R}^q$; the precise values of these vectors are immaterial. The Curie–Weiss–Potts model is a spin model defined on the complete graph on n vertices $1, 2, \ldots, n$. It is a mean-field approximation to the well-known Potts model (Wu 1982). In the Curie–Weiss–Potts model, the spin at site $j \in \{1, 2, \ldots, n\}$ is denoted by ω_j, a quantity taking values in Λ. The configuration space for the model is the set $\Omega_n = \Lambda^n$ containing all microstates $\omega = (\omega_1, \omega_2, \ldots, \omega_n)$ with each $\omega_j \in \Lambda$.

Let $\rho = (1/q)\sum_{i=1}^q \delta_{y_i}$ and let P_n denote the product measure on Ω_n with one-dimensional marginals ρ. Thus $P_n\{\omega\} = 1/q^n$ for each configuration $\omega = \{\omega_i, i =$

$1, \ldots, n\} \in \Omega_n$. We also denote by ρ the probability vector in $\mathbb{R}^q$ all of whose coordinates equal q^{-1}. For $\omega \in \Omega_n$, the Hamiltonian is defined by

$$H_n(\omega) = -\frac{1}{2n} \sum_{i,j=1}^{n} \delta(\omega_i, \omega_j),$$

where $\delta(\omega_i, \omega_j)$ equals 1 if $\omega_i = \omega_j$ and equals 0 otherwise. The probability of ω corresponding to an inverse temperature $\beta > 0$ is defined by the canonical ensemble

$$P_{n,\beta}\{\omega\} = \frac{1}{Z_n(\beta)} \cdot \exp[-\beta H_n(\omega)]\, P_n\{\omega\}, \tag{6.18}$$

where $Z_n(\beta)$ is the partition function

$$Z_n(\beta) = \int_{\Omega_n} \exp[-\beta H_n(\omega)]\, P_n(d\omega) = \sum_{\omega \in \Omega_n} \exp[-\beta H_n(\omega)]\, \frac{1}{2^n}.$$

In order to carry out a large deviation analysis of the Curie–Weiss–Potts model, we rewrite the sequence of Hamiltonians H_n as a function of a sequence of macroscopic variables, which is a sequence of random variables that satisfies a large deviation principle. This sequence is the sequence of empirical vectors

$$L_n = L_n(\omega) = (L_n(\omega, y_1), L_n(\omega, y_2), \ldots, L_n(\omega, y_q)),$$

the kth component of which is defined by

$$L_n(\omega, y_k) = \frac{1}{n} \sum_{j=1}^{n} \delta(\omega_j, y_k).$$

This quantity equals the relative frequency with which y_k appears in the configuration ω. The empirical vectors L_n take values in the set of probability vectors

$$\mathcal{P}_q = \left\{ \nu \in \mathbb{R}^q : \nu = (\nu_1, \nu_2, \ldots, \nu_q) : \nu_k \geq 0, \sum_{k=1}^{q} \nu_k = 1 \right\}.$$

Let $\langle \cdot, \cdot \rangle$ denote the inner product on $\mathbb{R}^q$. Since

$$\langle L_n(\omega), L_n(\omega) \rangle = \frac{1}{n^2} \sum_{i,j=1}^{n} \left(\sum_{k=1}^{q} \delta(\omega_i, y_k) \cdot \delta(\omega_j, y_k) \right) = \frac{1}{n^2} \sum_{i,j=1}^{n} \delta(\omega_i, \omega_j),$$

it follows that the Hamiltonian and the canonical ensemble can be rewritten as

$$H_n(\omega) = -\frac{1}{2n} \sum_{i,j=1}^{n} \delta(\omega_i, \omega_j) = -\frac{n}{2} \langle L_n(\omega), L_n(\omega) \rangle$$

and

$$P_{n,\beta}\{\omega\} = \frac{1}{\int_{\Omega_n} \exp[(n\beta/2)\langle L_n(\omega), L_n(\omega)\rangle]\, P_n\{\omega\}} \cdot \exp\left[\frac{n\beta}{2} \langle L_n(\omega), L_n(\omega) \rangle \right] P_n\{\omega\}.$$

We next establish a large deviation principle for L_n with respect to $P_{n,\beta}$. According to the special case of Sanov's Theorem given in Theorem 6.3.3, with respect to the product measures P_n, L_n satisfies the large deviation principle on $\mathcal{P}_\alpha$ with the rate function being the relative entropy I_ρ. Because of the form of $P_{n,\beta}$ given in the last equation, the large deviation principle for L_n with respect to $P_{n,\beta}$ is an immediate consequence of Theorem 6.5.9 with $\mathcal{X} = \mathcal{P}_q$ and $\psi(\nu) = -\frac{1}{2}\beta\langle\nu,\nu\rangle$ for $\nu \in \mathcal{P}_q$. We record the large deviation principle in the next theorem.

Theorem 6.6.3. *With respect to the canonical ensemble $P_{n,\beta}$ defined in eqn (6.18), the empirical vector L_n satisfies the large deviation principle on $\mathcal{P}_\alpha$ with rate function*

$$I_\beta(\nu) = I_\rho(\nu) - \tfrac{1}{2}\beta\langle\nu,\nu\rangle - \inf_{\gamma\in\mathcal{P}_q}\{I_\rho(\gamma) - \tfrac{1}{2}\beta\langle\gamma,\gamma\rangle\}\,.$$

As in the Curie–Weiss model, we define the set $\mathcal{E}_\beta$ of canonical equilibrium macrostates for the empirical vector L_n in the Curie–Weiss–Potts model to be the zero set of the rate function I_β or, equivalently, the set of $\nu \in \mathcal{P}_q$ at which $I_\rho(\nu) - \frac{1}{2}\beta\langle\nu,\nu\rangle$ attains its minimum. Thus

$$\begin{aligned}\mathcal{E}_\beta &= \{\nu \in \mathcal{P}_q : I_\beta(\nu) = 0\}\\ &= \left\{\nu \in \mathcal{P}_q : I_\rho(\nu) - \tfrac{1}{2}\beta\langle\nu,\nu\rangle \text{ is minimized}\right\}.\end{aligned}$$

The structure of $\mathcal{E}_\beta$ given in the next theorem is consistent with the entropy–energy competition underlying the phase transition. Let β_c be the critical inverse temperature given in the theorem. For $0 < \beta < \beta_c$, I_β has a unique global minimum point coinciding with the unique global minimum point of I_ρ at $\rho = (q^{-1}, q^{-1}, \ldots, q^{-1})$. For $\beta > \beta_c$, I_β has q global minimum points, a structure that is consistent with the fact that $-\frac{1}{2}\beta\langle\gamma,\gamma\rangle$ has q global minimum points in $\mathcal{P}_q$ at the vectors that equal 1 in the kth coordinate and 0 in the other coordinates for $i = 1, 2, \ldots, q$. As β increases through β_c, $\mathcal{E}_\beta$ bifurcates discontinuously from $\{\rho\}$ for $0 < \beta < \beta_c$ to a set containing $q+1$ distinct points for $\beta = \beta_c$ and to a set containing q distinct points for $\beta > \beta_c$. Because of this behavior, the Curie–Weiss–Potts model is said to have a discontinuous or first-order phase transition at β_c. The structure of $\mathcal{E}_\beta$ is given in terms of the function $\varphi : [0,1] \to \mathcal{P}_q$ defined by

$$\varphi(s) = (q^{-1}[1 + (q-1)s], q^{-1}(1-s), \ldots, q^{-1}(1-s))\,;$$

the last $(q-1)$ components all equal $q^{-1}(1-s)$. We note that $\varphi(0) = \rho$.

Theorem 6.6.4. *We fix a positive integer $q \geq 3$. Let $\beta_c = (2(q-1)/(q-2))\log(q-1)$ and, for $\beta > 0$, let $s(\beta)$ be the largest solution of the equation $s = (1-e^{-\beta s})/(1+(q-1)e^{-\beta s})$. The following conclusions hold.*

(a) *The quantity $s(\beta)$ is well defined. It is positive, strictly increasing, and differentiable in β on an open interval containing $[\beta_c, \infty)$, $s(\beta_c) = (q-2)/(q-1)$, and $\lim_{\beta\to\infty} s(\beta) = 1$.*

(b) *For $\beta \geq \beta_c$, define $\nu^1(\beta) = \varphi(s(\beta))$ and let $\nu^k(\beta)$, $k = 2, \ldots, q$, denote the points in $\mathcal{P}_q$ obtained by interchanging the first and kth coordinates of $\nu^1(\beta)$. Then*

$$\mathcal{E}_\beta = \begin{cases} \{\rho\} & \text{for } 0 < \beta < \beta_c\,, \\ \{\nu^1(\beta), \nu^2(\beta), \ldots, \nu^q(\beta)\} & \text{for } \beta > \beta_c\,, \\ \{\rho, \nu^1(\beta_c), \nu^2(\beta_c), \ldots, \nu^q(\beta_c)\} & \text{for } \beta = \beta_c\,. \end{cases}$$

For $\beta \geq \beta_c$, the points in $\mathcal{E}_\beta$ are all distinct, and $\nu^k(\beta)$ is a continuous function of $\beta \geq \beta_c$.

This theorem is proved in Ellis and Wang (1990) by replacing $I_\rho(\gamma) - \frac{1}{2}\beta\langle\gamma, \gamma\rangle$ by another function that has the same global minimum points but for which the analysis is much more straightforward. The two functions are related by Legendre–Fenchel transforms. Probabilistic limit theorems for the Curie–Weiss–Potts model are proved in Ellis and Wang (1990) and Ellis and Wang (1992). Having completed our discussion of the Curie–Weiss–Potts model, we next consider a third mean-field model that exhibits different features.

6.6.3 Mean-field Blume–Capel model

We end this section by considering a mean-field version of an important spin model due to Blume and Capel (Blume 1966; Capel 1966, 1967*a*,*b*). This mean-field model is one of the simplest models that exhibits the following intricate phase-transition structure: a curve of second-order points; a curve of first-order points; and a tricritical point, which separates the two curves. A generalization of the Blume–Capel model is studied in Blume *et al.* (1971).

The mean-field Blume–Capel model is defined on the complete graph on n vertices $1, 2, \ldots, n$. The spin at site $j \in \{1, 2, \ldots, n\}$ is denoted by ω_j, a quantity taking values in $\Lambda = \{-1, 0, 1\}$. The configuration space for the model is the set $\Omega_n = \Lambda^n$, containing all microstates $\omega = (\omega_1, \omega_2, \ldots, \omega_n)$ with each $\omega_j \in \Lambda$. In terms of a positive parameter K representing the interaction strength, the Hamiltonian is defined by

$$H_{n,K}(\omega) = \sum_{j=1}^n \omega_j^2 - \frac{K}{n}\left(\sum_{j=1}^n \omega_j\right)^2$$

for each $\omega \in \Omega_n$. Let P_n be the product measure on Λ^n with identical one-dimensional marginals $\rho = \frac{1}{3}(\delta_{-1} + \delta_0 + \delta_1)$. Thus P_n assigns the probability 3^{-n} to each $\omega \in \Omega_n$. The probability of ω corresponding to an inverse temperature $\beta > 0$ and an interaction strength $K > 0$ is defined by the canonical ensemble

$$P_{n,\beta,K}\{\omega\} = \frac{1}{Z_n(\beta, K)} \cdot \exp[-\beta H_{n,K}(\omega)]\, P_n\{\omega\}\,,$$

where $Z_n(\beta, K)$ is the partition function

$$Z_n(\beta, K) = \int_{\Omega_n} \exp[-\beta H_{n,K}(\omega)]\, P_n(d\omega) = \sum_{\omega \in \Omega_n} \exp[-\beta H_{n,K}(\omega)]\, \frac{1}{3^n}\,.$$

The large deviation analysis of the canonical ensemble $P_{n,\beta,K}$ is facilitated by absorbing the noninteracting component of the Hamiltonian into the product measure P_n, and we obtain

$$P_{n,\beta,K}\{\omega\} = \frac{1}{\widetilde{Z}_n(\beta,K)} \cdot \exp\left[n\beta K\left(\frac{S_n(\omega)}{n}\right)^2\right](P_\beta)_n\{\omega\}. \tag{6.19}$$

In this formula, $S_n(\omega)$ equals the total spin $\sum_{j=1}^n \omega_j$; $(P_\beta)_n$ is the product measure on Ω_n with identical one-dimensional marginals

$$\rho_\beta\{\omega_j\} = \frac{1}{Z(\beta)} \cdot \exp(-\beta\omega_j^2)\,\rho\{\omega_j\}; \tag{6.20}$$

$Z(\beta)$ is the normalization, equal to $\int_\Lambda \exp(-\beta\omega_j^2)\rho(d\omega_j) = (1+2e^{-\beta})/3$; and

$$\widetilde{Z}_n(\beta,K) = \frac{[Z(\beta)]^n}{Z_n(\beta,K)} = \int_{\Omega_n} \exp\left[n\beta K\left(\frac{S_n(\omega)}{n}\right)^2\right](P_\beta)_n\{\omega\}.$$

Comparing eqn (6.19) with (6.15), we see that the mean-field Blume–Capel model has the form of a Curie–Weiss model in which β and the product measure P_n in the latter are replaced by $2\beta K$ and the β-dependent product measure $(P_\beta)_n$ in the former. For each n, S_n/n takes values in $[-1,1]$. Hence the large deviation principle for S_n/n with respect to the canonical ensemble $P_{n,\beta,K}$ for the mean-field Blume–Capel model is proved exactly like the analogous large deviation principle for the Curie–Weiss model given in Theorem 6.6.1. By Cramér's Theorem (Theorem 6.5.4), with respect to the product measures $(P_\beta)_n$, S_n/n satisfies the large deviation principle with rate function

$$J_\beta(x) = \sup_{t\in\mathbb{R}}\{tx - c_\beta(t)\}, \tag{6.21}$$

where $c_\beta(t)$ is the cumulant-generating function

$$c_\beta(t) = \log\int_\Lambda \exp(t\omega_1)\,d\rho_\beta(d\omega_1) = \log\left(\frac{1+e^{-\beta}(e^t+e^{-t})}{1+2e^{-\beta}}\right).$$

The following large deviation principle for S_n/n with respect to $P_{n,\beta,K}$ is an immediate consequence of Theorem 6.5.9 with $\mathcal{X} = [-1,1]$ and $\psi(x) = \beta K x^2$ for $x \in [-1,1]$. In this context, the sequence S_n/n is called the sequence of macroscopic variables for the mean-field Blume–Capel model.

Theorem 6.6.5. *With respect to the canonical ensemble $P_{n,\beta,K}$ defined in eqn (6.19), the spin per site S_n/n satisfies the large deviation principle on $[-1,1]$ with rate function*

$$I_{\beta,K}(x) = J_\beta(x) - \beta K x^2 - \inf_{y\in[-1,1]}\{J_\beta(y) - \beta K y^2\},$$

where $J_\beta(x)$ is defined in eqn (6.21).

As in the Curie–Weiss model and the Curie–Weiss–Potts model, we define the set $\mathcal{E}_{\beta,K}$ of canonical equilibrium macrostates for the spin per site S_n/n in the mean-field Blume–Capel model to be the zero set of the rate function $I_{\beta,K}$ or, equivalently, the set of $x \in [-1,1]$ at which $J_\beta(x) - \beta K x^2$ attains its minimum. Thus

$$\begin{aligned}\mathcal{E}_{\beta,K} &= \{x \in [-1,1] : I_{\beta,K}(x) = 0\} \\ &= \left\{x \in [-1,1] : J_\beta(x) - \beta K x^2 \text{ is minimized}\right\}.\end{aligned}$$

In the case of the Curie–Weiss model, the rate function I in Cramér's Theorem for S_n/n is defined explicitly in eqn (6.13). This explicit formula greatly facilitates the analysis of the structure of the equilibrium macrostates for the spin per site in that model. By contrast, the analogous rate function J_β in the mean-field Blume–Capel model is not given explicitly. In order to determine the structure of $\mathcal{E}_{\beta,K}$, we use the theory of Legendre–Fenchel transforms to replace $J_\beta(x) - \beta K x^2$ by another function that has the same global minimum points but for which the analysis is much more straightforward.

The critical inverse temperature for the mean-field Blume–Capel model is $\beta_c = \log 4$. The structure of $\mathcal{E}_{\beta,K}$ is given first for $0 < \beta \leq \beta_c$ and second for $\beta > \beta_c$. The first theorem, proved in Theorem 3.6 of Ellis *et al.* (2005), describes the continuous bifurcation in $\mathcal{E}_{\beta,K}$ as K increases through a value $K(\beta)$. This bifurcation corresponds to a second-order phase transition.

Theorem 6.6.6. *For $0 < \beta \leq \beta_c$, we define*

$$K(\beta) = \frac{1}{2\beta c''_\beta(0)} = \frac{e^\beta + 2}{4\beta}. \tag{6.22}$$

For these values of β, $\mathcal{E}_{\beta,K}$ has the following structure.

(a) *For $0 < K \leq K(\beta)$, $\mathcal{E}_{\beta,K} = \{0\}$.*
(b) *For $K > K(\beta)$, there exists $m(\beta,K) > 0$ such that $\mathcal{E}_{\beta,K} = \{\pm m(\beta,K)\}$.*
(c) *$m(\beta,K)$ is a positive, increasing, continuous function for $K > K(\beta)$, and as $K \to (K(\beta))^+$, $m(\beta,K) \to 0^+$. Therefore, $\mathcal{E}_{\beta,K}$ exhibits a continuous bifurcation at $K(\beta)$.*

The next theorem, proved in Theorem 3.8 of Ellis *et al.* (2005), describes the discontinuous bifurcation in $\mathcal{E}_{\beta,K}$ for $\beta > \beta_c$ as K increases through a value $K_1(\beta)$. This bifurcation corresponds to a first-order phase transition.

Theorem 6.6.7. *For $\beta > \beta_c$, $\mathcal{E}_{\beta,K}$ has the following structure in terms of the quantity $K_1(\beta)$, denoted by $K_c^{(1)}(\beta)$ in Ellis* et al. *(2005) and defined implicitly for $\beta > \beta_c$ in Ellis* et al. *(2005, p. 2231).*

(a) *For $0 < K < K_1(\beta)$, $\mathcal{E}_{\beta,K} = \{0\}$.*
(b) *For $K = K_1(\beta)$, there exists $m(\beta, K_1(\beta)) > 0$ such that $\mathcal{E}_{\beta,K_1(\beta)} = \{0, \pm m(\beta, K_1(\beta))\}$.*
(c) *For $K > K_1(\beta)$, there exists $m(\beta,K) > 0$ such that $\mathcal{E}_{\beta,K} = \{\pm m(\beta,K)\}$.*

(d) *$m(\beta, K)$ is a positive, increasing, continuous function for $K \geq K_1(\beta)$, and as $K \to K_1(\beta)^+$, $m(\beta, K) \to m(\beta, K_1(\beta)) > 0$. Therefore, $\mathcal{E}_{\beta,K}$ exhibits a discontinuous bifurcation at $K_1(\beta)$.*

Because of the nature of the phase transitions expressed in these two theorems, we refer to the curve $\{(\beta, K(\beta)), 0 < \beta < \beta_c\}$ as the second-order curve and to the curve $\{(\beta, K_1(\beta)), \beta > \beta_c\}$ as the first-order curve. The point $(\beta_c, K(\beta_c)) = (\log 4, 3/2 \log 4)$ separates the second-order curve from the first-order curve and is called the tricritical point.

This completes our analysis of the mean-field Blume–Capel model. Probabilistic limit theorems for this model are proved in Costeniuc *et al.* (2007) and Ellis *et al.* (2005, 2009). In the next section, we discuss the equivalence and nonequivalence of ensembles for a general class of models in statistical mechanics. This discussion is based on a large deviation analysis that was inspired by the work in the present section.

6.7 Equivalence and nonequivalence of ensembles for a general class of models in statistical mechanics

Equilibrium statistical mechanics specifies two ensembles that describe the probability distribution of microstates in statistical-mechanical models. These are the microcanonical ensemble and the canonical ensemble. Particularly in the case of models of coherent structures in turbulence, the microcanonical ensemble is physically more fundamental because it expresses the fact that the Hamiltonian is a constant of the Euler dynamics underlying the model.

The introduction of two separate ensembles raises the basic problem of ensemble equivalence. As we will see in this section, the theory of large deviations and the theory of convex functions provide the perfect tools for analyzing this problem, which forces us to reevaluate a number of deep questions that have often been dismissed in the past as being physically obvious. These questions include the following. Is the temperature of a statistical-mechanical system always related to its energy in a one-to-one fashion? Are the microcanonical equilibrium properties of a system calculated as a function of the energy always equivalent to its canonical equilibrium properties calculated as a function of the temperature? Is the microcanonical entropy always a concave function of the energy? Is the heat capacity always a positive quantity? Surprisingly, the answer to each of these questions is in general no.

Starting with the work of Lynden-Bell and Wood (1968) and of Thirring (1970), physicists have come to realize in recent decades that systematic incompatibilities between the microcanonical and canonical ensembles can arise in the thermodynamic limit if the microcanonical entropy function of the system under study is nonconcave. The reason for this nonequivalence can be explained mathematically by the fact that, when applied to a nonconcave function, the Legendre–Fenchel transform is noninvolutive, i.e. performing it twice does not give back the original function but gives back its concave envelope (Ellis *et al.* 2005; Touchette *et al.* 2004). As a consequence of this property, the Legendre–Fenchel structure of statistical mechanics, traditionally used to establish a one-to-one relationship between the entropy and the free energy and between the energy and the temperature, ceases to be valid when the entropy is nonconcave.

From a more physical perspective, the explanation is even simpler. When the entropy is nonconcave, the microcanonical and canonical ensembles are nonequivalent because the nonconcavity of the entropy implies the existence of a nondifferentiable point of the free energy, and this, in turn, marks the presence of a first-order phase transition in the canonical ensemble (Ellis *et al.* 2000; Gross 1997). Accordingly, the ensembles are nonequivalent because the canonical ensemble jumps over a range of energy values at a critical value of the temperature and is therefore prevented from entering a subset of energy values that can always be accessed by the microcanonical ensemble (Ellis *et al.* 2000; Gross 1997; Thirring 1970). This phenomenon lies at the root of ensemble nonequivalence, which is observed in systems as diverse as the following. It is the typical behavior of systems, such as these, that are defined in terms of long-range interactions.

- Lattice spin models, including the Curie–Weiss–Potts model (Costeniuc *et al.* 2005*a*, 2006*a*), the mean-field Blume–Capel model (Barré *et al.* 2001, 2002; Ellis *et al.* 2004*b*, 2005), mean-field versions of the Hamiltonian model (Dauxois *et al.* 2002; Latora *et al.* 2001), and the XY model (Dauxois *et al.* 2000).
- Gravitational systems (Gross 1997; Hertel and Thirring 1971; Lynden-Bell and Wood 1968; Thirring 1970).
- Models of coherent structures in turbulence (Caglioti *et al.* 1992; Ellis *et al.* 2000, 2002; Eyink and Spohn 1993; Kiessling and Lebowitz 1997; Robert and Sommeria 1991).
- Models of plasmas (Kiessling and Neukirch 2003; Smith and O'Neil 1990).
- Model of the Lennard-Jones gas (Borges and Tsallis 2002).

Many of these models can be analyzed by the methods to be introduced in this section, which summarize the results presented in Ellis *et al.* (2000). Further developments in the theory are given in Costeniuc *et al.* (2005*b*). The reader is referred to these two papers for additional references to the large literature on ensemble equivalence for classical lattice systems and other models.

In the examples just cited, as well as in other cases, the microcanonical formulation gives rise to a richer set of equilibrium macrostates than does the canonical formulation, a phenomenon that occurs especially in the negative-temperature regimes of vorticity dynamics models (DiBattista *et al.* 1998, 2000; Eyink and Spohn 1993; Kiessling and Lebowitz 1997). For example, it has been shown computationally that the strongly reversing zonal-jet structures on Jupiter, as well as the Great Red Spot, fall into the nonequivalent range of the microcanonical ensemble with respect to the energy and circulation invariants (Turkington *et al.* 2001).

6.7.1 Large deviation analysis

The general class of models to be considered includes both spin models and models of coherent structures in turbulence. In order to simplify the presentation, we focus on spin models only. For models of coherent structures in turbulence, the definitions and theorems take slightly different forms (Ellis 2008, Section 10.1). The models to be considered are defined in terms of the following quantities.

- A sequence of probability spaces $(\Omega_n, \mathcal{F}_n, P_n)$ indexed by $n \in I\!N$, which typically represents a sequence of finite-dimensional systems. The Ω_n are the configuration spaces, $\omega \in \Omega_n$ are the microstates, and the P_n are the prior measures.
- For each $n \in I\!N$, the Hamiltonian H_n, a bounded, measurable function mapping Ω_n into $\mathbb{R}$.
- A sequence of positive scaling constants $a_n \to \infty$ as $n \to \infty$. In general, a_n equals the total number of degrees of freedom in the model. In many cases a_n equals the number of particles.

Models of coherent structures in turbulence often incorporate other dynamical invariants besides the Hamiltonian. In this case one replaces H_n in the second bullet point above by the vector of dynamical invariants and makes other corresponding changes in the theory, which are all purely notational. For simplicity, we work only with the Hamiltonian in this section.

A large deviation analysis of the general model is possible provided that there exist, as specified in the next four items, a space of macrostates, a sequence of macroscopic variables, and an interaction representation function, and provided that the macroscopic variables satisfy the large deviation principle on the space of macrostates. One can easily verify that this general setup applies to the three mean-field models considered in Section 6.6.

1. *Space of macrostates.* This is a complete, separable metric space $\mathcal{X}$, which represents the set of all possible macrostates.
2. *Macroscopic variables.* These are a sequence of random variables Y_n mapping Ω_n into $\mathcal{X}$. These functions associate a macrostate in $\mathcal{X}$ with each microstate $\omega \in \Omega_n$.
3. *Hamiltonian representation function.* This is a bounded, continuous function $\tilde{H}$ that maps $\mathcal{X}$ into $\mathbb{R}$ and enables us to write H_n, either exactly or asymptotically, as a function of the macrostate via the macroscopic variable Y_n. Namely, as $n \to \infty$,

$$H_n(\omega) = a_n \tilde{H}(Y_n(\omega)) + \mathrm{o}(a_n) \quad \text{uniformly for } \omega \in \Omega_n\,,$$

   i.e.

$$\lim_{n\to\infty} \sup_{\omega\in\Omega_n} |H_n(\omega)/a_n - \tilde{H}(Y_n(\omega))| = 0\,. \tag{6.23}$$

4. *Large deviation principle for the macroscopic variables.* There exists a function I mapping $\mathcal{X}$ into $[0, \infty]$ and having compact level sets such that, with respect to P_n, the sequence Y_n satisfies the large deviation principle on $\mathcal{X}$ with rate function I and scaling constants a_n. In other words, for any closed subset F of $\mathcal{X}$,

$$\limsup_{n\to\infty} \frac{1}{a_n} \log P_n\{Y_n \in F\} \le -\inf_{x\in F} I(x)\,,$$

   and for any open subset G of $\mathcal{X}$,

$$\liminf_{n\to\infty} \frac{1}{a_n} \log P_n\{Y_n \in G\} \ge -\inf_{x\in G} I(x)\,.$$

Here is a partial list of statistical-mechanical models to which the large deviation formalism has been applied. Further details are given in Costeniuc *et al.* (2005*b*, Example 2.1).

- The Miller–Robert model of fluid turbulence based on the two-dimensional Euler equations (Boucher *et al.* 2000).
- A model of geophysical flows based on equations describing barotropic, quasi-geostrophic turbulence (Ellis *et al.* 2002).
- A model of soliton turbulence based on a class of generalized nonlinear Schrödinger equations (Ellis *et al.* 2004*a*).
- Lattice spin models, including the Curie–Weiss model (Ellis 2006, Section IV.4), the Curie–Weiss–Potts model (Costeniuc *et al.* 2005*a*), the mean-field Blume–Capel model (Ellis *et al.* 2005), and the Ising model (Föllmer and Orey 1987; Olla 1988). The large deviation analysis of these models illustrates the three levels of the Donsker–Varadhan theory of large deviations, which are explained in Ellis (2006, Chapter 1).
  - *Level 1.* As we saw in Subsection 6.6.1, for the Curie–Weiss model the macroscopic variables are the sample means of i.i.d. random variables, and the large deviation principle with respect to the prior measures is the version of Cramér's Theorem given in Corollary 6.5.5. Similar comments apply to the mean-field Blume–Capel model considered in Subsection 6.6.3.
  - *Level 2.* As we saw in Subsection 6.6.2, for the Curie–Weiss–Potts model (Costeniuc *et al.* 2005*a*) the macroscopic variables are the empirical vectors of i.i.d. random variables, and the large deviation principle with respect to the prior measures is the version of Sanov's Theorem given in Theorem 6.3.3.
  - *Level 3.* For the Ising model, the macroscopic variables are an infinite-dimensional generalization of the empirical measure known as the empirical field, and the large deviation principle with respect to the prior measures is derived in Föllmer and Orey (1987) and Olla (1988). This is related to level 3 of the Donsker–Varadhan theory, which is formulated for a general class of Markov chains and Markov processes (Donsker and Varadhan 1983). A special case is treated in Ellis (2006, Chapter IX), which proves the large deviation principle for the empirical process of i.i.d. random variables taking values in a finite state space. The complicated large deviation analysis of the Ising model is outlined in Ellis (1995, Section 11).

Returning now to the general theory, we introduce the microcanonical ensemble, the canonical ensemble, and the basic thermodynamic functions associated with each ensemble: the microcanonical entropy and the canonical free energy. We then sketch the proofs of the large deviation principles for the macroscopic variables Y_n with respect to the two ensembles. As in the case of the Curie–Weiss model, the zeros of the corresponding rate functions define the corresponding sets of equilibrium macrostates, one for the microcanonical ensemble and one for the canonical ensemble. The problem of ensemble equivalence investigates the relationship between these two sets of equilibrium macrostates.

In general terms, the main result is that a necessary and sufficient condition for equivalence of ensembles to hold at the level of equilibrium macrostates is that it holds at the level of thermodynamic functions, which is the case if and only if the microcanonical entropy is concave. The necessity of this condition has the following striking formulation. If the microcanonical entropy is not concave at some value of its argument, then the ensembles are nonequivalent in the sense that the corresponding set of microcanonical equilibrium macrostates is disjoint from any set of canonical equilibrium macrostates. The reader is referred to Ellis *et al.* (2000, Section 1.4) for a detailed discussion of models of coherent structures in turbulence in which nonconcave microcanonical entropies arise.

We start by introducing the function whose support and concavity properties completely determine all aspects of ensemble equivalence and nonequivalence. This function is the microcanonical entropy, defined for $u \in \mathbb{R}$ by

$$s(u) = -\inf\{I(x) : x \in \mathcal{X}, \tilde{H}(x) = u\}. \tag{6.24}$$

Since I maps $\mathcal{X}$ into $[0, \infty]$, s maps $\mathbb{R}$ into $[-\infty, 0]$. Moreover, since I is lower semicontinuous and $\tilde{H}$ is continuous on $\mathcal{X}$, s is upper semicontinuous on $\mathbb{R}$. We define $\mathrm{dom}\, s$ to be the set of $u \in \mathbb{R}$ for which $s(u) > -\infty$. In general, $\mathrm{dom}\, s$ is nonempty, since $-s$ is a rate function (Ellis *et al.* 2000, Proposition 3.1(a)). For each $u \in \mathrm{dom}\, s$, $r > 0$, $n \in I\!N$, and $A \in \mathcal{F}_n$, the microcanonical ensemble is defined to be the conditioned measure

$$P_n^{u,r}\{A\} = P_n\{A \mid H_n/a_n \in [u-r, u+r]\}. \tag{6.25}$$

As shown in Ellis *et al.* (2000, p. 1027), if $u \in \mathrm{dom}\, s$, then for all sufficiently large n the conditioned measures $P_n^{u,r}$ are well defined.

A mathematically more tractable probability measure is the canonical ensemble. For each $n \in I\!N$, $\beta \in \mathbb{R}$, and $A \in \mathcal{F}_n$, we define the partition function

$$Z_n(\beta) = \int_{\Omega_n} \exp[-\beta H_n]\, dP_n\,,$$

which is well defined and finite; the canonical free energy

$$\varphi(\beta) = -\lim_{n\to\infty} \frac{1}{a_n} \log Z_n(\beta)\,;$$

and the probability measure

$$P_{n,\beta}\{A\} = \frac{1}{Z_n(\beta)} \cdot \int_A \exp[-\beta H_n]\, dP_n\,. \tag{6.26}$$

The measures $P_{n,\beta}$ are Gibbs states that define the canonical ensemble for the given model. Although for spin models one usually takes $\beta > 0$, in general $\beta \in \mathbb{R}$ is allowed; for example, negative values of β arise naturally in the study of coherent structures in two-dimensional turbulence.

Among other reasons, the canonical ensemble was introduced by Gibbs in the hope that in the limit $n \to \infty$ the two ensembles would be equivalent, i.e. all macroscopic properties of the model obtained via the microcanonical ensemble could be realized as macroscopic properties obtained via the canonical ensemble. However, as we will see, this is not in general the case.

The large deviation analysis of the canonical ensemble is summarized in the next theorem, Theorem 6.7.1. Part (a) of Theorem 6.7.1 shows that the limit defining $\varphi(\beta)$ exists and is given by a variational formula. Part (b) states the large deviation principle for the macroscopic variables with respect to the canonical ensemble. Part (b) is the analogue of Theorem 6.6.1 for the Curie–Weiss model. In part (c), we consider the set $\mathcal{E}_\beta$ consisting of points at which the rate function in part (b) attains its infimum of 0. The second property of $\mathcal{E}_\beta$ given in part (c) justifies calling this the set of canonical equilibrium macrostates. Part (c) is a special case of Theorem 6.5.3.

Theorem 6.7.1. *We assume that there exists a space of macrostates $\mathcal{X}$, macroscopic variables Y_n, and a Hamiltonian representation function $\tilde{H}$ satisfying*

$$\lim_{n\to\infty} \sup_{\omega\in\Omega_n} |H_n(\omega)/a_n - \tilde{H}(Y_n(\omega))| = 0\,, \tag{6.27}$$

where H_n is the Hamiltonian. We also assume that with respect to the prior measures P_n, Y_n satisfies the large deviation principle on $\mathcal{X}$ with some rate function I and scaling constants a_n. For each $\beta \in \mathbb{R}$, the following conclusions hold.

(a) *The canonical free energy $\varphi(\beta) = -\lim_{n\to\infty}(1/a_n)\log Z_n(\beta)$ exists and is given by*

$$\varphi(\beta) = \inf_{x\in\mathcal{X}}\{\beta\tilde{H}(x) + I(x)\}\,.$$

(b) *With respect to the canonical ensemble $P_{n,\beta}$ defined in eqn (6.26), Y_n satisfies the large deviation principle on $\mathcal{X}$ with scaling constants a_n and rate function*

$$I_\beta(x) = I(x) + \beta\tilde{H}(x) - \varphi(\beta)\,.$$

(c) *We define the set of canonical equilibrium macrostates*

$$\mathcal{E}_\beta = \{x \in \mathcal{X} : I_\beta(x) = 0\}\,.$$

Then $\mathcal{E}_\beta$ is a nonempty, compact subset of $\mathcal{X}$. In addition, if A is a Borel subset of $\mathcal{X}$ such that $\overline{A}\cap\mathcal{E}_\beta = \emptyset$, then $I_\beta(\overline{A}) > 0$ and, for some $C < \infty$,

$$P_{n,\beta}\{Y_n \in A\} \le C\exp[-nI_\beta(\overline{A})/2] \to 0 \quad \textit{as } n\to\infty\,.$$

Proof.

(a) Once we take into account the error between H_n and $a_n\tilde{H}(Y_n)$ expressed in eqn (6.27), the proof of (a) and (b) follows from the Laplace principle. Here are the details. By eqn (6.27),

$$\begin{aligned}
&\left|\frac{1}{a_n}\log Z_n(\beta) - \frac{1}{a_n}\log\int_{\Omega_n}\exp[-\beta a_n\tilde{H}(Y_n)]\,dP_n\right| \\
&= \left|\frac{1}{a_n}\log\int_{\Omega_n}\exp[-\beta H_n]\,dP_n - \frac{1}{a_n}\log\int_{\Omega_n}\exp[-\beta a_n\tilde{H}(Y_n)]\,dP_n\right| \\
&\le |\beta|\frac{1}{a_n}\sup_{\omega\in\Omega_n}|H_n(\omega) - a_n\tilde{H}(Y_n(\omega))| \to 0 \text{ as } n\to\infty\,.
\end{aligned}$$

Since $\tilde{H}$ is a bounded continuous function mapping $\mathcal{X}$ into $\mathbb{R}$, the Laplace principle satisfied by Y_n with respect to P_n yields part (a):

$$\begin{aligned}\varphi(\beta) &= -\lim_{n\to\infty} \frac{1}{a_n} \log Z_n(\beta) \\ &= -\lim_{n\to\infty} \frac{1}{a_n} \log \int_{\Omega_n} \exp[-\beta a_n \tilde{H}(Y_n)]\, dP_n \\ &= -\sup_{x\in\mathcal{X}} \{-\beta\tilde{H}(x) - I(x)\} \\ &= \inf_{x\in\mathcal{X}} \{\beta\tilde{H}(x) + I(x)\}\,.\end{aligned}$$

(b) This is an immediate consequence of Theorem 6.5.9 with $\psi = \tilde{H}$.
(c) This is proved in Theorem 6.5.3. The proof of the second equation in part (c) is based on the large deviation upper bound for Y_n with respect to $P_{n,\beta}$ (part (b) of this theorem). The proof of the theorem is complete. ■

We next present the large deviation analysis of the microcanonical ensemble $P_n^{u,r}$ defined in eqn (6.25). In this context, the microcanonical entropy s defined in eqn (6.24) has the following property: $-s$ is the rate function in the large deviation principles, with respect to the prior measures P_n, for both $\tilde{H}(Y_n)$ and H_n/a_n. In order to see this, we recall that with respect to P_n, Y_n satisfies the large deviation principle with rate function I. Since $\tilde{H}$ is a continuous function mapping $\mathcal{X}$ into $\mathbb{R}$, the large deviation principle for $\tilde{H}(Y_n)$ is a consequence of the contraction principle (Theorem 6.5.8). For $u \in \mathbb{R}$, the rate function is given by

$$\inf\{I(x) : x \in \mathcal{X}, \tilde{H}(x) = u\} = -s(u)\,.$$

In addition, since

$$\lim_{n\to\infty} \sup_{\omega\in\Omega_n} |H_n(\omega)/a_n - \tilde{H}(Y_n(\omega))| = 0\,,$$

H_n/a_n inherits from $\tilde{H}(Y_n)$ the large deviation principle with the same rate function. This follows from Dupuis and Ellis (1997, Theorem 1.3.3) or can be derived as in the proof of part (a) of Theorem 6.7.1 by using the equivalent Laplace principle. We summarize this large deviation principle by the notation

$$P_n\{H_n/a_n \in [u-r, u+r]\} \asymp \exp[a_n s(u)] \quad \text{as } n \to \infty, r \to 0\,. \tag{6.28}$$

For $x \in \mathcal{X}$ and $\alpha > 0$, $B(x, \alpha)$ denotes the open ball with center x and radius α. We next motivate the large deviation principle for Y_n with respect to the microcanonical ensemble $P_n^{u,r}$ by estimating the exponential-order contribution to the probability $P_n^{u,r}\{Y_n \in B(x,\alpha)\}$ as $n \to \infty$. Specifically, we seek a function I^u such that for all $u \in \operatorname{dom} s$, all $x \in \mathcal{X}$, and all $\alpha > 0$ sufficiently small,

$$P_n^{u,r}\{Y_n \in B(x,\alpha)\} \approx \exp[-a_n I^u(x)] \quad \text{as } n \to \infty, r \to 0, \alpha \to 0\,. \tag{6.29}$$

The calculation that we present shows both the interpretative power of the large deviation notation and the value of left-handed thinking.

We first work with $x \in \mathcal{X}$ for which $I(x) < \infty$ and $\tilde{H}(x) = u$. Such an x exists, since $u \in \text{dom}\, s$ and thus $s(u) > -\infty$. Because

$$\lim_{n \to \infty} \sup_{\omega \in \Omega_n} |H_n(\omega)/a_n - \tilde{H}(Y_n(\omega))| = 0,$$

for all sufficiently large n depending on r, the set of ω for which both $Y_n(\omega) \in B(x, \alpha)$ and $H_n(\omega)/a_n \in [u-r, u+r]$ is approximately equal to the set of ω for which both $Y_n(\omega) \in B(x, \alpha)$ and $\tilde{H}(Y_n(\omega)) \in [u-r, u+r]$. Since $\tilde{H}$ is continuous and $\tilde{H}(x) = u$, for all sufficiently small α compared with r this set reduces to $\{\omega : Y_n(\omega) \in B(x, \alpha)\}$. Hence, for all sufficiently small r, all sufficiently large n depending on r, and all sufficiently small α compared with r, the assumed large deviation principle for Y_n with respect to P_n and the large deviation principle for H_n/a_n summarized in eqn (6.28) yield

$$\begin{aligned} P_n^{u,r}\{Y_n \in B(x,\alpha)\} &= \frac{P_n\{\{Y_n \in B(x,\alpha)\} \cap \{H_n/a_n \in [u-r, u+r]\}\}}{P_n\{H_n/a_n \in [u-r, u+r]\}} \\ &\approx \frac{P_n\{Y_n \in B(x,\alpha)\}}{P_n\{H_n/a_n \in [u-r, u+r]\}} \\ &\approx \exp[-a_n(I(x) + s(u))]. \end{aligned}$$

On the other hand, if $\tilde{H}(x) \neq u$, then a similar calculation shows that for all sufficiently small r, all sufficiently small α, and all sufficiently large n, $P_n^{u,r}\{Y_n \in B(x,\alpha)\} = 0$. Comparing these approximate calculations with the desired asymptotic form (6.29) motivates the correct formula for the rate function (Ellis *et al.* 2000, Theorem 3.2):

$$I^u(x) = \begin{cases} I(x) + s(u) & \text{if } \tilde{H}(x) = u, \\ \infty & \text{if } \tilde{H}(x) \neq u. \end{cases} \tag{6.30}$$

We record the facts in the next theorem, which is proved in Ellis *et al.* (2000, Section 3). An additional complication occurs in part (b) because the large deviation principle involves the double limit $n \to 0$ followed by $r \to 0$. In part (c), we introduce the set of microcanonical equilibrium macrostates $\mathcal{E}^u$ and state a property of this set with respect to the microcanonical ensemble that is analogous to the property satisfied by the set $\mathcal{E}_\beta$ of canonical equilibrium macrostates with respect to the canonical ensemble. The proof, given in Ellis *et al.* (2000, Theorem 3.5), is similar to the proof of the analogous property of $\mathcal{E}_\beta$ given in part (c) of Theorem 6.7.1 and is omitted.

Theorem 6.7.2. *We assume that there exists a space of macrostates $\mathcal{X}$, macroscopic variables Y_n, and a Hamiltonian representation function $\tilde{H}$ satisfying eqn (6.23). We also assume that with respect to the prior measures P_n, Y_n satisfies the large deviation principle on $\mathcal{X}$ with scaling constants a_n and some rate function I. For each $u \in \text{dom}\, s$ and any $r \in (0,1)$, the following conclusions hold.*

(a) *With respect to P_n, $\tilde{H}(Y_n)$ and H_n/a_n both satisfy the large deviation principle with scaling constants a_n and rate function $-s$.*

(b) *We consider the microcanonical ensemble $P_n^{u,r}$ defined in eqn (6.25). With respect to $P_n^{u,r}$ and in the double limit $n \to \infty$ and $r \to 0$, Y_n satisfies the large deviation* *principle on $\mathcal{X}$ with scaling constants a_n and the rate function I^u defined in eqn (6.30). That is, for any closed subset F of $\mathcal{X}$,*

$$\lim_{r\to 0}\limsup_{n\to\infty}\frac{1}{a_n}\log P_n^{u,r}\{Y_n\in F\}\le -I^u(F)$$

and, for any open subset G of $\mathcal{X}$,

$$\lim_{r\to 0}\liminf_{n\to\infty}\frac{1}{a_n}\log P_n^{u,r}\{Y_n\in G\}\ge -I^u(G)\,.$$

(c) *We define the set of equilibrium macrostates*

$$\mathcal{E}^u=\{x\in\mathcal{X}:I^u(x)=0\}\,.$$

Then $\mathcal{E}^u$ is a nonempty, compact subset of $\mathcal{X}$. In addition, if A is a Borel subset of $\mathcal{X}$ such that $\overline{A}\cap\mathcal{E}^u=\emptyset$, then $I^u(\overline{A})>0$ and there exists $r_0>0$ and, for all $r\in(0,r_0]$, there exists $C_r<\infty$ such that

$$P_{n,\beta}\{Y_n\in A\}\le C_r\exp[-n\,I_\beta(\overline{A})/2]\to 0\quad \textit{as } n\to\infty\,.$$

This completes the large deviation analysis of the general spin model. In the next subsection, we investigate the equivalence and nonequivalence of the canonical and microcanonical ensembles.

6.7.2 Equivalence and nonequivalence of ensembles

The study of the equivalence and nonequivalence of the canonical and microcanonical ensembles involves the relationships between the two sets of equilibrium macrostates

$$\mathcal{E}_\beta=\{x\in\mathcal{X}:I_\beta(x)=0\}\ \text{ and }\ \mathcal{E}^u=\{x\in\mathcal{X}:I^u(x)=0\}\,.$$

The following questions will be considered.

1. Given $\beta\in\mathbb{R}$ and $x\in\mathcal{E}_\beta$, does there exists $u\in\mathbb{R}$ such that $x\in\mathcal{E}^u$? In other words, is any canonical equilibrium macrostate realized microcanonically?
2. Given $u\in\mathbb{R}$ and $x\in\mathcal{E}^u$, does there exist $\beta\in\mathbb{R}$ such that $x\in\mathcal{E}_\beta$? In other words, is any microcanonical equilibrium macrostate realized canonically?

As we will see in Theorem 6.7.4, the answer to question 1 is always yes, but the answer to question 2 is much more complicated, involving three possibilities.

2a. *Full equivalence.* There exists $\beta\in\mathbb{R}$ such that $\mathcal{E}^u=\mathcal{E}_\beta$.
2b. *Partial equivalence.* There exists $\beta\in\mathbb{R}$ such that $\mathcal{E}^u\subset\mathcal{E}_\beta$ but $\mathcal{E}^u\ne\mathcal{E}_\beta$.
2c. *Nonequivalence.* $\mathcal{E}^u$ is disjoint from $\mathcal{E}_\beta$ for all $\beta\in\mathbb{R}$.

One of the big surprises of the theory to be presented here is that we are able to decide on which of these three possibilities occurs by examining the support and concavity properties of the microcanonical entropy $s(u)$. This is remarkable because the sets $\mathcal{E}_\beta$ and $\mathcal{E}^u$ are, in general, infinite-dimensional while the microcanonical entropy is a function on $\mathbb{R}$.

In order to begin our study of ensemble equivalence and nonequivalence, we first recall the definitions of the corresponding rate functions:

$$I_\beta(x) = I(x) + \beta \tilde{H}(x) - \varphi(\beta)\,,$$

where $\varphi(\beta)$ denotes the canonical free energy

$$\varphi(\beta) = \inf_{x \in \mathcal{X}} \{\beta \tilde{H}(x) + I(x)\}\,,$$

and

$$I^u(x) = \begin{cases} I(x) + s(u) & \text{if } \tilde{H}(x) = u\,, \\ \infty & \text{if } \tilde{H}(x) \neq u\,, \end{cases}$$

where $s(u)$ denotes the microcanonical entropy

$$s(u) = -\inf\{I(x) : x \in \mathcal{X}, \tilde{H}(x) = u\}\,.$$

Using these definitions, we see that the two sets of equilibrium macrostates have the alternate characterizations

$$\mathcal{E}_\beta = \{x \in \mathcal{X} : I(x) + \beta \tilde{H}(x) \text{ is minimized}\}$$

and

$$\mathcal{E}^u = \{x \in \mathcal{X} : I(x) \text{ is minimized subject to } \tilde{H}(x) = u\}\,.$$

Thus $\mathcal{E}^u$ is defined by the following constrained minimization problem for $u \in \mathbb{R}$:

$$\text{minimize } I(x) \text{ over } \mathcal{X} \text{ subject to the constraint } \tilde{H}(x) = u\,. \tag{6.31}$$

By contrast, $\mathcal{E}_\beta$ is defined by the following related, unconstrained minimization problem for $\beta \in \mathbb{R}$:

$$\text{minimize } I(x) + \beta \tilde{H}(x) \text{ over } x \in \mathcal{X}\,. \tag{6.32}$$

In this formulation, β is a Lagrange multiplier dual to the constraint $\tilde{H}(x) = u$. The theory of Lagrange multipliers outlines suitable conditions under which the solutions of the constrained problem (6.31) lie among the critical points of $I + \beta \tilde{H}$. However, it does not give, as we will do in Theorem 6.7.4, necessary and sufficient conditions for the solutions of eqn (6.31) to coincide with the solutions of the unconstrained minimization problem (6.32). These necessary and sufficient conditions are expressed in terms of support and concavity properties of the microcanonical entropy $s(u)$.

Before we explain this, we point out a basic property relating the two thermodynamic functions $s(u)$ and $\varphi(\beta)$.

Theorem 6.7.3. *The microcanonical entropy $s(u)$ defined in eqn (6.24) and the canonical free energy $\varphi(\beta)$ defined in part (a) of Theorem 6.7.1 are related via the Legendre–Fenchel transform*

$$\varphi(\beta) = \inf_{u \in \mathbb{R}} \{\beta u - s(u)\}\,. \tag{6.33}$$

Proof. By the variational formula given in part (a) of Theorem 6.7.1,

$$\begin{aligned}\varphi(\beta) &= \inf_{x\in\mathcal{X}}\{\beta\tilde{H}(x) + I(x)\} \\ &= \inf_{u\in\mathbb{R}} \inf\{\beta\tilde{H}(x) + I(x) : x \in \mathcal{X}, \tilde{H}(x) = u\} \\ &= \inf_{u\in\mathbb{R}}\{\beta u + \inf\{I(x) : x \in \mathcal{X}, \tilde{H}(x) = u\} \\ &= \inf_{u\in\mathbb{R}}\{\beta u - s(u)\}\,.\end{aligned}$$

This completes the proof. ■

The formula (6.33), which exhibits $\varphi(\beta)$ as a concave function on $\mathbb{R}$, is related to a fundamental lack of symmetry between the microcanonical and canonical ensembles. Although one can obtain $\varphi(\beta)$ from $s(u)$ via a Legendre–Fenchel transform, in general one cannot obtain $s(u)$ from $\varphi(\beta)$ via the dual formula $s(u) = \inf_{\beta\in\mathbb{R}}\{\beta u - \varphi(\beta)\}$ unless s is concave on $\mathbb{R}$, which in general is not the case. In fact, the concavity of s on $\mathbb{R}$ depends on properties of I and $\tilde{H}$. For example, if I is convex on $\mathcal{X}$ and $\tilde{H}$ is affine, then s is concave on $\mathbb{R}$. On the other hand, microcanonical entropies s that are not concave on $\mathbb{R}$ arise in many models involving long-range interactions, including those listed in the five bullet points at the beginning of Section 6.7. This discussion indicates that of the two thermodynamic functions, the microcanonical entropy is the more fundamental, a state of affairs that is reinforced by the results on ensemble equivalence and nonequivalence to be presented in Theorem 6.7.4.

In order to state this theorem, we need several definitions. A function f on $\mathbb{R}$ is said to be concave on $\mathbb{R}$, or simply "concave," if f maps $\mathbb{R}$ into $\mathbb{R}\cup\{-\infty\}$, $f(u) > -\infty$ for some $u \in \mathbb{R}$, and, for all u and v in $\mathbb{R}$ and all $\lambda \in (0,1)$,

$$f(\lambda u + (1-\lambda)v) \geq \lambda f(u) + (1-\lambda)f(v)\,.$$

Let $f \not\equiv -\infty$ be a function mapping $\mathbb{R}$ into $\mathbb{R}\cup\{-\infty\}$. We define $\operatorname{dom} f$ to be the set of $u \in \mathbb{R}$ for which $f(u) > -\infty$. For β and u in $\mathbb{R}$, the Legendre–Fenchel transforms f^* and f^{**} are defined by (Rockafellar 1970, p. 308)

$$f^*(\beta) = \inf_{u\in\mathbb{R}}\{\beta u - f(u)\} \quad\text{and}\quad f^{**}(u) = (f^*)^*(u) = \inf_{\beta\in\mathbb{R}}\{\beta u - f^*(\beta)\}\,.$$

As in the case of convex functions (Ellis 2006, Theorem VI.5.3), f^* is concave and upper semicontinuous on $\mathbb{R}$, and for all $u \in \mathbb{R}$ we have $f^{**}(u) = f(u)$ if and only if f is concave and upper semicontinuous on $\mathbb{R}$. If f is not concave and upper semicontinuous on $\mathbb{R}$, then f^{**} is the smallest concave, upper semicontinuous function on $\mathbb{R}$ that satisfies $f^{**}(u) \geq f(u)$ for all $u \in \mathbb{R}$ (Costeniuc *et al.* 2005*b*, Proposition A.2). In particular, if for some u, $f(u) \neq f^{**}(u)$, then $f(u) < f^{**}(u)$. We say that f is concave at $u \in \operatorname{dom} f$ if $f(u) = f^{**}(u)$ and that f is nonconcave at $u \in \operatorname{dom} f$ if $f(u) < f^{**}(u)$. These definitions are reasonable because f^{**} is concave on $\mathbb{R}$.

We now state the main theorem concerning the equivalence and nonequivalence of the microcanonical and canonical ensembles. According to part (a), canonical equilibrium macrostates are always realized microcanonically. However, according to parts (b)–(d), the converse in general is false. The three possibilities given in parts (b)–(d) depend on concavity and support properties of the microcanonical entropy. The next theorem is proved in Ellis *et al.* (2000, Section 4), where it is formulated somewhat differently. The formulation given here specializes Theorem 3.1 in Costeniuc *et al.* (2005*b*) to dimension 1.

Theorem 6.7.4. *In parts (b), (c), and (d),* u *denotes any point in* $\operatorname{dom} s$.

(a) Canonical is always realized microcanonically. *We define* $\tilde{H}(\mathcal{E}_\beta)$ *to be the set of* $u \in \mathbb{R}$ *having the form* $u = \tilde{H}(x)$ *for some* $x \in \mathcal{E}_\beta$. *Then, for any* $\beta \in \mathbb{R}$, *we have* $\tilde{H}(\mathcal{E}_\beta) \subset \operatorname{dom} s$ *and*

$$\mathcal{E}_\beta = \bigcup_{u \in \tilde{H}(\mathcal{E}_\beta)} \mathcal{E}^u .$$

(b) Full equivalence. *There exists* $\beta \in \mathbb{R}$ *such that* $\mathcal{E}^u = \mathcal{E}_\beta$ *if and only if* s *has a strictly supporting line at* u *with tangent* β, *i.e.*

$$s(v) < s(u) + \beta(v-u) \text{ for all } v \neq u .$$

(c) Partial equivalence. *There exists* $\beta \in \mathbb{R}$ *such that* $\mathcal{E}^u \subset \mathcal{E}_\beta$ *but* $\mathcal{E}^u \neq \mathcal{E}_\beta$ *if and only if* s *has a nonstrictly supporting line at* u *with tangent* β, *i.e.*

$$s(v) \leq s(u) + \beta(v-u) \text{ for all } v, \text{ with equality for some } v \neq u .$$

(d) Nonequivalence. *For all* $\beta \in \mathbb{R}$, $\mathcal{E}^u \cap \mathcal{E}_\beta = \emptyset$ *if and only if* s *has no supporting line at* u, *i.e.*

$$\text{for all } \beta \in \mathbb{R} \text{ there exists } v \text{ such that } s(v) > s(u) + \beta(v-u) .$$

Here are some useful criteria for full or partial equivalence of ensembles and for nonequivalence of ensembles.

- *Full or partial equivalence.* Except possibly for boundary points of $\operatorname{dom} s$, s has a supporting line at $u \in \operatorname{dom} s$ if and only if s is concave at u (Costeniuc *et al.* 2005*b*, Theorem A.5(c)), and thus, according to parts (a) and (b) of Theorem 6.7.4, full or partial equivalence of ensembles holds.
- *Full equivalence.* Assume that $\operatorname{dom} s$ is a nonempty interval and that s is strictly concave on the interior of $\operatorname{dom} s$, i.e. for all $u \neq v$ in the interior of $\operatorname{dom} s$ and all $\lambda \in (0,1)$,

$$s(\lambda u + (1-\lambda)v) > \lambda s(u) + (1-\lambda)s(v) .$$

  Then, except possibly for boundary points of $\operatorname{dom} s$, s has a strictly supporting line at all $u \in \operatorname{dom} s$, and thus, according to part (a) of the theorem, full equivalence of ensembles holds (Costeniuc *et al.* 2005*b*, Theorem A.4(c)).
- *Nonequivalence.* Except possibly for boundary points of $\operatorname{dom} s$, s has no supporting line at $u \in \operatorname{dom} s$ if and only if s is nonconcave at u (Costeniuc *et al.* 2005*b*, Theorem A.5(c)).

The various possibilities in parts (b), (c), and (d) of Theorem 6.7.4 are illustrated in Ellis *et al.* (2004*b*) for the mean-field Blume–Capel spin model. In Ellis *et al.* (2002), the theory is applied to a model of coherent structures in two-dimensional turbulence. Numerical computations implemented for geostrophic turbulence over topography in a zonal channel demonstrate that nonequivalence of ensembles occurs over a wide range of the model parameters and that physically interesting equilibria seen microcanonically are often omitted by the canonical ensemble. The coherent structures observed in the model resemble the coherent structures observed in the midlatitude, zone-belt domains on Jupiter.

In Costeniuc *et al.* (2005*b*), the theory developed in Ellis *et al.* (2000) and summarized in Theorem 6.7.4 is extended. In Costeniuc *et al.* (2005*b*), it is shown that when the microcanonical ensemble is nonequivalent to the canonical ensemble on a subset of values of the energy, it is often possible to modify the definition of the canonical ensemble so as to recover equivalence with the microcanonical ensemble. Specifically, we give natural conditions under which one can construct a so-called Gaussian ensemble that is equivalent to the microcanonical ensemble when the canonical ensemble is not. This is potentially useful if one wants to work out the equilibrium properties of a system in the microcanonical ensemble, a notoriously difficult problem because of the equality constraint appearing in the definition of this ensemble. An overview of Costeniuc *et al.* (2005*b*) is given in Costeniuc *et al.* (2006*b*), and in Costeniuc *et al.* (2006*a*) it is applied to the Curie–Weiss–Potts model.

This completes our presentation of the theory of large deviations. The seed from which the theory blossomed was Boltzmann's discovery of a statistical interpretation of entropy. After examining his discovery and outlining a number of basic results in large deviation theory, in this paper we have brought the theory back to its roots by applying it to several problems in statistical mechanics. It is hoped that the reader will be inspired by this paper to further investigate the symbiotic and mutually invigorating relationship between these two fields.

Acknowledgment

The research of Richard S. Ellis is supported by a grant from the National Science Foundation (NSF-DMS-0604071).

References

Barré J., D. Mukamel, and S. Ruffo (2001). Inequivalence of ensembles in a system with long-range interactions. *Phys. Rev. Lett.* 87:030601.

Barré J., D. Mukamel, and S. Ruffo (2002). Ensemble inequivalence in mean-field models of magnetism. In T. Dauxois, S. Ruffo, E. Arimondo, and M. Wilkens (editors), *Dynamics and Thermodynamics of Systems with Long-Range Interactions*, Lecture Notes in Physics 602, pp. 45–67. New York: Springer.

Blume M. (1966). Theory of the first-order magnetic phase change in UO_2. *Phys. Rev.* 141:517–524.

Blume M., V. J. Emery, and R. B. Griffiths (1971). Ising model for the λ transition and phase separation in He^3–He^4 mixtures. *Phys. Rev. A* 4:1071–1077.

Boltzmann L. (1877). Über die Beziehung zwischen dem zweiten Hauptsatze der mechanischen Wärmetheorie und der Wahrscheinlichkeitsrechnung respecktive den Sätzen über das Wärmegleichgewicht. [On the relationship between the Second Law of the mechanical theory of heat and the probability calculus.] *Wiener Berichte* 2, no. 76, 373–435.

Borges E. P. and C. Tsallis (2002). Negative specific heat in a Lennard-Jones-like gas with long-range interactions. *Physica A* 305:148–151.

Boucher C., R. S. Ellis, and B. Turkington (2000). Derivation of maximum entropy principles in two-dimensional turbulence via large deviations. *J. Stat. Phys.* 98:1235–1278.

Caglioti E., P. L. Lions, C. Marchioro, and M. Pulvirenti (1992). A special class of stationary flows for two-dimensional Euler equations: A statistical mechanical description. *Commun. Math. Phys.* 143:501–525.

Capel H. W. (1966). On the possibility of first-order phase transitions in Ising systems of triplet ions with zero-field splitting. *Physica* 32:966–988.

Capel H. W. (1967*a*). On the possibility of first-order phase transitions in Ising systems of triplet ions with zero-field splitting II. *Physica* 33:295–331.

Capel H. W. (1967*b*). On the possibility of first-order phase transitions in Ising systems of triplet ions with zero-field splitting III. *Physica* 37:423–441.

Costeniuc M., R. S. Ellis, and H. Touchette (2005*a*). Complete analysis of phase transitions and ensemble equivalence for the Curie–Weiss–Potts model. *J. Math. Phys.* 46:063301 (25 pages).

Costeniuc M., R. S. Ellis, H. Touchette, and B. Turkington (2005*b*). The generalized canonical ensemble and its universal equivalence with the microcanonical ensemble. *J. Stat. Phys.* 119:1283–1329.

Costeniuc M., R. S. Ellis, and H. Touchette (2006*a*). Nonconcave entropies from generalized canonical ensembles. *Phys. Rev. E* 74:010105(R) (4 pages).

Costeniuc M., R. S. Ellis, H. Touchette, and B. Turkington (2006*b*). Generalized canonical ensembles and ensemble equivalence. *Phys. Rev. E* 73:026105 (8 pages).

Costeniuc M., R. S. Ellis, and P. T.-H. Otto (2007). Multiple critical behavior of probabilistic limit theorems in the neighborhood of a tricritical point. *J. Stat. Phys.* 127:495–552.

Dauxois T., P. Holdsworth, and S. Ruffo (2000). Violation of ensemble equivalence in the antiferromagnetic mean-field XY model. *Eur. Phys. J. B* 16:659.

Dauxois T., V. Latora, A. Rapisarda, S. Ruffo, and A. Torcini (2002). The Hamiltonian mean field model: From dynamics to statistical mechanics and back. In T. Dauxois, S. Ruffo, E. Arimondo, and M. Wilkens (editors), *Dynamics and Thermodynamics of Systems with Long-Range Interactions*, Lecture Notes in Physics 602, pp. 458–487. New York: Springer.

Dembo A. and O. Zeitouni (1998). *Large Deviations Techniques and Applications*, second edition. New York: Springer.

DiBattista M., A. Majda, and B. Turkington (1998). Prototype geophysical vortex structures via large-scale statistical theory. *Geophys. Astrophys. Fluid Dyn.* 89:235–283.

DiBattista M., A. Majda, and M. Grote (2000). Meta-stability of equilibrium statistical structures for prototype geophysical flows with damping and driving. *Physica D* 151:271–304.

Dobrushin R. L. and S. B. Shlosman (1994). Large and moderate deviations in the Ising Model. *Adv. Soviet Math.* 20:91–219.

Donsker M. D. and S. R. S. Varadhan (1983). Asymptotic evaluation of certain Markov process expectations for large time, IV. *Commun. Pure Appl. Math.* 36:183–212.

Dupuis P. and R. S. Ellis (1997). *A Weak Convergence Approach to the Theory of Large Deviations.* New York: Wiley.

Ellis R. S. (1995). An overview of the theory of large deviations and applications to statistical mechanics. *Scand. Actuarial J.* No. 1, 97–142.

Ellis R. S. (2006). *Entropy, Large Deviations, and Statistical Mechanics.* New York: Springer, 1985. Reprinted in *Classics of Mathematics* series.

Ellis R. S. (2008). The theory of large deviations and applications to statistical mechanics. Lecture notes for École de Physique Les Houches, August 5–8, 2008. Posted at http://www.math.umass.edu/~rsellis/pdf-files/Les-Houches-lectures.pdf.

Ellis R. S. and C. M. Newman (1978*a*). Limit theorems for sums of dependent random variables occurring in statistical mechanics. *Z. Wahrsch. verw. Geb.* 44:117–139.

Ellis R. S. and C. M. Newman (1978*b*). The statistics of Curie–Weiss models. *J. Stat. Phys.* 19:149–161.

Ellis R. S. and K. Wang (1990). Limit theorems for the empirical vector of the Curie–Weiss–Potts model. *Stoch. Proc. Appl.* 35:59–79.

Ellis R. S. and K. Wang (1992). Limit theorems for maximum likelihood estimators in the Curie–Weiss–Potts model. *Stoch. Proc. Appl.* 40:251–288.

Ellis R. S., C. M. Newman, and J. S. Rosen (1980). Limit theorems for sums of dependent random variables occurring in statistical mechanics, II: Conditioning, multiple phases, and metastability. *Z. Wahrsch. verw. Geb.* 51:153–169.

Ellis R. S., K. Haven, and B. Turkington (2000). Large deviation principles and complete equivalence and nonequivalence results for pure and mixed ensembles. *J. Stat. Phys.* 101:999–1064.

Ellis R. S., K. Haven, and B. Turkington (2002). Nonequivalent statistical equilibrium ensembles and refined stability theorems for most probable flows. *Nonlinearity* 15:239–255.

Ellis R. S., R. Jordan, P. Otto, and B. Turkington (2004*a*). A statistical approach to the asymptotic behavior of a generalized class of nonlinear Schrödinger equations. *Commun. Math. Phys.*, 244:187–208.

Ellis R. S., H. Touchette, and B. Turkington (2004*b*). Thermodynamic versus statistical nonequivalence of ensembles for the mean-field Blume–Emery–Griffiths model. *Physica A* 335:518–538.

Ellis R. S., P. Otto, and H. Touchette (2005). Analysis of phase transitions in the mean-field Blume–Emery–Griffiths model. *Ann. Appl. Probab.* 15:2203–2254.

Ellis R. S., J. Machta, and P. T. Otto (2008). Asymptotic behavior of the magnetization near critical and tricritical points via Ginzburg–Landau polynomials. *J. Stat. Phys.* 133:101–129.

Ellis R. S., J. Machta, and P. T. Otto (2009). Asymptotic behavior of the finite-size magnetization as a function of the approach to criticality. Submitted for publication.

Ethier S. N. and T. G. Kurtz (1986). *Markov Processes: Characterization and Convergence.* New York: Wiley.

Everdell W. R. (1997). *The First Moderns.* Chicago: The University of Chicago Press.

Eyink G. L. and H. Spohn (1993). Negative-temperature states and large-scale, long-lived vortices in two-dimensional turbulence. *J. Stat. Phys.* 70:833–886.

Föllmer H. and S. Orey (1987). Large deviations for the empirical field of a Gibbs measure. *Ann. Probab.* 16:961–977.

Gross D. H. E. (1997). Microcanonical thermodynamics and statistical fragmentation of dissipative systems: The topological structure of the n-body phase space. *Phys. Rep.* 279:119–202.

Hertel P. and W. Thirring (1971). A soluble model for a system with negative specific heat. *Ann. Phys. (NY)* 63:520.

Kiessling M. K.-H. and J. L. Lebowitz (1997). The micro-canonical point vortex ensemble: Beyond equivalence. *Lett. Math. Phys.* 42:43–56.

Kiessling M. K.-H. and T. Neukirch (2003). Negative specific heat of a magnetically self-confined plasma torus. *Proc. Natl. Acad. Sci. USA* 100:1510–1514.

Lanford O. E. (1973). Entropy and equilibrium states in classical statistical mechanics. In A. Lenard (editor), *Statistical Mechanics and Mathematical Problems*, Lecture Notes in Physics 20, pp. 1–113. Berlin: Springer.

Latora V., A. Rapisarda, and C. Tsallis (2001). Non-Gaussian equilibrium in a long-range Hamiltonian system. *Phys. Rev. E* 64:056134.

Lindley D. (2001). *Boltzmann's Atom: The Great Debate That Launched a Revolution in Physics.* New York: Free Press.

Lynden-Bell D. and R. Wood (1968). The gravo-thermal catastrophe in isothermal spheres and the onset of red-giant structure for stellar systems. *Mon. Not. R. Astron. Soc.* 138:495.

Olla S. (1988) Large deviations for Gibbs random fields. *Probab. Theor. Rel. Fields* 77:343–359.

Parisi G. (1988). *Statistical Field Theory.* Redwood City, CA: Addison-Wesley.

Robert R. (1991). A maximum-entropy principle for two-dimensional perfect fluid dynamics. *J. Stat. Phys.* 65:531–553.

Robert R. and J. Sommeria (1991). Statistical equilibrium states for two-dimensional flows. *J. Fluid Mech.* 229:291–310.

Rockafellar R. T. (1970) *Convex Analysis.* Princeton, NJ: Princeton University Press.

Smith R. A. and T. M. O'Neil (1990). Nonaxisymmetric thermal equilibria of a cylindrically bounded guiding center plasma or discrete vortex system. *Phys. Fluids B* 2:2961–2975.

Thirring W. (1970) Systems with negative specific heat. *Z. Phys.* 235:339–352.

Touchette H. (2009). The large deviation approach to statistical mechanics. *Phys. Rep.* 478:1–69.

Touchette H., R. S. Ellis, and B. Turkington (2004). An introduction to the thermodynamic and macrostate levels of nonequivalent ensembles. *Physica A* 340:138–146.

Turkington B., A. Majda, K. Haven, and M. DiBattista (2001). Statistical equilibrium

predictions of jets and spots on Jupiter. *Proc. Natl. Acad. Sci. USA* 98:12346–12350.
Varadhan S. R. S. (1966). Asymptotic properties and differential equations. *Commun. Pure Appl. Math.* 19:261–286.
Wightman A. S. (1979). Convexity and the notion of equilibrium state in thermodynamics and statistical mechanics. Introduction to R. B. Israel, *Convexity in the Theory of Lattice Gases.* Princeton: Princeton University Press.
Wu F. Y. (1982). The Potts model. *Rev. Mod. Phys.* 54:235–268.

7

Solving ordinary differential equations when the coefficients have low regularity: A kinetic point of view (after R. Di Perna and P. L. Lions)

François CASTELLA

IRMAR & IRISA, Université de Rennes 1, Campus de Beaulieu, 35042 Rennes Cedex, France

This document gathers together some known results on the solution of ordinary differential equations (ODEs) of the form

$$\frac{\mathrm{d}}{\mathrm{d}t}X(t) = b(t, X(t)), \qquad X(0) = x \in \mathbb{R}^d, \tag{7.1}$$

in the case when the vector field $b = b(t, x)$ possesses *low regularity* in the variable x. Our convention is that b depends on both a time variable $t \in \mathbb{R}$ and a space variable $x \in \mathbb{R}^d$, where d denotes the space dimension.

Equation (7.1) allows us to compute the *flow* induced by the vector field b. If $b(t, x)$ describes the velocity of a fluid particle sitting at $x \in \mathbb{R}^d$ at time $t \in \mathbb{R}$, say, then $X(t)$ is called the flow induced by the vector field b. It represents the trajectory of a fluid particle that has issued from x at time $t = 0$, as the particle is transported along the velocity field b.

While solving eqn (7.1) is a standard task when b is a smooth function, thanks to the well-known Cauchy–Lipschitz Theorem, which we briefly present in Section 7.1, solving eqn (7.1) when b lacks regularity is an old problem, and an essentially optimal result was discovered by R. Di Perna and P. L. Lions in 1989. The goal of these notes is to present the main ideas of the Di Perna–Lions theory. We try to insist on the key ideas, and to point out those aspects or techniques that are of more general interest in the mathematical analysis of evolution equations.

Note that the question of solving eqn (7.1) when b has low regularity in x is natural when one is considering real-life fluids, for which the velocity field $b(t, x)$ is given as the solution of, typically, the Euler equations, and therefore has limited regularity in x owing to strong, nonlinear, energy exchanges between small scales and large scales or, equivalently, between large Fourier modes and small Fourier modes. It is also a natural task in the modeling of various flows, such as polymer flows, and even in the modeling of random flows, for which the velocity fields again lack smoothness.

In order to stress the key result of the Di Perna–Lions theory, let us mention the following simple facts. In essence, the standard theory of ODEs (the Cauchy–Lipschitz Theorem) requires that b possesses Lipschitz regularity, namely

$$\frac{Db}{Dx}(t, x) \text{ is locally bounded in } x, \text{ for any time } t,$$

where Db/Dx denotes the Jacobian matrix of b. The Di Perna–Lions theory allows us to weaken the above requirement, and to ask only that Db/Dx possesses the following integrability property:

$$\frac{Db}{Dx}(t, x) \text{ is locally integrable in } x, \text{ for any time } t.$$

Besides, it is known that this requirement is roughly optimal: weaker regularities cannot be treated in general. More precise statements are provided below in this text.

To conclude, the situation is the following:

- The Di Perna–Lions theory relaxes the assumption that $Db/Dx(t, x)$ is locally bounded in x, and only requires the integral of $Db/Dx(t, x)$ over any bounded set in the variable x to be bounded, a considerably weaker requirement.

- Nevertheless, we stress that the Di Perna–Lions theory requires Db/Dx to be well defined (locally integrable). In particular, b is required to "possess one derivative." From this point of view, the "number of derivatives" of b that are needed is the same as in the standard theory of ODEs. In other words, solving $\mathrm{d}X/\mathrm{d}t = b(t, X)$ requires b to be once differentiable in any circumstance, a somewhat surprising statement at first glance (the first derivative of b does not enter the original ODE).
- We point out in this text the reason for requiring Db/Dx to be bounded (classical theory) or locally integrable (Di Perna–Lions). These assumptions are needed to ensure *stability* of the solutions to $\mathrm{d}X/\mathrm{d}t = b(t, X)$. Unstable solutions may otherwise be constructed when b is only continuous.

7.1 The Cauchy–Lipschitz Theorem

Throughout this section, we take a vector field $b(t, x) \in \mathbb{R}^d$, defined for any time $t \in \mathbb{R}$ and any space point $x \in \mathbb{R}^d$, where d is the space dimension. Our main smoothness assumption on b is the following. We assume that Db/Dx is *locally bounded* in x, for any time t. Using this assumption, we indicate how the ordinary differential equation

$$\frac{\mathrm{d}}{\mathrm{d}t} X(t) = b\,(t, X(t))$$

can be solved. We point out the (deep) role of the assumption "Db/Dx is locally bounded," concentrating mainly on stability issues.

7.1.1 Stability and uniqueness of solutions to $\dot{X} = b(t, X)$ when Db/Dx is globally bounded

Let us first assume that Db/Dx is *globally* bounded in x, in that there exists a constant $C_0 > 0$ such that

$$\forall t \in \mathbb{R}\,, \quad \forall x \in \mathbb{R}^d\,, \qquad \left| \frac{Db}{Dx}(t, x) \right| \leq C_0\,. \tag{7.2}$$

We say that b is *globally Lipschitz* in the variable x.

Let us assume for a while that for any $x \in \mathbb{R}^d$, there exists a function $X(t, x)$ that is a solution to the ODE

$$\frac{\mathrm{d}}{\mathrm{d}t} X(t, x) = b\,(t, X(t, x))\,, \qquad X(0, x) = x\,. \tag{7.3}$$

Under these circumstances, let us take two initial points x and y in $\mathbb{R}^d$, and estimate the difference $X(t, x) - X(t, y)$ as time evolves. In other words, let us estimate the *stability* of two nearby trajectories, starting from nearby points x and y.

We write

$$X(t, x) = x + \int_0^t \frac{\mathrm{d}}{\mathrm{d}s} X(s, x)\,\mathrm{d}s = x + \int_0^t b\,(s, X(s, x))\,\mathrm{d}s\,,$$

and similarly for $X(t, y)$. Therefore, we may estimate, for $t \geq 0$,

$$\begin{aligned}|X(t,x) - X(t,y)| &\le |x-y| + \int_0^t |b(s, X(s,x)) - b(s, X(s,y))| \, \mathrm{d}s \\ &\le |x-y| + \int_0^t |b(s, X(s,x)) - b(s, X(s,y))| \, \mathrm{d}s \\ &\le |x-y| + C_0 \int_0^t |X(s,x) - X(s,y)| \, \mathrm{d}s \,.\end{aligned}$$

Therefore, setting

$$\delta(t) := |X(t,x) - X(t,y)| \,,$$

we arrive at the estimate

$$\delta(t) \le |x-y| + C_0 \int_0^t \delta(s) \, \mathrm{d}s \,.$$

We now invoke the following standard Gronwall Lemma.

Lemma 7.1. (Gronwall Lemma.) *Let $\epsilon(t)$ be any function satisfying*

$$0 \le \epsilon(t) \le \alpha + \beta \int_0^t \epsilon(s) \, \mathrm{d}s \,,$$

for some constants $\alpha > 0$ and $\beta > 0$. Then, the function $\epsilon(t)$ also satisfies the estimate

$$\epsilon(t) \le \alpha \exp(\beta t) \,,$$

whenever $t \ge 0$.

Remark 7.2. *Note that the* equality

$$\epsilon(t) = \alpha + \beta \int_0^t \epsilon(s) \, \mathrm{d}s \,,$$

which also reads

$$\epsilon'(t) = \beta \epsilon(t) \,, \qquad \epsilon(0) = \alpha \,,$$

implies *the equality*

$$\epsilon(t) = \alpha \exp(\beta t) \,.$$

The Gronwall Lemma simply asserts that a similar assertion holds true when equalities are replaced by upper bounds everywhere.

Proof (of the Gronwall Lemma). The proof easily follows, observing that

$$\frac{1}{\beta} \frac{\mathrm{d}}{\mathrm{d}s} \left(\log \left(\alpha + \beta \int_0^t \epsilon(s) \, \mathrm{d}s \right) \right) = \frac{\epsilon(t)}{\alpha + \beta \displaystyle\int_0^t \epsilon(s) \, \mathrm{d}s} \le 1 \,,$$

from which we deduce

$$\alpha + \beta \int_0^t \epsilon(s)\,\mathrm{d}s \le \alpha \exp(\beta t)\,.$$

Finally, we recover

$$\epsilon(t) \le \alpha + \beta \int_0^t \epsilon(s)\,\mathrm{d}s \le \alpha \exp(\beta t)\,.$$

■

In our case, using the Gronwall Lemma with $\epsilon(t)$ replaced by $\delta(t)$ provides

$$\delta(t) \le |x-y| \exp(C_0\, t)\,.$$

Gathering the above estimates together, we have proved the following theorem.

Theorem 7.3. (Stability of solutions to $\dot{X} = b(t, X)$ when b is globally Lipschitz.) *Assume that the vector field b is globally Lipschitz (see above for the definition). Let $X(t, x)$ be the solution to the differential equation*

$$\frac{\mathrm{d}}{\mathrm{d}t} X(t,x) = b\,(t, X(t,x))\,, \qquad X(0,x) = x\,.$$

Then, initially close trajectories diverge at most exponentially, in that

$$|X(t,x) - X(t,y)| \le |x-y| \exp\left(C_0\, t\right)$$

whenever $t \ge 0$, where we use the notation

$$C_0 = \sup_{z\in\mathbb{R}^d, s\in\mathbb{R}} \left|\frac{Db}{Dz}(s,z)\right|.$$

This estimate is the key to all subsequent results, in at least two respects.

First, we shall admit that the above stability estimate implies roughly that solutions of the ODE $\dot{X} = b(t, X)$ *exist*, and therefore are uniquely determined (this second point being obvious in view of the—stronger—stability estimate). In other words, we shall admit that the above result, which states the uniqueness and stability of solutions, in fact also implies the existence of the latter.

Second, the above estimate and its proof clearly show the necessity of having a well-defined and bounded first derivative of b: this is the key to having *stable* trajectories.

7.1.2 Stability and uniqueness of solutions to $\dot{X} = b(t, X)$ when Db/Dx is locally bounded

Let us assume here that Db/Dx is only *locally* bounded, in that for any time $T > 0$, and any radius $R > 0$, there exists a constant $C(T, R) > 0$ such that

$$\forall 0 \le t \le T\,, \quad \forall x \in B(0,R)\,, \qquad \left|\frac{Db}{Dx}(t,x)\right| \le C(T,R)\,, \tag{7.4}$$

where $B(0, R) \subset \mathbb{R}^d$ denotes the ball centered at the origin with radius R.

In this particular case, it turns out that another assumption on b is needed, which is different in nature. It is not a smoothness assumption. To have trajectories that are well defined *for all values of time*, we need to assume that b has at most linear growth at infinity in x, namely

$$\text{For any } T > 0, \text{ there is a } C(T) > 0 \text{ such that } \forall x \in \mathbb{R}^d ,$$
$$\forall 0 \le t \le T , \quad |b(t,x)| \le C(T)\,(1 + |x|) .$$

The point is, without an auxiliary assumption of the type "b has at most linear growth at infinity in x," trajectories may not be defined for all times, but only for *short* times, as the following standard example shows. The solution to $\dot{X} = X^2$, $X(0) = x \in \mathbb{R}$, with $x > 0$, say, satisfies $X(t) = -1/(t - 1/x)$ whenever $0 \le t < 1/x$, so that $|X(t)| \to \infty$ as t approaches the critical time $1/x$, and $X(t)$ ceases to be defined for later times: blowup occurs. This blowup phenomenon is obviously created by too strong an amplification of the signal $X(t)$ through the ODE $\dot{X} = X^2$ as $X(t)$ increases. We shall not dwell on this aspect later in this text, and we shall always make the necessary assumptions which ensure the solution to the ODE $\dot{X} = b(t, X)$ is defined for all times.

Under these circumstances, we have the following theorem.

Theorem 7.4. (Stability of solutions to $\dot{X} = b(t, X)$ when b is locally Lipschitz.) *Assume that the vector field b is locally Lipschitz and has linear growth at infinity (see above). Let $X(t, x)$ be the solution to the differential equation*

$$\frac{\mathrm{d}}{\mathrm{d}t} X(t,x) = b\,(t, X(t,x))\,, \qquad X(0,x) = x\,.$$

Take an arbitrary time $T > 0$, and an arbitrary radius $R_0 > 0$.

Then, there exists a radius $R(T, R_0)$ such that given any $x \in B(0, R_0)$, and any $y \in B(0, R_0)$, the two trajectories $X(t, x)$ and $X(t, y)$ entirely belong to $B(0, R(T, R_0))$ whenever $0 \le t \le T$, namely

$$\forall |x| \le R_0\,, \qquad |X(t,x)| \le R(T, R_0)\,.$$

Also, the two trajectories $X(t, x)$ and $X(t, y)$ diverge at most exponentially, namely

$$\forall |x| \le R_0\,, \quad \forall |y| \le R_0\,, \qquad |X(t,x) - X(t,y)| \le |x - y| \, \exp\left(C_0(T, R_0)\, t\right),$$

whenever $0 \le t \le T$, where we have used the notation

$$C_0(T, R_0) = \sup_{\substack{|z| \le R(T, R_0) \\ 0 \le s \le T}} \left| \frac{Db}{Dz}(s, z) \right| .$$

Proof. The proof of the above theorem is essentially the same as the one we provided in the previous section.

To prove the first statement, we take a $T > 0$ and an $R_0 > 0$, take some $x \in B(0, R_0)$, and write, whenever $0 \le t \le T$,

$$|X(t,x)| \le |x| + \int_0^t |b(s, X(s,x))| \, \mathrm{d}s \le R_0 + C(T) \int_0^t (1 + |X(s,x)|) \, \mathrm{d}s \,,$$

where we have used the linear-growth assumption. The Gronwall Lemma then provides

$$|X(t,x)| \le (R_0 + T\, C(T)) \exp(t\, C(T)) \,,$$

and the first statement follows.

The second statement then becomes obvious. With the above notation, taking y in $B(0, R_0)$, we write

$$\begin{aligned} |X(t,x) - X(t,y)| &\le |x - y| + \int_0^t |b(s, X(s,x)) - b(s, X(s,y))| \, \mathrm{d}s \\ &\le |x - y| + C_0(T, R_0) \int_0^t |X(s,x) - X(s,y)| \, \mathrm{d}s \,, \end{aligned}$$

where we have used the local Lipschitz property together with the fact that the whole trajectories $X(t,x)$ and $X(t,y)$ belong to $B(0, R(T, R_0))$. The Gronwall Lemma provides

$$|X(t,x) - X(t,y)| \le |x - y| \exp(t\, C_0(T, R_0)) \,.$$

The proof is complete. ∎

7.1.3 Lack of stability/uniqueness when Db/Dx is not bounded: An example.

Up to now, we have seen that solutions to $\dot{X} = b(t, X)$ are well defined, unique, and stable whenever Db/Dx is bounded (globally or locally).

The following example shows a lack of stability when this boundedness assumption is removed.

Define the vector field $b(t,x) \equiv b(x)$ as

$$b(x) = 0 \text{ when } x \le 0 \,, \qquad = \sqrt{x} \text{ when } x \ge 0 \,.$$

Here, $d = 1$ and $x \in \mathbb{R}$. The above vector field satisfies

$$b'(x) = \frac{1}{2\sqrt{x}} \text{ when } x > 0 \,, \qquad = 0 \text{ when } x < 0 \,.$$

Hence b' is unbounded close to $x = 0$.

On the other hand, as an easy computation shows, the whole family of functions parameterized by $\lambda > 0$ and defined as

$$X_\lambda(t) = \left(\frac{t - \lambda}{2} \right)^2 \text{ when } t \ge \lambda \,, \qquad = 0 \text{ when } t \le \lambda \,,$$

satisfies

$$\dot{X}_\lambda(t) = b(X_\lambda(t)) \,, \qquad X_\lambda(0) = 0 \,.$$

In other words, the whole family X_λ corresponds to trajectories which start at the equilibrium point $x = 0$ of the vector field b, remain there for some time, and then suddenly leave this state at some instant $t = \lambda$ to reach positive values at later times. Clearly, stability and/or uniqueness of solutions to $\dot{X} = b(t, X)$ is violated in the present case. Such a situation is made possible by the fact that the equilibrium point $x = 0$ can be reached *in finite time*, which in turn is a consequence of the lack of smoothness of b close to $x = 0$.

This example is unacceptable, and the trajectory $X(t) \equiv 0$ should be somehow selected as "the" good trajectory in this specific situation. All other trajectories indeed somehow correspond to a particle at rest which suddenly leaves equilibrium (without any influx of energy into the system!) to reach nonequilibrium states at later times.

In any circumstance, this example shows how both quantitative *and* qualitative aspects are involved when dealing with ODEs with nonsmooth coefficients, and regularity here does not involve only analytical issues.

7.2 The method of characteristics and the transport equation: A link between the nonlinear, finite-dimensional ODE and a linear, infinite-dimensional system

One key ingredient in the Di Perna–Lions theory of ODEs with nonsmooth coefficients is given by the transport equation, a partial differential equation (PDE). To give only the result, the point is that it turns out that the ODE

$$\dot{X}(t) = b(t, X(t)),$$

whose unknown is the finite-dimensional vector $X(t) \in \mathbb{R}^d$, has a deep link with the following PDE, called the "transport equation along the field b," which reads

$$\partial_t f(t, x) + b(t, x) \cdot \nabla_x f(t, x) = 0.$$

Here, the unknown is the function $f(t, x) \in \mathbb{R}$, depending on a time variable $t \in \mathbb{R}$ and a space variable $x \in \mathbb{R}^d$, and ∂_t stands for $\partial/\partial t$ and ∇_x refers to the gradient in the x-direction. The function $f(t, x)$ may be seen, as in the conventional kinetic theory of gases, as the probability that a particle (in some gas, say) possesses a position x at time t. Note that, if we consider the PDE $\partial_t f + b \cdot \nabla_x f = 0$ as an evolution equation in time, where the spatial variable x plays the role of a parameter, then the PDE appears as a linear but infinite-dimensional system.

In other words, a key point of the analysis is that one may somehow "see" the original ODE, a nonlinear yet finite-dimensional system, as a linear yet infinite-dimensional system through the above transport equation. This point of view clearly makes us lose finite-dimensionality while gaining linearity. It turns out that many more tools are available at the level of the linear PDE, which therefore is easier to solve in the case when b is not smooth.

This procedure is very reminiscent of the standard "coarse graining" in conventional statistical mechanics: when modeling the behavior of large N-particles systems, one may either write down the large-dimensional but more or less "simple" (possibly linear) dynamics governing the N particles, or adopt a coarse-grained point of view, leading to the solution of a "complex" (and in any circumstance nonlinear) but low-dimensional dynamics, describing the behavior of the coarse-grained ensemble.

Let us now come to the details, at least when b is smooth to begin with (the nonsmooth case, which is the core of the Di Perna–Lions theory, is presented later).

7.2.1 Solving the transport equation with the help of the ODE $\dot{X} = b(t, X)$

Throughout this section, we take a smooth vector field b, such that for any time $T > 0$, there exists a $C(T) > 0$ such that

$$\forall 0 \leq t \leq T\,, \quad \forall x \in \mathbb{R}^d\,, \quad \left|\frac{Db}{Dx}(t, x)\right| \leq C(T)\,.$$

In other words, we assume here for simplicity that b is Lipschitz in x, globally in x and locally in t. Note that this assumption automatically ensures that b has at most linear growth at infinity, since $|b(t, x)| \leq |b(t, 0)| + C(T)|x|$ in this case, whenever $0 \leq t \leq T$.

We define, for later convenience, the flow $X_s^t(x)$ (we have changed the notation here in comparison with the previous paragraphs—the reason for this change will become clear later) as follows:

$$X_s^t(x) \text{ is the solution to } \frac{\mathrm{d}}{\mathrm{d}t} X_s^t(x) = b(t, X_s^t(x))\,, \qquad X_s^s(x) = x\,.$$

In other words, $X_s^t(x)$ is the unique solution to the ODE $\dot{X} = b(t, X)$ which starts at position x at the initial time $t = s$. Using the notation of the previous paragraphs, the previously defined $X(t, x)$ coincides with the new $X_0^t(x)$.

Now, let us try to solve the transport equation

$$\partial_t f(t, x) + b(t, x) \cdot \nabla_x f(t, x) = 0\,, \qquad \text{with initial value } f(0, x) = f^0(x), \forall x\,.$$

Here, $f^0(x)$ is a given, smooth function, describing the initial distribution of particles at time $t = 0$.

The key to the subsequent analysis lies in considering the equation satisfied by

$$g(t, x) \equiv f(t, X_0^t(x))$$

rather than the PDE satisfied by f. In other words, it turns out that it may be more convenient to consider the function f transported along the flow X_s^t, instead of f itself.

Assuming that the function $f(t, x)$ is smooth for the present time, we compute

$$\begin{aligned}
\frac{\partial}{\partial t}(g(t, x)) &= \frac{\partial}{\partial t}\left(f\left(t, X_0^t(x)\right)\right) \\
&= (\partial_t f)\left(t, X_0^t(x)\right) + \frac{\mathrm{d}}{\mathrm{d}t} X_0^t(x) \cdot (\nabla_x f)\left(t, X_0^t(x)\right) \\
&= (\partial_t f)\left(t, X_0^t(x)\right) + b\left(t, X_0^t(x)\right) \cdot (\nabla_x f)\left(t, X_0^t(x)\right) \\
&= (\partial_t f + b \cdot \nabla_x f)\left(t, X_0^t(x)\right) \\
&= 0\,.
\end{aligned}$$

Therefore, we arrive at

$$g(t, x) = g(0, x) = f(0, x) = f^0(x)\,,$$

and the function g coincides with the known function f^0 at any time.

In turn, coming back to the definition of g, we recover

$$f(t, X_0^t(x)) = f^0(x).$$

This means that the function $f(t,x)$ is constant along the flow $X_0^t(x)$.

To end the above computation, there remains to change the variable $X_0^t(x) = y$ in the relation $f(t, X_0^t(x)) = f^0(x)$. This is done by using the key relation

$$X_s^t(X_t^s(x)) = x, \quad \forall x.$$

This relation asserts that, since trajectories associated with the ODE $\dot{X} = b(t,X)$ are known to be unique (recall that b is taken smooth here), solving the ODE when starting at position x at time t, during the time interval $[t,s]$, and next solving the ODE backwards, starting from the final point $X_t^s(x)$ at time s and coming back to what happened at time t, eventually takes us back to the initial point x.

As a consequence, setting

$$X_0^t(x) = y,$$

we recover the inverse relation

$$x = X_t^0(y),$$

and the identity $f(t, X_0^t(x)) = f^0(x)$ turns out to be equivalent to

$$f(t,y) = f^0\left(X_t^0(y)\right), \quad \forall y.$$

In other words, we have proved the following theorem.

Theorem 7.5. (Solving the transport equation with smooth coefficients.) *Take a vector field b such that for any time $T > 0$ there exists a $C(T) > 0$ which satisfies*

$$\forall 0 \le t \le T, \quad \forall x \in \mathbb{R}^d, \quad \left|\frac{Db}{Dx}(t,x)\right| \le C(T).$$

Take a smooth function $f^0(x)$, say f^0 is once differentiable ($f^0(x) \in C^1(\mathbb{R}^d)$).

Then, the unique smooth solution $f(t,x)$ ($f(t,x) \in C^1(\mathbb{R} \times \mathbb{R}^d)$) to the transport equation

$$\partial_t f(t,x) + b(t,x) \cdot \nabla_x f(t,x) = 0$$

is given by

$$f(t,x) = f^0(X_t^0(x)), \quad \forall (t,x).$$

As we can see, there is a natural duality between the original ODE $\dot{X} = b(t,X)$ and the linear transport equation $\partial_t f + b \cdot \nabla_x f = 0$. More precisely, once the ODE is solved, i.e. once the flow X_s^t is known, the solution f is known as well. Conversely, once the function f is known "for sufficiently many initial data f^0," the function $f^0(X_t^0(x))$ is known for sufficiently many f^0 as well, and hence the flow X_s^t is essentially known (see below for more precise statements).

7.2.2 Stability estimates for the transport equation

Continuing in this direction, it is a natural question to ask what the stability estimate that holds at the level of trajectories becomes, when one is arguing at the level of the transport equation. In other words, what does the following stability upper bound become, which is valid for any $0 \le t \le T$,

$$\left|X_0^t(x) - X_0^t(y)\right| \le |x-y| \exp(C(T)\, t)$$

(where $C(T) = \sup_{s\in[0,T],z\in\mathbb{R}^d} |(Db/Dz)(s,z)|$), when one is arguing at the level of the solution $f(t,x)$ to the transport equation?

As we shall see, it turns out that the analogous estimate on f is physically more accurate, and it is best formulated using the following L^p norms.

Definition 7.6. (Definition of L^p norms.) *Let $u(x) \in \mathbb{R}$ be a function defined for all values of $x \in \mathbb{R}^d$. Take an index $1 \le p < +\infty$. We define the L^p norm of u as*

$$\|u\|_{L^p(\mathbb{R}^d)} := \left(\int_{\mathbb{R}^d} |u(x)|^p \,\mathrm{d}x\right)^{1/p}.$$

Analogously, if $p = +\infty$, we define

$$\|u\|_{L^\infty(\mathbb{R}^d)} := \sup_{x\in\mathbb{R}^d} |u(x)|\,.$$

Naturally, if u takes values in $\mathbb{R}^N$ for some N, the definition extends to $\|u\|_{L^p} = \left(\sum_{i=1}^N \|u_i\|^p_{L^p(\mathbb{R}^d)}\right)^{1/p}$, while if $K \subset \mathbb{R}^d$ is some subset of $\mathbb{R}^d$, this definition extends to $\|u\|_{L^p(K)} := \left(\int_K |u(x)|^p \,\mathrm{d}x\right)^{1/p}$.

With the above definition, we have the following theorem.

Theorem 7.7. (Stability of solutions to the transport equation with smooth coefficients.) *Take two smooth initial data points f^0 and g^0, say f^0 and $g^0 \in C^1(\mathbb{R}^d)$. Take a vector field such that for any T there is a $C(T)$ such that $|Db/Dx(t,x)| \le C(T)$ whenever $0 \le t \le T$ and $x \in \mathbb{R}^d$. Define $f(t,x)$ and $g(t,x)$ as the unique smooth (C^1) solutions to the transport equation $\partial_t f + b\cdot\nabla_x f = 0$ with initial data f^0 and g^0, respectively. Then, we have the following stability estimate, valid for any $1 \le p \le \infty$, and any $0 \le t \le T$:*

$$\|f(t,x) - g(t,x)\|_{L^p(\mathbb{R}^d)} \le \left\|f^0 - g^0\right\|_{L^p(\mathbb{R}^d)} \exp\left(\frac{t}{p} \times \|\mathrm{div}_x b\|_{L^\infty([0,T]\times\mathbb{R}^d)}\right).$$

Remark 7.8. *In other words, the distance between the two solutions $f(t)$ and $g(t)$ at time t, when measured in L^p norms, grows exponentially as time increases, as does the distance between nearby trajectories $X_0^t(x) - X_0^t(y)$.*

Yet the result is a bit more precise than what we get when dealing with the characteristics $X_0^t(x)$ themselves: while the growth rate of the distance between trajectories is *governed by the L^∞ norm of the full Jacobian matrix Db/Dx, here the growth rate of the distance between solutions of the transport equation is given by the L^∞ norm of the mere* divergence *of the vector field b. This fact translates the fact that the elementary volume $\mathrm{d}x$, when transported along the vector field b, grows or decays exponentially along the value of the divergence of b, a well-known fact in the context.*

We begin with a naive proof of the above theorem, based on a direct manipulation of the characteristics $X_0^t(x)$ themselves.

Proof (a naive proof). We may at once take advantage of the explicit formula

$$f(t,x) = f^0(X_t^0(x))\,, \quad g(t,x) = g^0(X_t^0(x))\,,$$

from which it is easily deduced that

$$\begin{aligned} \|f(t,x) - g(t,x)\|_{L^p(\mathbb{R}^d)}^p &= \int_{\mathbb{R}^d} |f(t,x) - g(t,x)|^p \, \mathrm{d}x \\ &= \int_{\mathbb{R}^d} \left|f^0(X_t^0(x)) - g^0(X_t^0(x))\right|^p \, \mathrm{d}x\,. \end{aligned}$$

Hence, performing the change of unknown $y = X_t^0(x)$ (i.e. $x = X_0^t(y)$) in the above integral, we recover

$$\|f(t,x) - g(t,x)\|_{L^p(\mathbb{R}^d)}^p = \int_{\mathbb{R}^d} |f^0(y) - g^0(y)|^p \left|\det\left(\frac{DX_0^t(y)}{Dy}\right)\right| \, \mathrm{d}y\,.$$

As a consequence, we finally obtain

$$\|f(t,x) - g(t,x)\|_{L^p(\mathbb{R}^d)}^p \leq \|f^0 - g^0\|_{L^p(\mathbb{R}^d)}^p \times \sup_{y\in\mathbb{R}^d} \left|\det\left(\frac{DX_0^t(y)}{Dy}\right)\right|\,.$$

It remains to estimate the Jacobian term DX/Dy.

Since $X_0^t(y)$ satisfies the ODE $\dot{X} = b(t,X)$, it is known that the Jacobian DX/dy satisfies the linearized system

$$\frac{\mathrm{d}}{\mathrm{d}t}\frac{DX_0^t(y)}{Dy} = \frac{Db}{Dx}\left(t, X_0^t(y)\right)\,\frac{DX_0^t(y)}{Dy}\,,$$

from which it follows, using the identity

$$\frac{\mathrm{d}\,(\det A(t))}{\mathrm{d}t} = \det A(t)\,\mathrm{Tr}\left(A(t)^{-1}\,\frac{\mathrm{d}A(t)}{\mathrm{d}t}\right)$$

(where Tr is the trace operator), that we have the relations

$$\begin{aligned} \frac{\mathrm{d}}{\mathrm{d}t}\left(\det\frac{DX_0^t(y)}{Dy}\right) &= \left(\det\frac{DX_0^t(y)}{Dy}\right)\,\mathrm{Tr}\left(\frac{DX_0^t(y)}{Dy}^{-1}\,\frac{Db}{Dx}\left(t,X_0^t(y)\right)\,\frac{DX_0^t(y)}{Dy}\right) \\ &= \left(\det\frac{DX_0^t(y)}{Dy}\right)\,\mathrm{Tr}\left(\frac{Db}{Dx}\left(t,X_0^t(y)\right)\right) \\ &= \left(\det\frac{DX_0^t(y)}{Dy}\right)\,\mathrm{div}_x b\left(t,X_0^t(y)\right)\,, \end{aligned}$$

by the definition of the divergence operator. We recognize here that the divergence of the vector field b controls the growth (when $\mathrm{div}_x b > 0$) or decay (when $\mathrm{div}_x b < 0$) of the volume measure $\det(DX_0^t(y)/Dy)$, as usual.

The above identity allows us to estimate, using the relation

$$\frac{DX_0^0(y)}{Dy} = \frac{Dy}{Dy} = \text{Identity}\,,$$

that whenever $0 \le t \le T$, we have

$$\begin{aligned}\left|\det \frac{DX_0^t(y)}{Dy}\right| &\le 1 + \int_0^t \left|\mathrm{div}_x b\,(s, X_0^s(y))\right| \left|\det \frac{DX_0^s(y)}{Dy}\right| \,\mathrm{d}s \\ &\le 1 + \underbrace{\left(\sup_{x\in\mathbb{R}, 0\le s\le T} |\mathrm{div}_x b(s,x)|\right)}_{=\|\mathrm{div}_x b\|_{L^\infty([0,T]\times\mathbb{R}^d)}} \int_0^t \left|\det \frac{DX_0^s(y)}{Dy}\right| \,\mathrm{d}s\,.\end{aligned}$$

From this it follows, using the Gronwall Lemma,

$$\sup_{y\in\mathbb{R}^d} \left|\det \frac{DX_0^t(y)}{Dy}\right| \le \exp\left(t \times \|\mathrm{div}_x b\|_{L^\infty([0,T]\times\mathbb{R}^d)}\right), \quad \forall 0 \le t \le T\,.$$

Coming back to the estimate of $\|f(t,x) - g(t,x)\|_{L^p}$, we recover for any time $t \in [0,T]$ the upper bound

$$\|f(t,x) - g(t,x)\|_{L^p(\mathbb{R}^d)}^p \le \left\|f^0 - g^0\right\|_{L^p(\mathbb{R}^d)}^p \times \exp\left(t\,\|\mathrm{div}_x b\|_{L^\infty([0,T]\times\mathbb{R}^d)}\right).$$

The stability estimate is proved whenever $1 \le p < \infty$. Since the result is obvious when $p = \infty$, the theorem is proved. ■

The above proof is perfectly correct, yet somehow unsatisfactory.

Indeed, remember our goal is to use the duality between the (linear) PDE $\partial_t f + b \cdot \nabla_x f = 0$ and the (nonlinear) ODE $\dot{X} = b(t,X)$, a duality that is valid when the coefficients are smooth at least, in order to solve the PDE when the coefficients are not smooth and to somehow deduce therefrom a reasonable notion of a solution for the ODE in the case when b lacks regularity.

In that perspective, the above proof has the obvious drawback that it establishes a stability estimate for the PDE by using known results on the ODE: one would expect a more direct argument, valid directly at the PDE level. Even more, our argument above requires us to consider the linearized ODE, obtained by differentiating once the relation $\dot{X} = b(t,X)$, a disaster when b lacks smoothness.

For these reasons, we now present a more direct argument. Roughly speaking, we shall now perform some kind of energy estimate, directly on the PDE.

Proof (obtaining stability as an energy estimate on the PDE). The argument is the following. It exploits the following two features of the PDE $\partial_t f + b \cdot \nabla_x f = 0$: namely, the equation is linear, and it involves only first-order derivatives.

Take two solutions f and g to the transport equation (as in the theorem).

Since the transport equation is linear, the difference $f - g$ satisfies

$$\partial_t(f-g)(t,x) + b(t,x)\cdot\nabla_x(f-g)(t,x) = 0\,. \tag{7.5}$$

Now, we want to estimate $|f-g|^p$. For that reason, we first observe the easy equality (here we drop the obvious (t,x) dependence for simplicity)

$$\partial_t\,|f-g|^p = p\,\mathrm{sgn}\,(f-g)\;(\partial_t(f-g))\;|f-g|^{p-1}\,. \tag{7.6}$$

This comes from the chain rule

$$\frac{\mathrm{d}}{\mathrm{d}z}\phi\,(\psi\,(z)) = \psi'(z)\,\phi'\,(\psi(z))$$

(where ϕ and ψ denote arbitrary smooth functions), together with the fact that[1]

$$\frac{\mathrm{d}}{\mathrm{d}z}|z|^p = p\,\mathrm{sgn}(z)\,|z|^{p-1} \qquad (z\in\mathbb{R})\,,$$

where

$$\mathrm{sgn}(z) = 0 \text{ when } z=0\,, \quad = 1 \text{ when } z>0\,, \quad = -1 \text{ when } z<0\,.$$

In a similar way, we have the relation

$$b\cdot\nabla_x\,|f-g|^p = p\,\mathrm{sgn}\,(f-g)\;(b\cdot\nabla_x(f-g))\;|f-g|^{p-1}\,. \tag{7.7}$$

Therefore, *since the transport equation is a first-order equation*, multiplying the relation $\partial_t(f-g) + b\cdot\nabla_x(f-g) = 0$ by

$$p\,\mathrm{sgn}\,(f-g)\;|f-g|^{p-1}$$

and using the above equalities, we successively recover

$$p\,\mathrm{sgn}\,(f-g)\;(\partial_t(f-g))\;|f-g|^{p-1} + p\,\mathrm{sgn}\,(f-g)\;(b\cdot\nabla_x(f-g))\;|f-g|^{p-1} = 0$$

and

$$\partial_t|f-g|^p(t,x) + b(t,x)\cdot\nabla_x|f-g|^p(t,x) = 0\,. \tag{7.8}$$

This is the key to many of the results we prove in the remainder of this text: the chain rule, combined with the structure of the transport equation, allows us to transform the original equation (7.5) on $f-g$ into eqn (7.8) on $|f-g|^p$, a *nonlinear function of* $f-g$. Besides, $f-g$ and $|f-g|^p$ actually satisfy the same transport equation.

[1] *Sensu stricto*, these relations are correct only when $p>1$, or when $p=1$ and $f-g\neq 0$—or $z\neq 0$. We sweep under the rug the truly harmless fact that these relations cannot be written when $p=1$ and $f=g$, or $p=1$ and $z=0$.

The proof now becomes a standard energy estimate. Integrating eqn (7.8) over $\mathbb{R}^d$ provides

$$\partial_t \left(\int_{\mathbb{R}^d} |(f-g)(t,x)|^p \, \mathrm{d}x \right) + \int_{\mathbb{R}^d} b(t,x) \cdot \nabla_x |(f-g)(t,x)|^p \, \mathrm{d}x = 0 \,.$$

Hence, performing the obvious integration by parts in the second integral, we recover

$$\int_{\mathbb{R}^d} b(t,x) \cdot \nabla_x |(f-g)(t,x)|^p \, \mathrm{d}x = - \int_{\mathbb{R}^d} \mathrm{div}_x b(t,x) \; |(f-g)(t,x)|^p \, \mathrm{d}x \,.$$

Hence

$$\partial_t \left(\int_{\mathbb{R}^d} |(f-g)(t,x)|^p \, \mathrm{d}x \right) = \int_{\mathbb{R}^d} \mathrm{div}_x b(t,x) \; |(f-g)(t,x)|^p \, \mathrm{d}x \,.$$

It readily follows that

$$\left| \partial_t \left(\int_{\mathbb{R}^d} |(f-g)(t,x)|^p \, \mathrm{d}x \right) \right| \leq \|\mathrm{div}_x b(t,x)\|_{L^\infty([0,T]\times\mathbb{R}^d)} \int_{\mathbb{R}^d} |(f-g)(t,x)|^p \, \mathrm{d}x \,,$$

whenever $0 \leq t \leq T$. Hence, writing

$$\int_{\mathbb{R}^d} |(f-g)(t,x)|^p \, \mathrm{d}x = \int_{\mathbb{R}^d} |(f-g)(0,x)|^p \, \mathrm{d}x + \int_0^t \partial_t \left(\int_{\mathbb{R}^d} |(f-g)(s,x)|^p \, \mathrm{d}x \right) \mathrm{d}s$$

and using the Gronwall Lemma provides

$$\|f(t,x) - g(t,x)\|^p_{L^p(\mathbb{R}^d)} \leq \left\|f^0 - g^0\right\|^p_{L^p(\mathbb{R}^d)} \times \exp\left(t \, \|\mathrm{div}_x b\|_{L^\infty([0,T]\times\mathbb{R}^d)}\right) ,$$

whenever $0 \leq t \leq T$.

The stability estimate is proved whenever $1 \leq p < \infty$.

When $p = \infty$, the argument is a slight modification of the previous one. We aim at proving that for any time $t \geq 0$ and any $x \in \mathbb{R}^d$, we have

$$-\|f^0 - g^0\|_{L^\infty} \leq f(t,x) - g(t,x) \leq \|f^0 - g^0\|_{L^\infty} \,.$$

In order to prove the inequality $f(t,x) - g(t,x) \leq \|f^0 - g^0\|_{L^\infty}$ (the other estimate follows along the same lines), we define

$$u(t,x) := f(t,x) - g(t,x) - \|f^0 - g^0\|_{L^\infty} \,,$$

and establish that $u(t,x) \leq 0$ for any (t,x).

In order to do so, we first observe that u satisfies the transport equation with nonpositive initial data

$$\partial_t u + b(t,x) \cdot \nabla_x u(t,x) = 0 \,, \quad u(0,x) \leq 0 \,.$$

Now, to prove that $u(t,x) \leq 0$, we establish

$$u(t,x)_+ = 0\,,$$

where for any $z \in \mathbb{R}$ we define

$$z_+ = z \text{ when } z \geq 0\,, \quad = 0 \text{ when } z < 0\,.$$

It is clear that

$$\frac{\mathrm{d}}{\mathrm{d}z} z_+ = \frac{z_+}{z}\,.$$

Therefore, multiplying the transport equation by u_+/u and using the chain rule provides

$$\partial_t u_+ + b \cdot \nabla_x u_+ = 0\,, \quad u_+(0,x) = 0\,.$$

Therefore, multiplying the above equation by u_+ and performing the natural integration by parts provides

$$\frac{\mathrm{d}}{\mathrm{d}t} \int_{\mathbb{R}^d} |u_+(t,x)|^2 \,\mathrm{d}x = - \int_{\mathbb{R}^d} \mathrm{div}_x b(t,x)\, |u_+(t,x)|^2 \,\mathrm{d}x\,.$$

From this it follows, using $u_+(0,x) = 0$, that for any time $0 \leq t \leq T$, with T fixed, we have

$$\int_{\mathbb{R}^d} |u_+(t,x)|^2 \,\mathrm{d}x \leq \|\mathrm{div}_x b\|_{L^\infty([0,T]\times\mathbb{R}^d)} \int_{\mathbb{R}^d} |u_+(t,x)|^2 \,\mathrm{d}x\,,$$

and the Gronwall Lemma provides

$$\int_{\mathbb{R}^d} |u_+(t,x)|^2 \,\mathrm{d}x \leq 0\,.$$

This means

$$u_+(t,x) = 0\,, \quad \forall (t,x)\,.$$

The proof of the theorem is complete. ■

7.2.3 Extending the above results when $f^0 \in L^p$ only

Up to now, we have solved the transport equation with the help of the ODE, for initial data f^0 that lie in C^1 at least. Since our purpose is to solve these equations for nonsmooth coefficients, it is desirable to at least go below this C^1 regularity for f^0. Besides, as we have seen, the distance between nearby solutions is best measured in L^p norms, so that dealing with f^0's having merely L^p regularity is clearly natural. Lastly, remember that in the context of statistical physics, $f(t,x)$ represents a probability of finding a particle at time t and position x: from this standpoint, an L^1 theory seems particularly natural.

Now, if we are to have even a rough idea of what could actually be a reasonable notion of a solution to the transport equation when $f(t,x)$ lies in L^p only, the key point is that the term $\partial_t f$ is well defined as a distribution, which can be shown by using the obvious integration by parts

$$\int (\partial_t f(t,x))\, \phi(t,x)\, \mathrm{d}t\, \mathrm{d}x = -\int f(t,x)\partial_t \phi(t,x)\, \mathrm{d}t\, \mathrm{d}x$$

whenever ϕ is smooth and compactly supported, while the transport term $b \cdot \nabla_x f$ is well defined as a distribution as well, which can be shown by writing

$$\int (b(t,x) \cdot \nabla_x f(t,x))\, \phi(t,x)\, \mathrm{d}t\, \mathrm{d}x = -\int f(t,x)\, \mathrm{div}_x\, (b(t,x)\phi(t,x))\, \mathrm{d}t\, \mathrm{d}x$$
$$= -\int f(t,x)\mathrm{div}_x\, (b(t,x))\, \phi(t,x)\, \mathrm{d}t\, \mathrm{d}x - \int f(t,x) b(t,x) \cdot \nabla_x \phi(t,x)\, \mathrm{d}t\, \mathrm{d}x\,, \tag{7.9}$$

the two last terms being well defined since $\mathrm{div}_x b$ and b belong to L^∞ locally in space. Hence writing the equation $\partial_t f + b \cdot \nabla_x f = 0$ when f has only L^p regularity makes perfect sense as an equality between distributions.

Based on these simple observations, it turns out that solving the transport equation with f^0's in L^p is fairly easy, so we can readily give the necessary results and ingredients right now.

Definition 7.9. (Weak solutions to the transport equation.) *Take a p such that $1 \le p \le \infty$. Take a function $f(t,x)$ such that for any time $T > 0$, we have*

$$\|f(t,x)\|_{L^\infty([0,T];L^p(\mathbb{R}^d))} := \sup_{t\in[0,T]} \|f(t,x)\|_{L^p(\mathbb{R}^d)} < \infty\,.$$

Take a function $f^0(x) \in L^p(\mathbb{R}^d)$.

We say that f is a weak solution to the transport equation $\partial_t f + b \cdot \nabla_x f = 0$ with initial data $f^0(x)$ whenever, for any smooth test function $\phi(t,x) \in C_c^\infty(\mathbb{R} \times \mathbb{R}^d)$, infinitely differentiable and with compact support, the following identity holds:

$$\int_0^{+\infty} \int_{\mathbb{R}^d} f(t,x)\partial_t \phi(t,x)\, \mathrm{d}t\, \mathrm{d}x =$$
$$-\int_{\mathbb{R}^d} f^0(x)\phi(0,x)\, \mathrm{d}x + \int_0^{+\infty} \int_{\mathbb{R}^d} f(t,x)\, \mathrm{div}_x\, (b(t,x)\phi(t,x))\, \mathrm{d}t\, \mathrm{d}x\,.$$

Remark 7.10. *The above definition extends the natural identity obtained when all functions are smooth by multiplying the transport equation by ϕ, integrating the result over $[0,+\infty[\times\mathbb{R}^d$, and performing the obvious integrations by parts to switch all derivatives from f to ϕ or b.*

This definition makes perfect sense whenever f has the required $L^\infty([0,T];L^p)$ regularity for all $T > 0$, since all derivatives of ϕ are smooth and $\mathrm{div}_x b$ is assumed to be bounded. To make things clear, the above assertion relies on the following estimates, obtained by using the Hölder inequality:

$$\left|\int_0^{+\infty}\int_{\mathbb{R}^d} f(t,x)\partial_t\phi(t,x)\,\mathrm{d}t\,\mathrm{d}x\right|$$
$$\leq \|f(t,x)\|_{L^\infty([0,T];L^p(\mathbb{R}^d))}\ \|\partial_t\phi(t,x)\|_{L^1([0,T];L^{p'}(\mathbb{R}^d))}$$
$$\leq \|f(t,x)\|_{L^\infty([0,T];L^p(\mathbb{R}^d))}\ \|\partial_t\phi(t,x)\|_{L^\infty([0,T]\times B(0,R))}\ T\,\mathrm{vol}\,(B(0,R))^{1/p'} < +\infty\,,$$

$$\left|\int_{\mathbb{R}^d} f^0(x)\phi(0,x)\,\mathrm{d}x\right| \leq \left\|f^0\right\|_{L^p(\mathbb{R}^d)}\ \|\phi(0,x)\|_{L^{p'}(\mathbb{R}^d)}$$
$$\leq \left\|f^0\right\|_{L^p(\mathbb{R}^d)}\ \|\phi(0,x)\|_{L^\infty(B(0,R))}\ \mathrm{vol}\,(B(0,R))^{1/p'} < +\infty\,,$$

$$\left|\int_0^{+\infty}\int_{\mathbb{R}^d} f(t,x)\,\mathrm{div}_x\,(b(t,x)\phi(t,x))\,\mathrm{d}t\,\mathrm{d}x\right|$$
$$\leq \|f(t,x)\|_{L^\infty([0,T];L^p(\mathbb{R}^d))}\ \Big(\|\mathrm{div}_x b(t,x)\|_{L^\infty([0,T]\times B(0,R))}\ \|\phi(t,x)\|_{L^\infty([0,T]\times B(0,R))}$$
$$+\ \|b(t,x)\|_{L^\infty([0,T]\times B(0,R))}\ \|\nabla_x\phi(t,x)\|_{L^\infty([0,T]\times B(0,R))}\Big)\,T\,\mathrm{vol}\,(B(0,R))^{1/p'} < +\infty\,,$$

where ϕ has support in $[0,T]\times B(0,R)$ (this defines T and R), $B(0,R)$ denotes the ball of radius R centered at the origin in $\mathbb{R}^d$, $\mathrm{vol}\,(B(0,R))$ is the volume of this ball, and p' is defined by

$$\frac{1}{p}+\frac{1}{p'}=1\,.$$

Recall that the Hölder inequality states that

$$\left|\int_{\mathbb{R}^d} u(x)v(x)\,\mathrm{d}x\right| \leq \|u\|_{L^p}\ \|v\|_{L^{p'}}\,,$$

which reduces to the Cauchy–Schwarz inequality when $p=2$ (in which case $p'=2$ as well).

Based on this definition, the following theorem is easily proved.

Theorem 7.11. (Existence and stability of weak solutions to the transport equation in L^p.) *Take a p such that $1\leq p\leq\infty$. Take two data points $f^0\in L^p(\mathbb{R}^d)$ and $g^0\in L^p(\mathbb{R}^d)$. Take a vector field b such that for any $T>0$ there is a $C(T)>0$ such that $\|Db/Dx(t,x)\|_{L^\infty(\mathbb{R}^d)}\leq C(T)$ whenever $0\leq t\leq T$.*

Then, there exist uniquely defined functions $f(t,x)\in L^\infty([0,T];L^p(\mathbb{R}^d))$ and $g(t,x)\in L^\infty([0,T];L^p(\mathbb{R}^d))$, which are weak solutions to the transport equation with initial data f^0 and g^0, respectively.

Also, $f(t,x)$ and $g(t,x)$ satisfy the following stability estimate, valid for any $0\leq t\leq T$:

$$\|f(t,x)-g(t,x)\|_{L^p(\mathbb{R}^d)} \leq \left\|f^0-g^0\right\|_{L^p(\mathbb{R}^d)}\ \exp\left(\frac{t}{p}\times\|\mathrm{div}_x b\|_{L^\infty([0,T]\times\mathbb{R}^d)}\right).$$

Proof (existence and stability of weak solutions to the transport equation). The proof is fairly easy, and it illustrates very well how much stability is the key ingredient for having both existence and uniqueness of solutions.

We first perform a regularization procedure, to take advantage of the previously proved theorem, valid for C^1 solutions to the transport equation.

Take a sequence $f_n^0(x)$ $(n \in \mathbb{N})$ of initial data such that

$$\forall n, \quad f_n^0(x) \in C^1(\mathbb{R}^d) \quad \text{and} \quad f_n^0(x) \underset{n\to\infty}{\to} f^0(x) \text{ in } L^p(\mathbb{R}^d)\,.$$

Similarly, take a sequence of C^1 functions g_n^0 such that g_n^0 tends to g^0 in L^p. Such sequences exist by standard arguments, which we detail later (see the appendix to this chapter).

Associated with f_n^0 and g_n^0, take the uniquely defined C^1 solutions $f_n(t,x)$ and $g_n(t,x)$ which solve the transport equation with initial data f_n^0 and g_n^0, respectively. The stability estimate derived before, which is valid for C^1 solutions, provides, for any integers n and m and for any time T,

$$\|f_n(t,x) - f_{n+m}(t,x)\|_{L^\infty([0,T];L^p(\mathbb{R}^d))} \leq C\left(T, \|\mathrm{div}_x b\|_{L^\infty([0,T]\times\mathbb{R}^d)}\right) \|f_n^0 - f_{n+m}^0\|_{L^p(\mathbb{R}^d)}\,,$$

where the constant $C\left(T, \|\mathrm{div}_x b\|_{L^\infty([0,T]\times\mathbb{R}^d)}\right)$ depends only on the arguments explicitly mentioned. On the other hand, since the sequence f_n^0 converges and therefore is a Cauchy sequence in L^p, we recover

$$\|f_n^0 - f_{n+m}^0\|_{L^p(\mathbb{R}^d)} \underset{n\to\infty}{\to} 0\,, \qquad \text{independently of } m\,.$$

Therefore,

$$\|f_n(t,x) - f_{n+m}(t,x)\|_{L^\infty([0,T];L^p(\mathbb{R}^d))} \underset{n\to\infty}{\to} 0\,, \qquad \text{independently of } m\,.$$

Hence the sequence $f_n(t,x)$ is a Cauchy sequence in $L^\infty([0,T];L^p(\mathbb{R}^d))$, which thus possesses a limit $f(t,x)$, namely

$$\|f_n(t,x) - f(t,x)\|_{L^\infty([0,T];L^p(\mathbb{R}^d))} \underset{n\to\infty}{\to} 0\,.$$

A similar argument holds for g_n, which provides a limit g. Obviously, by passing to the limit, f and g satisfy the stability estimate

$$\|f(t,x) - g(t,x)\|_{L^p(\mathbb{R}^d)} \leq \left\|f^0 - g^0\right\|_{L^p(\mathbb{R}^d)} \exp\left(\frac{t}{p} \times \|\mathrm{div}_x b\|_{L^\infty([0,T]\times\mathbb{R}^d}\right),$$

whenever $0 \leq t \leq T$.

It remains to prove that f is a solution to the transport equation. This is an easy task. Indeed, for any n, we know that f_n is a solution to the transport equation. Therefore, given any smooth and compactly supported function ϕ, we have

$$\int_0^{+\infty}\int_{\mathbb{R}^d} f_n(t,x)\partial_t\phi(t,x)\,\mathrm{d}t\,\mathrm{d}x =$$
$$-\int_{\mathbb{R}^d} f_n^0(x)\phi(0,x)\,\mathrm{d}x + \int_0^{+\infty}\int_{\mathbb{R}^d} f_n(t,x)\mathrm{div}_x\left(b(t,x)\phi(t,x)\right)\,\mathrm{d}t\,\mathrm{d}x\,.$$

Now, the estimates in Remark 7.10 readily allow us to pass to the limit $n \to \infty$ in the above expression, and to obtain

$$\int_0^{+\infty} \int_{\mathbb{R}^d} f(t,x)\partial_t\phi(t,x)\,\mathrm{d}t\,\mathrm{d}x = \\ -\int_{\mathbb{R}^d} f^0(x)\phi(0,x)\,\mathrm{d}x + \int_0^{+\infty} \int_{\mathbb{R}^d} f(t,x)\mathrm{div}_x\left(b(t,x)\phi(t,x)\right)\,\mathrm{d}t\,\mathrm{d}x\,.$$

The theorem is now proved. ■

Remark 7.12. *We stress the fact that the above theorem is essentially a trivial consequence of the stability estimate obtained for smooth solutions to the transport equation. A simple regularization procedure directly provides the existence of L^p solutions once stability is given.*

7.3 The Di Perna–Lions theory

Up to now, we have proved the existence, uniqueness, and stability of solutions to either the ODE $\dot{X} = b(t,X)$ or to the PDE $\partial_t f + b\cdot\nabla_x f = 0$, provided the vector field b has the Lipschitz property, namely that for any $T > 0$ we assume there is a $C(T)$ such that

$$\forall 0 \le t \le T\,, \quad \left\| \frac{Db}{Dx}(t,x) \right\|_{L^\infty(\mathbb{R}^d)} \le C(T)\,. \tag{7.10}$$

We also have seen that the ODE and the PDE are in a sense dual, through the formula

$$f(t,x) = f^0(X_t^0(x))\,. \tag{7.11}$$

Our aim in this section is to relax the Lipschitz condition (7.10) and merely assume that for any $T > 0$ and $R > 0$, there is a $C(T,R) > 0$ such that

$$\forall 0 \le t \le T\,, \quad \left\| \frac{Db}{Dx}(t,x) \right\|_{L^1(B(0,R))} \le C(T,R)\,. \tag{7.12}$$

In other words, we relax the requirement "Db/Dx should be (globally) bounded" into "Db/Dx should be (locally) integrable."

This means that we are about to somehow solve the ODE $\dot{X} = b(t,X)$ (or, more precisely, to solve the associated transport equation) for b's such that Db/Dx may present local singularities of the form $1/|x-x_0|^\alpha$ at some points x_0, where $\alpha < d$. One aspect is that, since we shall actually solve the transport equation instead of the ODE itself, the so-obtained (generalized) flow of the ODE will eventually be well defined for *almost every* initial data point only.

Some more technical comments and assumptions are in order.

Firstly, the fact that the first assumption (7.10) is made globally in x while the second (7.12) is only made locally is an artefact of our presentation. In any circumstance, the assumption (7.12) needs to be supplemented by the following assumption, which is more technical in nature:

$$\begin{aligned}&\text{For any } T>0, \text{ there exists a } C(T)>0 \text{ such that } \forall x\in\mathbb{R}^d\,,\\ &\forall 0\le t\le T\,,\quad |b(t,x)|\le C(T)\,(1+|x|)\,.\end{aligned} \tag{7.13}$$

As discussed already, this hypothesis roughly prevents possible blowup in the ODE $\dot{X}=b(t,X)$, since[2] any (formal) solution to $\dot{X}=b(t,X)$ satisfies, with this assumption, the estimate $|\dot{X}(t)|\le C(T)\,(1+|X(t)|)$ whenever $0\le t\le T$. This in turn provides, using the Gronwall Lemma, the upper bound $|X(t)|\le C(T)\exp(tC(T))$ whenever $0\le t\le T$, and $X(t)$ turns out to be well defined for all times.

Secondly, since the divergence of the vector field b controls the growth/decay of the elementary volume measure as it is transported along the flow of the ODE $\dot{X}=b(t,X)$, we shall need yet another assumption, this time about the divergence of b, namely

$$\text{For any } T>0, \text{ there exists a } C(T)>0 \text{ such that } \|\mathrm{div}_x b(t,x)\|_{L^\infty(\mathbb{R}^d)}\le C(T)\,. \tag{7.14}$$

The reader should not be confused about the meaning of this assumption: we are definitely not assuming that all first derivatives of b are bounded, and we again stress that first derivatives of b are only assumed to be locally integrable from now on. Yet the divergence of b is assumed to be bounded. Typically, one may imagine a divergence-free vector field whose components have possibly singular first derivatives. Again, and as in eqn (7.13), the assumption (7.14) merely prevents a possible blowup, this time a blowup of the elementary volume measure transported by the flow.

From now on, we assume that b is such that the key assumption (7.12), together with the auxiliary assumptions (7.13) and (7.14), is met. On this basis, we show that the transport equation $\partial_t f+b\cdot\nabla_x f=0$ may be solved under these assumptions. We next deduce information about the flow of $\dot{X}=b(t,X)$ in that case.

7.3.1 Existence of solutions to the transport equation when b is not smooth

Let b satisfy the assumptions (7.12), (7.13), and (7.14). Take an initial set of data $f^0\in L^p(\mathbb{R}^d)$ for some $1\le p\le\infty$. Let us first prove the *existence* of solutions to the transport equation $\partial_t f+b\cdot\nabla_x f=0$ with initial data f_0.

We proceed as in the proof of Theorem 7.11 above; namely, we perform a regularization procedure, and we pass to the limit in the regularization parameter. For the sake of simplicity, we assume $1<p<\infty$ to begin with, and indicate the necessary modifications to the argument in the cases $p=1$ and $p=\infty$ later.

Let us take a sequence $b_n(t,x)$ which converges (in a sense that we shall make precise below) towards b as $n\to\infty$, and is such that for any fixed n, b_n is as smooth

[2]We emphasize that this is only a formal discussion: the low regularity at hand on b does not allow us immediately to define $X(t)$ at this level of discussion!

as necessary (in particular, its first derivatives are bounded). With this sequence we associate the uniquely defined sequence of weak solutions $f_n(t,x)$ in $L^p(\mathbb{R}^d)$ to

$$\partial_t f_n(t,x) + b_n(t,x)\cdot\nabla_x f_n(t,x) = 0\,, \qquad f_n(0,x) = f^0(x) \in L^p(\mathbb{R}^d)\,.$$

In other words, if we take a fixed test function $\phi(t,x) \in C_c^\infty([0,+\infty[\times\mathbb{R}^d)$, infinitely differentiable with compact support, the following integral relation holds:

$$\begin{aligned}\int_{\mathbb{R}^+\times\mathbb{R}^d} f_n(t,x)\partial_t\phi(t,x)\,\mathrm{d}x\,\mathrm{d}t &= \int_{\mathbb{R}^d} f^0(x)\phi(0,x)\,\mathrm{d}x\,\mathrm{d}t \\ &+ \int_{\mathbb{R}^+\times\mathbb{R}^d} f_n(t,x)\,(\mathrm{div}_x b_n(t,x)\,\phi(t,x) + b_n(t,x)\cdot\nabla_x\phi(t,x))\,\mathrm{d}x\,\mathrm{d}t\,,\end{aligned} \tag{7.15}$$

where we have decomposed $\mathrm{div}_x(b_n\phi) = \mathrm{div}_x b_n\phi + b_n\cdot\nabla_x\phi$. In order to fix the notation, let us assume that the support of ϕ lies in $[0,T]\times B(0,R)$, which defines T and R. Our goal now is to pass to the limit in all of the terms of the above integrals.

To make the convergence of b_n to b a bit more precise, observe that we know from the assumptions (7.13) and (7.14) that $\mathrm{div}_x b$ lies in $L^\infty(\mathbb{R}^d)$ while b itself lies in $L^\infty(B(0,R))$. For this reason, it is a standard fact (see the appendix) that one may always assume that b_n goes to b in the following sense:

$$\begin{aligned}\forall 1 \le q < \infty, \quad &\|b_n - b\|_{L^q([0,T]\times B(0,R))} \underset{n\to\infty}{\to} 0\,, \\ \text{and } &\|\mathrm{div}_x b_n - \mathrm{div}_x b\|_{L^q([0,T]\times B(0,R))} \underset{n\to\infty}{\to} 0\,.\end{aligned} \tag{7.16}$$

Besides, one may equally well assume that there exists a constant $C > 0$ such that

$$\|\mathrm{div}_x b_n\|_{L^\infty([0,T]\times\mathbb{R}^d)} \le C\,.$$

On the other hand, for any fixed value of n, since the regularized vector field b_n is smooth, we know from the standard stability of solutions to a transport equation that

$$\|f_n\|_{L^\infty([0,T];L^p(\mathbb{R}^d))} \le C\,,$$

for some constant $C > 0$, independent of n. In particular,

$$\|f_n\|_{L^p([0,T]\times B(0,R))} \le C$$

as well. Hence one may assume, up to extracting a subsequence, that there exists a function $f \in L^p([0,T]\times B(0,R))$ such that

$$f_n(t,x) \underset{n\to\infty}{\to} f(t,x) \text{ weakly in } L^p([0,T]\times B(0,R))\,.$$

We are in position to pass to the limit in eqn (7.15). Indeed, arguing as in Remark 7.10, we have

$$\int f_n(t,x)\partial_t\phi(t,x) \underset{n\to\infty}{\to} \int f_0(t,x)\partial_t\phi(t,x)\,,$$

because $f_n \to f$ weakly and $\partial_t \phi$ is a smooth and compactly supported function. The next convergence is a bit more delicate. We write

$$\int f_n(t,x)\,\mathrm{div}_x b_n(t,x)\,\phi(t,x) \underset{n\to\infty}{\to} \int f(t,x)\,\mathrm{div}_x b(t,x)\,\phi(t,x)\,,$$

because $f_n \to f$ weakly in L^p and $\mathrm{div}_x b_n \to \mathrm{div}_x b$ strongly in $L^{p'}$, so that $f_n\,\mathrm{div}_x b_n \to f\,\mathrm{div}_x b$ weakly in the sense of measures (say). Similarly, we obtain

$$\int f_n(t,x)\,b_n(t,x)\cdot\nabla_x\phi(t,x) \underset{n\to\infty}{\to} \int f(t,x)\,b(t,x)\cdot\nabla_x\phi(t,x)\,.$$

The above argument is easily adapted in the cases $p=1$ and $p=\infty$, by replacing the above weak convergences in L^p by the appropriate weak convergences (see the appendix). We shall not provide the unnecessary technical details.

In any circumstance, we have now established the following theorem.

Theorem 7.13. (Existence of L^p solutions (possibly unstable) to the transport equation when b is not smooth.) *Take a p such that $1 \le p \le \infty$. Take an initial set of data $f^0 \in L^p(\mathbb{R}^d)$. Take a vector field b such that the assumptions (7.12), (7.13), and (7.14) are met.*

Then, there exists a function $f(t,x)$, defined over $\mathbb{R}^+ \times \mathbb{R}^d$, such that for any $T>0$ we have $f(t,x) \in L^\infty([0,T];L^p(\mathbb{R}^d))$, and which is a weak solution to the transport equation with initial data f^0.

Remark 7.14. Sensu stricto, *the above proof does not provide the regularity $f(t,x) \in L^\infty([0,T];L^p(\mathbb{R}^d))$ but rather $f(t,x) \in L^p([0,T]\times\mathbb{R}^d)$. This point is purely technical and harmless.*

Remark 7.15. *Note that the sole difficulty in the above proof lies in passing to the limit in the product $f_n b_n$ or $f_n\,\mathrm{div} b_n$. This is made possible since f_n is bounded (thanks to the stability estimate valid in the smooth case) and hence weakly convergent, and b_n is strongly convergent. Besides, it is a standard fact that the product of a weakly convergent sequence and a strongly convergent sequence is weakly convergent.*

Remark 7.16. *Not coincidentally, note that we did not use the assumption (7.12), according to which the first derivatives of b should be locally integrable, although we insisted that this is the key assumption in the present context.*

This point is extremely important. The hidden fact is, we have only proved the existence *of solutions to the transport equation; we did not prove the* stability *of the so-constructed solutions. At this level of analysis, we have proved that* some *solutions may be obtained by a regularizing procedure, but we have neither proved that* all *solutions can be obtained along these lines, nor proved the uniqueness and stability of* all *solutions to the transport equation. In a sense, the result presented in this section, whose proof is after all easy, is useless without a stability estimate.*

The proof of stability, performed in the next sections, uses precisely the assumption (7.12). This part is the core of the whole theory.

7.3.2 Obtaining a stability estimate: What is the difficulty?

Before coming to the actual proofs, let us point out the difficulty. How did we obtain stability in the smooth case?

We had actually two proofs. The first proof uses the trajectories themselves. We can forget this proof, since we are now dealing with a transport equation with nonsmooth coefficients, and trajectories are not even defined in this case. The second proof uses an energy estimate. The procedure is the following. We multiply the transport equation by $\operatorname{sgn}(f-g)\,|f-g|^{p-1}$, use the chain rule, and eventually obtain

$$\partial_t|f-g|^p + b\cdot\nabla_x|f-g|^p = 0\,.$$

Once it is established that $|f-g|^p$ satisfies a transport equation as well, it remains to integrate the equation over $\mathbb{R}^d$, and we come up with a stability estimate. So the key argument is: if $f-g$ satisfies a transport equation, then so does $|f-g|^p$, a nonlinear function of $f-g$. And the building block of this argument is the chain rule, according to which

$$\frac{\mathrm{d}}{\mathrm{d}z}\phi(\psi(z)) = \phi'(\psi(z))\,\psi'(z)\,,$$

or, to be more specific,

$$\partial_t\,|f-g|^p = p\operatorname{sgn}(f-g)\;(\partial_t(f-g))\;|f-g|^{p-1}\,, \tag{7.17}$$

and

$$b\cdot\nabla_x\,|f-g|^p = p\operatorname{sgn}(f-g)\;(b\cdot\nabla_x(f-g))\;|f-g|^{p-1}\,. \tag{7.18}$$

Naturally, all the above relations hold true when f, g, and b are smooth enough. Now, the question is: does the chain rule apply when the functions f, g, and b lack regularity?

Let us readily point out that this question is fairly natural, and important, in many applications, and the answer is by no means obvious. Indeed, the operator $f\mapsto b\cdot\nabla_x f$ is nothing but the derivative of f along the flow of b, an operator which is central in various contexts, such as fluid mechanics, plasma physics, and the kinetic theory of gases. Besides, in these contexts, considering vector fields b that lack regularity is fairly natural, including from the modeling point of view. To quote an example, in fluid mechanics, owing to exchange of energy between small scales and large scales, the velocity field $u(t,x)$ in a fluid, along which fluid particles are themselves transported, naturally has low regularity. Last, we may mention a simple, academic, but illustrative counterexample. Take

$$f(x) = 1 \text{ when } x>0\,, \quad = 0 \text{ when } x\le 0\,.$$

We stress that, with such a definition, the value of f at $x=0$ (here taken to be 0) is ambiguous, and any value would do. Recall that a locally integrable function is only well defined almost everywhere, and modifying such a function on countably many points does not modify the value of the function when seen, say, as a distribution or as a locally integrable function. Now, with this function f, we have

$$\partial_x |f(x)|^2 = \partial_x f(x) = \delta(x) \qquad \text{(Dirac mass)}\,,$$

while the *product*

$$2f(x)\,\partial_x f(x) = 2\delta(x)\,f(x) \text{ is not even well defined}\,,$$

since the Dirac mass picks out the value at $x = 0$, a value at which the definition of $f(x)$ is ambiguous. Hence the chain rule, clearly, can by no means apply here ....

7.3.3 Obtaining a stability estimate II: Validity of the chain rule for nonsmooth functions

One answer to the question of the validity of the chain rule in the case of nonsmooth functions is provided in the subsequent lemma. This lemma is the key technical tool for the Di Perna–Lions theory of transport equations with nonsmooth coefficients.

Lemma 7.17. (Validity of the chain rule for nonsmooth functions.) *Let the vector field $b(t,x)$ satisfy the assumptions (7.12), (7.13), and (7.14). Take a function $f(t,x)$ such that for any time $T > 0$, and for any radius $R > 0$, there is a $C(T,R) > 0$ such that*

$$\forall 0 \le t \le T\,, \quad \|f(t,x)\|_{L^\infty(B(0,R))} \le C(T,R)\,.$$

On the other hand, take a regularizing kernel $\chi(x)$ as in the appendix (namely $\chi(x) \in C_c^\infty(\mathbb{R}^d)$, $\int \chi(x)\,\mathrm{d}x = 1$, while $\chi(x) \ge 0$ for any x and $\chi(x) = 0$ whenever $|x| \ge 1$) and define, for any function $u(x)$, the notation[3]

$$u_n(x) := u * \chi_n(x) = \int_{\mathbb{R}^d} u\left(x - \frac{y}{n}\right)\,\chi(y)\,\mathrm{d}y\,.$$

Then the following hold.

(i) (A commutator estimate.) *For any choice of $T > 0$ and $R > 0$, we have*

$$\Big\|\,[(b\cdot\nabla_x f) * \chi_n]\,(t,x) - [b\cdot\nabla_x\,(f*\chi_n)]\,(t,x)\Big\|_{L^1([0,T]\times B(0,R))} \underset{n\to\infty}{\longrightarrow} 0\,.$$

(ii) (Regularization of the transport equation.) *Using the above notation, assume that the function f satisfies the transport equation*

$$\partial_t f(t,x) + b(t,x)\cdot\nabla_x f(t,x) = g(t,x)\,, \qquad f(0,x) = f^0(x)\,,$$

in the sense of L^∞ weak solutions to the transport equation (see Definition 7.9, and also the proof of this lemma), where $f^0(x) \in L^\infty(\mathbb{R}^d)$ and $g(t,x)$ is such that

[3] We recall from the appendix that u_n has C^∞ smoothness if the original u lies in some L^p, and that u_n goes strongly in L^p to u as $n \to \infty$ in this situation—hence u_n is a C^∞ approximation of u whenever u lies in some L^p.

for any time $T > 0$, and for any radius $R > 0$, there is a $C(T, R) > 0$ such that for any $0 \leq t \leq T$ we have $\|g(t,x)\|_{L^\infty(B(0,R))} \leq C(T, R)$.
Then, as a consequence of point (i) above, the function f_n satisfies the following transport equation:

$$\partial_t f_n(t,x) + b(t,x) \cdot \nabla_x f_n(t,x) = g_n(t,x) + r_n(t,x), \qquad f_n(0,x) = f_n^0(x),$$

where $f_n = f * \chi_n$, $f_n^0 = f^0 * \chi_n$, $g_n = g * \chi_n$, and the remainder term r_n satisfies

$$\|r_n(t,x)\|_{L^1([0,T]\times B(0,R))} \underset{n\to\infty}{\to} 0, \qquad \text{whenever } T > 0 \text{ and } R > 0.$$

(iii) (Validity of the chain rule for solutions to the transport equation with nonsmooth coefficients.) *Take a smooth nonlinear function*

$$\beta : \mathbb{R} \to \mathbb{R}.$$

Assume the function f satisfies the transport equation as in point (ii) above. Then the function $\beta(f)$ satisfies the transport equation as well, in that

$$\partial_t \beta(f(t,x)) + b(t,x) \cdot \nabla_x \beta(f(t,x)) = \beta'(f(t,x))\, g(t,x),$$
$$\beta(f(0,x)) = \beta\left(f^0(x)\right),$$

in the sense of L^∞ weak solutions to the transport equation. In other words, the chain rule applies, at least in the sense of weak solutions.

Remark 7.18. *The statement we have given here may be formulated in many ways, and the precise regularity that we require for f, g, and b may be somewhat modified. In that respect, we do not claim to give optimal statements here. The point is, however, that we claim the assumption "b should have locally integrable first derivatives in x" is optimal.*

Proof (of part (i)). In any circumstance, we may write

$$\begin{aligned}
&R_n(t,x) \\
&:= \left[(b \cdot \nabla_x f) * \chi_n\right](t,x) - \left[b \cdot \nabla_x (f * \chi_n)\right](t,x) \\
&= \int_{\mathbb{R}^d} \left[b\left(t, x - \frac{y}{n}\right) \cdot \nabla_x f\left(t, x - \frac{y}{n}\right) - b(t,x) \cdot \nabla_x f\left(t, x - \frac{y}{n}\right)\right] \chi(y)\, \mathrm{d}y \\
&= \int_{\mathbb{R}^d} \left[b\left(t, x - \frac{y}{n}\right) - b(t,x)\right] \cdot \nabla_x f\left(t, x - \frac{y}{n}\right) \chi(y)\, \mathrm{d}y \\
&= + \underbrace{\int_{\mathbb{R}^d} n \left[b\left(t, x - \frac{y}{n}\right) - b(t,x)\right] f\left(t, x - \frac{y}{n}\right) \cdot \nabla_y \chi(y)\, \mathrm{d}y}_{=:A_n(t,x)} \\
&\quad - \underbrace{\int_{\mathbb{R}^d} \mathrm{div}_x b\left(t, x - \frac{y}{n}\right) f\left(t, x - \frac{y}{n}\right) \chi(y)\, \mathrm{d}y}_{=:B_n(t,x)},
\end{aligned}$$

where we have performed the obvious integration by parts to let the gradient act on χ or on b, but not on f (recall that f is not assumed differentiable). Besides, it is readily seen that

$$B_n(t,x) = [(f\,\mathrm{div}_x b) * \chi_n]\,(t,x)\,,$$

where we recall that $\mathrm{div}_x b$ is assumed to belong to $L^\infty([0,T]\times B(0,R))$ for any $T>0$, $R>0$, while f belongs to $L^1([0,T]\times B(0,R))$, so that the product $f\,\mathrm{div}_x b$ belongs to $L^1([0,T]\times B(0,R))$. Hence, using the standard results that we recall in the appendix,

$$B_n(t,x) \underset{n\to\infty}{\to} (f\,\mathrm{div}_x b)\,(t,x), \text{ strongly in } L^1([0,T]\times B(0,R))\,.$$

Therefore, to establish that $R_n \to 0$ in the correct norm, it remains to prove that

$$A_n(t,x) \underset{n\to\infty}{\to} +(f\,\mathrm{div}_x b)\,(t,x), \text{ strongly in } L^1([0,T]\times B(0,R))\,. \tag{7.19}$$

The proof of eqn (7.19) is performed in two steps.

First step: Proving eqn (7.19) when b and f are smooth. In the case when b and f are as smooth as desired, say $C_c^\infty(\mathbb{R}\times\mathbb{R}^d)$, we may first observe that

$$A_n(t,x) = \int_{|y|\le 1} n\,\left[b\left(t,x-\frac{y}{n}\right) - b\,(t,x)\right]\,f\left(t,x-\frac{y}{n}\right)\cdot\nabla_y\chi(y)\,\mathrm{d}y\,,$$

so that $A_n(t,x)$ is compactly supported both in time t and in space x. More precisely, if f and b are assumed to be supported in $[0,T]\times B(0,R)$, then $A_n(t,x)$ is supported in the fixed compact set $[0,T]\times B(0,R+1)$.

On the other hand, we have

$$\begin{aligned}
A_n(t,x) &\underset{n\to\infty}{\to} -f\,(t,x)\int_{|y|\le 1}\frac{Db}{Dx}\,(t,x)\;(y)\cdot\nabla_y\chi(y)\,\mathrm{d}y\\
&= -f\,(t,x)\sum_{i,j=1}^d\frac{\partial b_i}{\partial x_j}\,(t,x)\qquad\underbrace{\int_{|y|\le 1} y_j\,\frac{\partial\chi}{\partial y_i}(y)\,\mathrm{d}y}_{\substack{=-1 \text{ if } i=j,\\ =0 \text{ otherwise, by an integration by parts}}}\\
&= -f\,(t,x)\sum_{i=1}^d\frac{\partial b_i}{\partial x_i}\,(t,x) \;= +f(t,x)\,\mathrm{div}_x b(t,x)\,.
\end{aligned}$$

Last, we may write the following easy uniform bound:

$$\begin{aligned}
|A_n(t,x)| &= \left|\int_{|y|\le 1}\int_0^1\frac{Db}{Dy}\left(t,x-s\frac{y}{n}\right)\;(y)\;f\left(t,x-\frac{y}{n}\right)\cdot\nabla\chi(y)\,\mathrm{d}y\,\mathrm{d}s\right|\\
&\le \int_{|y|\le 1}\int_0^1\left|\frac{Db}{Dy}\left(t,x-s\frac{y}{n}\right)\right|\left|f\left(t,x-\frac{y}{n}\right)\right|\,|y\cdot\nabla\chi(y)|\;\mathrm{d}y\,\mathrm{d}s\\
&\le \left\|\frac{Db}{Dx}\,(t,x)\right\|_{L^\infty([0,T]\times B(0,R))}\|f\|_{L^\infty([0,T]\times B(0,R+1))}\int_{|y|\le 1}|y\cdot\nabla\chi(y)|\;\mathrm{d}y\,\mathrm{d}s\,.
\end{aligned}$$

These three pieces of information (uniform boundedness of A_n, pointwise convergence of A_n, and the fact that A_n is supported in a fixed compact set) provide

$$A_n(t,x) \underset{n\to\infty}{\to} +f(t,x)\,\mathrm{div}_x b(t,x) \quad \text{strongly in } L^1([0,T]\times B(0,R+1)),$$

as desired.

Second step: Proving eqn (7.19) when f and b lack smoothness. In view of the above formula for A_n, we may always write down the following uniform estimate, valid whenever f has L^∞ smoothness locally, and Db/Dx only has L^1 smoothness locally, whenever $T>0$ and $R>0$ are fixed:

$$\begin{aligned}
&\|A_n(t,x)\|_{L^1([0,T]\times B(0,R))} \\
&\le \left\| \int_{|y|\le 1}\int_0^1 \left|\frac{Db}{Dy}\left(t,x-s\frac{y}{n}\right)\right| \left|f\left(t,x-\frac{y}{n}\right)\right| |y\cdot\nabla\chi(y)|\,\mathrm{d}y\,\mathrm{d}s \right\|_{L^1([0,T]\times B(0,R))} \\
&\le \|f\|_{L^\infty([0,T]\times B(0,R+1))} \left(\int_0^T \int_{|x'|\le R+1} \left|\frac{Db}{Dy}(t,x')\right| \mathrm{d}t\,\mathrm{d}x'\right) \left(\int_{\mathbb{R}^d} |y\cdot\nabla\chi(y)|\,\mathrm{d}y\right) \\
&= \|f\|_{L^\infty([0,T]\times B(0,R+1))} \left\|\frac{Db}{Dy}\right\|_{L^1([0,T]\times B(0,R+1))} \left(\int_{\mathbb{R}^d} |y\cdot\nabla\chi(y)|\,\mathrm{d}y\right). \qquad (7.20)
\end{aligned}$$

We are now in position to prove that $A_n \to f\,\mathrm{div}_x b$ in $L^1([0,T]\times B(0,R+1))$. To this end, we take an arbitrary positive number $\epsilon>0$ and prove that we can take an $N>0$ such that for any $n\ge N$, we have $\|A_n(t,x)-f(t,x)\,\mathrm{div}_x b(t,x)\|_{L^1([0,T]\times B(0,R))} \le C\,\epsilon$, for some constant $C>0$ which does not depend on ϵ nor on n.

Since the functions f and b and the number ϵ are given, we can always take two functions $f_\epsilon(t,x)$ and $b_\epsilon(t,x)$, having C_c^∞ smoothness, such that

$$\|f-f_\epsilon\|_{L^\infty([0,T]\times B(0,R+1))} \le \epsilon, \qquad \left\|\frac{D(b-b_\epsilon)}{Dy}\right\|_{L^1([0,T]\times B(0,R+1))} \le \epsilon,$$

$$\|\mathrm{div}_y(b-b_\epsilon)\|_{L^\infty([0,T]\times B(0,R+1))}\,.$$

Therefore, defining $A_{n,\epsilon}$ in the same way as that in which A_n is defined, but with f and b replaced by A_ϵ and b_ϵ, we recover

$$\begin{aligned}
&\|A_n(t,x)-f(t,x)\,\mathrm{div}_x b(t,x)\|_{L^1([0,T]\times B(0,R))} \\
&\le \|A_n - A_{n,\epsilon}\|_{L^1([0,T]\times B(0,R))} + \|A_{n,\epsilon} - f_\epsilon\,\mathrm{div}_x b_\epsilon\|_{L^1([0,T]\times B(0,R))} \\
&\quad + \|f_\epsilon\,\mathrm{div}_x b_\epsilon - f\,\mathrm{div}_x b\|_{L^1([0,T]\times B(0,R))} \\
&\le \left(\int_{\mathbb{R}^d} |y\cdot\nabla\chi(y)|\,\mathrm{d}y\right) \left(\|f-f_\epsilon\|_{L^\infty([0,T]\times B(0,R+1))} \left\|\frac{Db_\epsilon}{Dy}\right\|_{L^1([0,T]\times B(0,R+1))} \right. \\
&\quad \left. + \|f\|_{L^\infty([0,T]\times B(0,R+1))} \left\|\frac{D(b-b_\epsilon)}{Dy}\right\|_{L^1([0,T]\times B(0,R+1))} \right)
\end{aligned}$$

$$+ \|A_{n,\epsilon} - f_\epsilon \operatorname{div}_x b_\epsilon\|_{L^1([0,T]\times B(0,R))} + \|f_\epsilon \operatorname{div}_x b_\epsilon - f \operatorname{div}_x b\|_{L^1([0,T]\times B(0,R))}$$

$$\le C\,\epsilon + \underbrace{\|A_{n,\epsilon} - f_\epsilon \operatorname{div}_x b_\epsilon\|_{L^1([0,T]\times B(0,R))}}_{\to 0 \text{ as } n\to\infty, \text{ thanks to the previous step}},$$

where $C > 0$ does not depend on ϵ nor on n. Therefore, we can take an $N > 0$ such that for any $n \ge N$, we have $\|A_n(t,x) - f(t,x)\operatorname{div}_x b(t,x)\|_{L^1([0,T]\times B(0,R))} \le C\,\epsilon$, for some constant $C > 0$ which does not depend on ϵ nor on n.

The proof is complete. ■

Proof (of part (ii) of the lemma). With the notation of the lemma, we write

$$\partial_t f(t,x) + b(t,x)\cdot\nabla_x f(t,x) = g(t,x)\,.$$

Hence,

$$\partial_t\left(f * \chi_n\right)(t,x) + \left([b\cdot\nabla_x f] * \chi_n\right)(t,x) = \left(g * \chi_n\right)(t,x)\,.$$

In other words, using part (i) of the lemma, we recover

$$\partial_t f_n(t,x) + b(t,x)\cdot\nabla_x f_n(t,x) = g_n(t,x) + r_n(t,x)\,,$$

where

$$r_n(t,x) = \left(b(t,x)\cdot\nabla_x\left[f * \chi_n\right]\right)(t,x) - \left([b\cdot\nabla_x f] * \chi_n\right)(t,x) \to 0 \text{ as } n\to\infty\,,$$
$$\text{strongly in } L^1([0,T]\times B(0,R))\,,$$

whenever $T > 0$ and $R > 0$ are fixed.

Once we let $n\to\infty$, the proof is complete. ■

Proof (of part (iii) of the lemma). With the notation of the lemma, let us prove the identity

$$\partial_t\beta\left(f(t,x)\right) + b(t,x)\cdot\nabla_x\beta\left(f(t,x)\right) = \beta'\left(f(t,x)\right)g(t,x),$$

in the sense of distributions or in the sense of L^p solutions to the transport equation (see Definition 7.9). With this aim, we take a smooth test function $\phi(t,x)\in C_c^\infty([0,T]\times\mathbb{R}^d)$, and we prove below that the following identity holds:

$$\begin{aligned}
&\int_0^{+\infty} \mathrm{d}t \int_{\mathbb{R}^d} \mathrm{d}x\, \beta\left(f(t,x)\right)\partial_t\phi(t,x) \\
&\quad = -\int_{\mathbb{R}^d} \mathrm{d}x\,\beta\left(f^0(x)\right)\phi(0,x) + \int_0^{+\infty}\mathrm{d}t\int_{\mathbb{R}^d}\mathrm{d}x\,\beta'\left(f(t,x)\right)g(t,x)\,\phi(t,x) \\
&\qquad + \int_0^{+\infty}\mathrm{d}t\int_{\mathbb{R}^d}\mathrm{d}x\,\beta\left(f(t,x)\right)\left[\operatorname{div}_x b(t,x)\,\phi(t,x) + b(t,x)\cdot\nabla_x\phi(t,x)\right].
\end{aligned} \tag{7.21}$$

The proof of eqn (7.21) relies on an appropriate regularization. We take a large integer n. We already know from part (ii) of the lemma that

$$\partial_t f_n(t,x) + b(t,x)\cdot\nabla_x f_n(t,x) = g_n(t,x) + r_n(t,x)\,,$$

where $f_n = f * \chi_n$ and $g_n = g * \chi_n$ are smooth (C^∞ and bounded), while for any $T > 0$ and any $R > 0$ we have

$$\|r_n(t,x)\|_{L^1([0,T]\times B(0,R))} \to 0 \text{ as } n \to \infty\,.$$

Therefore, we may write

$$\partial_t \beta\left(f_n(t,x)\right) + b(t,x)\cdot\nabla_x\beta\left(f_n(t,x)\right) = \beta'\left(f_n(t,x)\right) g_n(t,x) + \beta'\left(f_n(t,x)\right) r_n(t,x)\,,$$

(because f_n is smooth and the chain rule applies for smooth functions) and, next,

$$\begin{aligned}
&\int_0^{+\infty} \mathrm{d}t \int_{\mathbb{R}^d} \mathrm{d}x\, \beta\left(f_n(t,x)\right)\, \partial_t\phi(t,x)\\
&\quad= -\int_{\mathbb{R}^d} \mathrm{d}x\, \beta\left(f^0(x)\right)\, \phi(0,x)\\
&\quad\quad+ \int_0^{+\infty} \mathrm{d}t \int_{\mathbb{R}^d} \mathrm{d}x\, \beta'\left(f_n(t,x)\right) \left[g_n(t,x) + r_n(t,x)\right] \phi(t,x)\\
&\quad\quad+ \int_0^{+\infty} \mathrm{d}t \int_{\mathbb{R}^d} \mathrm{d}x\, \beta\left(f_n(t,x)\right) \left[\mathrm{div}_x b(t,x)\,\phi(t,x) + b(t,x)\cdot\nabla_x\phi(t,x)\right]. \qquad (7.22)
\end{aligned}$$

Now let T and R be such that $\phi(t,x)$ is supported in $[0,T]\times B(0,R)$. Since $f \in L^\infty([0,T]\times B(0,R))$, we have (see appendix)

$$f_n \to_{n\to\infty} f \text{ in } L^q([0,T]\times B(0,R)) \quad \text{for any } 1 \leq q < \infty\,,$$

and hence

$$\begin{aligned}
\beta(f_n) &\underset{n\to\infty}{\to} \beta(f) \text{ in } L^1([0,T]\times B(0,R))\,,\\
\beta'(f_n) &\underset{n\to\infty}{\to} \beta'(f) \text{ in } L^1([0,T]\times B(0,R))\,.
\end{aligned}$$

Similarly, we know that

$$g_n \underset{n\to\infty}{\to} g \text{ in } L^1([0,T]\times B(0,R))\,.$$

Hence, passing to the limit $n \to \infty$ in eqn (7.22) provides the desired identity (7.21).
The proof is complete. ■

One important conclusion from the above computations is the following. Lemma 7.17 establishes that any solution f to the transport equation also satisfies the non-linear identity

$$\partial_t\beta(f) + b\cdot\nabla_x\beta(f) = 0\,,$$

whenever $\beta : \mathbb{R} \to \mathbb{R}$ is smooth. Conversely, if f is such that

$$\partial_t\beta(f) + b\cdot\nabla_x\beta(f) = 0$$

for any β, clearly f is a weak solution to the transport equation as well (take $\beta(z) \equiv z$).
This basic observation is the key to what follows.

7.3.4 Conclusion: Existence, uniqueness, and stability of solutions to the transport equation when b is not smooth

Armed with the tools developed in the previous paragraphs, we are now in position to complete the study of the transport equation when b is not smooth.

To begin with, we state the following definition.

Definition 7.19. (Renormalized solutions to the transport equation.) *Take a $1 \leq p \leq \infty$ and take a function $f(t,x)$ such that*

$$f(t,x) \in L^\infty([0,T]; L^p(\mathbb{R}^d))$$

for any $T > 0$. Take a function $f^0(x) \in L^p(\mathbb{R}^d)$.

We say that f is a renormalized solution *to the transport equation, with initial data f^0, whenever the equation*

$$\partial_t \beta(f) + b \cdot \nabla_x \beta(f) = 0\,, \qquad \beta(f(0,x)) = \beta(f^0(x))$$

is satisfied in the standard weak sense, for any *smooth function $\beta : \mathbb{R} \to \mathbb{R}$ such that $z \mapsto \beta(z)$ and $z \to z\beta'(z)$ are bounded over $\mathbb{R}$, and β vanishes close to the origin. Such functions β are called* admissible functions.

In other words, a renormalized solution is required to satisfy all *equations that are formally obtained from the transport equation by using the chain rule.*

Remark 7.20. *The reader's attention is drawn to the fact that being a renormalized solution is a nonlinear concept, even though the original equation is linear.*

Remark 7.21. *The reason for restricting the definition to functions β that are bounded, vanishing close to the origin, and such that $z\,\beta'(z)$ is bounded as well (namely, admissible functions) is a bit technical. Let us simply mention that when $p < \infty$, and when $f \in L^p$, the function $\beta(f)$ automatically belongs to $L^1 \cap L^\infty$ when β is admissible. Technically speaking, this fact allows us to manipulate the transport equation satisfied by $\beta(f)$ in an easy fashion.*

In order to become familiar with the notion of renormalized solutions, let us first investigate the link between renormalized solutions and standard weak solutions.

In the L^∞ setting, the situation is simple, as shown by the following proposition.

Proposition 7.22. *With the notation of Definition 7.19, when $p = \infty$, a renormalized solution to the transport equation is automatically a weak solution of the transport equation. Conversely, a weak solution is automatically a renormalized solution as well.*

Proof. The proof of this proposition is easy. Take a renormalized solution f, and write that $\beta(f)$ is a weak solution for any admissible β. Therefore, we have, for any test function $\phi(t,x) \in C_c^\infty(\mathbb{R}^+ \times \mathbb{R}^d)$, the relation

$$\begin{aligned}
&\int_{\mathbb{R}^+ \times \mathbb{R}^d} \beta(f(t,x))\, \partial_t \phi(t,x)\, \mathrm{d}t\, \mathrm{d}x \\
&= -\int_{\mathbb{R}^d} \beta(f^0(x))\, \phi(0,x)\, \mathrm{d}x + \int_{\mathbb{R}^+ \times \mathbb{R}^d} \beta(f(t,x))\, \mathrm{div}_x\, (b(t,x)\, \phi(t,x))\, \mathrm{d}t\, \mathrm{d}x\,.
\end{aligned}$$

It remains to take a sequence of β's which converge towards the identity for us to conclude, exploiting the relation $f \in L^\infty$, that f satisfies

$$\int_{\mathbb{R}^+ \times \mathbb{R}^d} f(t,x)\, \partial_t \phi(t,x)\, \mathrm{d}t\, \mathrm{d}x$$
$$= -\int_{\mathbb{R}^d} f^0(x)\, \phi(0,x)\, \mathrm{d}x + \int_{\mathbb{R}^+ \times \mathbb{R}^d} f(t,x)\, \mathrm{div}_x\, (b(t,x)\, \phi(t,x))\, \mathrm{d}t\, \mathrm{d}x\,.$$

Therefore, f is a weak solution to the transport equation.

Conversely, if $f \in L^\infty$ is a weak solution, Lemma 7.17, part (iii), immediately establishes that f is a renormalized solution as well.

The proof is complete. ■

In the L^p setting, the situation is different.

Proposition 7.23. *With the notation of Definition 7.19, when $p < \infty$, a renormalized solution to the transport equation is automatically a weak solution of the transport equation.*

The converse assertion may not be true if b is not Lipschitz: a weak solution may not be a renormalized solution.

However, if b is Lipschitz, being a weak solution implies being a renormalized solution.

The proof of the implication "renormalized solution $\Rightarrow$ weak solution" is obtained as above, by taking a sequence of admissible β's which converge towards the identity. The fact that the converse is true *when b is Lipschitz* comes from the validity of the chain rule when b has Lipschitz smoothness. Otherwise, a weak solution may differ from a renormalized solution in general.

In other words, the whole point in the notion of renormalized solutions lies in testing nonlinear functions of the unknown, so as to select those solutions which in some sense satisfy the chain rule.

With this definition at hand, we may state the key result of the whole theory.

Theorem 7.24. (Existence, uniqueness, and stability of renormalized solutions to the transport equation.) *Let the vector field $b(t,x)$ satisfy the assumptions (7.12), (7.13), and (7.14). Take a function $f^0 \in L^p(\mathbb{R}^d)$ for some $1 \le p \le \infty$.*

Then, there exists a unique function $f(t,x)$ such that for any time $T > 0$, we have $\sup_{0 \le t \le T} \|f(t,x)\|_{L^p(\mathbb{R}^d)} \le C(T)$ for some $C(T) > 0$, and such that it is a renormalized solution *to the transport equation*

$$\partial_t f(t,x) + b(t,x) \cdot \nabla_x f(t,x) = 0\,, \qquad f(0,x) = f^0(x)\,.$$

Also, the solution so obtained is stable, *in that for any two sets of initial data $f^0 \in L^p$ and $g^0 \in L^p$, if $f(t,x)$ and $g(t,x)$ denote the associated (uniquely defined) renormalized solutions to the transport equation with initial data f^0 and g^0, respectively, we have, for any time $T > 0$, the following estimate, valid whenever $0 \le t \le T$:*

$$\|f(t,x) - g(t,x)\|_{L^p(\mathbb{R}^d)} \le \left\|f^0 - g^0\right\|_{L^p(\mathbb{R}^d)} \exp\left(\frac{t}{p} \times \|\mathrm{div}_x b(t,x)\|_{L^\infty([0,T]\times\mathbb{R}^d)}\right).$$

Remark 7.25 (about the notion of a renormalized solution). *One important aspect of the whole theory is the following. We are definitely in a situation here where uniqueness is difficult to obtain, the difficulty being that uniqueness is obtained, in the smooth case, through some sort of energy estimate, the latter relying on the validity of the chain rule and hence on the smoothness of* b*. From that perspective, one controls both regularity and uniqueness simultaneously, in some way, by requiring that* f and nonlinear functions of f *satisfy the equation, which selects a unique solution.*

Let us mention here that this principle is in some sense general in the modern approach to PDEs, as in the theory of conservation laws in fluid mechanics, for instance: "good," unique solutions can be selected by considering both the equation satisfied by the unknown and the system satisfied by nonlinear functions of it. In the above-mentioned context, for instance, this principle is known under the name of the "entropy principle": a unique solution is selected by considering only those solutions which decrease some entropy functional, a nonlinear function of the unknown.

Proof (sketch of proof). Uniqueness of a solution is easy: if two renormalized solutions f_1 and f_2 exist, corresponding to the same initial data f^0, by the definition of a renormalized solution we obtain for any β two weak solutions $\beta(f_1)$ and $\beta(f_2)$ corresponding to the initial data $\beta(f^0)$. The point is, these solutions now have L^∞ smoothness, and hence the chain rule applies (thanks to Lemma 7.17). Therefore, the usual energy estimate provides uniqueness. As a consequence,

$$\beta(f_1) = \beta(f_2) .$$

Since the above relation holds *for any* β, we conclude that $f^1 = f^2$.

The existence of a solution has already been proved, by solving for any n the equation

$$\partial_t \tilde{f}_n + b_n \cdot \nabla_x \tilde{f}_n = 0 , \quad \tilde{f}_n(0, x) = f^0(x) ,$$

where

$$b_n = b * \chi_n .$$

We know that $\tilde{f}_n$ converges towards f, a solution to the transport equation. To prove that the so-obtained f is a renormalized solution as well, we take β as in the definition of a renormalized solution, and write, using Lemma 7.17,

$$\partial_t \beta(\tilde{f}_n) + b_n \cdot \nabla_x \beta(\tilde{f}_n) = r_n ,$$

where $r_n \to 0$ in $L^1([0, T] \times B(0, R))$ for any $T > 0$ and $R > 0$. Therefore $\beta(\tilde{f}_n)$ goes to a solution $f_\beta(t, x)$ to the transport equation. Uniqueness then asserts that the limit f_β necessarily coincides with $\beta(f)$.

The stability estimate is actually a consequence of a more general fact: whenever a sequence of initial data f_n^0 and vector fields b_n is given, which converges towards f^0 and b, respectively, in the natural topology, then the associated sequence $f_n(t, x)$ of renormalized solutions to the transport equation converges strongly to $f(t, x)$, which is the unique renormalized solution to the transport equation with initial data f^0 and vector field b. We shall not provide details on that point; we simply mention that this stability result comes roughly from considerations similar to the ones that provide uniqueness and existence, as above, combined with refined but standard facts from functional analysis.

As a consequence of this more general stability statement, we recover the stability claimed in the above statement. To do so, it is enough to take a sequence of solutions f_n and g_n to the transport equation, with the vector field b replaced by $b * \chi_n$. These solutions f_n and g_n satisfy the stability estimate claimed. Passing to the limit, f and g satisfy the same estimate since f_n and g_n go strongly to f and g, respectively. ■

We state, finally, the ODE version of the previous theorem.

Theorem 7.26. (Solving the ODE $\dot{X} = b(t, X)$ when b is not smooth.) *Let the vector field $b(t, x)$ satisfy the assumptions (7.12), (7.13), and (7.14).*

Then, there exists a function $X(t, x)$, which is C^1 in the time variable, defined for almost every initial data point $x \in \mathbb{R}^d$, such that $X(t, x)$ satisfies the differential equation $\dot{X}(t, x) = b(t, X(t, x))$ with initial data $X(0, x) = x$.

Remark 7.27. *We omit the proof of this theorem. Its proof is based mainly on defining X by duality, through the formula*

$$f_0(X(t, x)) = f(t, x),$$

which is valid when b is smooth at least, and which can be extended to the case when b is not smooth by defining f as the renormalized solution to the transport equation. In that case, the above formula may be seen in turn as a definition of X, so that we are allowed to define $f^0(X(t, x))$ for any f^0. It turns out that such an indirect definition provides a well-defined value of $X(t, x)$ for almost every x.

7.3.5 Recent extensions of the Di Perna–Lions theory

In essence, it was proved in the original paper by Di Perna and Lions that the condition $Db/Dx \in L^1([0, T] \times B(0, R))$ for any T and R cannot be weakened. Counterexamples were provided.

To be a bit more precise, recent work has shown that one may relax this assumption and require only that Db/Dx is a measure which has finite mass locally, as in the case of a Dirac mass or similar. Roughly speaking, this is essentially the same requirement as above, in that finite mass is required in any circumstance: the improvement lies in authorizing measure-valued derivatives instead of Lebesgue-integrable ones.

Some other extensions have been made for split systems, or systems having a special structure, such as Hamiltonian systems: in these cases, the particular structure of the underlying ODE allows us to relax the assumptions a little in specific directions. But again, the main requirement remains the same in spirit.

We shall not give more details in that direction.

7.4 Appendix

In this appendix we gather together some known facts in the field of analysis, which are used in the above proofs.

7.4.1 Convergence in L^p: The case $1 < p < \infty$

Proposition 7.28.

(i) For any given function $f(x) \in L^p(\mathbb{R}^d)$, there exists a sequence of functions $f_n(x)$ such that

$$\forall n\,, \quad f_n(x) \in C_c^\infty(\mathbb{R}^d) \qquad (f_n \textit{ is infinitely differentiable with compact support})$$

and

$$f_n \underset{n\to\infty}{\to} f \textit{ strongly in } L^p, \textit{ i.e. } \|f_n - f\|_{L^p} \underset{n\to\infty}{\to} 0\,.$$

(ii) Take a sequence $g_n(x)$ of functions in $L^p(\mathbb{R}^d)$ such that for some constant $C > 0$ we have

$$\forall n\,, \quad \|g_n\|_{L^p} \leq C\,.$$

Then, up to extracting a subsequence of the original sequence g_n, one may always assume that there exists a function $g(x) \in L^p(\mathbb{R}^d)$ such that $g_n \to_{n\to\infty} g$ weakly in L^p, i.e. for any function $\phi(x) \in L^{p'}(\mathbb{R}^d)$, we have

$$\int_{\mathbb{R}^d} (g_n(x) - g(x))\, \phi(x)\, \mathrm{d}x \underset{n\to\infty}{\to} 0\,.$$

Here, p' denotes the conjugate exponent of p, i.e. $1/p + 1/p' = 1$.

Remark 7.29. *Note that convergence is strong in (i) and weak in (ii). Obviously, strong convergence implies weak convergence, while the converse is false. To see that the converse is false, simply take the sequence*

$$g_n(x) = \sin(nx) \textit{ when } 0 \leq x \leq 1\,, \quad = 0 \textit{ otherwise}\,.$$

This sequence goes weakly to zero in $L^2(\mathbb{R})$ while the L^2 norm of g_n goes to $1/\sqrt{2}$.

Part (i) of the above proposition may be stated more precisely, as we do now.

Proposition 7.30. *Take a smooth function $\chi(x) \in C_c^\infty(\mathbb{R}^d)$ such that*

$$\forall x\,, \quad \chi(x) \geq 0\,,$$

$$\int_{\mathbb{R}^d} \chi(x)\, \mathrm{d}x = 1\,,$$

$$\chi(x) = 0 \textit{ whenever } |x| \geq 1\,.$$

Such a function is usually called a regularizing kernel. Define, for any $n \in \mathbb{N}^*$,

$$\chi_n(x) = n^d\, \chi(nx)\,.$$

Take a function $f(x) \in L^p(\mathbb{R}^d)$ and define

$$f_n(x) = (f * \chi_n)(x) = \int_{\mathbb{R}^d} f(y)\chi_n(y)\, \mathrm{d}y = \int_{\mathbb{R}^d} f\left(x - \frac{y}{n}\right) \chi(y)\, \mathrm{d}y\,.$$

Then, we have

$$f_n(x) \underset{n\to\infty}{\to} f(x) \textit{ strongly in } L^p(\mathbb{R}^d)\,.$$

Remark 7.31. *Note that*

$$\chi_n \to \delta$$

in the sense of distributions, where δ is the Dirac mass. This fact relies on the simple observation (valid when f is smooth at least)

$$\begin{aligned} f_n(x) &= \int_{\mathbb{R}^d} f\left(x - \frac{y}{n}\right) \chi(y)\,\mathrm{d}(y) \\ &= \left(f(x) + \mathcal{O}\left(\frac{1}{n}\right)\right) \int_{\mathbb{R}^d} \chi(y)\,\mathrm{d}(y) \\ &= f(x) + \mathcal{O}\left(\frac{1}{n}\right). \end{aligned}$$

Note that the second equality uses the fact that χ has compact support, while the third equality uses $\int \chi = 1$.

7.4.2 Convergence in L^1 and L^∞

Proposition 7.32.

(i) For any given function $f(x) \in L^1(\mathbb{R}^d)$, there exists a sequence of functions $f_n(x)$ such that

$$\forall n, \quad f_n(x) \in C_c^\infty(\mathbb{R}^d) \quad (f_n \text{ is infinitely differentiable with compact support})$$

and

$$f_n \underset{n\to\infty}{\to} f \text{ strongly in } L^1, \text{ i.e. } \|f_n - f\|_{L^1} \underset{n\to\infty}{\to} 0.$$

(ii) Take a sequence $g_n(x)$ of functions in $L^1(\mathbb{R}^d)$ such that for some constant $C > 0$ we have

$$\forall n, \quad \|g_n\|_{L^1} \le C.$$

Then, up to extracting a subsequence of the original sequence g_n, one may always assume that there exists a finite measure *g defined over $\mathbb{R}^d$ such that $g_n \to_{n\to\infty} g$ weakly in the sense of finite measures, i.e. for any function $\phi(x) \in C_b^0(\mathbb{R}^d)$ (ϕ is continuous and bounded), we have*

$$\int_{\mathbb{R}^d} g_n(x)\,\phi(x)\,\mathrm{d}x \underset{n\to\infty}{\to} \langle g, \phi\rangle.$$

Roughly speaking, the duality product $\langle g, \phi\rangle$ may be written $\int_{\mathbb{R}^d} g(x)\phi(x)\,\mathrm{d}x$, keeping in mind that $g(x)$ is a measure, i.e. g may typically be a Dirac mass or something similar.

Remark 7.33. *Note the difference, in point (ii), between the L^p situation $(1 < p < \infty)$ and the L^1 situation: here the test function ϕ is not taken in $L^{1'} = L^\infty$, but in the smaller space C^0 of continuous and bounded functions. Also, the limit g may not belong to L^1, but be in the larger space of finite measures.*

The function χ_n of the previous paragraph is the prototype of such a situation: χ_n is bounded in L^1 (its L^1 norm is one for any n), but the limit is the Dirac mass, a measure.

Proposition 7.34.

(i) For any given function $f(x) \in L^\infty(\mathbb{R}^d)$, there exists a sequence of functions $f_n(x)$ such that

$$\forall n, \quad f_n(x) \in C_c^\infty(\mathbb{R}^d) \quad (f_n \text{ is infinitely differentiable with compact support}),$$

and, for any compact set K, and for any $1 \le q < +\infty$, we have

$$\|f_n - f\|_{L^q(K)} \underset{n\to\infty}{\to} 0\,.$$

(ii) Take a sequence $g_n(x)$ of functions in $L^\infty(\mathbb{R}^d)$ such that for some constant $C > 0$ we have

$$\forall n, \quad \|g_n\|_{L^\infty} \le C\,.$$

Then, up to extracting a subsequence of the original sequence g_n, one may always assume that there exists a function $g(x) \in L^\infty(\mathbb{R}^d)$ such that $g_n \to_{n\to\infty} g$ weakly- in L^∞, i.e. for any function $\phi(x) \in L^1(\mathbb{R}^d)$, we have*

$$\int_{\mathbb{R}^d} (g_n(x) - g(x))\, \phi(x)\, \mathrm{d}x \underset{n\to\infty}{\to} 0\,.$$

Remark 7.35. *Here, point (ii) is essentially the same as in the L^p case $(1 < p < \infty)$. The difference is in point (i): the sequence f_n does not (in general) converge to f in L^∞, but does so in any smaller space $L^q(K)$ provided we restrict our attention to the compact set K, and the L^∞ norm over K is replaced by the weaker L^q norm over K for some $q < \infty$.*

Further reading

L. Ambrosio (2007). The flow associated to weakly differentiable vector fields: Recent results and open problems, *Boll. Unione Mat. Ital. Sez. B Artic. Ric. Mat. (8)*, **10**(1), 25–41.

L. Ambrosio (2008). Transport equation and Cauchy problem for nonsmooth vector fields, in *Calculus of Variations and Nonlinear Partial Differential Equations*, Lecture Notes in Mathematics, No. 1927, pp. 1–41, Springer, Berlin.

L. Ambrosio, F. Bouchut, and C. De Lellis (2004). Well-posedness for a class of hyperbolic systems of conservation laws in several space dimensions, *Commun. Partial Differ. Equ.*, **29**(9–10), 1635–1651.

L. Ambrosio, C. De Lellis, and J. Malý (2007). On the chain rule for the divergence of *BV*-like vector fields: Applications, partial results, open problems, *Contemp. Math.*, **446**, 31–67.

L. Ambrosio, M. Lecumberry, and S. Maniglia (2005). Lipschitz regularity and approximate differentiability of the Di Perna–Lions flow, *Rend. Sem. Mat. Univ. Padova*, **114**, 29–50.

F. Bouchut (2001). Renormalized solutions to the Vlasov equation with coefficients of bounded variation, *Arch. Ration. Mech. Anal.*, **157**(1), 75–90.

F. Bouchut and G. Crippa (2006). Uniqueness, renormalization, and smooth approximations for linear transport equations, *SIAM J. Math. Anal.*, **38**(4), 1316–1328.

F. Bouchut and L. Desvillettes (2001). On two-dimensional Hamiltonian transport equations with continuous coefficients, *Differ. Integral Equ.*, **14**(8), 1015–1024.

F. Bouchut and F. James (1998). One-dimensional transport equations with discontinuous coefficients, *Nonlinear Anal.*, **32**(7), 891–933.

F. Bouchut, F. Golse, and M. Pulvirenti (2000). *Kinetic Equations and Asymptotic Theory*, Series in Applied Mathematics, No. 4, Gauthier-Villars, Éditions Scientifiques et Médicales Elsevier, Paris.

F. Bouchut, F. James, and S. Mancini (2005). Uniqueness and weak stability for multi-dimensional transport equations with one-sided Lipschitz coefficient, *Ann. Sc. Norm. Super. Pisa Cl. Sci. (5)*, **4**(1), 1–25.

Y. Brenier (1997). A homogenized model for vortex sheets, *Arch. Ration. Mech. Anal.*, **138**(4), 319–353.

F. Colombini and N. Lerner (2005). Uniqueness of L^∞ solutions for a class of conormal BV vector fields, *Contemp. Math.*, **368**, 133–156.

G. Crippa and C. De Lellis (2006). Oscillatory solutions to transport equations, *Indiana Univ. Math. J.*, **55**(1), 1–13.

G. Crippa and C. De Lellis (2008). Estimates and regularity results for the Di Perna–Lions flow, *J. Reine Angew. Math.*, **616**, 15–46.

C. De Lellis (2008). ODEs with Sobolev coefficients: The Eulerian and the Lagrangian approach, *Discrete Contin. Dyn. Syst. Ser. S*, **1**(3), 405–426.

R. Di Perna, P.-L. Lions (1989). Équations différentielles ordinaires et équations de transport avec des coefficients irréguliers, Séminaire sur les Équations aux Dérivées Partielles, 1988–1989, École Polytechnique, Palaiseau.

R. Di Perna and P.-L. Lions (1989). Ordinary differential equations, transport theory and Sobolev spaces, *Invent. Math.*, **98**(3), 511–547.

F. Golse and L. Saint-Raymond (2004). The Navier–Stokes limit of the Boltzmann equation for bounded collision kernels, *Invent. Math.*, **155**(1), 81–161.

M. Hauray (2003). On two-dimensional Hamiltonian transport equations with L^p_{loc} coefficients, *Ann. Inst. H. Poincaré Anal. Non Linéaire*, **20**(4), 625–644.

C. Le Bris and P.-L. Lions (2004). Renormalized solutions of some transport equations with partially $W^{1,1}$ velocities and applications, *Ann. Mat. Pura Appl. (4)*, **183**(1), 97–130.

N. Lerner (2004). Transport equations with partially BV velocities, *Ann. Sc. Norm. Super. Pisa Cl. Sci. (5)*, **3**(4), 681–703.

N. Lerner (2004). Équations de transport dont les vitesses sont partiellement BV, Séminaire: Équations aux Dérivées Partielles, 2003–2004, Exp. No. X, 19, École Polytechnique, Palaiseau.

P.-L. Lions (1996). *Mathematical Topics in Fluid Mechanics*, Vol. 1, Oxford Lecture Series in Mathematics and Its Applications, No. 3, Oxford University Press.
S. Maniglia (2007). Probabilistic representation and uniqueness results for measure-valued solutions of transport equations, *J. Math. Pures Appl. (9)*, **87**(6), 601–626.

8
The mean-field limit for interacting particles

P. E. JABIN[1] and Antoine GERSCHENFELD[2]

[1] Equipe Tosca, Inria, 2004 route des Lucioles, BP 93, 06902 Sophia Antipolis
Laboratoire Dieudonné, Université de Nice, Parc Valrose, 06108 Nice Cedex 02, France
[2] École Normale Supérieure, Paris

8.1 Introduction

The validity of kinetic models as a limit of systems of many interacting particles is still an important open issue. The number of particles to take into account is so large in most applications (plasma physics, galaxy formation, ...) that the use of continuous models is absolutely required.

The same issues arise directly in the use of particle methods. These methods rely on the assumption that a large (but not too large) number of "meta-particles" correctly represents the dynamics of a much larger number of real particles. This assumption would be directly implied by the convergence of the system to the unique solution to some equation.

The scaling under consideration here leads to so-called mean-field limits. These limits were classically established under strong regularity assumptions for the interaction, which are not satisfied in many physical situations of interest. We aim at describing these classical approaches but also to present new ideas recently developed for singular cases.

We consider N identical particles with positions/velocities (X_i, V_i) in the phase space, interacting through the two-body interaction kernel $K(x)$, which leads to the evolution equations

$$\begin{cases} \dfrac{\mathrm{d}}{\mathrm{d}t} X_i = V_i\,, \\ \dfrac{\mathrm{d}}{\mathrm{d}t} V_i = \dfrac{1}{N} \displaystyle\sum_j K(X_i - X_j)\,. \end{cases} \tag{8.1}$$

The factor $1/N$ in the second equation is a scaling term so that positions, velocities, and accelerations are now of order 1.

The kernel K may take many different forms depending on the physical setting. The guiding example and the one with the most important physical applications is the Coulomb interaction, which reads, in dimension d,

$$K(x) = -\nabla\phi(x)\ ,\ \phi(x) = \frac{\alpha}{|x|^{d-2}} + \text{(regular terms)}\,,$$

where $\alpha > 0$ (or $\alpha < 0$) corresponds to the repulsive (or attractive, respectively) case.

In what follows, the dynamics will be considered on the torus $X_i \in \Pi^d$, $d \geq 2$, mainly to simplify the exposition. Note that even then the velocities are still in $\mathbb{R}^d$.

8.2 Well-posedness of the microscopic dynamics

The Cauchy–Lipschitz Theorem applies to eqn (8.1) if $K(x)$ is Lipschitz, in which case there exists a unique solution for any initial condition. In the repulsive Coulomb case, it is still possible to apply it by noting that energy conservation,

$$E(t) = \frac{1}{N} \sum_i |V_i|^2 + \frac{\alpha}{N^2} \sum_{i \neq j} \frac{1}{|X_i - X_j|} = E(0)\,,$$

implies that the $|X_i - X_j|$ have a time-independent lower bound depending on $1/N^2$: one may therefore consider K as Lipschitz on its attainable domain for a given set of initial conditions. One should note, however, that the form of this estimate makes it improper to use the $N \to \infty$ limit.

It is possible to assume less regularity of K by restricting the set of acceptable initial conditions. In particular, results by DiPerna and Lions (1989), Ambrosio (2004), and Hauray (2005) apply to almost every initial condition.

8.3 Existence of the macroscopic limit

Given a sequence of initial conditions $Z^{N0} = (X_1^{N0}, \ldots, X_N^{N0}, V_1^{N0}, \ldots, V_N^{N0})$ with corresponding solutions $Z^N(t)$, one expects the empirical density in phase space,

$$f_N(t,x,v) = \frac{1}{N} \sum_i \delta(x - X_i^N(t)) \otimes \delta(v - V_i^N(t)) \,,$$

to converge, in some sense, as $N \to \infty$, to a limit f satisfying an evolution equation, the "limiting dynamics," with initial conditions $f^0 = \lim_N f_N(0,\cdot)$.

If K is continuous or if $X_i^N(t) \neq X_j^N(t)$ for all t and $i \neq j$, then, assuming that $K(0) = 0$, one can write the N-body evolution in the form of a Vlasov equation

$$\begin{cases} \partial_t f_N + v \cdot \nabla_x f_N + (K \star_x \rho_N) \cdot \nabla_v f_N = 0 \,, \\ \rho_N(t,x) = \int \mathrm{d}v \, f_N(t,x,v) \,. \end{cases} \tag{8.2}$$

Then $f_N \to f$ in weak-$\star$ topology (for the space of Radon measures $M^1(\Pi^d \times \mathbb{R}^d)$), and f solves eqn (8.2) for the initial conditions $\lim_N f_N(0,\cdot)$.

Equation (8.2) cannot be obtained from eqn (8.1) by such an immediate method for any kind of singular interaction $K \notin C_0$. However, even for a Coulomb potential, eqn (8.2) is well posed provided some assumptions about the initial conditions are made, such as $f(0,\cdot) \in L^1 \cap L^\infty$ and that it has compact support on velocities (see Horst 1981; Lions and Perthame 1991; Pfaffelmoser 1992; Schaeffer 1991). However, the nonlinear term $(K \star_x \rho_N) \cdot \nabla_v f_N$ makes the $f_N \to f$ limit highly nontrivial for noncontinuous K.

8.4 Physical-space models

The above question is easier to solve in the case of hydrodynamics-related models, which evolve according to a first-order equation of the form

$$\frac{\mathrm{d}}{\mathrm{d}t} X_i = \frac{1}{N} \sum_{j \neq i} \mu_i \mu_j K(X_i - X_j) \,. \tag{8.3}$$

Using $\rho_N(t,x) = \sum_i \mu_i \delta(x - X_i(t))$, this equation can be rewritten as

$$\partial_t \rho + \nabla_x((K \star \rho)\rho) = 0 \,. \tag{8.4}$$

For instance, in dimension 2, the above yields the incompressible Euler equation for $\mu_i = \pm 1, K(x) = x_\perp / |x^2|$.

As a rule of thumb, the $N \to \infty$ limit is easier to take in this case than in eqn (8.1). A crucial ingredient of the study is a bound on $d_{\min}(t) = \inf_{i \neq j} |X_i(t) - X_j(t)|$. This offers direct control over the right-hand term in eqn (8.3), which becomes regular if $d^N_{\min} \sim N^{-1/d}$ for singular force terms K (up to a Coulombian singularity).

More precisely, assume that, up to time t and for $x \sim x_i$, there exists a locally bounded F such that

$$\left\| \frac{1}{N} \sum_{j \neq i} K(x - X_j(t)) \right\|_{W^{1,\infty}} \leq F\left(\frac{d_{\min}}{N^{1/d}}\right).$$

Let (k, l) be the particles such that $d_{\min}(t) = |X_k - X_l|$. If one also assumes that $\mu_k = \mu_l$, then

$$\begin{aligned}\frac{\mathrm{d}}{\mathrm{d}t} d_{\min} = \frac{\mathrm{d}}{\mathrm{d}t}|X_k - X_l| &\geq -\frac{1}{N}\left| \sum_{j \neq k,l} \mu_j (K(X_k - X_j) - K(X_l - X_j)) \right| + o(1) \\ &\geq -\left\| \frac{1}{N} \sum_{j \neq i} K(x - X_j(t)) \right\|_{W^{1,\infty}} d_{\min} \geq d_{\min} F\left(\frac{d_{\min}}{N^{1/d}}\right).\end{aligned}$$

Hence one can apply Gronwall's Lemma to $d_{\min}$, propagating $d_{\min} \sim N^{-1/d}$ and ensuring regularity in eqn (8.3). For more on this kind of limit and, particularly, point-vortex approximations to the 2D Euler equations, see for example Goodman *et al.* (1990) and Schochet (1996).

Note, however, that such an approach is inapplicable in phase space: in that case, the physical distance bound $d_{\min}$ still controls the regularity of the force terms, but the evolution equations control only the phase space distance

$$d^v_{\min} = \inf_{i \neq j} (|X_i - X_j| + |V_i - V_j|) \geq d_{\min}.$$

Thus one cannot obtain a closed estimate of $d_{\min}$.

8.5 Macroscopic limit in the regular case

8.5.1 Existence and weak solutions

If K is regular enough, it becomes possible to pass to the limit in the nonlinear term $(K \star \rho_N) \nabla_v f_N$, thus ensuring the existence of a solution to the limiting dynamics (8.2).

Theorem 8.1. *If K is continuous and if the initial conditions are uniformly bounded in velocity ($|V_i^N(0)| \leq R$ for some R), then there exists a subsequence $f_{\sigma(N)}$ of f_N such that:*

1. $f_{\sigma(N)} \overset{w-\star}{\longrightarrow} f$ *in* $L^\infty(\mathbb{R}_+, M^1(\Pi^d \times \mathbb{R}^d))$;
2. $\rho_{\sigma(N)} \overset{w-\star}{\longrightarrow} \rho = \int f \, \mathrm{d}v$ *in* $L^\infty(\mathbb{R}_+, M^1(\Pi^d))$;
3. *f is a solution to eqn (8.2) in the sense of a distribution.*

This theorem proves the existence of measure-valued solutions to eqn (8.2), but its assumptions about K are too weak to ensure their uniqueness (through the convergence of the full sequence f_N, for instance).

Proof. First, notice that, as eqn (8.2) conserves probability or total mass,

$$\int \mathrm{d}x\,\mathrm{d}v\, f(t,x,v) = 1 \qquad \forall t\,.$$

Hence $f_N \in L^\infty(\mathbb{R}_+, M^1(\Pi^d \times \mathbb{R}^d))$. This space, as the dual of $L^1(\mathbb{R}_+, \mathcal{C}^0(\Pi^d \times \mathbb{R}^d))$, is weak-$\star$ compact: therefore $f_{\sigma(N)} \overset{w-\star}{\longrightarrow} f$. By construction, it follows that one can take the $N \to \infty$ limit in both linear terms of eqn (8.2).

Then, using

$$|V_i(t)| \leq |V_i(0)| + \frac{1}{N}\sum_{j\neq i}\int_0^t |K(X_i - X_j)|\,ds \leq R + t\|K\|_\infty\,,$$

one obtains (by the compactness of compact-supported continuous functions) the result that $\rho_{\sigma(N)} = \int f_{\sigma(N)}\,\mathrm{d}v \overset{w-\star}{\longrightarrow} \rho = \int f\,\mathrm{d}v$ in $L^\infty(\mathbb{R}_+, M^1(\Pi^d))$.

Given that $\rho, \rho_N \in M^1(\Pi^d)$, $K \star \rho_N(t,\cdot)$ is equicontinuous in x for all fixed t (with the same continuity modulus as K). Moreover, integrating eqn (8.2) over v yields

$$\partial_t \rho_N + \nabla_x \cdot \left(\int v f_N\,\mathrm{d}v\right) = 0\,,$$

which yields $\partial_t \rho_N \in L^\infty(\mathbb{R}_+, W^{-s,1}(\mathbb{R}^d))$ $\forall s > 1$, so that, finally, $(K \star \rho_N)$ is equicontinuous in (x,t) over all $\Pi^d \times [0,T]$.

One may therefore apply Ascoli's Theorem, to obtain $(K \star \rho_{\sigma(N)}) \to (K \star \rho)$, uniformly over all $\Pi^d \times [0,T]$. Hence $(K \star \rho_{\sigma(N)}) f_{\sigma(N)} \to (K \star \rho) f$ in the sense of distributions, and f is a solution of eqn (8.2) in the same sense. ■

8.5.2 Stability and well-posedness

In order to obtain a stability estimate for the weak solutions derived above, it is necessary to strengthen K's regularity (Braun and Hepp 1977; Dobrushin 1979; Spohn 1991).

Theorem 8.2. *If $K \in W^{1,\infty}$ and $f^1, f^2 \in L^\infty(\mathbb{R}_+, M^1(\Pi^d \times \mathbb{R}^d))$ are two solutions to eqn (8.2) with compact support in velocity, then*

$$\|f^1(t) - f^2(t)\|_{W^{-1,1}(\Pi^d\times\mathbb{R}^d)} \leq C\|f^1_0 - f^2_0\|_{W^{-1,1}(\Pi^d\times\mathbb{R}^d)} \exp[C\|\nabla K\|_{L^\infty} t]\,. \tag{8.5}$$

This stability estimate yields the convergence of the f_N to the limiting dynamics, as well as the uniqueness and well-posedness of its solutions: in practice, it acts as a limit to the concentration of the corresponding measures. One should also note that, although the exponential growth of this estimate is certainly not optimal, the constant C depends only on the total masses of f^1 and f^2, as shown below.

Proof. For $\gamma \in \{1,2\}$, the characteristics of each solution, defined as

$$\begin{cases} \dfrac{\mathrm{d}}{\mathrm{d}t} X^\gamma(t,x,v) = V^\gamma(t,x,v)\,, \\ \\ \dfrac{\mathrm{d}}{\mathrm{d}t} V^\gamma(t,x,v) = (K \star \rho^\gamma)(t, X^\gamma)\,, \end{cases} \quad \text{for} \begin{cases} X(0,x,v) = x \\ V(0,x,v) = v \end{cases},$$

are well defined by the Cauchy–Lipschitz Theorem (since $K \in \mathcal{C}^1$, $\rho^\gamma \star K$ is Lipschitz), and satisfy the corresponding estimate

$$|\nabla X^\gamma| + |\nabla V^\gamma| \le C e^{C\|\nabla K\|_\infty t}\,. \tag{8.6}$$

Moreover, letting $\mathcal{L} = \{\phi \in \mathcal{C}^1(\Pi^d \times \mathbb{R}^d) : \|\phi\|_\infty \le 1, \|\nabla\phi\|_\infty \le 1\}$, one has

$$\begin{aligned}
\|f^1(t) - f^2(t)\|_{W^{-1,1}} &= \sup_{\phi\in\mathcal{L}} \int \mathrm{d}x\,\mathrm{d}v\,\phi(x,v)\left(f^1(t,x,v) - f^2(t,x,v)\right) \\
&= \sup_{\phi\in\mathcal{L}} \int \mathrm{d}x\,\mathrm{d}v\,\left(\phi(X^1(t),V^1(t))f^1(0,x,v) - \phi(X^2(t),V^2(t))f^2(0,x,v)\right) \\
&\le \sup_{\phi\in\mathcal{L}} \int \mathrm{d}x\,\mathrm{d}v\,\phi(X^1(t),V^1(t))\left|f^1(0,x,v) - f^2(0,x,v)\right| \\
&\quad + \sup_{\phi\in\mathcal{L}} \int \mathrm{d}x\,\mathrm{d}v\,\left|\phi(X^1(t),V^1(t)) - \phi(X^2(t),V^2(t))\right| f^2(0,x,v) \\
&\le \|\nabla(X^1,V^1)\|_\infty \|f_0^1 - f_0^2\|_{W^{-1,1}} + \|(X^1,V^1) - (X^2,V^2)\|_\infty \|f_0^2\|_{L^1}\,.
\end{aligned}$$

Since the first term in the above is bounded by eqn (8.6), it is sufficient to bound $\|(X^1,V^1) - (X^2,V^2)\|_\infty$. Given that $(\mathrm{d}/\mathrm{d}t)|X_1 - X_2| \le |V_1 - V_2|$ and

$$\begin{aligned}
\frac{\mathrm{d}}{\mathrm{d}t}|V_1 - V_2| &\le |(K\star\rho^1)(t,X^1) - (K\star\rho^2)(t,X^2)| \\
&\le |(K\star\rho^1)(t,X^1) - (K\star\rho^1)(t,X^2)| + |(K\star(\rho^1-\rho^2))(t,X^2)| \\
&\le |X^1 - X^2|\|\nabla K\|_\infty \int \rho^1\,\mathrm{d}x + \|\nabla K\|_\infty \|\rho^1 - \rho^2\|_{W^{-1,1}}\,,
\end{aligned}$$

one obtains

$$\frac{\mathrm{d}}{\mathrm{d}t}\|f^1(t) - f^2(t)\|_{W^{-1,1}} \le C\|\nabla K\|_\infty \|f^1(t) - f^2(t)\|_{W^{-1,1}}\,,$$

from which eqn (8.5) is derived by Gronwall's Lemma. ∎

8.6 Well-posedness for singular kernels

The above establishes the well-posedness of the limiting dynamics in the case $K \in \mathcal{C}^1$, while hinting that $K \in \mathcal{C}^0$ is likely to be insufficient to reach a similarly satisfactory conclusion. These kernels remain far from the Coulomb-like interactions that one would like to use in practice. It is, however, possible to obtain further results by exploiting the particular nature of these kernels, namely $K \in \mathcal{C}^1(\mathbb{R}^d \setminus \{0\})$.

8.6.1 The weakly singular case

For kernels less singular at 0 than $1/|x|$ (so this never contains the Coulombian case), it is still possible to derive well-posedness from Gronwall-type estimates (Hauray and Jabin 2007). Such a distinction between $K = o(1/|x|)$ and the rest makes sense physically, since if $K = -\nabla\phi$ it corresponds exactly to the cases of bounded vs. unbounded potentials ϕ.

Theorem 8.3. *Given a kernel $K \in \mathcal{C}^1(\mathbb{R}^d \setminus \{0\})$ such that $|K(x)| \underset{x\to 0}{\sim} 1/|x|^\alpha$ with $\alpha < 1$ and a sequence Z^{N0} of initial conditions with uniform compact support such that*

$$d_{\min}(0) = \min_{i\neq j}\left(|X_i^{N0} - X_j^{N0}| + |V_i^{N0} - V_j^{N0}|\right) \geq cN^{-1/2d},$$

then there exists $c' > 0$ such that $d_{\min}(t) \geq c'N^{-1/2d}$ for any $t > 0$. Then the sequence of N-body solutions f_N converges weakly towards the unique solution $f \in L^1 \cap L^\infty$ of eqn (8.2), which is compactly supported.

The condition $\alpha < 1$ above is probably close to optimal, although it remains far from the Coulomb case. This stems from the need to bound the integrals of the force along the trajectories of close particles, which take the form

$$\int \frac{\mathrm{d}t}{|X + Vt|^\alpha} < \infty.$$

Similarly, the condition on $d_{\min}(0)$ is quite remote from physical reality, as the probability of having it satisfied vanishes to 0 for sets of particles with random initial positions and velocities. However, this does not preclude its practical use, for instance for numerical purposes. It is a strong assumption on its own, since

$$f_N^0 \overset{w}{\longrightarrow} f^0 \text{ and } d_{\min}(0) \geq \frac{c}{N^{1/2d}} \Rightarrow f_N^0 \overset{\text{weak-}*\ M^1}{\longrightarrow} f^0 \in L^1 \cap L^\infty.$$

A similar, stronger theorem can be proven for physical-space models (Hauray 2008):

Theorem 8.4. *If $|K(x)| \sim 1/|x|^\alpha$ with $\alpha < d - 1$, then, for any sequence of initial data X^{N0} such that*

$$d_{\min}(0) = \min_{i\neq j}|X_i^{N0} - X_j^{N0}| \geq cN^{-1/d},$$

the dynamics (8.3) with $\mu_i = 1$ and $(\mathrm{d}/\mathrm{d}t)X_i = (1/N)\sum_{j\neq i} K(X_i - X_j)$ satisfies $d_{\min}(t) \geq c'N^{-1/d}$ for all $t \leq T$, and for all $t \geq 0$ if $\nabla \cdot K = 0$. Thus the sequence (ρ_N) converges towards the unique solution ρ to eqn (8.4).

The condition $\alpha < d - 1$ is a noticeable improvement on the Vlasov case, as the 2D Euler equation ($\alpha = 1$) is now the limiting case; however, the assumption about $d_{\min}$ remains just as strong.

Proof. As shown in Section 8.4, the uniqueness of the limit results from the estimate of $d_{\min}$. Since $K \sim |x|^{-\alpha}$,

$$\begin{aligned}\frac{\mathrm{d}}{\mathrm{d}t}|X_i - X_k| &\geq -\frac{1}{N}(|K(X_i - X_k)| + |K(X_k - X_i)|) \\ &\quad - \frac{1}{N}\sum_{j\neq i,k}(|K(X_i - X_j)| + |K(X_k - X_j)|) \\ &\geq -\frac{|X_i - K_k|}{N}\sum_{j\neq i,k}\left(\frac{C}{|X_i - X_j|^{\alpha+1}} + \frac{C}{|X_k - X_j|^{\alpha+1}}\right) - \frac{2C}{N d_{\min}^{\alpha}}.\end{aligned}$$

In order to bound $(1/N)\sum_{j\neq i}(C/|X_i - X_j|^{\alpha+1})$, let $N_k = |\{j \neq i \text{ such that } |X_i - X_j| \in [2^k d_{\min}, 2^{k+1} d_{\min}]\}|$. By the definition of $d_{\min}$, $N_k = 0$ for any $k < 0$; furthermore, as Π has diameter 1, $N_k = 0$ as well for $k > k_0 = -\log_2 d_{\min}$. Hence

$$\frac{1}{N}\sum_{j\neq i}\frac{C}{|X_i - X_j|^{\alpha+1}} \leq \frac{C_d}{N}\sum_{k=0}^{k_0}\frac{N_k}{2^{k(\alpha+1)} d_{\min}^{\alpha+1}}.$$

By the definition of $d_{\min}$, $N_k \leq C_d 2^{kd}$ as all particles are farther than $d_{\min}$ from each other; therefore, as $\alpha + 1 < d$,

$$\frac{1}{N}\sum_{j\neq i}\frac{C}{|X_i - X_j|^{\alpha+1}} \leq \frac{C_d}{N}\sum_{k=0}^{k_0}\frac{2^{k(d-\alpha-1)}}{d_{\min}^{\alpha+1}} \leq \frac{C_d}{N}\frac{2^{k_0(d-\alpha-1)}}{d_{\min}^{\alpha+1}} \leq \frac{C_d}{N d_{\min}^{d}}.$$

Gathering the estimates together yields

$$\frac{\mathrm{d}}{\mathrm{d}t}|X_i - X_k| \geq -|X_i - X_k|\frac{C_d}{N d_{\min}^{d}} - \frac{C}{N d_{\min}^{\alpha}}.$$

Hence, choosing (i,k) such that $d_{\min} = |X_i - X_k|$, one gets

$$\frac{\mathrm{d}}{\mathrm{d}t}d_{\min} \geq -d_{\min}\frac{C_d}{N d_{\min}^{d}} - \frac{C}{N d_{\min}^{\alpha}} = -d_{\min}\frac{N^{-1}}{d_{\min}^{d}}\left(C_d + C d_{\min}^{d-\alpha}\right),$$

from which we may bound $d_{\min}$ by Gronwall's Lemma. The rest of the theorem follows easily from here. ∎

8.7 An almost-everywhere approach

If one attempts to derive a stability estimate, for example a bound on $|X_i^N(t, Z^{N0}) - X_i^N(t, Z^{N0} + \delta)|$ for small δ, while avoiding Cauchy–Lipschitz/Gronwall-like methods, one can only succeed for almost all initial data, for some definition of "almost all." As the dimension N of the system of differential equations changes, quantitative estimates are needed, unlike the more traditional approaches to ordinary differential equations. Such an estimate was derived by Crippa and De Lellis (2008) in a finite-dimensional framework, and a natural idea that will be followed in future work by Barré, Hauray, and Jabin is to try to extend it like

$$\int_{(\Pi^d\times\mathbb{R}^d)^N} \mathrm{d}\mathbb{P}[Z^{N0}] \log\left(1+\frac{1}{N\|\delta\|_\infty}\sum_{i=1}^N \left(|X_i^N(t,Z^{N0})-X_i^N(t,Z^{N0}+\delta)|\right.\right.$$
$$\left.\left.+|V_i^N(t,Z^{N0})-V_i^N(t,Z^{N0}+\delta)|\right)\right) \leq C(1+t)\,, \quad (8.7)$$

where the measure $\mathbb{P}$ on initial configurations determines the meaning of "almost everywhere," and thus must be chosen carefully.

In order to prove eqn (8.7), one has to differentiate the integral with respect to t, and then perform the change of variables $Z^{N0} \to Z^N(t)$ (which has Jacobian 1). One then needs to estimate integrals of the form

$$\int_{(\Pi^d\times\mathbb{R}^d)^N} \frac{\mathrm{d}\mathbb{P}_t(Z^{N0})}{|X_1^0 - X_2^0|^{\alpha+1}}\,,$$

where $\mathbb{P}_t$ is the image of $\mathbb{P}$ created by the flow at time t, which should remain finite for $\alpha < d-1$.

Note that one has to be careful in doing the estimate as a simple direct computation would require a bound on

$$\int \mathrm{d}\mathbb{P}_t(Z^{N0}) \max_i \left(\frac{1}{N}\sum_{j\neq i}\frac{1}{|X_i^0 - X_j^0|^{\alpha+1}}\right) = +\infty\,.$$

The proof involves the following conditions on $\mathbb{P}$:

$$\forall t\,, \int_{\Pi^{d(N-k)}\times\mathbb{R}^d} \mathbb{P}_t(Z^{N0})\mathrm{d}X_{N-k}^0\ldots\mathrm{d}X_N^0\mathrm{d}V_1^0\ldots\mathrm{d}V_N^0 \leq C^k\,,$$

which can be checked if $\mathbb{P}$ is flow-invariant, but would be very hard to investigate otherwise. If K derives from a potential, such that $K = -\nabla\phi$ with $\phi \geq 0$, one such invariant is the energy

$$H_N = \frac{1}{N}\sum_{i=1}^N |V_i^N|^2 + \frac{1}{N^2}\sum_{i\neq j}\phi(X_i^N - X_j^N)\,,$$

and one can choose $\mathbb{P}(Z^{N0}) \propto e^{-H_N(Z^{N0})}$ or $\mathbb{P}(Z^{N0}) \propto e^{-NH_N(Z^{N0})}$. The first choice leads to an easier proof, but the corresponding sequences (f_N^0) of initial conditions are such that $f_N^0 \xrightarrow{\text{a.s.}} 0$, and this choice is therefore very restrictive; the second implies that $f_N(0) \xrightarrow{\text{a.s.}} \rho(x)e^{-|v|^2}$, where ρ minimizes

$$\int_{\Pi^{2d}} \frac{1}{2}\phi(x-y)\rho(x)\rho(y)\,\mathrm{d}x\,\mathrm{d}y + \int_\Pi \rho(x)\log\rho(x)\,\mathrm{d}x\,.$$

In both cases, eqn (8.7) can be used to perturb the dynamics instead of the initial conditions. For instance, by letting $f_N^{(\delta)}$ be the evolution according to a regularized kernel $K_\delta = \mathbb{1}_{\{|x|>\delta\}} K$ with the same initial conditions, one gets

$$\int_{(\Pi^d \times \mathbb{R}^d)^N} d\mathbb{P}(Z^{N0}) \log\left(1 + \frac{1}{|\delta|} \|f_N(t) - f_N^{(\delta)}(t)\|_{W^{-1,1}}\right) \leq C(1+t),$$

which yields the convergence of (f_N) towards a (unique) solution f of eqn (8.2) with $f^0 = \lim_N f_N(0)$. Hence the two choices of $\mathbb{P}$ prove the stability of the problem's two stationary solutions (zero and thermal equilibrium) involving Dirac masses.

Further application of this approach would require considering noninvariant measures for $\mathbb{P}$.

References

Ambrosio L. (2004). Transport equation and Cauchy problem for BV vector fields. *Invent. Math.*, **158**, 227–260.

Braun W. and Hepp K. (1977). The Vlasov dynamics and its fluctuations in the 1/N limit of interacting classical particles. *Commun. Math. Phys.*, **56**(2), 101–113.

Crippa G. and De Lellis C. (2008). Estimates and regularity results for the DiPerna–Lions flow. *J. Reine Angew. Math.*, **616**, 15–46.

DiPerna R. J. and Lions P. L. (1989). Ordinary differential equations, transport theory and Sobolev spaces. *Invent. Math.*, **98**(3), 511–547.

Dobrushin R. L. (1979). Vlasov equations. *Funct. Anal. Appl.*, **13**(2), 115–123.

Goodman J., Hou, T. Y., and Lowengrub J. (1990). Convergence of the point vortex method for the 2D Euler equations. *Commun. Pure Appl. Math.*, **43**(3), 415–430.

Hauray M. (2005). On Liouville transport equation with force field in BV_{loc}. *Commun. Part. Differ. Equ.*, **29**(1), 207–217.

Hauray M. (2008). Approximation of Euler and quasi Euler equations by systems of vortices. To appear in *Math. Models Methods Appl. Sci.*

Hauray M. and Jabin P. E. (2007). N-particles approximation of the Vlasov equations with singular potential. *Arch. Rat. Mech. Anal.*, **183**(3), 489–524.

Horst E. (1981). On the classical solutions of the initial value problem for the unmodified nonlinear Vlasov equation. I. General theory. *Math. Methods Appl. Sci.*, **3**(2), 229–248.

Lions P. L. and Perthame B. (1991). Propagation of moments and regularity for the 3-dimensional Vlasov–Poisson system. *Invent. Math.*, **105**(1), 415–430.

Pfaffelmoser K. (1992). Global classical solutions of the Vlasov–Poisson system in three dimensions for general initial data. *J. Differ. Equ.*, **95**(2), 281–303.

Schaeffer J. (1991). Global existence of smooth solutions to the Vlasov–Poisson system in three dimensions. *Commun. Part. Differ. Equ.*, **16**(8), 1313–1335.

Schochet S. (1996). The point-vortex method for periodic weak solutions of the 2D Euler equations. *Commun. Pure Appl. Math.*, **49**(9), 911–965.

Spohn H. (1991). *Large Scale Dynamics of Interacting Particles.* Springer-Verlag, Berlin.

9
On the origin of phase transitions in long- and short-range interacting systems

Michael KASTNER

Physikalisches Institut, Universität Bayreuth, 95440 Bayreuth, Germany

9.1 Introduction

Almost regardless of the field of physics you are interested in, phase transitions are a phenomenon of great importance. Phase transitions occur in both equilibrium and nonequilibrium, but in this lecture our concern will be exclusively the equilibrium situation. Typical examples of such equilibrium phase transitions are the transitions between different states of matter (solid, liquid, gaseous, ...) and the transition from normal conductivity to superconductivity. In the vicinity of a phase transition point, a small change in some external control parameter (such as pressure or temperature) results in a dramatic change of certain physical properties (such as specific heat or electrical resistance) of the system under consideration.

9.1.1 Thermodynamic description of phase transitions

In a theoretical description of phase transitions in the framework of thermodynamics, the abrupt changes of physical properties motivate the following definition.

Definition 9.1. *An equilibrium phase transition is defined as a nonanalyticity of the free-energy density* $\bar{f}$.

To capture a phase transition of interest with the above definition, the free-energy density has to be considered as a function of the relevant control parameters, i.e. those which, upon variation, give rise to the phase transition. For the phase transitions between the aggregate states of, say, water, the (Gibbs) free-energy density as a function of temperature and pressure is a suitable choice. The examples we will discuss later are typically spin systems where we have at most two such relevant control parameters, the temperature T and an external magnetic field h, and therefore we will consider the free-energy density $\bar{f}(\beta, h)$ as a function of the inverse temperature $\beta = 1/(k_{\mathrm{B}}T)$ and the magnetic field h, where Boltzmann's constant k_{B} is set to unity in the following. Feel free to replace β and h by any other parameters you prefer.

Quantities such as the specific heat or caloric curves which are typically measured in an experiment are then given in terms of derivatives of the free-energy density. Nonanalyticities of $\bar{f}$ may hence lead to discontinuities or divergences in these quantities, which are experimental hallmarks of phase transitions.

Various aspects of phase transitions have been studied extensively in recent decades. Arguably the biggest success was the invention of renormalization group techniques, which led to an understanding of the universal behavior of physically very different systems in the vicinity of a critical point (Kadanoff 1966; Wilson 1971; see also Binney *et al.* 1992). In this lecture, we want to take one step back and ask an even more basic, more fundamental question.

Question 9.2. *What is the microscopic "origin" of a (macroscopic) phase transition?*

9.1.2 Statistical-physical description of phase transitions

Statistical physics was developed in the late nineteenth and early twentieth centuries by Boltzmann, Gibbs, and others to provide the microscopic foundations of thermodynamics. The starting point for such a description is a Hamiltonian function (in classical mechanics) or a Hamiltonian operator (in quantum mechanics), characterizing the interactions between all the elementary constituents of a physical system. In what follows, we will use the language of classical mechanics, characterized by a Hamiltonian function

$$H(p;q) = E(p;q) - hM(p;q) \tag{9.1}$$

defined on phase space, where $p = (p_1, \dots, p_N)$ is the vector of momenta and $q = (q_1, \dots, q_N)$ is the vector of position coordinates. h is the second control parameter that we considered in the thermodynamic description of Section 9.1.1. We will comment briefly on a generalization to quantum mechanical systems in the outlook (Section 9.4.2).

Contact with the thermodynamic description is made by defining the canonical free-energy density of a system with N degrees of freedom as

$$\bar{f}_N(\beta, h) = -\frac{1}{N\beta} \ln \int \mathrm{d}p\, \mathrm{d}q\, \mathrm{e}^{-\beta H(p;q)}, \tag{9.2}$$

where the integration is over phase space. Accordingly, we have to modify Definition 9.1 for a phase transition in this context.

Definition 9.3. *An equilibrium phase transition is defined as a nonanalyticity of the canonical free-energy density* $\bar{f}_N$.

This is one of the two commonly used definitions of an equilibrium phase transition in statistical physics. The other one is based on the (non)uniqueness of translationally invariant Gibbs measures (see Lebowitz (1999) for an introductory discussion, and Georgii (1988) for a comprehensive treatment). The two definitions apparently coincide in most but not all cases.

Now we are able to reformulate Question 9.2 in the language of statistical mechanics.

Question 9.4. *Under what conditions on the number of degrees of freedom* N *and on the Hamiltonian function* H *is the canonical free-energy density* $\bar{f}_N$ *nonanalytic?*

A partial answer to this question is well known: the canonical free-energy density $\bar{f}_N(\beta, h)$ is analytic in β and h for all finite N (see Griffiths 1972), and a nonanalyticity may be observed only in the thermodynamic limit of an infinite number of degrees of freedom,

$$\bar{f}(\beta, h) = \lim_{N\to\infty} \bar{f}_N(\beta, h). \tag{9.3}$$

Still, it would be desirable to establish criteria on the Hamiltonian function H which guarantee, or exclude, that a phase transition will occur in the thermodynamic limit. Unfortunately, this appears to be notoriously difficult, and a general answer is too much to even hope for. Nonetheless, it might be rewarding to pursue such a line of thinking.

9.1.3 Different statistical ensembles

Concerning phase transitions in the framework of statistical physics, there is a somewhat irritating fact I would like to bring to your attention: when the important and ubiquitous phenomenon of a phase transition was defined in Definition 9.3, explicit reference was made to the canonical ensemble. This does not cause much of a problem as long as we are considering systems with short-range[1] interactions in the thermodynamic limit, since under these conditions the various statistical ensembles are known to yield equivalent results (Ruelle 1969). But as soon as we are interested in long-range systems, we may run into trouble: nonequivalence of statistical ensembles may occur in these circumstances. Accordingly, we have to work in the statistical ensemble which reflects the physical situation we are interested in, for example the canonical ensemble when the system is coupled to a heat bath, or the microcanonical ensemble when the system is energetically isolated. In the latter case, it would not make sense to define a phase transition in terms of the canonical free-energy density $\bar{f}$. Instead, the microcanonical entropy density

$$\bar{s}_N(\varepsilon, m) = \frac{1}{N} \ln \int \mathrm{d}p\, \mathrm{d}q\, \delta(E(p;q) - N\varepsilon)\delta(M(p;q) - Nm) \tag{9.4}$$

is the starting point of statistical physics in the microcanonical ensemble, where δ denotes the Dirac distribution and, again, the integration is over phase space. From these considerations it appears natural to ask, in the spirit of Question 9.4, the following.

Question 9.5. *Under what conditions on the number of degrees of freedom N and on the Hamiltonian function H is the microcanonical entropy density $\bar{s}_N$ nonanalytic?*

Remarkably, the answer differs significantly from the canonical situation. In particular, we will see below that the microcanonical entropy density can be nonanalytic even for finite N. Whether it is reasonable to interpret nonanalyticities of $\bar{s}_N$ as phase transitions is doubtful (and, of course, a matter of definition), and we will come back to this question in Section 9.3.4.

9.1.4 Outline

We have seen in the preceding sections that, inspired by the canonical definition of a phase transition, there is an interest in the analyticity properties of other thermodynamic functions apart from the canonical free-energy density. In particular, when we study finite systems or systems with long-range interactions, nonequivalence of ensembles may occur and quantities such as the microcanonical entropy density may be of physical interest.

In Section 9.2, we investigate nonanalyticities of the microcanonical entropy from a thermodynamic point of view. To this purpose, the concavity properties of entropy functions in the thermodynamic limit are discussed for both long-range and short-range systems. These properties will provide some intuition about the various mechanisms giving rise to phase transitions. In the long-range case, a simple maximization procedure on the macroscopic level may generate a nonanalyticity of the entropy. The short-range case is more intricate and is discussed in Section 9.3 from a statistical-physical point of view: here the nonanalytic behavior (if it occurs) is generated when

[1] The precise technical conditions are *stability* and *temperedness* of the potential, but we will use the sloppy term "short-range" as a synonym; see Ruelle (1969) for details.

we switch over from a microscopic to a macroscopic description. We work out necessary conditions on the Hamiltonian function H for such nonanalytic behavior to occur: stationary points (i.e. points of vanishing gradient) of H are necessary for a phase transition to occur in short-range systems. Moreover, the curvature at those points plays an important role in determining which stationary points remain relevant even in the thermodynamic limit. Having gained some understanding of the analyticity properties of the microcanonical entropy, we may interpret some of our findings in the canonical context also. Finally, we comment on the consequences of these findings for long-range and short-range interacting systems and comment on an extension to quantum systems in Section 9.4.

9.2 Nonanalyticities in short- and long-range systems

9.2.1 Concavity properties of entropy functions

The concavity properties of the microcanonical entropy play a crucial role in the equivalence or nonequivalence of the microcanonical and the canonical ensemble: whenever the entropy is a concave function, equivalence of the microcanonical and the canonical ensemble holds (Ellis *et al.* 2000). Concavity of the entropy

$$\bar{s}(\varepsilon, m) = \lim_{N \to \infty} \bar{s}_N(\varepsilon, m) \tag{9.5}$$

in the thermodynamic limit is known to hold true if both E and M in eqn (9.1) are of short-range (Lanford 1973). In contrast, the microcanonical entropy has been shown to be nonconcave for several examples of long-range interacting systems which undergo a discontinuous (first-order) phase transition (see Touchette (2008) for some simple examples, as well as a list of further references). Typical examples of entropy functions $s(\varepsilon, m)$ of ferromagnetic spin systems with long- and short-range interactions, one being nonconcave, the other being concave, are plotted in Fig. 9.1. We will see in the subsequent sections that the different concavity properties of $\bar{s}(\varepsilon, m)$ have immediate implications for the occurrence of phase transitions.

9.2.2 Contraction of $\bar{s}(\varepsilon, m)$

The entropy $\bar{s}(\varepsilon, m)$ does not properly describe the typical physical situation when an isolated system is studied. Usually, the total energy is a conserved quantity, but the magnetization m is not. Therefore the entropy

$$s(u) = \lim_{N \to \infty} \frac{1}{N} \ln \int \mathrm{d}p \, \mathrm{d}q \, \delta(H(p;q) - Nu) \tag{9.6}$$

is a suitable choice to describe this physical situation. In terms of $\bar{s}$, this quantity can be expressed as

$$\begin{aligned} s(u) &= \lim_{N \to \infty} \frac{1}{N} \ln \int \mathrm{d}p \, \mathrm{d}q \, \delta(E(p;q) - hM(p;q) - Nu)) \int \mathrm{d}(Nm) \, \delta(M(p;q) - Nm) \\ &= \lim_{N \to \infty} \frac{1}{N} \ln \int \mathrm{d}(Nm) \int \mathrm{d}p \, \mathrm{d}q \, \delta(E(p;q) - N(u + hm))\delta(M(p;q) - Nm) \\ &= \lim_{N \to \infty} \frac{1}{N} \ln \int \mathrm{d}(Nm) \exp\{N \bar{s}_N(u + hm, m)\} = \max_m \bar{s}(u + hm, m) \, , \end{aligned} \tag{9.7}$$

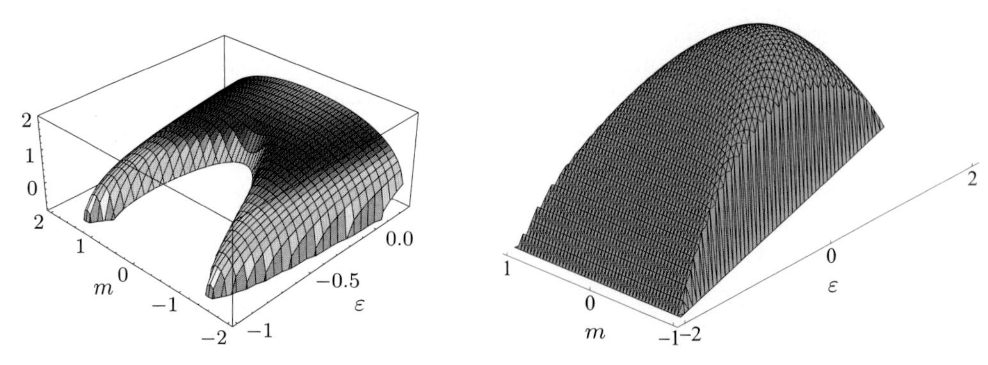

Fig. 9.1 Plots of typical graphs of entropy functions $\bar{s}(\varepsilon, m)$ of ferromagnetic spin systems. Left: in the presence of long-range interactions, s is not necessarily a concave function. In this example, this is obvious from the fact that the domain in the (ε, m) plane is not even a convex set. Adapted from Hahn and Kastner (2006). Right: for a ferromagnetic system with short-range interactions, the entropy is concave, but not strictly concave in the coexistence region. Adapted from Kastner (2002).

where Laplace's method for the asymptotic evaluation of integrals has been used. Of particular interest for the study of phase transitions is the symmetric case of vanishing external magnetic field $h = 0$, where we obtain

$$s(u) \equiv s(\varepsilon) = \max_{m} \bar{s}(\varepsilon, m) \,. \tag{9.8}$$

This simple contraction formula can provide some interesting insights into how a thermodynamic function may have, or may develop, a nonanalyticity in the thermodynamic limit.

9.2.3 Contraction of analytic, nonconcave entropy functions

Assume that, for some long-range interacting system, you find that the entropy $\bar{s}(\varepsilon, m)$ is *analytic* and *nonconcave* with a shape qualitatively similar to that plotted in Fig. 9.1 (left). Then, although $\bar{s}(\varepsilon, m)$ is analytic, its contraction $s(\varepsilon)$ shows a nonanalyticity for some value of ε. This becomes evident from the following simple calculation.

The function

$$\bar{s}_1(\varepsilon, m) = \varepsilon - \varepsilon^2 - 2\varepsilon m^2 - m^4 \tag{9.9}$$

is, as is easily verified, an analytic and nonconcave function. Maximizing with respect to the second variable of $\bar{s}_1$, one obtains

$$s_1(\varepsilon) = \max_{m} \bar{s}_1(\varepsilon, m) = \begin{cases} \varepsilon & \text{for } \varepsilon < 0 \,, \\ \varepsilon(1 - \varepsilon) & \text{for } \varepsilon \geqslant 0 \,, \end{cases} \tag{9.10}$$

which is nonanalytic at $\varepsilon = 0$.

The above conditions—analyticity and nonconcavity of the entropy $\bar{s}(\varepsilon, m)$—are believed to be not uncommon for ferromagnetic, long-range-interacting spin systems, and they may be proved explicitly for some simple models.

Example 9.6. The mean-field φ^4 model is characterized by the Hamiltonian function[2]

$$H(\varphi) = -\frac{J}{2N}\left(\sum_{i=1}^{N} \varphi_i\right)^2 + \sum_{i=1}^{N}\left(-\frac{1}{2}\varphi_i^2 + \frac{1}{4}\varphi_i^4\right), \qquad \varphi_i \in \mathbb{R}, \tag{9.11}$$

with coupling constant $J > 0$. An exact expression for the entropy $\bar{s}(\varepsilon, m)$ has been computed by large-deviation techniques (Hahn and Kastner 2005, 2006; Campa *et al.* 2007). The result is indeed analytic and nonconcave, and the graph of $\bar{s}(\varepsilon, m)$ of the mean-field φ^4 model is the one plotted in Fig. 9.1 (left) as an example of a typical entropy in the presence of long-range interactions. It may be verified that the contraction $s(\varepsilon)$ is in fact nonanalytic at a certain critical energy $\varepsilon = \varepsilon_c$ (see Hahn and Kastner (2005) for details). ■

What is remarkable about these observations? The above considerations show that, for long-range systems in the case of nonconcave entropy functions, it is possible to go from the microscopic to the macroscopic level by performing the thermodynamic limit $N \to \infty$ and still retain analyticity. A nonanalyticity arises in this case only *on the macroscopic level* when the contraction of $\bar{s}(\varepsilon, m)$ to $s(\varepsilon)$ is performed. We will see in the next section that this stands in stark contrast to the situation for short-range systems, where the entropy is necessarily concave.

9.2.4 Contraction of concave entropy functions

Assume that you find an entropy function $\bar{s}(\varepsilon, m)$ which is *analytic* and *concave*. Let us exclude the trivial case where $\bar{s}$ is linear in one or both of its arguments (which should be an irrelevant special case in statistical physics). Then, the contraction $s(\varepsilon)$ is also an analytic function. (This should be obvious from geometric considerations.) Similarly to the previous section, this may be illustrated by a simple calculation.

Consider the analytic, concave function

$$\bar{s}_2(\varepsilon, m) = \varepsilon - \varepsilon^2 - 2m^2 - m^4 \,. \tag{9.12}$$

A maximization of this function with respect to m yields

$$s_2(\varepsilon) = \max_m \bar{s}_2(\varepsilon, m) = \varepsilon - \varepsilon^2 \,, \tag{9.13}$$

which is again analytic.

For a phase transition to occur in a short-range system, we need a concave entropy function of the type plotted in Fig. 9.1 (right), with a region in the (ε, m) plane where $\bar{s}$ is concave, but not strictly concave. However, there is no way of realizing such a behavior by means of an analytic entropy.

[2]The mean-field φ^4 model, like other classical spin models, does not have a kinetic-energy term, and hence the "energy function" H in eqn (9.11) is not a Hamiltonian in a strict sense.

Example 9.7. The Ising model on a two-dimensional square lattice with nearest-neighbor interactions is characterized by the Hamiltonian function

$$H(\sigma) = -J \sum_{\langle i,j \rangle} \sigma_i \sigma_j - h \sum_{i=1}^{N} \sigma_i \,, \qquad \sigma_i \in \{-1, +1\} \,, \tag{9.14}$$

where the angle brackets denote a summation over pairs of nearest neighbors on the lattice and $J > 0$ is a coupling constant. An exact analytic solution for the entropy $\bar{s}(\varepsilon, m)$ in the thermodynamic limit is not known, but numerical data from Monte Carlo simulations of large, finite systems allow us to approximate this function pretty well. An approximate graph of the entropy $\bar{s}(\varepsilon, m)$ of the two-dimensional Ising model is shown in the right plot of Fig. 9.1. The function is nonanalytic and concave, but not strictly concave, and its contraction $s(\varepsilon)$ is nonanalytic at a certain critical energy $\varepsilon = \varepsilon_c$ (see Kastner (2002) for details). ■

9.2.5 Origin of nonanalyticities of entropy functions

Summarizing the observations made in the preceding two sections, we note that non-analyticities of the entropy may be generated on two different levels of description: either by a maximization over one variable of a macroscopic, analytic entropy function, which has to be nonconcave, or when going from the microscopic to the macroscopic level of description by means of the thermodynamic limit. The second "mechanism" may occur in both short-range and long-range systems, but this is not so for the first which is a genuine long-range phenomenon.

What is the consequence of this difference? Since nonanalyticities in the entropy of short-range systems have their origin on the microscopic level, we may hope to find hallmarks or precursors of a phase transition on the microscopic level. For long-range systems, however, we cannot necessarily expect such signatures. These statements about "precursors" on a "microscopic level" are a bit vague and more of an intuitive kind. However, we will see in the following section what such precursors may look like. Moreover, their presence will indeed be seen to be *necessary* for phase transitions in short-range interacting systems, but not in the long-range case.

9.3 Phase transitions, configuration space topology, and energy landscapes

We have argued that, at least for short-range interacting systems, we expect to find signatures of a phase transition even on the microscopic level. But where should we look for such signatures? What quantities should we study? One possible line of reasoning goes like this.

9.3.1 Topology changes in configuration space

The fundamental quantity of the microcanonical ensemble is the entropy

$$s_N(u) = \frac{1}{N} \ln \Omega_N(u) \,, \tag{9.15}$$

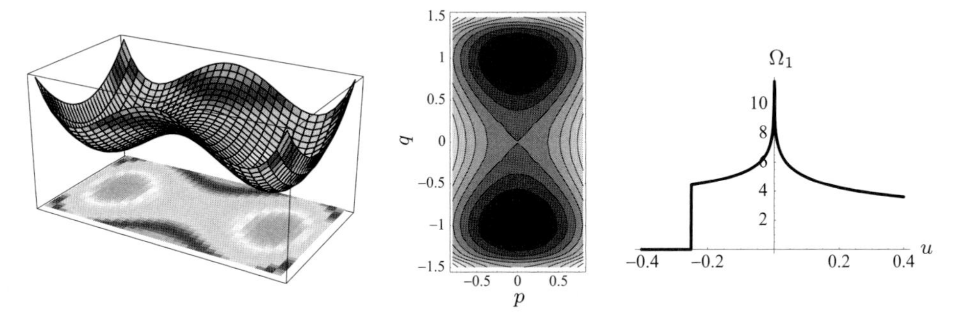

Fig. 9.2 Plot of the Hamiltonian function (9.18) as a three-dimensional plot (left) and as a contour plot (middle). The corresponding density of states $\Omega_1(u)$ is nonanalytic at the energy $u = -1/4$ of the minima and at the energy $u = 0$ of the saddle point of H, at which the contours also change their topology (right).

where

$$\Omega_N(u) = \int \mathrm{d}p\, \mathrm{d}q\, \delta(H(p;q) - Nu) \tag{9.16}$$

is the density of states. The latter expression can be interpreted as the volume of the constant-energy surface

$$\Sigma_u = H^{-1}(Nu) = \big\{(p;q) \,\big|\, H(p;q) = Nu\big\} \tag{9.17}$$

in phase space. Upon variation of the value of u, one may expect that this volume will typically vary in a smooth way. At certain special values of u, however, the *topology* of Σ_u changes, and here the volume will change in a nonanalytic fashion.[3] These ideas were pioneered in the late 1990s by Pettini and coworkers; for a review on the subject, see Kastner (2008).

Example 9.8. Consider the Hamiltonian function

$$H(p;q) = \tfrac{1}{2}p^2 + \tfrac{1}{4}q^4 - \tfrac{1}{2}q^2, \qquad p, q \in \mathbb{R}. \tag{9.18}$$

This double well has two minima at $(p,q) = (0, \pm 1)$ with energy $u = H(0;\pm 1) = -1/4$, and a saddle point at $(p,q) = (0,0)$ with energy $u = H(0;0) = 0$ (see Fig. 9.2, left). Precisely at these values of u, the surfaces of constant energy Σ_u change topology: from the empty set to two 1-spheres at $u = -1/4$, and to a single 1-sphere at $u = 0$ (see the contour plot in Fig. 9.2, middle). As expected, the density of states $\Omega_1(u)$ displays nonanalyticities at the very same values of u (Fig. 9.2, right). ■

It is convenient to restrict the following discussions to Hamiltonian functions of the standard form

$$H(p;q) = \sum_{i=1}^{N} \frac{p_i^2}{2m_i} + V(q), \tag{9.19}$$

where N is the number of degrees of freedom and m_i is the mass associated with the ith degree of freedom. The potential V is a function of all position coordinates,

[3]We speak of a topology change at u when $\Sigma_{u-\delta}$ and $\Sigma_{u+\delta}$ are not homeomorphic for arbitrarily small $|\delta| > 0$.

and it may include external potentials, two-body interactions, or arbitrary n-body interactions. Since the kinetic-energy term in eqn (9.19) is quadratic in the momenta, it is harmless as regards the occurrence of phase transitions.[4] Hence we can focus our attention in the following on the potential V and study the configurational density of states

$$\Omega_N(v) = \int \mathrm{d}q \, \delta\left(V(q) - Nv\right) \tag{9.20}$$

and the configurational microcanonical entropy

$$s_N(v) = \frac{1}{N} \ln \Omega_N(v) \,. \tag{9.21}$$

The corresponding surfaces of constant potential energy are given by

$$\Sigma_v = V^{-1}(vN) = \left\{ q \,\middle|\, V(q) = Nv \right\} . \tag{9.22}$$

Often, it may be convenient to study the topology of the related subsets

$$M_v = V^{-1}(-\infty, Nv] = \left\{ q \,\middle|\, V(q) \leqslant Nv \right\} . \tag{9.23}$$

Since Σ_v is the border of M_v, we expect to find topology changes of M_v whenever Σ_v changes topology.

9.3.2 Morse theory and energy landscapes

The ideas outlined in the previous section lead us to deal with topology changes, i.e. in principle we need to determine whether a homeomorphism that maps some subset M_a of configuration space onto another subset M_b exists or not. Fortunately, Morse theory—a branch of differential topology—helps us to rephrase this problem in more familiar terms (for introductory textbooks on Morse theory, see Milnor (1963) or Matsumoto (2002)).

Morse theory, in general, establishes a relation between the topology of a manifold and the stationary points of an analytic function on this manifold, i.e. points with vanishing gradient. In the notation of the previous section, Morse theory allows us to characterize the topology of M_v by studying the stationary points q_s of an analytic potential V in configuration space, i.e. those points for which

$$\mathrm{d}V(q_{\mathrm{s}}) = 0 \tag{9.24}$$

holds. The only restriction we have to impose on V is that it has the so-called *Morse property*, meaning that for all stationary points q_s of V, the Hessian $\mathfrak{H}_V$ of V has a nonvanishing determinant, i.e.

$$\det \mathfrak{H}_V(q_{\mathrm{s}}) \neq 0 \,. \tag{9.25}$$

One may argue that this is an insignificant restriction, since Morse functions form an open dense subset of the space of smooth functions (Demazure 2000) and are

[4]Such quadratic terms give rise to Gaussian integrals in the partition function or density of states which can be solved, resulting in analytic contributions to the canonical free energy or microcanonical entropy.

therefore generic. This means that, if the potential V of the Hamiltonian system we are interested in is not a Morse function, we can transform it into one by adding an arbitrarily small perturbation. One important consequence of the Morse property is that all stationary points of a Morse function are isolated.

Under these conditions, Morse theory has the following to say about topology changes of the configuration space subsets M_v:

1. If the interval $[a, b]$ is free of stationary values, i.e. for all stationary points q_{s} of V we have
$$v_{\mathrm{s}} = \frac{V(q_{\mathrm{s}})}{N} \notin [a, b]\,, \tag{9.26}$$
then M_a and M_b are homeomorphic.
2. If there is a single stationary point q_{s} with stationary value $v_{\mathrm{s}} = V(q_{\mathrm{s}})/N$, then the topology of M_v changes at $v = v_{\mathrm{s}}$ in a way which is determined by the index i of the stationary point, i.e. by the number of negative eigenvalues of the Hessian $\mathfrak{H}_V$ at q_{s} (for details, see Matsumoto (2002)).

In summary, we can obtain all information about the topology changes of M_v by determining the stationary points of the potential V and their indices. In this way, we make contact with what is often termed the concept of *energy landscapes*. When speaking about energy landscapes, one typically refers to the study of stationary points of an energy function, such as the potential V in our case. This concept has proved particularly useful for the investigation of *dynamical* properties such as reaction pathways in chemical physics and conformational changes in biophysics: after one has computed local minima of the energy landscape and transition states (i.e. stationary points of index $i = 1$), reaction pathways and reaction rates can be predicted from these data (see Wales (2004) for a comprehensive textbook on this subject). We will see in the following that the study of stationary points proves useful also in the context of statistical physics.

9.3.3 Stationary points of V and nonanalyticities of the finite-system entropy

We can now pursue the idea outlined in Section 9.3.1 that a nonanalyticity of the entropy s_N should have its origin in a topology change of the configuration space subsets M_v. With the results from Morse theory sketched in Section 9.3.2, this idea can be rephrased in terms of stationary points of the potential-energy landscape:

Question 9.9. *What is the effect of a stationary point of the potential V on the analyticity properties of the configurational entropy s_N?*

An answer to this question has been given by Kastner *et al.* (2007):

Theorem 9.10. *Let $V : G \to \mathbb{R}$ be a Morse function with a single critical point q_s of index i in an open region $G \subset \mathbb{R}^N$. Without loss of generality, we assume $V(q_s) = 0$. Then there exists a polynomial P of degree less than $N/2$ such that at $v = 0$ the configurational density of states (9.20) can be written in the form*

$$\Omega_N(v) = P(v) + \frac{h_{N,i}(v)}{\sqrt{|\det[\mathfrak{H}_V(q_s)]|}} + o(v^{N/2-\epsilon}) \tag{9.27}$$

for any $\epsilon > 0$. *Here* Θ *is the Heaviside step function,* o *denotes Landau's little-o symbol for asymptotic negligibility, and*

$$h_{N,i}(v) = \frac{(N\pi)^{N/2}}{\Gamma(N/2)} \begin{cases} (-1)^{(N-i)/2}(-v)^{(N-2)/2}\,\Theta(-v) & \text{for } N, i \text{ odd}, \\ (-1)^{i/2}\,v^{(N-2)/2}\,\Theta(v) & \text{for } i \text{ even}, \\ (-1)^{(i+1)/2}\,v^{(N-2)/2}\,\pi^{-1}\ln|v| & \text{for } N \text{ even}, i \text{ odd}, \end{cases} \tag{9.28}$$

is a universal function which is nonanalytic at $v = 0$.

For a proof of this theorem, the density of states is calculated separately below and above the critical value $v = 0$. By complex continuation, it is possible to subtract the two contributions and to evaluate the leading order of the difference. A detailed proof of an even stronger result (including higher-order terms) has been given by Kastner *et al.* (2008).

The content of Theorem 9.10 can be summarized as follows:

1. Every stationary point q_s gives rise to a nonanalyticity of the configurational entropy $s_N(v)$ at the corresponding stationary value $v = v_s = V(q_s)/N$.
2. The order of this nonanalyticity is $\lfloor (N-3)/2 \rfloor$, i.e. $s_N(v)$ is precisely $\lfloor (N-3)/2 \rfloor$ times differentiable at $v = v_s$.

At this point, it is interesting to note that numerical studies and heuristic arguments indicate that the number of stationary points of a generic potential V increases *exponentially* with the number of degrees of freedom N (Doye and Wales 2002). As a consequence, for large (but finite) N we can expect to find a very large number of nonanalyticities of s_N. Such behavior shows a pronounced difference from the properties of *canonical* thermodynamic functions of finite systems, which are known to be always analytic (see Section 9.1.2).

The first of the above observations can be rewritten in the language of configuration space topology:

1.′ Every topology change of M_v at some value $v = v_s$ gives rise to a nonanalyticity of the configurational entropy $s_N(v)$ at this value.

Moreover, as the number N of degrees of freedom increases, the number of values of the potential energy v at which M_v changes topology typically increases exponentially with N.

9.3.4 Phase transitions in finite systems?

In agreement with the intuition produced in Section 9.3.1, we have found a relation between topology changes in configuration space and nonanalyticities of s_N. But what is the physical significance of these nonanalyticities?

In Section 9.1.3, we described the motivation for the study of nonanalyticities of thermodynamic functions in general, and of the microcanonical entropy in particular, by reference to the conceptual similarity to Definition 9.3, where a phase transition was defined as a nonanalyticity of the canonical free-energy density. But does it make sense to consider the nonanalyticities of the entropy as phase transitions of finite systems in the microcanonical ensemble?

I don't think so, the reason being that these nonanalyticities by no means show the remarkable properties which account for the interest of physicists in phase transitions, namely the dramatic changes of physical properties of large systems due to collective effects. In contrast, the nonanalyticities of the microcanonical entropy that we have found typically occur at a huge number of energy values, and they become weaker and weaker the larger the number of degrees of freedom of the system under consideration becomes. This is simply not the physical phenomenon which we would like to call a phase transition!

But there is no need for disappointment. On the one hand, we have learned that it would not be reasonable to generalize the canonical Definition 9.3 of a phase transition to the microcanonical ensemble in a naive fashion by reference to nonanalyticities of the microcanonical entropy. On the other hand, we should keep in mind that we started out with the goal of understanding the origin of a phase transition in the usual (thermodynamic limit) sense. We have one more step to go.

9.3.5 Phase transitions and configuration space topology

Now that we have observed that topology changes can be numerous and that the corresponding nonanalyticities of the microcanonical entropy become weaker with increasing N, it may appear doubtful whether they are related at all to the occurrence of phase transitions in the thermodynamic limit. However, for several models, calculations of quantities characterizing the topology changes of M_v have been performed, and the results give strong indications that a relation between topology changes and the occurrence of phase transitions in the thermodynamic limit *does* exist.

Example 9.11. One of those models for which topological quantities can be computed analytically is the mean-field k-trigonometric model, characterized by the potential

$$V_k(q) = \frac{\Delta}{N^{k-1}} \sum_{i_1,\ldots,i_k=1}^{N} \left[1 - \cos\left(q_{i_1} + \cdots + q_{i_k}\right)\right], \tag{9.29}$$

where $\Delta > 0$ is a coupling constant and the position coordinates $q_i \in [0, 2\pi)$ are angular variables. The potential describes a k-body interaction, where $k \in \mathbb{N}$, and the model is known to undergo a phase transition for $k \geqslant 2$ at a transition potential energy $v = \Delta$, whereas no phase transition takes place for $k = 1$. In Angelani *et al.* (2003), all stationary points of V and the corresponding indices have been calculated, and these results can be used to compute the *Euler characteristic* χ of M_v. For our purposes it is sufficient to know that χ is a topological invariant, i.e. it changes its value at most when M_v changes topology. A plot of

$$\sigma = \lim_{N\to\infty} \frac{1}{N} \ln |\chi(M_v)| \tag{9.30}$$

is shown in Fig. 9.3 (left) to illustrate the thermodynamic-limit behavior of the Euler characteristic. Remarkably, even this purely topological quantity signals the absence or presence of a phase transition, being smooth for $k = 1$, in the absence of a phase transition, and nonanalytic precisely at the phase transition energy $v = \Delta$ when $k \geqslant 2$. ■

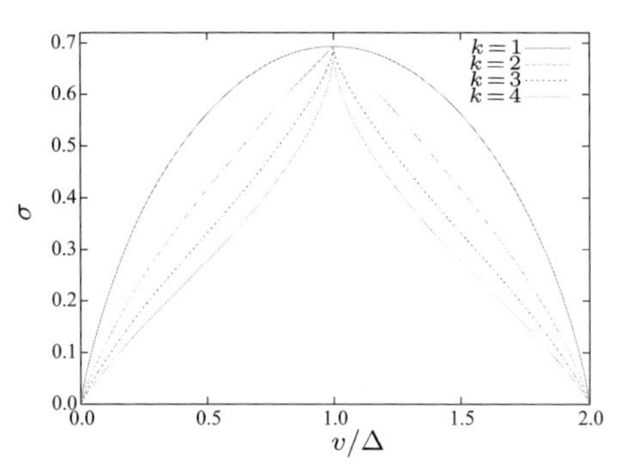

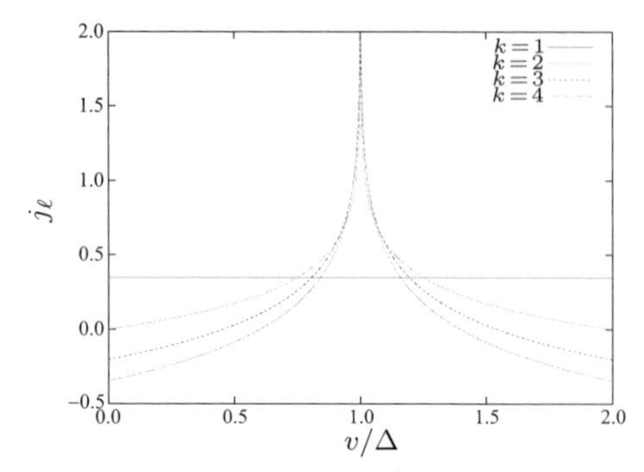

Fig. 9.3 Logarithmic modulus of the Euler characteristic of M_v (left) and the "flatness indicator" j_ℓ (right) as functions of v/Δ for the mean-field k-trigonometric model in the thermodynamic limit. See Section 9.3.6 for further details of j_ℓ.

Other models for which a relation between topology changes and the occurrence of phase transitions has been found are listed in Table 1 of Kastner (2008). Further evidence of such a relation is provided by a theorem due to Franzosi and Pettini (2004), stating that topology changes are a *necessary* condition for a phase transition to take place. Here we give only a very sloppy formulation of this result:

Sloppy theorem 9.12. *Let V be the potential of a system with N degrees of freedom and short-range interactions. If some interval $[a, b]$ of potential energy per degree of freedom remains, for any large enough N, free of stationary values of V, then the configurational entropy $s(v)$ does not show a phase transition in this interval.*

Note that a precise formulation of this theorem requires further technical conditions on the potential V (see Franzosi *et al.* (2007) for details).

9.3.6 Flat stationary points may give rise to phase transitions

Although topology changes are necessary for a phase transition to occur, their presence is by no means sufficient. This is evident from Example 9.11, where topology changes were found to lie densely in an interval of the potential-energy (per degree of freedom) axis, but a phase transition takes place at the single value of $v = \Delta$ only. To better understand the microscopic origin of a phase transition, we might now like to ask the following question:

Question 9.13. *Which of the stationary points of V may give rise to a nonanalyticity of the microcanonical entropy $s(u)$ in the thermodynamic limit?*

An answer to this question was given by Kastner and Schnetz (2008). The idea behind this result is to sum the nonanalytic contributions to the entropy as given in eqns (9.27) and (9.28) for all stationary points and to take the thermodynamic limit of this sum. Then a bound on the magnitude of this sum is derived, which can be interpreted in the following way:

Sloppy theorem 9.14. *The sum of the nonanalytic contributions of the stationary points to the entropy cannot induce a phase transition at the potential energy per particle $v = v_t$ if, in a neighborhood of v_t, the following are true:*

1. *The number of critical points is bounded by* $\exp(CN)$ *for some* $C > 0$.
2. *The stationary points do not become "asymptotically flat" in the thermodynamic limit.*

"Asymptotically flat" here means that the determinant of the Hessian of V at the stationary points goes to zero in the thermodynamic limit in some suitable sense (see Kastner and Schnetz (2008) for details). This result qualifies a subset of all stationary points of V as "harmless" as regards phase transitions and leaves only the (hopefully few!) asymptotically flat ones as candidates for being the origin of a phase transition. To illustrate the power of this theorem, we reconsider the mean-field k-trigonometric model of Example 9.11.

Example 9.15. Without going into the details, we consider a function $j_\ell(v)$ which can be computed from the stationary points of V and has the property of being divergent whenever stationary points with stationary value v become asymptotically flat in the thermodynamic limit, and finite otherwise (see Kastner *et al.* (2008) for details). For the mean-field k-trigonometric model (9.29), a plot of this function is shown in Fig. 9.3 (right). In agreement with Theorem 9.14, j_ℓ is finite for $k = 1$, where no phase transition takes place, but shows a divergence at the transition potential energy $v = \Delta$ for $k \geqslant 2$. Despite the superficial similarity of j_ℓ and the Euler characteristic in Fig. 9.3, it is worth pointing out that, by virtue of Theorem 9.14, j_ℓ has a predictive power with respect to the occurrence of phase transitions which the Euler characteristic lacks. ■

With the results of the present section we have finally established conditions on the *microscopic level*, i.e. local properties in configuration space, which are relevant to the occurrence of a phase transition on the *macroscopic level*. We may consider such conditions as a way of better understanding the origin of a phase transition.

9.4 Conclusions and outlook

9.4.1 Conclusions

In Section 9.3, we have studied nonanalyticities of the microcanonical entropy of finite and infinite systems, and, in particular, their relation to topology changes in configuration space (or, equivalently, to stationary points of the potential V). In summary, we have seen the following:

1. Stationary points of V (or topology changes of M_v) cause *nonanalytic points* of order $\lfloor (N-3)/2 \rfloor$ in the entropy of *finite* systems; this result shows a pronounced difference from the analytic behavior of the canonical free-energy density of finite systems.
2. The number of stationary points is believed to generically increase exponentially with N.
3. Owing to the large number of finite-system nonanalyticities, their relation to phase transitions is clearly not one-to-one, but has to be somewhat more subtle.
4. Stationary points of V are *necessary* for a phase transition to occur in a short-range system.

5. Stationary points of *asymptotically vanishing curvature* may cause a phase transition in the thermodynamic limit.

At this stage it is instructive to return to the considerations of Section 9.2.5, where we have argued that, for short-range interacting systems in the thermodynamic limit, nonanalyticities of the entropy necessarily have to be generated when we switch over from the microscopic to the macroscopic level of description. Therefore we expected to find a precursor of this behavior on the microscopic level. This is in contrast to the case of long-range interactions, where, owing to the possibility of generating nonanalyticities entirely on the macroscopic level, the presence of such a precursor is possible but not necessary.

Remarkably, the results of Section 9.3 reflect these presuppositions very well: by virtue of Theorem 9.12, stationary points of the potential V have been shown to be necessary for a phase transition only in short-range interacting systems. In fact, certain long-range systems (such as the mean-field φ^4 model of Example 9.6) have been shown to display a phase transition also in the absence of stationary points of V (Baroni 2002; Garanin *et al.* 2004; Hahn and Kastner 2005). Still, although not necessary, an asymptotically flat stationary point *may* be the origin of a phase transition in a long-range system, as observed for the mean-field k-trigonometric model in Example 9.15.

Note that, in the case of concave entropy functions in the thermodynamic limit, nonanalyticities of the microcanonical entropy always correspond to nonanalyticities of the canonical free-energy density $\bar{f}$, and therefore to phase transitions according to the standard Definition 9.3.

9.4.2 Quantum outlook

The concepts of topology changes of the configuration space subsets M_v and of stationary points of the potential-energy function V are of purely classical nature. However, *geometric quantum mechanics*[5] (Kibble (1979); see Ashtekar and Schilling (1999) for an introduction) provides a suitable framework to take over all the key features of these concepts to quantum mechanics. In this geometric framework, starting from a Hamiltonian operator $\hat{H}$ on Hilbert space $\mathcal{H}$, an energy expectation value function h is defined on the quantum phase space $\mathcal{P}(\mathcal{H})$, i.e. on the *complex projective space* corresponding to $\mathcal{H}$. The function h can be shown to have a number of stationary points which, remarkably, correspond to the eigenstates of the operator $\hat{H}$. Analogously to the classical-mechanical case, the constant-energy subsets of quantum phase space $\mathcal{P}(\mathcal{H})$ change topology precisely at the stationary values of h. Further elaboration and application of these concepts in a quantum context is currently in progress.

9.4.3 Outlook on applications

Up to now we have mostly emphasized the conceptual aspects of an analysis of configuration space topology or, equivalently, stationary points of the potential energy V: we have gained insights into the origin of phase transitions and have found differences in

[5]Note that geometric quantum mechanics is not a different theory, but a reformulation of standard quantum mechanics in symplectic geometric language.

how nonanalyticities are generated in short-range and long-range systems. Additionally, an analysis of the stationary points of V is of interest in applications as well. For dynamical properties, as mentioned in Section 9.3.2, such energy landscape methods have already been used extensively. Together with the link between stationary points and statistical-physical properties presented in this lecture, it might be promising to investigate both the dynamical and the statistical properties of a given system simultaneously on the basis of stationary points of the potential-energy landscape. Such an approach should prove particularly useful for the study of protein folding and the glass transition, where both dynamical and statistical features play a role.

References

Angelani L., Casetti L., Pettini M., Ruocco G., and Zamponi F. (2003). Topological signature of first-order phase transitions in a mean-field model. *Europhys. Lett.*, **62**, 775–781.

Ashtekar A. and Schilling T. A. (1999). Geometrical formulation of quantum mechanics. In *On Einstein's Path: Essays in Honor of Engelbert Schucking* (ed. A. Harvey), pp. 23–65. Springer.

Baroni F. (2002). Transizioni di fase e topologia dello spazio delle configurazioni di modelli di campo medio. Master's thesis, Università degli Studi di Firenze.

Binney J. J., Dowrick N. J., Fisher A. J., and Newman, M. E. J. (1992). *The Theory of Critical Phenomena: An Introduction to the Renormalization Group.* Clarendon Press.

Campa A., Ruffo S., and Touchette H. (2007). Negative magnetic susceptibility and nonequivalent ensembles for the mean-field φ^4 spin model. *Physica A*, **385**, 233–248.

Demazure M. (2000). *Bifurcations and Catastrophes: Geometry of Solutions to Nonlinear Problems.* Springer.

Doye J. P. K. and Wales D. J. (2002). Saddle points and dynamics of Lennard-Jones clusters, solids, and supercooled liquids. *J. Chem. Phys.*, **116**, 3777–3788.

Ellis R. S., Haven K., and Turkington B. (2000). Large deviation principles and complete equivalence and nonequivalence results for pure and mixed ensembles. *J. Stat. Phys.*, **101**, 999–1064.

Franzosi R. and Pettini M. (2004). Theorem on the origin of phase transitions. *Phys. Rev. Lett.*, **92**, 060601.

Franzosi R., Pettini M., and Spinelli L. (2007). Topology and phase transitions I. Preliminary results. *Nucl. Phys. B*, **782**, 189–218.

Garanin D. A., Schilling R., and Scala A. (2004). Saddle index properties, singular topology, and its relation to thermodynamic singularities for a ϕ^4 mean-field model. *Phys. Rev. E*, **70**, 036125.

Georgii H.-O. (1988). *Gibbs Measures and Phase Transitions*, de Gruyter Studies in Mathematics 9. de Gruyter.

Griffiths R. B. (1972). Rigorous results and theorems. In *Phase Transitions and Critical Phenomena* (ed. C. Domb and M. S. Green), Volume 1, pp. 7–109. Academic Press.

Hahn I. and Kastner M. (2005). The mean-field φ^4 model: Entropy, analyticity, and configuration space topology. *Phys. Rev. E*, **72**, 056134.

Hahn I. and Kastner M. (2006). Application of large deviation theory to the mean-field φ^4-model. *Eur. Phys. J. B*, **50**, 311–314.

Kadanoff L. P. (1966). Scaling laws for Ising models near T_c. *Physics*, **2**, 263–272.

Kastner M. (2002). Existence and order of the phase transition of the Ising model with fixed magnetization. *J. Stat. Phys.*, **109**, 133–142.

Kastner M. (2008). Phase transitions and configuration space topology. *Rev. Mod. Phys.*, **80**, 167–187.

Kastner M. and Schnetz O. (2008). Phase transitions induced by saddle points of vanishing curvature. *Phys. Rev. Lett.*, **100**, 160601.

Kastner M., Schreiber S., and Schnetz O. (2007). Phase transitions from saddles of the potential energy landscape. *Phys. Rev. Lett.*, **99**, 050601.

Kastner M., Schnetz O., and Schreiber S. (2008). Nonanalyticities of the entropy induced by saddle points of the potential energy landscape. *J. Stat. Mech. Theory Exp.*, **2008**, P04025.

Kibble T. W. B. (1979). Geometrization of quantum mechanics. *Commun. Math. Phys.*, **65**, 189–201.

Lanford O. E. (1973). Entropy and equilibrium states in classical statistical mechanics. In *Statistical Mechanics and Mathematical Problems* (ed. A. Lenard), Lecture Notes in Physics 20, pp. 1–113. Springer.

Lebowitz J. L. (1999). Statistical mechanics: A selective review of two central issues. *Rev. Mod. Phys.*, **71**, S346–S357.

Matsumoto Y. (2002). *An Introduction to Morse Theory.* Translations of Mathematical Monographs 208. American Mathematical Society.

Milnor J. (1963). *Morse Theory.* Annals of Mathematical Studies 51. Princeton University Press.

Ruelle D. (1969). *Statistical Mechanics: Rigorous Results.* Benjamin.

Touchette H. (2008). Simple spin models with non-concave entropies. *Am. J. Phys.*, **76**, 26–30.

Wales D. J. (2004). *Energy Landscapes.* Cambridge University Press.

Wilson K. G. (1971). Renormalization group and critical phenomena. I. Renormalization group and the Kadanoff scaling picture. *Phys. Rev. B*, **4**, 3174–3183.

Part IV

Gravitational interaction

10
Statistical mechanics of gravitating systems: An overview

T. PADMANABHAN

Inter-University Centre for Astronomy and Astrophysics,
Post Bag 4, Ganeshkhind,
Pune-411 007, India
email: nabhan@iucaa.ernet.in

10.1 Overview of the key issues and results

The statistical mechanics of systems dominated by gravity has close connections with the areas of condensed matter physics, fluid mechanics, renormalization groups, etc. and poses an incredible challenge as regards the basic formulation. The ideas of this field also find application in many different areas of astrophysics and cosmology, especially in the study of globular clusters, galaxies, and gravitational clustering in the expanding universe. (For an overall review of the statistical mechanics of gravitating systems, see Padmanabhan 1990, 2000*a*, 2002*a* and Binney and Tremaine 1987; reviews of gravitational clustering in an expanding background are available in Padmanabhan 2000*b* and in several textbooks on cosmology (Peebles 1993; Padmanabhan 1993, 1996*a*, 2002*d*); for a sample of different attempts to understand these phenomena by different groups, see Chavanis 2002, de Vega *et al.* 1998, de Vega and Siebert 2002, Valageas 2001*a*,*b*, Follana and Laliena 2000, Bottaccio *et al.* 2002, Scoccimarro 2000, and the references cited therein.) It will be useful to begin with a broad overview and a description of the issues which will be addressed in this article.

In Newtonian theory, the gravitational force can be described as the gradient of a scalar potential and the evolution of a set of particles under the action of gravitational forces can be described by the equations

$$\ddot{\mathbf{x}}_i = -\nabla\phi(\mathbf{x}_i, t)\,, \quad \nabla^2\phi = 4\pi G\sum_i m_i\delta_D(\mathbf{x}-\mathbf{x}_i)\,, \tag{10.1}$$

where $\mathbf{x}_i$ is the position of the ith particle and m_i is its mass. (Obviously, the existence of point particles, indicated by the Dirac delta function, is an idealization; but this does not affect anything once we ignore self-interaction.) For an isolated system with a sufficiently large number of particles, it is useful to investigate whether some kind of statistical description of such a system is possible. Such a description, however, is complicated by the long-range, unscreened nature of the gravitational force. If a self-gravitating system is divided into two parts, the total energy of the system cannot be expressed as the sum of the gravitational energy of the components. The conventional results in statistical physics are based on the extensivity of the energy, which is clearly invalid for gravitating systems. To construct a statistical description of such a system, one must begin with the construction of the microcanonical ensemble describing such a system. If the Hamiltonian of the system is $H(p_i, q_i)$, then the volume $g(E)$ of the constant-energy surface $H(p_i, q_i) = E$ will be of primary importance in the microcanonical ensemble. The logarithm of this function will give the entropy $S(E) = \ln g(E)$ and the temperature of the system will be $T(E) \equiv \beta(E)^{-1} = (\partial S/\partial E)^{-1}$. [Of course, the fundamental ensemble to use for *any* statistical-mechanical system—say, even an ideal gas—is the microcanonical ensemble. In the case of systems without long-range interactions, for which extensivity of energy holds, this is just a pedantic issue. But in the case of gravitating systems, for example, this distinction has clear practical implications. Our discussion should be viewed against this backdrop.]

In the case of systems for which a description based on the canonical ensemble is possible, the Laplace transform of $g(E)$ with respect to the variable β will give the partition function $Z(\beta)$. It is, however, trivial to show that the gravitating systems of interest in astrophysics cannot be described by a canonical ensemble (Padmanabhan 1990, 2002*a*; Lynden-Bell 1998; Lynden-Bell and Lynden-Bell 1977). The virial theorem holds for such systems and we have $2K + U = 0$, where K and U are the total kinetic and potential energies of the system. This leads to $E = K + U = -K$; since the temperature of the system is proportional to the total kinetic energy, the specific heat will be negative: $C_V \equiv (\partial E/\partial T)_V \propto (\partial E/\partial K) < 0$. On the other hand, the specific heat of any system described by a canonical ensemble, $C_V = \beta^2 \langle (\Delta E)^2 \rangle$, must be positive definite. Thus one cannot describe self-gravitating systems of the kind we are interested in by a canonical ensemble. (The version of the virial theorem $2K + U = 0$ is rigorously true only for point particles without any confinement. Such a system, as we shall see, is pathological in any ensemble. The argument described above, however, holds even when we avoid this idealization. We shall say more about this later on.)

One may attempt to find the equilibrium configuration for self-gravitating systems by maximizing the entropy $S(E)$ or the phase volume $g(E)$. (In a more pedantic approach, one would make a distinction between variables such as entropy and temperature defined for an ensemble of systems and the corresponding quantities for a particular member of the ensemble. We will not do this, because it is of no relevance to our discussion.) It is again easy to show that no global maximum of the entropy exists for classical point particles interacting via Newtonian gravity. To prove this, we only need to construct a configuration with arbitrarily high entropy, which can be achieved as follows. Consider a system of N particles initially occupying a region of finite volume in phase space and with total energy E. We now move a small number of these particles (in fact, a pair of them; particles 1 and 2, say, will do) arbitrarily close to each other. The potential energy of interaction of these two particles, $-Gm_1m_2/r_{12}$, will become arbitrarily high as $r_{12} \to 0$. By transferring some of this energy to the rest of the particles, we can increase their kinetic energy without limit. This will clearly increase the phase volume occupied by the system without bound. This argument can be made more formal by dividing the original system into a small, compact core and a large, diffuse halo and allowing the core to collapse and transfer the energy to the halo.

The absence of a global maximum for the entropy—as argued above—depends on the idealization that there is no short-distance cutoff in the interaction of the particles, so that we could take the limit $r_{12} \to 0$. If we assume, instead, that each particle has a minimum radius a, then the typical lower bound on the gravitational potential energy contributed by a pair of particles will be $-Gm_1m_2/2a$. This will put an upper bound on the amount of energy that can be made available to the rest of the system.

We have also assumed that part of the system can expand without limit—in the sense that any particle with sufficiently large energy can move to arbitrarily large distances. In real life, no system is completely isolated, and eventually one has to assume that a meandering particle is better treated as a member of another system. One way of obtaining a truly isolated system is to confine the system inside a spherical region of radius R with, say, a reflecting wall. (Most of our discussion is confined to three dimensions, and the situation is different in two dimensions; see e.g. Engineer *et al.* (1999) and Padmanabhan and Kanekar (2000).)

The two cutoffs a and R will make the upper bound on the entropy finite, but even with the two cutoffs, the primary nature of the gravitational instability cannot be avoided. The basic phenomenon described above (namely, the formation of a compact core and a diffuse halo) will still occur since this is the direction of increasing entropy. Particles in the hot, diffuse component will permeate the entire spherical cavity, bouncing off the walls and having a kinetic energy which is significantly larger than the potential energy. The compact core will exist as a gravitationally bound system with very little kinetic energy. A more formal way of understanding this phenomenon is based on the virial theorem for a system with a short-distance cutoff confined to a sphere of volume V. In this case, the virial theorem will read as (Padmanabhan 2000*a*)

$$2T + U = 3PV + \Phi\,, \tag{10.2}$$

where P is the pressure on the walls and Φ is the correction to the potential energy arising from the short-distance cutoff. This equation can be satisfied in essentially three different ways. If T and U are significantly higher than $3PV$, then we have $2T+U \approx 0$, which describes a self-gravitating system in standard virial equilibrium but not in the state of maximum entropy. If $T \gg U$ and $3PV \gg \Phi$, one can have $2T \approx 3PV$, which describes an ideal gas with no potential energy confined to a container of volume V; this will describe the hot, diffuse component at late times. If $T \ll U$ and $3PV \ll \Phi$, then one can have $U \approx \Phi$, describing the compact potential-energy-dominated core at late times. In general, the evolution of the system will lead to the production of a core and a halo and each component will satisfy the virial theorem in the form (10.2). Such an asymptotic state with two distinct phases (Aaronson and Hansen 1972; Padmanabhan 1989*a*; Padmanabhan and Narasimha, unpublished work) is quite different from what would have been expected for systems with only short-range interaction. Considering its importance, I shall briefly describe in Section 10.2 a toy model which captures the essential physics of the above system in an exactly solvable context.

The above discussion focused on the existence of global maximum of the entropy and we proved that it does not exist in the absence of two cutoffs. It is, however, possible to have *local* extrema of the entropy which are not global maxima. Intuitively, one would have expected the distribution of matter in a configuration which is a local extremum of the entropy to be described by a Boltzmann distribution, with the density given by $\rho(\mathbf{x}) \propto \exp[-\beta\phi(\mathbf{x})]$, where ϕ is the gravitational potential related to ρ by Poisson's equation. This is indeed true; for a formal proof, see Padmanabhan (1990, 2002*a*). This configuration is usually called the isothermal sphere (because it can be shown that, among all solutions to this equation, the one with spherical symmetry maximizes the entropy) and it is a local maximum of the entropy. The second (functional) derivative of the entropy with respect to the configuration variables will determine whether the local extremum of the entropy is a local maximum or a saddle point (Antonov 1962; Padmanabhan 1989*b*).

The relevance of the long-range of gravity in all the above phenomena can be understood by studying model systems with an attractive potential varying as $r^{-\alpha}$ with different values for α. Such studies confirm the results and interpretation given above (see Ispolatov and Cohen (2001) and references cited therein).

Let us now consider the situation in the context of an expanding background. There is considerable amount of observational evidence to suggest that one of the dominant energy densities in the universe is contributed by self-gravitating point particles. The smooth average energy density of these particles drives the expansion of the universe, while any small deviation from the homogeneous energy density will cluster gravitationally. (For a review of cosmology from a contemporary perspective, see e.g. Padmanabhan (2005*b*, 2006*a*,*b*, 2008).) One of the central problems in cosmology is to describe the nonlinear phases of this gravitational clustering starting from a initial spectrum of density fluctuations. It is often enough (and necessary) to use a statistical description and to relate different statistical indicators (such as the power spectra and nth-order correlation functions) of the resulting density distribution to the statistical parameters (usually the power spectrum) of the initial distribution. The relevant scales at which gravitational clustering is nonlinear are less than about 10 Mpc (where 1 Mpc $= 3 \times 10^{24}$ cm is the typical separation between galaxies in the universe), while the expansion of the universe has a characteristic scale of about few thousand Mpc. Hence, nonlinear gravitational clustering in an expanding universe can be adequately described by Newtonian gravity provided the rescaling of lengths due to the background expansion is taken into account. This is easily done by introducing a *proper* coordinate for the ith particle $\mathbf{r}_i$, related to the *comoving* coordinate $\mathbf{x}_i$ by $\mathbf{r}_i = a(t)\mathbf{x}_i$, with $a(t)$ describing the stretching of length scales due to cosmic expansion. The Newtonian dynamics works with the proper coordinates $\mathbf{r}_i$, which can be translated to the behavior of the comoving coordinates $\mathbf{x}_i$ by this rescaling. (This implies that, for all practical purposes, we are still in the domain of Newtonian gravity. There is a far deeper connection between thermodynamics and gravity (Padmanabhan 2002*b*,*c*, 2005*a*; Paranjape *et al.* 2006) in the general-relativistic domain, which we will not discuss in these lectures.)

As is to be expected, cosmological expansion completely changes the nature of the problem because of several new factors which come in. (a) The problem has now become time-dependent and it is pointless to look for equilibrium solutions in the conventional sense of the word. (b) On the other hand, the expansion of the universe has a “civilizing” influence on the particles and acts counter to the tendency of gravity to make systems unstable. (c) In any small local region of the universe, one would assume that the conclusions describing a finite gravitating system will still hold true approximately. In that case, particles in any small subregion will be driven towards configurations of local extrema of the entropy (say, isothermal spheres) and towards global maxima of the entropy (say, core–halo configurations).

An extra feature comes into play as regards the expanding halo from any subregion. The expansion of the universe acts as a damping term in the equations of motion and drains the particles of their kinetic energy—this is essentially the lowering of the temperature of any system participating in cosmic expansion. This, in turn, helps gravitational clustering since the potential wells of nearby subregions can capture particles in the expanding halo of one region when the kinetic energy of the expanding halo has been sufficiently reduced.

The actual behavior of the system will, of course, depend on the form of $a(t)$. However, for understanding the nature of clustering, one can take $a(t) \propto t^{2/3}$, which describes a matter-dominated universe with the critical density. Such a power law has the advantage that there is no intrinsic scale in the problem. Since the Newtonian gravitational force is also scale-free, one would expect some scaling relations to exist in the pattern of gravitational clustering. Converting this intuitive idea into a concrete mathematical statement turns out to be nontrivial and difficult.

It is clear that cosmological expansion introduces several new factors into the problem when compared with the study of the statistical mechanics of isolated gravitating systems. (For a general review of the statistical mechanics of gravitating systems, see Padmanabhan (1990, 2002*a*). For a sample of different approaches, see Follana and Laliena (2000), Cooray and Sheth (2002), Scoccimarro and Frieman (1996), de Vega *et al.* (1998), Padmanabhan (1989*b*), Buchert and Dominguez (2005), and the references cited therein. Review of gravitational clustering in an expanding background is also available in several textbooks on cosmology (Peebles 1993; Padmanabhan 1993, 1996*a*, 2002*d*; Peebles 1980).) Though this problem can be tackled in a "practical" manner using high-resolution numerical simulations (for a review, see Bagla (2004) and Bagla and Padmanabhan (1997*b*)), such an approach hides the physical principles which govern the behavior of the system. To understand the physics, it is necessary to attack the problem from several directions using analytic and semianalytic methods. Several such attempts exist in the literature, based on Zeldovich(like) approximations (Zeldovich 1970; Gurbatov *et al.* 1989; Brainerd *et al.* 1993; Matarrese *et al.* 1992; Bagla and Padmanabhan 1994; Padmanabhan and Engineer 1998; Padmanabhan and Kanekar 2000; Engineer *et al.* 2000; Tatekawa 2004), path integral and perturbative techniques (Buchert 1994; Valageas 2001*a*,*b*), nonlinear scaling relations (Hamilton *et al.* 1991; Padmanabhan *et al.* 1996; Munshi *et al.* 1997; Bagla *et al.* 1998; Kanekar *et al.* 2001; Nityananda and Padmanabhan 1994; Padmanabhan 1996*b*; Ray *et al.* 2005), and many other techniques. In spite of all these, it is probably fair to say that we still do not have a clear analytic grasp of this problem, mainly because each of these approximations has a different domain of validity and does not arise from a central paradigm.

I propose to attack the problem from a different angle, which has not received much attention in the past. The approach begins from the dynamical equation for the density contrast in Fourier space and casts it as an integro-differential equation. Though this equation is known in the literature (see e.g. Peebles 1980), it has received very little attention because it is not "closed" mathematically; that is, it involves variables which are not natural to the formalism and thus further progress is difficult. I will, however, argue that there exists a natural closure condition for this equation based on a Zeldovich approximation, thereby allowing us to write down a *closed* integro-differential equation for the gravitational potential in Fourier space.

It turns out that this equation can form the basis for several further investigations, some of which are described in Padmanabhan (2002*a*, 2005*c*, 2006*c*). Here I will concentrate on just two specific features, centered around the following issues:

- If the initial power spectrum is sharply peaked in a narrow band of wavelengths, how does the evolution transfer the power to other scales? In particular, does the nonlinear evolution in the case of gravitational interactions lead to a universal power spectrum (like the Kolmogorov spectrum in fluid turbulence)?

- What is the nature of the time evolution at late stages? Does the gravitational clustering at late stages wipe out the memory of initial conditions and evolve in a universal manner?

A fair amount of progress can be made as regards these questions using the integro-differential equation mentioned above, and some of these aspects will be discussed in detail.

10.2 Phases of a self-gravitating system

As described in Section 10.1, the statistical mechanics of finite, self-gravitating systems has the following characteristic features. (a) These systems exhibit negative specific heat while in virial equilibrium. (b) They are inherently unstable to the formation of a core–halo structure, and a global maximum of the entropy does not exist without cutoffs at short and large distances. (c) They can be broadly characterized by two phases, one of which is compact and dominated by potential energy while the other is diffuse and behaves more or less like an ideal gas. The purpose of this section is to describe a simple toy model which exhibits all these features and mimics a self-gravitating system (Padmanabhan 1990, 2002*a*).

Consider a system with two particles described by a Hamiltonian of the form

$$H(\mathbf{P}, \mathbf{Q}; \mathbf{p}, \mathbf{r}) = \frac{\mathbf{P}^2}{2M} + \frac{\mathbf{p}^2}{2\mu} - \frac{Gm^2}{r}, \tag{10.3}$$

where $(\mathbf{Q}, \mathbf{P})$ are the coordinates and momenta of the center of mass, $(\mathbf{r}, \mathbf{p})$ are the relative coordinates and momenta, $M = 2m$ is the total mass, $\mu = m/2$ is the reduced mass, and m is the mass of the individual particles. This system may be thought of as consisting of two particles (each of mass m) interacting via gravity. We shall assume that the quantity r varies in the interval (a, R). This is equivalent to assuming that the particles are hard spheres of radius $a/2$ and that the system is confined to a spherical box of radius R. We will study the "statistical mechanics" of this simple toy model. It may appear strange that we are attempting a statistical description of a system with $N = 2$ particles when usually thermodynamics works in the $N \to \infty$ limit. This issue, however, is actually irrelevant. As we have stressed above, given a particular Hamiltonian one can construct the microcanonical ensemble and—when the Laplace transform of the density of states exists—a canonical ensemble. The real question is not whether it is possible but whether it is of any use. As we shall see, in this case, it *is* and this simple system actually acts as a good proxy for much more complex gravitating many-body systems.

To do this, we start with the volume $g(E)$ of the constant-energy surface $H = E$. Straightforward calculation gives

$$g(E) = AR^3 \int_a^{r_{\max}} r^2\, dr \left[E + \frac{Gm^2}{r}\right]^2, \tag{10.4}$$

where $A = 64\pi^5 m^3/3$. The range of integration in eqn (10.4) should be limited to the region in which the expression in the square brackets is positive. So we should use

$r_{\rm max} = Gm^2/|E|$ if $-Gm^2/a < E < -Gm^2/R$, and $r_{\rm max} = R$ if $-Gm^2/R < E < +\infty$. Since $H \geq -Gm^2/a$, we trivially have $g(E) = 0$ for $E < -Gm^2/a$. The constant A is unimportant for our discussions and hence will be omitted from the formulae hereafter. The integration in eqn (10.4) gives the following result:

$$\frac{g(E)}{(Gm^2)^3} = \begin{cases} \dfrac{R^3}{3}(-E)^{-1}\left(1+\dfrac{aE}{Gm^2}\right)^3, & -Gm^2/a < E < -Gm^2/R, \\ \dfrac{R^3}{3}(-E)^{-1}\left[\left(1+\dfrac{RE}{Gm^2}\right)^3 - \left(1+\dfrac{aE}{Gm^2}\right)^3\right], & -Gm^2/R < E < \infty. \end{cases} \tag{10.5}$$

This function $g(E)$ is continuous and smooth at $E = -Gm^2/R$. We define the entropy $S(E)$ and the temperature $T(E)$ of the system by the relations

$$S(E) = \ln g(E), \quad T^{-1}(E) = \beta(E) = \frac{\partial S(E)}{\partial E}. \tag{10.6}$$

All the interesting thermodynamic properties of the system can be understood from the $T(E)$ curve.

Consider first the case of low energies with $-Gm^2/a < E < -Gm^2/R$. Using eqns (10.5) and (10.6), one can easily obtain $T(E)$ and write it in dimensionless form as

$$t(\epsilon) = \left[\frac{3}{1+\epsilon} - \frac{1}{\epsilon}\right]^{-1}, \tag{10.7}$$

where we have defined $t = aT/Gm^2$ and $\epsilon = aE/Gm^2$.

This function exhibits the peculiarities characteristic of gravitating systems. At the lowest energy admissible for our system, which corresponds to $\epsilon = -1$, the temperature t vanishes. This describes a tightly bound, low-temperature phase of the system with negligible random motion. The function $t(\epsilon)$ is clearly dominated by the first term of eqn (10.7) for $\epsilon \simeq -1$. As we increase the energy of the system, the temperature *increases*, which is the normal behavior for a system. This trend continues up to

$$\epsilon = \epsilon_1 = -\frac{1}{2}(\sqrt{3}-1) \simeq -0.36, \tag{10.8}$$

at which point the $t(\epsilon)$ curve reaches a maximum and turns around. As we increase the energy further, the temperature *decreases*. The system *exhibits negative specific heat in this range.*

Equation (10.7) is valid from the minimum energy $-Gm^2/a$ all the way up to the energy $-Gm^2/R$. For realistic systems, $R \gg a$ and hence this range is quite wide. For a small region in this range (from $-Gm^2/a$ to $-0.36Gm^2/a$), we have a positive specific heat; for the rest of the region, the specific heat is negative. *The positive-specific-heat region owes its existence to the nonzero short-distance cutoff.* If we set $a = 0$, the first term in eqn (10.7) will vanish; we will have $t \propto (-\epsilon^{-1})$ and negative specific heat in this entire domain.

For $E \geq -Gm^2/R$, we have to use the second expression in eqn (10.5) for $g(E)$. In this case, we get

$$t(\epsilon) = \left[\frac{3\left[(1+\epsilon)^2 - (R/a)(1+(R/a)\epsilon)^2\right]}{(1+\epsilon)^3 - (1+(R/a)\epsilon)^3} - \frac{1}{\epsilon} \right]^{-1} . \tag{10.9}$$

This function, of course, matches smoothly with eqn (10.7) at $\epsilon = -a/R$. As we increase the energy, the temperature continues to decrease for a little while, exhibiting negative specific heat. However, this behavior is soon halted at some $\epsilon = \epsilon_2$, say. The $t(\epsilon)$ curve reaches a minimum at this point, turns around, and starts increasing with increasing ϵ. We thus enter another (high-temperature) phase with positive specific heat. From eqn (10.9) it is clear that $t \simeq (1/2)\epsilon$ for large ϵ. (Since $E = (3/2)NkT$ for an ideal gas, we might have expected to find $t \simeq (1/3)\epsilon$ for our system with $N = 2$ at high temperatures. This is indeed what we would have found if we had defined our entropy as the logarithm of the volume of the phase space with $H \leq E$. With our definition, the energy of the ideal gas is actually $E = [(3/2)N - 1]kT$; hence we get $t = (1/2)\epsilon$ when $N = 2$.) The form of $t(\epsilon)$ for $a/R = 10^{-4}$ is shown in Fig. 10.1 by the dashed curve. The specific heat is positive along the portions AB and CD and is negative along BC.

The overall picture is now clear. Our system has two natural energy scales, $E_1 = -Gm^2/a$ and $E_2 = -Gm^2/R$. For $E \gg E_2$, gravity is not strong enough to keep $r < R$ and the system behaves like a gas confined by the container; we have a high-temperature phase with positive specific heat. As we lower the energy to $E \simeq E_2$, the effects of gravity begin to be felt. For $E_1 < E < E_2$, the system is unaffected by either the box or the short-distance cutoff; this is the domain dominated entirely by gravity and we have negative specific heat. As we go to $E \simeq E_1$, the hard-core nature of the particles begins to be felt and the gravity is again resisted. This gives rise to a low-temperature phase with positive specific heat.

We can also consider the effect of increasing R, keeping a and E fixed. Since we imagine the particles to be hard spheres of radius $a/2$, we should consider only $R > 2a$. It is amusing to note that if $2 < R/a < \sqrt{3} + 1$, there is no region of negative specific heat. As we increase R, the negative-specific-heat region appears and increasing R increases the range over which the specific heat is negative. Suppose a system is originally prepared with some E and R values such that the specific heat is positive. If we now increase R, the system may find itself in a region of negative specific heat. *This suggests the possibility that an instability may be triggered in a constant-energy system if its radius increases beyond a critical value.* We will see later that this is indeed true.

Since systems described by the canonical distribution cannot exhibit negative specific heat, it follows that the canonical distribution will lead to a very different physical picture for this range of (mean) energies $E_1 < E < E_2$. It is, therefore, of interest to look at our system from the point of view of the canonical distribution by computing the partition function. In the partition function

$$Z(\beta) = \int d^3P\, d^3p\, d^3Q\, d^3r \exp(-\beta H) , \tag{10.10}$$

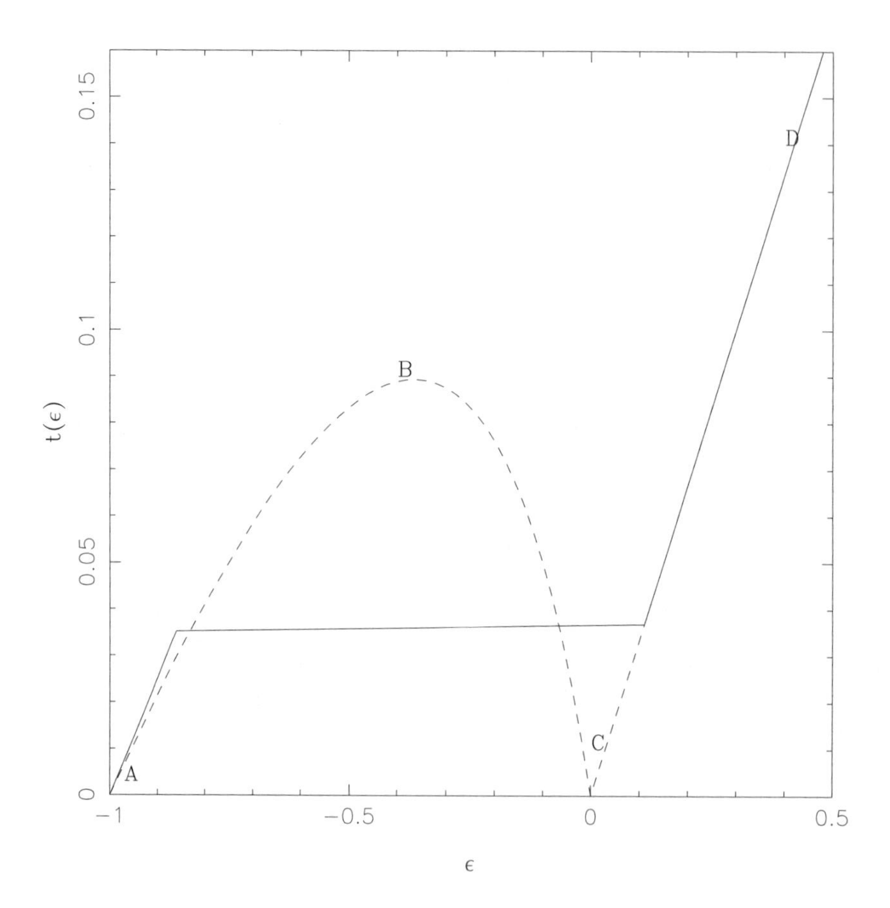

Fig. 10.1 The relation between temperature and energy for a model mimicking self gravitating systems. The dashed line is the result for the microcanonical ensemble and the solid line is for the canonical ensemble. The negative-specific-heat region, BC, in the microcanonical description is replaced by a phase transition in the canonical description. See text for more details.

the integrations over P, p, and Q can be performed trivially. Omitting an overall constant which is unimportant, we can write the answer in dimensionless form as

$$Z(t) = t^3 \left(\frac{R}{a}\right)^3 \int_1^{R/a} dx\, x^2 \exp\left(\frac{1}{xt}\right), \tag{10.11}$$

where t is the dimensionless temperature defined in eqn (10.7). Though this integral cannot be evaluated in closed form, all the limiting properties of $Z(\beta)$ can be easily obtained from eqn (10.11).

The integrand in eqn (10.11) is large for both large and small x and reaches a minimum for $x = x_m = 1/2t$. At high temperatures, $x_m < 1$ and hence the minimum falls outside the domain of integration. The exponential contributes very little to the integral and we can approximate Z adequately by

$$Z \approx t^3 \left(\frac{R}{a}\right)^3 \int_1^{R/a} dx\, x^2 \left[1 + \frac{2x_m}{x}\right] = \frac{t^3}{3} \left(\frac{R}{a}\right)^6 \left(1 + \frac{3a}{2Rt}\right). \tag{10.12}$$

On the other hand, if $x_m > 1$, the minimum lies between the limits of the integration and the exponential part of the curve dominates the integral. We can easily evaluate this contribution by a saddle point approach, and obtain

$$Z \approx \left(\frac{R}{a}\right)^3 t^4 (1-2t)^{-1} \exp\left(\frac{1}{t}\right). \tag{10.13}$$

As we lower the temperature, making x_m cross 1 from below, the contribution switches over from eqn (10.12) to eqn (10.13). The transition is exponentially sharp. The critical temperature at which the transition occurs can be estimated by finding the temperature at which the two contributions are equal. This occurs at

$$t_c = \frac{1}{3} \frac{1}{\ln(R/a)}. \tag{10.14}$$

For $t < t_c$, we should use eqn (10.13), and for $t > t_c$ we should use eqn (10.12).

Given $Z(\beta)$, all thermodynamic functions can be computed. In particular, the mean energy of the system is given by $E(\beta) = -(\partial \ln Z/\partial \beta)$. This relation can be inverted to give $T(E)$, which can be compared with the $T(E)$ obtained earlier using the microcanonical distribution. From eqns (10.12) and (10.13), we get

$$\epsilon(t) = \frac{aE}{Gm^2} = 4t - 1 \tag{10.15}$$

for $t < t_c$ and

$$\epsilon(t) = 3t - \frac{3a}{2R} \tag{10.16}$$

for $t > t_c$. Near $t \approx t_c$, there is a rapid variation of the energy and we cannot use either asymptotic form. The system undergoes a phase transition at $t = t_c$, absorbing a large amount of energy

$$\Delta\epsilon \approx \left(1 - \frac{1}{3\ln(R/a)}\right). \tag{10.17}$$

The specific heat is, of course, positive throughout the range. This is to be expected because the canonical ensemble cannot lead to negative specific heats.

The T–E curves obtained from the canonical (unbroken line) and microcanonical (dashed line) distributions are shown in Fig. 10.1. (For convenience, we have rescaled the T–E curve of the microcanonical distribution so that $\epsilon \simeq 3t$ asymptotically.) At both very low and very high temperatures, the canonical and microcanonical descriptions match. The crucial difference occurs at intermediate energies and temperatures. The microcanonical description predicts a negative specific heat and a reasonably slow variation of the energy with temperature. The canonical description, on the other hand, predicts a phase transition with a rapid variation of the energy with temperature. Such phase transitions are accompanied by large fluctuations in the energy, which is the main reason for the disagreement between the two descriptions (Padmanabhan 1990, 2002*a*; Lynden-Bell 1998; Lynden-Bell and Lynden-Bell 1977).

Numerical analysis of more realistic systems confirms all these features. Such systems exhibit a phase transition from the diffuse virialized phase to a core-dominated phase when the temperature is lowered below a critical value (Aaronson and Hansen 1972; Padmanabhan 1989*a*; Padmanabhan and Narasimha, unpublished work). The transition is very sharp and occurs at nearly constant temperature. The energy released by the formation of the compact core heats up the diffuse halo component.

10.3 Isothermal sphere

While a global maximum of the entropy does not exist in the absence of two cutoffs, it is, however, possible to have *local* extrema of the entropy which are not global maxima. Such a configuration is described by a Boltzmann distribution, with the density given by $\rho(\mathbf{x}) \propto \exp[-\beta\phi(\mathbf{x})]$, where ϕ is the gravitational potential related to ρ by Poisson's equation (for a formal proof see Padmanabhan (1990, 2002*a*)). Among all solutions to this equation, the one with spherical symmetry maximizes the entropy, and this configuration is usually called the isothermal sphere. The extremum condition for the entropy is equivalent to the following differential equation for the gravitational potential:

$$\nabla^2\phi = 4\pi G\rho_c e^{-\beta[\phi(\mathbf{x})-\phi(0)]} . \tag{10.18}$$

Given the solution to this equation, all other quantities can be determined. As we shall see, this system shows several peculiarities.

It is convenient to introduce length, mass, and energy scales by the definitions

$$L_0 \equiv (4\pi G\rho_c\beta)^{1/2} , \quad M_0 = 4\pi\rho_c L_0^3, \quad \phi_0 \equiv \beta^{-1} = \frac{GM_0}{L_0} , \tag{10.19}$$

where $\rho_c = \rho(0)$. All other physical variables can be expressed in terms of the dimensionless quantities

$$x \equiv \frac{r}{L_0}, \quad n \equiv \frac{\rho}{\rho_c}, \quad m = \frac{M(r)}{M_0}, \quad y \equiv \beta[\phi - \phi(0)] . \tag{10.20}$$

In terms of $y(x)$, the isothermal equation (10.18) becomes

$$\frac{1}{x^2}\frac{d}{dx}\left(x^2\frac{dy}{dx}\right) = \mathrm{e}^{-y} , \tag{10.21}$$

with the boundary condition $y(0) = y'(0) = 0$. Let us consider the nature of the solutions to this equation.

By direct substitution, we see that $n = 2/x^2$, $m = 2x$, $y = 2\ln x$ satisfies these equations. This solution, however, is singular at the origin and hence is not physically admissible. The importance of this solution lies in the fact that other (physically admissible) solutions tend to this solution (Padmanabhan 1990, 2002*a*; Chandrasekhar 1939) for large values of x. This asymptotic behavior of all solutions shows that the density decreases as $1/r^2$ for large r, implying that the mass contained inside a sphere of radius r increases as $M(r) \propto r$ at large r. To find physically useful solutions, it is necessary to assume that the solution is cut off at some radius R. For example, one may assume that the system is enclosed in a spherical box of radius R. In what follows, it will be assumed that the system has some cutoff radius R.

Equation (10.21) is invariant under the transformation $y \to y + a$, $x \to kx$ with $k^2 = \mathrm{e}^a$. This invariance implies that, given a solution with some value of $y(0)$, we can obtain a solution with any other value of $y(0)$ by simple rescaling. Therefore, only one of the two integration constants in eqn (10.21) is really nontrivial. Hence it must be possible to reduce the degree of the equation from two to one by a judicious choice of variables (Chandrasekhar 1939). One such set of variables is

$$v \equiv \frac{m}{x}, \quad u \equiv \frac{nx^3}{m} = \frac{nx^2}{v}. \tag{10.22}$$

In terms of v and u, eqn (10.18) becomes

$$\frac{u}{v}\frac{dv}{du} = -\frac{u-1}{u+v-3}. \tag{10.23}$$

The boundary conditions $y(0) = y'(0) = 0$ translate into the following: v is zero at $u = 3$, and $dv/du = -5/3$ at $(3, 0)$. The solution $v(u)$ has to be obtained numerically: it is plotted in Fig. 10.2 as the spiraling curve. The singular points of this differential equation are given by the intersection of the straight lines $u = 1$ and $u + v = 3$, on which the numerator and denominator of the right-hand side of eqn (10.23) vanish; that is, the singular point is at $u_s = 1$, $v_s = 2$, corresponding to the solution $n = 2/x^2$, $m = 2x$. It is obvious from the nature of the equations that the solutions will spiral around the singular point.

The nature of the solution shown in Fig. 10.2 allows us to put interesting bounds on some physical quantities, including the energy. To see this, we shall compute the total energy E of the isothermal sphere. The potential and kinetic energies are

$$\begin{aligned} U &= -\int_0^R \frac{GM(r)}{r}\frac{dM}{dr}dr = -\frac{GM_0^2}{L_0}\int_0^{x_0} mnx\,dx\,, \\ K &= \frac{3}{2}\frac{M}{\beta} = \frac{3}{2}\frac{GM_0^2}{L_0}m(x_0) = \frac{GM_0^2}{L_0}\frac{3}{2}\int_0^{x_0} nx^2\,dx\,, \end{aligned} \tag{10.24}$$

where $x_0 = R/L_0$. The total energy is, therefore,

$$\begin{aligned} E = K + U &= \frac{GM_0^2}{2L_0}\int_0^{x_0} dx\,(3nx^2 - 2mnx) \\ &= \frac{GM_0^2}{2L_0}\int_0^{x_0} dx\frac{d}{dx}\{2nx^3 - 3m\} = \frac{GM_0^2}{L_0}\left\{n_0x_0^3 - \frac{3}{2}m_0\right\}, \end{aligned} \tag{10.25}$$

where $n_0 = n(x_0)$ and $m_0 = m(x_0)$. The dimensionless quantity RE/GM^2 is given by

$$\lambda = \frac{RE}{GM^2} = \frac{1}{v_0}\left\{u_0 - \frac{3}{2}\right\}. \tag{10.26}$$

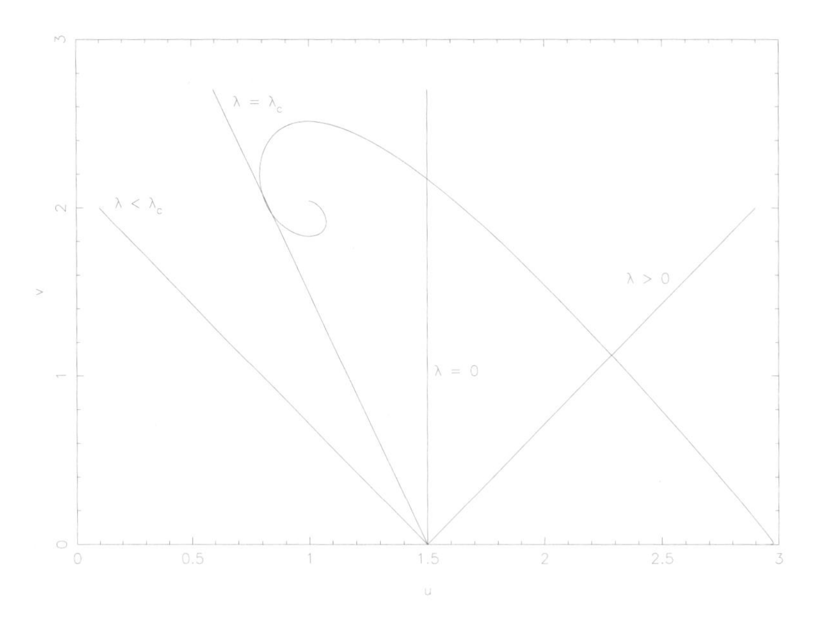

Fig. 10.2 Bound on RE/GM^2 for the isothermal sphere.

Note that the combination RE/GM^2 is a function of (u, v) alone. Let us now consider the constraints on λ. Suppose we specify some value for λ by specifying R, E, and M. Then such an isothermal sphere *must* lie on the curve

$$v = \frac{1}{\lambda}\left(u - \frac{3}{2}\right), \qquad \lambda \equiv \frac{RE}{GM^2}, \tag{10.27}$$

which is a straight line through the point $(1.5, 0)$ with a slope λ^{-1}. On the other hand, since *all* isothermal spheres must lie on the u–v curve, *an isothermal sphere can exist only if the line in eqn (10.27) intersects the u–v curve.*

For large positive λ (positive E), there is just one intersection. When $\lambda = 0$ (zero energy), we still have a unique isothermal sphere. (For $\lambda = 0$, eqn (10.27) is a vertical line through $u = 3/2$.) When λ is negative (negative E), the line can cut the u–v curve at more than one point; thus more than one isothermal sphere can exist with a given value of λ. (Of course, specifying M, R, E individually will remove this nonuniqueness.) But as we decrease λ (more and more negative E), the line in eqn (10.27) will slope more and more to the left; and when λ is smaller than a critical value λ_c, the intersection will cease to exist. *Thus no isothermal sphere can exist if RE/GM^2 is below a critical value λ_c.*[1] This fact follows immediately from the nature

[1]This approach to the critical value of λ is due to the author (Padmanabhan 1989*b*). Chandrasekhar (1939), who discussed the isothermal sphere in u–v coordinates, obtained the u–v curve but did not overlay lines of constant λ to obtain something like Fig. 10.2. If he had investigated the situation with confinement at finite R, he might have discovered the Antonov instability decades before Antonov did (Antonov 1962).

of the u–v curve and eqn (10.27). The value of λ_c can be found from the numerical solution shown in the figure. It turns out to be about -0.335.

The isothermal sphere has a special status as a solution to the mean-field equations. Isothermal spheres, however, cannot exist if $RE/GM^2 < -0.335$. Even when $RE/GM^2 > -0.335$, the isothermal solution need not be stable. The stability of this solution can be investigated by studying the second variation of the entropy. Such a detailed analysis shows that the following results are true (Antonov 1962; Lynden-Bell and Wood 1968; Padmanabhan 1989*b*). (i) Systems with $RE/GM^2 < -0.335$ cannot evolve into isothermal spheres. The entropy has no extremum for such systems. (ii) Systems with $RE/GM^2 > -0.335$ and $\rho(0) > 709\,\rho(R)$ can exist in a metastable (saddle point state) isothermal-sphere configuration. Here $\rho(0)$ and $\rho(R)$ denote the densities at the center and edge, respectively. The entropy extrema exist but they are not local maxima. (iii) Systems with $RE/GM^2 > -0.335$ and $\rho(0) < 709\,\rho(R)$ can form isothermal spheres which are a local maximum of the entropy.

10.4 An integral equation to describe nonlinear gravitational clustering

Let us next consider the gravitational clustering of a system of collisionless point particles *in an expanding universe*, which poses several challenging theoretical questions. Though the problem can be tackled in a "practical" manner using high-resolution numerical simulations, such an approach hides the physical principles which govern the behavior of the system. To understand the physics, it is necessary that we attack the problem from several directions using analytic and semianalytic methods. These sections will describe such attempts and will emphasize the semianalytic approach and outstanding issues, rather than more well-established results.

The expansion of the universe sets a natural length scale (called the Hubble radius) $d_H = c(\dot{a}/a)^{-1}$, which is about 4000 Mpc in the current universe. In any region which is small compared with $d_{\rm H}$, one can set up an unambiguous coordinate system in which the *proper* coordinate of a particle $\mathbf{r}(t) = a(t)\mathbf{x}(t)$ satisfies the Newtonian equation $\ddot{\mathbf{r}} = -\nabla_{\mathbf{r}}\Phi$, where Φ is the gravitational potential. The Lagrangian for such a system of particles is given by

$$L = \sum_i \left[\frac{1}{2} m_i \dot{\mathbf{r}}_i^2 + \frac{G}{2} \sum_j \frac{m_i m_j}{|\mathbf{r}_i - \mathbf{r}_j|} \right]. \tag{10.28}$$

In the term

$$\frac{1}{2}\dot{\mathbf{r}}_i^2 = \frac{1}{2}\left[a^2 \dot{\mathbf{x}}_i^2 + \dot{a}^2 \mathbf{x}_i^2 + a\dot{a}\frac{d\mathbf{x}_i^2}{dt} \right] = \frac{1}{2}\left[a^2 \dot{\mathbf{x}}_i^2 - a\ddot{a}\mathbf{x}_i^2 + \frac{d a\dot{a}\mathbf{x}_i^2}{dt} \right], \tag{10.29}$$

we note that (i) the total time derivative can be ignored, and (ii) using $\ddot{a} = -(4\pi G/3)\rho_b a$, the term $\Phi_{FRW} = -(1/2)a\ddot{a}\mathbf{x}_i^2 = (2\pi G/3)\rho_0(x_i^2/a)$ can be identified as the gravitational potential due to the uniform Friedmann background of density $\rho_b = \rho_0/a^3$. Hence the Lagrangian can be expressed as

$$L = \sum_i m_i \left[\frac{1}{2} a^2 \dot{\mathbf{x}}_i^2 + \phi(t, \mathbf{x}_i) \right] = T - U \,, \tag{10.30}$$

where

$$\phi = -\frac{G}{2a} \sum_j \frac{m_j}{|\mathbf{x}_i - \mathbf{x}_j|} - \frac{2\pi G \rho_0}{3} \frac{x_i^2}{a} \tag{10.31}$$

is the difference between the total potential and the potential for the background Friedmann universe Φ_{FRW}. By varying the Lagrangian in eqn (10.30) with respect to $\mathbf{x}_i$, we find the equation of motion to be

$$\ddot{\mathbf{x}} + 2\frac{\dot{a}}{a}\dot{\mathbf{x}} = -\frac{1}{a^2}\nabla_x \phi \,. \tag{10.32}$$

Since ϕ is the gravitational potential generated by the *perturbed* mass density, it satisfies an equation with a source $\rho - \rho_b \equiv \rho_b \delta$:

$$\nabla_x^2 \phi = 4\pi G \rho_b a^2 \delta \,. \tag{10.33}$$

Equations (10.32) and (10.33) govern the nonlinear gravitational clustering in an expanding background.

Usually one is interested in the evolution of the density contrast $\delta(t, \mathbf{x})$ rather than in the trajectories. Since the density contrast can be expressed in terms of the trajectories of the particles, it should be possible to write down a differential equation for $\delta(t, \mathbf{x})$ based on the equations for the trajectories $\mathbf{x}(t)$ derived above. It is, however, somewhat easier to write down an equation for $\delta_{\mathbf{k}}(t)$, which is the spatial Fourier transform of $\delta(t, \mathbf{x})$. To do this, we begin with the fact that the density $\rho(\mathbf{x}, t)$ due to a set of point particles, each of mass m, is given by

$$\rho(\mathbf{x}, t) = \frac{m}{a^3(t)} \sum_i \delta_D[\mathbf{x} - \mathbf{x}_i(t)] \,, \tag{10.34}$$

where $\mathbf{x}_i(t)$ is the trajectory of the ith particle and δ_D is the Dirac delta function. The density contrast $\delta(\mathbf{x}, t)$ is related to $\rho(\mathbf{x}, t)$ by

$$1 + \delta(\mathbf{x}, t) \equiv \frac{\rho(\mathbf{x}, t)}{\rho_b} = \frac{V}{N} \sum_i \delta_D[\mathbf{x} - \mathbf{x}_i(t)] = \int d\mathbf{q}\, \delta_D[\mathbf{x} - \mathbf{x}_T(t, \mathbf{q})] \,. \tag{10.35}$$

In arriving at the last equality, we have taken the continuum limit by doing the following. (i) Replacing $\mathbf{x}_i(t)$ by $\mathbf{x}_T(t, \mathbf{q})$, where $\mathbf{q}$ stands for a set of parameters (such as the initial position and velocity) of a particle; for simplicity, we shall take this to be the initial position. The subscript T is just to remind ourselves that $\mathbf{x}_T(t, \mathbf{q})$ is the *trajectory* of the particle. (ii) Replacing V/N by $d^3\mathbf{q}$ since both represent the volume per particle. Fourier transforming both sides, we get

$$\delta_{\mathbf{k}}(t) \equiv \int d^3\mathbf{x}\, \mathrm{e}^{-i\mathbf{k}\cdot\mathbf{x}} \delta(\mathbf{x}, t) = \int d^3\mathbf{q}\, \exp[-i\mathbf{k}.\mathbf{x}_T(t, \mathbf{q})] - (2\pi)^3 \delta_D(\mathbf{k}) \,. \tag{10.36}$$

By differentiating this expression and using eqn (10.32) for the trajectories, one can obtain an equation for $\delta_{\mathbf{k}}$ (see e.g. Padmanabhan 2005*c*, 2006*c*, eqn 10). The structure of this equation can be simplified if we use the perturbed gravitational potential (in Fourier space) $\phi_{\mathbf{k}}$, related to $\delta_{\mathbf{k}}$ by

$$\delta_{\mathbf{k}} = -\frac{k^2\phi_{\mathbf{k}}}{4\pi G\rho_b a^2} = -\left(\frac{k^2 a}{4\pi G\rho_0}\right)\phi_{\mathbf{k}} = -\left(\frac{2}{3H_0^2}\right)k^2 a\phi_{\mathbf{k}}\,. \tag{10.37}$$

In terms of $\phi_{\mathbf{k}}$, the *exact* evolution equation reads as

$$\ddot{\phi}_{\mathbf{k}} + 4\frac{\dot{a}}{a}\dot{\phi}_{\mathbf{k}} = -\frac{1}{2a^2}\int\frac{d^3\mathbf{p}}{(2\pi)^3}\phi_{\frac{1}{2}\mathbf{k}+\mathbf{p}}\phi_{\frac{1}{2}\mathbf{k}-\mathbf{p}}\left[\left(\frac{k}{2}\right)^2 + p^2 - 2\left(\frac{\mathbf{k}.\mathbf{p}}{k}\right)^2\right]$$
$$+\left(\frac{3H_0^2}{2}\right)\int\frac{d^3\mathbf{q}}{a}\left(\frac{\mathbf{k}.\dot{\mathbf{x}}}{k}\right)^2 e^{i\mathbf{k}.\mathbf{x}}\,, \tag{10.38}$$

where $\mathbf{x} = \mathbf{x}_T(t,\mathbf{q})$. Of course, this equation is not "closed." It contains the velocities of the particles $\dot{\mathbf{x}}_T$ and their positions explicitly in the second term on the right and one cannot—in general—express them in simple form in terms of $\phi_{\mathbf{k}}$. As a result, it might seem that we are in no better position than when we started. I will now suggest a strategy to tame this term in order to close this equation. This strategy depends on two features:

1. First, extremely nonlinear structures do not contribute to the right-hand side of eqn (10.38) though, of course, they contribute individually to the two terms. More precisely, the right-hand side of eqn (10.38) will lead to a density contrast that will scale as $\delta_{\mathbf{k}} \propto k^2$ if originally—in the linear theory—$\delta_k \propto k^n$ with $n > 2$ as $k \to 0$. This leads to a $P \propto k^4$ tail in the power spectrum (Peebles 1980; Padmanabhan 2005*c*, 2006*c*). (We will provide a derivation of this result in the next section.)
2. Second, we can use the Zeldovich approximation to evaluate this term, once the above fact is realized. It is well known that, when the density contrasts are small, this term grows as $\delta_{\mathbf{k}} \propto a$ in the linear limit. One can easily show that such a growth corresponds to particle displacements of the form

$$\mathbf{x}_T(a,\mathbf{q}) = \mathbf{q} - a\,\nabla\psi(\mathbf{q})\,, \qquad \psi \equiv (4\pi G\rho_0)^{-1}\phi\,. \tag{10.39}$$

A useful approximation to describe the quasi-linear stages of clustering is obtained by using the trajectory in eqn (10.39) as an ansatz valid *even at quasi-linear epochs*. In this approximation (called the Zeldovich approximation), the velocities $\dot{\mathbf{x}}$ can be expressed in terms of the initial gravitational potential.

We now combine the two results mentioned above to obtain a closure condition for our dynamical equation. At any given moment of time, we can divide the particles in the system into three different sets. First, there are particles which are already a part of the virialized cluster in the nonlinear regime. The second set is made of particles which are completely unbound and are essentially contributing to the power spectrum at the linear scales. The third set is made of all particles which cannot be put into either of these two baskets. Of these three, we know that the first two sets of particles do not contribute significantly to the right-hand side of eqn (10.38), so we will not incur any serious error in ignoring these particles in computing the right-hand side. For the description of the particles in the third set, the Zeldovich approximation should be fairly good. In fact, we can do slightly better than the standard Zeldovich approximation. We note that in eqn (10.39) the velocities were taken to be proportional to the gradient of the *initial* gravitational potential. We can improve on this ansatz by taking the velocities to be given by the gradient of the *instantaneous* gravitational potential, which has the effect of incorporating the influence of particles in bound clusters on the rest of the particles to a certain extent.

Given this ansatz, it is straightforward to obtain a *closed* integro-differential equation for the gravitational potential. Direct calculation shows that the gravitational potential is described by the following closed integral equation (the details of this derivation can be found in Padmanabhan (2005*c*, 2006*c*) and will not be repeated here):

$$\ddot{\phi}_{\mathbf{k}} + 4\frac{\dot{a}}{a}\dot{\phi}_{\mathbf{k}} = -\frac{1}{3a^2}\int\frac{d^3\mathbf{p}}{(2\pi)^3}\phi_{\frac{1}{2}\mathbf{k}+\mathbf{p}}\phi_{\frac{1}{2}\mathbf{k}-\mathbf{p}}\left[\frac{7}{8}k^2+\frac{3}{2}p^2-5\left(\frac{\mathbf{k}\cdot\mathbf{p}}{k}\right)^2\right]. \tag{10.40}$$

This equation provides a powerful method for analyzing nonlinear clustering since estimating eqn (10.38) by the Zeldovich approximation has a very large domain of applicability. In the next two sections, I will use this equation to study the transfer of power in gravitational clustering.

10.5 Inverse cascade in nonlinear gravitational clustering: The k^4 tail

There is an interesting and curious result that is characteristic of gravitational clustering and can be obtained directly from eqn (10.40). Consider an initial power spectrum which has very little power at large scales; more precisely, assume that $P(k) \propto k^n$ with $n > 4$ for small k (i.e. the power dies away faster than k^4 for small k). If these large spatial scales are described by the linear theory—as one would have normally expected—then the power at these scales can only grow as $P \propto a^2 k^n$ and will always be subdominant to k^4. It turns out that this conclusion is incorrect. As the system evolves, small-scale nonlinearities will develop in the system and—if the large scales have too little power intrinsically—then the long-wavelength power will soon be dominated by the "k^4 tail" of the short-wavelength power arising from the nonlinear clustering. This is a purely nonlinear effect, which we shall now describe. (This result is known in the literature (Peebles 1980; Padmanabhan 2005*c*, 2006*c*) but we will derive it from the formalism developed in the last section, which adds fresh insight.)

A formal way of obtaining the k^4 tail is to solve eqn (10.40) for long wavelengths, i.e. near $\mathbf{k} = 0$. Writing $\phi_{\mathbf{k}} = \phi^{(1)}_{\mathbf{k}} + \phi^{(2)}_{\mathbf{k}} + \ldots$, where $\phi^{(1)}_{\mathbf{k}} = \phi^{(L)}_{\mathbf{k}}$ is the time-*independent* gravitational potential in the linear theory and $\phi^{(2)}_{\mathbf{k}}$ is the next-order correction, we get from eqn (10.40) the equation

$$\ddot{\phi}^{(2)}_{\mathbf{k}} + 4\frac{\dot{a}}{a}\dot{\phi}^{(2)}_{\mathbf{k}} \cong -\frac{1}{3a^2}\int\frac{d^3\mathbf{p}}{(2\pi)^3}\phi^L_{\frac{1}{2}\mathbf{k}+\mathbf{p}}\phi^L_{\frac{1}{2}\mathbf{k}-\mathbf{p}}\mathcal{G}(\mathbf{k},\mathbf{p}), \tag{10.41}$$

where $\mathcal{G}(\mathbf{k},\mathbf{p}) \equiv [(7/8)k^2 + (3/2)p^2 - 5(\mathbf{k}\cdot\mathbf{p}/k)]^2$. The solution to this equation is the sum of a solution to the homogeneous part (which decays as $\dot{\phi} \propto a^{-4} \propto t^{-8/3}$, giving

$\phi \propto t^{-5/3}$) and a particular solution which grows as a. Ignoring the decaying mode at late times and taking $\phi_{\mathbf{k}}^{(2)} = aC_{\mathbf{k}}$, one can determine $C_{\mathbf{k}}$ from the above equation. Plugging it back, we find the lowest-order correction to be

$$\phi_{\mathbf{k}}^{(2)} \cong -\left(\frac{2a}{21H_0^2}\right)\int \frac{d^3\mathbf{p}}{(2\pi)^3}\phi^L_{\frac{1}{2}\mathbf{k}+\mathbf{p}}\phi^L_{\frac{1}{2}\mathbf{k}-\mathbf{p}}\mathcal{G}(\mathbf{k},\mathbf{p})\,. \tag{10.42}$$

Near $\mathbf{k} \simeq 0$, we have

$$\begin{aligned}\phi_{\mathbf{k}\simeq 0}^{(2)} &\cong -\frac{2a}{21H_0^2}\int \frac{d^3\mathbf{p}}{(2\pi)^3}|\phi^L_{\mathbf{p}}|^2\left[\frac{7}{8}k^2 + \frac{3}{2}p^2 - \frac{5(\mathbf{k}\cdot\mathbf{p})^2}{k^2}\right]\\ &= \frac{a}{126\pi^2 H_0^2}\int_0^\infty dp\, p^4 |\phi_{\mathbf{p}}^{(L)}|^2\,,\end{aligned} \tag{10.43}$$

which is independent of $\mathbf{k}$ to the lowest order. Correspondingly, the power spectrum for the density $P_\delta(k) \propto a^2k^4P_\varphi(k) \propto a^4k^4$ in this order.

The generation of a long-wavelength k^4 tail is easily seen in simulations if one starts with a power spectrum that is sharply peaked in $|\mathbf{k}|$. Figure 10.3 (adapted from Bagla and Padmanabhan (1997a)) shows the results of such a simulation. The y-axis is $\Delta(k)/a(t)$, where $\Delta^2(k) \equiv k^3P/2\pi^2$ is the power per logarithmic band in k. In the linear theory, $\Delta \propto a$ and this quantity should not change. The curves labeled by

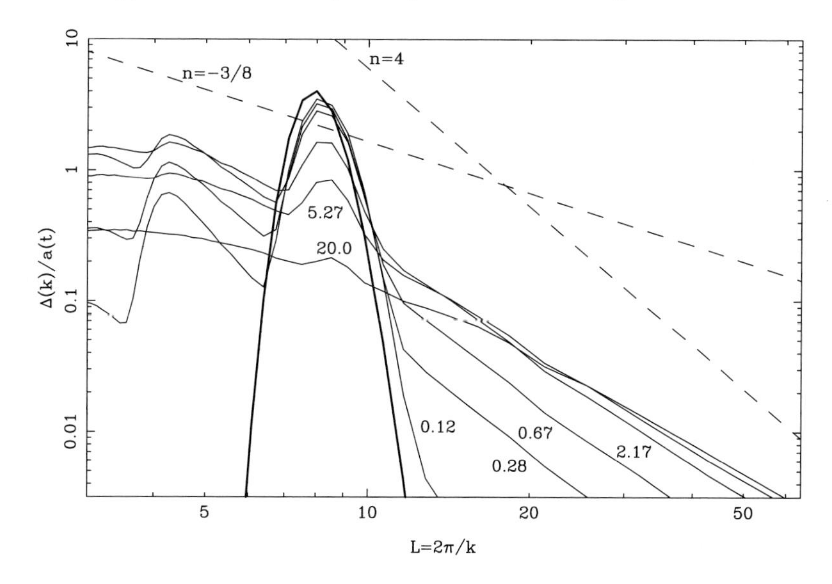

Fig. 10.3 The transfer of power to long wavelengths so as to form a k^4 tail, illustrated using simulation results. Power is injected in the form of a narrow peak at $L = 8$. Note that the y-axis is Δ/a so that there will be no change in shape of the power spectrum under linear evolution with $\Delta \propto a$. As time goes on, a k^4 tail is generated owing purely to nonlinear coupling between the modes. (Figure adapted from Bagla and Padmanabhan (1997a).)

$a = 0.12$ to $a = 20.0$ show the *effects of nonlinear evolution*, especially the development of the k^4 tail. (Actually, one can do better. The formulation can also reproduce the subharmonic at $L \simeq 4$ seen in Fig. 10.3 and other details; see Padmanabhan (2005*c*, 2006*c*).)

10.6 Analogue of Kolmogorov spectrum for gravitational clustering

If power is injected at some scale L into an ordinary viscous fluid, it cascades down to smaller scales because of the nonlinear coupling between different modes. The resulting power spectrum, for a wide range of scales, is well approximated by the Kolmogorov spectrum, which plays a key role in the study of fluid turbulence. It is possible to obtain the form of this spectrum from fairly simple, general considerations, though the actual equations of fluid turbulence are intractably complicated. Let us now consider the corresponding question for nonlinear gravitational clustering. If power is injected at a given length scale very early on, how does the dynamical evolution transfer power to other scales at late times? In particular, does the nonlinear evolution lead to an analogue of the Kolmogorov spectrum with some level of universality, in the case of gravitational interactions?

Surprisingly, the answer is "yes," even though normal fluids and collisionless self-gravitating particles constitute very different physical systems. If power is injected at a given scale $L = 2\pi/k_0$, then the gravitational clustering transfers the power to both larger and smaller spatial scales. At large spatial scales, the power spectrum goes as $P(k) \propto k^4$ as soon as nonlinear coupling becomes important. We have already seen this result in the previous section. More interestingly, the cascading of power to smaller scales leads to a *universal pattern* at late times just as in the case of fluid turbulence. This is because eqn (10.40) admits solutions for the gravitational potential of the form $\phi_{\mathbf{k}}(t) = F(t)D(\mathbf{k})$ at late times when the initial condition is irrelevant; here $F(t)$ satisfies a nonlinear differential equation and $D(\mathbf{k})$ satisfies an integral equation. One can analyze the relevant equations analytically as well as verify the conclusions by numerical simulations. Such a study (the details of which can be found in Padmanabhan and Ray (2006)) confirms that nonlinear gravitational clustering does lead to a universal power spectrum at late times if the power is injected at a given scale initially. (In cosmology, there is very little motivation to study the transfer of power by itself, and most of the numerical simulations in the past have concentrated on evolving the broadband initial power spectrum. So this result was missed.)

Our aim is to look for *late-time* scale-free evolution of the system, exploiting the fact that eqn (10.40) allows self-similar solutions of the form $\phi_{\mathbf{k}}(t) = F(t)D(\mathbf{k})$. Substituting this ansatz into eqn (10.40), we obtain two separate equations for $F(t)$ and $D(\mathbf{k})$. It is also convenient at this stage to use the expansion factor $a(t) = (t/t_0)^{2/3}$ of the matter-dominated universe as the independent variable rather than the cosmic time t. Then simple algebra shows that the governing equations are

$$a\frac{d^2F}{da^2} + \frac{7}{2}\frac{dF}{da} = -F^2 \tag{10.44}$$

and

$$H_0^2 D_{\mathbf{k}} = \frac{1}{3}\int \frac{d^3\mathbf{p}}{(2\pi)^3} D_{\frac{1}{2}\mathbf{k}+\mathbf{p}} D_{\frac{1}{2}\mathbf{k}-\mathbf{p}} \mathcal{G}(\mathbf{k},\mathbf{p})\,. \tag{10.45}$$

Equation (10.44) governs the time evolution, while eqn (10.45) governs the shape of the power spectrum. (The separation ansatz, of course, has the scaling freedom $F \to \mu F, D \to (1/\mu)D$, which will change the right-hand side of eqn (10.44) to $-\mu F^2$ and the left-hand side of eqn (10.45) to $\mu H_0^2 D_{\mathbf{k}}$. But, as expected, our results will be independent of μ; so we have set it to unity.) Our interest lies in analyzing the solutions of eqn (10.44) subject to the initial conditions $F(a_i) =$ constant, $(dF/da)_i = 0$ at some small enough $a = a_i$.

Inspection shows that eqn (10.44) has the exact solution $F(a) = (3/2)a^{-1}$. This, of course, is a special solution and will not satisfy the relevant initial conditions. However, eqn (10.44) fortunately belongs to a class of nonlinear equations which can be mapped to a homologous system. In such cases, the special power-law solutions will arise as an asymptotic limit. (The example well known to astronomers is that of the isothermal sphere (Chandrasekhar 1939). Our analysis below has a close parallel.) To find the general behavior of the solutions to eqn (10.44), we will make the substitution $F(a) = (3/2)a^{-1}g(a)$ and change the independent variable from a to $q = \log a$. Then eqn (10.44) reduces to the form

$$\frac{d^2g}{dq^2} + \frac{1}{2}\frac{dg}{dq} + \frac{3}{2}g(g-1) = 0\,. \tag{10.46}$$

This represents a particle moving in a potential $V(g) = (1/2)g^3 - (3/4)g^2$ under friction. For our initial conditions, the motion will lead the "particle" to asymptotically come to rest at the stable minimum at $g = 1$ with damped oscillations. In other words, $F(a) \to (3/2)a^{-1}$ for large a, showing that this is indeed the asymptotic solution. From the Poisson equation, it follows that $k^2\phi_{\mathbf{k}} \propto \delta_{\mathbf{k}}/a$ so that $\delta_{\mathbf{k}}(a) \propto g(a)k^2 D(\mathbf{k})$, giving a direct physical meaning to the function $g(a)$ as the growth factor for the density contrast. The asymptotic limit ($g \simeq 1$) corresponds to a rather trivial case of $\delta_{\mathbf{k}}$ becoming independent of time. What will be more interesting—and accessible in simulations—will be the approach to this asymptotic solution. To obtain this, we introduce the variable

$$u = g + 2\left(\frac{dg}{dq}\right) \tag{10.47}$$

so that our system becomes homologous. It can be easily shown that we now find the first-order form of the autonomous system to be

$$\frac{du}{dg} = -\frac{6g(g-1)}{u-g}\,. \tag{10.48}$$

The critical points of the system are at $(0,0)$ and $(1,1)$. Standard analysis shows that (i) the point $(0,0)$ is an unstable critical point and the second point $(1,1)$ is the stable critical point; (ii) for our initial conditions, the solution spirals around the stable critical point.

Figure 10.4 (from Padmanabhan and Ray 2006) describes the solution in the g–a plane. The $g(a)$ curves clearly approach the asymptotic value of $g \approx 1$ with superposed

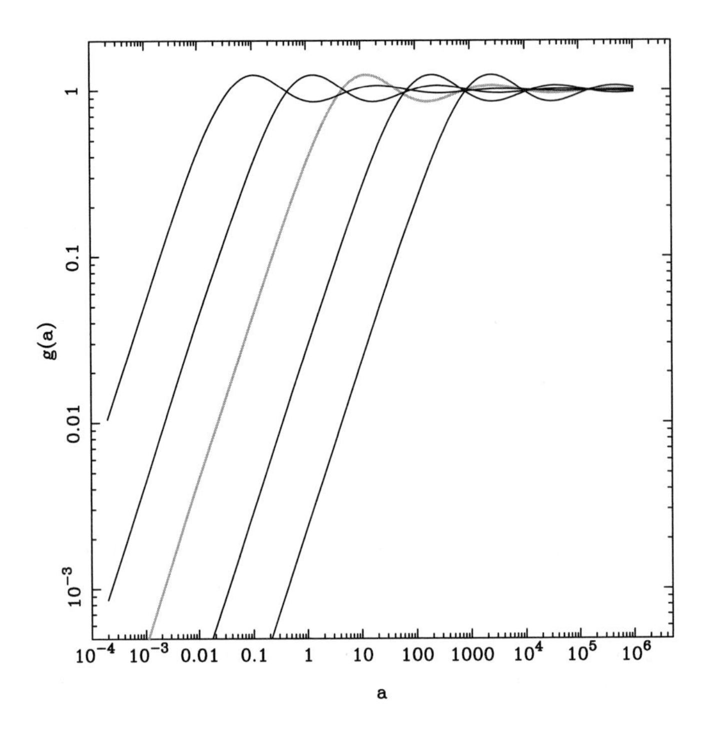

Fig. 10.4 Solutions to eqn (10.46) plotted in the g–a plane. The function $g(a)$ asymptotically approaches unity with oscillations. The different curves correspond to the rescaling freedom in the initial conditions. For more details, see Padmanabhan and Ray (2006).

oscillations. The different curves in Fig. 10.4 are for different initial values, which arise from the scaling freedom mentioned earlier. (The thick red line corresponds to the initial conditions used in the simulations described below.) The solution $g(a)$ describes the time evolution and solves the problem of determining the asymptotic time evolution.

To test the correctness of these conclusions, we performed a high-resolution simulation using the TreePM method (Bagla 2002; Bagla and Ray 2003) and its parallel version (Ray and Bagla 2004) with 128^3 particles on a 128^3 grid. Details of the code parameters can be found in Bagla and Ray (2003). The initial power spectrum $P(k)$ was chosen to be a Gaussian peaked at the scale of $k_p = 2\pi/L_p$, with $L_p = 24$ grid lengths and with a standard deviation $\Delta k = 2\pi/L_{\rm box}$, where $L_{\rm box} = 128$ is the size of one side of the simulation volume. The amplitude of the peak was taken such that $\Delta_{\rm lin}\,(k_p = 2\pi/L_p, a = 0.25) = 1$.

The late-time evolution of the power spectrum (in terms of $\Delta_k^2 \equiv k^3 P(k)/2\pi^2$, where $P = |\delta_k|^2$ is the power spectrum of density fluctuations) obtained from the simulations is shown in Fig. 10.5 (left panel). In the right panel, we have rescaled the Δ_k, using the appropriate solution $g(a)$. The fact that the curves fall on top of each other shows that the late-time evolution indeed scales as $g(a)$ within numerical accuracy. A reasonably accurate fit for $g(a)$ at the late times used in this figure is given

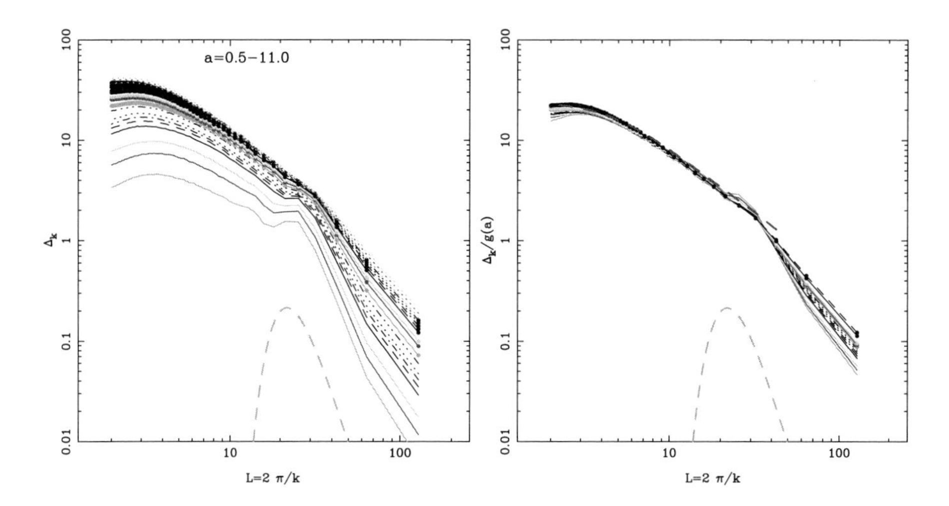

Fig. 10.5 Left panel: results of a numerical simulation with an initial power spectrum which is a Gaussian peaked at $L = 24$. The y-axis gives Δ_k, where $\Delta_k^2 = k^3P/2\pi^2$ is the power per logarithmic band. The evolution generates the well-known k^4 tail at large scales and leads to cascading of power to small scales. Right panel: the simulation data is reexpressed by factoring out the time evolution function $g(a)$ obtained by integrating eqn (10.46). The fact that the curves fall nearly on top of each other shows that the late-time evolution is scale-free and described by the ansatz discussed in the text. The rescaled spectrum is very well described by $P(k) \propto k^{-0.4}/(1+(k/k_0)^{2.6})$, which is shown by the completely overlapping broken curve. For more details, see Padmanabhan and Ray (2006).

by $g(a) \propto a(1 - 0.3\ln a)$. The key point to note is that the asymptotic time evolution is essentially $\delta(a) \propto a$ except for a logarithmic correction, *even on highly nonlinear scales.* (This was first noticed from somewhat lower-resolution simulations in Bagla and Padmanabhan (1997*a*).) Since the evolution at *linear* scales is always $\delta \propto a$, this allows for a form-invariant evolution of the power spectrum at all scales. Gravitational clustering evolves towards this asymptotic state.

To the lowest order of accuracy, the power spectrum at this range of scales is approximated by the mean index $n \approx -1$ with $P(k) \propto k^{-1}$. A better fit to the power spectrum in Fig. 10.5 is given by

$$P(k) \propto \frac{k^{-0.4}}{1+(k/k_0)^{2.6}}, \qquad \frac{2\pi}{k_0} \approx 4.5. \tag{10.49}$$

This fit is shown by the broken blue line in the figure, which completely overlaps with the data and is barely visible. (Note that this fit is applicable only at $L < L_p$, since the k^4 tail will dominate scales to the right of the initial peak; see the discussion in Bagla and Padmanabhan (1997*a*).) At nonlinear scales, $P(k) \propto k^{-3}$, making Δ_k flat, as seen in Fig. 10.5. (This is *not* a numerical artifact, and we have sufficient dynamic range in the simulation to ascertain this.) At quasi-linear scales, $P(k) \propto k^{-0.4}$. The effective index of the power spectrum varies between -3 and -0.4 in this range of scales.

What interpretation of this behavior could be possible? It is difficult to provide a simple but precise answer, but one possible line of reasoning is as follows. In the case of viscous fluids, the energy is dissipated at the smallest scales as heat. In the steady state, energy cannot accumulate at any intermediate scale and hence the rate of flow of energy from one scale to the next (lower) scale must be a constant. This constancy immediately leads to the Kolmogorov spectrum. In the case of gravitating particles, there is no dissipation and each scale will evolve towards virial equilibrium. At any given time t, the power will have cascaded down only up to some scale $l_{\rm min}(t)$, which itself, of course, is a decreasing function of time. So, at time t we expect very little power for $1 < kl_{\rm min}(t)$ and a k^4 tail for $kL_p < 1$, say. The really interesting band is between $l_{\rm min}$ and L_p.

To understand this band, let us recall that the Lagrangian in eqn (10.30) leads to the time-dependent Hamiltonian $H(\mathbf{p}, \mathbf{x}, t) = \sum[p^2/2ma^2 + U]$. The evolution of the energy in the system is governed by the equation $dH/da = (\partial H/\partial a)_{\mathbf{p},\mathbf{x}}$. It is clear from eqn (10.31) that $(\partial U/\partial a)_{\mathbf{p},\mathbf{x}} = -U/a$, while $(\partial T/\partial a)_{\mathbf{p},\mathbf{x}} = -2T/a$. Hence the time evolution of the total energy $H = E$ of the system is described by

$$\frac{dE}{da} = -\frac{1}{a}(2T + U) = -\frac{2E}{a} - \frac{U}{a} = -\frac{E}{a} - \frac{T}{a}. \tag{10.50}$$

In the continuum limit, ignoring the infinite self-energy term, the potential energy can be written as

$$U = -\frac{G\rho_0^2}{2a}\int d^3\mathbf{x}\int d^3\mathbf{y}\frac{\delta(\mathbf{x}, a)\delta(\mathbf{y}, a)}{|\mathbf{x} - \mathbf{y}|}. \tag{10.51}$$

Hence

$$\frac{d(a^2E)}{da} = -a^2U = \frac{G\rho_0^2 a}{2}\int d^3\mathbf{x}\int d^3\mathbf{y}\frac{\delta(\mathbf{x}, a)\delta(\mathbf{y}, a)}{|\mathbf{x} - \mathbf{y}|}. \tag{10.52}$$

The ensemble average of the right-hand side, per unit proper volume, will be

$$\mathcal{E} = -\frac{G\rho_0^2}{2Va^2}\int d^3\mathbf{x}\, d^3\mathbf{y}\frac{\langle\delta(\mathbf{x}, a)\delta(\mathbf{y}, a)\rangle}{|\mathbf{x} - \mathbf{y}|} \propto \int d^3k\frac{|\delta_k|^2}{a^2k^2} \propto \int_0^\infty \frac{dk}{k}\frac{kP(k)}{a^2}, \tag{10.53}$$

where V is the comoving volume.

When a particular scale is virialized, we expect $\mathcal{E} \approx$ constant at that scale in comoving coordinates. That is, we would expect that the energy density in eqn (10.53) would have reached equipartition and contribute the same amount per logarithmic band of scales in the intermediate scales between $l_{\rm min}$ and $L_{\rm peak}$. *This requires* $P(k) \propto a^2/k$, *which is essentially what we have found from simulations.* The time dependence of P is essentially $P \propto a$ except for a logarithmic correction. Similarly, the scale dependence is $P \propto k^{-1}$, which is indeed a good fit to the simulation results. The flattening of the power at small scales, modeled by the more precise fitting function in eqn (10.49), can be understood from the fact that equipartition has not yet been achieved at smaller scales. The same result holds for the kinetic energy if the motion is dominated by scale-invariant radial flows (Bagla and Padmanabhan 1997*a*; Klypin and Melott 1992). Our result suggests that gravitational power transfer evolves towards this equipartition.

Acknowledgments

I thank the organizers of the Les Houches School for inviting me to give these lectures.

References

Aaronson E. B. and Hansen C. J. (1972). *Astrophys. J.*, **177**, 145.
Antonov V. A. (1962). *Vest. Leningrad Univ.*, **7**, 135; translation available in *IAU Symp.*, **113**, 525 (1985).
Bagla J. S. (2002). *J. Astrophys. Astron.*, **23**, 185.
Bagla J. S. (2004). astro-ph/0411043.
Bagla J. S. and Padmanabhan T. (1994). *Mon. Not. R. Astron. Soc.*, **266**, 227.
Bagla J. S. and Padmanabhan T. (1997*a*). *Mon. Not. R. Astron. Soc.*, **286**, 1023.
Bagla J. S. and Padmanabhan T. (1997*b*). *Pramana*, **49**, 161–192.
Bagla J. S. and Ray S. (2003). *New Astron.*, **8**, 665.
Bagla J. S. *et al.* (1998). *Astrophys. J.*, **495**, 25.
Binney J. and Tremaine S. (1987). *Galactic Dynamics*, Princeton University Press.
Bottaccio M. *et al.* (2002). *Europhys. Lett.*, **57**, 315–321.
Brainerd T. G. *et al.* (1993). *Astrophys. J.*, **418**, 570.
Buchert T. (1994). *Mon. Not. R. Astron. Soc.*, **267**, 811–820.
Buchert T. and Dominguez A. (2005). *Astron. Astrophys.*, **438**, 443–460.
Chandrasekhar S. (1939). *An Introduction to the Study of Stellar Structure*, Dover.
Chavanis P.-H. (2002). "Statistical mechanics of two-dimensional vortices and three-dimensional stellar systems," in *Dynamics and Thermodynamics of Systems with Long Range Interactions*, Eds. T. Dauxois, S. Ruffo, E. Arimondo, and M. Wilkens, pp. 208–289, Lecture Notes in Physics 206, Springer.
Cooray A. and Sheth R. (2002). *Phys. Rep.*, **372**, 1–129.
de Vega H. J. and Siebert J. (2002). *Phys. Rev. E*, **66**, 016112.
de Vega H. J. , Sánchez N., and Combes F. (1998). "Fractal structures and scaling laws in the universe: Statistical mechanics of the self-gravitating gas," in Special issue of *J. Chaos, Solitons and Fractals, Superstrings, M, F, S...theory*, Eds. M. S. El Naschie and C. Castro, Vol. 10, pp. 329–343, astro-ph/9807048.
Engineer S. *et al.* (1999). *Astrophys. J.*, **512**, 1.
Engineer S. *et al.* (2000). *Mon. Not. R. Astron. Soc.*, **314**, 279.
Follana E. , Laliena V. (2000). *Phys. Rev. E*, **61**, 6270.
Gurbatov, S. N. *et al.* (1989). *Mon. Not. R. Astron. Soc.*, **236**, 385.
Hamilton A. J. S. *et al.* (1991). *Astrophys. J.*, **374**, L1.
Ispolatov I. and Cohen E. G. D. (2001). Collapse in $1/r^{\alpha}$ interacting systems, cond-mat/0106381.
Kanekar N. *et al.* (2001). *Mon. Not. R. Astron. Soc.*, **324**, 988.
Klypin A. A. and Melott A. L. (1992). *Astrophys. J.*, **399**, 397.
Lynden-Bell D. (1998). "Negative specific heat in astronomy, physics and chemistry," *Proceedings of XXth IUPAP International Conference on Statistical Physics*, Paris, July 20–24, 1998, *Physica A*, **263**, 293.
Lynden-Bell D. and Lynden-Bell R. M. (1977). *Mon. Not. R. Astron. Soc.*, **181**, 405.
Lynden-Bell D. and Wood R. (1968). *Mon. Not. R. Astron. Soc.*, **138**, 495.
Matarrese S. *et al.* (1992). *Mon. Not. R. Astron. Soc.*, **259**, 437–452.

Munshi D. *et al.* (1997). *Mon. Not. R. Astron. Soc.*, **290**, 193.
Nityananda R. and Padmanabhan T. (1994). *Mon. Not. R. Astron. Soc.*, **271**, 976.
Padmanabhan T. (1989*a*). *Phys. Lett. A*, **136**, 203.
Padmanabhan T. (1989*b*). *Astrophys. J. Suppl.*, **71**, 651.
Padmanabhan T. (1990). *Phys. Rep.*, **188**, 285.
Padmanabhan T. (1993). *Structure Formation in the Universe*, Cambridge University Press.
Padmanabhan T. (1996*a*). *Cosmology and Astrophysics through Problems*, Cambridge University Press.
Padmanabhan T. (1996*b*). *Mon. Not. R. Astron. Soc.*, **278**, L29.
Padmanabhan T. (2000*a*). *Theoretical Astrophysics*, Vol. I: *Astrophysical Processes*, Chapter 10, Cambridge University Press.
Padmanabhan P. (2000*b*). "Aspects of gravitational clustering," in *Large Scale Structure Formation*, Eds. R. Mansouri and R. Brandenberger, pp. 97–168, Astrophysics and Space Science Library 247, Kluwer Academic.
Padmanabhan T. (2002*a*). "Statistical mechanics of gravitating systems in static and cosmological backgrounds," in *Dynamics and Thermodynamics of Systems with Long Range Interactions*, Eds. T. Dauxois, S. Ruffo, E. Arimondo, and M. Wilkens, Lecture Notes in Physics 602, pp. 165–207, Springer.
Padmanabhan T. (2002*b*). *Class. Quantum Grav.*, **19**, 5387.
Padmanabhan T. (2002*c*). *Gen. Rel. Grav.*, **34**, 2029.
Padmanabhan T. (2002*d*). *Theoretical Astrophysics*, Vol. III: *Galaxies and Cosmology*, Cambridge University Press.
Padmanabhan T. (2005*a*). *Phys. Rep.*, **406**, 49.
Padmanabhan T. (2005*b*). "Understanding our universe: Current status and open issues," in *100 Years of Relativity: Space–time Structure: Einstein and Beyond*, Ed. A. Ashtekar, pp. 175–204, World Scientific.
Padmanabhan T. (2005*c*). astro-ph/0511536.
Padmanabhan T. (2006*a*). *AIP Conf. Proc.*, **843**, 111.
Padmanabhan T. (2006*b*). *AIP Conf. Proc.*, **861**, 179.
Padmanabhan T. (2006*c*). *C. R. Phys.*, **7**, 350.
Padmanabhan T. (2008). *Gen. Rel. Grav.*, **40**, 529.
Padmanabhan T. and Engineer S. (1998). *Astrophys. J.*, **493**, 509.
Padmanabhan T. and Kanekar N. (2000). *Phys. Rev. D*, **61**, 023515.
Padmanabhan T. and Ray S. (2006). *Mon. Not. R. Astron. Soc. Lett.*, **372**, L53–L57.
Padmanabhan T. *et al.* (1996). *Astrophys. J.*, **466**, 604.
Paranjape A. *et al.* (2006). *Phys. Rev. D*, **74**, 104015.
Peebles P. J. E. (1980). *Large Scale Structure of the Universe*, Princeton University Press.
Peebles P. J. E. (1993). *Principles of Physical Cosmology*, Princeton University Press.
Ray S. and Bagla J. S. (2004). astro-ph/0405220.
Ray S. *et al.* (2005). *Mon. Not. R. Astron. Soc.*, **360**, 546.
Scoccimarro R. (2000). "A new angle on gravitational clustering," in *Proceedings of the 15th Florida Workshop in Nonlinear Astronomy and Physics, "The Onset of Nonlinearity"*, astro-ph/0008277.

Scoccimarro R. and Frieman J. (1996). *Astrophys. J.*, **473**, 620.
Tatekawa T. (2004). astro-ph/0412025.
Valageas P. (2001*a*). *Astron. Astrophys.*, **379**, 8.
Valageas P. (2001*b*). *Astron. Astrophys.*, **382**, 477.
Zeldovich Ya. B. (1970). *Astron. Astrophys.*, **5**, 84.

11

Statistical mechanics of the cosmological many-body problem and its relation to galaxy clustering

William C. SASLAW[1,2] and Abel YANG[1]

[1] Department of Astronomy, University of Virginia, USA

[2] Institute of Astronomy, Cambridge, UK

11.1 Introduction

E pluribus unus
(One composed of many: Virgil, *Moretum*, 1, 104)

Imagine an expanding universe filled with objects moving under their mutual gravitational forces. What decides the distribution of these objects in space, and their velocities, after the memory of their initial state has faded? Richard Bentley, England's leading seventeenth-century classicist, essentially posed the spatial part of this question to his friend Isaac Newton. Their ensuing letters may be found today in the library of Trinity College, Cambridge, and are discussed in some detail elsewhere (Saslaw 2000), along with the subject's subsequent history. Briefly, Newton surmised that if the universe were finite the objects would eventually all fall together into one large cluster. But if the universe were infinite, they "could never convene into one mass; but some of it would convene into one mass and some into another, so as to make an infinite number of great masses, scattered at great distances from one to another throughout all that infinite space." And there the question rested for almost 300 years.

Today we restate this question by replacing "finite" and "infinite" by "static" and "expanding" and talking about galaxies instead of stars. The cosmological many-body problem appears to be a major component of the clustering of galaxies, although a detailed analysis of the astronomical observations also involves the roles of dark matter, galaxy formation and evolution, and perhaps dark energy.

Early investigations of the gravitational many-body problem used a fluid approach when Jeans (1902) explored the stability of a self-gravitating gas. In the cosmological problem, the fundamental particles are usually galaxies. Jeans's results described gravitational collapse, and gave a timescale, proportional to $(G\rho)^{-1/2}$, for the collapse.

Jeans's solutions were developed for a static universe, since expansion had not been discovered. However, Eddington (1930) showed that a static universe is unstable and a slight perturbation will cause it to start either expanding or collapsing. Hubble's (1929) discovery of expansion and the possible existence of dark energy (Riess *et al.* 1998; Perlmutter *et al.* 1999) provide a framework for the expanding universe and suggest that the universe will continue expanding so that structure formation and growth may eventually cease. This leads to interesting properties that we will explore below.

In addition to expansion, the other major discovery that simplifies the cosmological many-body problem is that the two-point correlation function of galaxies, $\xi_2(r)$, has a power-law form $\xi_2(r) \propto r^{-\gamma}$ (Totsuji and Kihara 1969). This two-point galaxy correlation function is defined by the relation

$$P(r|N_1 = 1)\,dr = 4\pi r^2 \overline{n}(1 + \xi_2(r))\,dr \tag{11.1}$$

for a statistically homogeneous system of average number density $\overline{n}$. $P(r|N_1 = 1)$ is the conditional probability, given a galaxy in an infinitesimal volume at an arbitrary coordinate origin, that there is another galaxy at a distance r. Conventionally, spherical volumes are used, although this can easily be generalized to volumes of arbitrary shape which provide further information. Two-point correlation functions therefore represent an excess over a random Poisson probability. In our case this excess is caused by the galaxies' mutual attraction. For positive γ, $\xi_2(r)$ decreases significantly at large r. Observationally there is a correlation length, R_1, beyond which the correlation function decreases even faster than a power law, and may be neglected. Therefore, over sufficiently large scales, the spatial distribution of galaxies is uncorrelated, and has a Poisson distribution, modified by its clustering on smaller scales. Modern surveys such as the 2DFGRS (Hawkins *et al.* 2003) have shown that

$$\xi_2(r) \approx \left(\frac{r}{5.05h^{-1}\ \mathrm{Mpc}}\right)^{-1.67}, \tag{11.2}$$

with $R_1 \approx 20h^{-1}$ Mpc and $H_0 = 100h$, where H_0 is the Hubble constant. Amusingly, the exponent and amplitude in eqn (11.2) are very close to those found in Totsuji and Kihara's analysis of the Lick survey, 40 years ago.

This result implies that on sufficiently large scales, the average density and the gravitational mean field are constant and isotropic. The resultant net force on a galaxy from the distant universe is negligible. On local scales, galaxies are subject to the long-range effects of gravity from their neighbors within R_1.

For an expanding universe, Saslaw and Fang (1996) showed that the expansion of the universe cancels the effect of the long-range gravitational field for distances greater than R_1. This result is valid for Einstein–Friedmann models and for models that incorporate a cosmological constant, including ΛCDM cosmology. Because of expansion, the infinite-range gravitational force has a finite effective range. Large-scale structures that form in the universe may be essentially stable as long as the universe is statistically homogeneous at scales larger than R_1.

These results simplify the problem. Since the expansion of the universe effectively cancels the long-range gravitational field, we can calculate the potential acting on a galaxy by integrating over a finite region of space instead of over the entire universe. Although the problem still involves an infinite volume, the gravitational field now has a finite range and we can consider a finite subvolume caused by this cancelation.

Gravitational systems with more than two interacting particles are essentially unstable and not in equilibrium. While thermodynamics describes equilibrium systems, the cosmological case is characterized by quasi-equilibrium. This means that macroscopic quantities such as average temperature, pressure, and density satisfy equilibrium relations whose variables change slowly compared with local relaxation timescales. For example, the average density of the universe (including dark matter) is approximately $\overline{\rho} = 3.5 \times 10^{10} m_\odot \mathrm{Mpc}^{-3}$ (Spergel *et al.* 2007), and an average galaxy has a mass of approximately $10^{11} m_\odot$. Hence the dynamical timescale of the universe is $\tau_{\mathrm{universe}} \approx 25$ Gyr, and there are about 0.35 galaxies per cubic megaparsec. The average peculiar velocity of a galaxy is ~ 1000 km s^{-1}, or 1 Mpc Gyr^{-1}. Therefore a cube with sides on the order of R_1 will have about 3000 galaxies, and the time for a galaxy to cross the cube is about 20 Gyr, or 1.5 times the age of the universe.

Local timescales are much shorter. A typical cluster similar to our local group of galaxies has a density (including dark matter) $\rho_{LG} \approx 1.5 \times 10^{12} m_\odot \mathrm{Mpc}^{-3}$. This is ~ 40 times greater than $\overline{\rho}$, and hence the local group has a dynamical timescale

$\tau_{LG} \approx 40^{-1/2}\tau_{\text{universe}} \approx 4$ Gyr. Rich clusters which are much denser than the local group will have correspondingly shorter dynamical timescales. Most clusters have an average diameter of 2–4 Mpc, and hence the time taken to cross from one end to the other is on the order of 3 Gyr. These numbers mean that "microscopic" perturbations on local scales will relax significantly faster than perturbations on the "macroscopic" scale of a cube of side R_1, so that the system generally evolves from one equilibrium state to another in quasi-equilibrium.

11.1.1 The GQED

The preceding considerations frame the problem. We have a gravitational field with an effective range R_1 in a system that is in quasi-equilibrium, for which we need the counts-in-cells distribution $f(N)$, i.e. the probability that a given cell located randomly in space contains N galaxies. Thermodynamics and statistical mechanics provide two approaches to this problem. The thermodynamic solution was the first to be investigated (Saslaw and Hamilton 1984) and gives

$$f(N;\overline{N},b) = \frac{\overline{N}(1-b)}{N!}\left[\overline{N}(1-b)+Nb\right]^{N-1} e^{-(\overline{N}(1-b)+Nb)}\,, \tag{11.3}$$

where $\overline{N}$ is the average number of galaxies in a cell, and b is a clustering parameter equal to the ratio between the correlation potential energy and twice the kinetic energy:

$$b = -\frac{W}{2K} = \frac{b_0\overline{n}T^{-3}}{1+b_0\overline{n}T^{-3}}\,, \tag{11.4}$$

with $0 \leq b \leq 1$. Here T is the kinetic temperature of the system, with kinetic energy $K = 3NT/2$ (taking the Boltzmann constant $= 1$), and $\overline{n} = \overline{N}/V$ is the average number of galaxies per unit volume, where V is the cell volume. The distribution given by eqn (11.3) describes the clustering of galaxies in quasi-equilibrium, and hence is known as the gravitational quasi-equilibrium distribution (GQED).

The galaxy spatial distribution function $f(N)$ is a simple but powerful statistic which characterizes the locations of galaxies in space. It includes statistical information on voids and other underdense regions, on clusters of all shapes and sizes, on the probability of finding a given number of neighbors around randomly located positions, on counts of galaxies in cells of arbitrary shapes and sizes randomly located, and on galaxy correlation functions of all orders. These are just some of its representations (Saslaw 2000). Moreover, it is also closely related to the distribution function of the peculiar velocities of galaxies around the Hubble flow (Saslaw *et al.* 1990; Leong and Saslaw 2004).

While the form of b in the second equality of eqn (11.4) was originally taken as an ansatz by Saslaw and Hamilton (1984), a physical reason for the form was given by Saslaw and Fang (1996) using constraints from thermodynamics and the boundary conditions of the problem. Writing the correlation potential energy in terms of the two-point correlation function gives

$$b = -\frac{W}{2K} = \frac{2\pi G m^2 \overline{n}}{3T}\int_V \xi_2(r) r\, dr\,, \tag{11.5}$$

where m is the mass of an individual galaxy.

The integral in eqn (11.5) indicates that b depends on the shape and size of the cell, often taken to be spherical for simplicity. More generally, the shape of a cell affects the correlation potential energy so that (Saslaw 2000)

$$b = \frac{Gm^2\bar{n}^2}{6\bar{N}T} \int_V \int_V \frac{\xi_2(|\mathbf{r}_1 - \mathbf{r}_2|)}{|\mathbf{r}_1 - \mathbf{r}_2|} d\mathbf{r}_1 \, d\mathbf{r}_2 \,. \tag{11.6}$$

11.1.2 The statistical-mechanical approach

The statistical-mechanical approach to this problem was originated by Ahmad *et al.* (2002), who treated the cells as a grand canonical ensemble with the universe as an energy and particle reservoir and the galaxies as particles in the ensemble. The statistical-mechanical results are essentially the same as those in the thermodynamic analysis of point particles, but can be generalized to extended particles. In this section we consider the simplest case with the following assumptions:

- Galaxies behave like point masses, and the two-galaxy interaction potential is $\phi(\mathbf{r}) = -Gm^2/|\mathbf{r}|$. This will be generalized to include a softening parameter, ϵ, at small scales in eqn (11.23). But since the lower limits of the integrals in eqns (11.7)–(11.9) converge uniformly in the limit $\epsilon = 0$, we first use this limit to give the simple formulae in eqns (11.10)–(11.22) of the original thermodynamic results.
- Only two-body interactions are considered as the dominant type of interactions.
- $\phi(\mathbf{r})T^{-1}$ is small.

All three assumptions can be relaxed, and we will discuss the resulting modifications in Section 11.2.

In order to compute the partition function, we note that since expansion cancels the long-range mean-field force, the integral over physical space has a finite cutoff and the integral

$$\int_0^{R_1} \exp\left[-\phi(\mathbf{r})T^{-1}\right] d^3\mathbf{r} = \int_0^{R_1} \exp\left[\frac{Gm^2}{|\mathbf{r}|T}\right] d^3\mathbf{r} \tag{11.7}$$

for a galaxy mass potential has no singularities if the third assumption above holds.

In order to compute the partition function, we start with the normalized phase space integral

$$\begin{aligned} Z_N(T,V) &= \frac{1}{\Lambda^{3N} N!} \int \exp\left[-\left(\sum_{i=1}^{N} \frac{\mathbf{p}_i^2}{2m} + \phi(\mathbf{r}_1, \ldots, \mathbf{r}_N)\right) T^{-1}\right] d^{3N}\mathbf{p}\, d^{3N}\mathbf{r} \\ &= \frac{1}{N!} \left(\frac{2\pi m T}{\Lambda^2}\right)^{3N/2} Q_N(T,V)\,, \end{aligned} \tag{11.8}$$

where Λ is a normalization constant and V is the volume of the ensemble. From the observation that most gravitational interactions are dominated by two-body interactions, and by taking the first-order expansion of the exponential, the potential-energy term of the configuration integral $Q_N(T,V)$ becomes

$$
\begin{aligned}
Q_N(T,V) &= \int \exp\left[-\left(\phi(\mathbf{r}_1,\ldots,\mathbf{r}_N)\right)T^{-1}\right] d^{3N}\mathbf{r} \\
&\approx \int \prod_{1\le i<j\le N} \exp\left[-\phi(\mathbf{r}_{ij})T^{-1}\right] d^{3N}\mathbf{r} \\
&\approx \int \prod_{1\le i<j\le N} \left[1+\frac{Gm^2}{T|\mathbf{r}_{ij}|}\right] d^{3N}\mathbf{r}\,,
\end{aligned}
\tag{11.9}
$$

assuming that galaxies at large distances behave like point masses (although their finite extent is important in close interactions and mergers), that two-body interactions dominate, and that $\phi(\mathbf{r}_{ij})T^{-1}$ is small. This gives the partition function for a canonical ensemble, the Helmholtz free energy, and the corresponding thermodynamic variables for the entropy, pressure, internal energy, and chemical potential:

$$
Z_N(T,V) = \frac{1}{N!}\left(\frac{2\pi mT}{\Lambda^2}\right)^{3N/2} V^N (1-b)^{1-N}\,,
\tag{11.10}
$$

$$
F = -T\ln Z_N(T,V) = NT\ln\left(\frac{N}{VT^{3/2}}\right) + NT\ln(1-b) - NT - \frac{3}{3}NT\ln\left(\frac{2\pi m}{\Lambda^2}\right),
\tag{11.11}
$$

$$
S = -\left(\frac{\partial F}{\partial T}\right)_{V,N} = -N\ln\left(\frac{N}{VT^{3/2}}\right) - N\ln(1-b) - 3Nb + \frac{5}{2}N + \frac{3}{2}N\ln\left(\frac{2\pi m}{\Lambda^2}\right),
\tag{11.12}
$$

$$
P = -\left(\frac{\partial F}{\partial V}\right)_{T,N} = \frac{NT}{V}(1-b)\,,
\tag{11.13}
$$

$$
U = F + TS = \frac{3}{2}NT(1-2b)\,,
\tag{11.14}
$$

$$
\mu = \left(\frac{\partial F}{\partial N}\right)_{T,V} = T\ln\left(\frac{N}{VT^{3/2}}\right) + T\ln(1-b) - Tb - \frac{3}{2}T\ln\left(\frac{2\pi m}{\Lambda^2}\right).
\tag{11.15}
$$

In these derivations, the functional form of b occurs naturally as

$$
b = \frac{(3/2)G^3m^6\bar{n}T^{-3}}{1+(3/2)G^3m^6\bar{n}T^{-3}}\,.
\tag{11.16}
$$

Comparing the coefficients of eqns (11.16) and (11.4), we see that

$$
b_0 = (3/2)G^3m^6\,,
\tag{11.17}
$$

which quantitatively relates the correlation potential energy to the mass of an individual galaxy and confirms the original ansatz in eqn (11.4).

The distribution function is obtained by summing over all energy states for given N:

$$
f(N) = \frac{e^{N\mu/T}Z_N(T,V)}{Z_G(T,V)}\,.
\tag{11.18}
$$

By using eqns (11.10)–(11.15) and the relation between the grand canonical partition function Z_G and the pressure equation of state, we then again obtain the distribution

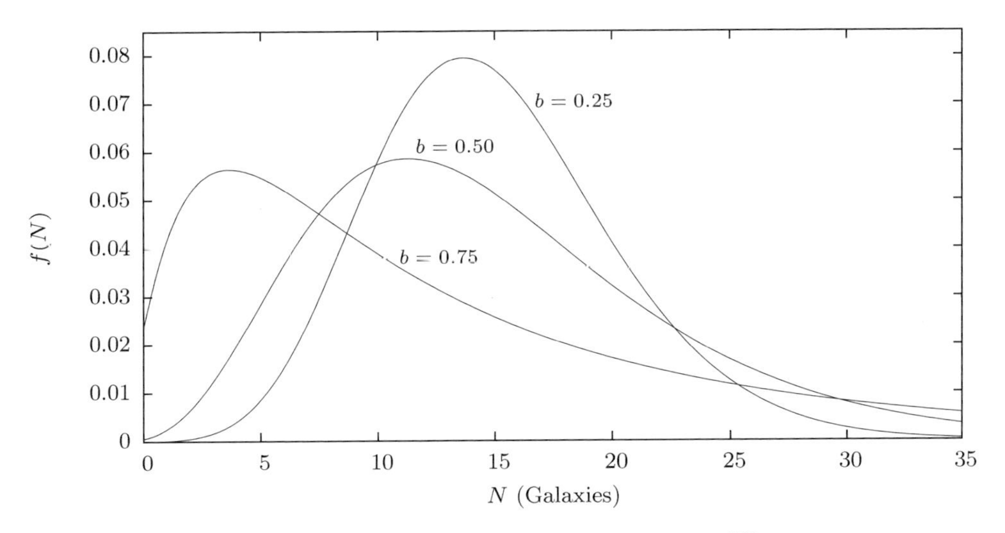

Fig. 11.1 The GQED obtained from eqn (11.3) for $\overline{N} = 15$.

function given by eqn (11.3). Figure 11.1 illustrates the counts-in-cells distribution for different values of b. For $b = 0$, there is no interaction and eqn (11.3) becomes a Poisson distribution.

11.1.3 The peculiar-velocity distribution

The GQED also contains a consistent distribution function for the peculiar velocities of galaxies. Since the partition function separates into kinetic- and potential-energy factors, the phase space distribution is separable such that (Saslaw *et al.* 1990)

$$f(N, v) \to f(N)h(v)\,. \tag{11.19}$$

To relate the counts-in-cells distribution to the peculiar-velocity distribution, we need a relation between the number density N and the velocity v. This arises from the proportionality between the fluctuations in potential energy (due to correlations) over a given volume and the local kinetic-energy fluctuations:

$$GmN\left\langle \frac{1}{r} \right\rangle = \alpha v^2\,. \tag{11.20}$$

Here α is a constant of proportionality such that $\alpha = \langle 1/r \rangle \left\langle v^2 \right\rangle^{-1}$, v is the peculiar velocity of a galaxy, and r is the separation between two galaxies (Leong and Saslaw 2004).

With the relation between N and v in eqn (11.20), the distribution in eqn (11.3) can be transformed to a distribution in v (Saslaw *et al.* 1990):

$$f(v)\,dv = \frac{2\alpha^2\beta(1-b)}{\Gamma\left(\alpha v^2 + 1\right)} \left[\alpha\beta(1-b) + \alpha b v^2\right]^{\alpha v^2 - 1} e^{-\left(\alpha\beta(1-b) + \alpha b v^2\right)} v\,dv\,, \tag{11.21}$$

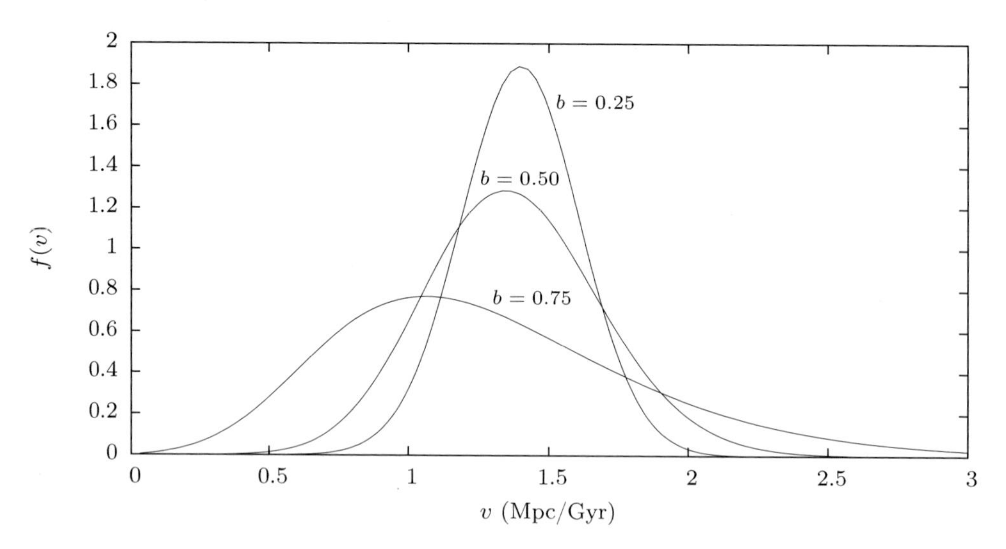

Fig. 11.2 The peculiar-velocity distribution obtained from eqn (11.21) for $\alpha = 10$ and $\beta = 2.0$, with units of velocity in Mpc Gyr^{-1}.

where $\beta \equiv \langle v^2 \rangle$, Γ is the standard gamma function, and averages are taken over the grand canonical ensemble. Figure 11.2 illustrates the peculiar-velocity distribution for different values of b.

Since the proper motions of galaxies are too small to be observed, we can only measure the radial component of their peculiar velocity along our line of sight. We can generally write the velocity as a component parallel to our line of sight and a component perpendicular to our line of sight such that $v^2 = v_{\|}^2 + v_{\perp}^2$. Then, to obtain a form of the velocity distribution directly comparable with observations, we integrate over all perpendicular velocities to get the radial velocity distribution function (Inagaki *et al.* 1992)

$$f(v_{\|}) = \alpha^2 \beta (1-b) e^{-\alpha\beta(1-b)} \int_0^\infty \frac{v_\perp}{\sqrt{v_{\|}^2 + v_\perp^2}} \times \frac{\left[\alpha\beta(1-b) + \alpha b \left(v_{\|}^2 + v_\perp^2\right)\right]^{\alpha\left(v_{\|}^2+v_\perp^2\right)+1}}{\Gamma\left[\alpha\left(v_{\|}^2 + v_\perp^2\right) + 1\right]} e^{-\alpha b\left(v_{\|}^2+v_\perp^2\right)} \, dv_\perp . \quad (11.22)$$

Figures 11.9 and 11.11 (shown later) compare this with simulations and observations.

11.2 Modifications and the general form of the GQED

Various modifications and extensions exist for the GQED. The simplest case involves a modification of the potential of a galaxy so that instead of treating galaxies as point masses, galaxies are treated as extended objects with a potential having the form (Ahmad *et al.* 2002)

$$\phi = -\frac{Gm^2}{(r^2 + \epsilon^2)^{1/2}}, \tag{11.23}$$

where ϵ is a softening parameter related to the radius of a galaxy. Such a modified potential is commonly used in N-body simulations to model extended galaxies, possibly with dark-matter halos, and to avoid the singularity in the point mass potential when the separation between galaxies is small. The result of such a softening parameter is a modification of b to b_ϵ such that

$$b_\epsilon = \frac{(3/2)G^3 m^6 \zeta(\epsilon/R_1)\,\overline{n}T^{-3}}{1 + (3/2)G^3 m^6 \zeta(\epsilon/R_1)\,\overline{n}T^{-3}}, \tag{11.24}$$

where $\zeta(\epsilon/R_1)$ is a term that depends on the interaction potential between a pair of galaxies. For $x \equiv \epsilon/R_1$,

$$\zeta(x) = \sqrt{1+x^2} + x^2 \ln \frac{x}{1+\sqrt{1+x^2}}. \tag{11.25}$$

In the case of a point mass, $\epsilon \to 0$ and $\zeta(\epsilon/R_1) \to 1$ so the partition function and its consequent thermodynamic properties converge to the solution for a point mass potential. Figure 11.3 illustrates $\zeta(\epsilon/R_1)$.

To generalize this result, we can modify the point mass potential so that

$$\phi = -\frac{Gm^2}{r}\kappa(r), \tag{11.26}$$

where $\kappa(r)$ is a dimensionless modification factor. It can then be shown that $\zeta(x)$ depends on the form of $\kappa(r)$ such that $\zeta(x) \to 1$ as $\kappa(r) \to 1$. This modification of the potential will affect only b_0. Therefore many other forms of the potential, such as multipole moments, are possible.

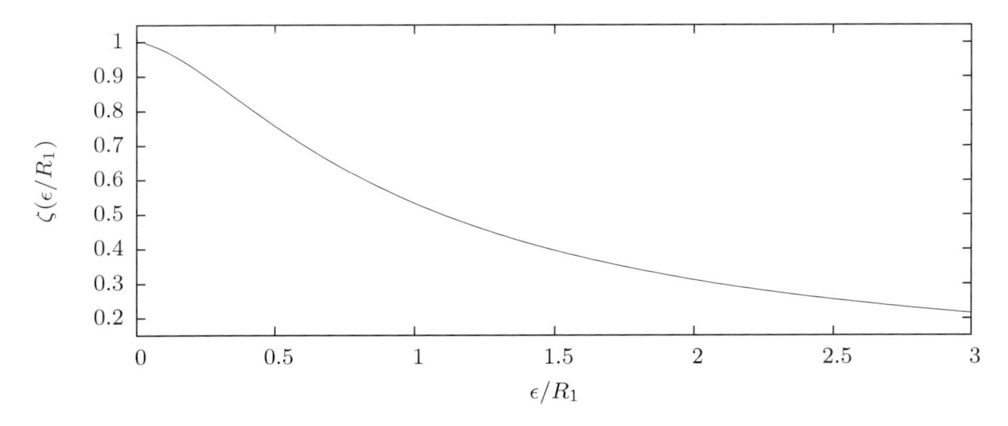

Fig. 11.3 $\zeta(\epsilon/R_1)$ for the softened point mass potential given by eqn (11.23).

11.2.1 Triplet interaction

Ahmad *et al.* (2006*b*) investigated the contribution of the irreducible triplet interactions, which are the cases where three-body interactions occur very close together. In such a case, the modified partition function for a canonical ensemble of $N \geq 3$ galaxies is

$$Z_N(T,V) = \frac{V^N}{N!}\left(\frac{2\pi mT}{\Lambda^2}\right)^{3N/2}\left[\left(\frac{1}{1-b}\right)^{N-1} + \frac{(N-1)(N-2)}{2}\frac{4}{9}\left(\frac{b}{1-b}\right)^3\right], \tag{11.27}$$

where we note that at large values of N, the $(b/(1-b))^3$ term becomes very small compared with the $(1/(1-b))^{N-1}$ term. This indicates that the contribution from irreducible triplets is negligible compared with that from pairwise interactions for large N.

From the partition function, the resulting distribution function taking into account irreducible triplets is

$$f(N;\overline{N},b) = \frac{\overline{N}(1-b)\left[\overline{N}(1-b)+Nb\right]^{N-1} + N^3\overline{N}^{N-3}L(N)}{N!(1+L(N))}e^{-(\overline{N}(1-b_t)+Nb_t)}, \tag{11.28}$$

where we write $a_N = 2(N-1)(N-2)/9$, and define $L(N)$ and b_t as follows:

$$L(N) = \begin{cases} a_N b^3(1-b)^{N-3} & \text{for } N \geq 3\,, \\ 0 & \text{for } N < 3\,, \end{cases} \tag{11.29}$$

$$b_t = \begin{cases} b\left[\dfrac{N+3a_N b^2(1-b)^{N-3}}{N+Na_N b^3(1-b)^{N-3}}\right] & \text{for } N \geq 3\,, \\ b & \text{for } N < 3\,. \end{cases} \tag{11.30}$$

Figure 11.4 compares the contributions of triplet interactions to the original GQED for different values of b.

11.2.2 Two species

In order to help understand the effects of a distribution of galaxy masses, Ahmad *et al.* (2006*a*) derived the distribution function for a system consisting of two different galaxy masses and pointed out how this can be generalized further to a distribution of masses. We let the total number of galaxies in a cell be $N = N_1 + N_2$, of which N_1 and N_2 have individual masses m_1 and m_2, respectively.

When we consider interactions between the different species, the partition function for a canonical ensemble of N galaxies becomes

$$\begin{aligned} Z_N(T,V) = {} & \frac{V^N}{\Lambda^{3N}N!}(2\pi m_1 T)^{3N_1/2}(2\pi m_2 T)^{3N_2/2} \\ & \times\left[1+b_0\overline{n}T^{-3}\right]^{N_1-1}\left[1+b_{12}\overline{n}T^{-3}\right]^{N_2}, \end{aligned} \tag{11.31}$$

where, for the case of a Newtonian point mass potential, $b_0 = (3/2)(Gm_1^2)^3$ and $b_{12} = (3/2)(Gm_1m_2)^3$.

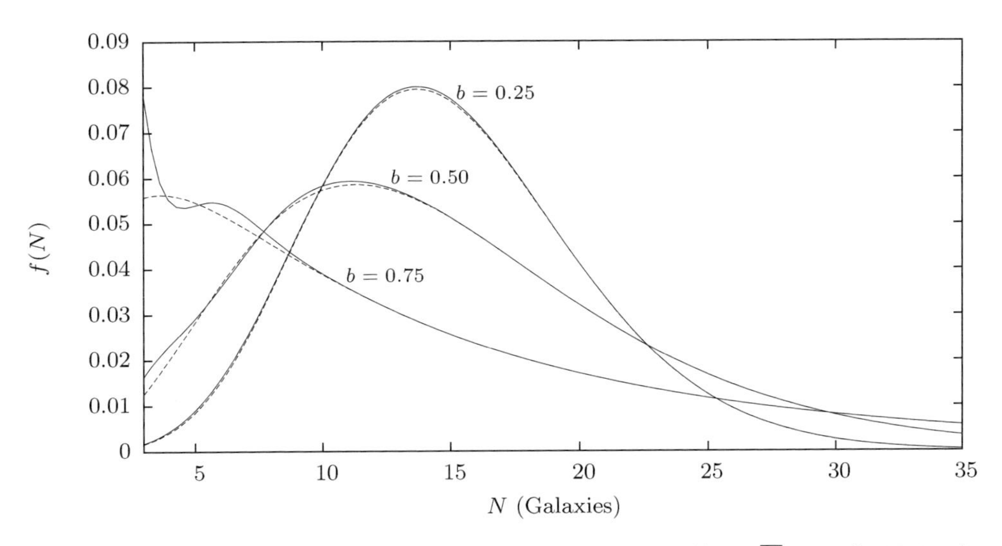

Fig. 11.4 The GQED including irreducible triplets (eqn (11.28)) for $\overline{N} = 15$ (solid lines). The dashed lines, for comparison, show the GQED (eqn (11.3)) obtained without including irreducible triplets.

The distribution function is then

$$f(N; \overline{N}, b) = \frac{\overline{N}(1-b)}{N!} \left[\overline{N}(1-b) + Nb\right]^{N_1 - 1} \times \left[\frac{\overline{N}(1-b) + (m_2/m_1)^3 Nb}{1 - b + (m_2/m_1)^3 b}\right]^{N_2} e^{-(\overline{N}(1-B)+NB)}, \qquad (11.32)$$

where b is given, as always, by eqns (11.4) and (11.5) and B is given by

$$B = \frac{b}{(1 + N_2/N_1)} \left[1 + \frac{(m_2/m_1)^3 (N_2/N_1)}{1 - b + (m_2/m_1)^3 b}\right]. \qquad (11.33)$$

We compare the two-species distribution function with the original GQED in Fig. 11.5.

11.2.3 Higher-order expansions of $\exp(-\phi(r))/T$

A higher-order expansion of the exponential in eqn (11.9) is considered by Saslaw and Ahmad (2009). Again writing $b_0 = (3/2)(Gm^2)^3$, the canonical partition function is now

$$Z_N(T, V) = \frac{V^N}{N!} \left(\frac{2\pi m T}{\Lambda^2}\right)^{3N/2} \left[1 + b_0 \overline{n} T^{-3} \zeta_1 + \left(b_0 \overline{n} T^{-3}\right)^2 \zeta_2\right]^{N-1}, \qquad (11.34)$$

where ζ_1 and ζ_2 are factors that arise from the second-order expansion. In the case of point masses, $\zeta_1 = 1$ and $\zeta_2 = 2/3$.

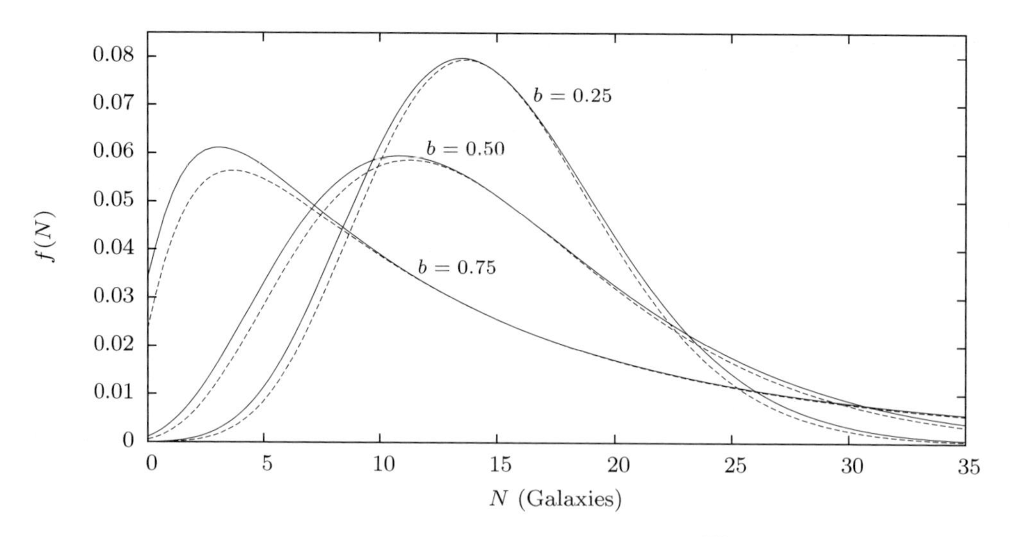

Fig. 11.5 The two-species GQED obtained from eqn (11.32) for $\overline{N} = 15$ (solid lines) with a number ratio of $N_2/N = 0.1$ and mass ratio of $m_2/m_1 = 10$. The dashed lines, for comparison, show the GQED (eqn (11.3)) with mass m_1 and $\overline{N} = 15$.

A quantity $b_\star$, representing the modification of b by the second-order expansion, can be defined as

$$b_\star = \frac{b_0 \bar{n} T^{-3} \zeta_1 + 2\left(b_0 \bar{n} T^{-3}\right)^2 \zeta_2}{1 + b_0 \bar{n} T^{-3} \zeta_1 + \left(b_0 \bar{n} T^{-3}\right)^2 \zeta_2} = \frac{b(1-b)\zeta_1 + 2b^2\zeta_2}{(1-b)^2 + b(1-b)\zeta_1 + b^2\zeta_2}. \tag{11.35}$$

The distribution function in this case is

$$f(N; \overline{N}, b) = \frac{\overline{N}(1-b)}{N!}\left[\frac{\overline{N}(1-b) + Nb\zeta_1 + (N^2b^2)/(N(1-b))\zeta_2}{(1-b) + b\zeta_1 + (b^2/(1-b))\zeta_2}\right]^{N-1}$$
$$\times \frac{e^{-(\overline{N}(1-b_\star) + Nb_\star)}}{(1-b) + b\zeta_1 + (b^2/(1-b))\zeta_2}. \tag{11.36}$$

Normally the effects of the high-order terms in this expansion are small, thereby confirming the essential form of $f(N)$ in eqn (11.3) and its consequences.

11.3 Properties of the GQED

The GQED has a number of properties that are discussed in detail in Saslaw (2000). We summarize some of them here.

11.3.1 Correlation functions

For the simplest case, the two-point correlation function enters $f(N)$ through eqn (11.6) since b depends on the volume integral of ξ_2. However, in general the form of the GQED also contains information about the volume integrals of all the correlation functions, where the average volume integral is defined as

$$\overline{\xi}_N = \frac{1}{V^N} \int_V \xi_N(\mathbf{r}_1, \ldots, \mathbf{r}_N)\, d^3\mathbf{r}_1 \ldots \mathbf{r}_N \, . \tag{11.37}$$

For the general N-point correlation function, Zhan and Dyer (1989) derived a relation between $\overline{\xi}_N$, $\overline{N}$, and b. The general form for ξ_N is

$$\overline{\xi}_N = \frac{1-b}{b\overline{N}^{N-1}} \sum_{M=N}^{\infty} \frac{M^{M-1}}{(M-N)!} b^M e^{-Mb} \, . \tag{11.38}$$

For $N = 1$, 2, and 3 this gives

$$\overline{\xi}_1 = 1 \, , \tag{11.39}$$

$$\overline{\xi}_2 = \frac{b}{(1-b)^2} \frac{2-b}{\overline{N}} \, , \tag{11.40}$$

$$\overline{\xi}_3 = \frac{b^2}{(1-b)^4} \frac{9 - 8b + 2b^2}{\overline{N}^2} \, . \tag{11.41}$$

Other quantities such as average moments of fluctuations in the thermodynamic variables have also been derived.

11.3.2 Specific heat

From the internal energy given by eqn (11.14), using eqn (11.4), the specific heat per galaxy at constant volume is

$$C_V = \frac{1}{N} \frac{\partial U}{\partial T}\bigg|_{V,N} = \frac{3}{2}(1 + 4b - 6b^2) \, , \tag{11.42}$$

which at $b = 0$ is 3/2, representing the monatomic ideal gas. As galaxies cluster, binaries start to dominate and C_V reaches its maximum of 5/2 at $b = 1/3$. At a critical value of b_{crit} such that

$$b_{\text{crit}} = \frac{2 + \sqrt{10}}{6} = 0.8604 \, , \tag{11.43}$$

$C_V = 0$. As b increases beyond b_{crit}, C_V decreases until C_V reaches $-3/2$ at $b = 1$, representing the specific heat of a fully virialized cluster. The transition from positive to negative specific heat, illustrated in Fig. 11.6, occurs at a rather high value of b, and does not involve a discontinuity, unlike some laboratory systems.

While negative specific heat is frequently encountered in the gravitational context, the system we are discussing is nominally a grand canonical ensemble. Grand canonical ensembles can only have positive specific heat (Tolman 1938, eqn 141.12). However, for a microcanonical ensemble, the specific heat can be negative (Thirring 1970). (For a review, see Dauxois *et al.* (2002).)

To resolve this "paradox," we consider clustering for increasing b. As b increases, galaxies become increasingly clustered and large voids are created. These voids have

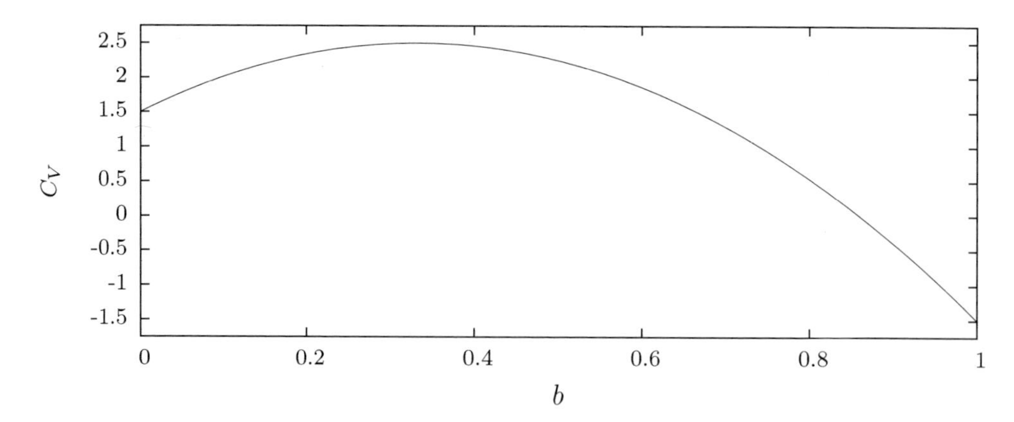

Fig. 11.6 The specific heat per galaxy at constant volume for varying b obtained from eqn (11.42).

the effect of insulating clusters from the energy and galaxy reservoir that is the rest of the universe. The net effect is that the universe breaks up into a collection of galaxy clusters. Each cluster is a microcanonical ensemble that does not exchange significant energy or galaxies with the rest of the universe.

The measured value of b for the local universe of redshifts $z \lesssim 0.1$ is about 0.86 for large 8° angular cells (Sivakoff and Saslaw 2005), which indicates that $b \approx b_{\text{crit}}$ and suggests that clustering is fairly advanced in the local universe. The average matter density (including dark matter) of the universe is $3.5 \times 10^{10} m_\odot \text{Mpc}^{-3}$ (Spergel *et al.* 2007). For comparison, the mass and radius of a typical rich cluster are about $10^{15} m_\odot$ and 2 Mpc, respectively, making the cluster approximately 1000 times denser than the average universe. To find a lower bound on the intercluster spacing, consider a simple model in which rich clusters are each surrounded by a void where no galaxies exist. Then each void would be approximately a sphere that would contain roughly $10^{15} m_\odot$ if the matter were initially distributed uniformly throughout the universe. The radius of such a sphere is approximately 19 Mpc, which suggests that the average separation of massive-cluster centers is at least 19 Mpc.

For an average galaxy with a peculiar velocity of about 1000 km/s to move from one cluster to another over a distance of 19 Mpc would require about 19 Gyr, which is longer than the age of the universe. The long transit times involved suggest that such clusters are no longer likely to exchange galaxies. At separations of 19 Mpc, the interaction potential energy of a pair of galaxies of $\sim 10^{12} m_\odot$ each is approximately four orders of magnitude lower than the kinetic energy of a single galaxy with a peculiar velocity of 1000 km/s, which indicates that at large values of b, clusters could exchange energy more effectively by exchanging galaxies rather than by long-range gravitational interactions. But galaxy exchange now takes too long. Hence at $b \gtrsim b_{\text{crit}}$, clusters are approximately microcanonical ensembles that are likely to be virialized (Leong and Saslaw 2004) and have a negative heat capacity. When most galaxies are bound to a cluster in the case of large b, the total heat capacity is negative, and hence the specific heat per galaxy is negative.

To see this, we can use the multiplicity function derived from eqn (11.3), which gives the probability that a physical cluster contains N galaxies (see Saslaw 2000, eqn 28.97):

$$\eta(N,b) = \frac{(Nb)^{N-1}}{N!}e^{-Nb} \text{ for } N = 1,2,3,\dots, \tag{11.44}$$

which is a truncated Borel distribution. From eqn (11.42), the total heat capacity of these clusters is therefore

$$\frac{3}{2}\left(1+4b-6b^2\right)\sum_{N=1}^{\infty}\overline{N}N_0\eta(N,b) = \frac{3}{2}\overline{N}N_0\left(1+4b-6b^2\right), \tag{11.45}$$

where N_0 is the total number of clusters having an average number $\overline{N}$ of galaxies per cluster. The multiplicity function sums to unity and we can divide the total heat capacity by $\overline{N}N_0$ to obtain the average specific heat of the collection of clusters, each of which is a microcanonical ensemble. This gives the same result as eqn (11.42), so the effective specific heat of this ensemble also becomes negative for $b > b_{\text{crit}}$ in eqn (11.43).

When $b \to 1$, $f(N) \to 0$ for $N > 0$. The void probability goes to 1 and all galaxies will be bound to a single cluster that can be represented as an isothermal sphere (Baumann*et al.* 2003). In that limit, the energy and galaxy reservoir is effectively depleted, and the entire system effectively becomes a single microcanonical ensemble.

This transition, as b increases, from a single grand canonical ensemble where the average cluster has a positive specific heat per galaxy, through a collection of microcanonical ensembles with negative specific heat per galaxy, to a single virialized microcanonical ensemble is an effect of increasing gravitational inhomogeneity. It is rather remarkable that eqn (11.42) can describe this entire ensemble transition in a smooth manner. Perhaps the reason is that the properties of the system for $b > b_{\text{crit}}$ are essentially determined by its properties for $b < b_{\text{crit}}$.

11.3.3 Robustness to mergers

Galaxies are known to interact and merge with each other. Over time, we can expect the average number of galaxies in a given comoving volume to decrease owing to mergers, and hence galaxy mergers will change the distribution of galaxies. We can classify galaxy mergers by the masses of the progenitors m_1 and m_2. Minor mergers are mergers where one galaxy is much less massive than the other galaxy such that $m_1 \ll m_2$ or $m_2 \ll m_1$. In such mergers, the resulting galaxy will have a position that is close to the center of mass of the larger of the two galaxies and the only change to the distribution is a reduction of the number of low-mass galaxies. The other class of mergers is that of major mergers, where both progenitors are of comparable mass such that $m_1 \approx m_2$, and the position of the resulting galaxy is approximately at the midpoint between the two. Major mergers will change the spatial distribution of galaxies more drastically. The average number density $\overline{N}$ decreases with time and the average mass of each galaxy increases with time. This affects the rate at which b changes.

To follow the change in the distribution of galaxies due to mergers, we consider the distribution of nearest-neighbor galaxy pairs whose members are separated by a distance $2\mathbf{r}$. We denote the separation between the midpoints of two such pairs by $\mathbf{R}$. Hence the interaction between two galaxy pairs, with separations $2\mathbf{r}_1$ and $2\mathbf{r}_2$, is given by

$$\phi(\mathbf{r}) = -\frac{Gm^2}{|\mathbf{R}|}\left(\frac{|\mathbf{R}|}{|2\mathbf{r}_1|} + \frac{|\mathbf{R}|}{|2\mathbf{r}_2|} + \frac{|\mathbf{R}|}{|\mathbf{R}+\mathbf{r}_1+\mathbf{r}_2|} + \frac{|\mathbf{R}|}{|\mathbf{R}-\mathbf{r}_1+\mathbf{r}_2|} + \frac{|\mathbf{R}|}{|\mathbf{R}+\mathbf{r}_1-\mathbf{r}_2|} + \frac{|\mathbf{R}|}{|\mathbf{R}-\mathbf{r}_1-\mathbf{r}_2|}\right), \tag{11.46}$$

where the term in brackets can be viewed as a modification to the point mass potential. By averaging over possible values of $\mathbf{r}_1$ and $\mathbf{r}_2$, we can use eqn (11.46) to define a modification $\kappa(|\mathbf{R}|, \langle\mathbf{r}\rangle)$ to the potential, where $|\mathbf{R}|$ is the separation between the two pairs. From Section 11.2, the modification due to such a potential will enter into the form of b_0 and hence the resulting distribution will also follow the GQED. By assuming that all galaxies will eventually merge, we note that when all galaxies have merged once, the spatial distribution of the resulting galaxies also follows the GQED. Hence the distribution of galaxies is robust to mergers.

11.3.4 The time evolution of b

Since there is no final equilibrium state for a classical infinite gravitating system, its clustering increases as described by the increase of b with time. From the thermodynamic variables in Section 11.1, we can write (cf. Saslaw 1992)

$$b = b_0(\overline{N})PT^{-4}. \tag{11.47}$$

Then, by taking the expansion of the universe to be an adiabatic process,

$$0 = dU + P\,dV = \frac{3}{2}(1-2b)\left[T\,d\overline{N}|_{T,P} + \overline{N}\,dT|_{\overline{N},P}\right] - 3\overline{N}T\,db + \overline{N}T(1-b)\frac{dV}{V}. \tag{11.48}$$

If galaxy mergers do not occur, $d\overline{N} = 0$. With the relation $dV/V = 3\,da/a$, we arrive at a relation relating the scale length, a, of the universe with b and $\overline{N}$ such that

$$\frac{1+6b}{8b}\frac{db}{da} = \frac{1-b}{a}. \tag{11.49}$$

Integrating eqn (11.49) we get (Saslaw 1992)

$$\frac{b^{1/8}}{(1-b)^{7/8}} = \frac{a}{a_*}, \tag{11.50}$$

where a_* is a constant of integration given by the initial state.

Galaxies, however, will merge, and hence we can expect that $d\overline{N} \neq 0$. This gives

$$\frac{1+6b}{8b}\frac{db}{da} = \frac{1-b}{a} + \frac{1-2b}{2\overline{N}}\left.\frac{d\overline{N}}{da}\right|_{T,P}. \tag{11.51}$$

From the definition of b given in eqn (11.4), we note that b depends on $\overline{n}$ and hence $\overline{N}$ and V. Denoting $\overline{m}$ as the average mass of a galaxy, by conservation of mass we have $\overline{m}\overline{N} =$ constant for an ensemble of comoving volumes. This gives

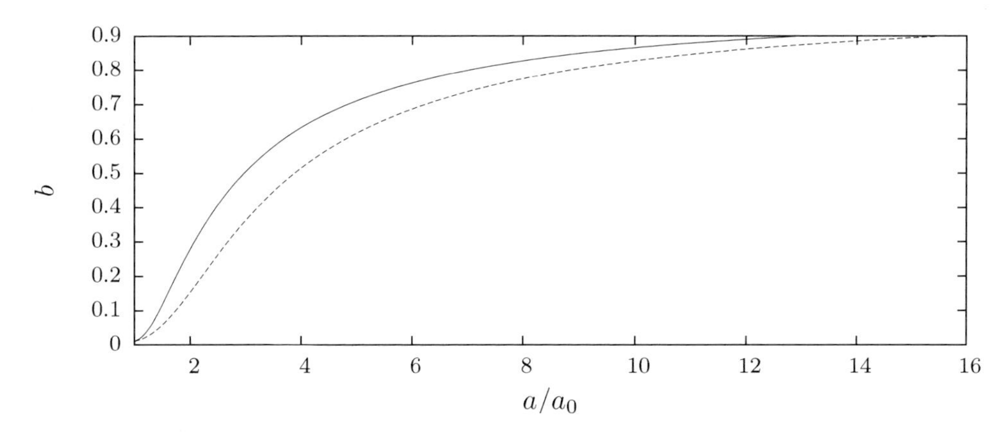

Fig. 11.7 The evolution of b with respect to scale length for an initial value of $b = 0.01$ at $a/a_0 = 1$. The solid line shows the case without mergers (eqn (11.50)) and the dashed line shows the case with mergers (eqn (11.55)).

$$\frac{db}{b} = -6\zeta_\star \left. \frac{d\overline{N}}{\overline{N}} \right|_{T,P} . \tag{11.52}$$

Then, from eqns (11.51) and (11.52), the rate of change of b is given by

$$\left(\frac{1+6b}{8b} + \frac{1-2b}{12b\zeta_\star(\epsilon/R_1)} \right) \frac{db}{da} = \frac{1-b}{a}, \tag{11.53}$$

where $\zeta_\star$ is a term that is given by

$$\zeta_\star(\epsilon/R_1) = 1 + \frac{1}{18} \frac{\epsilon}{R_1} \frac{\zeta'(\epsilon/R_1)}{\zeta(\epsilon/R_1)} \tag{11.54}$$

for the case of the isothermal halo, with $\zeta(\epsilon/R_1)$ given by eqn (11.25). Here $\zeta'(\epsilon/R_1)$ is the derivative of $\zeta(\epsilon/R_1)$ with respect to ϵ/R_1. In this case $\zeta_\star$ is a factor of order unity that varies between 1 and 17/18.

In the limit of a small halo, $\epsilon/R_1 \to 0$ and $\zeta_\star(\epsilon/R_1) \to 1$ and eqn (11.53) integrates to become

$$\frac{b^{5/24}}{(1-b)^{19/24}} = \frac{a}{a_*} . \tag{11.55}$$

In both the nonmerging and the merging case, we note that b increases as a increases, with b increasing faster in the case with no mergers than in the case with mergers. We compare the two cases in Fig. 11.7.

11.4 Simulations and observations

The results for the GQED have been compared with N-body simulations and various observations, and there is very good agreement between predictions, simulations, and observations without having to introduce additional (e.g. nongravitational) parameters.

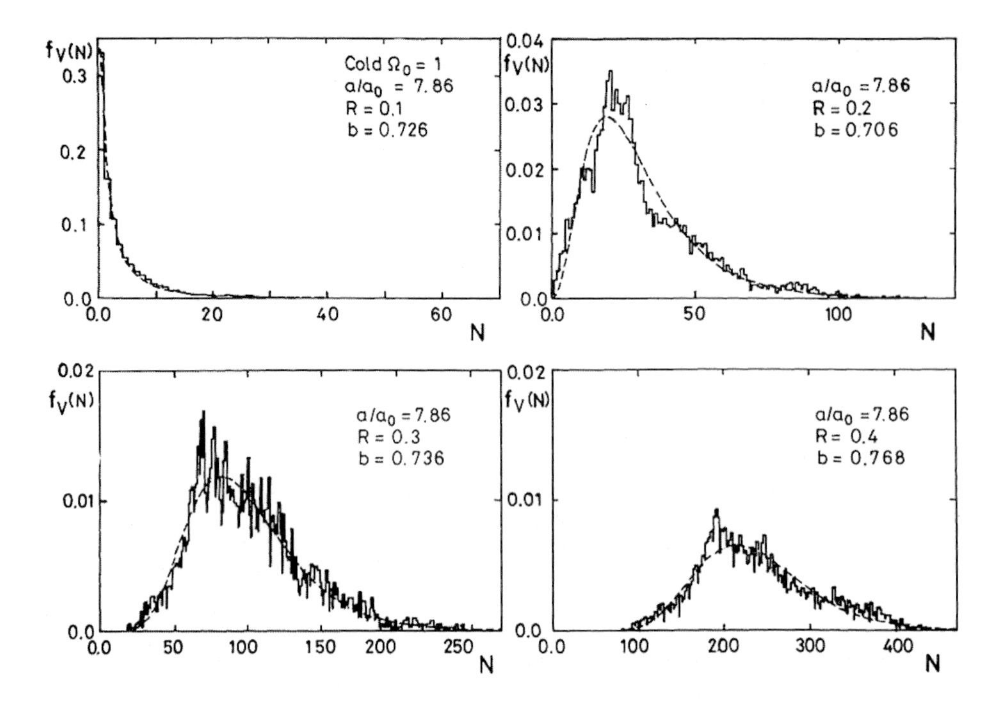

Fig. 11.8 $f_V(N)$ for a single simulation with $\Omega_0 = 1$, cold initial conditions, and an expansion factor $a/a_0 = 7.86$, from Itoh *et al.* (1988). Results for various cell radii R between 0.1 and 0.4 are plotted. The solid lines are from the simulation and the dashed lines are from eqn (11.3).

11.4.1 N-body simulations

A series of computer simulations (Itoh *et al.* 1988, 1990, 1993; Inagaki *et al.* 1992) examined various aspects of the GQED in a comoving volume with N between 4000 and 10 000. The spatial distribution and peculiar-velocity distributions were examined for cases where galaxies had the same mass and for various mass spectra. The primary results of these simulations were that the distributions of galaxies followed the GQED very well for identical masses, and even better for a more realistic case where galaxies have different masses. Figure 11.8 shows a typical example for identical masses, and Fig. 11.9 shows examples of velocity distributions.

11.4.2 Observations of $f_V(N)$

There have been many observations of $f_V(N)$ using various galaxy catalogs. Some recent ones (which also give earlier references) are the following:

1. An analysis of two-dimensional angular cells (Sivakoff and Saslaw 2005), shown in Fig. 11.10, using data from the 2MASS survey, an all-sky survey in the infrared with up to 439 754 galaxies. A good fit to predictions was found, with b approaching $b_{\rm crit}$ at large cell sizes.

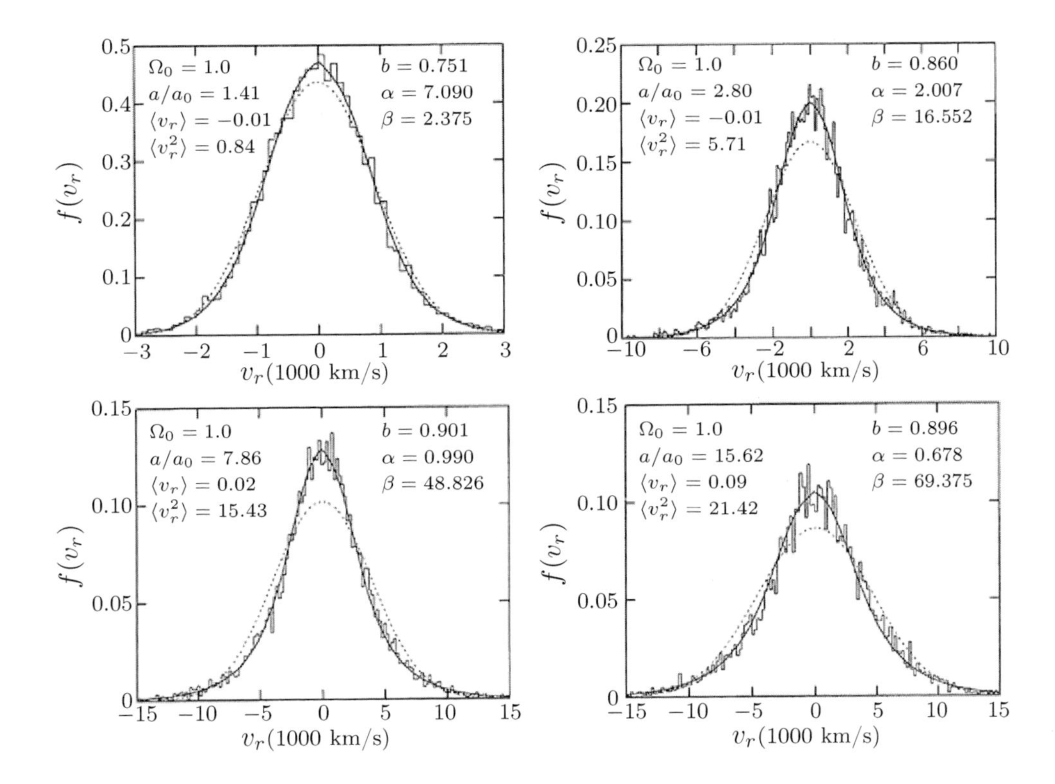

Fig. 11.9 Radial velocity distributions $f(v_r)$ for a single simulation with $\Omega_0 = 1.0$, from Inagaki *et al.* (1992). The histogram shows data from the simulation. Results for various expansion factors a/a_0 are plotted. The solid lines are from eqn (11.22) and the dashed lines show Maxwell–Boltzmann distributions with the same $\langle v_r^2 \rangle$.

2. A three-dimensional analysis of the Pisces–Perseus supercluster with 4501 galaxies (Saslaw and Haque-Copilah 1998) found $\overline{N} = 3.38$ and $b = 0.80$ for cells of $10h^{-1}$ Mpc, where h is the reduced Hubble constant such that $h = H_0/100$. The χ^2 value was found to be 0.089, indicating a relatively good fit to the GQED.
3. At high redshifts, Rahmani *et al.* (2009) found that the projected spatial distribution followed the GQED out to a redshift of $z = 1.5$ using the GOODS survey catalog. Owing to the small sample size and limited sky coverage, large differences between the north and south fields in the GOODS catalog were evident. Nevertheless, there was good agreement between the form of the GQED and the observed spatial distribution of galaxies even at these high redshifts.

11.4.3 Observations of $f(v)$

The radial velocity distribution was analyzed observationally (Raychaudhury and Saslaw 1996) using the Matthewson catalog of spiral galaxies. Since radial velocities of galaxies require a secondary distance indicator and are not easily obtained, the

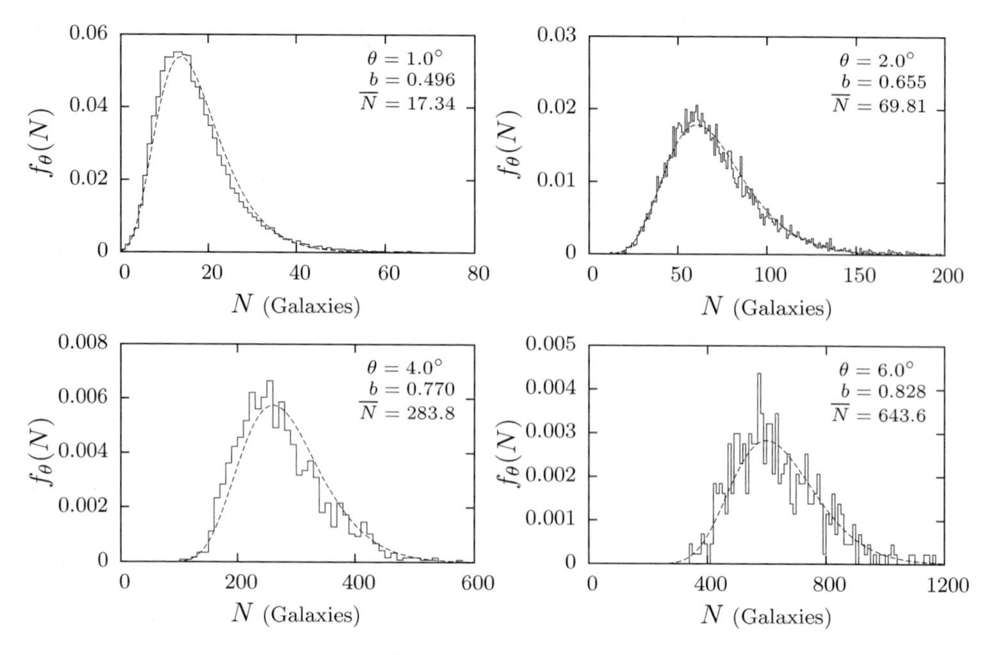

Fig. 11.10 The angularly projected GQED $f_\theta(N)$ for square cells with different values of θ, from Sivakoff and Saslaw (2005). The solid histogram shows data from observations. The dashed curve is from eqn (11.3). Values of b and $\overline{N}$ were found directly from the observations.

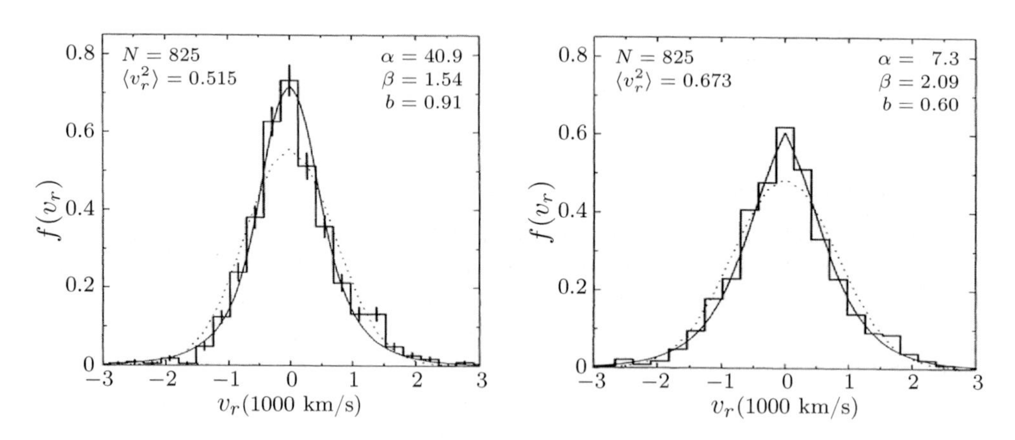

Fig. 11.11 The observed radial velocity distribution $f(v_r)$, from Raychaudhury and Saslaw (1996). The histogram shows data from observations. The solid lines are from eqn (11.22) and the dotted lines are best-fit Maxwell–Boltzmann distributions. The left panel shows the radial velocity distribution corrected for a bulk motion of 599 km/s, while the right panel shows the radial velocity distribution corrected for an extra Hubble expansion with an effective local Hubble constant of $H_0 = 92$ km/s/Mpc.

catalog is relatively small, with 1353 galaxies. Despite the small sample, a relatively good fit to the radial velocity distribution of eqn (11.22) was found, as illustrated in Fig. 11.11.

11.5 Conclusion

Although the cosmological many-body problem is essentially a long-range gravitating system, the expansion of the universe cancels the long-range gravitational mean field so that interactions effectively have a finite range. Using this property and its consequences, we have described the spatial and velocity distributions of particles in such a system. Its fundamental particles are galaxies. In the simplest case, the spatial distribution of galaxies can be characterized by just the average number of galaxies in a cell, $\overline{N}$, and a clustering parameter b, which is essentially the ratio of the interaction and kinetic energies. This assumes that galaxies are equal point masses, and that pairwise interactions are the dominant form of interactions. Relaxing these assumptions introduces minor higher-order corrections into b. However, the form of the spatial distribution does not change, suggesting it is very robust to a wide range of physical conditions.

The velocity distribution function can be derived consistently from the spatial distribution, which can also be used to obtain the volume integrals of the N-point correlation functions. From the thermodynamic quantities, the time evolution of b can be derived, describing the rate at which clusters form. With increasing b, the grand canonical ensemble gradually breaks up into a collection of microcanonical ensembles, each of which is a virialized cluster that individually has negative specific heat such that the average specific heat goes negative when b is greater than a critical value $b_{\text{crit}} = 0.8604$.

Finally, comparisons with N-body simulations and observations show that the GQED does indeed describe the physical world, at both low and high redshifts and for both spatial and velocity distributions.

Acknowledgments

We are both grateful to the organizers of the Les Houches Summer School for an environment of stimulating discussion. We particularly thank Thierry Dauxois and Michael Kastner for their comments on negative specific heat.

References

Ahmad F., Saslaw W. C., and Bhat N. I. (2002). *Astrophysical Journal*, **571**, 576–584.

Ahmad F., Malik M. A., and Masood S. (2006*a*). *International Journal of Modern Physics D*, **15**, 1267–1282.

Ahmad F., Saslaw W. C., and Malik M. A. (2006*b*). *Astrophysical Journal*, **645**, 940–949.

Baumann D., Leong B., and Saslaw W. C. (2003). *Monthly Notices of the Royal Astronomical Society*, **345**, 552–560.

Dauxois T., Ruffo S., Arimondo E., and Wilkens M. (2002). "Dynamics and thermodynamics of systems with long-range interactions: An introduction," in *Dynamics*

and Thermodynamics of Systems with Long-Range Interactions (ed. T. Dauxois, S. Ruffo, E. Arimondo, and M. Wilkens), Lecture Notes in Physics 602, pp. 1–19. Berlin: Springer-Verlag.
Eddington A. S. (1930). *Monthly Notices of the Royal Astronomical Society*, **90**, 668–678.
Hawkins E. *et al.* (2003). *Monthly Notices of the Royal Astronomical Society*, **346**, 78–96.
Hubble E. (1929). *Proceedings of the National Academy of Science*, **15**, 168–173.
Inagaki S., Itoh M., and Saslaw W. C. (1992). *Astrophysical Journal*, **386**, 9–18.
Itoh M., Inagaki S., and Saslaw W. C. (1988). *Astrophysical Journal*, **331**, 45–63.
Itoh M., Inagaki S., and Saslaw W. C. (1990). *Astrophysical Journal*, **356**, 315–331.
Itoh M., Inagaki S., and Saslaw W. C. (1993). *Astrophysical Journal*, **403**, 476–496.
Jeans J. H. (1902). *Royal Society of London Philosophical Transactions Series A*, **199**, 1–53.
Leong B. and Saslaw W. C. (2004). *Astrophysical Journal*, **608**, 636–646.
Perlmutter S. *et al.* (1999). *Astrophysical Journal*, **517**, 565–586.
Rahmani H., Saslaw W. C., and Tavasoli S. (2009). *Astrophysical Journal*, **695**, 1121–1126.
Raychaudhury S. and Saslaw W. C. (1996). *Astrophysical Journal*, **461**, 514–524.
Riess A. G. *et al.* (1998). *Astronomical Journal*, **116**, 1009–1038.
Saslaw, W. C. (1992). *Astrophysical Journal*, **391**, 423–428.
Saslaw W. C. (2000). *The Distribution of the Galaxies: Gravitational Clustering in Cosmology.* Cambridge: Cambridge University Press.
Saslaw W. C. and Ahmad F. (2009). In preparation.
Saslaw W. C. and Fang F. (1996). *Astrophysical Journal*, **460**, 16–27.
Saslaw W. C. and Hamilton A. J. S. (1984). *Astrophysical Journal*, **276**, 13–25.
Saslaw W. C. and Haque-Copilah S. (1998). *Astrophysical Journal*, **509**, 595–607.
Saslaw W. C., Chitre S. M., Itoh M., and Inagaki S. (1990). *Astrophysical Journal*, **365**, 419–431.
Sivakoff G. R. and Saslaw W. C. (2005). *Astrophysical Journal*, **626**, 795–808.
Spergel D. N. *et al.* (2007). *Astrophysical Journal Supplement*, **170**, 377–408.
Thirring W. (1970). *Zeitschrift für Physik*, **235**, 339–352.
Tolman R. C. (1938). *The Principles of Statistical Mechanics.* Oxford: Clarendon Press.
Totsuji H. and Kihara T. (1969). *Publications of the Astronomical Society of Japan*, **21**, 221–229.
Zhan Y. and Dyer C. C. (1989). *Astrophysical Journal*, **343**, 107–112.

12
A lecture on the relativistic Vlasov–Poisson equations

Michael K.-H. KIESSLING

Department of Mathematics,
Rutgers, The State University of New Jersey,
110 Frelinghuysen Rd., Piscataway, NJ 08854, USA

Preface

When the organizers invited me to their Les Houches Summer School on Long-Range Interacting Systems, they suggested that I lecture on Coulomb and Newton interactions. But the topic of Newtonian self-gravitating systems was already being addressed by two other lecturers at the summer school, and also Coulomb interactions were covered in other lectures, though not explicitly highlighted in their titles. To be sure, I expected to seriously disagree with some things that would be said in some of these lectures, but I decided rather to lecture on some physics not covered otherwise. The relativistic theories of electromagnetism and gravity came to my mind, but, precisely because relativity was not covered in the previous lectures, in my 1 1/2 hour lecture I had to focus on a single aspect that could easily be accommodated. The perfect topic—spherically symmetric relativistic dynamics—basically suggested itself: since there are no spherically symmetric electromagnetic or gravitational waves, the interactions are then governed by an elliptic equation, Poisson's equation for electromagnetic systems, and a more complicated set of equations which, in the weak-field limit reduces to Poisson's equation for gravitating systems. Happily, this combined the organizers' wishes that I lecture on Coulomb and Newton interactions with my own wishes to address relativistic problems.

The results presented in this lecture are extracted from a joint paper with my colleague A. Shadi Tahvildar-Zadeh at Rutgers. For this reason, this writeup of my lecture is only a summary of the things I said. It is meant as an appetizer, and anyone who feels hungry for the mathematical details is encouraged to read our paper, which at the time of writing of these lines is going to appear in the *Indiana University Mathematics Journal.*

12.1 The basic equations

At this summer school quite some time has been devoted already to the discussion of some Vlasov model or other, so that I can assume a certain familiarity with the type of equations that constitute such a model. In particular, a Vlasov–Poisson system of equations for a single-species N-body system in all space features a kinetic equation for the relative density function $f_t : \mathbb{R}^3 \times \mathbb{R}^3 \to \mathbb{R}_+$, given by

$$\left(\partial_t + v \cdot \nabla_q + \sigma \nabla_q \phi_t(q) \cdot \nabla_p\right) f_t(p, q) = 0\,, \tag{12.1}$$

in which the velocity $v \in \mathbb{R}^3$ and momentum $p \in \mathbb{R}^3$ of a (point) particle of unit mass are related either by Newton's formula

$$v = p \tag{12.2}$$

or by Einstein's formula (with the speed of light $c = 1$)

$$v = \frac{p}{\sqrt{1 + |p|^2}}\,, \tag{12.3}$$

and in which the scalar field $\phi_t : \mathbb{R}^3 \to \mathbb{R}_-$ satisfies Poisson's equation

$$\Delta_q \phi_t(q) = 4\pi \int_{\mathbb{R}^3} f_t(p,q)\,\mathrm{d}p\,; \tag{12.4}$$

finally, $\sigma \in \{-1,+1\}$ decides whether the gradient force field generated by ϕ is attractive ($\sigma = -1$) or repulsive ($\sigma = +1$). Equation (12.1) is supplemented by smooth initial data $f_0(p,q) \geq 0$, which are normalized thus, $\iint f_0(p,q)\,\mathrm{d}p\,\mathrm{d}q = 1$, and eqn (12.4) by some asymptotic conditions "at infinity"; here we choose

$$\phi_t(q) \asymp -|q|^{-1} \tag{12.5}$$

when $|q| \to \infty$, so that

$$\phi_t(q) = -\int_{\mathbb{R}^3} \left(|q-q'|^{-1} \int_{\mathbb{R}^3} f_t(p,q')\,\mathrm{d}p \right) \mathrm{d}q'\,. \tag{12.6}$$

Note that $\phi_t = -|\mathrm{id}|^{-1} * \int f_t\,\mathrm{d}p$ does *not* represent dynamical degrees of freedom beyond those of f_t. An overview of Vlasov–Poisson results up to 1997 can be found in Rein (1997).

12.2 Physical interpretation

In any Vlasov model, the function (or functions, if more than a single species is modeled) f_t is properly interpreted as a continuum approximation to the *actual microscopic particle density of an underlying N-body system.* By scaling, one can normalize f_t as I have done, so that f_t becomes a relative particle density. Thus, at any time t, and for any Borel set $B \subset \mathbb{R}^3 \times \mathbb{R}^3$ of macroscopic size, the actual fraction of particles in B is given by $\iint_B f_t(p,q)\,\mathrm{d}p\,\mathrm{d}q$. In order to be conceptually well conceived, this continuum approximation has to be a very good approximation. This implies that the number of particles of our N-body system is huge.

Unfortunately, there is a lot of confusion and misunderstanding in the literature about the physical meaning of f_t, so I shall add some words of caution. Namely, a relative particle density formally satisfies the criteria for a probability density function, but there is nothing nontrivially probabilistic about f_t which would be of any relevance in the Vlasov model. For if f_t had merely the probabilistic significance of conveying the incomplete information about an underlying random (vector) variable (P,Q) of a particle that at time t the probability for (P,Q) being in $\mathrm{d}p\,\mathrm{d}q$ around (p,q) is $f_t(p,q)\,\mathrm{d}p\,\mathrm{d}q$, then eqn (12.1) coupled with eqn (12.4) would make little sense. Indeed, this probabilistic interpretation in itself does not distinguish between a particle in an N-body system and a *single-particle system*, but in the latter case the dynamics of f_t cannot be determined by the gradient of a potential generated by this incomplete amount of information: the particle evolves according to its own dynamics, so why should it care about what I know about it? The correct equation for such a probability density is not the Vlasov kinetic equation but the linear Liouville equation, which would be eqn (12.1), where ϕ_t is a given (externally generated) potential field. It is important not to confuse these superficially similar objects.

After these words of caution, I return to the Vlasov–Poisson models (12.1) and (12.6), with either eqn (12.2) or eqn (12.3) in place, and ask: what are the underlying N-body systems? The integral representation (12.6) seems to make it plain that the underlying N-body system effectively interacts with either Newtonian (attractive) or Coulombian (repulsive) $1/r^2$ forces. With eqn (12.2) in place, the physical validity of this interpretation is well established, but with eqn (12.3) in place it is very questionable: relativistic particle motions influenced by instantaneous long-range pair interactions are not known to constitute a realistic physical model, although a mathematician may want to investigate them for other reasons. However, there is a physically viable interpretation of eqns (12.1) and (12.6) with eqn (12.3), but it is not in terms of an N-body system with Coulomb or Newton $1/r^2$ forces between the particles, and it requires restricting the model to spherically symmetric solutions.

Namely, while this so-called "relativistic Vlasov–Poisson system" (rVP) is not truly relativistic in the sense of proper Lorentz or general covariance, for spherical solutions rVP can actually be obtained from truly relativistic and physically relevant Vlasov models in certain limiting physical regimes; hence, rVP does have some physical significance. Its version with $\sigma = +1$ (rVP$^+$) is obtained directly from the relativistic Vlasov–Maxwell system (rVM) for a single species of electrically charged physical particles just by imposing spherically symmetric initial data—no further conditions are required; see Horst (1990), and see Glassey and Strauss (1986, 1987) and Klainerman and Staffilani (2002) for state-of-the-art results on rVM more generally. The reason is that there are no spherically symmetric magnetic fields, and therefore no electromagnetic waves with this symmetry. So, in the repulsive case, $\phi_t(q)$ can be thought of as Coulomb's electrical potential. The version with $\sigma = -1$ (rVP$^-$) is expected to emerge in a "weak-field limit" of the physically relevant general covariant Vlasov–Einstein system (VE) with spherically symmetric data. The reason is that there are no spherically symmetric gravitational waves in general relativity; the weak-field regime is necessary because the general-relativistic field equations for spherically symmetric potentials are not Poisson. So far, the only work we know (Rendall 1994) is on the combined weak-field plus low-velocity limit, which leads to the familiar nonrelativistic Vlasov–Poisson system (12.1), (12.6), with eqn (12.2) in place, and with $\sigma = -1$ (VP$^-$). In the attractive case, $\phi_t(q)$ may therefore be thought of as Newton's gravitational potential at the space point q at time t.

Unfortunately, this gravitational interpretation has to be taken with a grain of salt. Some interesting phenomena such as stationary bound states (Batt 1989; Hadžić and Rein 2007) and finite-time blowup (Glassey and Schaeffer 1985) (associated with gravitational collapse to a singularity; see Lemou *et al.* (2008*b*)) of rVP$^-$ occur in the *strong-field* regime, where rVP$^-$ can no longer be expected to be a legitimate approximation to VE.

In Kiessling and Tahvildar-Zadeh (2009), we give an unconventional physical interpretation of rVP$^-$ with spherical symmetry in terms of distributional solutions of rVM for a neutral two-species plasma which is not restricted to weak fields. In a nutshell, the basic idea behind this electromagnetic interpretation can be summarized thus. Consider an overall neutral system of N positive and N negative charges of equal magnitude, locally neutral and spherically distributed independently and identically.

Then each particle is acted on mainly by a net attractive central force, for from each particle's perspective the rest of the system is singly oppositely charged, as the rest of the system always contains one more of the oppositely charged than the equally charged particles. The force is (approximately) central by the (approximate) spherical symmetry of the $2N$-body system. In the limit $N \to \infty$ this is expected to become exact, but the mathematically rigorous vindication of this electrical interpretation of rVP$^-$ is an important open problem!

If mathematically confirmed, our interpretation would mean that the strong-field results for rVP$^-$ have interesting applications in space physics after all. Namely, while the electric force effectively due to a single charge would not seem to amount to much, the enormous difference between the gravitational and electric coupling constants implies that "electrical collapse in finite time" might actually be contributing to the dynamical formation of cosmic bodies with fewer than 10^{36} or so particles.

In the remainder of these notes, I summarize the main rigorous results for rVP$^-$ proved in Kiessling and Tahvildar-Zadeh (2009) and set them briefly into perspective.

12.3 Mathematical results

I begin by stipulating some notation. Thus, by $\mathfrak{P}(\mathrm{d}p\,\mathrm{d}q)$ we denote the probability measures on $\mathbb{R}^3 \times \mathbb{R}^3$ (momentum $\times$ physical space), by $\mathfrak{P}_n(\mathrm{d}p\,\mathrm{d}q)$ those having a finite nth moment, and by $(\mathfrak{P}_n \cap \mathfrak{L}^\alpha)(\mathrm{d}p\,\mathrm{d}q)$, $\alpha \geq 1$, those measures which are absolutely continuous with respect to the Lebesgue measure $\mathrm{d}p\,\mathrm{d}q$ on $\mathbb{R}^3 \times \mathbb{R}^3$ with density in $\mathfrak{L}^\alpha(\mathrm{d}p\,\mathrm{d}q)$; similarly, by $(\mathfrak{P}_n \cap \mathfrak{C}^1)(\mathrm{d}p\,\mathrm{d}q)$ we denote those whose density is in $\mathfrak{C}^1(\mathrm{d}p\,\mathrm{d}q)$, which are the functions with one continuous classical derivative.

Our main results presented in Kiessling and Tahvildar-Zadeh (2009) can be summarized as follows.

Theorem 12.3.1. *A unique global classical solution of rVP$^-$ exists for all spherically symmetric initial data $f_0 \in \mathfrak{P}_1 \cap \mathfrak{C}^1(\mathrm{d}p\,\mathrm{d}q)$ which are compactly supported in momentum space, vanish for $p \times q = 0$, and satisfy $\|f_0\|_{3/2} < C_{3/2}$, with $C_{3/2} = \frac{3}{8}\left(\frac{15}{16}\right)^{1/3} \approx 0.367$. The $\mathfrak{L}^{3/2}$ bound $C_{3/2}$ on f_0 is optimal in the sense that initial data f_0 exist which satisfy all the hypotheses except that $\|f_0\|_{3/2} > C_{3/2}$, and which lead to a blowup in finite time.*

Remark 12.3.2. *The critical case $\|f_0\|_{3/2} = \frac{3}{8}\left(\frac{15}{16}\right)^{1/3}$ is not covered by our theorem.*

The following comments put our Theorem 12.3.1 into perspective.

Some of the assumptions in the above global existence and uniqueness theorem are adapted from Glassey and Schaeffer (1985), in particular the compact-p-support condition and the condition that f_0 vanishes for $p \times q = 0$. We expect that the compact-support conditions can be replaced by sufficiently rapid decay "at infinity," and that $f_0 = 0$ for $p \times q = 0$ is not necessary.

All global-in-time solutions covered by Theorem 12.3.1 have positive energy, and those that blow up in finite time nonpositive energy. Among the initial data that lead to finite-time blowup there are indeed some sets of data with zero energy. This improves on Glassey and Schaeffer's result (Glassey and Schaeffer 1985) that negative-energy data will lead to finite-time blowup.

In Kiessling and Tahvildar-Zadeh (2009) we also prove a global existence and uniqueness theorem analogous to Theorem 12.3.1 with the sharp $\mathfrak{L}^{3/2}$ condition on f_0 replaced by a sharp $\mathfrak{L}^{\beta}$ condition on f_0 for any $\beta > 3/2$, with $C_{3/2}$ replaced by a corresponding sharp constant $C_\beta \geq C_{3/2}^{3(1-1/\beta)}$, and we give a variational principle for C_β. Besides the sharp $C_{3/2}$ given in Theorem 12.3.1, a lower bound on C_β for $\beta > 3/2$ follows from the interpolation inequality: $f_0 \in \mathfrak{P}_1 \cap \mathfrak{L}^{\beta}$ with $\beta > 3/2$ implies $f_0 \in P_1 \cap \mathfrak{L}^{3/2}$, with $\|f_0\|_\beta < C_{3/2}^{3(1-1/\beta)}$ implying $\|f_0\|_{3/2} < C_{3/2}$. We also give an explicit upper bound on C_β for $\beta > 3/2$ based on our variational principle. Meanwhile, Brent Young has evaluated this variational principle for C_β numerically.

When restricted to $\beta > 3/2$, our variational principle is equivalent to the variational principle (1.14) of Lemou *et al.* (2008*a*) for polytropes with a certain interval of possibilities for the polytropic index, mapped one-to-one into β. Polytropes have shown up in gravitational physics in many occasions, and their stability has been investigated in many works. Of particular relevance in the context of our work is Lemou *et al.* (2008*b*), where the dynamical collapse of supercritical solutions initially close to a polytrope was investigated.

Since $f_0 \in \mathfrak{P}_1 \cap \mathfrak{L}^{3/2}$ with $\|f_0\|_{3/2} < C_{3/2}$ does not imply any bound on $\|f_0\|_\beta$ for $\beta > 3/2$, our $\mathfrak{L}^{3/2}$ condition is weaker than any of the possible $\mathfrak{L}^{\beta}$ conditions with $\beta > 3/2$. In fact, our $\mathfrak{L}^{3/2}$ bound is the weakest possible $\mathfrak{L}^{\beta}$ condition for which an analog of Theorem 12.3.1 can be formulated, in the sense that our $\mathfrak{L}^{3/2}$ bound on f_0 cannot be replaced by an $\mathfrak{L}^{\beta}$ bound with $\beta \in (1, 3/2)$, whatever that bound. Indeed, among the sets of initial data f_0 satisfying any such $\mathfrak{L}^{\beta}$ bound with $\beta < 3/2$ are some with negative energy, and which lead to a blowup in finite time by Glassey and Schaeffer's blowup theorem (evidently, $\|f_0\|_{3/2} > C_{3/2}$ for those sets of data).

By the previous two remarks, everything else being equal, our sharp $\mathfrak{L}^{3/2}$ condition on f_0 improves on the (nonsharp) $\mathfrak{L}^{\infty}$ condition on f_0 given in Glassey and Schaeffer (1985). For a follow-up on Glassey and Schaeffer (1985), see Glassey and Schaeffer (2001).

In contrast to rVP$^-$, classical solutions of nonrelativistic VP$^-$ do not blow up but stay smooth forever (Pfaffelmoser 1992; Schaeffer 1991). Unfortunately, this does not seem to help in settling the still open Newtonian N-body problem, because the Vlasov time scale is “infinitely” short compared with the N-body time scale. More precisely, in the Vlasov limit for a sequence of N-body systems with fixed particle parameters, one follows more and more particles over shorter and shorter times. Thus, arbitrarily long “Vlasov times” do not correspond to infinite time spans in the underlying N-body problems.

We wonder how much of our other results generalizes to nonspherical solutions. This is not just of mathematical interest. Although the sphericity assumption essentially defines the realm of physical validity of rVP$^-$, spherical symmetry is never a perfect symmetry of nature. Therefore it is important to show that significant qualitative results do not depend sensitively on having exact spherical symmetry.

Acknowledgments

I thank A. Shadi Tahvildar-Zadeh at Rutgers for a wonderful collaboration, and my student Brent Young for the numerical results shown in my lecture. The first partial results were obtained while visiting CNRS-Université de Provence in 2001 upon the invitation of Yves Elskens, supported in part by CNRS through a *poste rouge* and in part by NSF grant DMS-0103808. The work was finalized while I was supported by NSF grant DMS-0406951. The NSF nowadays requests the following disclaimer from all their grantees: "Any opinions, conclusions, or recommendations expressed in this material are those of the author and do not necessarily reflect those of the NSF." I immensely enjoyed my week at the Les Houches summer school and thank Leticia Cugliandolo, Thierry Dauxois, and Stefano Ruffo for their superb organization and their support. My thanks go also to their secretaries for their efficient help and friendly assistance in all matters small and large.

References

Batt J. (1989). Steady state solutions of the relativistic Vlasov–Poisson system, in *Proceedings of the Fifth Marcel Grossmann Meeting on General Relativity*, Perth, 1988, pp. 1235–1247, World Scientific.

Glassey R. T. and Schaeffer J. (1985). On symmetric solutions to the relativistic Vlasov–Poisson system. *Commun. Math. Phys.*, **101**, 459–473.

Glassey R. T. and Schaeffer J. (2001). On global symmetric solutions to the relativistic Vlasov–Poisson equation in three dimensions. *Math. Methods Appl. Sci.*, **24**, 143–157.

Glassey R. and Strauss W. (1986). Singularity formation in a collisionless plasma could occur only at high velocities. *Arch. Rat. Mech. Anal.*, **92**, 59–90.

Glassey R. and Strauss W. (1987). Absence of shocks in an initially dilute collisionless plasma. *Commun. Math. Phys.*, **113**, 191–208.

Hadžić M. and Rein G. (2007). Global existence and nonlinear stability for the relativistic Vlasov–Poisson system in the gravitational case. *Indiana Univ. Math. J.*, **56**, 2453–2488.

Horst E. (1990). Symmetric plasmas and their decay. *Commun. Math. Phys.*, **126**, 613–633.

Kiessling M. K.-H. and Tahvildar-Zadeh A. S. (2009). On the relativistic Vlasov–Poisson system. *Indiana Univ. Math. J.*, published online Jan. 5th, 2009.

Klainerman S. and Staffilani G. (2002). A new approach to study the Vlasov–Maxwell system. *Commun. Pure Appl. Anal.*, **1**, 103–125.

Lemou M., Méhats F., and Raphaël P. (2008*a*). Structure of the linearized gravitational Vlasov–Poisson system close to a polytropic ground state. *SIAM J. Math. Anal.*, **39**, 1711–1739.

Lemou M., Méhats F., and Raphaël P. (2008*b*). Stable self-similar blow-up dynamics for the three-dimensional relativistic Vlasov–Poisson system. *J. Am. Math. Soc.*, **21**, 1019–1063.

Pfaffelmoser K. (1992). Global classical solutions of the Vlasov–Poisson system in three dimensions with generic initial data. *J. Differ. Equ.*, **95**, 281–303.
Rein G. (1997). Self-gravitating systems in Newtonian theory – the Vlasov–Poisson system, in *Mathematics of Gravitation*, Part I, Banach Center Publications **41**, pp. 179–194, Warsaw.
Rendall A. (1994). The Newtonian limit for asymptotically flat solutions of the Vlasov–Einstein system. *Commun. Math. Phys.*, **163**, 89–112.
Schaeffer J. (1991). Global existence of smooth solutions to the Vlasov–Poisson system in three dimensions. *Commun. PDE*, **16**, 1313–1335.

Part V

Coulomb and wave–particle interaction

13
Plasma collisional transport

Daniel H. E. Dubin

University of California at San Diego, Physics Department 0319, 9500 Gilman Drive, La Jolla, CA 92093, USA

13.1 Estimates

The question addressed in the following lectures dates from the earliest days of plasma physics research: how do collisions in a plasma affect the transport (i.e. redistribution) of energy, momentum, and particles? Surprisingly, several aspects of this venerable problem remain incompletely understood. As befits lectures at a school on long-range interactions, many of these thorny issues relate to the long-range nature of the Coulomb interaction.

We will first review some basic concepts relating to collisions in plasmas. Consider a particle of mass m with velocity $\mathbf{v} = (v_x, v_y, v_z)$. The probability density associated with the velocity is, in thermal equilibrium, given by a Maxwellian distribution,

$$f_{\max}(\mathbf{v}) = \frac{e^{-mv^2/2T}}{(\sqrt{2\pi T/m})^3}, \tag{13.1}$$

where T is the temperature. The thermal speed $\bar{v}$ of a collection of particles with temperature T is the r.m.s. velocity associated with this distribution,

$$\begin{aligned}\bar{v}^2 &= \int d^3v\, v_x^2 f_{\max}(v) = \int d^3v\, v_y^2 f_{\max} = \int d^3v\, v_z^2 f_{\max} \\ &= T/m. \end{aligned} \tag{13.2}$$

Exercise 13.1 For two particles taken from the distribution (13.1), show that the distribution of their relative velocities, $\mathbf{v}_r = \mathbf{v}_1 - \mathbf{v}_2$, is also Maxwellian but at twice the temperature:

$$f_{\mathrm{rel}}(v_r) = \frac{e^{-mv_r^2/4T}}{(\sqrt{4\pi T/m})^3}. \tag{13.3}$$

To estimate the rate of collisions ν between particles, assuming they all have charge e, we rather arbitrarily define this rate as the mean rate at which the relative velocity of two colliding particles scatters by 90° or more. Two particles separated by an impact parameter ρ (see Fig. 13.1) and with an initial relative velocity $\mathbf{v}_r = \mathbf{v}_1 - \mathbf{v}_2$ require, for scattering by 90° or more, that

$$\rho \lesssim \frac{e^2}{mv_r^2}. \tag{13.4}$$

The rate ν at which such collisions occur is

$$\nu = \int d^3v_r\, n f_{\mathrm{rel}}(v_r) v_r \times \int_0^{e^2/mv_r^2} 2\pi\rho\, d\rho, \tag{13.5}$$

where n is the particle density. The first integral in this expression gives the flux of particles impacting on a given charge, and the second integral, over impact parameters,

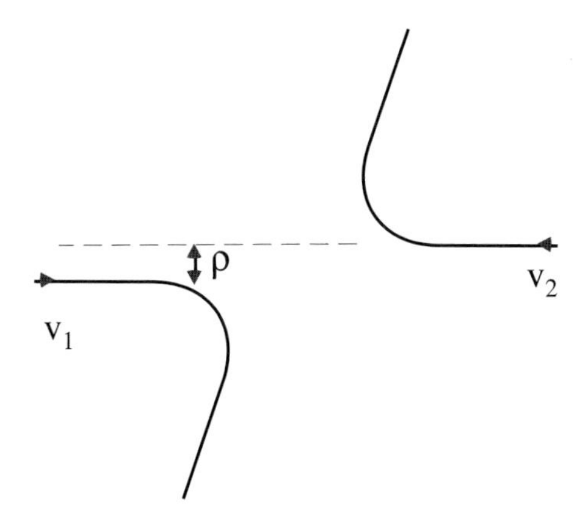

Fig. 13.1 Collision between two particles in an unmagnetized plasma.

estimates the cross section for 90° scattering. Performing the ρ integral and using eqn (13.3) yields

$$\nu = \frac{\sqrt{\pi}}{2} n \bar{v} b^2 \int_{v_{\min}}^{\infty} \frac{dv_r \, e^{-v_r^2/4\bar{v}^2}}{v_r} , \tag{13.6}$$

where $b \equiv e^2/T$ is the distance of closest approach, and we have introduced a lower cutoff $v_{\min}$ into the relative-velocity integral since otherwise the integral diverges logarithmically. Such divergences are a common problem in plasma kinetic theory, and extra physical effects must be added to the model to determine the cutoffs. In this case, we note that $v_r \to 0$ is equivalent to the maximum impact parameter approaching ∞ (see eqn (13.4)), so $v_{\min}$ is set by the maximum reasonable ρ. This is usually taken to be the Debye length λ_D, defined as

$$\lambda_D \equiv \sqrt{\frac{T}{4\pi e^2 n}} , \tag{13.7}$$

which implies that

$$v_{\min} = \sqrt{\frac{e^2}{m \lambda_D}} . \tag{13.8}$$

The Debye length is the length over which potentials are shielded out in a plasma. If a stationary test charge Q is placed in a plasma consisting of two species with charges $\pm e$ at densities $n_+ = n_- = n$, in thermal equilibrium the density is perturbed by the presence of Q according to Boltzmann distributions

$$n_{\substack{+ \\ (-)}} = n \, e^{\mp e\phi/T} , \tag{13.9}$$

where ϕ is the potential, which satisfies Poisson's equation,

$$\nabla^2 \phi = -4\pi Q \delta(\mathbf{r}) - 4\pi e n_+ + 4\pi e n_- . \tag{13.10}$$

Substituting for n_+ and n_-, linearizing in ϕ (assuming $e\phi/T \ll 1$), and solving yields

$$\phi(r) = Q\,\frac{e^{-\sqrt{2}r/\lambda_D}}{r}. \tag{13.11}$$

The factor of $\sqrt{2}$ arises because two species do the shielding: like charges are repelled from Q and opposite charges are attracted, and each shielding effect is additive. Returning to our expression for the collision rate, we note that $v_{\min} \ll \bar{v}$ for typical plasmas for which $\lambda_D/b \gg 1$, in which case, to "logarithmic accuracy," the velocity integral in eqn (13.6) can be approximated by

$$\int_{v_{\min}}^{\bar{v}} \frac{dv_r}{v_r} \cong \ln\left(\frac{\bar{v}}{v_{\min}}\right) = \frac{1}{2}\ln\left(\frac{\lambda_D}{b}\right), \tag{13.12}$$

where, in the second form, we have substituted for $v_{\min}$ from eqn (13.8). This approximation to the actual integral in eqn (13.6) neglects an additive constant of order unity, which is reasonable provided that $\ln(\lambda_D/b) \gg 1$. Typical values of $\ln(\lambda_D/b)$ are of order 10 in the experiments that we will consider later. In other words, a change in our estimate of $v_{\min}$ by a factor of 2 causes only a small change in ν. That is why only an estimate of $v_{\min}$ is needed. Thus, the collision rate in a plasma scales as (Spitzer 1956; Montgomery and Tidman 1964)

$$\nu \sim n\bar{v}\,b^2 \ln\left(\frac{\lambda_D}{b}\right) \tag{13.13}$$

$$\equiv \nu_0 \ln\left(\frac{\lambda_D}{b}\right). \tag{13.14}$$

As the density increases, ν increases since collisions become more likely; and as the temperature increases, ν decreases because higher-velocity particles are harder to deflect. The rate $\nu_0 \equiv n\bar{v}b^2$ will appear often in the following analyses.

Exercise 13.2 Estimate ν for a virialized globular cluster of 10^6 solar-mass stars, with mean density 1 star/cubic light year. Note that Debye shielding does not occur in self-gravitating systems, so λ_D must be replaced by the system radius.

13.1.1 Transport in unmagnetized plasmas

The mean free path ℓ of a particle is the mean distance the particle travels between collisions, given by

$$\ell \simeq \frac{\bar{v}}{\nu}. \tag{13.15}$$

After each collision time ν^{-1}, the particle's velocity is randomized, so that its path resembles a "random walk" with a step size of order ℓ. This leads to diffusion of a distribution of such particles: the mean square change in position increases linearly with time t, as

$$\langle[\mathbf{r}(t) - \mathbf{r}(0)]^2\rangle = 6Dt\,. \tag{13.16}$$

Here, the average $\langle\ \rangle$ is over the distribution of initial conditions and D is the particle diffusion coefficient, which may be estimated as

$$D \cong \nu \ell^2 \sim \frac{\bar{v}^2}{\nu}\,. \tag{13.17}$$

Equation (13.17) also holds in gases with short-range interactions; the long-range nature of the Coulomb interaction does not affect this estimate. A smaller collision frequency implies larger diffusion, owing to the increase in the mean free path.

The particle density $n(x,t)$ for a diffusive process follows the diffusion equation

$$\frac{\partial n}{\partial t} = \frac{\partial}{\partial x} D \frac{\partial n}{\partial x}\,. \tag{13.18}$$

Similar equations govern the transport of energy and momentum in a plasma. A plasma with a nonuniform temperature $T(x,t)$ evolves through collisions according to the heat equation,

$$C\frac{\partial T}{\partial t} = \frac{\partial}{\partial x}\kappa\frac{\partial T}{\partial x}\,, \tag{13.19}$$

where C is the specific heat and κ is the thermal conductivity. The ratio $\chi = \kappa/C$ has units of a diffusion coefficient and is termed the thermal diffusivity. Similarly, momentum transport in a system in which the fluid velocity is sheared, $\mathbf{V} = V_y(x,t)\hat{y}$, is also governed by a diffusion equation,

$$mn\frac{\partial V_y}{\partial t} = \frac{\partial}{\partial x}\eta\frac{\partial V_y}{\partial x}\,, \tag{13.20}$$

where η is the shear viscosity (we do not consider flows for which the bulk viscosity is important). The ratio $\lambda = \eta/mn$ also has dimensions of a diffusion coefficient and is called the kinematic viscosity.

In an unmagnetized plasma,

$$\chi \sim \lambda \sim D\,, \tag{13.21}$$

since particles carry their energy and momentum with them as they diffuse (Spitzer 1956; Simon 1955; Spitzer and Harm 1952).

13.1.2 Magnetized plasma: Classical theory of collisional transport

When a uniform magnetic field $\mathbf{B} = B\hat{z}$ is applied, collisional transport is reduced in the directions transverse to the field. This is because an isolated charge no longer travels in a straight-line trajectory, but rather executes circular cyclotron motion with radius $r_c = v_\perp/\Omega_c$, where $v_\perp = \sqrt{v_x^2 + v_y^2}$ is the perpendicular particle speed and $\Omega_c = eB/mc$ is the cyclotron frequency, which is the frequency (in radians/sec) of the circular motion. For an isolated particle, the center of the circular orbit, termed the "guiding center" position, is fixed (except for the uniform motion along $\mathbf{B}$). However,

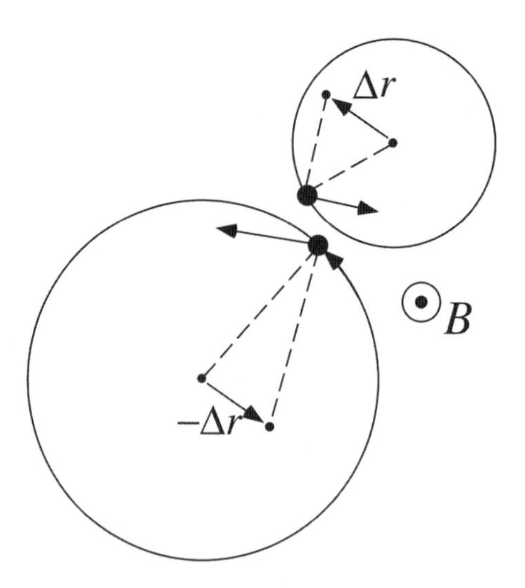

Fig. 13.2 Typical collision between two like particles in a magnetized plasma.

when two particles collide, the guiding centers of the two particles step across the magnetic field.

In classical transport theory, this cross-field step happens because of the velocity scattering that we described previously in an unmagnetized plasma. In the collision, the perpendicular velocity $\mathbf{v}_\perp$ for each particle changes, leading to a change $\Delta\mathbf{r}$ in the guiding-center position; see Fig. 13.2. The mean size of such steps is of order

$$\bar{r}_c = \bar{v}/\Omega_c \,, \tag{13.22}$$

the mean cyclotron radius. The rate of these steps is the collision rate ν, so we can estimate the diffusion coefficient as

$$D = \nu \bar{r}_c^2 \,. \tag{13.23}$$

Comparing this with eqn (13.17), we see that the mean free path has been replaced by the cyclotron radius because the magnetic field limits the cross-field motion to a distance of order $\bar{r}_c$. Again, the thermal diffusivity and kinematic viscosity are of order D. These coefficients were worked out rigorously in the early days of plasma physics (Spitzer 1956; Longmire and Rosenbluth 1956; Simon 1955; Rosenbluth and Kaufman 1958; Braginskii 1958, 1965), and are given in Table 13.1.[1] Interestingly, to my knowledge none of these coefficients have ever been measured experimentally to better than order-of-magnitude accuracy, mainly because of the difficulty of producing a quiescent magnetized plasma where collisional transport is not swamped by competing effects such as instabilities or turbulence.

[1]Not all authors agree on the values of the numerical coefficients for χ and λ. The coefficients quoted in Table 13.1 are those given in Braginskii (1965) for ion–ion collisions.

Table 13.1 Classical theory of collisional transport[a]

D	χ	λ
$\frac{4}{3}\sqrt{\pi}\nu_0\bar{r}_c^2 \ln\left(\frac{\rho_{\max}}{b}\right)$	$\frac{16}{9}\sqrt{\pi}\nu_0\bar{r}_c^2 \ln\left(\frac{\rho_{\max}}{b}\right)$	$\frac{2}{5}\sqrt{\pi}\nu_0\bar{r}_c^2 \ln\left(\frac{\rho_{\max}}{b}\right)$

[a] $\nu_0 = n\bar{v}b^2, \quad \rho_{\max} = \begin{cases} \lambda_D\,, & \lambda_D < \bar{r}_c\,, \\ \bar{r}_c\,, & \lambda_D > \bar{r}_c\,. \end{cases}$

13.1.3 Long-range collisions

The classical theory has been used for decades to evaluate transport due to collisions in a magnetized plasma. However, the validity of the theory is limited by the assumption that transport is due primarily to collisions that scatter the perpendicular velocity vector. In fact, collisions with large impact parameters ("long-range" collisions) that do not lead to appreciable velocity scattering can dominate the collisional transport.

For example, consider the situation where the magnetic field is sufficiently strong that $\lambda_D > \bar{r}_c$. The velocity-scattering collisions pictured in Fig. 13.2 occur only for guiding centers separated across $\mathbf{B}$ by a distance of order $\bar{r}_c$; but many more collisions occur with larger impact parameters ρ, of order λ_D. These collisions do not look like those in Fig. 13.2; such a collision is shown in Fig. 13.3.

No appreciable perpendicular velocity scattering occurs in these collisions, but momentum and energy are still transported across the magnetic field. For instance, consider energy transport. Through such long-range collisions, a particle on one magnetic field line can transfer parallel energy to a particle on a field line separated by a Debye length (by exchange of parallel velocities). The distance over which energy is transferred is now λ_D, with exchanges occurring at a rate ν, so the thermal diffusivity is

$$\chi \sim \nu\lambda_D^2\,. \tag{13.24}$$

This is much larger than that given by the classical theory (Table 13.1) when $\lambda_D > \bar{r}_c$ (Psimopoulos and Li 1992; Dubin and O'Neil 1997). A similar argument for momentum transport yields the kinematic viscosity

$$\lambda \sim \nu\lambda_D^2 \tag{13.25}$$

as well (O'Neil 1985).

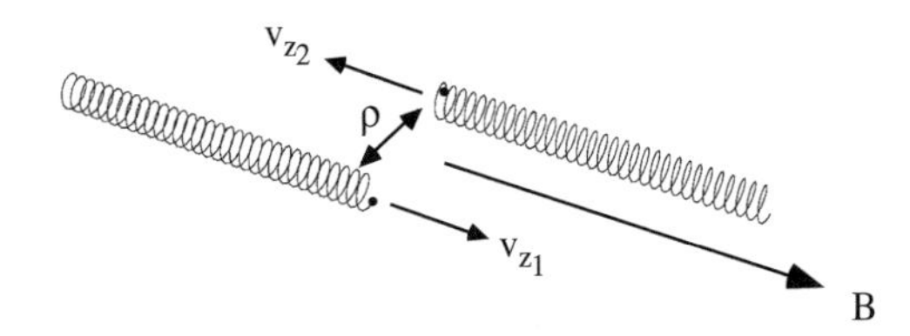

Fig. 13.3 Typical collision with impact parameter $\rho \sim \lambda_D$ between two particles in a magnetized plasma where $\bar{r}_c \ll \lambda_D$.

Furthermore, the exchange of energy and momentum is not limited to interactions between particles separated by only λ_D. Particles can transfer energy by emitting and absorbing weakly damped waves that travel large distances across the plasma (Rosenbluth and Liu 1976; Dubin and O'Neil 1997). While the impact parameter for such "collisions" is limited only by the plasma size, the effective rate of collisions is smaller than ν because the fluctuation energy in lightly damped plasma waves is a small fraction of the total fluctuation energy in the plasma. Even so, we will see that detailed calculations predict that lightly damped waves can dominate the energy and momentum transport in plasmas that are sufficiently large.

Particle diffusion also occurs owing to the long-range collisions shown in Fig. 13.3. The origin of the diffusion is the $\mathbf{E} \times \mathbf{B}$ drift that occurs as charges encounter one another (Lifshitz and Pitaevskii 1981). To understand $\mathbf{E} \times \mathbf{B}$ drifts, consider the dynamics of a particle in a uniform magnetic field $B\hat{z}$ *and* a uniform electric field $E\hat{x}$. The particle's motion is given by

$$m\frac{d\mathbf{v}}{dt} = e\left(\mathbf{E} + \frac{\mathbf{v} \times \mathbf{B}}{c}\right). \tag{13.26}$$

The solution to this linear ordinary differential equation for $\mathbf{v}(t)$ is a sum of a particular and a homogeneous solution. The homogeneous solution is simply the circular cyclotron motion described previously. The particular solution (for the case where $\mathbf{E} \perp \mathbf{B}$) is

$$\mathbf{v}_{E\times B} = \frac{\mathbf{E} \times \mathbf{B}}{B^2}c = -\frac{E}{B}c\hat{y}\,. \tag{13.27}$$

This velocity, the $\mathbf{E} \times \mathbf{B}$ drift, is superimposed on the cyclotron motion.

Now consider a collision between two particles, shown in Fig. 13.3. Assuming that $\Omega_c \gg v_r/\lambda_D$ and $\lambda_D \gg r_c$, the electric field between the charges varies slowly compared with the cyclotron motion and can be taken to be nearly uniform in space and time. This immediately leads to an $\mathbf{E} \times \mathbf{B}$ drift velocity due to the perpendicular component of $\mathbf{E}$, given by eqn (13.27), which for an electric field of order e/λ_D^2 yields

$$v_{E\times B} \sim \frac{e}{\lambda_D^2 B}c\,. \tag{13.28}$$

This drift acts for a time $t_d \sim \lambda_D/v_r \sim \lambda_D/\bar{v}$ as particles pass one another, implying a cross-magnetic-field step $\Delta r = v_{\mathbf{E}\times\mathbf{B}}t_d$, or

$$\Delta r \sim \frac{e}{\lambda_D B}\frac{c}{\bar{v}}\,. \tag{13.29}$$

The rate of these collisions is roughly $n\bar{v}\lambda_D^2$, so the particle diffusion coefficient D is

$$\begin{aligned} D &\sim n\bar{v}\lambda_D^2\,\Delta r^2 \\ &= \nu_0\bar{r}_c^2\,, \end{aligned} \tag{13.30}$$

neglecting constants of order unity and logarithmic factors. Thus $\mathbf{E} \times \mathbf{B}$ drift diffusion due to long-range collisions has roughly the same scaling as the diffusion due to the velocity scattering described by the classical theory (Lifshitz and Pitaevskii 1981; Anderegg *et al.* 1997; Dubin 1997). In lecture 2 we will see that in one set of experiments the $\mathbf{E} \times \mathbf{B}$ diffusion is about 10 times the diffusion predicted by the classical theory.

It should be noted that velocity-scattering collisions still occur when $\lambda_D > \bar{r}_c$, owing to collisions with impact parameter $\rho \lesssim \bar{r}_c$. The classical transport coefficients describe these collisions, except that the maximum impact parameter appearing in the Coulomb logarithm is no longer λ_D, but rather $\bar{r}_c$ (Montgomery *et al.* 1974). When $\lambda_D > \bar{r}_c$, the total transport is a sum of classical transport due to collisions with impact parameters ρ less than $\bar{r}_c$, and long-range transport due to collisions with $\rho > \bar{r}_c$.

13.2 Kinetic theory of $\mathbf{E} \times \mathbf{B}$ drift diffusion, and experiments

13.2.1 Integration along unperturbed orbits

In this lecture we will rigorously calculate the diffusion due to $\mathbf{E} \times \mathbf{B}$ drifts that we estimated in the previous lecture, and we will compare the result with experimental measurements (Anderegg *et al.* 1997). As before, we assume an infinite uniform plasma in the regime $\lambda_D > \bar{r}_c$, and concentrate on the motion of the guiding centers only, since cyclotron motion is not important in this process.

We consider the interaction of two like particles, labeled 1 and 2, following the motion of the guiding centers. The equations of motion for the guiding center of particle 1, at position $\mathbf{r}_1 = (x_1, y_1, z_1)$ with axial velocity v_{z_1}, are

$$\frac{dx_1}{dt} = -\frac{c}{eB}\frac{\partial \phi_{12}}{\partial y_1}, \quad \frac{dy_1}{dt} = \frac{c}{eB}\frac{\partial \phi_{12}}{\partial x_1}, \quad \frac{dz_1}{dt} = v_{z_1}, \quad \frac{dv_{z_1}}{dt} = -\frac{1}{m}\frac{\partial \phi_{12}}{\partial z_1}, \tag{13.31}$$

where $\phi_{12} = e^2/|\mathbf{r}_1 - \mathbf{r}_2|$ is the Coulomb potential energy. For particle 2, the equations of motion are obtained by interchanging labels 1 and 2 in eqn (13.31).

As the particles pass one another, they step across the magnetic field. The step in the x-direction for particle 1, δx_1, is given by integrating dx_1/dt:

$$\delta x_1 = \frac{ce}{B}\int_{-\infty}^{\infty} dt \frac{y_1 - y_2}{[(x_1 - x_2)^2 + (y_1 - y_2)^2 + (z_1 - z_2)^2]^{3/2}} . \tag{13.32}$$

In order to evaluate this integral, we assume that the interaction between the particles only weakly perturbs the particle orbits, so we use the unperturbed orbits in the integrand, taking $z_r \equiv z_1 - z_2 = v_r t$, $x_1 - x_2 = \text{const.}$, $y_1 - y_2 = \text{const.}$ This well-known approximation method is called the method of integration along unperturbed orbits (IUO). Then eqn (13.32) yields

$$\delta x_1 = \frac{2ce}{B|v_r|\rho^2}(y_1 - y_2), \tag{13.33}$$

where $\rho = \sqrt{(x_1 - x_2)^2 + (y_1 - y_2)^2}$ is the impact parameter. This result has the same scaling as the previous estimate, eqn (13.29).

The diffusion of particle 1 can now be determined as a series of uncorrelated steps δx_1 due to collisions with an incident flux of particles 2 streaming past 1 along the magnetic field:

$$D = \frac{1}{2}\frac{\langle \Delta x^2 \rangle}{\Delta t} = \frac{1}{2}\int_{-\infty}^{\infty} dv_r \, n f_{\rm rel}(v_r) \int_0^{\infty} \rho \, d\rho \int_0^{2\pi} d\theta \, |v_r| (\delta x_1)^2 \,. \tag{13.34}$$

By substituting for $(\delta x_1)^2$ from eqn (13.33) and for the distribution of relative z-velocities $f_{\rm rel}(v_r) = e^{-mv_r^2/4T}/\sqrt{4\pi T/m}$, the integral over θ can be performed but the integrals over v_r and ρ must be cut off owing to logarithmic divergences:

$$D = \frac{1}{2}\left(\frac{2ce}{B}\right)^2 \frac{n}{\sqrt{4\pi T/m}} 2 \int_{\rho_{\rm min}}^{\rho_{\rm max}} \frac{d\rho}{\rho} \int_{v_{\rm min}}^{\infty} \frac{dv_r}{|v_r|} e^{-mv_r^2/4T} \,. \tag{13.35}$$

We deal with the logarithmically divergent ρ integral by positing that Debye shielding acts to cut off the long ranges, giving $\rho_{\rm max} = \lambda_D$, and that the $\mathbf{E} \times \mathbf{B}$ drift approximation breaks down at short ranges, giving $\rho_{\rm min} = \bar{r}_c$. For $\rho < \bar{r}_c$, the velocity-scattering collisions described by the classical theory dominate. The classical diffusion arising from this range of impact parameters, given in Table 13.1, must be added to eqn (13.35).

The divergence in v_r is a bit more subtle. It comes about because when $v_r \to 0$, particles may interact for a long time and take a correspondingly large drift step—see eqn (13.33). Here we note that, in reality, particle velocities do not remain constant forever—collisions with surrounding particles cause v_r to diffuse, so even if it is initially zero it does not remain so. On average, particles with $v_r = 0$ initially will obey

$$\langle v_r^2 \rangle = 2D_v t \,, \tag{13.36}$$

where D_v is the velocity diffusion coefficient due to collisions, $D_v \sim \nu \bar{v}^2$. Since, on average, $\sqrt{\langle v_r^2 \rangle}$ increases like $\sqrt{D_v t}$, the r.m.s. relative z position $\sqrt{\langle z_r^2 \rangle}$ will also increase, like $\sqrt{D_v}\, t^{3/2}$. The interaction between the particles is reduced by order unity once z_r has increased from 0 to ρ, which requires a time of order $(\rho/\sqrt{D_v})^{2/3}$. The mean relative speed over this time is $v_{\rm min} = \rho(\sqrt{D_v}/\rho)^{2/3} = (D_v \rho)^{1/3}$. Using this result in eqn (13.35) yields the following expression for the diffusion coefficient across the magnetic field due to $\mathbf{E} \times \mathbf{B}$ drifts in the regime $\lambda_D > \bar{r}_c$:

$$D_{\rm IUO} = 2\sqrt{\pi}\nu_0 r_c^2 \ln\left(\frac{\bar{v}}{|D_v \sqrt{\lambda_D \bar{r}_c}|^{1/3}}\right) \ln\left(\frac{\lambda_D}{\bar{r}_c}\right), \tag{13.37}$$

where $\nu_0 \equiv n\bar{v}b^2$.

We have labeled the diffusion coefficient as $D_{\rm IUO}$ because integration along unperturbed orbits was used in the derivation. We will find that this method needs to be modified, because integration along unperturbed orbits contains a subtle error.

Another effect, shear in the background plasma flow, can also set an effective minimum relative velocity. Particles separated by ρ would then have $v_{\rm min} \sim S\rho$, where $S = |\nabla \mathbf{V}|$ is the shear rate of the flow. Equation (13.37) assumes that S is sufficiently small that $v_{\rm min}$ is set by collisions, not shear.

13.2.2 Experiments

We will now compare this prediction with experimental measurements, performed in a pure ion plasma by our group at UCSD (Anderegg *et al.* 1997). The plasma was a

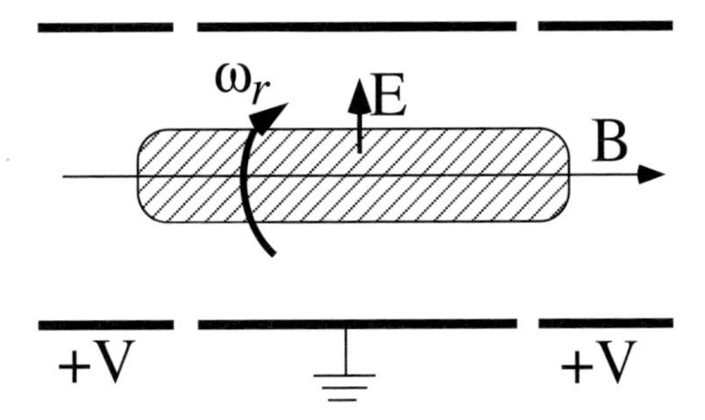

Fig. 13.4 Nonneutral plasma confined in a Malmberg–Penning trap.

collection of roughly 10^{10} Mg^+ ions confined in a Malmberg–Penning trap geometry, shown in Fig. 13.4. The cylindrical electrodes were biased so as to provide an axial potential well for the ions. Radial confinement was provided by an axial magnetic field. In such an apparatus, the plasma rotates about the axis of symmetry at a frequency ω_r, providing a $\mathbf{v} \times \mathbf{B}$ force to balance the radial electric field E_r: radial force balance implies

$$\frac{e\omega_r r B}{c} = eE_r + m\omega_r^2 r\,, \tag{13.38}$$

and axial force balance implies $E_z = 0$ in the plasma. Solving eqn (13.38) for E_r and applying Poisson's equation $\nabla \cdot \mathbf{E} = 4\pi e n$, we obtain the following relation between rotation frequency and density:

$$n = \frac{m\omega_r}{2\pi e^2}(\Omega_c - \omega_r)\,. \tag{13.39}$$

Pure ion plasmas can be confined in a quiescent near-thermal-equilibrium state for arbitrarily long time periods (Dubin and O'Neil 1999). In the experiments, $n \sim 10^7\,\mathrm{cm}^{-3}$, B ran from 1 to 4 T, and T could be varied over the range $0.03\,\mathrm{eV} < T < 3\,\mathrm{eV}$. For such plasmas, the Debye length is greater than the cyclotron radius, so transport due to long-range collisions is an important effect.

Exercise 13.3 Show that the maximum possible density (called the Brillouin density, n_B) is related to the magnetic field by

$$n_B = m\Omega_c^2/8\pi e^2\,,$$

and find n_B for Mg^+ ions in a 1 T magnetic field.

In order to measure diffusion across the magnetic field, lasers were used to tag some of the ions via their spins. First, a laser directed across the column intersected all ions as the plasma rotated and pumped them all into the $s_z = +1/2$ spin state. This beam was then turned off and a second "tagging" beam, directed along the trap axis, pumped ions in the beam into the $s_z = -1/2$ state. This could be done very quickly compared with the rate at which particles diffused out of the beam. After this tagging beam was turned off, the $s_z = -1/2$ ions (the "test particles") diffused across the magnetic field. A weak probe beam that did not change the spin was then used to measure the density $n_t(r, t)$ of the test particles, through their fluorescence in the probe beam. Given $n_t(r, t)$, one can obtain the radial test particle flux Γ_r through the continuity equation,

$$\frac{\partial n_t}{\partial t} = -\frac{1}{r}\frac{\partial}{\partial r}(r\Gamma_r), \tag{13.40}$$

which implies

$$\Gamma_r(r, t) = \int_\infty^r r'\,dr'\frac{\partial n_t}{\partial t}(r', t). \tag{13.41}$$

We can compare Γ_r with Fick's law, which states that in a diffusive process the flux of test particles is proportional to the gradient of their concentration:

$$\Gamma_r = -Dn\frac{\partial}{\partial r}\left(\frac{n_t}{n}\right). \tag{13.42}$$

Note that while n_t diffused, n remained fixed, since the plasma was in equilibrium. Rather, the concentration of test particles eventually became uniform. Also, the experiments were performed on near-thermal-equilibrium plasmas which rotated rigidly, so we assume that $v_{\min}$ is set by collisions, not flow shear.

The experimentally determined values of Γ_r were found to be proportional to $n\,\partial/\partial r\;(n_t/n)$, and their ratio yields D. Results for D using this method are displayed in Fig. 13.5 versus plasma temperature for a range of magnetic field strengths and densities. The figure shows that the measured diffusion is roughly 10 times the classical-theory prediction from Table 13.1, but also about 3 times the prediction of eqn (13.37) for $\mathbf{E} \times \mathbf{B}$ drift collisions. Evidently, the theory presented so far requires modification.

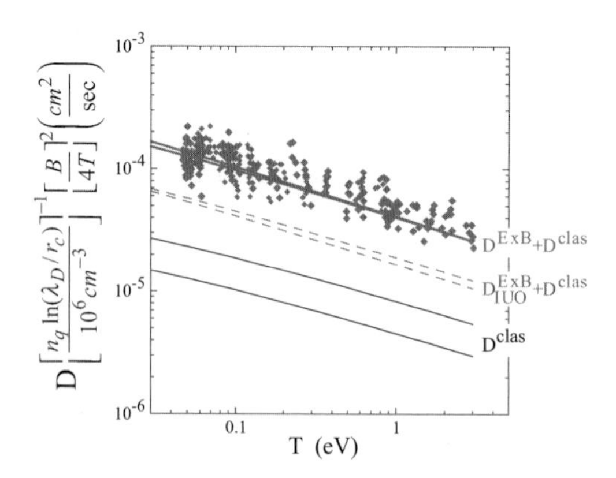

Fig. 13.5 Measured test particle diffusion compared with eqn (13.37) (dashed lines), the classical theory (lower solid lines), and the improved theory discussed in Section 13.2.3 (upper solid lines) (from Anderegg *et al.* (1997)).

13.2.3 Correction to integration along unperturbed orbits

The IUO technique used in deriving eqn (13.37) gives a diffusion coefficient which is up to three times too small because of a subtle "collisional caging" effect. To understand this, we will start the calculation over again using a slightly different approach, based on the Green–Kubo expression for $\mathbf{E} \times \mathbf{B}$ diffusion,

$$D = \frac{1}{2}\left(\frac{c}{B}\right)^2 \int_{-\infty}^{\infty} dt \, \langle E_y(t) E_y(0) \rangle \,, \tag{13.43}$$

where $E_y(t)$ is the fluctuating electric field acting on a test particle, labeled "1," due to its collisions with other particles as they stream by. This fluctuating field creates a fluctuating velocity $v_x(t) = cE_y/B$ that is responsible for the diffusion in x. Equation (13.43) can be derived from the equation of motion $dx/dt = v_x(t)$, which implies

$$\langle x^2 \rangle(t) = \int_0^t dt' dt'' \langle v_x(t') v_x(t'') \rangle \,. \tag{13.44}$$

The derivation relies on three assumptions: (1) $\langle v_x(t') v_x(t'') \rangle$ is a function only of $t' - t''$ (the fluctuating velocities are stationary); (2) this correlation function approaches zero for $|t' - t''| > \tau$ (the "autocorrelation time"); and (3) times of interest satisfy $t \gg \tau$. Using these assumptions, it is not difficult to show that $\langle x^2 \rangle = 2Dt$, with D given by eqn (13.43) (Reif 1965).

The electric field in eqn (13.43) can be expressed as a Fourier transform,

$$E_y(t) = -\sum_{j=2}^{N} e \int \frac{d^3k}{(2\pi)^3} \frac{4\pi e i k_y}{k^2} e^{i\mathbf{k}\cdot\Delta\mathbf{r}_j(t)} \,, \tag{13.45}$$

where $\Delta\mathbf{r}_j(t) = \mathbf{r}_1(t) - \mathbf{r}_j(t)$ is the difference between the position of the test particle and that of particle j. Their relative position evolves according to

$$\Delta\mathbf{r}_j(t) = \Delta\mathbf{r}_j(0) + \hat{z} v_r t + \tilde{z}_r(t) \,, \tag{13.46}$$

where v_r is the initial relative velocity, and $\tilde{z}_r(t)$ is the fluctuating relative-position change caused by collisions:

$$\frac{d^2 \tilde{z}_r}{dt^2} = \frac{eE_z(t)}{m} \,, \tag{13.47}$$

where $E_z(t)$ is a fluctuating electric field due to interactions with the plasma.

Exercise 13.4 Show that

$$\langle e^{ik_z \tilde{z}_r(t)} \rangle = e^{-k_z^2 D_v |t|^3/3} \,, \tag{13.48}$$

where $D_v = (e/m)^2 \int_0^\infty dt' \langle E_z(t') E_z(0) \rangle$ is the velocity diffusion coefficient, assuming that $t \gg \tau$, where τ is the autocorrelation time for the fluctuations in $E_z(t)$.

Equation (13.43) then becomes

$$D = \frac{1}{2}\left(\frac{ce}{B}\right)^2 \sum_{j=2}^{N}\sum_{\ell=2}^{N}\int_{-\infty}^{\infty} dt \int \frac{d^3k\, d^3k'}{(2\pi)^6}\frac{(4\pi i)^2 k_y k_y'}{k^2 k'^2}\langle e^{i\mathbf{k}\cdot\Delta\mathbf{r}_j(t)+i\mathbf{k}'\cdot\Delta r_\ell(0)}\rangle . \quad (13.49)$$

In order to evaluate the average, we assume that the initial conditions for the particles are uncorrelated, so that the probability distribution for these initial conditions is a product of the distribution functions for each separate charge. Also, we *assume* that the fluctuations in $E_z(t)$ are uncorrelated from the initial conditions, so that

$$\langle e^{i\mathbf{k}\cdot\Delta\mathbf{r}_j(t)+i\mathbf{k}'\cdot\Delta r_\ell(0)}\rangle = \langle e^{i\mathbf{k}\cdot(\Delta r_j(0)+v_r t)+i\mathbf{k}'\cdot\Delta\mathbf{r}_\ell(0)}\rangle \times e^{-k_z^2 D_v |t|^3/3} . \quad (13.50)$$

(See Exercise 13.4.) It is then not difficult to show that

$$\left\langle \sum_j \sum_\ell e^{i\mathbf{k}\cdot\Delta\mathbf{r}_j(0)+i\mathbf{k}'\cdot\Delta\mathbf{r}_\ell(0)}\right\rangle = \frac{N-1}{V}\int d^3\Delta r_j(0)\int dv_r f_{\mathrm{rel}}(v_r) \times e^{i(\mathbf{k}+\mathbf{k}')\cdot\Delta\mathbf{r}_j(0)} = n(2\pi)^3\delta(\mathbf{k}+\mathbf{k}')\int dv_r f_{\mathrm{rel}}(v_r) . \quad (13.51)$$

Applying this expression to eqn (13.49) and performing the $\mathbf{k}'$ integral, we can express the diffusion coefficient as

$$D = \frac{(4\pi)^2}{2}\left(\frac{ce}{B}\right)^2 n\int_{-\infty}^{\infty} dv_r f_{\mathrm{rel}}(v_r)\int \frac{d^3k}{(2\pi)^3}\frac{k_y^2}{k^4}\int_{-\infty}^{\infty} dt\, e^{ik_z v_r t - k_z^2 D_v |t|^3/3} . \quad (13.52)$$

We write d^3k in cylindrical coordinates $(k_\perp, \theta, k_z)$, and define $\bar{k}_z = k_z/k_\perp$ and $\bar{t} = k_\perp v_r t$. Then eqn (13.52) can be written as

$$D = \left(\frac{ce}{B}\right)^2 n\int_{-\infty}^{\infty} dv_r \frac{e^{-mv_r^2/4T}}{\sqrt{4\pi T/m}}\int_0^{\infty} k_\perp dk_\perp \frac{1}{k_\perp^2 |v_r|}\int_{-\infty}^{\infty} d\bar{t}\, J(\bar{D}_v, \bar{t}) , \quad (13.53)$$

where $\bar{D}_v = D_v/(3k_\perp v_r^3)$ and

$$J(\bar{D}_v, \bar{t}) = \int_{-\infty}^{\infty} \frac{d\bar{k}_z}{(1+\bar{k}_z^2)^2} e^{i\bar{k}_z\bar{t} - \bar{k}_z^2 \bar{D}_v |\bar{t}|^3} . \quad (13.54)$$

This function is plotted versus $\bar{t}$ in Fig. 13.6.

We require the integral over $\bar{t}$ of J in the expression for D, eqn (13.53). The result of this integration depends on the value of $\bar{D}_v$. For $\bar{D}_v = 0$, it is clear that

$$\int_{-\infty}^{\infty} d\bar{t}\, J(0, \bar{t}) = 2\pi \quad (13.55)$$

since $\int_{-\infty}^{\infty} d\bar{t}\, e^{i\bar{k}_z \bar{t}} = 2\pi\delta(\bar{k}_z)$. This result, together with imposition of the appropriate cutoffs, leads back to eqn (13.37), as expected since $\bar{D}_v = 0$ is equivalent to integration

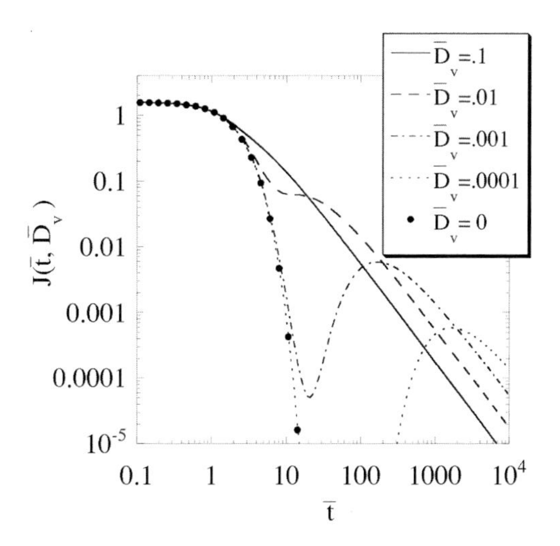

Fig. 13.6 The function $J(\bar{D}_v, \bar{t})$.

along unperturbed orbits. On the other hand, when $\bar{D}_v \neq 0$ but $\bar{D}_v \to 0$, one can show the following:

$$\lim_{\bar{D}_v \to 0+} \int_{-\infty}^{\infty} d\bar{t}\, J(\bar{D}_v, \bar{t}) = 6\pi\,. \tag{13.56}$$

This is *three times* the result one obtains for $\bar{D}_v$ identically equal to zero (Dubin 1997).

Exercise 13.5 Verify eqn (13.56). (Hint: transform variables. See Dubin (1997) for details.)

Even an infinitesimal amount of velocity diffusion makes a large change to the result for D, increasing it by a factor of 3. Equation (13.56) can be verified by numerical integration, and the result is displayed in Fig. 13.7. Note that as $v_r \to 0$, $\bar{D}_v \to \infty$ because $\bar{D}_v = D_v/3k_\perp v_r^3$. Also, Fig. 13.7 shows that for $\bar{D}_v \gtrsim 1$, $\int_{-\infty}^{\infty} J(\bar{D}_v, t)\, dt \to 0$. This provides a natural cutoff to the logarithmically divergent v_r integral in eqn (13.53) at $v_{\min} \sim (D_v/3k_\perp)^{1/3}$. This is the same minimum velocity as was used in deriving eqn (13.37), but expressed in k-space. Then, after the v_r integral is performed to logarithmic order, eqn (13.53) becomes

$$D = \frac{6\sqrt{\pi}}{\bar{v}} \left(\frac{ce}{B}\right)^2 n \int_0^{\infty} \frac{dk_\perp}{k_\perp} \ln\left(\frac{\bar{v}}{(D_v/3k_\perp)^{1/3}}\right). \tag{13.57}$$

Next, the logarithmic divergence in $k_\perp$ is cut off at λ_D^{-1} and r_c^{-1} for the same reasons as we cut off the ρ integral in eqn (13.35), resulting in

$$D = 3D_{\mathrm{IUO}}\,, \tag{13.58}$$

where D_{IUO} is the result of integration along unperturbed orbits ($\bar{D}_v = 0$), given by eqn (13.37).

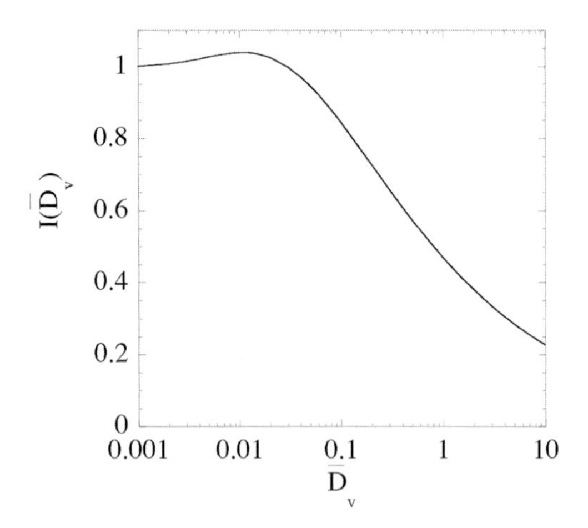

Fig. 13.7 The function $I(\bar{D}_v) \equiv (1/6\pi)\int_{-\infty}^{\infty} J(\bar{D}_v, \bar{t})\, d\bar{t}$.

Integration along unperturbed orbits fails to capture the extra factor of 3 because when $\bar{D}_v = 0$, as assumed in IUO, particles only collide once as they pass one another along the magnetic field. However, when $\bar{D}_v$ is small but finite, particles encounter one another many times as their relative velocities diffuse: eventually, particle velocities reverse and the pair collides again. Effectively, collisions with surrounding particles (responsible for the velocity diffusion) cause a colliding pair to interact over a longer period of time than would occur in the absence of collisions (this is the "collisional caging" effect referred to at the beginning of this section: surrounding particles "cage" the interacting pair).

This longer interaction time can be seen directly in the plot of $\bar{J}(\bar{D}_v, \bar{t})$ versus time in Fig. 13.6. For $\bar{D}_v = 0$, $\bar{J}$ displays a single peak at $\bar{t} = 0$, due to the single collision. However, for $\bar{D}_v$ small but finite, a second peak appears, caused by further collisions between the pair as their diffusing relative velocity changes sign. This peak occurs at $\bar{t} \sim 0.1/\bar{D}_v$. These multiple collisions are responsible for the increase (by a factor of 3) of the diffusion D.

Another way to think of the factor of 3 increase is that our estimate $v_{\min} \sim (D_v\rho)^{1/3}$ was incorrect. One can rewrite $3\ln(\bar{v}/(D_v\rho)^{1/3})$ as $\ln(\bar{v}/v'_{\min})$, where $v'_{\min} = \nu\rho$ is an improved estimate for the minimum relative velocity. This is much smaller than $(D_v\rho)^{1/3}$, and comes about because particles interact for a much longer time than our previous estimate suggested—see Fig. 13.6.

Returning now to the effect of fluid shear on the transport, we note that shear will supercede collisions when $S > \nu$, so that $v_{\min} = S\rho$ rather than $\nu\rho$. Accounting for fluid shear then yields

$$D = 2\sqrt{\pi}\,\nu_0 r_c^2 \ln\left(\frac{\lambda_D}{r_c}\right) \ln\left(\frac{\bar{v}}{\mathrm{Max}(S, \nu)\sqrt{\lambda_D r_c}}\right). \tag{13.59}$$

Equation (13.59) can be verified by including fluid shear from the beginning in the previous diffusion calculation; however, the details are too complex to include here. One portion of the calculation is considered in Exercise 13.6. Also, in a previous description of the effect of fluid shear on diffusion (Driscoll *et al.* 2002), a multiplicative parameter α was introduced to account for the enhancement factor. Equation (13.59) is equivalent to that description since the enhancement is included through the use of our improved estimate for the minimum relative velocity.

In comparing theory with experiment we assume $S < \nu$, since the plasmas were near thermal equilibrium, rotating rigidly. Figure 13.6 shows the experimental results compared with the improved theory of eqn (13.58) or, equivalently, (13.59).

Exercise 13.6 Derive eqn (13.59) for the case $\bar{D}_v = 0$ (i.e. $v_{\min}$ determined by shear) by repeating the analysis that leads to eqn (13.58), but setting $\tilde{z}_r = 0$ in eqn (13.46) and keeping the shear, replacing the equation by $\Delta\mathbf{r}_j(t) = \Delta\mathbf{r}_j(0) + \hat{z}v_r t + \hat{y}S\,\Delta x\,t$.

13.3 Heat conduction across B

In this lecture we consider the problem of heat conduction across a strong magnetic field, in a plasma for which $\bar{r}_c \ll \lambda_D$. Following Psimopoulos and Li (1992), we employ an approach based on an ad hoc generalization of the Boltzmann collision operator that describes isolated two-particle collisions. Weaknesses in this approach will become apparent, but it has the advantage of being quite straightforward, and it points out the necessity of deriving a new collision operator that is capable of properly handling long-range interactions.

Based on the estimate discussed in Lecture 1, we consider a picture of collisions as shown in Fig. 13.3: two particles on field lines separated by $\rho \sim \lambda_D$ interact as they pass by one another. Initially, their parallel velocities are v_{z_1} and v_{z_2}, respectively. After their interaction, their velocities are v'_{z_1} and v'_{z2}. The change in kinetic energies of the two particles results in transfer of energy across **B**.

In such a one-dimensional collision, the relative energy E_r of the two particles is conserved:

$$E_r = \frac{mv_r^2}{4} + \frac{e^2}{(\rho^2 + z_r^2)^{1/2}}\,, \tag{13.60}$$

where $z_r = z_1 - z_2$ is the relative position. This implies that the initial and final relative speeds are the same: $|v_r| = |v'_r|$. However, the sign of v_r can change. If, initially (when $z_r \gg \rho$),

$$\frac{mv_r^2}{4} < \frac{e^2}{\rho}\,, \tag{13.61}$$

the repulsive Coulomb potential will cause the particles to reflect from one another so that $v'_r = -v_r$. If, on the other hand, the inequality (13.61) is not satisfied, the relative energy is sufficient to overcome the repulsion and the particles do not reflect, so that $v'_r = v_r$. Combining these results with momentum conservation, $v_{z_1} + v_{z_2} = v'_{z_1} + v'_{z_2}$, we find that when the inequality (13.61) is satisfied,

$$v'_{z_1} = v_{z2} \text{ and } v'_{z_2} = v_{z1}\,, \tag{13.62}$$

i.e. the particles exchange velocities; if eqn (13.61) is not satisfied, their velocities are unchanged by the interaction.

Now, consider the effect that these interactions have on a distribution of particles $f(x_1, v_{z_1}, t)$, described by a Maxwellian with varying temperature $T(x_1, t)$:

$$f(x_1, v_{z_1}, t) = \frac{n\, e^{-mv_{z_1}^2/2T(x_1,t)}}{\sqrt{2\pi T(x_1,t)/m}}\,. \tag{13.63}$$

In a Boltzmann picture (Lifshitz and Pitaevskii 1981), the number of particles in an element dv_{z_1} at position x_1, $f(x_1, v_{z_1}, t)\, dv_{z_1}$, varies in time as collisions remove particles from this element and other collisions introduce particles into the element. The rate of removal is the number of particles in dv_{z_1}, $f(x_1, v_{z_1})\, dv_{z_1}$, multiplied by the total number of collisions per unit time,

$$\int dx_2\, dy_2 \int dv_{z_2}\, |v_r| f(x_2, v_{z_2}, t)\,, \tag{13.64}$$

where $v_r = v_{z_1} - v_{z_2}$, and the integral over x_2 and y_2 must satisfy eqn (13.61); otherwise there is no change in the velocities due to the interaction. Similarly, the rate at which collisions introduce particles into the element is

$$dv_{z_1} \int dx_2\, dy_2 \int dv_{z_2}\, |v_r| f(x_1, v_{z_2}, t) f(x_2, v_{z_1}, t)\,, \tag{13.65}$$

since particles that begin at velocity v_{z_2} will end up with velocity v_{z_1} in a collision with a particle moving at that velocity. Taking the difference between these two rates yields the overall rate of change of $f(x_1, v_{z_1})$:

$$\begin{aligned} \frac{d}{dt} f(x_1, v_{z_1}, t) = \int dx_2\, dy_2 \int_{mv_r^2/4<e^2/\rho} & dv_{z_2} |v_r| \\ & \times \left(f(x_1, v_{z_2}, t) f(x_2, v_{z_1}, t) - f(x_1, v_{z_1}, t) f(x_2, v_{z_2}, t)\right). \end{aligned} \tag{13.66}$$

The right-hand side is similar in form to the Boltzmann collision operator, except that the distribution functions are evaluated at *different* points in space, x_1 and x_2. This is important in order to describe heat conduction caused by energy-transferring collisions.

The rate of change of temperature may now be found by integrating eqn (13.66) over $mv_{z_1}^2$, and substituting for f from eqn (13.63):

$$\begin{aligned} \frac{\partial T}{\partial t}(x_1, t) &= \frac{1}{n} \int dv_{z_1} m v_{z_1}^2 \frac{\partial f}{\partial t}(x_1, v_{z_1}, t) \\ &= \int dx_2\, dy_2 \int_{mv_r^2/4<e^2/\rho} dv_{z_1} dv_{z_2} \frac{|v_r| m^2 n v_{z_1}^2}{2\pi\sqrt{T_1 T_2}} \\ &\qquad \times \left[e^{-m(v_{z_2}^2/T_1 + v_{z_1}^2/T_2)/2} - e^{-m(v_{z_1}^2/T_1 + v_{z_2}^2/T_2)/2}\right], \end{aligned} \tag{13.67}$$

where $T_1 = T(x_1, t)$ and $T_2 = T(x_2, t)$.

By interchanging the dummy variables v_{z_1} and v_{z_2}, eqn (13.67) can be expressed as

$$\frac{\partial T_1}{\partial t} = \int dx_2 \, F(x_1, x_2, t) \,, \tag{13.68}$$

where the function F is odd under interchange of x_1 and x_2:

$$\begin{aligned} F(x_1, x_2, t) = \int dy_2 \int_{mv_r^2/4<e^2/\rho} dv_{z_1} dv_{z_2} \frac{|v_r| m^2 n}{4\pi\sqrt{T_1 T_2}} (v_{z_1}^2 - v_{z_2}^2) \\ \times \left[e^{-m(v_{z_2}^2/T_1 + v_{z_1}^2/T_2)/2} - e^{-m(v_{z_1}^2/T_1 + v_{z_2}^2/T_2)/2} \right]. \end{aligned} \tag{13.69}$$

Since $F(x_1, x_2, t)$ is odd under interchange of x_1 and x_2, we will write it as

$$F(x_1, x_2, t) = \bar{F}(X, x_r, t) \,, \tag{13.70}$$

where $X = (x_1 + x_2)/2$, $x_r = x_2 - x_1$, and $\bar{F}$ is odd in x_r (i.e. $\bar{F} \to -\bar{F}$ as $x_r \to -x_r$). This transformation implies that

$$x_1 = X - x_r/2 \,, x_2 = X + x_r/2 \,. \tag{13.71}$$

Then

$$\int dx_2 \, F(x_1, x_2, t) = \int dx_2 \, \bar{F}\left(\frac{x_1 + x_2}{2}, x_2 - x_1, t \right) \tag{13.72}$$

and, converting the integration variable to x_r, we obtain

$$\int dx_2 \, F = \int dx_r \, \bar{F}\left(x_1 + \frac{x_r}{2}, x_r \right). \tag{13.73}$$

We now assume that the values of x_r required in eqn (13.68) are small (of order λ_D) compared with the scale of variation of $T(x_1, t)$, so that we may Taylor expand $\bar{F}$ in the first argument, obtaining

$$\int dx_2 \, F = \int dx_r \, \bar{F}(x_1, x_r, t) + \frac{\partial}{\partial x_1} \int dx_r \frac{x_r}{2} \bar{F}(x_1, x_r, t) \,. \tag{13.74}$$

The first integral vanishes because $\bar{F}$ is odd in x_r. Converting $\bar{F}$ back to F using eqns (13.70) and (13.71) yields

$$\frac{\partial T_1}{\partial t} = \frac{\partial}{\partial x_1} \int dx_r \frac{x_r}{2} F\left(x_1 - \frac{x_r}{2}, x_1 + \frac{x_r}{2}, t \right). \tag{13.75}$$

Substituting for F from eqn (13.69) and Taylor expanding $T(x_1 \pm x_r/2, t)$ to first-order in x_r, we obtain the heat equation:

$$\frac{\partial T_1}{\partial t} = \frac{\partial}{\partial x_1} \left(\chi \frac{\partial T_1}{\partial x_1} \right), \tag{13.76}$$

where the thermal diffusivity χ is given by the expression

$$\chi = \int dx_2\, dy_2 \int_{mv_r^2/4<e^2/\rho} dv_{z_1}\, dv_{z_2} \frac{x_r^2}{2} \frac{|v_r| m^3 n}{8\pi T_1^3} (v_{z_1}^2 - v_{z_2}^2)^2 e^{-m(v_{z_1}^2+v_{z_2}^2)/2T_1} \,. \quad (13.77)$$

The integrals over velocity can be most easily performed by converting to relative and center-of-mass variables, and the result, assuming $b \ll \rho$, is

$$\chi = \frac{1}{\sqrt{\pi}} n\bar{v}b^2 \int dx_2\, dy_2 \frac{x_r^2}{\rho^2} \,. \quad (13.78)$$

Exercise 13.7 Derive eqn (13.78) from eqn (13.77).

Converting to polar coordinates, for which $dx_2\, dy_2 = \rho\, d\rho\, d\theta$ and $x_r = \rho \cos\theta$, we can perform the θ integral in eqn (13.78), yielding

$$\chi = \sqrt{\pi} n\bar{v}b^2 \int_0^\infty \rho\, d\rho \,. \quad (13.79)$$

This result is clearly divergent at large impact parameters. The range of impact parameters must be cut off owing to Debye shielding, so we take $\rho_{\max} = \lambda_D$, yielding

$$\chi = \frac{\sqrt{\pi}}{2} n\bar{v}b^2 \lambda_D^2 \,. \quad (13.80)$$

Note, however, that unlike the previous logarithmic divergences, a change in $\rho_{\max}$ by a factor of 2 causes a large change in χ, by a factor of 4. Thus, eqn (13.80) can only be regarded as an estimate; the numerical coefficient is unknown. Furthermore, our treatment of collisions as isolated two-body events breaks down at impact parameters of order λ_D, since many other particles surrounding the colliding pair intervene. This approach, while a useful exercise, points to the need for a proper derivation of a collision operator to replace eqn (13.66); one which can rigorously describe the effect of long-range interactions without assuming that they consist of isolated two-body collisional events. This will be the topic of Lecture 4.

13.4 Collision operator for long-range interactions

In this lecture, we will derive a new collision operator that describes collisional interactions between particles separated by a Debye length or more in a plasma for which $\bar{r}_c \ll \lambda_D$.

13.4.1 Plasma response to a moving charge

As a first step, we will examine the response of a magnetized plasma to a charge Q moving with velocity $v_0\hat{z}$ along the magnetic field $B\hat{z}$. Previously, we analyzed how a plasma responds to a stationary charge by Debye-shielding the charge. When the charge is moving, the plasma response is more complicated: the possibility exists that the charge will emit plasma waves. This response should be an integral component of our collision operator, since the operator must describe interactions between moving charges, including the self-consistent plasma response.

The moving charge creates a potential disturbance $\delta\phi(\mathbf{r},t)$ that follows from Poisson's equation:

$$\nabla^2\delta\phi = -4\pi e\int \delta f\, dv_z - 4\pi Q\delta(\mathbf{r} - v_0\hat{z}t)\,, \tag{13.81}$$

where $\delta f(\mathbf{r}, v_z, t)$ is the perturbation away from the equilibrium distribution function $f_0(v_z)$. This perturbation can be obtained by solving the guiding-center Vlasov equation

$$\frac{\partial f}{\partial t} + v_z\frac{\partial f}{\partial z} + \frac{e}{m}E_z\frac{\partial f}{\partial v_z} = 0\,. \tag{13.82}$$

Here we assume for simplicity that B is very large, so only motion along z is needed, and we neglect cyclotron motion and $\mathbf{E}\times\mathbf{B}$ drifts (Chen 1974). Linearizing the equation in $\delta\phi$ and taking $\delta f = f - f_0$, we obtain the linearized Vlasov equation

$$\frac{\partial\,\delta f}{\partial t} + v_z\frac{\partial\,\delta f}{\partial z} - \frac{e}{m}\frac{\partial\,\delta\phi}{\partial z}\frac{\partial f_0}{\partial v_z} = 0\,. \tag{13.83}$$

We will solve eqns. (13.81) and (13.83) for $\delta\phi(\mathbf{r},t)$ via a Fourier–Laplace transform, writing

$$\delta\phi(\mathbf{r},t) = \int\frac{d^3k}{(2\pi)^3}\int_C\frac{dp}{2\pi i}e^{i\mathbf{k}\cdot\mathbf{r}+pt}\delta\phi_{\mathbf{k}p}\,, \tag{13.84}$$

where $\delta\phi_{\mathbf{k}p}$ is the Fourier–Laplace amplitude, and the contour C runs from $-i\infty$ to $i\infty$, to the right of any poles in $\delta\phi_{\mathbf{k}p}$. Applying the Fourier–Laplace transform to eqn (13.83) yields

$$(p + ik_zv_z)\delta f_{\mathbf{k}p} - \frac{e}{m}ik_z\delta\phi_{\mathbf{k}p}\frac{\partial f_0}{\partial v_z} = \delta f_{\mathbf{k}}(t=0)\,, \tag{13.85}$$

where $\delta f_{\mathbf{k}}(t=0)$ is the Fourier transform of $\delta f(\mathbf{r}, t=0)$. We assume that initially the plasma is unperturbed, so $\delta f_{\mathbf{k}}(t=0) = 0$. Then, solving eqn (13.85) for $\delta f_{\mathbf{k}p}$ and substituting the result into the Fourier–Laplace transform of eqn (13.81),

$$-k^2\,\delta\phi_{\mathbf{k}p} = -4\pi e\int dv_z\,\delta f_{\mathbf{k}p} - \frac{4\pi Q}{p + ik_zv_0}\,, \tag{13.86}$$

we obtain the following expression for $\delta\phi_{\mathbf{k}p}$:

$$\delta\phi_{\mathbf{k}p} = \frac{4\pi Q}{k^2D_{\mathbf{k}p}}\frac{1}{p + ik_zv_0}\,, \tag{13.87}$$

where $D_{\mathbf{k}p}$ is the *linear plasma dielectric function*,

$$D_{\mathbf{k}p} = 1 - \frac{4\pi e^2 ik_z}{mk^2}\int dv_z\frac{\partial f_0/\partial v_z}{p + ik_zv_z}\,. \tag{13.88}$$

This function describes the shielding response of the plasma. As written, eqn (13.88) is correct only for $\mathrm{Re}(p) > 0$, since the Laplace transform contour C must run to the right of all singularities, and the v_z integral in eqn (13.88) is singular at $\mathrm{Re}(p) = 0$. For $\mathrm{Re}(p) \leq 0$, eqn (13.88) must be analytically continued: the v_z integration contour must be deformed below the real line in the complex v_z plane, so that p remains above the v_z integration contour (Krall and Trivelpiece 1986).

Equation (13.87) must now be substituted into eqn (13.84) to obtain $\delta\phi(\mathbf{r}, t)$. The inverse Laplace transform in eqn (13.84) has two types of poles in the integrand: zeros in $D_{\mathbf{k}p}$ and the pole at $p = -ik_z v_0$. We will assume that all zeros in $D_{\mathbf{k}p}$ are damped, i.e. these zeros are at $\mathrm{Re}\, p < 0$, so that they produce a time-dependent response in $\delta\phi$ that is damped away, giving no contribution at large times. The undamped pole, at $p = -ik_z v_0$, yields

$$\delta\phi(\mathbf{r}, t) = 4\pi Q \int \frac{d^3k}{(2\pi)^3} \frac{e^{i\mathbf{k}\cdot\mathbf{r} - ik_z v_0 t}}{k^2 D_{\mathbf{k}, -ik_z v_0 + \varepsilon}}, \tag{13.89}$$

where ε is a positive infinitesimal used to ensure that the integration in p has passed to the right of singularities in $D_{\mathbf{k}p}$ (this is equivalent to deforming the v_z integration contour below the pole at $v_z = v_0$). Equation (13.89) implies that $\delta\phi = \delta\phi(x, y, z - v_0 t)$, i.e. the potential is a stationary perturbation as seen in the frame of the moving charge.

In order to complete our examination of the response of the plasma to the moving charge, we need to understand the plasma dielectric function $D_{\mathbf{k}, ik_z v_0 + \varepsilon}$. This function clearly depends on the velocity v_0 of the charge. If $v_0 = 0$, eqn (13.88) implies

$$D_{\mathbf{k},0} = 1 - \frac{4\pi e^2}{mk^2} \int \frac{dv_z}{v_z} \frac{\partial f_0}{\partial v_z}. \tag{13.90}$$

If f_0 is given by a Maxwellian distribution, $\partial f_0 / \partial v_z = -v_z f_0 / \bar{v}^2$, and eqn (13.90) yields

$$D_{\mathbf{k},0} = 1 + \frac{1}{k^2 \lambda_D^2}, \tag{13.91}$$

the dielectric response for static Debye shielding. Using this in eqn (13.89) leads to $\delta\phi(\mathbf{r}, t) = Q e^{-r/\lambda_D}/r$, as expected for Debye shielding of a stationary charge due to a single plasma species.

For $v_0 \neq 0$, the plasma response is more complicated. If we define $\omega = k_z v_0$, and again assume that f_0 is given by a Maxwellian, the dielectric function can be written as

$$D_{\mathbf{k}, -i\omega + \varepsilon} = 1 + \frac{k_z}{k^2 \lambda_D^2} \int_{-\infty}^{\infty} dv_z\, f_{\max}(v_z) \frac{v_z}{k_z v_z - \omega - i\varepsilon}. \tag{13.92}$$

Since $\omega = k_z v_0$ is real and ε is a positive infinitesimal we can use the Plemelj formula to break the integral into a principal part and an imaginary contribution from the pole at $\omega = k_z v_z - i\varepsilon$:

$$\begin{aligned} D_{\mathbf{k}, -i\omega + \varepsilon} &= 1 + \frac{k_z}{k^2 \lambda_D^2} \left[P\!\!\int_{-\infty}^{\infty} dv_z \frac{f_{\max}(v_z)}{k_z v_z - \omega} + i\pi \int_{-\infty}^{\infty} dv_z f_{\max}(v_z) v_z \delta(k_z v_z - \omega) \right] \\ &= 1 + \frac{1}{k^2 \lambda_D^2} \left[a\left(\frac{v_0}{\bar{v}}\right) + ib\left(\frac{v_0}{\bar{v}}\right) \mathrm{sgn}(k_z) \right], \end{aligned} \tag{13.93}$$

where the real functions $a(x)$ and $b(x)$ are defined as

$$a(x) \equiv \oint_{-\infty}^{\infty} \frac{ds}{\sqrt{2\pi}} \frac{s e^{-s^2/2}}{s - x} , \quad b(x) \equiv \sqrt{\frac{\pi}{2}} x e^{-x^2/2} . \tag{13.94}$$

Then the potential surrounding the charge is found by substituting eqn (13.93) into eqn (13.89).

Exercise 13.8 Show that $\delta\phi(x, y, z - v_0 t)$ can be written as

$$\delta\phi = \frac{Q}{r} g\left(\frac{\rho}{\lambda_D}, \frac{z - v_0 t}{\lambda_D}\right), \tag{13.95}$$

where $\rho^2 = x^2 + y^2$, $r^2 = z^2 + \rho^2$, and the shielding function $g(\rho, z)$ is

$$g(\rho, z) = \frac{2}{\pi} r \,\mathrm{Re}\left[\int_0^\infty k_\perp dk_\perp J_0(k_\perp \rho) \int_0^\infty dk_z \frac{e^{ik_z z}}{k_z^2 + k_\perp^2 + a(v_0/\bar{v}) + ib(v_0/\bar{v})}\right]. \tag{13.96}$$

(Hint: Evaluate eqn (13.89) in cylindrical coordinates, scaling **k** to the Debye length.)

The shielding function is plotted in Figs. 13.8 and 13.9 for several values of the particle speed v_0. At $v_0 = 0$, the static Debye-shielding response is evident; $g(\rho, z) = e^{-\sqrt{\rho^2+z^2}}$. However, for $v_0 \neq 0$ the dynamical shielding is incomplete, and as v_0 increases beyond $v_0/\bar{v} \sim$ 2–3 a damped wave develops behind the moving charge. This wave is due to weakly damped plasma waves that are resonant with the particle, i.e. waves whose phase velocity in the z-direction, ω/k_z, matches the speed v_0 of the particle. Such resonant waves are strongly excited by the passage of the particle.

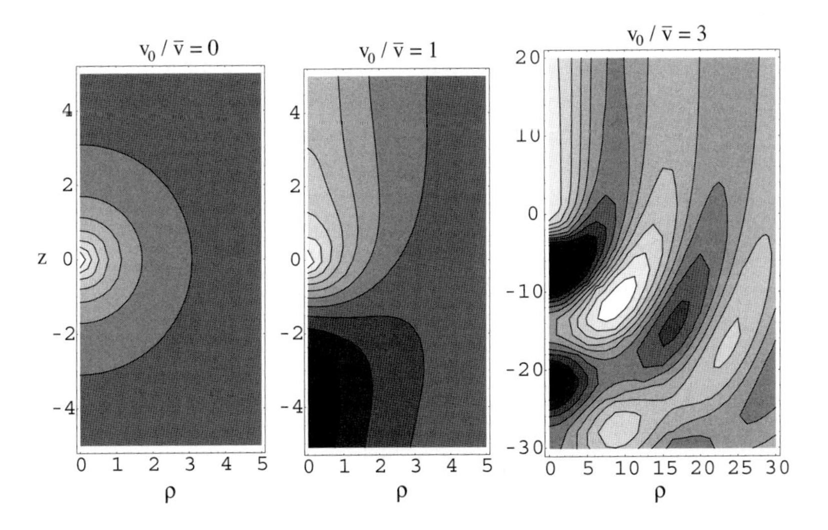

Fig. 13.8 Contour plots of the shielding function $g(\rho, z)$ at $v_0/\bar{v} = 0$, 1, and 3.

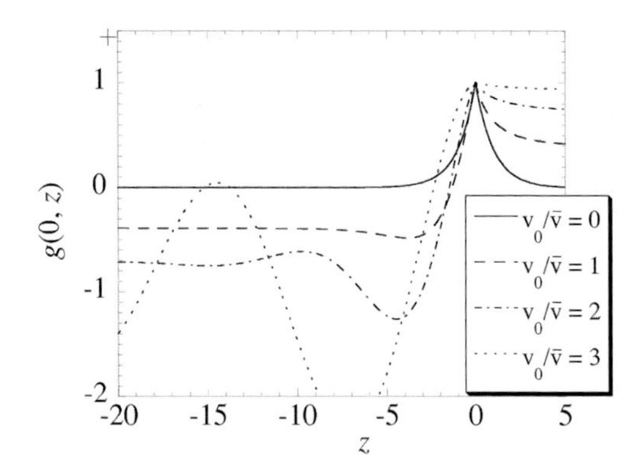

Fig. 13.9 Plot of the shielding function $g(0, z)$ for $v_0/\bar{v} = 0$, 1, 2, and 3.

Exercise 13.9 Taking $v_0 = \omega/k_z$ in eqn (13.93) and assuming $v_0 \gg \bar{v}$, show that D is approximately

$$D_{\mathbf{k},-i\omega+\varepsilon} \simeq 1 - \frac{\omega_p^2 k_z^2}{\omega^2 k^2} + \frac{ib\,\mathrm{sgn}(k_z)}{k^2 \lambda_D^2}. \tag{13.97}$$

Mathematically, the waves are excited by near-zeros in $D_{\mathbf{k},-i\omega+\varepsilon}$, which provide a large contribution to the integral in eqn (13.89). A near-zero in D exists for $\omega/k_z \gg \bar{v}$, at

$$\omega^2 \cong \frac{\omega_p^2 k_z^2}{k^2} \tag{13.98}$$

(see Exercise 13.9), which is the dispersion relation for magnetized plasma waves. Note that the small imaginary term in eqn (13.97) yields a negative imaginary part of ω, causing the waves to slowly damp. This is the origin of the wavelike response to the passage of the particle when $v_0 \gg \bar{v}$. These waves, a form of Cerenkov radiation, are similar to the wake observed behind a moving boat. They carry energy and momentum away from the particle, and this energy and momentum can be reabsorbed by particles a great distance away. This has important consequences for the thermal conduction and viscosity of the plasma, as we will see.

Equation (13.95) can also be used to determine the drag force on the moving charge due to the plasma. By subtraction of the bare Coulomb potential $Q/|r|$ from eqn (13.95), the remaining potential $\delta\phi_p(\rho, z - v_0 t)$ can be obtained. This potential, due to the plasma only, creates a force $F = -\partial\,\delta\phi_p(0, z)/dz|_{z=0}$ that acts on the moving charge to slow it down. One can show that the force scales as $F \sim -m\nu v_0$ for $v_0/\bar{v} \ll 1$, as one might expect. (The logarithmic divergence in ν (see eqn (13.13)) arises from a large-k logarithmic divergence in the derivative with respect to z of the wavenumber integral in eqn (13.96), which must be cut off at $k = b^{-1}$.)

13.4.2 Collision operator for long-range collisions

We now use what we have learned about the shielding response of a plasma to a moving charge in order to derive the collision operator, including self-consistent plasma-shielding effects. We assume a plasma in a uniform magnetic field $B\hat{z}$ with several species of charges with mass m_α, charge e_α, and density $n_\alpha(x)$, where α is a species label (e.g. $\alpha =$ electron or ion). Each species has a distribution function $f_\alpha(x, v_z, t)$. In the absence of collisions, this distribution would satisfy the guiding-center Vlasov equation

$$\frac{\partial f_\alpha}{\partial t} + v_z \frac{\partial f_\alpha}{\partial z} + \frac{\mathbf{E}_0 \times \hat{z}}{B} c \cdot \nabla f_\alpha + \frac{e_\alpha}{m_\alpha} E_{0_z} \frac{\partial f_\alpha}{\partial v_z} = 0\,, \tag{13.99}$$

where $\mathbf{E}_0 = -\nabla \phi_0$ is the plasma electric field, given by Poisson's equation,

$$\nabla^2 \phi_0 = -4\pi \sum_\infty e_\alpha \int f_\alpha \, dv_z\,. \tag{13.100}$$

Equation (13.99) keeps the $\mathbf{E} \times \mathbf{B}$ drift that was neglected in eqn (13.82), as it is important in determining the viscosity and diffusion due to long-range collisions.

The assumption that $f_\alpha = f_\alpha(x, v_z, t)$ and $\mathbf{E}_0 = E_0(x, t)\hat{x}$ implies that eqn (13.99) reduces to

$$\frac{\partial f_\alpha}{\partial t} = 0\,, \tag{13.101}$$

implying that any function of x and v_z alone is an equilibrium solution of eqn (13.99).

Normally, one chooses as an initial condition for eqn (13.99) a smooth function of x and v_z. However, in reality, the plasma consists of a set of discrete particles. If one therefore chooses as the initial condition

$$f_\alpha(\mathbf{r}, v_z, t = 0) = \sum_{i=1}^{N_\alpha} \delta(\mathbf{r} - \mathbf{r}_i(0))\delta(v_z - v_{z_i}(0))\,, \tag{13.102}$$

i.e. a series of δ-functions at the N_α discrete positions and velocities of the charges for each species α, the solution to the Vlasov–Poisson system can still be found in principle, but it will contain all the detailed information concerning the microscopic interactions between individual charges.

Exercise 13.10 Show that the solution to eqn (13.99) with the initial conditions (13.102) is given by the *Klimontovitch density* $\eta_\alpha(\mathbf{r}, v_z, t)$,

$$\eta_\alpha(\mathbf{r}, v_z, t) = \sum_{i=1}^{N_\alpha} \delta(\mathbf{r} - \mathbf{r}_i(t))\delta(v_z - v_{z_i}(t))\,, \tag{13.103}$$

where $(\mathbf{r}_i(t), v_{z_i}(t))$ is the phase-space trajectory of the ith particle of species α. [Hint: substitute eqn (13.103) into eqn (13.99).]

As shown in Exercise 13.10, the Klimontovitch density satisfies eqn (13.99) with the initial condition (13.102), i.e.

$$\frac{\partial \eta_\alpha}{\partial t} + v_z \frac{\partial \eta_\alpha}{\partial z} + \frac{\mathbf{E} \times \hat{z}}{B} c \cdot \nabla \eta_\alpha + \frac{e_\alpha}{m_\alpha} E_z \frac{\partial \eta_\alpha}{\partial v_z} = 0 , \tag{13.104}$$

where

$$\mathbf{E} = -\nabla \Phi \tag{13.105}$$

and

$$\nabla^2 \Phi = -4\pi \sum_\alpha e_\alpha \int dv_z \, \eta_\alpha . \tag{13.106}$$

Equation (13.104) is often referred to as the Klimontovitch equation (Krall and Trivelpiece 1986), although the only difference between it and the Vlasov equation is the form of the initial condition. This is a big difference, however, since a smooth Vlasov distribution has a different microscopic form from the N_α particles moving along chaotic trajectories described by the Klimontovitch density $\eta_\alpha(\mathbf{r}, v_z, t)$. The connection between these two pictures of the plasma (smoothed and discrete) may be made by averaging over an ensemble of initial conditions for the discrete particle positions and velocities, *assuming that* each individual particle i of species α has its position $\mathbf{r}_i$ and velocity v_{z_i} initially distributed according to a smooth probability density $f_\alpha(x_i, v_{z_i}, t = 0)/N_\alpha$.

Then, at later times, the following function $f_\alpha(x, v_z, t)$ is *defined* by the average over initial conditions:

$$\langle \eta_\alpha(\mathbf{r}, v_z, t) \rangle \equiv f_\alpha(x, v_z, t) . \tag{13.107}$$

One may easily show using eqn (13.103) that $\langle \eta_\alpha(\mathbf{r}, v_z, t = 0) \rangle = f_\alpha(x, v_z, t = 0)$, so eqn (13.107) is consistent with the initial conditions.

The averaged distribution function f_α does not satisfy the Vlasov equation; rather, it evolves slowly in time owing to collisions. An equation for f_α can be obtained by applying the above-described averaging procedure to the Klimontovitch equation itself:

$$\frac{\partial f_\alpha}{\partial t} + v_z \frac{\partial f_\alpha}{\partial z} + \left\langle \frac{\mathbf{E} \times \hat{z}}{B} c \cdot \nabla \eta_\alpha \right\rangle + \left\langle \frac{e_\alpha}{m_\alpha} E_z \frac{\partial \eta_\alpha}{\partial v_z} \right\rangle = 0 . \tag{13.108}$$

It is useful to break $\mathbf{E}$ and η_α into a smooth, averaged part and a fluctuation due to discreteness:

$$\eta_\alpha = f_\alpha(x, v_z, t) + \delta\eta_\alpha(\mathbf{r}, v_z, t) \; , \; \mathbf{E} = E_0(x, t)\hat{x} + \delta\mathbf{E}(\mathbf{r}, t) . \tag{13.109}$$

Then eqn (13.108) can be written as

$$\frac{\partial f_\alpha}{\partial t} = C_\alpha[\mathbf{f}] , \tag{13.110}$$

where the collision operator $C_\alpha[\mathbf{f}]$ is

$$C_\alpha[\mathbf{f}] = -\nabla \cdot \left\langle \frac{c}{B} \delta\mathbf{E} \times \hat{z} \delta\eta_\alpha \right\rangle - \frac{\partial}{\partial v_z} \left\langle \frac{e_\alpha}{m_\alpha} \delta E_z \delta\eta_\alpha \right\rangle \tag{13.111}$$

and the notation $C_\alpha[\mathbf{f}]$ denotes a functional dependence on f_β for all β. This dependence is implicit in eqn (13.111) through the dependence of the fluctuations $\delta\eta_\alpha$ and

$\delta\mathbf{E}$ on f_β, and may be uncovered by subtracting eqn (13.108) from eqn (13.104) to obtain equations for the fluctuations $\delta\eta_\alpha$ and $\delta\mathbf{E}$. If one assumes that the fluctuations are small, one may linearize the resulting equation in the fluctuations to obtain

$$\frac{\partial\,\delta\eta_\alpha}{\partial t} + v_z\frac{\partial\,\delta\eta_\alpha}{\partial z} + V_y\frac{\partial\,\delta\eta_\alpha}{\partial y} + \frac{\delta E_y c}{B}\frac{\partial f_\alpha}{\partial x} + \frac{e_\alpha}{m_\alpha}\delta E_z\frac{\partial f_\alpha}{\partial v_z} = 0\,, \tag{13.112}$$

where $V_y(x,t) = -cE_0/B$ is the mean $\mathbf{E}\times\mathbf{B}$ drift of the plasma in the y-direction, $\delta\mathbf{E} = -\nabla\,\delta\phi$, and

$$\nabla^2\,\delta\phi = -4\pi\sum_\alpha e_\alpha\int dv_z\,\delta\eta_\alpha\,. \tag{13.113}$$

We solve eqn (13.112) using a Laplace transform in time and Fourier transforms in z and y. Also, we assume that the fluctuations evolve rapidly in time compared with the slow time variation of f_α so that we may assume f_α and V_y are time-independent in eqn (13.112). Therefore, in what follows, we take $f_\alpha = f_\alpha(x,v_z)$ and $V_y = V_y(x)$. The solution to eqn (13.112) is then

$$\delta\eta_\alpha(\mathbf{r},v_z,t) = \int_C\frac{dp}{2\pi i}e^{pt}\int\frac{dk_y\,dk_z}{(2\pi)^2}e^{ik_yy+ik_zz}\delta\hat{\eta}_\alpha(x,k_y,k_z,v_z,p)\,, \tag{13.114}$$

where $\delta\hat{\eta}_\alpha$ satisfies

$$(p+ik_zv_z+ik_yV_y)\delta\hat{\eta}_\alpha = \left(ik_y\frac{c}{B}\frac{\partial f_\alpha}{\partial x} + ik_z\frac{e_\alpha}{m_\alpha}\frac{\partial f_\alpha}{\partial v_z}\right)\delta\hat{\phi} + \delta\hat{\eta}_{\alpha_0} \tag{13.115}$$

where $\delta\hat{\phi}(x,k_y,k_z,p)$ is the Fourier–Laplace transform of $\delta\phi$, and $\delta\hat{\eta}_{\alpha_0}(x,k_y,k_z,v_z)$ is the Fourier transform of the initial condition for $\delta\eta_\alpha$, $\delta\eta_{\alpha_0}(\mathbf{r},v_z)$.

Exercise 13.11 Show that, if one assumes particles are *uncorrelated*,

$$\langle\delta\eta_{\alpha_0}(\mathbf{r},v_z)\,\delta\eta_{\beta_0}(\mathbf{r}',v_z')\rangle = f_\alpha(x,v_z)\delta_{\alpha\beta}\delta(\mathbf{r}-\mathbf{r}')\delta(v_z-v_z') \tag{13.116}$$

and

$$\begin{aligned}\langle\delta\hat{\eta}_{\alpha_0}(x,k_y,k_z,v_z)\,\delta\hat{\eta}_{\beta_0}(x',k_y',k_z',v_z')\rangle &= f_\alpha(x,v_z)\delta_{\alpha\beta}\delta(x-x')(2\pi)^2\\ &\times\delta(k_y+k_y')\delta(k_z+k_z')\delta(v_z-v_z')\,.\end{aligned} \tag{13.117}$$

Equation (13.115), when combined with Poisson's equation for the fluctuating potential $\delta\hat{\phi}$, yields the following expression for $\delta\hat{\phi}$:

$$\delta\hat{\phi}(x,k_y,k_z,p) = -4\pi\sum_\alpha e_\alpha\int dx'\,\psi(x,x',k_y,k_z,p)\int\frac{dv_z'\,\delta\hat{\eta}_{0_\alpha}(x',k_z,k_z,v_z')}{p+ik_zv_z'+ik_yV_y(x')}\,, \tag{13.118}$$

where ψ is a Green's function for the potential, satisfying

$$\frac{\partial^2\psi}{\partial x^2}-\left[k_y^2+k_z^2-4\pi i\sum_\alpha\frac{e_\alpha^2}{m_\alpha}\int dv_z\frac{(k_y/\Omega_\alpha)(\partial f_\alpha/\partial x)+k_z(\partial f_\alpha/\partial v_z)}{p+ik_zv_z+ik_yV_y}\right]\psi=\delta(x-x'), \tag{13.119}$$

and where $\Omega_\alpha = e_\alpha B/m_\alpha c$ is the cyclotron frequency for species α.

Equation (13.118) says that the potential fluctuations at position x with wavenumbers (k_y, k_z) and frequency $\omega = ip$ are driven by the discreteness in the particle distribution described by $\delta\hat{\eta}_{0\alpha}$, through the shielded Green's function ψ. Note that if $f_\alpha(x, v_z)$ were independent of x we could Fourier transform eqn (13.119) in x to obtain

$$\psi=-\int\frac{dk_x}{2\pi}\frac{e^{ik_x(x-x')}}{k^2D_{\mathbf{k}p}}, \tag{13.120}$$

where $D_{\mathbf{k}p}$ is a generalization of the plasma dielectric function discussed in the previous section:

$$D_{\mathbf{k}p}=1-\frac{1}{k^2}\sum_\alpha\frac{4\pi ie_\alpha^2}{m_\alpha}k_z\int\frac{dv_z\,\partial f_\alpha/\partial v_z}{p+ik_zv_z+ik_yV_y}. \tag{13.121}$$

Comparing eqn (13.120) with eqn (13.89) reveals that ψ is a function similar to the function $\delta\phi$ discussed previously; it describes a Debye-shielded plasma response when $p + ik_yV_y \to 0$, and a wavelike response for $|\mathrm{Im}\, p| \gg k_z\bar{v}$.

In order to evaluate the collision operator, we require the two averages $\langle\delta E_y\,\delta\eta_\alpha\rangle$ and $\langle\delta E_z\,\delta\eta_\alpha\rangle$. (One can show by symmetry that $\langle\delta E_x\,\delta\eta_\alpha\rangle$ is zero.) These can be evaluated by first expressing δE_y (or δE_z) and $\delta\eta_\alpha$ in terms of their Fourier–Laplace transforms and then using eqn (13.115):

$$\begin{aligned}\langle\delta E_{\substack{y\\(z)}}\,\delta\eta_\alpha\rangle&=-i\int\frac{dk_y'\,dk_z'\,dk_y\,dk_z}{(2\pi)^4}\int_C\frac{dp\,dp'}{(2\pi i)^2}e^{(p+p')t+i(k_y+k_y')y+i(k_z+k_z')z}\\&\times k'_{\substack{y\\(z)}}\left[i\frac{e_\alpha}{m_\alpha}\left(\frac{k_y}{\Omega_\alpha}\frac{\partial f_\alpha}{\partial x}+k_z\frac{\partial f_\alpha}{\partial v_z}\right)\langle\delta\hat{\phi}(x,k_y',k_z',p')\,\delta\hat{\phi}(x,k_y,k_z,p)\rangle\right.\\&\left.+\langle\delta\hat{\phi}(x,k_y',k_z',p')\delta\hat{\eta}_{0\alpha}(x,k_y,k_z,v_z)\rangle\right]\Big/(p+ik_yV_y+ik_zv_z).\end{aligned} \tag{13.122}$$

In turn, the two averages appearing in eqn (13.122) can be expressed in terms of the averages over initial fluctuations, given by eqn (13.117), by substituting eqn (13.118) for $\delta\hat{\phi}$:

$$\begin{aligned}&\langle\delta\hat{\phi}(x,k_y',k_z',p')\,\delta\hat{\eta}_{0\alpha}(x,k_y,k_z,v_z)\rangle\\&=-4\pi\sum_\beta e_\beta\int dx'\,dv_z'\frac{\psi(x,x',k_y',k_z',p')}{p'+ik_z'v_z'+ik_y'V_y'}\langle\delta\hat{\eta}_{0\beta}(x',k_y',k_z',v_z')\,\delta\hat{\eta}_{0\alpha}(x,k_y,k_z,v_z)\rangle\\&=-4\pi e_\alpha\frac{\psi(x,x,-k_y,-k_z,p')}{p'-ik_zv_z-ik_yV_y}(2\pi)^2\delta(k_y+k_y')\delta(k_z+k_z')f_\alpha(x,v_z)\end{aligned} \tag{13.123}$$

and

$$
\begin{aligned}
&\langle \delta\hat{\phi}(x, k'_y, k'_z, p')\, \delta\hat{\phi}(x, k_y, k_z, p)\rangle \\
&\quad = (4\pi)^2 \sum_\alpha \sum_\beta e_\alpha e_\beta \int dx'\, dv'_z\, dx''\, dv''_z \frac{\psi(x, x', k'_y, k'_z, p')\psi(x, x'', k_y, k_z, p)}{(p' + ik'_z v'_z + ik'_y V'_y)(p + ik_z v''_z + ik_y V''_y)} \\
&\qquad \times \langle \delta\hat{\eta}_{0\alpha}(x', k'_y, k'_z, v'_z)\, \delta\hat{\eta}_{0\beta}(x'', k_y, k_z, v''_z)\rangle \\
&\quad = (4\pi)^2 \sum_\beta e_\beta^2 \int dx' dv'_z \frac{\psi(x, x', k'_y, k'_z, p')\psi(x, x', k_y, k_z, p)}{(p' + ik'_z v'_z + ik'_y V'_y)(p + ik_z v'_z + ik_y V'_y)} \\
&\qquad \times (2\pi)^2 \delta(k_y + k'_y)\delta(k_z + k'_z) f_\beta(x', v'_z)\,,
\end{aligned}
\tag{13.124}
$$

where $V'_y \equiv V_y(x')$.

We must now perform the inverse Laplace transforms in eqn (13.122). In so doing, we note that poles in ψ are damped, with $\mathrm{Re}\, p < 0$, so that at large times these poles do not contribute to the fluctuations. The only contributions that remain at large times arise from the resonant denominators, and lead to a time-independent result, just as in the case discussed in Section 13.4.1. It is useful to consider the contributions to eqn (13.122) of eqns (13.123) and (13.124) separately. Equation (13.123) contributes the term

$$
-4\pi i \int \frac{dk_y\, dk_z}{(2\pi)^2} \underset{(z)}{k_y}\, e_\alpha \psi(x, x, -k_y, -k_z, i(k_y V_y + k_z v_z)) f_\alpha(x, v_z) \tag{13.125}
$$

and eqn (13.124) contributes the term

$$
\begin{aligned}
&-(4\pi)^2 \frac{e_\alpha}{m_\alpha} \sum_\beta e_\beta^2 \int \frac{dk_y\, dk_z}{(2\pi)^2} \int dx'\, dv'_z \underset{(z)}{k_y} \left(\frac{k_y}{\Omega_\alpha} \frac{\partial f_\alpha}{\partial x}(x, v_z) + k_z \frac{\partial f_\alpha}{\partial v_z}(x, v_z) \right) \\
&\times \left|\psi(x, x', k_y, k_z, -i(k_y V_y + k_z v_z))\right|^2 f_\beta(x', v'_z) \pi\delta(k_y(V_y - V'_y) + k_z(v_z - v'_z))\,.
\end{aligned}
\tag{13.126}
$$

Here, in evaluating the inverse Laplace transform, we have used the identity

$$
\lim_{t\to\infty} \mathrm{Re} \frac{e^{i(\omega' - \omega)t} - 1}{i(\omega' - \omega)} = \pi\delta(\omega - \omega')\,. \tag{13.127}
$$

(Only the real part contributes, by symmetry of the integrand.) Also, we have used the identity

$$
\psi(x, x', -k_y, -k_z, i\omega) = \psi^*(x, x', k_y, k_z, -i\omega)\,, \tag{13.128}
$$

which follows from eqn (13.119).

The expression (13.125) can be rewritten in a form that looks more like eqn (13.126) by means of an identity that follows from eqn (13.119). Multiplying this equation by $\psi^*(x, k_y, k_z, -i\omega)$, integrating over x, and taking the imaginary part yields

$$\begin{aligned}
&\operatorname{Im}\psi^*(x', x', k_y, k_z, -i\omega) \\
&\quad = \operatorname{Im}\int dx \left(\psi^*(x, x', k_y, k_z, -i\omega)\frac{\partial^2 \psi}{\partial x^2}(x, x', k_y, k_z, -i\omega) \right. \\
&\qquad \left. - 4\pi i \sum_\beta \frac{e_\beta^2}{m_\beta} \int dv_z \frac{(k_y/\Omega_\beta)(\partial f_\beta/\partial x) + k_z(\partial f_\beta/\partial v_z)|\psi(x, x', k_y, k_z - i\omega)|^2}{ik_z v_z + ik_y V_y - i\omega + \varepsilon} \right),
\end{aligned} \tag{13.129}$$

where the infinitesimal ε in the denominator arises from the fact that the values of the Laplace transform variable $p = -i\omega$ must always lie to the right of poles. The first term in the integral over x vanishes because, on integration by parts, it has no imaginary part. Application of the Plemelj formula to the second term allows one to extract the imaginary part analytically, yielding

$$\begin{aligned}
\operatorname{Im}\psi^*(x', x', k_y, k_z, -i\omega) &= -4\pi \sum_\beta \frac{e_\beta^2}{m_\beta} \int dx\, dv_z |\psi(x, x', k_y, k_z, -i\omega)|^2 \\
&\quad \times \left(\frac{k_y}{\Omega_\beta}\frac{\partial f_\beta}{\partial x} + k_z \frac{\partial f_\beta}{\partial v_z} \right) \pi\delta(k_y v_z + k_y V_y - \omega)\,.
\end{aligned} \tag{13.130}$$

If we apply this identity to the expression (13.125) and combine it with eqn (13.126) we obtain the result

$$\begin{aligned}
\langle \delta \underset{(z)}{E_y}\, \delta\eta_\alpha \rangle &= e_\alpha \sum_\beta (4\pi e_\beta)^2 \int \frac{dk_y\, dk_z}{(2\pi)^2} \underset{(z)}{k_y} \int dx'\, dv_z' \\
&\quad \times |\psi(x, x', k_y, k_z, -i(k_y V_y + k_z v_z))|^2 \pi\delta(k_y(V_y - V_y') + k_z(v_z - v_z')) \\
&\quad \times \left[\frac{1}{m_\beta}\left(\frac{k_y}{\Omega_\beta}\frac{\partial f_\beta'}{\partial x'} + k_z \frac{\partial f_\beta'}{\partial v_z'} \right) f_\alpha - \frac{1}{m_\alpha}\left(\frac{k_y}{\Omega_\alpha}\frac{\partial f_\alpha}{\partial x} + k_z \frac{\partial f_\beta}{\partial v_z} \right) f_\beta' \right],
\end{aligned} \tag{13.131}$$

where $f_\alpha = f_\alpha(x, v_z)$ and $f_\beta' = f_\beta(x', v_z')$. This result, along with eqn (13.111), provides the collision operator for long-range interactions in a magnetized plasma.

This operator describes shielded interactions between particles on different field lines at x and x'. The interaction is moderated by the Green's function ψ, which describes a Debye-shielded potential when the particles move slowly compared with $\bar{v}$, and describes wave emission and absorption when they move rapidly. The δ function implies that the important interactions are resonant, so that a given Fourier component of the interaction provides a steady, time-independent force. The product of distribution functions in the square bracket is somewhat similar in form to the previous Boltzmann operator, eqn (13.66), representing the rate at which particles at (x, v_z) enter and leave the phase space element $dx\, dv_z$ owing to collisions with particles at (x', v_z').

The collision operator satisfies several conservation laws. First, it clearly conserves particle number for each separate species; $\dot{N}_\alpha = \int dx\, dv_z(\partial f_\alpha/\partial t) = 0$. Second, total momentum along the field is conserved:

$$\dot{P}_z = \sum_\alpha \int dx\, dv_z\, m_\alpha v_z \frac{\partial f_\alpha}{\partial t} = -\sum_\alpha \int dx\, dv_z\, m_\alpha v_z C_\alpha[\mathbf{f}] = 0\,. \tag{13.132}$$

This follows by integration by parts of eqn (13.132), which yields

$$\dot{P}_z = \sum_{\alpha\beta}(4\pi e_\alpha e_\beta)^2 \int \frac{dk_y\,dk_z}{(2\pi)^2} k_z \int dx\,dv_z\,dx'\,dv_z'|\psi|^2$$
$$\times \pi\delta(k_v(V_y - V_y') + k_z(v_z - v_z'))[\ldots]\,, \tag{13.133}$$

where $[\ldots]$ stands for the square bracket in eqn (13.131). However, this bracket is antisymmetric under interchange of the dummy variables (x, v_z) and (x', v_z') and therefore the integral vanishes by symmetry, proving eqn (13.132). (Note that this requires $|\psi|^2$ to be symmetric on an interchange of x and x'. Proof of this is left as an exercise.)

Similar arguments imply that the canonical momentum

$$P_y = \sum_\alpha \frac{e_\alpha B}{c} \int dx\,dv_z\,x f_\alpha \tag{13.134}$$

is conserved, as well as the energy

$$E = \sum_\alpha \int dx\,dv_z \left(\frac{m_\alpha v_z^2}{2} + e_\alpha \phi_0\right) f_\alpha\,. \tag{13.135}$$

Exercise 13.12 Prove $\dot{P}_y = \dot{E} = 0$.

Finally, one can show that the collision operator maximizes the entropy functional $S[\mathbf{f}]$, where

$$S = -\sum_\alpha \int dx\,dv_z\,f_\alpha \ln f_\alpha\,. \tag{13.136}$$

Exercise 13.13 Show that $\dot{S} \geq 0$.

13.5 Heat conduction, viscosity, and diffusion due to long-range collisions

In this lecture, we will use the collision operator derived in the previous lecture to determine the cross-magnetic-field heat conduction and viscosity of a plasma due to long-range collisions. We will also revisit the diffusion problem considered in Lecture 2.

13.5.1 Heat conduction

As discussed in Lectures 1 and 3, the mechanism for heat conduction due to long-range collisions involves only motion parallel to the magnetic field $B\hat{z}$, and yields a result for the thermal conductivity that is essentially magnetic-field-independent. Therefore, in what follows, we simplify the derivations by taking $B \to \infty$ so that we can neglect $\mathbf{E} \times \mathbf{B}$ drift corrections of $O(1/B)$. Also, we assume that there is only a single species. Furthermore, we assume that close velocity-scattering collisions not described by our collision operator keep the distribution function f a local Maxwellian with temperature $T(x,t)$:

$$f(x, v_z, v_\perp, t) = \frac{n}{(\sqrt{2\pi T(x,t)/m})^3} e^{-m(v_z^2+v_\perp^2)/2T(x,t)} . \tag{13.137}$$

Here we have explicitly kept the dependence of f on $v_\perp \equiv \sqrt{v_x^2 + v_y^2}$ because it will play a role in what follows. (We dropped this dependence in the previous two lectures because it was not needed.)

An evolution equation for the plasma temperature follows from the $B \to \infty$ version of eqn (13.110):

$$\frac{\partial f}{\partial t} = -\frac{e}{m}\frac{\partial}{\partial v_z}\langle \delta E_z \, \delta\eta \rangle . \tag{13.138}$$

This equation implies that the kinetic energy density of the plasma, $K = \frac{3}{2}nT(x,t)$, evolves according to

$$\frac{3}{2}n\frac{\partial T}{\partial t} = \int d^3v \frac{m(v_z^2 + v_\perp^2)}{2}\frac{\partial f}{\partial t} = e\int dv_z \, v_z \langle \delta E_z \, \delta\eta \rangle . \tag{13.139}$$

The last expression is merely the Joule heating due to the parallel-current fluctuations. This heats the parallel kinetic energy, but the velocity-scattering collisions are assumed to be sufficiently rapid that the energy is quickly shared with the perpendicular degrees of freedom, implying that the specific heat per particle is 3/2. This is why we have kept the perpendicular energy in f.

Using eqn (13.131) for $\langle \delta E_z \, \delta\eta \rangle$ and eqn (13.137) for f, we have

$$\begin{aligned}\frac{3}{2}n\frac{\partial T}{\partial t} &= -\frac{e^2}{m}(4\pi e)^2 \int \frac{dk_y \, dk_z}{(2\pi)^2} \int d^3v \, d^3v' \, dx' \, k_z^2 v_z \\ &\times |\psi|^2(x, x', k_y, k_z, -ik_z v_z)\pi\delta(k_z(v_z - v_z'))ff'm\left(\frac{v_z'}{T(x')} - \frac{v_z}{T(x)}\right),\end{aligned} \tag{13.140}$$

which implies

$$\begin{aligned}\frac{\partial T}{\partial t} &= \frac{2}{3}me^2 n(4\pi e)^2\pi \int \frac{dk_z \, dk_z}{(2\pi)^2} \int dv_z \, dx' \, |k_z| v_z^2 \\ &\times |\psi|^2 \frac{(x, x', k_y, k_z, -ik_z v_z)}{2\pi\sqrt{T(x)T(x')}} e^{-mv_z^2(1/T(x)+1/T(x'))/2}\left(\frac{1}{T(x)} - \frac{1}{T(x')}\right).\end{aligned} \tag{13.141}$$

This is very similar in form to eqn (13.68), obtained using the approximate Boltzmann formalism. Just as in that analysis, the integrand is odd under interchange of x and x'. If we (incorrectly) assume that the Green's function ψ is *local*, i.e. a sharply peaked function of $x - x'$, we may obtain a local heat conduction equation using the same argument as in Lecture 3:

$$\frac{\partial T}{\partial t} = \frac{\partial}{\partial x}\chi\frac{\partial T}{\partial x}\,, \tag{13.142}$$

where

$$\chi = \frac{2}{3}me^2 n\frac{(4\pi e)^2\pi}{T^3(x)}\int\frac{dk_y\,dk_z}{(2\pi)^3}\int dv_z\,dx'\,|k_z|v_z^2 e^{-mv_z^2/T(x)} \times |\psi|^2(x,x',k_y,k_z,-ik_zv_z)\frac{(x-x')^2}{2}\,. \tag{13.143}$$

Also, when $|\psi|^2$ is sharply peaked in $x - x'$, we can use the Fourier-transform form, eqn (13.120), for ψ, which implies that

$$\int |\psi|^2\frac{(x-x')^2}{2}dx' = 2\int\frac{dk_x}{2\pi}\frac{k_x^2}{k^8|D_{\mathbf{k},-ik_zv_z}|^4}\,, \tag{13.144}$$

where $D_{\mathbf{k}p}$ is given by eqn (13.88) (or eqn (13.93) with $v_0 = v_z$).

Exercise 13.14 Using eqn (13.120), prove eqn (13.144).

Then the local form of the thermal diffusivity due to long-range collisions is

$$\chi = \frac{2}{3}(4\pi e^2)^2\frac{mn}{T^3}\int\frac{d^3k}{(2\pi)^3}\int\frac{dv_z\,e^{-mv_z^2/T}v_z^2|k_z|k_x^2}{|k^2D_{\mathbf{k},-ik_zv_z}|^4}\,. \tag{13.145}$$

If we approximate the dielectric function by a Debye-shielded form,

$$D_{\mathbf{k},-ik_zv_z} = 1 + \frac{1}{k^2\lambda_D^2}\,, \tag{13.146}$$

then the integrals can be performed, yielding

$$\chi^{\text{Debye}} = \frac{\sqrt{\pi}}{18}n\bar{v}b^2\lambda_D^2 = \frac{e^2}{72\sqrt{\pi}m\bar{v}}\,. \tag{13.147}$$

However, we have seen in Lecture 4 that eqn (13.146) is correct only for $v_z = 0$. For $v_z > \bar{v}$, the dielectric exhibits near-zeros at $k_zv_z = \pm k_z\omega_p/k$, which cause the integrand in eqn (13.145) to blow up. This can be seen by using scaled variables $\bar{\mathbf{k}} = \mathbf{k}\lambda_D$ and $\bar{v}_z = v_z/\bar{v}$ in eqn (13.145), and noting that D is a function only of $|\bar{\mathbf{k}}|$, $\bar{v}_z$, and the sign of $\bar{k}_z$ (see eqn (13.93)). Then, writing d^3k in spherical coordinates and integrating over solid angles yields

$$\chi = \frac{4}{3\pi}n\bar{v}b^2\lambda_D^2\int_0^\infty d\bar{k}\,g(\bar{k})\,, \tag{13.148}$$

where

$$g(\bar{k}) = \frac{\pi}{2}\bar{k}^5 \int_{-\infty}^{\infty} d\bar{v}_z \frac{\bar{v}_z^2 e^{-\bar{v}_z^2}}{[(\bar{k}^2 + a(\bar{v}_z))^2 + b^2(\bar{v}_z)]^2}, \tag{13.149}$$

with the functions a and b given by eqn (13.94). Note that for small $\bar{k}$, there is a near-zero in the denominator of the integrand where $\bar{k}^2 + a = 0$ and $|b| \ll 1$, owing to weakly damped waves.

The function $g(\bar{k})$ is plotted in Fig. 13.10, and has a nonintegrable singularity at small $\bar{k}$. Our expression for the thermal conduction diverges because weakly damped waves are excited, and these waves transfer energy large distances across the plasma. This causes a breakdown in the local approximation used to obtain eqn (13.145) from eqn (13.141).

In order to obtain a nondivergent heat equation we must go beyond the local approximation. Returning to eqn (13.141), we instead assume only that $T(x)$ is nearly uniform,

$$T(x) = T + \delta T(x), \tag{13.150}$$

and we expand eqn (13.141) for small δT, obtaining

$$\begin{aligned}\frac{\partial T}{\partial t} &= \frac{2}{3} m e^2 n \frac{(4\pi e)^2 \pi}{T^3} \int \frac{dk_y\, dk_z}{(2\pi)^3} \int dv_z\, dx'\, |k_z| v_z^2 e^{-m v_z^2/T} \\ &\times |\psi|^2(x, x', k_y, k_z, -ik_z v_z)(\delta T(x') - \delta T(x)).\end{aligned} \tag{13.151}$$

Also, to lowest order in δT we can treat the plasma as a uniform slab of width L, and we expand ψ in Fourier modes of the slab:

$$\psi(x, x', k_y, k_z, -ik_z v_z) = \frac{2}{L} \sum_{k_x} \frac{\sin(k_x x)\sin(k_x x')}{k^2 D_{\mathbf{k}, -ik_z v_z}}, \tag{13.152}$$

where $k_x = n\pi/L$, $n = 1, 2, \ldots, \infty$, and $k^2 = k_x^2 + k_y^2 + k_z^2$.

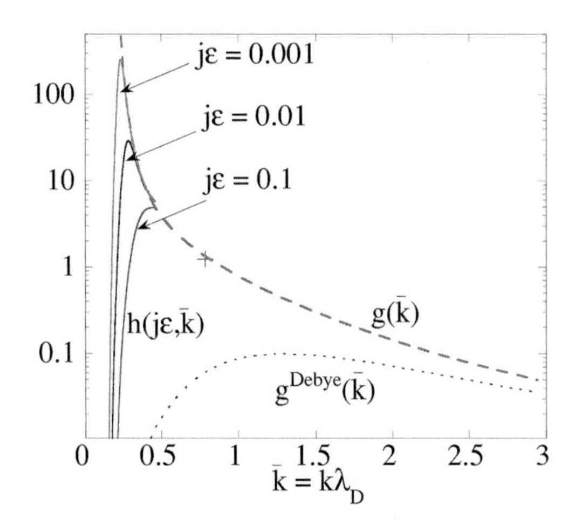

Fig. 13.10 The functions $g(\bar{k})$, $g^{\mathrm{Debye}}(\bar{k})$, and $h(j\varepsilon, \bar{k})$ for three values of $j\varepsilon$.

For large plasmas with $L \gg \lambda_D$, we evaluate the integrals in eqn (13.151) asymptotically in the small parameter $\varepsilon \equiv \pi\lambda_D/L$. Since $|\psi|^2$ appears in eqn (13.151), a double sum over k_x appears, as $\sum_{k_x}\sum_{k'_x}$. However, the flux is dominated by $k_x \simeq k'_x$; otherwise, the integral phase-mixes away upon integration over x' (because $\delta T(x)$ varies slowly but $\sin k_x x$ varies rapidly for k_x of $O(\lambda_D^{-1})$). Writing $\sin(k_x x')\sin(k'_x x') = (1/2)[\cos(k_x - k'_x)x' - \cos(k_x + k'_x)x']$ and dropping the second term because of phase-mixing, we obtain

$$\begin{aligned}\frac{\partial T}{\partial t} &= \frac{2}{3}me^2 n\frac{(4\pi e)^2\pi}{T^3}\int\frac{dk_y\,dk_z}{(2\pi)^3}\int dv_z\,dx' \\ &\quad\times|k_z|v_z^2 e^{-mv_z^2/T}\frac{1}{L^2}\sum_{k_x}\sum_{\Delta k_x}\frac{1}{k^2 D_{\mathbf{k},-ik_zv_z}k'^2 D^*_{\mathbf{k}',-ik_zv_z}} \\ &\quad\times\cos(\Delta k_x x)\cos(\Delta k_x x')(\delta T(x') - \delta T(x))\,, \end{aligned} \tag{13.153}$$

where $\Delta k_x = k'_x - k_x$.

As $\varepsilon \to 0$ the integrand takes two asymptotic forms, depending on the size of $k\lambda_D$. These forms can be asymptotically matched at $k\lambda_D = 0.4$. For $k\lambda_D > 0.4$, there are no lightly damped waves and $1/D$ varies slowly with k. Then, by Taylor expanding $1/k'^2 D_{\mathbf{k}',-ik_zv_z}$ in Δk_x, one finds that the $(\Delta k_x)^0$ term in eqn (13.153) vanishes because

$$\sum_{\Delta k_x}\cos\Delta k_x\, x\int_0^L dx'\cos\Delta k_x\, x'(T(x') - T(x)) = 0\,. \tag{13.154}$$

One may show this directly by writing $T(x)$ and $T(x')$ as Fourier cosine series, and performing the integral in eqn (13.154). Also, the $(\Delta k_x)^1$ term in eqn (13.153) vanishes because it is odd in Δk_x, and the $O(\Delta k_x^2)$ term leads back to the local form for the heat equation with χ given by eqn (13.145), except that the integral over k in the expression for χ is limited to $k\lambda_D > 0.4$.

For $k\lambda_D < 0.4$, lightly damped waves provide the main contribution to the integral. When v_z approaches a zero of the dielectric function, $1/D_{\mathbf{k},-ik_zv_z}D^*_{\mathbf{k}',-ik_zv_z}$ becomes sharply peaked at the points $v_z = v_0$ and $v_z = v'_0$, where v_0 and v'_0 satisfy

$$k^2\lambda_D^2 + a(v_0/\bar{v}) = 0 \tag{13.155}$$

and

$$k'^2\lambda_D^2 + a(v'_0/\bar{v}) = 0\,; \tag{13.156}$$

see eqn (13.93). Note that for $k\lambda_D \ll 1$ each equation has two solutions, since a is an even function and $a < 0$. When $k\lambda_D \ll 1$, v_0 is such that $b(v_0/\bar{v}) \ll 1$, and similarly for $b' \equiv b(v'_0/\bar{v})$. Then, upon integrating over the sharp peaks, we obtain

$$\int \frac{dv_z}{k^2 k'^2 D_{\mathbf{k}-ik_z v_z} D^*_{\mathbf{k}-ik_z v_z}}$$
$$\simeq \int \frac{dv_z\, \lambda_D^4}{[(\partial a/\partial v_0)(v_z - v_0) + ib\,\mathrm{sgn}(k_z)][(\partial a/\partial v_0')(v_z - v_0') - ib'\,\mathrm{sgn}(k_z)]}$$
$$= \frac{2\pi\lambda_D^4}{(\partial a/\partial v_0)(\partial a/\partial v_0')} \frac{\bar{\gamma} - i\,\Delta v\,\mathrm{sgn}(k_z)}{\bar{\gamma}^2 + \Delta v^2}, \tag{13.157}$$

where $\bar{\gamma} = b/(\partial a/\partial v_0) + b'/(\partial a/\partial v_0')$ and $\Delta v = v_0 - v_0'$. (This result must be doubled to account for the second negative solution of eqns (13.155) and (13.156).) When we substitute this expression into eqn (13.153), the imaginary part vanishes because it is odd both in k_z and under interchange of k_x and k_x'. Also, we note that one can add to the right-hand side of eqn (13.157) any function that is independent of Δk_x (this follows from eqn (13.154)). We therefore subtract $\pi\lambda_D^4/(b\,\partial a/\partial v_0)$, which causes the integrand to vanish at $\Delta k_x = 0$. Turning the sum over $k_x \geq 0$ into an integral over all k_x, dividing by two, and multiplying by two because eqn (13.155) has two roots for v_0 yields

$$\frac{\partial T^{\mathrm{waves}}}{\partial t} = \frac{2}{3} m e^2 n \frac{(4\pi e)^2 \pi}{T^3} \lambda_D^4 \int_{k\lambda_D < 0.4} \frac{d^3k}{(2\pi)^3} \int dx'\, |k_z| v_0^2 e^{-mv_0/T}$$
$$\times \frac{1}{\pi L} \sum_{\Delta k_x} \left(\frac{2\pi}{(\partial a/\partial v_0)(\partial a/\partial v_0')} \frac{\bar{\gamma}}{\bar{\gamma}^2 + \Delta v^2} - \frac{\pi}{b(\partial a/\partial v_0)} \right)$$
$$\times \cos \Delta k_x\, x\, \cos \Delta k_x\, x' (\delta T(x') - \delta T(x)) \,. \tag{13.158}$$

The main contribution arises from $\Delta k_x \ll k_x$, allowing us to take $\partial a/\partial v_0 \simeq \partial a/\partial v_0'$, and $\bar{\gamma} \simeq 2b/(\partial a/\partial v_0)$. This also implies $\Delta v = -2k_x\,\Delta k_x/(\partial a/\partial v_0)$ (see eqns (13.155) and (13.156)).

If we now integrate by parts with respect to x' and define $\Delta k_x = j\pi/L$, we can write eqn (13.158) as

$$\frac{\partial T^{\mathrm{waves}}}{\partial t} = \frac{2}{3} m e^2 n \frac{(4\pi e)^2 \pi \lambda_D^4}{T^3} \int_{k\lambda_D < 0.4} \frac{d^3k}{(2\pi)^3} |k_z| v_0^2 e^{-mv_0^2/T}$$
$$\times \sum_{j=1}^{\infty} \hat{T}_j \frac{\partial}{\partial x} (\sin \Delta k_x\, x) \frac{\partial a/\partial v_0}{b} \frac{\Delta v^2/\Delta k_x^2}{4b^2 + \Delta v^2 (\partial a/\partial v_0)^2}, \tag{13.159}$$

where $\hat{T}_j \equiv 2/L \int dx' \sin(j\pi x'/L)\, \partial T/\partial x'$ is the Fourier coefficient of the temperature gradient. Finally, we add to eqn (13.159) the contribution from $k\lambda_D > 0.4$ to obtain the total rate of change of T (Dubin and O'Neil 1997):

$$\frac{\partial T}{\partial t} = \frac{\partial}{\partial x} \sum_{j=1}^{\infty} (\chi^{\mathrm{local}} + \chi_j^{\mathrm{waves}}) \hat{T}_j \sin \frac{j\pi x}{L} \,. \tag{13.160}$$

Here,

$$\chi^{\text{local}} = \frac{4}{3\pi} n\bar{v}b^2\lambda_D^2 \int_{0.4}^{\infty} d\bar{k}\, g(\bar{k}) \tag{13.161}$$

is the local contribution to the thermal diffusivity, with g given by eqn (13.149). Also,

$$\chi_j^{\text{waves}} = \frac{4}{3\pi} n\bar{v}b^2\lambda_D^2 \int_{0}^{0.4} d\bar{k}\, h(\bar{k}, j\varepsilon) \tag{13.162}$$

is the wave contribution to the thermal diffusivity, where

$$h(\bar{k}, \varepsilon) = \pi \int d\Omega\, \bar{k}^2 |\bar{k}_z| \frac{\bar{v}_0^2 e^{-\bar{v}_0^2}}{(\partial a/\partial \bar{v}_0) b(\bar{v}_0)} \frac{\bar{k}_x^2}{b^2(\bar{v}_0) + \bar{k}_x^2 j^2 \varepsilon^2}, \tag{13.163}$$

$\bar{v}_0 = v_0/\bar{v}$, and $d\Omega$ is the element of solid angle that arises in $d^3\bar{k} = \bar{k}^2\, d\bar{k}\, d\Omega$. When evaluating eqn (13.163), we must recall that $\bar{v}_0$ is a function of $k\lambda_D$ through the solution of eqn (13.155); it is the (scaled) phase velocity of weakly damped plasma waves.

The functions $g(\bar{k})$ and $h(\bar{k}, j\varepsilon)$ are plotted in Fig. 13.10. For $j\varepsilon \ll 1$, they can be matched at $\bar{k} \sim 0.4$. Note that h is not divergent at small $\bar{k}$, so the heat flux is now finite, depending on the scale length $L_j = L/j\pi$ of the temperature gradient through the parameter $j\varepsilon$.

A numerical integration yields

$$\chi^{\text{local}} = \frac{0.0652 e^2}{m\bar{v}} = 0.819 \nu_0 \lambda_D^2 . \tag{13.164}$$

Equation (13.164) is an order of magnitude larger than the value obtained from simple Debye shielding, eqn (13.147). This is because the interaction cannot be accurately characterized by the simple Debye-shielding dielectric response, even when $k\lambda_D > 0.4$. Off-resonant plasma waves greatly increase $g(\bar{k})$ compared with the form $g(\bar{k})$ would take using simple Debye shielding, i.e. $g^{\text{Debye}}(\bar{k})$ obtained by setting $a = 1$ and $b = 0$ in eqn (13.149),

$$g^{\text{Debye}}(\bar{k}) = \frac{\pi^{3/2}}{4} \frac{\bar{k}^5}{(1+\bar{k}^2)^4} . \tag{13.165}$$

Only for $\bar{k} \gtrsim 3$ do the functions approach one another—see Fig. 13.10.

Values for χ_j^{waves} are provided in Table 13.2. These values differ slightly from those in Dubin and O'Neil (1997), because here we have used a more accurate solution to eqn (13.155) in determining the function $h(j\varepsilon, \bar{k})$. Table I of Dubin and O'Neil (1997) provides values of $\kappa_{\text{waves}}^j = (3/2) n x_{\text{waves}}^j$. Some other references (e.g. Dubin (1998) and Hollman *et al.* (2000)) quote a different value of χ based on a "neutral-plasma" heat capacity of 5/2 rather than 3/2 (i.e. the heat capacity at constant pressure rather than at constant volume). The resulting factor of 3/5 must be accounted for when comparing formulae for χ in those papers with those quoted here.

The table shows that when the scale length of the temperature gradient L_j is greater than about $100\lambda_D$, emission and absorption of lightly damped waves is the

Table 13.2 Wave contribution to heat flux

$j\varepsilon$	$\chi^j_{\mathrm{waves}}/(e^2/m\bar{v})$
0.1	0.0115
0.05	0.0262
0.01	0.102
0.005	0.167
0.001	0.518
0.0005	0.859
10^{-4}	2.93
10^{-5}	19.0

dominant heat transport mechanism. For L_j less than this, a local theory of thermal conduction suffices, with χ^{local} given by eqn (13.164).

Experiments have tested the theory in the regime where $\lambda_D > \bar{r}_c$ and $L_j < 100\lambda_D$, where the transport is local (Hollmann *et al.* 2000). In these experiments, the laser-diagnosed Mg^+ plasma described in Lecture 2 was again used. Now, the lasers were used both to manipulate the initial plasma temperature and to diagnose the resulting temperature evolution. First, the plasma was brought to a near-uniform temperature with a slight peak at $r = 0$. Then the laser was turned off, and the evolution of $T(r,t)$ was recorded. Using the same basic methodology for measuring test particle transport, the temperature data was compared with the prediction of Fourier's Law, that the radial heat flux is proportional to the local temperature gradient. The coefficient of proportionality is the local thermal diffusivity χ^{local}. Results of several such measurements are shown in Fig. 13.11 and compared with both eqn (13.164) and the classical theory (Table 13.1), versus temperature. Here n_7 is the density in units of $10^7\,\mathrm{cm}^{-3}$, and B is in tesla. The results are independent of magnetic field and density, as predicted by eqn (13.164), and are in quantitative agreement with the theory to within a factor of 2.

13.5.2 Viscosity

We now consider viscous relaxation due to long-range collisions in a plasma for which there is a shear flow with $\mathbf{V} = V_y(x)\hat{y}$. For simplicity, we assume a single species with uniform density and temperature. The distribution function is then

$$f(v_z) = \frac{n e^{-mv_z^2/2T}}{\sqrt{2\pi T/m}}\,. \tag{13.166}$$

Viscous relaxation in such a plasma manifests itself through a cross-field particle flux Γ_x. This flux arises because the shear flow creates a shear stress in the y-direction (i.e. a force density):

$$F_y = \frac{\partial}{\partial x}\eta\frac{\partial V_y}{\partial x} \tag{13.167}$$

(see eqn (13.20)). In a magnetized plasma, this force density creates an $\mathbf{F}\times\mathbf{B}$ drift (the equivalent to an $\mathbf{E}\times\mathbf{B}$ drift except for a general force rather than an electric field):

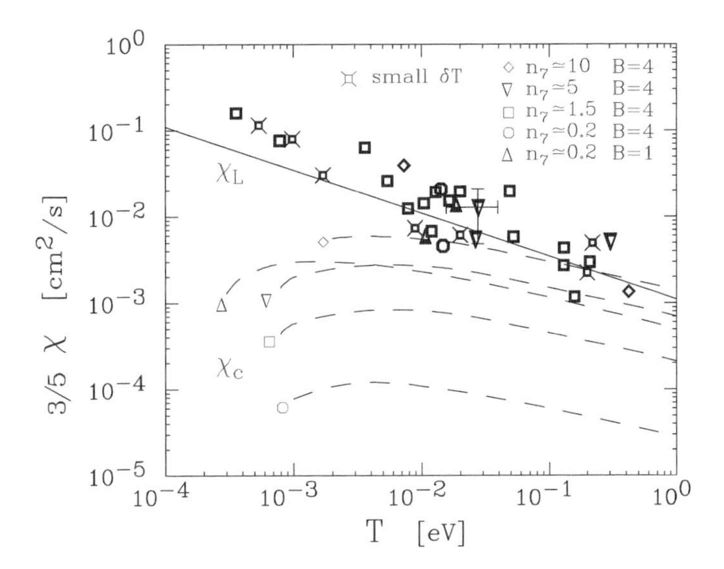

Fig. 13.11 Comparison of experimentally measured thermal diffusivity with theory. The solid line shows eqn (13.164); the dashed lines are for the classical theory at the various densities and magnetic fields used in the experiment. The factor of 3/5 arises from a different definition of χ used in Hollmann *et al.* (2000).

$$\Gamma_x = \frac{c}{eB} F_y = \frac{c}{eB} \frac{\partial}{\partial x} \eta \frac{\partial V_y}{\partial x}. \tag{13.168}$$

Using eqns (13.110), (13.111), and (13.131) we can obtain the following expression for the particle flux due to long-range collisions, which should be equivalent to eqn (13.168):

$$\begin{aligned}\Gamma_x &= \int dv_z \frac{c}{B} \langle \delta E_y \, \delta\eta \rangle \\ &= \frac{ec}{mB} (4\pi e)^2 \int dx' \, dv_z' \, dv_z \int \frac{dk_y \, dk_z}{(2\pi)^2} k_y \pi \delta(k_z(v_z - v_z') + k_y(V_y - V_y')) \\ &\quad \times |\psi|^2(x, x', k_y, k_z, -ik_y V_y - ik_z v_z) \frac{m^2 n^2}{2\pi T^2} k_z (v_z - v_z') e^{-m(v_z^2 + v_z'^2)/2T}. \end{aligned} \tag{13.169}$$

Using the δ-function, we may replace the $k_z(v_z - v_z')$ term in the integrand by $k_y(V_y' - V_y)$. Hence, the flux is proportional to the difference in the fluid velocity between points x and x'.

If $|\psi|^2$ is *assumed* to be a sharply peaked function of $x - x'$, a local approximation to the flux can be made, just as was done previously for heat transport. We then obtain, in a manner entirely analogous to the derivation of the local form of the heat conduction,

$$\Gamma_x = \frac{c}{eB} \frac{\partial}{\partial x} \left(mn\lambda \frac{\partial V_y}{\partial x} \right), \tag{13.170}$$

with the kinematic viscosity λ given by

$$\lambda = \frac{(4\pi e^2)^2 n}{T^2} \int dv_z \, e^{-mv_z^2/T} \int \frac{d^3k}{(2\pi)^3} \frac{k_x^2 k_y^2}{|k_z||k^2 D_{\mathbf{k},-ik_z v_z}|^4} \,. \tag{13.171}$$

Here we have simplified the expression for λ by assuming B is large so that we can neglect $k_y V_y$ compared with $k_z v_z$. This expression for the local kinematic viscosity due to long-range collisions can be easily evaluated if one assumes a Debye-shielded response, $D = 1 + 1/k^2\lambda_D^2$. Noting that the logarithmic divergence at small k_z must be cut off at $k_{z\,\min} = \mathrm{Max}(S, \nu)/\bar{v}$ in order to properly account for velocity diffusion (and include the factor-of-3 enhancement effect due to collisional caging discussed in Lecture 2), we can write eqn (13.171) to logarithmic order as (O'Neil 1985)

$$\begin{aligned} \lambda &= \frac{(4\pi e^2)^2 n}{T^2} \sqrt{\pi}\bar{v} \int \frac{dk_\perp k_\perp d\theta}{(2\pi)^3} \frac{k_\perp^4 \sin^2\theta \cos^2\theta}{(k_\perp^2 + \lambda_D^{-2})^4} \, 2 \int_{k_{z_{\min}}}^{\lambda_D^{-1}} \frac{dk_z}{|k_z|} \\ &= \frac{\sqrt{\pi}}{6} n\bar{v} b^2 \lambda_D^2 \ln\left(\frac{\omega_p}{\mathrm{Max}(S,\nu)}\right). \end{aligned} \tag{13.172}$$

However, if we use the full dielectric response we again find a nonlocal contribution due to lightly damped waves that cannot be neglected. A calculation analogous to that given previously for thermal conduction yields

$$\Gamma_x = \frac{n}{\Omega_c} \frac{\partial}{\partial x} \sum_{j=1}^{\infty} (\lambda^{\mathrm{local}} + \lambda_j^{\mathrm{waves}}) \, \hat{V}_j \frac{\sin j\pi x}{L} \,, \tag{13.173}$$

where $\hat{V}_j$ is the jth Fourier coefficient of $\partial V_y/\partial x$,

$$\hat{V}_j = \frac{2}{L} \int_0^L dx \frac{\partial V_y}{\partial x} \frac{\sin j\pi x}{L} \,, \tag{13.174}$$

and λ^{local} is the local contribution due to wavenumbers $k\lambda_D > 0.4$, given by

$$\lambda^{\mathrm{local}} = 0.0465 \frac{e^2}{m\bar{v}} \ln\left(\frac{\omega_p}{\mathrm{Max}(S,\nu)}\right) = 0.585 \nu_0 \lambda_D^2 \ln\left(\frac{\omega_p}{\mathrm{Max}(S,\nu)}\right), \tag{13.175}$$

roughly twice the result given by simple Debye shielding, eqn (13.172). The wave contribution is

$$\lambda_j^{\mathrm{waves}} = n\bar{v} b^2 \lambda_D^2 \ln\left(\frac{\omega_p}{\mathrm{Max}(S,\nu)}\right) \int_0^{0.4} d\bar{k}_\perp h_\lambda(\bar{k}_\perp, j\varepsilon) \,, \tag{13.176}$$

with

$$h_\lambda = \frac{4\bar{k}_\perp^5 e^{-\bar{v}_0^2}}{(\partial a/\partial \bar{v}_0) b(\bar{v}_0)} \int_0^{2\pi} d\theta \frac{\cos^2\theta \sin^2\theta}{b^2(\bar{v}_0) + (\bar{k}_\perp j\varepsilon)^2 \cos^2\theta} \,, \tag{13.177}$$

where the functions a and b are given by eqn (13.94), $\bar{v}_0 = v_0/\bar{v}$ is the solution to eqn (13.155) taking $k_z = 0$, and $\bar{k}_\perp = k_\perp \lambda_D$.

Table 13.3 Wave contribution to viscosity

$j\varepsilon$	λ_j^{waves} $(\ln(\omega_p/\text{Max}(S,\nu))e^2/m\bar{v})$
0.1	1.45×10^{-3}
0.05	3.19×10^{-3}
0.01	1.06×10^{-2}
0.005	1.61×10^{-2}
0.001	4.11×10^{-2}
0.0005	6.29×10^{-2}
10^{-4}	0.181
10^{-5}	0.955

Table 13.4 Transport coefficients due to long-range collisions in a single-species plasma with $\lambda_D > \bar{r}_c$. In this regime, these coefficients must be added to the classical results (Table 13.1).[a]

D	χ	λ
$2\sqrt{\pi}\nu_0\bar{r}_c^2 \ln\left(\frac{\lambda_D}{\bar{r}_c}\right)\ln\left(\frac{\bar{v}}{\gamma\sqrt{\lambda_D\bar{r}_c}}\right)$	$0.0652\, e^2/m\bar{v}$ + wave contribution important for $L_j > 100\,\lambda_D$	$0.0465(e^2/m\bar{v})\ln\left(\frac{\omega_p}{\gamma}\right)$ + wave contribution important for $L_j > 10^3\lambda_D$

[a] $S \equiv |\partial V_y/\partial x|$, $\nu_0 \equiv n\bar{v}b^2$, $\gamma \equiv \text{Max}(S,\nu)$

The wave contribution is tabulated in Table 13.3. One can see that waves are important to viscous transport only when the scale length of the shear flow $L_j = L/j\pi$ exceeds roughly 1000 Debye lengths.

The wave contribution to the viscosity is an order of magnitude smaller than the corresponding contribution to the thermal conduction. Evidently, weakly damped plasma waves transmit energy more efficiently than momentum. So far, there have been no experiments to test these results in detail. However, some experiments have been performed that measured viscosities consistent with eqn (13.175) (Kriesel and Driscoll 2001). More experiments need to be done to test the density, temperature, and magnetic-field dependence of the viscosity coefficient.

Table 13.4 summarizes our results for the flux of particles, energy, and momentum caused by long-range collisions.

13.5.3 Diffusion revisited, and the Ludwig–Soret effect

As a final application of the new collision operator, we reanalyze particle diffusion due to long-range collisions where there are several species and temperature gradients. We also consider the effects of lightly damped waves on diffusion. We will find that such effects are negligible, justifying our neglecting them in Lecture 2. By including temperature gradients, we will predict a new transport coefficient related to the Ludwig–Soret effect, where a temperature gradient induces a particle flux in a multispecies system.

The system is assumed to be at temperature $T(x)$, and, as always, velocity-scattering collisions keep the distribution function for each species in a local Maxwellian form,

$$f_\alpha(x, v_z) = \frac{n_\alpha(x)}{\sqrt{2\pi T(x)/m_\alpha}} e^{-m_\alpha v_z^2/2T(x)} . \tag{13.178}$$

Equation (13.131) then implies the following expression for the particle flux of species α:

$$\begin{aligned}
\Gamma_{\alpha_x}(x) &= \int dv_z \frac{c}{B} \langle \delta E_y \, \delta \eta_\alpha \rangle \\
&= \frac{ce_\alpha}{B} \sum_\beta (4\pi e_\beta)^2 \int dx' \, dv'_z \, dv_z \int \frac{dk_y \, dk_z}{(2\pi)^2} k_y |\psi|^2 (x, x', k_y, k_z, -ik_y V_y - ik_z v_z) \\
&\quad \times \pi \delta(k_z(v_z - v'_z) + k_y(V_y - V'_y)) \frac{\sqrt{m_\alpha m_\beta}}{2\pi \sqrt{TT'}} e^{-(m_\alpha v_z^2/2T + m_\beta v_z'^2/2T')} \\
&\quad \times \left\{ \frac{n_\alpha}{m_\beta} \left(\frac{k_y}{\Omega_{c\beta}} \left[\frac{\partial n'_\beta}{\partial x'} + \frac{n'_\beta}{T'} \frac{\partial T'}{\partial x'} \left(\frac{m_\beta v_z'^2}{2T'} - \frac{1}{2} \right) \right] - \frac{m_\beta v'_z}{T'} k_z n'_\beta \right) \right. \\
&\quad \left. - \frac{n'_\beta}{m_\alpha} \left(\frac{k_y}{\Omega_{c\alpha}} \left[\frac{\partial n_\alpha}{\partial x} + \frac{n_\alpha}{T} \frac{\partial T}{\partial x} \left(\frac{m_\alpha v_z^2}{2T} - \frac{1}{2} \right) \right] - \frac{m_\alpha v_z}{T} k_z n_\alpha \right) \right\} .
\end{aligned} \tag{13.179}$$

We assume that $|\psi|^2$ is strongly peaked in $x - x'$, so that everywhere in the integrand except in $|\psi|^2$ itself we may take $x' = x$. The δ-function then implies $v_z = v'_z$, and eqn (13.179) becomes

$$\Gamma_{\alpha_x} = -\sum_\beta \left(D_{\alpha\beta} \frac{\partial n_\alpha}{\partial x} - \bar{D}_{\alpha\beta} \frac{\partial n_\beta}{\partial x} \right) + \gamma_\alpha \frac{\partial T}{\partial x} , \tag{13.180}$$

where

$$D_{\alpha\beta} = \left(\frac{4\pi e_\beta c}{B} \right)^2 \frac{n_\beta}{2T} \sqrt{m_\alpha m_\beta} \int \frac{dk_y \, dk_z}{(2\pi)^2} \frac{k_y^2}{|k_z|} \int dx' \, dv_z |\psi|^2 e^{-(m_\alpha + m_\beta) v_z^2/2T} \tag{13.181}$$

is the diffusion coefficient of species α due to collisions with species β and $\bar{D}_{\alpha\beta}$ is an "off-diagonal" diffusion coefficient, producing a flux of species α due to gradients in other species. The diffusion coefficient $\bar{D}_{\alpha\beta}$ is related to $D_{\alpha\beta}$ via

$$e_\beta n_\beta \bar{D}_{\alpha\beta} = e_\alpha n_\alpha D_{\alpha\beta} . \tag{13.182}$$

Note that when there is only a single species, eqns (13.180) and (13.182) imply that the diffusive flux vanishes, as expected. For a single species the flux is viscous, not diffusive. When there are two species with identical mass and charge that have some other property that discriminates them, such as their spin, eqns (13.180) and (13.182) are equivalent to eqn (13.42) for the flux of test particles through a background plasma of like particles.

The Ludwig–Soret coefficient γ_α provides the particle flux of species α due to a temperature gradient:

$$\gamma_\alpha = \frac{ce_\alpha}{2B}\sum_\beta (4\pi e_\beta)^2 n_\alpha n_\beta \int dx'\, dv_z \int \frac{dk_y\, dk_z}{(2\pi)^2}\frac{k_y^2}{|k_z|}|\psi|^2 \frac{\sqrt{m_\alpha m_\beta}}{T^2} e^{-((m_\alpha+m_\beta)/2T)v_z^2}$$
$$\times\left[\frac{1}{m_\beta \Omega_{c_\beta}}\left(\frac{m_\beta v_z^2}{2T}-\frac{1}{2}\right)-\frac{1}{m_\alpha\Omega_{c_\alpha}}\left(\frac{m_\alpha v_z^2}{2T}-\frac{1}{2}\right)\right]. \tag{13.183}$$

The above expressions for $D_{\alpha\beta}$ and γ_α are evaluated under the assumption that $|\psi|^2$ is sharply peaked. Spatial variation in the coefficients of eqn (13.119) can then be neglected, from which Parseval's theorem implies

$$\int dx'\, |\psi|^2 = \int \frac{dk_x}{2\pi}\frac{1}{|k^2 D_{\mathbf{k},-ik_z v_z}|^2}. \tag{13.184}$$

Thus, eqn (13.181) may be written as

$$D_{\alpha\beta} = \left(\frac{4\pi e_\beta c}{B}\right)^2 \frac{n_\beta}{2T}\sqrt{m_\alpha m_\beta}\int \frac{d^3k\, dv_z}{(2\pi)^3}\frac{k_y^2}{|k_z||k^2 D_{\mathbf{k},-ik_z v_z}|^2} e^{-(m_\alpha+m_\beta)v_z^2/2T}. \tag{13.185}$$

When we note the log divergence at small k_z and cut it off in the usual manner at $k_{z\,\min} = \mathrm{Max}(S,\nu)/\bar{v}$ and $k_{z\,\max} = k_\perp$, eqn (13.185) becomes

$$D_{\alpha\beta} = \left(\frac{4\pi e_\beta c}{B}\right)^2 \frac{n_\beta}{T}\sqrt{m_\alpha m_\beta}$$
$$\times \int \frac{dv_z\, d\bar{k}_\perp}{(2\pi)^3}\frac{\pi \bar{k}_\perp^3 e^{-(m_\alpha+m_\beta)v_z^2/2T}}{(\bar{k}_\perp^2 + a(v_z/\bar{v}))^2 + b^2(v_z/\bar{v}))}\log\left(\frac{\bar{k}_\perp \omega_p}{\mathrm{Max}(S,\nu)}\right), \tag{13.186}$$

where $\bar{k}_\perp \equiv k_\perp \lambda_D$. While the denominator does exhibit near-zeros due to lightly damped waves, these occur only for $\bar{k}_\perp < 1$. However, the integrand is dominated by the logarithmic divergence at large $\bar{k}_\perp$, which we cut off at $\bar{k}_\perp \simeq \lambda_D/\bar{r}_c$, where here $\bar{r}_c$ is the mean cyclotron radius for species α and β.

In this range of $\bar{k}_\perp$ the integrand can be approximated by neglecting a and b (i.e. using unshielded interactions), yielding

$$D_{\alpha\beta} = \left(\frac{4\pi e_\beta c}{B}\right)^2 \frac{n_\beta}{T}\sqrt{m_\alpha m_\beta}\frac{\pi}{(2\pi)^3}\int dv_z\, e^{-(m_\alpha+m_\beta)v_z^2/2T}\int_1^{\lambda_D/\bar{r}_c}\frac{d\bar{k}_\perp}{\bar{k}_\perp}\ln\left(\frac{\bar{k}_\perp\omega_p}{\mathrm{Max}(S,\nu)}\right)$$
$$= 2\sqrt{\pi}\left(\frac{e_\beta c}{B}\right)^2 n_\beta\sqrt{\frac{2m_\alpha m_\beta}{(m_\alpha+m_\beta)T}}\ln\left(\frac{\lambda_D}{\bar{r}_c}\right)\ln\left(\frac{\bar{v}}{\mathrm{Max}(S,\nu)\sqrt{\lambda_D \bar{r}_c}}\right). \tag{13.187}$$

Thus, lightly damped waves have a negligible effect on the diffusion coefficient, justifying the analysis of Lecture 2. For a single species, eqn (13.187) is the same as eqn (13.59).

Applying the same analysis to the Ludwig–Soret coefficient given by eqn (13.183) yields

$$\gamma_\alpha = \sum_\beta \frac{n_\alpha D_{\alpha\beta}}{2T}\frac{e_\beta m_\beta - e_\alpha m_\alpha}{e_\beta(m_\alpha+m_\beta)}. \tag{13.188}$$

As expected, this coefficient vanishes in a single-species system.

13.6 Enhanced transport in nearly 2D plasmas

The theory of transport due to long-range collisions, summarized in Table 13.4, assumes plasmas of infinite extent along the magnetic-field direction $\hat{z}$. However, new effects come into play for plasmas that are of finite length L_z when both the collision frequency ν and the shear rate S are small compared with the frequency $\omega_B = \pi\bar{v}/L_z$ at which particles bounce from end to end in the plasma. When this axial bounce frequency is large, particles encounter one another many times as they bounce, just as in the collisional-caging phenomenon discussed in Lecture 2, except that now the ends of the plasma (rather than surrounding particles) act to cage the particles. The effect of these multiple encounters is to increase the correlation time of the interactions, leading to larger particle diffusion and viscosity (we will see that thermal diffusion is largely unaffected).

This effect is already contained in our collision operator. Let us apply it to a system of finite length with *periodic* boundary conditions in z of length L_z; the argument can be generalized to more realistic boundary conditions (Dubin and O'Neil 1998). The imposition of periodic boundary conditions has the effect of replacing the integral over k_z in eqn (13.179) by a sum, $\int dk_z \to (2\pi/L_z)\sum_{k_z}$. This sum includes a $k_z = 0$ term, representing the z-averaged (or "bounce-averaged") interaction. In our previous derivations, this term did not enter. The contribution of $k_z = 0$ to the flux is

$$\Gamma_\alpha^{k_z=0} = \frac{ce_\alpha}{B}\sum_\beta (4\pi e_\beta)^2 \int dx' \frac{dk_y}{2\pi L_z} k_y^2 |\psi|^2(x, x', k_y, 0, -ik_yV_y)$$
$$\times \pi\delta(k_y(V_y - V_y'))\left[\frac{n_\alpha}{m_\beta\Omega_{c_\beta}}\frac{\partial n_\beta'}{\partial x'} - \frac{n_\beta'}{m_\alpha\Omega_{c_\alpha}}\frac{\partial n_\alpha}{\partial x}\right], \tag{13.189}$$

where we have dropped temperature gradients for simplicity.

If $V_y(x)$ is not a monotonic function of x, the δ-function allows resonant interactions between separate x and x' points that obey $V_y = V_y'$, leading to enhanced viscosity (Dubin and O'Neil 1988). However, if $V_y(x)$ is monotonic, only $x = x'$ satisfies $V_y = V_y'$ and eqn (13.189) predicts a diffusive contribution to the flux with a diffusion coefficient $D_{\alpha\beta}^{2\text{D}}$ given by

$$D_{\alpha\beta}^{2\text{D}} = \sum_\beta \left(\frac{4\pi e_\beta c}{B}\right)^2 n_\beta \int \frac{dk_y}{2\pi L_z}|k_y|\frac{\pi|\psi|^2(x, x, k_y, 0, -ik_yV_y)}{|\partial V_y/\partial x|}. \tag{13.190}$$

The integral over k_y is logarithmically divergent at large k_y. In this range, the Green's function is unshielded, eqn (13.119) becoming

$$\frac{\partial^2\psi}{\partial x^2} - k_y^2\psi = \delta(x - x'), \tag{13.191}$$

which implies $\psi(x, x) = 1/(2|k_y|)$ at large k_y. Substituting this form into eqn (13.190) yields, to logarithmic accuracy,

$$D_{\alpha\beta}^{2\text{D}} = \sum_\beta \left(\frac{4\pi e_\beta c}{B}\right)^2 \frac{n_\beta}{4L_z|S|}\ln\left(\frac{k_{y\,\max}}{k_{y\,\min}}\right), \tag{13.192}$$

where $k_{y\,\mathrm{max}}$ and $k_{y\,\mathrm{min}}$ are cutoffs determined, as always, by physics outside of the model equations used to obtain the result (Chavanis 2000; Dubin and Jin 2001; Dubin 2003). We neglect for the moment the rather subtle effects that enter into these cutoffs, and instead compare eqn (13.192) with our previous expression for $D_{\alpha\beta}$, eqn (13.187). This expression included only $k_z \neq 0$ terms, and will now be referred to as $D^{\mathrm{3D}}_{\alpha\beta}$. The ratio of $D^{\mathrm{2D}}_{\alpha\beta}$ to $D^{\mathrm{3D}}_{\alpha\beta}$ is roughly

$$\frac{D^{\mathrm{2D}}_{\alpha\beta}}{D^{\mathrm{3D}}_{\alpha\beta}} \sim \frac{\omega_B}{S}\,, \tag{13.193}$$

where we have neglected constants of order unity, and the logarithmic factors.

Thus, when $\omega_B > S$, the $k_z = 0$ (two-dimensional) contribution to the diffusion dominates. Also, note that the 2D contribution scales with B as B^{-1} (since $S \propto 1/B$) rather than B^{-2}.

The effect of a large axial bounce frequency on transport can be understood from the following simple argument. The rapid axial bounce motion can be averaged out, which effectively replaces the particles by rods of charge with charge per unit length $q \equiv e/L_z$. The charged rods then $\mathbf{E}\times\mathbf{B}$ drift in the electric field created by the other rods. This model is also called a point vortex gas, for reasons that will soon become apparent. The equations of motion for this two-dimensional system of charged rods are

$$\begin{aligned}\frac{dx_i}{dt} &= -\frac{c}{B}\frac{\partial}{\partial y_i}\phi(x_i, y_i, \ldots, x_N, y_N)\,,\\ \frac{dy_i}{dt} &= \frac{c}{B}\frac{\partial}{\partial x_i}\phi(x_1, y_1, \ldots, x_N, y_N)\,.\end{aligned} \tag{13.194}$$

As first noted by Taylor and McNamara (1971), the magnetic field can be scaled out of these equations by scaling time as $\bar{t} = ct/B$. Thus, a change in B simply changes the time scale of the dynamics without affecting the orbits in any other way. An increase in B by a factor of 2 slows the dynamics down by this factor. The particle flux due to diffusion or viscosity must therefore scale as B^{-1}. For the case of diffusion, the generalized Fick's law (eqn (13.180)) then implies that diffusion coefficients scale as $1/B$ for this system, as we obtained in eqn (13.192). Also, for viscosity, a particle flux proportional to $1/B$ implies that $\lambda \propto B^1$ (see eqn (13.170) and recall that $V_y \propto 1/B$), which is larger than the B^0 scaling obtained previously. The viscosity and particle diffusion are increased because the correlation time of the density and potential fluctuations is now set by slow $\mathbf{E}\times\mathbf{B}$ drift motion rather than relatively fast motion along the magnetic field.

Also, this 2D plasma model implies a heat flux scaling as $1/B$ as well, but this flux is only a small correction to the previous $O(B^0)$ results of Lectures 3 and 5.

We can use the intuition gained from the model of the plasma as a collection of charged rods to understand eqn (13.192) in more detail. Rods with charge per unit length $q = e/L_z$ drift past one another owing to the flow shear, $S = \partial V_y/\partial x$, and as they interact they take an $\mathbf{E}\times\mathbf{B}$ drift step across the magnetic field. The $\mathbf{E}\times\mathbf{B}$ drift motion of two rods in a shear flow S is an integrable problem in Hamiltonian mechanics (Dubin and Jin 2001), described by the Hamiltonian

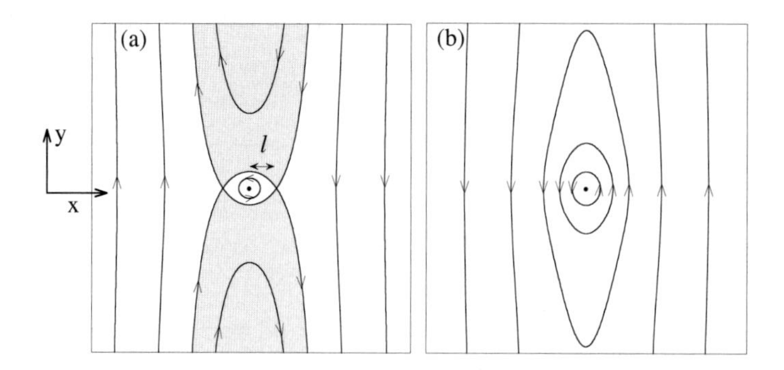

Fig. 13.12 Streamlines for rods in a shear flow. (a) $qS < 0$ (retrograde flow); (b) $qS > 0$ (prograde flow).

$$H(x, y) = -\frac{1}{2}\frac{qB}{c}Sx^2 + q^2 \ln(2\sqrt{x^2 + y^2}), \tag{13.195}$$

where $\pm(x, y)$ is the location of each rod relative to the center of charge, and the dynamics is viewed in a frame where the center of charge is stationary. The equations of motion for one of the two rods are

$$\frac{dx}{dt} = \frac{c}{qB}\frac{\partial H}{\partial y}, \quad \frac{dy}{dt} = -\frac{c}{qB}\frac{\partial H}{\partial x}, \tag{13.196}$$

and the other rod is at $-(x, y)$. The rods move along surfaces of constant H, shown in Fig. 13.12(a) and 13.12(b) for the two cases $qS < 0$ and $qS > 0$, respectively. When $qS < 0$ (termed "retrograde flow"), a separatrix exists in the flow, whose width is

$$\ell = \sqrt{-qc/BS} \tag{13.197}$$

at $y = 0$. Rods in the shaded region of the figure take an $\mathbf{E} \times \mathbf{B}$ drift step $\Delta x \sim \ell$. The number of such interactions per unit time is roughly $S\ell^2 n_{2\mathrm{D}}$ as rods are swept past one another by the flow, where $n_{2\mathrm{D}} \equiv nL_z$ is the number of rods per unit area. Thus, the diffusion coefficient is roughly

$$\begin{aligned} D^{2\mathrm{D}} &\sim S\ell^2 n_{2\mathrm{D}} \cdot \ell^2 \\ &= \left(\frac{qc}{B}\right)^2 \frac{n_{2\mathrm{D}}}{|S|}, \end{aligned} \tag{13.198}$$

in agreement with the scaling of eqn (13.192).

In fact, collisions between rods with impact parameters $\rho \lesssim \ell$, which result in reflections as depicted in Fig. 13.12(a), are not included in eqn (13.192) (which was derived through linearization, i.e. integration along unperturbed orbits). The linearization requires that the orbits are only slightly perturbed, and so eqn (13.192) applies only for impact parameters larger than ℓ. This sets one estimate for $k_{y\,\max}$; $k_{y\,\max} \sim \ell^{-1}$.

For impact parameters less than ℓ, a detailed Boltzmann calculation of the diffusion (Dubin and Jin 2001) yields

$$D^{2D}_{\text{Boltzmann}} = 4\left(\frac{qc}{B}\right)^2 \frac{n_{2D}}{|S|} \ln^2(0.17 n_{2D}\ell^2)\,. \tag{13.199}$$

This result must be added to eqn (13.192) to obtain the total diffusion. Equation (13.199) is only valid if $n_{2D}\ell^2 < 1$, which requires strong shear or low density. For $n_{2D}\ell^2 \gtrsim 1$ (large densities, low shear), the two-particle picture of collisions shown in Fig. 13.12(a) is incorrect because other particles intervene. Only eqn (13.192) is required to describe the diffusion in this high-density, low-shear regime.

Note however, that if $qS > 0$ (termed "prograde flow"), there is no separatrix in the flow (see Fig. 13.12(b)); instead, two rods trap one another in their mutual interaction, circulating indefinitely. Diffusive steps are taken only when these trapped rods interact with others, disrupting the circulation. No rigorous theory for this situation has been developed, and we do not expect eqn (13.192) to accurately describe this case. Thus, eqn (13.192) applies only to the case $qS < 0$ (or for multiple species, when $(q_\alpha + q_\beta)S < 0$; see Dubin (2003)).

Another issue that must be addressed is the physics of the cutoffs of the logarithmic divergence of eqn (13.192), $k_{y\,\min}$ and $k_{y\,\max}$. These wavenumbers are set, respectively, by maximum and minimum distance scales along the flow direction. The maximum distance scale is determined by the system size. In a rotating plasma with a circular cross section transverse to the magnetic field, the maximum distance at radius r is the circumference $2\pi r$, which sets $k_{y\,\min} \sim 1/r$. The minimum distance scale is set by two possibilities: the distance ℓ (see Fig. 13.12(a)) or a distance $\delta = \sqrt{D^{2D}/|S|}$ determined by the diffusion itself. Particles separated by a distance less than δ diffuse apart before they shear apart, contradicting the assumption of integration along unperturbed orbits used in deriving eqn (13.192). Thus, we take $k_{y\,\max} = 1/\max(\delta, \ell)$.

One more issue must be addressed before we compare eqn (13.192) with results from experiments and computer simulations. Equation (13.192) indicates that as the flow shear S decreases, transport increases. This is because adjacent rods have less relative velocity and interact for a longer time as S decreases. However, as $S \to 0$ this breaks down: the relative motion of rods is no longer set by the shear flow, but rather by the diffusion process itself. Rods diffuse apart by a distance of the order of the Debye length λ_D at a rate of order D^{2D}/λ_D^2. If we simply replace S by this rate in eqn (13.192), we obtain (assuming $\lambda_D > \ell$)

$$\begin{aligned}(D^{2D})^2 &\sim \left(\frac{4\pi ec}{B}\right)^2 \frac{n\lambda_D^2}{4L_z} \ln\left(\frac{r}{\lambda_D}\right) \\ &= \left(\frac{4\pi qc}{B}\right)^2 \frac{n_{2D}\lambda_D^2}{4} \ln\left(\frac{r}{\lambda_D}\right).\end{aligned} \tag{13.200}$$

This is an estimate of the diffusion coefficient in a shear-free plasma in the two-dimensional limit, first put forward by Taylor and McNamara (1971) and Dawson *et al.* (1971). If the temperature is sufficiently high, so that the Debye length is greater than the plasma radius r_p, one should replace λ_D by r_p, and set the logarithm to unity

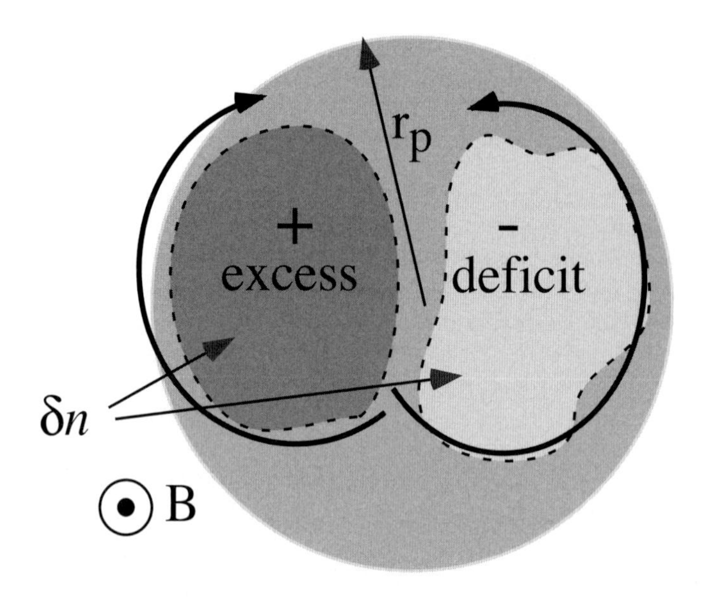

Fig. 13.13 Sketch of Dawson–Okuda vortices.

since its argument is only an estimate. In this case the result is usually referred to as Taylor–McNamara diffusion, and can be written as

$$D^{\mathrm{TM}} = \frac{qc}{2B}\sqrt{\frac{N}{\pi}}, \tag{13.201}$$

where $N = \pi r_p^2 n_{\mathrm{2D}}$ is the number of charges in the plasma. Another way to understand diffusion in this limit is through the action of long-wavelength density/potential fluctuations (see Fig. 13.13). A density fluctuation with a size scale of order r_p would, on average, have a magnitude of order $\delta n_{\mathrm{2D}} \sim \sqrt{N}\, n_{\mathrm{2D}}/N$ and cause an $\mathbf{E}\times\mathbf{B}$ drift velocity of order $(c/B)\, q\, \delta n_{\mathrm{2D}}\, r_p$. The "diffusive step" is of order r_p in such a fluctuation, and the rate of the steps is the circulation rate in the fluctuation, $(c/B)\, q\, \delta n_{\mathrm{2D}}$. Thus,

$$\begin{aligned} D^{\mathrm{TM}} &\sim \frac{c}{B} q\, \delta n_{\mathrm{2D}}\, r_p^2 \\ &= \frac{qc}{\pi B}\sqrt{N}, \end{aligned} \tag{13.202}$$

which has the same scaling as eqn (13.201). If we replace the spatial size r_p of these fluctuations by λ_D, we recover the scaling of eqn (13.200). These fluctuations are often referred to as "convective cells" or "Dawson–Okuda vortices." For nonzero shear, the Dawson–Okuda vortices are pulled apart by the flow shear, reducing the radial scale of the diffusive steps. This is why D^{2D} given by eqn (13.192) decreases with increasing S.

Experiments and simulations have been performed on plasmas in the 2D regime to test eqn (13.192). In the simulations, N identical charged rods were placed randomly

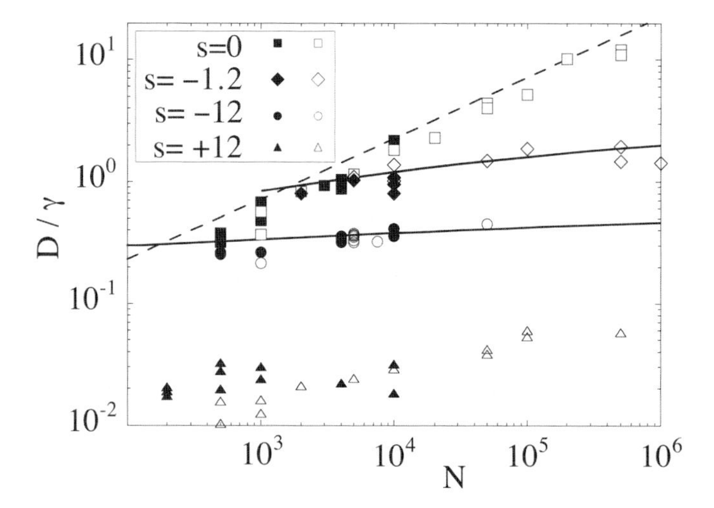

Fig. 13.14 Diffusion measured in simulations for four different shear rates, versus number of rods N. Here $s = SB/(2\pi qcn_{2D})$ is a scaled shear rate and $\gamma = 4\pi qc/B$. The solid lines show eqn (13.192) added to eqn (13.199). The dashed line shows eqn (13.201).

in a plasma with a circular cross section, and their 2D $\mathbf{E}\times\mathbf{B}$ drift motion was followed using eqn (13.194). To add flow shear, an external radial electric field was also applied, and the radial diffusion of the rods was measured. The resulting diffusion coefficient is plotted versus N for four different shear rates in Fig. 13.14. In the figure, $\gamma = 4\pi qc/B$ in the "circulation" associated with a rod (which happens to have units of a diffusion coefficient), and $s = 2S/n_{2D}\gamma$ is a scaled shear rate. For $s < 0$, eqn (13.192) (with eqn (13.199) added) accurately predicts the diffusion. For $s = 0$, eqn (13.201) works well. (Note that random placement of the rods corresponds to infinite temperature and hence $\lambda_D \gg r_p$, so eqn (13.201) should be used rather than eqn (13.200).)

For $s > 0$, corresponding to the case $qS > 0$, for which eqn (13.192) is not expected to apply, the measured diffusion is an order of magnitude less than eqn (13.192) predicts (triangles). As yet, there is no theory available for this case.

Experiments on Mg^+ ion plasmas have also measured diffusion, as discussed in Lecture 2, and when $\omega_B > |S|$ the experiments do observe the expected scaling of D with shear (Fig. 13.15). When shear is large enough that $\omega_B/|S| < 1$, the plasma is in the 3D regime described by eqn (13.59); but when the shear is smaller, the results show the expected increase of the diffusion as the shear decreases, although the magnitude is off by about a factor of 2 compared with eqn (13.192) (Driscoll *et al.* 2002). For comparison, the Taylor–McNamara zero-shear result of eqn (13.202) is also shown.

13.6.1 Continuum limit and 2D Euler flow

So far, these lectures have focused on collisional-transport theory. However, it is worth mentioning that the model introduced in this lecture, of a magnetized plasma as a collection of charged rods moving via $\mathbf{E}\times\mathbf{B}$ drift dynamics in two dimensions, has

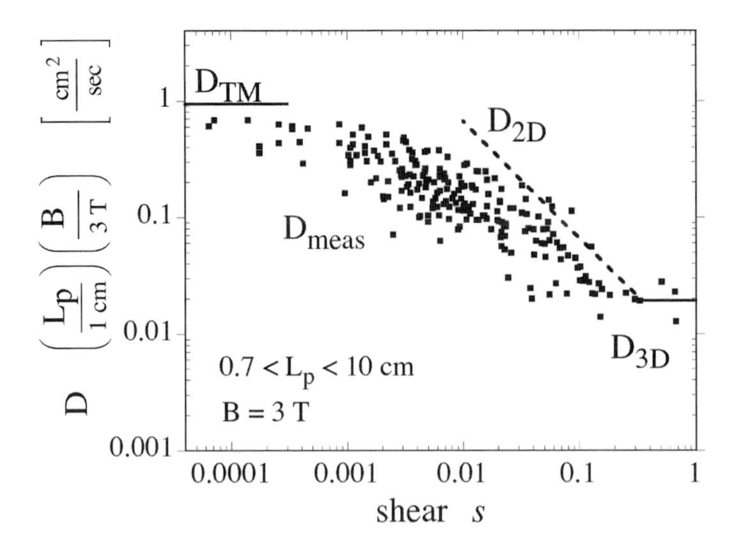

Fig. 13.15 Experimental measurement of diffusion in the 2D regime compared with eqn (13.192). Here s is the scaled shear rate (see Fig. 13.14), and the plasma length is referred to as L_p.

interesting implications beyond the area of transport theory. We shall briefly review some aspects of the extensive research in this area. In the continuum limit of the model (i.e. neglecting collisions), the system is described by the Vlasov equation for the 2D rod density $n_{2\text{D}}(x, y, t)$,

$$\frac{\partial n_{2\text{D}}}{\partial t} - \frac{c}{B}\nabla\phi \times \hat{z} \cdot \nabla n_{2\text{D}} = 0\,, \tag{13.203}$$

where the electrostatic potential $\phi(x, y, t)$ is related to $n_{2\text{D}}$ via the 2D Poisson's equation

$$\nabla_\perp^2 \phi = -4\pi q n_{2\text{D}}\,. \tag{13.204}$$

This plasma model is isomorphic to a two-dimensional neutral inviscid fluid, which is described by Euler's equations for the vorticity $\zeta(x, y, t) = \hat{z} \cdot \nabla \times \mathbf{V}(x, y, t)$, where $\mathbf{V}$ is the flow velocity in the x–y plane. Euler's equations are

$$\frac{\partial \zeta}{\partial t} + \mathbf{V} \cdot \nabla \zeta = 0\,, \tag{13.205}$$

where the fluid velocity $\mathbf{V}$ is related through the incompressibility condition $\nabla \cdot \mathbf{V} = 0$ to a streamfunction $\psi(x, y, t)$,

$$\mathbf{V} = \hat{z} \times \nabla\psi\,, \tag{13.206}$$

and ψ is given in terms of ζ by

$$\nabla_\perp^2 \psi = \zeta\,. \tag{13.207}$$

Comparing eqns (13.203)–(13.204) with eqns (13.205)–(13.207), we see that the vorticity ζ is related to the density $n_{2\text{D}}$ through

$$\zeta = -4\pi q n_{2\mathrm{D}} \frac{c}{B}, \tag{13.208}$$

and the streamfunction is related to the electrostatic potential via

$$\psi = \frac{c\phi}{B}. \tag{13.209}$$

Using these relations, a 2D plasma can be employed to study 2D Euler flow. The 2D Euler equations are a useful paradigm for various fluid flows occurring in nature, from turbulence in soap films to large-scale atmospheric flows such as hurricanes or Jupiter's Great Red Spot. Also, eqns (13.194), describing the $\mathbf{E} \times \mathbf{B}$ motion of charged rods, are the same as the equations of motion for a point vortex gas consisting of N point vortices with identical circulation $\gamma = 4\pi qc/B$. (The circulation is the area integral of the vorticity.) The point vortex gas has been employed as a model for turbulent flow for many years.

Experiments on pure electron plasmas, carried out in the 2D regime where $\omega_B \gg \nu, S$, have investigated several aspects of 2D Euler flow, and have observed some striking new phenomena. In these experiments, the electron plasma was trapped in a Penning trap configuration similar to those used in the previously-described transport experiments using Mg^+ ion plasmas; see Fig. 13.4. One must change the sign of the confinement voltages to contain electrons rather than ions.

To diagnose the plasma, the end confinement electrode voltage was rapidly brought to ground, allowing the electrons to stream out along the magnetic field. An accelerating voltage increased their parallel energy to around 10 kV and they impacted on a phosphor screen (not shown in the figure). A camera image of the glowing phosphor provided a direct measurement of $n_{2\mathrm{D}}(x, y, t)$. Although the measurement was destructive, by repeatedly creating the same initial condition and allowing it to evolve for different times, a time history for $n_{2\mathrm{D}}(x, y, t)$ could be built up, providing a complete record of the dynamics of the 2D flow. Movies starting from various initial conditions can be found on the "Nonneutral Plasmas" website, http://nnp.ucsd.edu.

Figure 13.16 shows a sequence of images in which two vorticity patches (electron columns) undergo a merger (Fine *et al.* 1991). Well-separated vortices would simply rotate around one another owing to their self-interaction. However, if the vortices are sufficiently closely spaced, the flow field of one distorts the other. This distortion lowers their self-energy, and in order to conserve total energy the vortices must therefore move closer together. (This energy argument is more intuitive if one thinks of the vortices as electron columns; the energy is just the repulsive potential energy of the columns, including their self- and mutual interaction.) This results in a merger when the vortices are sufficiently close together.

Figure 13.17 shows the merger time as a function of $2D/2R_v$, where $2D$ is the initial separation and $2R_v$ is the initial vortex diameter. The time for the merger increases by a factor of 10^4 as D/R_v passes from 1.5 to 1.8. These results are in good agreement with analytical theory and numerical computations for 2D ideal fluids (Moore and Saffman 1975; Saffman and Szeto 1980; Rossow 1977; Melander *et al.* 1988). The eventual merger for $D/R_v \gtrsim 2$ is caused by slow expansion of the vortices due to the weak viscous effects analyzed previously in these lectures.

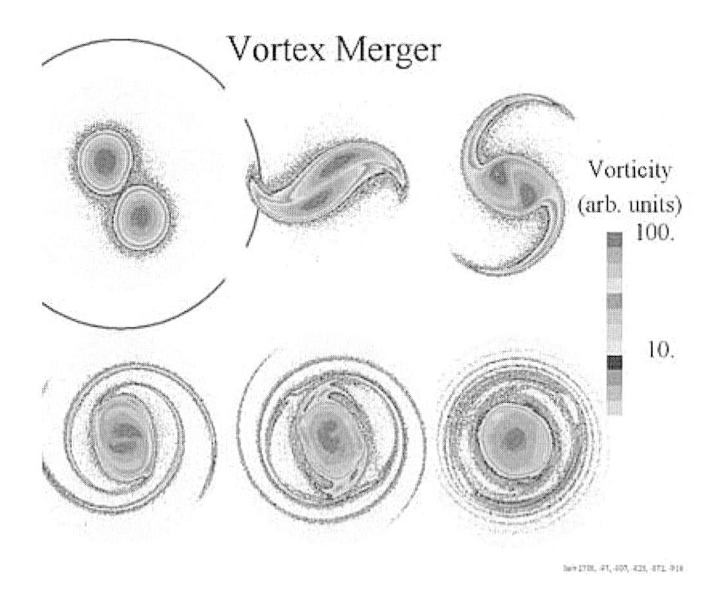

Fig. 13.16 Merger of two like-sign vortices.

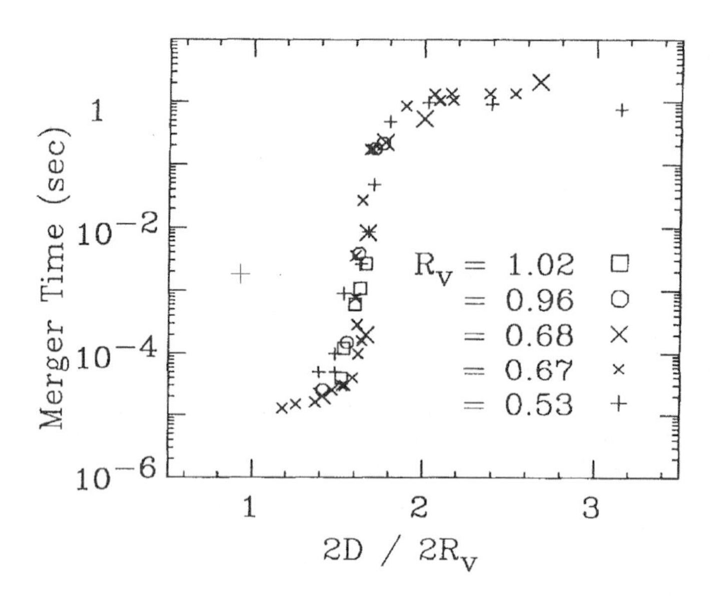

Fig. 13.17 Merger time versus vortex separation.

This merger process is an important element in the dynamics of more complex 2D flows. An example is shown in Fig. 13.18 (Fine *et al.* 1995). Starting from a highly unstable filamented initial condition, the filaments of vorticity quickly form many in-

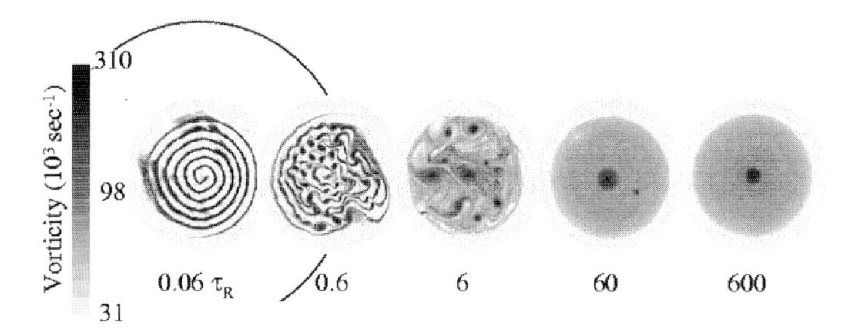

Fig. 13.18 Free relaxation of a turbulent flow.

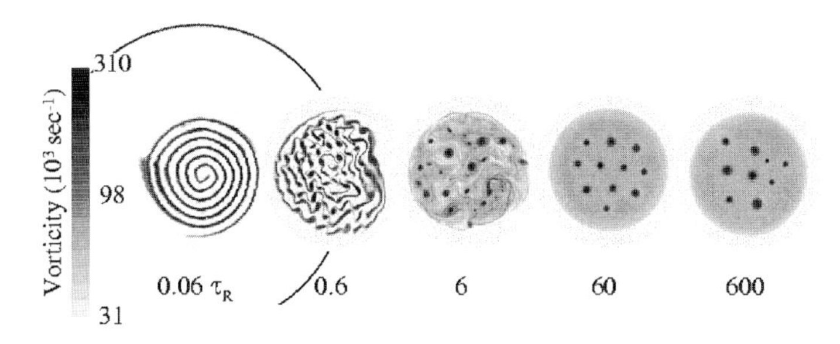

Fig. 13.19 Formation of a vortex crystal from relaxation of 2D turbulence. Here τ_R is the overall rotation time of the vorticity patch.

tense vortices. The vortices subsequently advect chaotically, and merge when they are advected within the merger radius. The mergers are a manifestation of the "inverse cascade" (Kraichnan 1967) of energy in 2D turbulence: mergers take filamented vorticity and clump it into larger-scale vorticity patches.

The relaxation of 2D turbulent flow to an equilibrium state has been considered by many authors. Some theories describe the relaxation process as ergodic mixing of the vorticity, resulting in maximization of an entropy functional $S[\zeta]$ (Onsager 1949; Joyce and Montgomery 1973; Smith 1991; Lynden-Bell 1967; Miller *et al.* 1992; Robert and Sommeria 1992). Other theories extremize other functionals, such as the enstrophy $Z_2(\zeta) = \int d^2r\, \zeta^2$ (Bretherton and Haidvogel 1976; Leith 1984). Such theories predict smoothly varying (though not necessarily monotonic) equilibrium states, unlike the rather sharply peaked final vorticity observed in Fig. 13.18. Evidently, maximum-entropy theories do not explain the final state in these experiments because the turbulent mixing is not fully ergodic—vorticity trapped in the cores of strong vortices does not ergodically mix throughout the flow.

Other theories (McWilliams 1990; Carnevale *et al.* 1991; Weiss and McWilliams 1993) describe the turbulent relaxation process as a series of merger events, resulting ultimately in one final vortex. For the flow depicted in Fig. 13.18, such theories describe the evolution rather well (Fine *et al.* 1995). However, a slightly different unstable initial condition results in a very different final state (Fig. 13.19). In this evolution, merger events occur initially, but then stop, and the final state consists of several strong vortices in a rigidly rotating regular pattern. These patterns are referred to as vortex

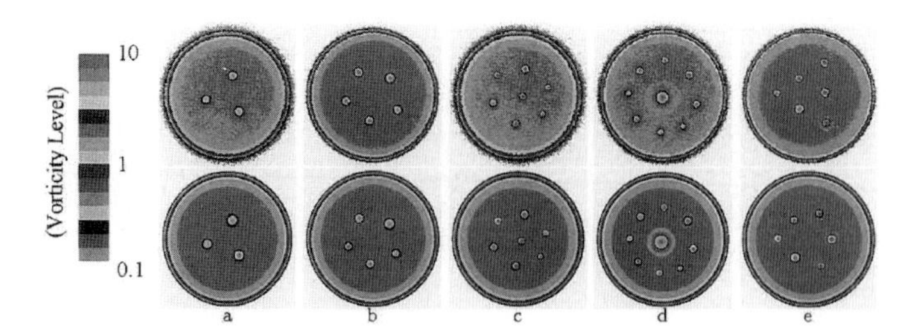

Fig. 13.20 Comparison of experimental crystal patterns (upper) with regional maximum entropy theory (lower).

crystal states, and were not predicted by any previous theory of relaxing 2D inviscid turbulence.

It is now understood that vortex crystal states occur because the strong vortices interact with a low-vorticity background that surrounds them. This background may be present initially or, as in the case of Fig. 13.19, may be produced by the merger process as filaments of vorticity are thrown off (see Fig. 13.16) and subsequently mixed. As the strong vortices mix the background, they maximize its entropy and this causes the strong vortices to "cool," falling into a minimum-energy state. In fact, by maximizing the entropy $S[\zeta_b]$ of the background vorticity ζ_b, subject to the constraint that there are a given number of strong vortices, the observed vortex crystal patterns can be predicted (Fig. 13.20). This is referred to as "regional maximum-entropy theory," since only the background is ergodically mixed to a maximum-entropy state, while the vorticity trapped in the strong vortices is not mixed (Jin and Dubin 1998).

The number of strong vortices in the final state is determined by a competition between the merger process and the cooling process. The rate at which mergers occur slows as the number of vortices decreases, and the rate of cooling increases as the background vorticity increases. Both effects act to eventually arrest mergers and form vortex crystals. However, since both the merger and the cooling processes involve chaotic dynamics, we can only estimate the number of strong vortices in the final state (Jin and Dubin 2000).

The interaction of strong vortices with a lower-vorticity background is important in the formation of vortex crystals, and is also important in a number of other applications. For instance, the motion of hurricanes on a rotating planet is influenced by the north–south gradient in the Coriolis parameter, which can be thought of as a (potential) vorticity gradient (Rossby 1948; Liu and Ting 1987; Carnevale *et al.* 1991; Reznik 1992; Smith 1993; Sutyrin and Flierl 1994). The location of Jupiter's Great Red Spot, and of other storms, can also be understood as being due to the interaction of the storm with the background vorticity gradients due to Jupiter's strong zonal winds (Schecter *et al.* 2000).

Figure 13.21 shows that clumps (strong vorticity excesses) ascend a background vorticity until they reach a peak, whereas holes (strong vorticity deficits) descend the gradient (Rossby 1948; Liu and Ting 1987). This gradient-driven separation helps organize storms into bands of like-sign vortices on planets (such as Jupiter) with strong

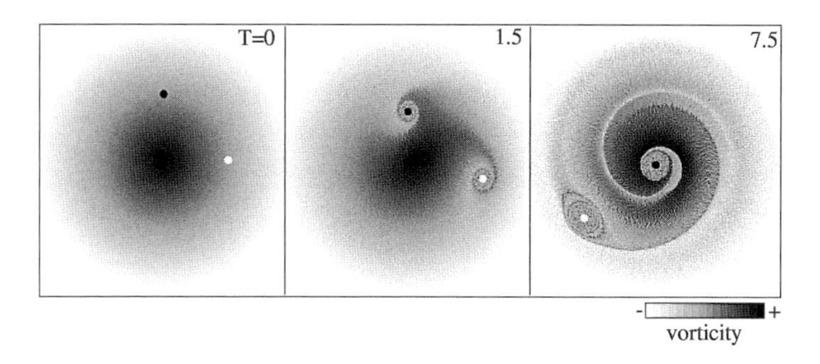

Fig. 13.21 Motion of a clump (and a hole) up (and down) a background vorticity gradient.

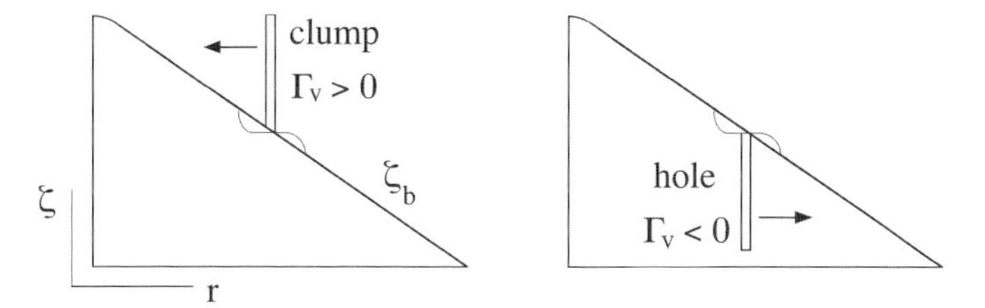

Fig. 13.22 Local mixing of the background vorticity increases $\langle r^2 \rangle_b$. By conservation of P_θ, clumps and holes react oppositely.

zonal winds, with holes in vorticity troughs and clumps at vorticity peaks (Schecter *et al.* 2000). The opposite motion of clumps and holes can be understood by momentum conservation. When there is just one strong vortex in an isolated background vorticity patch, the conserved angular momentum consists of two pieces: a background contribution and a vortex contribution:

$$P_\theta = \gamma_b \langle r^2 \rangle_b + \gamma_v r_v^2 , \tag{13.210}$$

where $\gamma_b > 0$ is the circulation of the background, $\langle r^2 \rangle_b$ is the mean square radius of the background patch, γ_v is the circulation of the strong vortex (positive for clumps and negative for holes), and r_v is the radial location of the vortex. The strong vortex mixes and flattens the θ-averaged background vorticity, as is visible in Fig. 13.21, and is shown schematically in Fig. 13.22. As the background is leveled, it is evident from Fig. 13.22 that $\langle r^2 \rangle_b$ increases. To conserve P_θ, eqn (13.210) implies that r_v^2 must decrease for clumps ($\gamma_v > 0$) and increase for holes ($\gamma_v < 0$).

The radial speed of clumps moving up a vorticity gradient can also be predicted. In fact we have already calculated it, in Lecture 5. There, the flux of a species α due to a density gradient in a *different* species β was given by

$$\Gamma_\alpha = \bar{D}_{\alpha\beta} \frac{\partial n_\beta}{\partial r} \tag{13.211}$$

(changing from planar to cylindrical coordinates). This flux is $\Gamma_\alpha = n_\alpha v_r$, where v_r is the radial velocity of species α up the density gradient due to species β. However, in

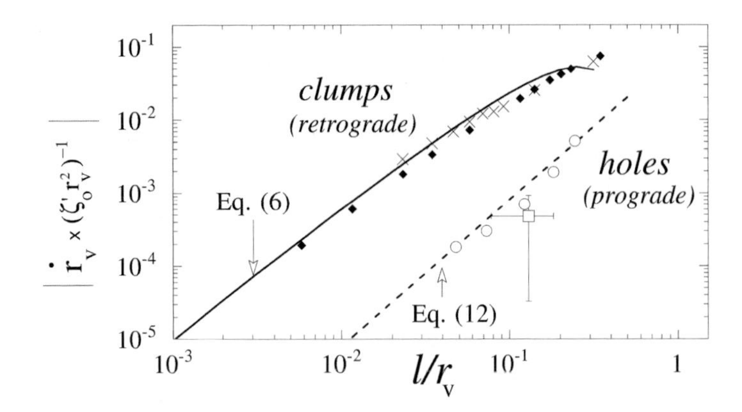

Fig. 13.23 Radial velocity of clumps and holes versus ℓ/r_v, where $\ell = \sqrt{|\gamma_v/2\pi S|}$. Here, "Eq. (6)" and "Eq (12)" refer to equations in Schecter and Dubin (1999). The solid line is eqn (13.212). The dashed line is a mix and move estimate.

the 2D regime we have seen that vorticity and density are the same, so eqn (13.211) predicts the velocity of a species α vortex up the vorticity gradient due to another species β. Using eqns (13.192) and (13.182), we then obtain

$$v_r = \frac{\gamma_v}{4|S|} \ln\left(\frac{r_v}{\ell}\right) \frac{\partial \zeta_b}{\partial r_v}(r_v), \tag{13.212}$$

where we have converted from plasma units to fluid units using eqn (13.208), and the circulation of the strong vortex (species α) is related to the charge q_α per unit length by $\gamma_v = -4\pi c q_\alpha / B$. Equation (13.212) works well in predicting the speed of strong vortices up or down a background gradient, *provided that* $\gamma_v S < 0$, the case of retrograde flow discussed previously (see Fig. 13.12a). For retrograde flow, most of the flow is only slightly perturbed by the strong vortex, and so a quasi-linear calculation using integration along unperturbed orbits is valid. However, for prograde flow, with $\gamma_v S > 0$ (Fig. 13.12(b)), the flow around the strong vortex is trapped and integration along unperturbed orbits does not work. For the background vorticity patch shown in Fig. 13.21, clumps are retrograde ($\gamma_v S < 0$) and holes are prograde ($\gamma_v S > 0$), so eqn (13.212) works for clumps but not holes. Holes are observed to move an order of magnitude slower than the clumps—see Fig. 13.23. This same nonlinear effect was responsible for the observed decrease in diffusion in simulations where $qS > 0$—see Fig. 13.14. As yet, no rigorous theory describes the velocity of prograde holes, although a "mix and move" estimate does appear to work (Schecter and Dubin 1999; Schecter *et al.* 2000).

Acknowledgments

The author thanks the organizers of the Les Houches lectures for the opportunity to contribute to the workshop. The author also thanks Jo Ann Christina for typing and proofreading the manuscript, and Professors C. F. Driscoll, T. M. O'Neil, and Dr. F.

Anderegg for useful scientific discussions. The author's research is currently supported by NSF grant 0354979 and NSF/DOE grant 0613740.

References

Anderegg, F., Huang, X.-P., Driscoll, C. F., Hollmann, E. M., O'Neil, T. M., and Dubin, D. H. E. (1997). Test particle transport due to long range interactions. *Phys. Rev. Lett.*, **78**, 2128–31.

Braginskii, S. I. (1958). Transport phenomena in a completely ionized two-temperature plasma. *Sov. Phys. JETP*, **6**, 358–69.

Braginskii, S. I. (1965). Transport processes in a plasma. In *Review of Plasma Physics*, Vol. 1 (edited by M. A. Leontovitch) pp. 205–311. Consultants Bureau, New York.

Bretherton, F. P. and Haidvogel, D. B. (1976). Two-dimensional turbulence above topography. *J. Fluid Mech.*, **78**, 129–54.

Carnevale, G. F., Kloosterziel. R. C., and Van Heijst, G. J. F. (1991). Propagation of barotropic vortices over topography in a rotating tank. *J. Fluid Mech.*, **233**, 119–39 (1991).

Chavanis, P.-H. (2000). Quasilinear theory of the 2D Euler equation. *Phys. Rev. Lett.*, **84**, 5512–5.

Chen, F. (1974). *Introduction to Plasma Physics*. Plenum Press, New York.

Dawson, J. M., Okuda, H., and Carlile, R. N. (1971). Numerical simulation of plasma diffusion across a magnetic field in two dimensions. *Phys. Rev. Lett.*, **27**, 491–4.

Driscoll, C. F., Anderegg, F., Dubin, D. H. E., Jin, D.-Z., Kriesel, J. M., Hollmann, E. M., and O'Neil, T. M. (2002). Shear reduction of collisional transport: Experiments and theory. *Phys. Plasmas*, **9**, 1905–1914.

Dubin, D. H. E. (1997). Test particle diffusion and the failure of integration along unperturbed orbits. *Phys. Rev. Lett.*, **79**, 2678–81.

Dubin, D. H. E. (1998). Collisional transport in nonneutral plasmas. *Phys. Plasmas*, **5**, 1688–94.

Dubin, D. H. E. (2003). Collisional diffusion in a two-dimensional point vortex gas or a two-dimensional plasma. *Phys. Plasmas*, **10**, 1338–50.

Dubin, D. H. E. and Jin, D.-Z. (2001). Collisional diffusion in a 2-dimensional point vortex gas. *Phys. Lett. A*, **284**, 112–7.

Dubin, D. H. E. and O'Neil, T. M. (1988). Two-dimensional guiding-center transport of a pure electron plasma. *Phys. Rev. Lett.*, **60**, 1286–89.

Dubin, D. H. E. and O'Neil, T. M. (1997). Cross-magnetic field heat conduction in nonneutral plasmas. *Phys. Rev. Lett.*, **78**, 3868–71.

Dubin, D. H. E. and O'Neil, T. M. (1998). Two-dimensional bounce-averaged collisional particle transport in a single species non-neutral plasma. *Phys. Plasmas*, **5**, 1305–14.

Dubin, D. H. E. and O'Neil, T.M. (1999). Trapped nonneutral plasmas, liquids, and crystals (the thermal equilibrium states). *Rev. Mod. Phys.*, **71**, 87–172.

Fine, K. S., Driscoll, C. F., Malmberg, J. H., and Mitchell, T. B. (1991). Measurements of symmetric vortex merger. *Phys. Rev. Lett.*, **67**, 588–91.

Fine, K. S., Cass, A. C., Flynn, W. G., and Driscoll, C. F. (1995). Relaxation of 2D turbulence to vortex crystals. *Phys. Rev. Lett.*, **75**, 3277–80.

Hollmann, E. M., Anderegg, F., and Driscoll, C. F. (2000). Measurement of cross-magnetic-field heat transport due to long-range collisions. *Phys. Plasmas*, **7**, 1767–73.

Jin, D.-Z. and Dubin, D. H. E. (1998). Regional maximum entropy theory of vortex crystal formation. *Phys. Rev. Lett.*, **80**, 4434–37.

Jin, D.-Z. and Dubin, D. H. E. (2000). Characteristics of two-dimensional turbulence that self-organizes into vortex crystals. *Phys. Rev. Lett.*, **84**, 1443–6.

Joyce, G. and Montgomery, D. (1973). Temperature states for the two-dimensional guiding-centre plasma. *J. Plasma Phys.*, **10**, 107–21.

Kraichnan, R. (1967). Inertial ranges in two-dimensional turbulence. *Phys. Fluids*, **10**, 1417–23.

Krall, N. and Trivelpiece, A. W. (1986). *Principles of Plasma Physics*. San Francisco Press, San Francisco.

Kriesel, J. M. and Driscoll, C. F. (2001). Measurements of viscosity in pure-electron plasmas. *Phys. Rev. Lett.*, **87**, 135003.

Leith, C. E. (1984). Minimum enstrophy vortices. *Phys. Fluids*, **27**, 1388–95.

Lifshitz, E. M. and Pitaevskii, L. P. (1981). *Physical Kinetics*, Section 60. Pergamon Press, Oxford.

Liu, P. L.-F. and Ting, L. (1987). Interaction of decaying trailing vortices in spanwise shear-flow. *Comput. Fluids*, **15**, 77–92.

Longmire, C. L. and Rosenbluth, M. N. (1956). Diffusion of charged particles across a magnetic field. *Phys. Rev.*, **103**, 507–10.

Lynden-Bell, D. (1967). Statistical mechanics of violent relaxation in stellar systems. *Mon. Not. R. Astron. Soc.*, **136**, 101–21.

McWilliams, J. C. (1990). The vortices of two-dimensional turbulence. *J. Fluid Mech.*, **219**, 361–85.

Melander, M. V., Zabusky, N. J., and McWilliams, J. C. (1988). Symmetric vortex merger in two dimensions: Causes and conditions. *J. Fluid Mech.*, **195**, 303–40.

Miller, J., Weichman, P. B., and Cross, M. C. (1992). Statistical mechanics, Euler's equation, and Jupiter's Red Spot. *Phys. Rev. A*, **45**, 2328–59.

Montgomery, D. C. and Tidman, D. A. (1964). *Plasma Kinetic Theory*. McGraw-Hill, New York.

Montgomery, D., Joyce, G., and Turner, L. (1974). Magnetic field dependence of plasma relaxation times. *Phys. Fluids*, **17**, 2201–04.

Moore, D. W. and Saffman, P. G. (1975). The density of organized vortices in a turbulent mixing layer. *J. Fluid Mech.*, **69**, 465–73.

O'Neil, T. M. (1985). New theory of transport due to like-particle collisions. *Phys. Rev. Lett.*, **55**, 943–6.

Onsager, L. (1949). Statistical hydrodynamics. *Nuovo Cimento Suppl.*, **6**, 279–87.

Psimopoulos, M. and Li, D. (1992). Cross field thermal transport in highly magnetized plasmas. *Proc. R. Soc.: Math. Phys. Sci.*, **437**, 55–65.

Reif, F. (1965). *Fundamentals of Statistical and Thermal Physics*. McGraw-Hill, New York.

Reznik, G. M. (1992). Dynamics of singular vortices on a beta-plane. *J. Fluid Mech.*, **240**, 405–32.

Robert, R. and Sommeria, J. (1992). Relaxation towards a statistical equilibrium state in two-dimensional perfect fluid dynamics. *Phys. Rev. Lett.*, **69**, 2776–9.

Rosenbluth, M. N. and Kaufman, A. N. (1958). Plasma diffusion in a magnetic field. *Phys. Rev.*, **109**, 1–5.

Rosenbluth, M. N. and Liu, C. S. (1976). Cross-field energy transport by plasma waves. *Phys. Fluids*, **19**, 815–18.

Rossby, C. G. (1948). On displacements and intensity changes of atmospheric vortices. *J. Marine Res.*, **7**, 175–87.

Rossow, V. J. (1977). Convective merging of vortex cores in lift generated wakes. *J. Aircraft*, **14**, 283–90.

Saffman, P. G. and Szeto, R. (1980). *Phys. Fluids*, **23**, 2339–42.

Schecter, D. A. and Dubin, D. H. E. (1999). Vortex motion driven by a background vorticity gradient. *Phys. Rev. Lett.*, **83**, 2191–4.

Schecter, D. A., Dubin, D. H. E., Cass, A. C., Driscoll, C. F., Lansky, I. M., and O'Neil, T. M. (2000). Inviscid damping of asymmetries on a two-dimensional vortex. *Phys. Fluids*, **12**, 2397–12.

Simon, A. (1955). Diffusion of like-particles across a magnetic field. *Phys. Rev.*, **100**, 1557–9.

Smith, R. A. (1991). Maximization of vortex entropy as an organizing principle in intermittent, decaying, two dimensional turbulence. *Phys. Rev. A*, **43**, 1126–9.

Smith, R. K. (1993). On the theory of tropical cyclone motion. In *Tropical Cyclone Disasters*, pp. 264–79. Peking University Press, Beijing.

Spitzer, L. (1956). *Physics of Fully Ionized Gases*. Interscience, New York.

Spitzer, L. and Harm, R. (1952). Transport phenomena in a completely ionized gas. *Phys. Rev.*, **89**, 977–81.

Sutyrin, G. G. and Flierl, G. R. (1994). Intense vortex motion on the beta plane: Development of the beta gyres. *J. Atmos. Sci.*, **51**, 773–90.

Taylor, J. B. and McNamara, B. (1971). Plasma diffusion in two dimensions. *Phys. Fluids*, **14**, 1492–9.

Weiss, J. B. and McWilliams, J. C. (1993). Temporal scaling behavior of decaying two-dimensional turbulence. *Phys. Fluids A*, **5**, 608–21.

14
Wave–particle interaction in plasmas: A qualitative approach

Dominique F. Escande

Physique des Interactions Ioniques et Moléculaires, UMR 6633, CNRS/Aix-Marseille Université, France

To the memory of Boris Chirikov (June 6, 1928–February 12, 2008)

14.1 Outlook

14.1.1 Introduction

Teaching in six hours modern wave–particle interaction in the Les Houches Summer School on Long-Range Interacting Systems was a very challenging task. Indeed, the basics had to be recalled to a non-plasma-physics audience, but they take a lot of time in conventional approaches. How to perform this necessary but somewhat lengthy step, while conveying some important building blocks of the present state of the art? This twofold constraint led me to design a course where concepts would be introduced by appealing to intuition, and by minimizing the use of mathematical formalism. I had to forget about my initial dream of teaching some parts of the book *Microscopic Dynamics of Plasmas and Chaos* (Elskens and Escande 2003) that conveys my view on the topic. I had to give up the book, but keep the message. As a result, during the preparation of my course this book had the fate of the Cheshire cat in *Alice in Wonderland*, which progressively disappears, except for its grin! This forced me into coming closer to the essence of the phenomena, and to introduce nonstandard views about reactive instabilities (Section 14.3.1), a new calculation of the Landau effect (Section 14.4.2), and a presentation of the saturation of the bump-on-tail instability, avoiding diffusion as a prerequisite (Section 14.4.6).

Like the Earth's surface, knowledge has a fractal structure. Before starting a trip to visit a new region, it is useful to look at a map, and to try and figure out how its geography looks as a whole. Then, during the subsequent journey, the successive landscapes can be integrated into a global view, without waiting for the end of the trip. As a result, the first chapter of this writeup is a large-scale presentation of the important concepts of the course, and of the theoretical views underlying it. In particular, it provides a qualitative description of the saturation of the instability produced by a beam of electrons streaming inside a plasma. Section 14.2, the first half of Section 14.3, and most of Section 14.4 may be viewed as an intermediate-scale presentation of the same concepts. Elements of a third scale are provided in the remainder of the manuscript. They were chosen while bearing in mind the need to provide general-purpose tools to the broad audience of the School.

14.1.2 Large-scale presentation of the main results

A plasma is a globally neutral set of free positive and negative charges, whose size is large enough with respect to the Debye length $\lambda_{\mathrm{D}} = (\epsilon_0 k_{\mathrm{B}} T/(ne^2))^{1/2}$, where n, T, and e are the electron density, temperature, and absolute value of the charge, respectively, k_{B} is the Boltzmann constant, and ϵ_0 is the vacuum permittivity. Over scales much larger than λ_{D} the plasma particles interact mainly through collective electric fields. A ubiquitous aspect of such fields is that of plasma waves, which brings about the issue of particle–wave interaction in plasmas, the topic of this course. Since, a priori, all plasma particles are involved in the wave propagation, it is important to make more precise from the outset that the interaction of interest is that of waves with particles whose velocity is close to the wave phase velocity.

We will consider very simple plasmas:

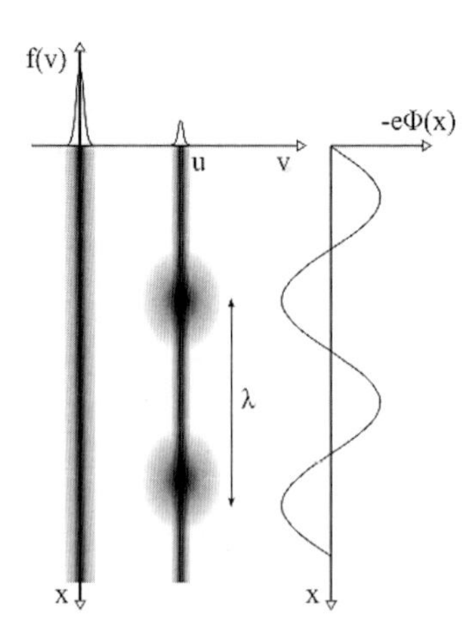

Fig. 14.1 Saturation of the cold beam–plasma instability. The shaded disks correspond to the saturation phase of the beam, and sketch beam–particle trapping in the wave troughs. In reality, the initial distributions of the plasma and beam are Dirac functions.

- They are collisionless; thus we forget about the noncollective effects occurring over scales smaller than λ_{D}, which were considered at length in Dan Dubin's lectures.
- The ions are nothing but a neutralizing background at rest: only electrons are active.
- The electron motion is considered as one-dimensional; this assumption is correct for the motion along the direction of a strong magnetic field, and it forgets about any dynamical effect due to this field.
- In harmony with the previous point, interactions between electrons are purely electrostatic: only a part of the particle–field self-consistency in plasmas is kept.

However, this simplistic description of a plasma already brings about a lot of interesting phenomena in this long-range-interaction medium. This course focuses on two extreme nonequilibrium cases, corresponding to the *coherent* and the *incoherent* limits of the beam–plasma interaction. In the *coherent limit*, we forget about the possibly thermal spreading of particle velocities. The upper part of Fig. 14.1 displays a velocity distribution function of electrons that is made up of two Dirac functions: one with zero velocity, corresponding to a cold plasma, and one with a positive velocity u and a lower amplitude, corresponding to a cold beam. The lower part of Fig. 14.1 displays the position of particles in the single-particle phase space (x, v) (i.e. in Boltzmann's μ space). Initially the plasma and beam particles are distributed uniformly in space. We will see that such a state is unstable, and that an electrostatic wave develops, which travels at a velocity close to u. The electrostatic potential of this wave is displayed in the lower right part of Fig. 14.1. We will show that beam particles end up trapped in the troughs of the growing wave, as balls would do on a rippled ground. This leads to a time-periodic bunching (or clustering) of the beam particles; it also suppresses

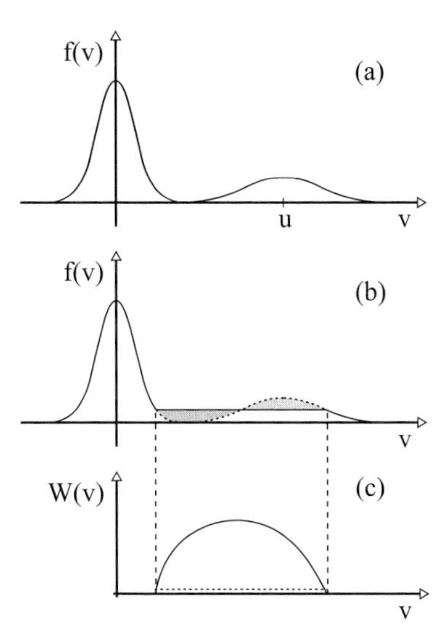

Fig. 14.2 Saturation of the bump-on-tail instability. (a) Initial distribution, (b) Final distribution with plateau, (c) Wave intensity spectrum; the dotted and continuous lines show the initial and final shapes, respectively.

the so-called *cold beam–plasma instability*, and brings about a concomitant oscillation of the wave amplitude about a constant level: the wave amplitude saturates owing to trapping, exchanging momentum periodically with the trapped particles.[1]

In the *incoherent limit* of beam–plasma interaction, the spreading of particle velocities will be shown to matter. Figure 14.2(a) displays an initial velocity distribution function of electrons that is made up of two Gaussians: one with zero mean velocity corresponding to the plasma, and one with a positive mean velocity u and a lower amplitude corresponding to the beam. Here again, the plasma and beam particles are distributed uniformly in space. We will see that such a state is also unstable, and that electrostatic waves develop with phase velocities covering the whole domain where there is a positive slope of the beam distribution function $f_{\mathrm{b}}(v)$ (Landau effect). As in the previous case, each wave aims to trap particles in its own potential troughs, but we will show that the competition between neighboring waves leads to a diffusion of particle velocities. This progressively broadens to the left the positive-slope part of $f_{\mathrm{b}}(v)$, while decreasing its maximum amplitude in order to keep the number of particles constant. The unstable-wave domain broadens concomitantly. Both broadenings stop when $f_{\mathrm{b}}(v)$ hits the plasma distribution function, but the diffusive behavior of particle velocities forces $f_{\mathrm{b}}(v)$ to flatten until a plateau is formed in the beam–plasma velocity distribution, as shown in Fig. 14.2(b). Figure 14.2(c) displays the correspond-

[1]A more precise description of the saturation is provided in Figs. 14.7 and 14.8.

ing phase velocity spectrum of the waves.[2] Since there is no longer any zone of $f_b(v)$ with a positive slope, waves are no longer unstable. Therefore Figs 14.2(b) and (c) display the eventual saturated state of the beam–plasma instability considered here, called the *bump-on-tail instability*. At variance with the previous case, at saturation $f_b(v)$ is static and spatially uniform. It is worth noticing that, since the plateau builds up both on the initial beam and on the bulk plasma velocity distributions, the actual dynamical object of interest is not simply $f_b(v)$, but rather the tail distribution function, which incorporates a part of the bulk distribution. At saturation, the transfer of momentum from the particles to the waves stops, and leads to a stationary wave spectrum.[3]

In describing the bump-on-tail instability, we introduced the Landau effect for the case of a positive slope of the beam distribution function. This effect exists as well for a negative slope of the particle distribution function, and then corresponds to a damping of the plasma wave. Both cases will be related to an average synchronization of particles with the wave.

14.1.3 Rigorous background to the intuitive approach

The above description of the incoherent case was longer than that of the coherent one. This is not by chance. Indeed, a thorough presentation of both cases involves more exotic physics, such as Hamiltonian chaos, in the latter. In order to avoid putting hot matter in the hands of beginners right away, we will start the intermediate-scale part of this course with the cold beam–plasma instability (hereafter called the "cold instability" for short). Indeed, while studying the coherent case of the beam–plasma interaction, we will introduce concepts useful for dealing with the incoherent case, in particular locality in velocity of the wave–particle interaction, and synchronization of particles with a wave. Furthermore, it will turn out to be very useful to consider any plasma as a set of cold beams.

This course uses an intuitive approach to wave–particle interaction, but it is important for the reader to be sure about the existence of a rigorous description of the same topic. Following Popper's paradigm, science is always in the making, and this is all the more true for plasma physics, which deals with a complex physical system, and which started to develop rapidly only half a century ago. Since it came after fluid mechanics and the kinetic theory of gases, and since it had similarities with both of these fields, plasma physics naturally borrowed fluid and kinetic models to describe its basic phenomena, even though the self-consistent mechanical approach described below was already possible for the linear theory. A posteriori, it turns out that the Ockham's razor principle was the underlying prescription for choosing out of those models the one to be used to describe a given phenomenon. Indeed, textbooks used the simplest model compatible with the actual physics, sometimes making sure that a more complete description confirmed the results of the simpler model. A metonymic process came with this empirical approach; in particular, according to the description

[2] This quantity will be precisely defined in Section 14.4.6.

[3] As will be explained in Section 14.4.6, this is followed on much longer time scales by a relaxation toward a hotter Maxwellian.

used for the linear stage, beam–plasma instabilities were termed "fluid" in the cold-beam case because a fluid description was sufficient, and "kinetic" in the bump-on-tail case, since a kinetic description was mandatory. Even when fluid models were sufficient, there was the implicit idea that the best possible description was the kinetic one, as provided by the mean-field Vlasov–Maxwell description.

This idea is still alive, even though plasma physicists who tackled the nonlinear stage of the cold instability were forced from the outset down a quite different alley (Onishchenko *et al.* 1970, O'Neil *et al.* 1971). Indeed, they kept from the kinetic description of the bulk plasma nothing but the degrees of freedom related to its collective motion: those corresponding to the above-mentioned electrostatic waves, which are called *Langmuir waves*; as to the beam, it was described as a set of particles in self-consistent interaction with the waves.[4] This approach led to the discovery of the above-described bunching mechanism, and was later on generalized to the bump-on-tail case (Escande 1989, Tennyson *et al.* 1994). At this point, we feel compelled to make a remark: the kinetic description of plasmas is a mean-field approximation of the classical N-body description (Spohn 1991); so why go through such a description to land on a model with a finite number of degrees of freedom incorporating beam particles? Indeed, it is possible to start with the classical-mechanics N-body description of the plasma to derive the same model through a sequence of controlled approximations. This model is summarized in a Hamiltonian

$$\mathcal{H}_{\text{self-consistent}} = \mathcal{H}_{\text{particles}} + \mathcal{H}_{\text{waves}} + \mathcal{H}_{\text{interaction}} , \tag{14.1}$$

which is the sum of three parts: one for the tail particle degrees of freedom, one for the wave degrees of freedom, and one for their mutual interaction. This description puts waves and particles on an equal footing, as occurs in modern field theory. It enables a unified derivation of all results obtained with the classical fluid and kinetic approaches, while shedding light on the actual physical mechanisms at work, and providing new handles on nonlinear aspects of wave–particle dynamics. This is the compass guiding our intuitive journey inside wave–particle interaction. The reader is referred to Elskens and Escande (2003) for an account of this approach. Those who are aware of it will recognize for sure the Cheshire cat's grin in the following sections! In magnetized plasmas, there are waves for which the Larmor precession plays an important role. A self-consistent Hamiltonian description is available also for this case (Krafft *et al.* 2005; see also Zaslavsky *et al.* 2008 and references therein).

Before providing more details about our topic, let us take an aerial view of the path described in the preceding paragraph. We notice that a grand dream of the nineteenth century comes true: the irreversible evolution of a macroscopic many-body system, the beam–plasma system, is described through classical mechanics only. Classical mechanics for sure, but its modern version, the one incorporating dynamical chaos! As a prelude to our study of Hamiltonian chaos, two classical paradigms of classical mechanics will be instrumental: the harmonic oscillator and the nonlinear pendulum.

Before starting our journey into wave–particle interaction, we shall take a look at the corresponding landscape. Section 14.2 introduces the harmonic oscillation of

[4] A Vlasovian description of the saturation of the cold instability is still not available.

electrons at the plasma frequency, and the corresponding Langmuir waves; it also shows the analogy between the phase space portraits of the nonlinear pendulum and of the motion of an electron in such a wave. Section 14.3 is devoted to the cold beam–plasma instability. It shows how the instability is related to the forcing of the electron harmonic oscillation, and its saturation is shown to occur by particle trapping. Then the mechanism of the average synchronization of particles with a wave, and the transfer of momentum between the wave and the particles are introduced, which provides a firmer foundation for the saturation process. Section 14.4 is devoted to the hot beam–plasma instability. It involves the Landau effect, a transfer of momentum between the wave and particles close to being resonant, which is due to the average wave–particle synchronization. Hamiltonian chaos is then introduced, and is shown to play an essential role in the saturation of this instability. Section 14.5 provides a further description of Hamiltonian chaos, which shows that this dynamics is not of hyperbolic character, but that appropriate averages enable its description as a diffusion.

14.2 Basics of collective and single-particle motion

14.2.1 Plasma frequency

We now consider a plasma without any beam, and make a further simplifying assumption to start with: the electrons are assumed to be cold. Figure 14.3(a) displays a plasma slab of width L whose electrons have been shifted by a small amount $\xi \ll L$ from the ion background. This produces the classical picture of an infinite plane capacitor. Gauss's theorem yields a uniform electric field inside the slab, $E = ne\xi/\epsilon_0$. Then Newton's law yields the acceleration of the electrons, $m\ddot{\xi} = -eE = -ne^2\xi/\epsilon_0$, where m is the electron mass. This equation is that of a harmonic oscillator with an angular frequency ω_p, called the *plasma frequency*, such that $\omega_\text{p}^2 = ne^2/(m\epsilon_0)$. Therefore, when released in the state of Fig. 14.3(a), the electron cloud sloshes back and forth between

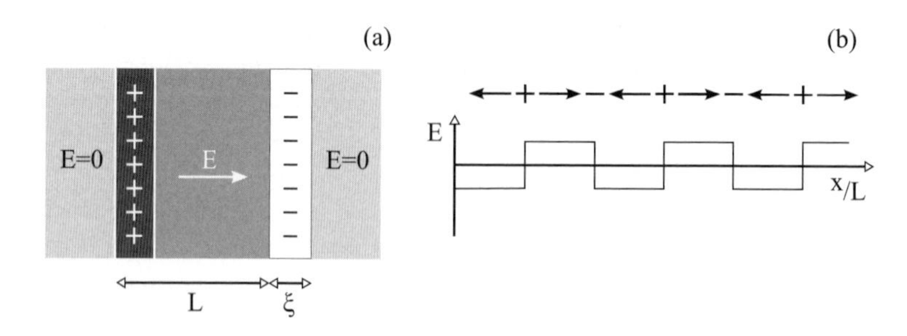

Fig. 14.3 Collective electron motion. (a) Plane electric capacitor created by a displacement ξ to the right of an electron slab of width L. (b) Upper part: localized charge distribution, and a series of slices with uniform electric field obtained by setting slabs of the kind shown in part (a) side by side, with alternating positive and negative ξ's. Lower part: spatial distribution of the electric field for $\xi \ll L$.

this state and the one obtained by exchanging the + and − signs. This is a collective motion where electrons lose their individuality.

In solid state physics, since $\hbar\omega_{\rm p} \gg k_{\rm B}T$, it is very hard to excite oscillations at the plasma frequency; these oscillations are quantized and are called plasmons. This would be also the case in a plasma with a vanishing absolute temperature. However, in most plasmas of interest the density and temperature are such that the opposite ordering holds. Therefore there are many plasmons even at a thermal level, and they may be described through their coherent states, i.e. classical electrostatic waves. We introduce them now.

14.2.2 Langmuir waves

By flipping Fig. 14.3(a) about its vertical axis, one gets another perturbed state of the previous cold plasma slab (the one related to $-\xi$). When Fig. 14.3(a) and its flipped version are set side by side in a periodic manner, one gets an infinite plasma with alternating uniform positive and negative electric fields, as shown in Fig. 14.3(b). The electric field has a periodicity $2L$. It oscillates again at the plasma frequency defined above, which creates a standing (square) wave pattern. Fourier analyzing this field in space yields a series of components $\sin(k_n x \pm \omega_{\rm p} t)$, where $k_n = 2\pi n/(2L)$. In the limit of small-amplitude oscillations, these components are uncoupled electrostatic waves called *Langmuir waves*. These waves are collective oscillations of the electrons with respect to the ions, which may be considered again as individual harmonic oscillators. Their wavenumber k may take any value (indeed, L is arbitrary). Their dispersion relation is shown in Fig. 14.4(a); it is made up of the two lines $\omega = \pm\omega_{\rm p}$. The phase velocity $v_\varphi = \omega/k$ of the Langmuir waves in a cold plasma goes from $-\infty$ to $+\infty$. Since Langmuir waves may be considered as individual harmonic oscillators, it is natural for the $\mathcal{H}_{\rm waves}$ part of $\mathcal{H}_{\rm self\text{-}consistent}$ to be a sum of harmonic-oscillator Hamiltonians. In contrast, $\mathcal{H}_{\rm particles}$ is a sum of free-particle Hamiltonians.

A neutralized beam of cold electrons with velocity u may be viewed as a plasma moving with this velocity. It supports Langmuir waves as well, whose dispersion (Fig. 14.4(b)) is merely Doppler shifted from that of a plasma at rest: $\omega = \pm\omega_{\rm b} + ku$, where $\omega_{\rm b}$ is the beam–plasma frequency, i.e. an $\omega_{\rm p}$ computed with the beam density. For a given wavenumber $k > k_{\rm b} \equiv \omega_{\rm b}/u$, there are two beam waves: a slow one and a fast one (slower and faster than the beam).

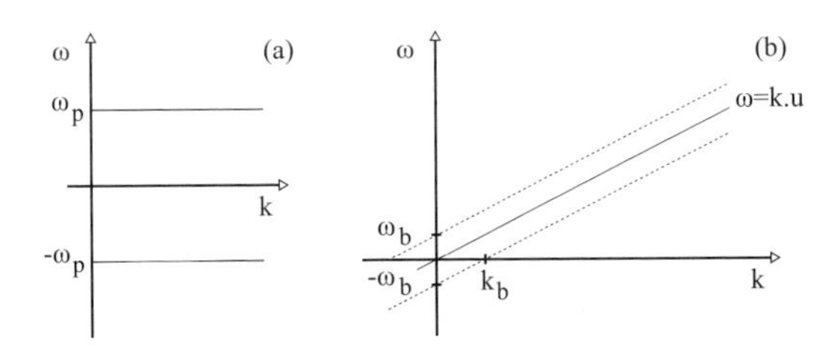

Fig. 14.4 Cold-plasma and cold-beam dispersion relations.

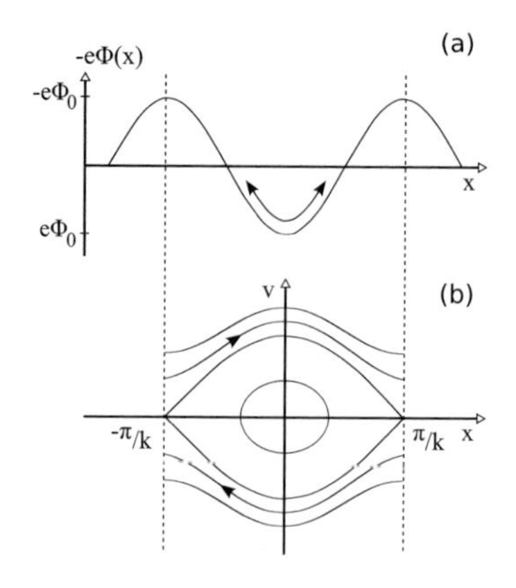

Fig. 14.5 Electron in an electric potential $\Phi(x) = \Phi_0 \cos(kx)$.

14.2.3 Motion of an electron in an electrostatic wave

In order to understand the saturation mechanism of the cold instability, as was sketched in Fig. 14.1, we need to understand the dynamics of particles in the presence of an electrostatic wave, which is one of the building blocks for this course. In the frame moving at the phase velocity of the wave, the wave electrostatic potential is static, as shown in Fig. 14.5(a). This potential imposes two kinds of motions on the electrons. Either they pass through, and their velocity is only modulated by the potential troughs and hills, or they are trapped inside the potential troughs. Figure 14.5(b) displays the phase portrait of this dynamics, which is exactly that of a nonlinear pendulum when two successive maxima of the mechanical potential $-e\Phi(x)$ are identified.[5] A separatrix with the shape of a cat's eye separates trapped orbits and passing ones with positive and negative velocities. The electron energy[6] is $\mathcal{E} = mv^2/2 - e\Phi_0 \cos(kx)$. Close to the origin in Fig. 14.5(b), the dynamics is that of a harmonic oscillator (another one!) whose angular frequency, called the bounce frequency, is $\omega_{\mathrm{B}} = k(e\Phi_0/m)^{1/2}$. Therefore the separatrix turns out to be the boundary between distorted free-particle orbits and distorted harmonic-oscillator orbits. At $x = 0$, the separatrix has a half-width in velocity $\Delta v = 2(e\Phi_0/m)^{1/2}$. If $\mathcal{E}$ is large, v has only small modulations about $(2\mathcal{E}/m)^{1/2}$. Particles inside the resonant domain are said to be resonant with the wave, and Δv is the typical width of this resonance process. It is important to notice that trapped particles have a time-averaged vanishing velocity. Therefore the set of particles with a time-averaged vanishing velocity covers a set of dimension 2 in phase space for $\Phi_0 \neq 0$, while it covers a set of dimension 1 whenever $\Phi_0 = 0$. This will provide an intuitive basis for understanding the Poincaré–Birkhoff theorem of Section 14.4.5. The

[5]When two successive maxima of the mechanical potential $-e\Phi(x)$ are not identified, the cat's eye structure repeats periodically in space, and creates what is called an island chain; this will be useful later.

[6]For future reference, let us note that the Hamiltonian $H(v, x)$ of the above dynamics is written exactly like $\mathcal{E}$ whenever $m = 1$.

above phase portrait shows that an electrostatic wave has a strong influence only on particles with a velocity close to its phase velocity: *wave–particle interaction is local in velocity.* This course focuses on this local interaction. Both resonant and nearly resonant aspects of this interaction will be important.

In a more general frame of reference, the wave–particle resonance condition for a particle with velocity v and a Langmuir wave with angular frequency ω and wavenumber k is $\omega = kv$. For a magnetized plasma and a wave for which Larmor precession plays an important role (e.g. lower hybrid waves), wave–particle resonance corresponds to $\omega - n\omega_\mathrm{L} = kv$, where n is any integer, ω_L is the Larmor frequency, and k and v are the components of the wavenumber and of the particle velocity parallel to the magnetic field.

14.3 Cold beam–plasma instability

Sections 14.3.1, 14.3.2, and 14.3.3 provide an intuitive description of the cold instability. Sections 14.3.4–14.3.6 use elementary calculations to perform a linear analysis of the cold instability, and to further substantiate its saturation mechanism.

14.3.1 Origin of the instability

By letting L become very small in Fig. 14.3(a), one gets a strongly localized harmonic oscillator. Then the plasma may be considered as a continuous set of such oscillators.[7] Assume that a cold electron beam with velocity u is flowing in the plasma. If the beam density is modulated with a wavenumber k, all the localized harmonic oscillators of the plasma are driven at an angular frequency $\omega = ku$ by the electric field resulting from this modulation. As is well known, the response of each oscillator is proportional to the excitation force multiplied by $1/(\omega_\mathrm{p}^2 - \omega^2)$. When ω crosses ω_p, the sign of the density response changes with respect to the forcing density perturbation. If the plasma response has the same sign as the forcing, this feeds back on the beam density perturbation, and leads to the cold instability. For $\omega \gg \omega_\mathrm{p}$, the electrons react weakly owing to their inertia, which rules out the possibility of an instability for such ω's and, by continuity, for $\omega > \omega_\mathrm{p}$. Then the plasma behaves like a classical dielectric, which screens the perturbing charge. As a result, the unstable forcing must correspond to $\omega = ku < \omega_\mathrm{p}$. Then the plasma amplifies the perturbation, which leads to a so-called reactive instability.

14.3.2 Intuitive description of linear aspects

We now consider the cold beam–plasma system described in Fig. 14.1. It is made up of a cold plasma with plasma frequency ω_p and of a weak cold beam (velocity u) with plasma frequency $\omega_\mathrm{b} \ll \omega_\mathrm{p}$. As a first attempt to build the dispersion relation of this beam–plasma system, we may draw in the same (k, ω) diagram the dispersion relations $\omega = \pm\omega_\mathrm{p}$ for the cold plasma and $\omega = \pm\omega_\mathrm{b} + ku$ for the cold beam. Figure 14.6 displays the $\omega > 0$ part of this diagram. There are two intersection points between

[7] Since the electric field vanishes outside a plane capacitor, the plasma is completely free there, and a series of such capacitors may be juxtaposed. The localization of the oscillators is consistent with the fact that Langmuir waves have a vanishing group velocity in a cold plasma.

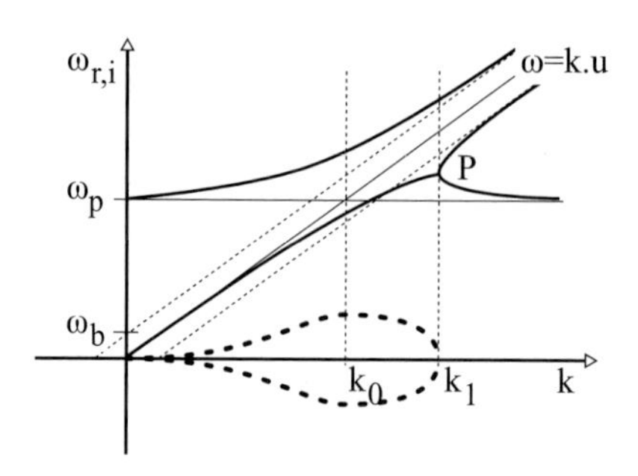

Fig. 14.6 Cold beam–plasma dispersion relation. The branch corresponding to ω close to $-\omega_{\mathrm{p}}$ is not displayed. For $k > k_1$ three real solutions are present, while for $k < k_1$ two solutions are complex conjugates; their imaginary parts are shown by dashed lines.

the uncoupled beam and plasma dispersion curves. There, we may expect the actual coupling of the two electron populations through the electric field to modify our naive diagram.

As will be shown later, there is in fact a frequency-crossing avoidance, which is analogous to the avoided crossings in quantum mechanics.[8] Figure 14.6 displays as thick continuous lines the modified dispersion relation to be derived in Section 14.3.4. More exactly, the real part of $\omega(k)$ is displayed. The branch with $0 < k < k_1$ has an imaginary part as well, which can take two opposite values; it is shown by thick dashed lines. One of these two conjugate roots is unstable, and corresponds to the cold instability. As was predicted in Section 14.3.1, the unstable domain corresponds to $\omega < \omega_{\mathrm{p}}$, except for a small extension due to frequency-crossing avoidance.[9] As could be expected from the response of a periodically forced harmonic oscillator, the largest growth rate occurs for a forcing with an angular frequency $\omega = ku$ close to ω_{p}. The real part of the dispersion relation corresponds to a modified degenerate ($\omega = ku$) beam mode, and not to one of the beam waves, in agreement with the beam density modulation picture of the previous section. To the right of point P, the dispersion relation takes on a hyperbolic shape, which corresponds to the coupling of the plasma wave and the slow wave. This hyperbolic branch joins at point P with the beam mode branch located to the left of this point. Classical textbooks invoke the "negative energy" of the slow wave as the origin of this reactive instability. This would rather suggest that the real part of $\omega(k)$ for the unstable mode would be like $-\omega_{\mathrm{b}} + ku$. This classical picture does not tell us either why the instability occurs instead for ku smaller than ω_{p}, or why it is strongest for $ku \simeq \omega_{\mathrm{p}}$. Therefore the forced-harmonic-oscillator

[8]This analogy is deeply mathematically rooted. Indeed, the cold instability is describable by a (linear) fluid model, which can be encapsulated in a variational principle involving a self-adjoint operator on a Hilbert space

[9]The weak difference between ω_{p} and ω_{b} in Fig. 14.6 exaggerates the relative distance between k_1 and $k_0 = \omega_{\mathrm{p}}/u$.

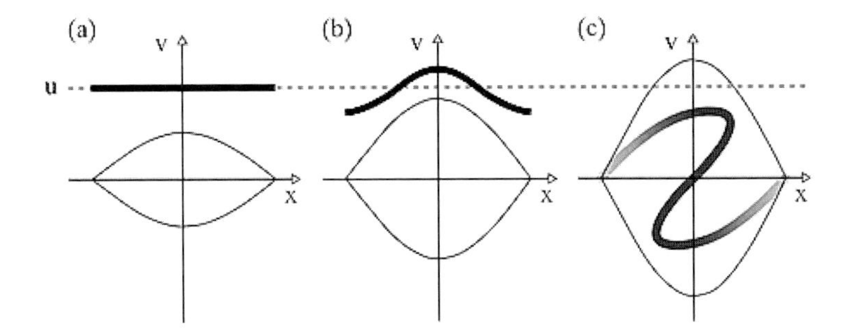

Fig. 14.7 Cold beam–plasma saturation: single-particle phase space represented over one wavelength only: (a) initial monokinetic beam and wave with a small amplitude; (b) distorted beam slightly before trapping in the wave; (c) trapped beam particles.

approach provides a more accurate, and possibly new, explanation of this reactive instability.

For further reference, it is important to notice that the phase velocity of the unstable modes is lower than the beam velocity u. This kind of instability can be observed experimentally when a cold beam is launched into a plasma column along a strong magnetic field parallel to its axis. Then, instead of having an instability developing in time, as in the theoretical model, one finds an instability growing spatially from the beam injection point.

A similar instability may occur in a magnetized plasma, the fan instability. It involves lower hybrid waves, which are electrostatic waves too. The instability occurs for particles that are slightly faster than the wave, and satisfy the resonance condition $v = (\omega + \omega_{\mathrm{L}})/k$, where ω_{L} is the Larmor frequency, ω is the angular frequency of the wave, and k and v are the components of the wavenumber and of the particle velocity parallel to the magnetic field (Kadomtsev 1992).

14.3.3 Saturation of the instability

Figure 14.6 shows that the growth rate of the instability has a maximum. Therefore exponential growth of the electric field leads to a peaking of the spectrum of the unstable wave about the most unstable mode.[10] This enables us to deal with the saturation of the instability by making a single-wave approximation. Figure 14.7 shows three successive phase portraits of the beam and of the trapping domain of the most unstable wave, inspired after Fig. 14.5(b). In agreement with our previous discussion of the dispersion relation, the wave has a phase velocity smaller than the initial beam velocity. Figure 14.7(a) corresponds to the linear growth stage, where the beam particle

[10]When the mode with the maximum growth rate is multiplied by a factor of e N times, the mode with half this growth rate exponentiates only $N/2$ times, which causes the wave intensity to peak about the most unstable mode.

velocities are almost unperturbed by the wave. Figure 14.7(b) corresponds to a moment where the wave amplitude is high enough for the beam particles to have their velocities strongly modified by the wave, and the beam trace to follow the upper part of the wave separatrix.

Further increase of the wave amplitude forces beam particles to enter the trapping domain. Figure 14.7(c) shows the beginning of the particle-trapping motion. The central part of the beam trace rotates as a rigid rod because motion at the bottom of the wave trough is harmonic. In contrast, particles that cross the instantaneous separatrix close to the unstable point "stick" to it; this creates lagging, but low-density, tails, which wind around the "rigid" central rod, as an elastic string would do. This produces a long, low-density spiraling filament as time goes on. This dressed rotating rod was indicated by the shaded disks in Fig. 14.1, and brings about the clustering introduced in Section 14.1.2; this clustering is maximum each time the rod is vertical in Fig. 14.7(c). The trapping of the beam particles by the wave destroys the monokinetic beam structure, which was the origin of the instability. Therefore we may expect this to produce a saturation of the instability. This is indeed so, as shown by numerical simulations (Onishchenko *et al.* 1970; O'Neil *et al.* 1971; Elskens and Escande 2003, Section 8) and experiments (Gentle and Lohr 1973). We will see later that there is a global momentum conservation between the wave and particles (it is at the heart of the Hamiltonian model (14.1)). The growth of the wave intensity means a growth of its momentum. This occurs at the expense of the momentum of the beam particles. Beam particle trapping sets an upper bound to their loss of momentum; indeed, their time-averaged velocity cannot go below the wave phase velocity. Their periodic bouncing implies an oscillation of their momentum, and accordingly of the wave amplitude with the opposite phase, as shown in Fig. 14.8. This is indeed so, as shown again by numerical simulations (Onishchenko *et al.* 1970; O'Neil *et al.* 1971; Elskens and Escande 2003, Section 8) and experiments (Gentle and Lohr 1971).

This first approach to the saturation of the cold instability already explains the bunching (clustering) mechanism of Fig. 14.1. Even though there exists a Debye screening, we see that an out-of-equilibrium plasma can experience a spatial clustering process, which is germane to the Jeans instability in gravitation. Though less

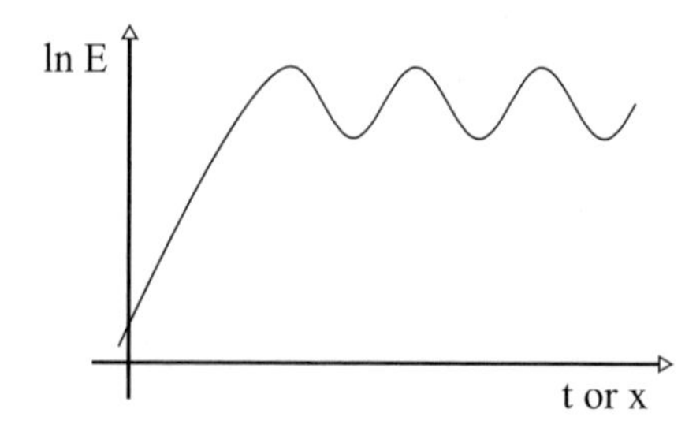

Fig. 14.8 Cold beam–plasma saturation: logarithm of the wave field amplitude.

dramatic, it is a lot more affordable, since it can be realized in small academic experiments! Finally, we notice that kinetic aspects are present during the saturation of the cold instability. Therefore only the linear phase of this instability may be termed fluid, as is often done in textbooks in a metonymic way. The existence of a filamentary particle distribution in the cold instability makes the Vlasov equation an awkward description of the nonlinear evolution of the dynamics. This motivated the finite-degree-of-freedom approach introduced in Section 14.1.3, and the description of the beam as a set of particles. The latter approach made possible the first numerical simulations of the saturation of the cold instability (Onishchenko *et al.* 1970; O'Neil *et al.* 1971; Elskens and Escande 2003, Chapter 8).

In the cold instability, the bulk plasma plays only the role of a wave-supporting medium: only the plasma character of the beam matters! This makes possible the production of the same instability by shining electron beams into other wave-supporting systems. This is done in traveling-wave tubes (TWTs) and free-electron lasers (FELs), which are used as sources of electromagnetic waves. FELs are used to produce waves over a large range of frequencies, up to those corresponding to X-rays. In this case, the absence of mirrors for X-rays forces these devices to work in the single-pass mode. Maximum power output is obtained when the first maximum is reached in Fig. 14.8. TWTs are used a lot as amplifiers in telecommunication satellites, but they also enable us to mimic beam–plasma interaction without the intrinsic noise of plasmas in academic laboratories (Dimonte and Malmberg 1977). We will see that this approach has enabled a series of experiments proving the reality of several of the concepts discussed in this course.

After this intuitive presentation of the cold instability, we now introduce a more rigorous one (the "third scale of the fractal") in the next three sections.

14.3.4 Linear theory

We start with the derivation of the linear theory of the cold instability. To this end we need an equation relating, for a cold plasma, the electronic displacement ξ to the density modulation when both are small. The following table provides the notation for the unperturbed plasma and for the perturbation:

	ξ	n	E
Unperturbed plasma	0	n_0	0
Perturbation	ξ	n_1	E_1

The amount of particles of the unperturbed plasma inside a small interval $[x, x+\delta x]$ must be conserved in the perturbed plasma. When the electron population is displaced by the distributed amount $\xi(x)$, this interval becomes $[x+\xi(x), x+\delta x+\xi(x+\delta x)]$. Simultaneously, the density goes from n_0 to $n_0+n_1(x)$. At the lowest order, density conservation yields $n_0\,\delta x = (n_0+n_1(x))[\delta x+\xi(x+\delta x)-\xi(x)]$, which provides the required expression $n_1 + n_0\,\partial\xi/\partial x = 0$. Together with Newton equation for the electrons $m\ddot{\xi} + eE_1(x,t) = 0$, this yields

$$\ddot{n}_1 = \frac{en_0}{m}\frac{\partial E_1}{\partial x}\,. \tag{14.2}$$

Fourier transforming in both space and time yields

$$n_{1,\mathrm{p}} = -\frac{en_0}{m\omega^2}\mathrm{i}kE_1\,, \tag{14.3}$$

where the index p indicates the cold plasma. For a cold beam with velocity u, this formula becomes

$$n_{1,\mathrm{b}} = -\frac{en_\mathrm{b}}{m(\omega - ku)^2}\mathrm{i}kE_1\,, \tag{14.4}$$

by taking into account the Doppler shift of the frequency in the beam frame. Poisson's equation brings self-consistency into the density-field dynamics. For the beam–plasma system, it reads

$$\frac{\partial E}{\partial x} = -\frac{en_1}{\varepsilon_0} = -\frac{e}{\varepsilon_0}[n_{1,\mathrm{p}} + n_{1,\mathrm{b}}]\,. \tag{14.5}$$

Fourier transforming this equation as above, and setting $n_{1,\mathrm{p}}$ and $n_{1,\mathrm{b}}$ equal to their values given by eqns (14.3) and (14.4), gives the following dispersion relation:

$$1 = \frac{\omega_\mathrm{p}^2}{\omega^2} + \frac{\omega_\mathrm{b}^2}{(\omega - ku)^2}\,, \tag{14.6}$$

where ω_p and ω_b are the plasma and beam–plasma frequencies. For $\omega \gg \omega_\mathrm{p}$, this dispersion reduces to the cold beam dispersion relation, which explains why we recover in Fig. 14.4 the slow and fast beam modes in this domain. Let $k_0 = \omega_\mathrm{p}/u$, $\delta k = k - k_0$, and $\delta\omega = \omega - \omega_\mathrm{p}$. Substituting these quantities in eqn (14.6) yields a fourth-order polynomial in $\delta\omega$. If both δk and $\delta\omega$ are small, it reduces to third order. It is easy to verify that the most unstable wavenumber corresponds to $\delta k = 0$. In this case one finds

$$\delta\omega = 2^{-1/3}\left(\frac{n_\mathrm{b}}{n_\mathrm{p}}\right)^{2/3}\omega_\mathrm{p}\exp\left(\frac{\mathrm{i}2\pi n}{3}\right), \qquad n = 0, 1, 2\,. \tag{14.7}$$

These three roots correspond to the three branches of the complex dispersion relation in Fig. 14.6 close to (k_0, ω_p). Since eqn (14.6) is second-order in k, the calculation of the cold instability occurring experimentally, i.e. ω real and k complex, is simpler. Finally, it is worth noticing that, in contrast to classical textbooks, the linear stage of this instability is not described by using a fluid description of the cold beam–plasma system, but by using a rigorous elementary kinetic description hereof.

14.3.5 Average synchronization of particles with a wave

In order to provide a more rigorous description of the saturation scenario for the cold instability, we show that particles released at $t = 0$ with a velocity u, and a *uniform initial spatial distribution*, have an average velocity which comes closer to the wave phase velocity over a bounded time. This is what we call average synchronization of particles with an electrostatic wave. Let $E(x,t) = E_\mathrm{w}\cos(kx - \omega t)$ be the electric field of the wave. We now compute the average change of velocity due to this wave for the above set of particles. When we go to the wave reference frame and choose $k = 1$, the particle equation of motion becomes $\ddot{x} = \varepsilon\cos x$, where $\varepsilon = -eE_\mathrm{w}/m$. The unperturbed orbit is $X_0(t) = x_0 + ut$. First let us complement this expression with arbitrary polynomial terms up to the power 5 in t. When this expansion is substituted into both sides of the equation of motion, $\ddot{x}$ introduces terms of degree up to 3. When these terms are equated with similar terms on the left-hand side, one gets the beginning of the Taylor expansion of the orbit. Averaging over x_0 yields

$$\Delta u(t) \equiv \langle \dot{x}(t) \rangle - u = -\frac{\varepsilon^2 u t^4}{24} . \tag{14.8}$$

Since $u\,\Delta u(t) < 0$, for t small there is an average synchronization of the particles with the wave, whatever the sign of their velocity relative to the wave. We notice that the effect vanishes for small $|u|$. Therefore this effect is not related at all to trapping inside the wave troughs.

This calculation is short, but does not tell us how long the average synchronization process lasts. A better calculation is a perturbative one to second-order in ε for the same unperturbed orbit $X_0(t)$. A very efficient way to perform the calculation is Picard's fixed-point method, which uses the fact that the exact solution is the fixed point of the iterative process starting from $n = 0$

$$\begin{aligned} X_n \quad &\to X_{n+1} , \\ \ddot{X}_{n+1} &= \varepsilon \cos X_n , \end{aligned}$$

with the boundary conditions $X_n(0) = x_0$ and $\dot{X}_n(0) = u$. This yields $X_1 = x_0 + ut - \varepsilon \sin(x_0)t/u - \varepsilon[\cos(X_0(t)) - \cos(x_0)]/u^2$, which is identical to the expression given by first-order perturbation theory, but it is obtained without needing to write a linearized version of the equation of motion. Similarly, by computing X_2 to second-order, we get an expression for the particle velocity to this order without needing to write a second-order expression for the equation of motion. This finally yields

$$\Delta u(t) = \varepsilon^2 \frac{\cos ut - 1 + \frac{1}{2} ut \sin (ut)}{u^3} . \tag{14.9}$$

$u\,\Delta u(t)$ is even in t and is negative from $t = 0$ up to $t = T \equiv 2\pi/|u|$, which means an average synchronization of the particles with the wave within this time interval whatever the sign of their velocity relative to the wave. Since over a time T the quantity $\Delta u(t)$ scales like $1/u^3$, the average synchronization is small for large $|u|$: it is a local effect in velocity. The effect is maximum for $|t| \simeq 3T/4$. For t small, eqn (14.9) becomes eqn (14.8). This average synchronization effect was proved to exist in an experiment with a TWT (Doveil *et al.* 2005).

The average synchronization of particles with a wave can be understood intuitively as follows. Consider two passing particles with the same velocity v located, at $t = 0$, symmetrically in the wave potential trough of Fig. 14.5(a). As a first approximation, they have exactly opposite accelerations over a small time Δt, as predicted by first-order perturbation theory in ε. However, over interval $[\Delta t, 2\,\Delta t]$ the particle whose velocity is moving away from the wave phase velocity (the desynchronized particle) experiences typically a smaller acceleration than the one coming closer (the synchronized particle), which yields an average synchronization, as predicted by second-order perturbation theory in ε. This calculation makes sense only if u is not too close to 0 (more precisely, if $|u|$ is large with respect to the trapping half-width of the wave). If the wave amplitude varies appreciably over T, the periodicity is lost, and the average synchronization may last longer.

14.3.6 A leading fact: Momentum conservation

We now make more precise the mechanism of the cold instability by using momentum conservation. For a cold plasma, the electron displacement $\xi(x,t)$ in an electrostatic wave is related to the electric field $E(x,t)$ through $\ddot{\xi} + eE(x,t)/m = 0$. Using the notation of the previous subsection, this yields the sloshing velocity $\dot{\xi}(x,t) = eE_{\mathrm{w}} \sin(kx - \omega t)/(m\omega)$. Poisson's equation yields the density perturbation due to the wave, $n_1(x,t) = -(\epsilon_0/e)\partial E/\partial x = (k\epsilon_0/e)E_{\mathrm{w}} \sin(kx - \omega t)$. As a result, the average over a wave period $2\pi/\omega$ of the momentum density of the sloshing electron cloud is

$$P_{\mathrm{w}} = \langle n_1(x,t) m \dot{\xi}(x,t) \rangle = \frac{\epsilon_0 k E_{\mathrm{w}}^2}{2\omega}. \tag{14.10}$$

P_{w} is also called the momentum density of the electrostatic wave, but it corresponds to the momentum of the bulk plasma particles.

Since the momentum of the plasma–beam system is conserved, $P_{\mathrm{w}} + P_{\mathrm{beam}} =$ constant, where P_{beam} is the momentum density of the beam particles. Therefore if the amplitude E_{w} of the wave increases, P_{beam} must decrease. Since there occurs an average synchronization of particles with the wave, in the cold instability only a wave with a phase velocity smaller that the beam velocity can be unstable. This explains why the unstable branch in Fig. 14.6 has phase velocities smaller than u. Now the mechanical origin of the cold instability is explained. It is clear that separatrix crossing is mandatory: beam particles tend to synchronize with a wave whose trapping domain increases. This leads to the particle trapping of Fig. 14.7(c), which stops the one-way transfer of beam momentum to the wave of Figs. 14.7(a) and (b). After trapping, the momentum transfer becomes oscillatory and brings about the wave amplitude oscillations of Fig. 14.8.

The Hamiltonian $\mathcal{H}_{\mathrm{ho}}$ of a harmonic oscillator written in terms of its action angle variables (I,θ) is $\mathcal{H}_{\mathrm{ho}} = \omega I$; then $\dot{\theta} = \partial\mathcal{H}_{\mathrm{ho}}/\partial I = \omega$. The $\mathcal{H}_{\mathrm{interaction}}$ part of $\mathcal{H}_{\mathrm{self\text{-}consistent}}$ is the sum of wave-potential-like terms proportional to $\cos(kx - \theta)$ (see Elskens and Escande 2003, Chapter 2). Since $\mathcal{H}_{\mathrm{self\text{-}consistent}}$ depends on particle positions and wave angles only through such cosines, it is easy to show that the sum of the wave momenta kI and the particle momenta is a constant of motion. This is essentially due to translational invariance.

14.4 Hot beam and plasma

14.4.1 Hot beam

Up to now we have focused on beam–plasma systems made up of two cold distributions of particles. We now will consider hotter matter, and we start by considering a small beam with a broad distribution of velocities. We saw before that the wave–particle average synchronization mechanism operates over a finite range of particle velocities for a given wave. Therefore we may expect a beam with a broad range of velocities

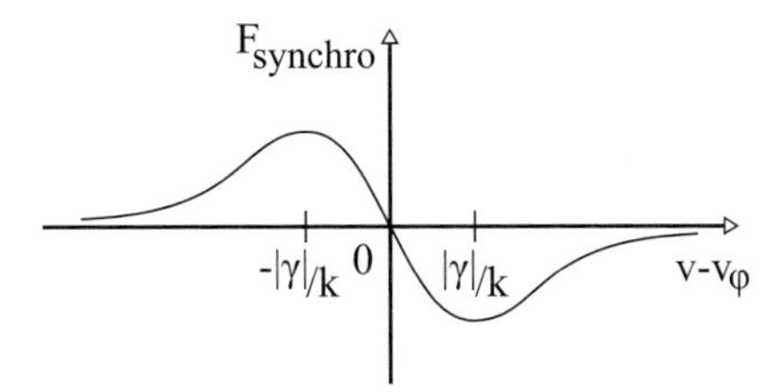

Fig. 14.9 Typical force that synchronizes particles with a wave.

to experience an average synchronization effect due to a given wave over some finite part of its velocity distribution function close to the wave phase velocity, leading to a change of its momentum. Conservation of wave–particle momentum should then relate this effect to a change in the wave amplitude.

Imagine a small wave varying with a growth rate $\gamma > 0$ or a damping rate $\gamma < 0$. Then our estimate of the average synchronization of a spatially uniform set of monokinetic particles with a wave having a constant amplitude makes sense only over a time of order $1/|\gamma|$. The average synchronization effect has been confirmed in this case by rigorous calculations (see Elskens and Escande 2003, Sections 3.9 and 4.1.3). Particles with velocity v such that $|v - v_\phi| \gg |\gamma|/k$ cannot synchronize over a time of order[11] $1/|\gamma|$. Particles with $|v - v_\phi| \ll |\gamma|/k$ either are trapped in the wave and no further average synchronization is possible, or move so slowly over a time $1/|\gamma|$ with respect to the wave that the synchronization effect is weak. Therefore this effect is maximum for particles with $|v - v_\phi| \sim |\gamma|/k$. Figure 14.9 displays the corresponding typical average synchronizing force acting on particles.

As a result, through wave–particle momentum conservation, the wave amplitude evolution is not sensitive to the full velocity distribution, but to a finite range of velocities that does not depend on the wave amplitude. The action of particles on the wave and vice versa is a pseudo-resonant effect: $|\gamma|/k$ is much larger than the trapping width of the wave Δv of Section 14.5, since small waves are considered. If the beam velocity width is much larger than $|\gamma|/k$, the wave does not know it is interacting with a beam. It feels only the effect of the particle velocity distribution close to its phase velocity. Finally, what matters is the quasi-resonant part of the tail particle distribution. In the next subsection we use the local (in velocity) balance of momentum exchange between a wave and the particles.

14.4.2 Landau effect

We now combine the average synchronization of particles with a wave with wave–particle momentum conservation to estimate the evolution of the wave amplitude. We consider a plasma with a velocity distribution function $f(v)$ whose tail has a typical

[11]This is analogous to the inefficient drive produced by a perturbation with a high angular frequency ω for a harmonic oscillator. An analogy can be proved by going to action-angle variables.

scale of variation Δv, and a Langmuir wave with phase velocity v_φ and an exponential growth rate γ. We define

$$u = v - v_\varphi \tag{14.11}$$

and let $F(u) = f(u + v_\varphi)$. It is convenient to consider $\Delta u(t)$, as defined by eqn (14.9), as a function of u as well, $\Delta u(u,t)$. The global change of particle momentum is

$$\Delta u_{\text{global}}(t) = \int F(u)\, \Delta u(u,t)\, du\,. \tag{14.12}$$

Provided $|\gamma|/k \ll \Delta v$, the antisymmetry of $\Delta u(u,t)$ and its fast oscillations in u provide, through a Taylor expansion in u, the good approximation

$$\Delta u_{\text{global}}(t) \simeq \int F'(0) u\, \Delta u(u,t)\, du\,. \tag{14.13}$$

The $(\cos ut - 1)/u^2$ term of $\Delta u(u,t)$ yields a contribution to this integral which may be integrated by parts. This leads to an integral of $\sin ut/(ut)$ over u, as occurs for the contribution of the other term. If $t \gg 1/(k\,\Delta v)$, we find $\Delta u_{\text{global}}(t) \simeq -\pi\varepsilon^2 F'(0)t/2$. If $t \ll |\gamma|^{-1}$ too, we may consider $\Delta u_{\text{global}}(t)/t$ as the time derivative of $\Delta u_{\text{global}}(t)$. When we take into account that $\int f(v)\, dv$ is the plasma density, we find that the derivative of the particle momentum density is

$$\frac{dP_{\text{part}}}{dt} = -\frac{\pi}{2}\frac{e^2 E_{\text{w}}^2}{mk} f'(v_\varphi)\,. \tag{14.14}$$

In this expression, we have introduced the explicit dependence on k which was hidden by our previous choice of $k = 1$. Through eqn (14.10), the conservation of the total wave–particle momentum implies

$$\frac{dE_{\text{w}}^2}{dt} = \gamma_{\text{L}} E_{\text{w}}^2\,, \tag{14.15}$$

and a growth of the wave amplitude like $\exp \gamma_{\text{L}} t$, where

$$\gamma_{\text{L}} = \frac{\pi}{2}\omega_{\text{p}}^3 \frac{f'(v_\varphi)}{k^2 n_{\text{p}}}\,. \tag{14.16}$$

In the case of a hot beam, there is a domain in velocity where $f'(v) > 0$. Therefore, waves with a phase velocity in this domain are unstable: this is the *Landau instability*. However, we notice that the above calculation predicts as well a damping of the wave if $f'(v_\varphi) < 0$, which is called *Landau damping*; γ_{L} is called the *Landau growth or damping rate* (Landau 1946; see also Elskens and Escande 2003, Sections 3.8.1 and 4.1.2). We notice that the evolution of the Langmuir wave amplitude is governed by the derivative of the velocity distribution function at the wave phase velocity. The hot-beam case corresponds to the bump-on-tail instability (hereafter called the "hot instability" for short). Notice that in this case the unstable modes are Langmuir waves, in contrast to the cold instability, where they were degenerate beam modes. The Landau-damping case can be understood from the wave point of view as a phase-mixing effect, as explained in Section 14.4.7. It is important to notice that the Landau phenomenon works for a distribution of velocities whose contour lines do not correspond to the orbits of particles in the initial wave.[12] In the opposite case, one gets a so-called Bernstein–Greene–Kruskal equilibrium (Bernstein *et al.* 1957).

14.4.3 Modification of the cold plasma dispersion relation

The above qualitative picture of synchronization, and the analytical calculations of this effect and of the Landau effect used implicitly the fact that $|\gamma_L| \ll \omega_p$, i.e. that the evolution of the wave amplitude is slow with respect to its period. When the phase velocity of a wave in a thermal plasma is progressively decreased, one finds that $|\gamma_L| \sim \omega_p$ for v_φ of the order of the thermal velocity v_{th}. This shows that Langmuir waves cannot propagate with low phase velocities, and that some lower $v_\varphi \sim v_{th}$ must exist. This is confirmed by analytical calculations performed within the Hamiltonian model (14.1) or the Vlasovian mean-field model (see Elskens and Escande 2003, Section 2.2.3). For small Landau damping, the Langmuir wave dispersion relation becomes $\omega^2 = \omega_p^2 + 3k^2 v_{th}^2$; this is called the Bohm–Gross dispersion relation. This implies $v_\varphi > \sqrt{3} v_{th}$. In reality, to avoid too strong a Landau damping, v_φ / v_{th} must be at least a few units for a Langmuir wave to exist. This introduces a practical upper bound on the wavenumber $k < k_D \equiv 1/\lambda_D$.

14.4.4 Transition from cold to hot beam

Consider a small hot beam with mean velocity u and width Δv, inducing a hot instability. When we take into account that $\omega_p / k \simeq v_\varphi$ and that $f(v)$ is typically $n_p / \Delta v$, then eqn (14.16) shows the maximum Landau growth rate to be of the order of $\gamma_M = (u/\Delta v)^2 \omega_p$. Now imagine we progressively decrease Δv while keeping the beam density constant. Then γ_M progressively increases. When γ_M becomes of the order of the growth rate of the cold instability $\gamma \sim (n_b/n_p)^{2/3} \omega_p$ as given by eqn (14.7), we may expect a qualitative change to occur, and the hot instability to become a cold instability. This occurs for $u/\Delta v \sim (n_b/n_p)^{1/3}$. A correct calculation shows that this is indeed the case (O'Neil and Malmberg 1968; Self *et al.* 1971). The transition from the cold instability to the hot instability is in fact a bifurcation, since the nature of the unstable mode changes from beam modes to Langmuir waves. In both cases the unstable modes have a velocity smaller than the beam mean velocity (for a symmetric beam distribution function). However, for the cold instability, the beam particles are initially nonresonant with the unstable wave.

14.4.5 An essential phenomenon: Hamiltonian chaos

We will see that during the saturation of the hot instability, the particle dynamics becomes chaotic. This requires a small amount of knowledge about Hamiltonian chaos to study this saturation. Consider a plasma with a spatial periodicity L. Then the wavenumbers of Langmuir waves are quantized, $k_n = n2\pi/L$, and so are their phase velocities. Since the phase space has a spatial periodicity L, we may identify $x = 0$ and

[12]The traditional Landau calculation and the synchronization calculation in Section 14.3.5 assume that at $t = 0$ the distribution function is uniform spatially.

$x = L$. This means we wrap the phase space around a cylinder. Consider now a set S of a finite number of Langmuir waves. For simplicity, assume the frequencies of these waves to be pairwise commensurate with a common period T. Since we are interested in the long-time behavior of particle orbits in the presence of these waves, we may look at the dynamics with a stroboscope of period T. This yields a stroboscopic plot, which corresponds to the more general Poincaré map of dynamical-system theory. In this plot, if we wait long enough, the successive points of orbits when S reduces to a single wave draw a picture like the phase portrait of a pendulum in Fig. 14.5. For a general S, since wave–particle interaction is strong only close to the wave phase velocity, it is natural to consider that the global dynamics is a kind of superposition of the dynamics in the various waves if their amplitudes are small enough. This leads to the naive phase space portrait shown in Fig. 14.10(a), obtained by superposing the phase space diagrams of each single wave. Each wave shows up through its unperturbed separatrix bounding the domain of orbits trapped in its troughs. Orbits fall into two categories: those trapped in one of the waves, and those passing in between two neighboring trapping domains. This naive picture corresponds to the natural idea that if waves have small amplitudes, the free orbits are weakly perturbed.

Mathematics confirms this idea, but in a nontrivial way that we briefly describe now. First consider the case of passing orbits. An orbit with mean velocity u can be characterized by its velocity $v_u(x_0, t_0)$ at position x_0 and time t_0. When the waves have a vanishing amplitude, the orbits are free-streaming. Since the velocity is then a constant of motion, $v_u(x_0, t_0) = u$. Since x_0 may be chosen anywhere in the interval $[0, L]$, the set of orbits with mean velocity u is continuous. It is a circle in our cylindrical phase space. Since the dynamics is periodic with period T, we may "wrap" time as

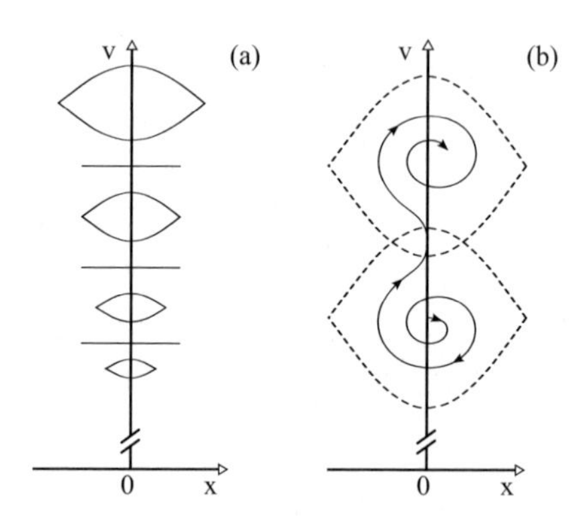

Fig. 14.10 Particle in a set of waves: (a) low-amplitude case, where there are almost free-streaming orbits intermingled with orbits trapped in the various trapping domains of the waves; (b) resonance overlap: the dashed curves are the separatrices of the waves considered individually, and the continuous spiraling line displays the beginning of successive, intermittent trapping motions in each wave. In both cases, a single trapping domain per wave is plotted.

well: we identify times differing by an integer multiple of T. The initial time t_0 for the orbit may be chosen anywhere in the interval $[0, T]$ of this wrapped time. Because of the wrapping in space and time, the manifold of $v_u(x_0, t_0)$'s for a given u is called a torus with velocity u by mathematicians; we denote it by T_u. Such a torus is defined whatever u is. The trace of T_u in the stroboscopic plot corresponds to the line $u =$ constant.

When the wave spectrum has a nonvanishing but small amplitude, it would be natural to think that the orbits with velocity u still belong to a torus, whose trace in the stroboscopic plot is slightly perturbed with respect to T_u. Mathematics tells us that this idea is correct for most u's, but wrong in general. Really, we were already aware of this, because of the phase portrait of the single-wave dynamics (Fig. 14.5). There, the torus with the wave phase velocity that exists for a vanishing amplitude of the wave breaks up into a pair of stable and unstable fixed points. In our stroboscopic plot, we may expect that for each wave, the corresponding unperturbed torus breaks up into a pair of stable and unstable periodic orbits. The Poincaré–Birkhoff theorem provides a mathematical justification for these breakups. It tells us that the T_u's with uT/L rational ("rational tori") are structurally unstable to perturbations, and are substituted with one or more pairs of stable and unstable periodic orbits. This backs up our naive superposition of cats' eyes: the trapped orbits correspond to orbits rotating about a stable periodic orbit. This is a sophisticated way of understanding the libration of the nonlinear pendulum! However, this theorem tells us a lot more: we should expect all T_u's with uT/L rational to break up. Indeed, a numerical calculation of orbits reveals this is so, as will be described in Section 14.5.4.

Fortunately, most numbers are irrational, and we may still expect the T_u's with uT/L irrational to be preserved by a small-amplitude perturbation. The Kolmogrov–Arnold–Moser (KAM) theorem tells us that this is correct for most irrational uT/L's only (those "irrational enough," i.e. satisfying a Diophantine condition). The KAM theorem may also be applied to justify the existence of trapped orbits in our naive stroboscopic plot. In this case we take as the unperturbed system one with a single wave. As was done for the passing particles, we can define unperturbed tori for the trapped particles. Then such a torus, corresponding to a trapping period τ such that τ/T is irrational enough, is preserved if all other waves have a small enough amplitude.

If we let all the wave amplitudes be multiplied by a factor F, and if we let F grow, at a certain moment two of the neighboring trapping domains are bound to touch each other. For a larger value of F, one would have overlap of these two trapping domains. This is obviously nonsense. Indeed, since a particle trapped in a wave has an average velocity equal to that of the wave phase velocity, simultaneous trapping in two waves with different phase velocities is ruled out. Obviously, something new must occur in the particle dynamics when separatrix overlap occurs. A numerical calculation of orbits starting in the intersection domain of two neighboring unperturbed separatrices shows that trapping into either of these waves is intermittent, and occurs alternately in each (see Fig. 14.10(b)). Transitions between these alternating phases occur in a seemingly erratic way. Two orbits starting at nearby points are seen to diverge exponentially in time. These facts are typical of dynamical chaos and, more precisely, of Hamiltonian chaos.

In this mechanical frame, orbits trapped in a wave are said to be in resonance with its sinusoidal potential. As a result, the above-mentioned separatrix overlap is called *resonance overlap* in the literature. For nearby waves with similar amplitudes, the value of F such that their unperturbed separatrices touch each other is a good approximation for the threshold of chaotic motion over the domain covered by the two trapping domains. This estimate is termed the Chirikov *resonance overlap criterion* (Chirikov 1979). It may be applied to a large class of chaotic Hamiltonian systems that can be written as an integrable part plus two or more resonant terms. However, this criterion underestimates the large-scale chaos threshold when the nearby waves considered have amplitudes that are too different (Escande 1985). The resonance overlap picture has been shown to describe exactly the merging of particle velocities in an experiment performed with a TWT (Doveil *et al.* 2005).

If F grows further, then more of the waves in Fig. 14.10(a) experience resonance overlap, and this overlap may become global when it occurs for all nearby waves (resonances) in a given velocity domain. Then the intermittent trapping described above of an orbit inside any of these waves ends in trapping in either of two nearby waves, and so on. Since the next nearby trapping wave is chosen at random, the orbit performs a kind of random walk in velocity. More concepts of Hamiltonian chaos will be introduced in Section 14.5.

14.4.6 Saturation of the bump-on-tail instability

With these concepts of Hamiltonian chaos, we can understand the saturation of the hot instability displayed in Fig. 14.2. Let us call $V(t)$ the velocity interval where $\partial f(v,t)/\partial v > 0$ (here $f(v,t)$ stands for the spatial average of the velocity distribution function). Particles with a velocity in $V(0)$ are acted upon by a broad spectrum of Langmuir waves. Since these waves are unstable, their trapping domains start overlapping at some moment during their growth, which destroys KAM tori in the domain $V(0)$. This erodes $f(v,t)$ toward lower velocities, and broadens $V(t)$ accordingly. This in turn increases the phase velocity domain of unstable Langmuir waves. This process carries on until the eroded beam distribution function hits the velocity distribution function of the bulk plasma. Then the spreading of the unstable phase velocity domain stops,[13] but chaos forces particles to fill the chaotic domain ergodically, which leads to the plateau formation. This picture has been confirmed experimentally (Roberson and Gentle 1971). The final distribution is spatially uniform, a quite different feature with respect to the clustering of the cold instability. This stationary spatial uniformity prevents any wave–wave coupling from occurring. Therefore the saturated Langmuir modes behave like linear waves, while the particle dynamics is chaotic (Escande and Elskens 2008).

This qualitative description calls for a mathematical formulation that would describe the genuine relaxation of the distribution function toward a plateau. A diffusion equation may be expected to hold after proper averaging, because of the random walk related to global resonance overlap introduced in the previous section. However, the correct choice of the averaging is a difficult issue, as will be explained in Section 14.5.1.

[13]Figure 14.2(c) displays the final wave intensity spectrum $W(v) = 2\pi|kE_k|^2/(\omega L)$, where E_k is the Fourier transform of the fluctuating electric field, and $k = \omega_{\rm p}/v$.

Finally, let us notice that the above growth of $V(t)$ corresponds to a direct cascade in wavenumber of the unstable wave spectrum (i.e. extension toward larger k values, as may occur in fluid turbulence).

If more physics is included, the plateau formation is nothing but the first stage of the relaxation of the hot instability toward a thermal equilibrium. Over larger time scales, because of spontaneous emission of waves by particles,[14] the plateau becomes progressively slanted toward positive velocities. Eventually, collisions bring the whole distribution toward a Maxwellian. Experimentally, these last two stages are too slow to be observed. After the first saturation phase described in Section 14.3.3, the cold instability involves chaotic motion also, because the unstable spectrum is not a single mode, and the system may switch to the hot-instability saturation mechanism.

In the case of the fan instability introduced in Section 14.3.2, diffusion may be superseded by a dynamical merging of resonances where a particle switches from a resonance with one wave to a resonance with a nearby one through a parametric instability involving both waves (Krafft *et al.* 2005). In the case of a dense, continuous wave spectrum this even leads to a trapping process in several waves (Krafft *et al.* 2006). Collective motion might play an important role in other media as well in the process of reaching thermal equilibrium. For instance, when there is an explosion in air, the emitted sound might have an important role, before the celebrated Boltzmann two-body collision mechanism becomes the dominant relaxation process.

14.4.7 Interpretation of Landau damping and growth

Our description of the Landau effect in Section 14.4.2 might have induced the reader into thinking that Landau growth and damping are symmetrical phenomena. This is not quite so! First consider the case of Landau damping, as could be measured along the x-axis of a magnetized plasma column. Imagine there is a grid perpendicular to the x-axis at $x = 0$ inside the plasma. This grid is made of very thin threads and has holes larger than the Debye length, and lets most of the electrons flow through its holes without feeling the grid potential. The grid potential is driven in such a way that it alternately blocks totally all electrons close enough to its threads during a time $\Delta t/2$, and lets them flow through the grid during the same amount of time. Since the plasma has a finite temperature, the electrons flowing though the grid have a finite velocity dispersion. Figure 14.11 is inspired by a similar figure in Baker *et al.* (1968). It shows a finite series of beamlets, which stream freely periodically after crossing the grid. For $x > 0$, a temporal modulation of the plasma density due to the fans produced by these beamlets is clearly visible. However, it is progressively smeared out as x grows, owing to the broadening of the positions of beamlet particles passing simultaneously through the grid, which produces an overlap of successive fans. Eventually, the perturbation completely vanishes. This is the very simple reason for Landau damping first introduced by van Kampen (van Kampen 1955; Elskens and Escande 2003, Section 3.8.3). Even for wave damping, it is useful to think of the plasma as a series of monokinetic beams! For further reference, it is important to notice that the density modulation produced by the grid exists also if the electron

[14]This phenomenon is analogous to the Cherenkov emission of electromagnetic waves by fast particles in a dielectric medium.

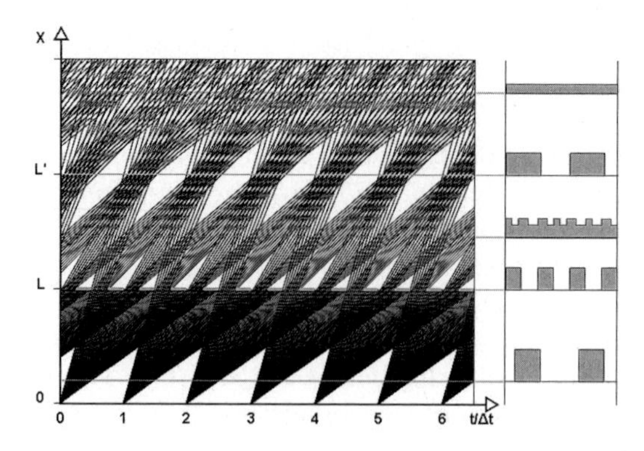

Fig. 14.11 Landau damping and plasma echo. Left part: in order to avoid displaying too busy a figure, the grid at $x = 0$ has been assumed to let particles flow in a time-periodic impulsive way, while the second grid lets particles flow over half of the modulation period. Right part: density modulation in time; for a series of values of x, the local density modulation in time is displayed by shaded boxes.

velocity distribution function has a positive slope (the relative density of the beamlets is not an issue).

Figure 14.11 has an upper part where a second grid is present at $x = L > 0$. This grid is excited like the first one, but with a shorter period. The previous beamlets go through this second grid and undergo a second periodic modulation that is smeared out like the previous one (again, Landau damping). However, we notice that at a larger distance L' the two grid modulations of the beamlets induce a cooperative effect: the so-called plasma echo[15] (Baker *et al.* 1968). This gedankenexperiment shows that Landau damping corresponds to the damping of the perturbation due to the collective action of the individual beamlets, but that each beamlet perturbation carries on forever. This shows that Landau damping is not a dissipation. Such an experiment has actually been performed in a plasma column (Baker *et al.* 1968).

Instead of a square modulation of the beamlet densities by the first grid, assume that a sinusoidal modulation $\sin(\omega t)$ is applied. Then, at a position x in between the two grids, the density modulation due to a beamlet with velocity v scales like $\sin[\omega(t - x/v)]$. The summation over the various beamlets provides a phase mixing that increases when x grows. A similar process occurs for the initial-value problem where the plasma is excited with a wavenumber k at $t = 0$. Therefore Landau damping is nothing but a phase-mixing mechanism of the beamlet density modulations (van Kampen 1955). The exponential Landau damping occurs only for a typical excitation

[15] Assume that the two grids are excited in a sinusoidal way with angular frequencies ω_1 and ω_2. If $f(v)$ is the velocity distribution function before the first grid is crossed, the radial part of the plasma perturbed by both grids has a distribution $f(v)\cos[\omega_1(t - x/v)]\cos[\omega_2(t - (x - L)/v)]$, where each cosine describes the modulation of one grid. This leads to a cosine whose argument does not depend on v if $x = L' \equiv \omega_2 L/(\omega_2 - \omega_1)$, which shows the coherence of the modulation for all v's, and thus the plasma echo at $x = L'$.

of the plasma density, which provides a typical amplitude distribution to the beamlet density modulations. A lot of Langmuir wave amplitude evolutions can be tailored by choosing appropriate amplitudes of the beam modes, in particular a non-Landau exponential damping (Belmont *et al.* 2008). It is important to notice that the Hamiltonian character of the wave–particle dynamics rules out the possibility that Landau damping might be related to a damped eigenmode. Indeed, this character requires a damped eigenvalue with damping rate γ to come with an unstable one with growth rate $-\gamma$. If Landau damping were related to a damped eigenmode, the unstable mode would show up for a typical excitation of the plasma.

It is clear that the hot instability cannot be understood as an inverse phase mixing. However, if a bump-on-tail velocity distribution flows through the first grid, as stressed before, it will also experience a Landau damping mechanism together with the development of the hot instability. How to reconcile these two views? This is one of the capabilities of the Hamiltonian model (14.1). A careful analysis (Elskens and Escande 2003, Section 3.8.3) shows that an initial perturbation with wavenumber k and amplitude 1 excites the perturbation $\exp(\gamma_{\rm L} t) + \exp(-\gamma_{\rm L} t) - \exp(-\gamma_{\rm L}|t|)$, where we recognize in the first two terms the above-mentioned pair of unstable and damped eigenmodes, and recognize the third term to be the Landau damping term coming from the spatial modulation of the above beamlets. Consistently with the reversibility of Hamiltonian mechanics, the system is unstable for t both growing and decreasing.

The original 1946 derivation of the Landau effect is a beautiful mathematical linear calculation starting from the Vlasovian mean-field description of the electron population. This mathematical technique can be applied to a wealth of other plasma waves, and works for a boundary value problem as well. Unfortunately, it hides completely the true nature of the underlying physics. In particular, it tells us nothing about the particle dynamics. The phenomenon was intuitive to so little an extent that it was debated for almost two decades, until its experimental proof by Malmberg and Wharton in 1964! After this discovery, people developed a series of small models in order to explain the Landau effect from the particle dynamics point of view. Some of these models, even in present textbooks, should still be considered with care.[16]

Finally, it is interesting to notice that the Landau effect is not a unique feature of plasmas, but that it occurs in many media with a continuous oscillation spectrum. It is at the root of wind-generated water waves (Vekstein 1998), but also exists in liquids with gas bubbles, in high-energy particle beams, in superfluids, and in quark–gluon plasmas. Analogs of Landau damping have also been discovered in biology in

[16]Three cases may be pointed out. The first one is the surfer model (Chen 1984); it comes with the explicit caveat that trapping is not involved in Landau damping, but the surfer image is so strongly suggestive of trapping that students are often misled. The second one is a popular calculation of the Landau effect by stating that the energy lost or gained by particles is gained or lost by the waves; as is made explicit by the Hamiltonian (14.1), there may be a third component in the total energy: the wave–particle coupling energy. But often calculations do not prove that the conservation law they use has only two terms (see, for instance, Chen 1984). A similar calculation can be made with momentum conservation where only two terms are to be considered (Vekstein 1998). The third case is a calculation made with a wave of constant amplitude, and using the Plemelj formula for a small positive parameter, which gives the Landau instability (Lifshitz and Pitaevskiĭ 1981); the final result is then claimed to provide Landau damping for a negative value of the parameter, which is not justified mathematically.

connection with the flashing of fireflies, and the pacemaker cells controlling the beating of the heart (see Vekstein 1998 for the corresponding references).

14.5 Hamiltonian chaos and diffusion

In Section 14.4.6, we anticipated that chaotic transport might be described as diffusion. This section aims at a better understanding of the way chaos develops for the dynamics of a particle in a series of waves. We will see that, though chaotic, the dynamics is not hyperbolic[17] at all. However, diffusion is a correct description provided that adequate averages are performed on the dynamics. The concepts and techniques described in this chapter apply to issues as different as chaos of magnetic field lines, the heating of particles by cyclotronic waves, the chaotic dispersion of pollutants in a nonstationary flow, chaos of rays in geometrical optics, etc.

14.5.1 Diffusion

Provided adequate averages are performed on the dynamics, diffusion is found numerically to be a correct description of chaotic transport (Bénisti and Escande 1997; Elskens and Escande 2003, Section 6.2; Escande and Sattin 2007). One mathematical result backs up this fact: diffusion has been proved for the dynamics of a particle in a Gaussian electrostatic potential in the limit of infinite resonance overlap (Elskens and Pardoux 2008). More precisely, this potential generates a field of the type $E(x,t) = \sum_m [A_m \cos(x - mt) + B_m \sin(x - mt)]$, where the sum runs over all integers, and where the A_m's and B_m's are Gaussian random variables with zero expectation and the same finite variance. However, the corresponding proof does not use at all the fact that the orbits are chaotic!

The classical quasi-linear approach to diffusion in wave–particle interaction does not use chaotic motion either. Let $v(t)$ be the velocity of a particle subjected to a set of M Langmuir waves with angular frequencies ω_j and wavenumbers k_j, $j = 1, M$; the numbering is done such that ω_j/k_j is a growing function of j. Let $\Omega_j = k_j v(0) - \omega_j$ be the Doppler wave frequency as seen by the particle at $t = 0$, and let $\Delta\Omega$ be the typical value of $|\Omega_{j+1} - \Omega_j|$. Let $\tau_{\text{cor}} = |\Omega_M - \Omega_1|^{-1}$ be the correlation time of the waves seen initially by the particle, and let $\tau_{\text{discr}} = \Delta\Omega^{-1}$; the latter time plays the role of a recurrence time of the wave electric field as seen by a particle with a fixed velocity $v(0)$. The quasi-linear approach makes the following simple calculation. One computes $\Delta v(t) = v(t) - v(0)$ by perturbation theory to first-order in the wave amplitudes. Then $\Delta v(t)$ is squared and *one averages over the phases of the waves*, assumed to be pairwise independent and uniformly distributed on the circle. In the limit where $\tau_{\text{cor}} \ll t \ll \tau_{\text{discr}}$, one finds $\langle \Delta v^2(t) \rangle/(2t)$ to be equal to the quasi-linear diffusion coefficient D_{QL}, which scales like the square of the wave amplitudes. More precisely, D_{QL} is a function of the initial velocity of the particle, and its value depends only on the amplitude of the waves close to being resonant with the particle, in agreement with the local character of wave–particle interaction. This shows that over the time

[17] A dynamics is hyperbolic if the phase space can be foliated by two transverse families of manifolds, such that in the corresponding coordinate system each trajectory is attracting along one family and repelling along the other family. The linearized dynamics around a typical trajectory behaves as if the reference motion were a saddle or X point.

interval $[\tau_{\rm cor}, \tau_{\rm discr}]$ there is a stochastic diffusion, even if the dynamics is nonchaotic. Numerical simulations back up this fact (Cary *et al.* 1990; Elskens and Escande 2003, Section 6.2). This can be related to the fact that the orbit feels an almost white-noise force during this time interval (Elskens and Escande 2003, Section 6.8.1).

Beyond the time $\tau_{\rm discr}$, if large-scale chaos does not rule the dynamics, $\langle \Delta v^2(t) \rangle$ saturates because all orbits are bounded by quasiperiodic KAM tori. If the chaos is strong enough for the exponentiation (Lyapunov) time of nearby orbits to be smaller than $\tau_{\rm discr}$, diffusion is found numerically to exist over large times; the diffusion coefficient is larger than $D_{\rm QL}$, but converges to this value when the resonance overlap goes to infinity. This can be approximately proved by taking into account that the effect of two phases of the dynamics is felt only after a long time in this limit (Bénisti and Escande 1997; Elskens and Escande 2003, Section 6.8.2). This is consistent with the fact that $D_{\rm QL}$ is the value of the diffusion coefficient in the theorem of Elskens and Pardoux (2008). In the general case, the diffusion picture can be backed up for waves with random phases by a theorem showing that in the chaotic regime, only a finite range of phase velocities around the particle velocity participates in the chaotic motion (Bénisti and Escande 1998; Elskens and Escande 2003, Section 6.3). The diffusive properties of the self-consistent wave–particle case is still an area of active research (Elskens and Escande 2003, Chapter 7).

14.5.2 Adiabatic description of Hamiltonian chaos

Consider the dynamics of an electron in three electrostatic waves defined by the Hamiltonian $H(v, x, t) = v^2/2 - A\cos(x) - B\cos(x - \Omega t) - B\cos(x + \Omega t)$. The waves have phase velocities Ω, 0, and $-\Omega$. For Ω large enough, the system is in the regime where the KAM theorem applies for most tori, and the corresponding stroboscopic plot looks like Fig. 14.10(a). When Ω decreases, resonance overlap sets in. It is interesting to consider the limit of infinite resonance overlap, i.e. $\Omega \to 0$. Then it is convenient to write the Hamiltonian as $H(v, x, t) = v^2/2 - A(1 + \eta\cos(\Omega t))\cos(x)$, where $\eta = 2B/A$. This Hamiltonian may be interpreted as describing the dynamics of a nonlinear pendulum in a slowly pulsating gravity field. An adiabatic picture of the system describes its phase space similarly to the picture in Fig. 14.5 but where the separatrix is slowly pulsating. Numerical simulations reveal that the domain swept by the separatrix in the Poincaré map looks like a chaotic sea where no island is visible (Menyuk 1985; Elskens and Escande 1993, 2003, Section 5.5.2). As a result, one might think that the limit of infinite overlap corresponded to some "pure" chaos. A fact pushing us in this direction is a theorem that says that in the domain swept by the separatrix, the homoclinic tangle is tight, which forces islands to have a size of order at most Ω (Elskens and Escande 1991). However, another theorem states that the total area covered by islands in the same domain remains finite in the limit (Neishtadt *et al.* 1997)! This shows that the dynamics is not hyperbolic at all; on the contrary, a lot of stable periodic orbits with their islands are lurking in the chaotic sea to wreck any too naive diffusive description: chaotic does not mean stochastic! This also shows that the chaos is not pure at all, and that the numerical simulation of orbits provides misleading information. This statement is right as far as the original mathematical model is concerned. However, if this model is thought of as an approximation of a true physical system, its dynamics undergoes real perturbations such as noise. These perturbations are likely to smear out the many minuscule islands of Neishtadt *et al.*'s theorem when Ω is small enough. Then the above numerical simulation gives the right physical picture! This raises the important issue of the structural stability of mathematical models when embedded in more realistic ones: numerical simulations might be more realistic than the mathematical model they approximate!

14.5.3 Mathematical view of resonance overlap

Mathematically speaking, in a Poincaré map (stroboscopic plot) a separatrix is a set of points which go to an unstable fixed point (X-point) for $t \to \pm\infty$. For a typical small perturbation of a dynamics exhibiting such a separatrix, the sets of points going toward the X-point for $t \to +\infty$ (stable manifold) and that for $t \to -\infty$ (unstable manifold) split and intersect (forming the so-called homoclinic intersections). They wiggle a lot and form an intricate trellis (the so-called homoclinic tangle; see Elskens and Escande 2003, Section 5.5.1). This is the structure underlying chaos. In the case of the particle-wave system of Section 14.4.5, when the wave amplitudes increase, chaos first occurs close to the unperturbed separatrices, but KAM tori still separate these local chaotic domains. Large-scale chaos related to resonance overlap corresponds to the breakup of the last KAM torus in between the two trapping domains. It also corresponds to the existence of (so-called heteroclinic) intersections between the stable manifold of one wave and the unstable manifold of the other one, and vice versa. Since the unperturbed separatrix of a wave is an approximation of the part close to the X-point of its stable and unstable manifolds, the Chirikov criterion may be viewed as an approximate rule for the heteroclinic intersection of the manifolds of nearby waves.

14.5.4 Renormalization group for Hamiltonian order

Numerical microscope. After exhibiting the structure underlying chaos, we now look for that underlying order in the many-wave system of Section 14.4.5. Since wave–particle interaction is a local effect in velocity, when we consider particles whose velocity falls between the phase velocities of two nearby waves, it is natural to approximate the dynamics by that due to these two waves only; this is called the *two-resonance approximation.* This idea has already been implicitly used when we computed the threshold for large-scale chaos by Chirikov's overlap criterion, which involves only the two nearest waves (or resonances). Therefore we now focus on a two-wave system. If we choose appropriate normalizations for time, space, and mass, and appropriate origins for space and time, the equation of motion of this system reads

$$\ddot{x} = -A \sin x - kB \sin k(x - t)\,. \tag{14.17}$$

The first and second waves have phase velocities 0 and 1, respectively, and k is the ratio of the original wavenumbers of the two waves considered. The force contains two sines, which lead to resonant effects in the dynamics, and are also called *resonances* by metonymy. The velocity of a resonance is defined as the value of $\dot{x}$ that makes the argument of the corresponding sine stationary. In wave language, this is the phase velocity. For simplicity we consider k to be a rational number r/s, where r and s are two mutually prime integers. This makes the phase space periodic in x with spatial period $L = 2s\pi$.

Consistently with the two-resonance approximation, we focus on orbits with average velocities u such that $0 < u < 1$. We now consider A and B to be proportional to some number ε. A numerical calculation of orbits in the stroboscopic plot reveals that, for ε small enough, there are chains of cats' eyes, or islands, corresponding to orbits with a mean velocity

$$u_{m,n}(k) = \frac{mk}{mk+n}, \tag{14.18}$$

where m and n are any positive integers. This was expected owing to the Poincaré–Birkhoff theorem. In reality, owing to the finite precision of computer calculations, one can find these chains only for bounded values of m and n. For a given small but finite value of ε, the island widths are quite inhomogeneous. When one looks at the whole range $0 < u < 1$, only a finite number of them are visible. Figure 14.12(a) displays a partial sketch of such a stroboscopic plot for $k = 1$ and $A = B$, displaying island chains with mean velocities $u_{0,1}(1) = 0$, $u_{1,1}(1) = 1/2$, and $u_{1,0}(1) = 1$. The other island chains were deemed too small to be plotted. Since $r = s = 1$, the chains with mean velocities $u_{0,1}(1) = 0$ and $u_{1,0}(1) = 1$ have only one island. More generally, the first one has s islands and the second one r islands over the spatial period $2s\pi$.

We now focus on orbits with mean velocities u such that $u_{1,1}(1) < u < u_{1,0}(1) = 1$. To this end, we blow up the previous stroboscopic plot, so that $u_{1,0}(1) - u_{1,1}(1)$ is rescaled to 1. For reasons which will become clearer later on, on top of this blowup of the velocity axis, we also flip the plot upside down. Then the new velocity is $v_1 = 2(1 - v)$. Figures 14.12(a) and (b) show this process for $k = 1$. Thanks to the blowup, the chain of three islands with the previous mean velocity $u_{2,1}(1) = 2/3$ is now visible, but the other island chains were again deemed too small to be plotted. As far as the chains with new mean velocities 0 and 1 are concerned, the second frame has exactly the structure of the first one, but for $k = k_1 \equiv 2/1$, and other values of A and B. It turns out that the mean new velocity of the three-island chain is $u_{1,1}(k_1) = 2/3$. Therefore Fig. 14.12(b) has exactly the structure of the first one, but for $k = 2$, and other values A_1 and B_1 of A and B.

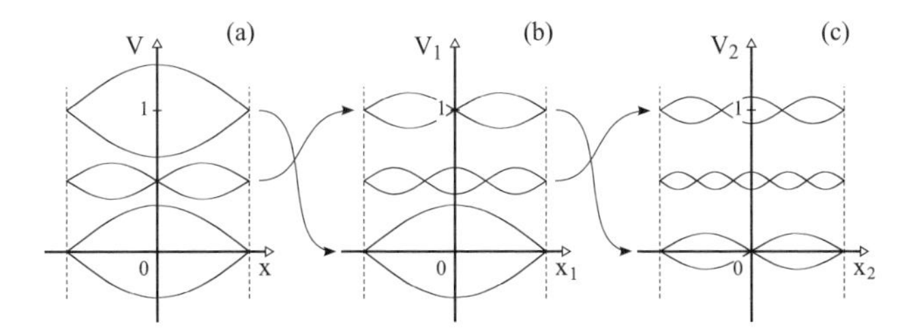

Fig. 14.12 Renormalization scheme. The finite resolution of the numerical microscope exhibits only three chains of islands at each successive step of the renormalization process.

We can iterate this process and focus on orbits with new mean velocities u such that $u_{1,1}(k_1) = 2/3 < u < u_{1,0}(k_1) = 1$ by blowing up the velocity axis and flipping it upside down as before. Figures 14.12(b) and (c) show this process. They reveal a chain of five islands. The new plot is again like the first one, but for $k = k_2 \equiv 3/2$, and new values A_2 and B_2 of A and B. Iterating this flipping and rescaling process (increasing the resolution of our numerical microscope) again and again turns out to be possible for ε small enough. For an initial $k = 1$, the successive k_l's turn out to be equal to F_{l+1}/F_l, where F_l is the Fibonacci number of order l, defined by $F_{l+1} = F_l + F_{l-1}$ and $F_0 = F_1 = 1$. For $l \to \infty$, $k_l \to g$, where $g = (\sqrt{5}+1)/2$ is the golden mean. In the meantime, A_l and B_l go to 0. In the original velocity frame, this iterative process converges toward an infinitely thin object with velocity $1/g$: a KAM torus. If ε is progressively increased, A_l and B_l go to 0 in a progressively slower way. Above some threshold in ε, A_l and B_l are seen to increase, and rapidly the numerical microscope reveals that the island structure is smeared out, and that motion is chaotic in the blown-up domain.

Secondary resonances. We now consider ε to be small, and we apply perturbation theory to second-order in ε to eqn (14.17) for an orbit with initial position x_0 belonging to the KAM torus with velocity u. At zeroth order, the orbit is $x^{(0)}(t) = x_0 + ut$. We may use Picard's technique again to compute the first-order contribution $x^{(1)}(t)$ to $x(t)$. The boundary conditions are $x(0) = x_0$ and a time-averaged value of $\dot{x}(t)$ equal to u. We get $\ddot{x}^{(1)}(t) = -A\sin(x_0+ut) - kB\sin k[x_0+(u-1)t]$. Therefore the first-order contribution is $x^{(1)}(t) = \xi^{(1)}(t) - \xi^{(1)}(0)$, where

$$\xi^{(1)}(t) = -\frac{A\sin(x_0+ut)}{u^2} - \frac{kB\sin k[x_0+(u-1)t]}{(u-1)^2}. \tag{14.19}$$

We notice that each wave yields a contribution that diverges when u is equal to its resonant velocity $u_{0,1}(k) = 0$ or $u_{1,0}(k) = 1$. This is the classical issue of small denominators related to the resonances in eqn (14.17). This makes the KAM theorem very hard to prove, and causes its statement to involve subtle number theory. The motion of a particle in a wave or, equivalently, of the nonlinear pendulum, makes the reason for this divergence obvious. The existence of a separatrix implies that there is a class of orbits, the trapped ones, whose motion is not a perturbation of the dynamics corresponding to a vanishing amplitude of the wave or of the gravity field. Indeed, trapped orbits correspond to a qualitative change in the topology of the phase space portrait.

The second iteration of Picard's technique corresponds to the calculation of the approximation $X_2(t)$ to the true orbit such that

$$\ddot{X}_2(t) = -A\sin[x^{(0)}(t) + x^{(1)}(t)] - kB\sin k[x^{(0)}(t) + x^{(1)}(t) - t]. \tag{14.20}$$

The second-order terms of $X_2(t)$ in ε lead to the second-order contribution $x^{(2)}(t)$ to $x(t)$. They include contributions involving sinusoidal functions of the sum of the arguments of the two sines in eqn (14.19), such as

$$\xi^{(2)}(t) \sim -\frac{AB\sin[(k+1)x_0 + [(k+1)u - k]t]}{(u-u_{1,1})^2}. \tag{14.21}$$

We notice that this term diverges for $u = u_{1,1}(k)$. This divergence looks analogous to that already found at the first order, and indicates the existence of a new resonance, called the *secondary resonance*, with velocity $u_{1,1}(k)$. Since the argument of the sine in eqn (14.21) depends on time like $[(k+1)u - k]t$, we can infer that this secondary resonance has a wavenumber $k+1$. A numerical calculation of the orbits and the Poincaré–Birkhoff theorem have already revealed that this is so. Indeed, for ε small enough, in the frame moving with velocity $u = u_{1,1}(k)$ there are static chains of cats' eyes containing orbits rotating about stable orbits predicted by this theorem. The other frames of Fig. 14.12 progressively reveal other signatures of the same theorem.

Renormalization group. We now make progressively more rigorous the mathematical meaning of the above numerical microscope. When focusing on orbits with average velocities u such that $u_{1,1}(k) < u < u_{1,0}(k) = 1$, we introduce a new position x_1 such that $x_1 = kx - kt$, and a new time $t_1 = -(k+1)t/k$ such that $u_{1,1}(k_1)$ becomes 1. This gives $v_1 = k(1-v)$ and a wavenumber $k_1 = (1+k)/k$ to the new resonance with velocity $u_{1,0}(k_1) = 1$. Implicitly, we are giving the resonance with velocity $u_{1,0}(k) = 1$ the previous role of the resonance with velocity $u_{0,1} = 0$, and we are giving the resonance with velocity $u_{1,1}(k)$ the previous role of that with velocity $u_{1,0}(k) = 1$. This already helps us to make more precise the axis of the successive frames in Fig. 14.12. We may expect the dynamics to be governed by an equation close to eqn (14.17) with new amplitudes A_1 and B_1 but, as yet, we have no way to prove it.

The correct way of implementing this idea is to consider the Hamiltonian formulation of the dynamics (14.17). Equation (14.17) comes from the Hamiltonian

$$H(v,x,t) = \frac{v^2}{2} - A\cos x - B\cos k(x-t) \tag{14.22}$$

through the compound use of the canonical equations $\dot{x} = \partial H/\partial v$ and $\dot{v} = -\partial H/\partial x$. For $B = 0$, this is the pendulum Hamiltonian introduced in Section 14.2.3. Then the above rescaling and flipping procedure is preceded by an important step: one performs a canonical transformation, "killing" the resonance with velocity $u_{0,1} = 0$ in the Hamiltonian (14.22) (Escande 1985). This procedure amounts to taking away most of the orbit distortion due to this resonance: after this procedure, in the new conjugate variables (x', v') the trace $v'(x')$ of any KAM torus in the stroboscopic plot has fewer wiggles than in the original variables. The correct version of the previous blowup and flipping procedure involves in reality the definitions $x_1 = kx' - kt$ and $v_1 = k(1-v')$, where (x', v') are substituted for (x, v) in the previous formulation. When we focus on the mean-velocity domain $[u_{1,1}(k), u_{1,0}(k)]$, it is natural to make again a two-resonance approximation, retaining only those resonances with velocities $u_{1,1}(k_1)$ and $u_{1,0}(k_1)$. The corresponding amplitudes of these resonances turn out to depend on v_1. Each amplitude is substituted by its value for the resonant velocity. One is then left with a Hamiltonian of the type (14.22) with new parameters (k_1, A_1, B_1). The mapping of the old Hamiltonian (14.22) into a new one describing another volume of phase space is called a renormalization group procedure.

Whatever the initial value of k is, k_l converges toward a single fixed point k_* when $l \to \infty$. Figure 14.13 shows the dynamics of (A_l, B_l) for $k = k_*$. It exhibits a hyperbolic fixed point whose stable manifold separates two domains. If the initial

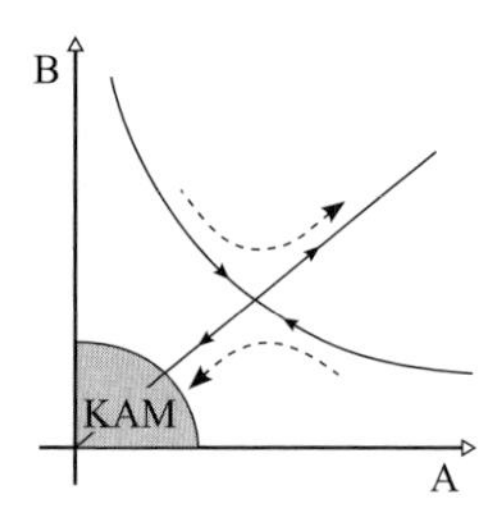

Fig. 14.13 Mapping of the two resonance amplitudes.

value of (A, B) is chosen in the lower domain, (A_l, B_l) converges toward zero. After a finite number of iterations, (A_l, B_l) lands in the shaded domain, where the KAM theorem applies for the torus with initial mean velocity $k/(k+g-1)$. This means that this torus is preserved. If the initial value of (A, B) is chosen in the upper domain, A_l and B_l increase indefinitely. Rapidly there is resonance overlap, which means the previous KAM torus is destroyed. When the KAM torus T_u breaks, the orbits with mean velocity u do not all disappear. Instead, their trace in the Poincaré map forms a Cantor set called a cantorus (MacKay *et al.* 1984; Elskens and Escande 2003, Sections 5.2.1 and 5.4.4).

Up to now we have focused on the range $[u_{1,1}(k_l), u_{1,0}(k_l)]$ for all l's. In fact, we may shift our focus at each l, defining a series of n_l's such that at step l we focus on the range $[u_{1,n_l+1}(k_l), u_{1,n_l}(k_l)]$. This in turn modifies the successive changes of variables. The series of n_l's defines the KAM torus that the renormalization procedure is focusing on. The interested reader is referred to Escande and Doveil (1981), Escande (1985), and Section 5.4 of Elskens and Escande (2003) for more details.

There are several instances of this approximate renormalization procedure. One of them involves killing the two resonances with velocities $u_{0,1} = 0$ and $u_{1,0} = 1$ at each step. In Fig. 14.12, this amounts to taking two steps at once. In this presentation, we avoided presenting this faster-converging scheme in order not to plot chains with too large a Fibonacci number of islands!

Though approximate, this renormalization procedure gives accurate estimates for the threshold in ε for breakup of the KAM torus (Escande and Doveil 1981). Furthermore, as in other renormalization group approaches, the hyperbolic fixed point enables the calculation of critical exponents; they turn out to be quite precise too (Escande 1985). One of them is useful for the description of chaotic transport. When the KAM torus T_u breaks, chaotic orbits sneak through the holes of the corresponding cantorus. The exponent of interest characterizes the flux of chaotic orbits through these holes. The above "golden mean" fixed point corresponds to the most robust KAM tori, as seen numerically. Therefore the corresponding exponent rules the large-scale chaotic transport close to the threshold for torus breakup. Close to this threshold, the time for the setting up of the random walk introduced in Section 14.4.5 may be very long. Furthermore, chains of islands and cantori look rather like coherent structures, which prevent the chaotic motion being stochastic. This confirms the need for appropriate averages to wash out these structures and to justify a diffusive description, as indicated in Section 14.5.1.

14.6 Conclusion

This presentation of wave–particle interaction in plasmas showed that two paradigms of classical mechanics are essential: the harmonic oscillator and the nonlinear pendulum. We found that wave–particle interaction is local in velocity, and that particles synchronize in an average sense with waves. Chaotic dynamics rules this interaction as soon as more than one wave is present. This dynamics is not of hyperbolic character, but appropriate averages enable its description as a diffusion. Chaos plays an essential role in the bump-on-tail instability, the incoherent limit of beam–plasma interaction, but may be neglected in the coherent limit related to a cold beam.

At this point there are many avenues to enter deeper into wave–particle interaction. Let us now quote a few. We might enter the description of wave–particle interaction by a mechanical approach using eqn (14.1) defining $\mathcal{H}_{\text{self-consistent}}$ (Elskens and Escande 2003, Chapters 2–4). Since thinking of the plasma as a set of monokinetic beams is useful, we would first decompose our plasma into such beams. Each of them would be an array of particles, in order to kill the spontaneous emission of waves by particles. Linear theory would lead to a high-dimensional Floquet problem that would turn out to be explicitly solvable. Instead of this approach to a single mechanical realization of the plasma, we also could perform a perturbative calculation in the wave amplitudes. An average over the initial positions of the particles and the phases of the waves would provide both the Landau effect and the quasi-linear diffusion of particles. The same calculation would include the spontaneous emission of Langmuir waves by particles, revealing that Landau damping is the exponential relaxation of Langmuir waves toward their thermal level. Therefore the mechanical approach using $\mathcal{H}_{\text{self-consistent}}$ brings about a synthesis and an extension of the descriptions of the linear aspects of wave–particle interaction provided in this course.

As far as the nonlinear regime of wave–particle interaction is concerned, we might consider the case of a single Langmuir wave interacting with a broad distribution of particles. In particular, an interesting second-order phase transition occurs in this dynamics (Firpo and Elskens 2000; Elskens and Escande 2003, Chapter 9). Furthermore, the mean-field Vlasovian description has been shown to miss the actual saturation of an unstable wave: the limits of infinite time and particle number do not commute (Doveil *et al.* 2001; Firpo *et al.* 2001). Landau damping can be computed for waves with a finite amplitude that depends slowly on both time and space. In particular, nonlinear Landau damping can be shown to decrease when the amplitude of a forced plasma wave grows (Bénisti and Gremillet 2007).

We might consider magnetized plasmas, where the Larmor precession plays an important role for several waves. The concepts introduced in this course apply to this case as well. When a charged particle is acted upon by an electrostatic wave propagating across a uniform magnetic field, the wave–particle interaction is no longer local in phase velocity. This allows nonchaotic particle energization (Bénisti *et al.* 1998).

Acknowledgments

R. Paškauskas processed his notes into a useful LaTeX input for my manuscript. F. Doveil and Y. Elskens made useful comments about a preliminary version of the manuscript. Y. Elskens also provided valuable help with correct typographical choices and made useful remarks about the second version. This version benefited from useful comments by T. Dauxois, Y. Peysson, and S. Ruffo, and from extensive and deep remarks by G. Belmont, D. Bénisti, and C. Krafft. Two PhD students, A. Lejeune and F. de Solminihac, spent hours with me discussing the first version. D. Guyomarc'h kindly agreed to draw the figures for this course. I express my gratitude to all of them for their valuable contributions.

References

Baker, D. R., Ahern, N. R., and Wong, A. Y. (1968). Ion-wave echoes. *Phys. Rev. Lett.*, **20**, 318–21.

Belmont, G., *et al.* (2008). Existence of non-Landau solutions for Langmuir waves. *Phys. Plasmas*, **15**, 052310.

Bénisti, D., and Escande, D. F. (1997). Origin of diffusion in Hamiltonian dynamics. *Phys. Plasmas*, **4**, 1576–81.

Bénisti, D., and Escande, D. F. (1998). Finite range of large perturbations in Hamiltonian dynamics. *J. Stat. Phys.*, **92**, 909–72.

Bénisti, D., and Gremillet, L. (2007). Nonlinear plasma response to a slowly varying electrostatic wave, and application to stimulated Raman scattering. *Phys. Plasmas*, **14**, 042304.

Bénisti, D., *et al.* (1998). Ion dynamics in multiple electrostatic waves in a magnetized plasma. *Phys. Plasmas*, **5**, 3224–32; 3233–41.

Bernstein, I. B., Greene, J. M., and Kruskal, M. D. (1957). Exact nonlinear plasma oscillations. *Phys. Rev.*, **108**, 546–50.

Cary, J. R., Escande, D. F., and Verga, A. D. (1990). Non quasilinear diffusion far from the chaotic threshold. *Phys. Rev. Lett.*, **65**, 3132–5.

Chen, F. F. (1984). *Introduction to Plasma Physics and Controlled Fusion.* New York: Plenum Press.

Chirikov, B. V. (1979). A universal instability of many-dimensional oscillator systems. *Phys. Rep.*, **52**, 263–379.

Dimonte, G., and Malmberg, J. H. (1977). Destruction of trapped-particle oscillations. *Phys. Rev. Lett.*, **38**, 401–4.

Doveil, F., *et al.* (2001). Trapping oscillations, discrete particle effects and kinetic theory of collisionless plasma. *Phys. Lett. A*, **284**, 279–85.

Doveil, F., *et al.* (2005*a*). Experimental observation of resonance overlap responsible for Hamiltonian chaos. *Phys. Plasmas*, **12**, 010702(L).

Doveil, F., Escande, D. F., and Macor, A. (2005*b*). Experimental observation of nonlinear synchronization due to a single wave. *Phys. Rev. Lett.*, **94**, 085003.

Elskens, Y., and Escande, D. F. (1991). Slowly pulsating separatrices sweep homoclinic tangles where islands must be small: An extension of classical adiabatic theory. *Nonlinearity*, **4**, 615–67.

Elskens, Y., and Escande, D. F. (1993). Infinite resonance overlap: A natural limit of Hamiltonian chaos. *Physica D*, **62**, 66–74.

Elskens, Y., and Escande, D. F. (2003). *Microscopic Dynamics of Plasmas and Chaos*. Bristol: Institute of Physics.

Elskens, Y., and Pardoux, E. (2008). Diffusion limit for many particles in a periodic stochastic acceleration field, arXiv:0811.0801[math.PR], submitted.

Escande, D. F. (1985). Stochasticity in classical Hamiltonian systems: Universal aspects. *Phys. Rep.*, **121**, 165–261.

Escande, D. F. (1989). Description of Landau damping and weak Langmuir turbulence through microscopic dynamics. In *Nonlinear World*, Vol. 2, eds. Bar'yakhtar, V. G., Chernousenko, V. M., Erokhin, N. S., Sitenko, A. G., and Zakharov, V. E., pp. 817–36. Singapore: World Scientific.

Escande, D. F., and Doveil, F. (1981). Renormalization method for computing the threshold of large-scale stochastic instability in two degrees of freedom Hamiltonian systems. *J. Stat. Phys.*, **26**, 257–84.

Escande, D. F., and Elskens, Y. (2008). Self-consistency vanishes in the plateau regime of the bump-on-tail instability. http://hal.archives-ouvertes.fr/hal-00295896/fr/.

Escande, D. F., and Sattin, F. (2007). When can the Fokker–Planck equation describe anomalous or chaotic transport? *Phys. Rev. Lett.*, **99**, 185005.

Firpo, M.-C., and Elskens, Y. (2000). Phase transition in the collisionless damping regime for wave–particle interaction. *Phys. Rev. Lett.*, **84**, 3318–21.

Firpo, M.-C., *et al.* (2001). Long-time discrete particle effects versus kinetic theory in the self-consistent single-wave model. *Phys. Rev. E*, **64**, 026407.

Gentle, K. W., and Lohr, J. (1971). Observations of the beam–plasma instability. *Phys. Fluids*, **14**, 2780–2.

Gentle, K. W., and Lohr, J. (1973). Experimental determination of the nonlinear interaction in a one dimensional beam–plasma system. *Phys. Fluids*, **16**, 1464–71.

Kadomtsev, B. B. (1992). *Tokamak Plasma: A Complex Physical System*, p. 117. Bristol: Institute of Physics.

Krafft, C., *et al.* (2005). Saturation of the fan instability: Nonlinear merging of resonances. *Phys. Plasmas*, **12**, 112309.

Krafft, C., *et al.* (2006). Stabilization of the fan instability: Electron flux relaxation. *Phys. Plasmas*, **13**, 122301.

Landau, L. D. (1946). On the vibrations of the electronic plasma. *Zh. Eksp. Teor. Fiz.*, **16**, 574–86; transl. *J. Phys. USSR*, **10**, 25.

Lifshitz, E. M., and Pitaevskiĭ, L. P. (1981). *Landau and Lifshitz's Course of Theoretical Physics*, Vol. 10, *Physical Kinetics*, transl. Sykes, J. B. and Franklin, R. N. Oxford: Pergamon.

MacKay, R. S., Meiss, J. D., and Percival, I. C. (1984). Transport in Hamiltonian-systems. *Physica D*, **13D**, 55–81.

Menyuk, C. R. (1985). Particle motion in the field of a modulated wave. *Phys. Rev. A*, **31**, 3282–90.

Neishtadt, A. I., Sidorenko, V. V., and Treschev, D. V. (1997). Stable periodic motion in the problem on passage through a separatrix. *Chaos*, **7**, 1–11.

O'Neil, T. M., and Malmberg, J. H. (1968). Transition of the dispersion roots from beam-type to Landau-type solutions. *Phys. Fluids*, **11**, 1754–60.

O'Neil, T. M., Winfrey, J. H., and Malmberg, J. H. (1971). Nonlinear interaction of a small cold beam and a plasma. *Phys. Fluids*, **14**, 1204–12.

Onishchenko, I. N., Linetskiĭ, A. R., Matsiborko, N. G., Shapiro, V. D., and Shevchenko, V. I. (1970). Contribution to the nonlinear theory of excitation of a monochromatic plasma wave by an electron beam. *ZhETF Pis. Red.*, **12**, 407–11, transl. *JETP Lett.*, **12**, 281–5.

Roberson, C., and Gentle, K. W. (1971). Experimental test of the quasilinear theory of the gentle bump instability. *Phys. Fluids*, **14**, 2462–9.

Self, S. A., Shoucri, M. M., and Crawford, F. W. (1971). Growth rates and stability limits for beam–plasma interaction. *J. Appl. Phys.*, **42**, 704–13.

Spohn, H. (1991). *Large Scale Dynamics of Interacting Particles.* Berlin: Springer.

Tennyson, J. L., Meiss, J. D., and Morrison, P. J. (1994). Self-consistent chaos in the beam–plasma instability. *Physica D*, **71**, 1–17.

van Kampen, N. G. (1955). On the theory of stationary waves in plasmas. *Physica*, **21**, 949–63.

Vekstein, G. E. (1998). Landau resonance mechanism for plasma and wind generated waves. *Am. J. Phys.*, **66**, 886–92.

Zaslavsky, A., *et al.* (2008). Wave–particle interaction at double resonance. *Phys. Rev. E*, **77**, 056407.

15
Long-range interaction in cold-atom optics

Philippe W. Courteille

Physikalisches Institut, Eberhard-Karls-Universität Tübingen,
Auf der Morgenstelle 14, D-72076 Tübingen, Germany

15.1 Introduction

Long-range effects play a role in a variety of systems, ranging from condensed matter physics and hydrodynamics to astrophysical systems, and a precise knowledge of their long-range interactions is often crucial to understanding their behavior. Long-range interacting systems are peculiar in many respects. The presence of long-range forces between microscopic particles can have a critical impact on the system's bulk behavior and lead to the emergence of instabilities and self-organization phenomena, i.e. the spontaneous generation of long-range order. Theoretical modeling of real physical systems is often complicated by dominating short-range interactions, inhomogeneities, or impurities. Furthermore, important quantities (such as density and energy) are often not amenable to direct measurement, and relevant parameters (size, temperature, interaction strength, etc.) cannot be tuned. For example, neither is it practicable to manipulate stars and galaxies or tune the gravitational constant, nor is it easy to shield the weak gravitational force from perturbations in a laboratory environment. On the other hand, the experimental verification of theories describing the statistical mechanics of long-range interactions requires simple, controllable, microscopic systems.

Atom optics provides ideal toy systems. Atoms are small, simple, and easy to manipulate in all their internal *and* external degrees of freedom. Cold-atom optics has seen incredible progress over 20 years. The invention of the laser permitted precise spectroscopic investigations of the atoms' internal structure. Powerful trapping and cooling techniques for dilute atomic gases and the possibility of tuning the interatomic forces paved the way to perfectly pure, absolutely controllable particle samples. The relevant degrees of freedom and their interaction with the environment can be controlled to such an extent that quantum mechanical effects not only emerge but become dominant, as in the case of Bose–Einstein condensation or Fermi degeneracy. Cold-atom optics is today providing idealized many-body model systems for condensed matter physics (e.g. Mott insulators and quantum phases in reduced dimensions), as well as for astrophysics. For example, it is possible to emulate gravitational forces under various circumstances, including white dwarfs and Bose stars. Sonic analogues of black holes and Hawking radiation are being studied in quantum optics labs.

Long-range interaction as a topic of research in atom optics is very new, which is partly due to the fact that cold-atom optics is a very recent field itself. A more prominent reason is, however, that ultralow temperatures are necessary for giving forces beyond binary collisions a chance to emerge at densities low enough to prohibit inelastic processes. Nevertheless, there are already numerous examples of long-range effects ruling the dynamics of cold and ultracold atoms: the influence of dipolar interactions on the global stability of a Bose–Einstein condensate (BEC), self-organization and self-synchronization in Wigner crystals of trapped ions, magneto-optical traps, and atomic clouds interacting with optical ring cavities. The latter system is particularly interesting for its feature of providing uniform coupling. In a ring cavity, the positions of all atoms are, independent of their distance, coupled by the scattering and rescattering of photons. This collective coupling can lead to instability and self-organization of the cloud. This phenomenon, termed *collective atomic-recoil lasing* (CARL), has been discussed theoretically for years, but was observed only recently. The CARL effect is closely linked to superradiant Rayleigh scattering (SRyS), which has been intensely studied with BECs in free space. However, the presence of a resonator dramatically enhances the coherence time.

The present chapter is organized as follows. Section 15.2 can be read as a eulogy of atom optics. Indeed, since most attendees of this Les Houches Summer School may not be familiar with cold-atom optics, I take the opportunity to advertise this field of physics to a broad community of statistical physicists and scientists studying long-range phenomena. I will give a very brief survey of the most important techniques, specify relevant orders of magnitude, and point to the observables and controllable parameters. In view of the many excellent review articles available on Bose–Einstein condensation and ultracold-matter-wave optics (Parkins and Wallis 1998; Dalfovo *et al.* 1999; Stamper-Kurn and Ketterle 2000; Courteille *et al.* 2001), I will concentrate on the few aspects which are particular to long-range phenomena.

In Section 15.3, I will briefly discuss a few atom-optical systems exhibiting long-range effects. Electromagnetic forces have massless exchange particles and hence infinite range. An example of a system entirely ruled by electrostatic forces is the cold charged plasma, a gaseous cloud of charged ions confined by an external potential where the ions repel each other by the Coulomb force. But even neutral particles can exert long-range forces on each other if they possess dipolar moments. Another type of long-range force is light-induced, although it also derives from the electromagnetic interaction. As first shown by Frisch (1933), the absorption and emission of radiation of frequency ω is always accompanied by a momentum transfer $\hbar\omega/c$. Examples of this are radiation pressure in magneto-optical traps (MOTs) and the collective atomic-recoil laser.

Section 15.4 is devoted to the two closely linked phenomena of CARL and SRyS. In both cases atoms are coupled via the scattering of photons. Both systems show self-organization leading to the spontaneous formation of a density grating of atoms, and demonstrate in a nice way that the existence of a long-range interaction is a prerequisite for the spontaneous formation of a global instability. Finally, Section 15.5 gives a conclusion and a brief outlook into the future.

15.2 Cold-atom-optics toolbox

Atom optics has a number of advantages. First of all, atom-optical experiments are table-top experiments. Even though making Bose–Einstein condensates can be a little tricky, all experiments can in principle be handled by a single person. Second, all important control parameters, such as size, temperature, and even the interaction strength, are tunable. Thermodynamic potentials such as the internal energy, the chemical potential, and the heat capacity can, in principle, be measured. Third, energy and particle exchange with the environment can be made very weak over the time scale of an experiment. Fourth, the degrees of freedom involved in an interaction can be cooled and controlled to such an extent that their quantum nature determines the dynamics.

The experimental challenge of atom optics is to control and manipulate the motional state of atoms by electromagnetic fields or light and to infer their response to the manipulations from images of their density distribution. In atom optics, all contributions to the total energy E can be controlled independently: the kinetic energy $\sum_i \mathbf{p}_i^2/2m$ by cooling, the potential energy $\sum_i V(\mathbf{r}_i)$ by appropriate trapping potentials, and even the self-energy, as we will see below. A particularity is, however, the fact that all experiments use trapped atoms, i.e. inhomogeneous density distributions. This is important to bear in mind when investigating thermodynamic properties such as phase transitions.

15.2.1 Trapping and cooling to low and ultralow temperatures

Atoms (or in some cases ions) are either confined and studied individually or trapped in large clouds of billions of atoms by external potentials. These potentials hold the atoms against gravity, isolate them from the surroundings, and compress the clouds to low or high densities, which are about ten orders of magnitude below atmospheric pressure. Hence, all experiments must be performed in ultrahigh-vacuum chambers. The confinement technique depends on the type of trapped atoms. Ions are stored in combinations of stationary electric and magnetic fields, called Penning traps, or time-dependent electric fields, called Paul traps. Owing to the strong interaction of charged particles with electromagnetic fields, ion traps can be made very deep. Arrays of ions stored in linear Paul traps are today the most promising system for quantum computational applications.

Spin-polarized paramagnetic atoms are trapped by their magnetic dipole moment in inhomogeneous magnetic fields near field minima. It is also possible to induce an electric dipole moment in the atoms by exciting them near an electronic resonance with a laser beam. The interaction of this dipole moment with intensity gradients in the laser beam induces forces, giving rise to a conservative trapping potential around intensity extrema such as foci or the antinodes of standing light waves. Both types of trapping forces are generally weak and work only for cold atoms having kinetic energies lower than about 1 mK.

Trapping potentials are often harmonic and cylindrically symmetric (i.e. cigar- or pancake-shaped) with secular frequencies ω_{trap} on the order of several tens or hundreds of Hz. However, the external electric, magnetic, and electromagnetic optical fields can be given almost arbitrary shapes and time dependences. Particularly interesting are periodic potentials, where perfect 1D, 2D, or 3D arrays of atoms can be formed without any defects.

The major breakthroughs for atom optics date from the 1980s and 1990s and come from the invention of cooling techniques capable of reducing the temperature of atomic clouds to the range of microkelvins and even nanokelvins. The most important cooling technique is laser cooling. Here, the atoms are exposed to laser beams, which are red-detuned from an electronic resonance. Because of the Doppler effect, the frequencies of the laser beams are shifted out of resonance for copropagating atoms. The atoms only absorb light from counterpropagating laser beams, whose frequencies are Doppler-shifted into resonance (see Fig. 15.1, right). The absorption process decelerates an atom by one photonic recoil momentum, i.e. $\hbar k$, where $\lambda = 2\pi/k = 780$ nm for the rubidium D_2 line; the subsequent photon reemission occurs isotropically. Arrangements of laser beams oriented in all six directions of space, called optical molasses, are able to cool large samples of atoms within milliseconds to temperatures of $T \simeq 10$ µK. When

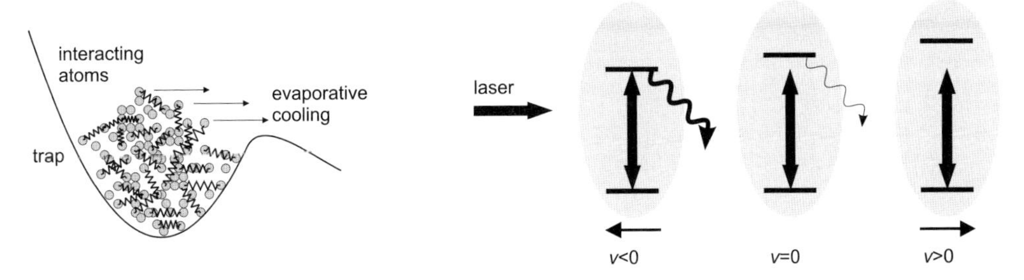

Fig. 15.1 Left: interacting atoms trapped in a harmonic potential. The potential has an spout, over which energetic atoms with large oscillation amplitudes may escape. Right: the idea of Doppler cooling.

talking about *cold* atoms, we normally mean Doppler-limited temperatures, i.e. when the cooling limit is set by a dynamical balance between cooling forces, which can be understood as friction, and diffusion processes linked to the stochastic nature of the photon reemission.

Even lower temperatures are obtained by evaporative cooling, where hot atoms with large oscillation amplitudes are allowed to escape from the trap (see Fig. 15.1, left). Provided the remaining atoms rethermalize fast enough via binary collisions, evaporation can be used for efficient temperature reduction down to well below 100 nK. The atoms are called *ultracold* if the kinetic energy of the atoms is below the photon recoil limit, i.e. for rubidium, $T \lesssim \hbar^2 k^2/2mk_B \simeq 180$ nK. At such low temperatures the atomic motion is not only quantized but also often dominated by quantum statistical effects. For example, ultracold bosonic atoms tend to form a Bose–Einstein condensate. The necessary condition for this is that the atomic de Broglie wavelength is longer than the mean interparticle distance, i.e. $T \lesssim 2\pi\hbar^2 n^{2/3}/m \simeq 750$ nK for typical densities of $n = 10^{14}$ cm^{-3}. In contrast, fermions form a degenerate Fermi gas and have even been made to form Cooper pairs and ultracold molecules.

15.2.2 Tuning contact interactions with Feshbach resonances

Collisions play an important role in the dynamics of atomic clouds. By means of *elastic* collisions, atoms exchange only kinetic energy, a process which is important for the thermalization of atomic clouds (e.g. during evaporative cooling). At short interatomic distances R, the Born–Oppenheimer interaction potentials are governed by exchange interactions. But at ranges exceeding $10a_B$, the interaction potentials are dominated by van der Waals forces, so that they rapidly decay as $V(R) \propto R^{-6}$, thus being clearly short-ranged. Since at low temperatures $T \lesssim 10$ µK, the atoms interact exclusively via s-wave collisions (higher partial waves are frozen out behind the centrifugal barrier), the collisions are described by a single parameter, the scattering length a_s, which is defined as the phase slip that the relative de Broglie wave acquires inside the interaction potential during the collision in the limit of zero temperature. For most practical cases, we may understand the collisions as contact interactions between hard spheres with radii a_s. The cross section is quadratic in the scattering length, $\sigma = 4\pi a_s^2$. In the mean-field approximation, the self-energy of a cloud can be written as the sum $E_{\text{self}} = \sum_{m<n}(4\pi\hbar^2/m)a_s\delta(\mathbf{r}_n - \mathbf{r}_m)$.

The mean-field approximation dramatically simplifies the description of BECs via a single macroscopic wavefunction $\psi(\mathbf{r})$ satisfying the so-called *Gross–Pitaevskii equation*, which resembles a Schrödinger equation except for an additional nonlinear term appearing in the system's Hamiltonian, $4\pi\hbar^2 a_s n(\mathbf{r})/m$, which depends only on the density of the gas $n = |\psi|^2$ and the scattering length a_s. The appearance of the condensate self-energy in the Gross–Pitaevskii equation emphasizes its impact on the shape, stability, and dynamics of the condensate. The interatomic potential decides the value of the scattering length. A repulsive potential corresponds to a positive scattering length. For a purely attractive potential without bound vibrational states, the scattering length is negative. For an attractive potential which supports bound states, the scattering length can be positive or negative. This depends on the closeness of the last vibrational state to the dissociation limit. Homogeneous condensates with a negative scattering length are unstable, because the condensate attempts to lower its self-energy by increasing its density until it collapses under the influence of inelastic two- or three-body collisions.[1]

Feshbach resonances were first predicted for nuclear systems (Feshbach 1958). Their recent revival in the context of cold atomic collisions is due to the prospect of their use for manipulating the scattering length in real time over huge ranges and thus controlling the mean-field energy of BECs. Furthermore, Feshbach resonances are interesting for their capacity to provide a coherent free–bound coupling between an unbound state of colliding atoms and a molecular bound state of the same atoms. This coupling has been used for the creation of molecular condensates.

Feshbach collision resonances arise when the energy of a collisional state coincides with the energy of a molecular vibrational level belonging to a higher asymptote (Courteille *et al.* 1998). This coincidence heavily perturbs the collisional channel, because the colliding atoms may tunnel into the molecular state for a short period of time, whose duration is, according to Heisenberg's uncertainty relation, inversely proportional to the energy gap. This time delay modifies the collisional phase shift and hence the scattering length. If the sum of the atomic magnetic dipole moments is different from the dipole moment of the molecule, the energy levels of the atoms and of the molecule can be tuned in opposite directions via external magnetic fields, exploiting the Zeeman effect, and hence can be brought into resonance. When a Feshbach resonance is crossed, the scattering length traverses a singularity (see Fig. 15.2, left). Note that in this simple picture, which only holds within the mean-field approximation, the Feshbach resonance influences the strength of the contact interaction, but not its range.

15.2.3 Imaging and measuring thermodynamic potentials

Gathering information on cold atoms is in most cases done via a technique called absorption imaging. According to the Lambert–Beer law, the intensity of a laser beam

[1] If the condensate is confined in a trap, the finite energy of the trap's ground state exerts a kinetic pressure which succeeds in counterbalancing the destabilizing influence of the interactions provided the condensate is not too big (Dalfovo *et al.* 1999).

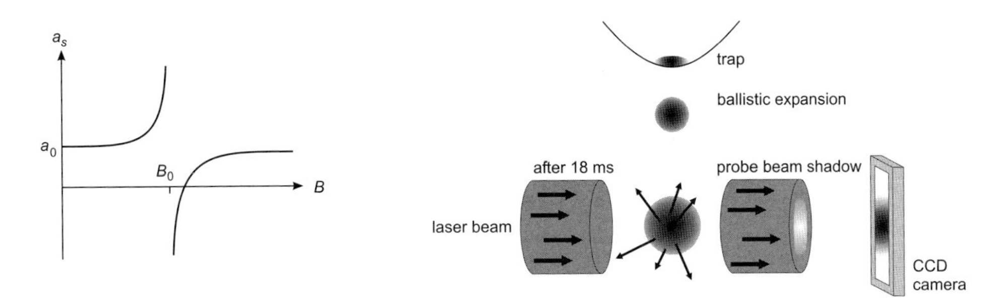

Fig. 15.2 Left: magnetic-field dependence of the interatomic scattering length near a Feshbach collision resonance. Right: time-of-flight absorption imaging of an atomic cloud suddenly released from a trap. The shadow imprinted by the atoms on a laser beam is recorded by a CCD camera.

traversing an atomic cloud decreases exponentially with the column density, i.e. the density integrated over the beam path (e.g. along the z-direction) weighted by the optical cross section, i.e. $I_{\text{out}}(x, y) = I_{\text{in}}(x, y) \exp\left[-\sigma_{\text{oc}} \int n(\mathbf{r})\, dz\right]$. Hence, imaging the shadow imprinted by the atomic cloud on an irradiating probe laser beam yields two-dimensional images (in x and y) of the density distribution.

The size of trapped clouds is often too small to be resolved by optical microscopes. Also, the cloud is optically dense and thus opaque for resonant laser light. This means that $n(\mathbf{r})$ is difficult to measure. However, we can suddenly switch off the trap, wait a few milliseconds, and then take a picture (see Fig. 15.2, right). In the course of the ballistic expansion, the potential energy and the self-energy are transformed into kinetic energy, the size of the cloud increases, and the optical density rapidly drops. The density distribution $n(\mathbf{r}, t_f)$ recorded after a time of flight t_f maps the initial momentum distribution $n(\mathbf{p})$. Note that the imaging process destroys the sample.

Below a critical temperature, which in a harmonic trap is given by the expression $k_B T_c = \hbar\omega_{\text{trap}}(N/1.2)^{1/3}$ (Dalfovo *et al.* 1999), many atoms condense to the lowest vibrational state of the trapping potential. Therefore, the Bose-condensed cloud is about 10 times denser than the thermal cloud. When the cloud is suddenly liberated from its trap, its explosion is accelerated by the self-energy, which is much higher for the condensed fraction, and the acceleration is faster in those directions in which the trapping potential was stronger (Castin and Dum 1996). Thus the two fractions separate during ballistic expansion, which allows one to determine the condensed fraction and hence the temperature of the cloud.

Dilute alkali gases are almost ideal systems. Their weak nonideal features become apparent in the behavior of the *thermodynamic quantities* as a function of temperature near the Bose–Einstein phase transition. A repulsive interatomic interaction (Bagnato *et al.* 1987) and finite-size effects, $N < \infty$, slightly reduce the critical temperature (Grossmann and Holthaus 1995). In contrast, for strongly interacting systems such as liquid ^{4}He, it is as difficult to measure the condensed fraction as to calculate the critical temperature.

The classification of Bose–Einstein condensation as a phase transition depends a lot on the behavior of the thermodynamic potentials near the critical point (Landau 1937; Huang 1987). In fact, measurements of the temperature dependence of some thermodynamic quantities such as the energy and the heat capacity (Ensher *et al.* 1996) have shown significant deviations from the ideal-gas behavior and exhibited the occurrence of interaction effects. For ideal confined gases, the equipartition theorem ensures that $E_{\text{kin}} + E_{\text{pot}} = 2E_{\text{kin}}$. For real gases, the repulsive energy of the mean field is added to this energy, i.e. $E = E_{\text{kin}} + E_{\text{pot}} + E_{\text{self}}$. A sudden extinction of the trapping potential takes away nonadiabatically the potential energy E_{pot}. The kinetic energy and the condensate self-energy are totally converted into kinetic energy, the so-called *release energy*, $U = E_{\text{kin}} + E_{\text{self}}$, during ballistic expansion (Holland *et al.* 1997). Hence, the release energy can be measured by calculating the second moments of the density distributions $n(\mathbf{r})$ obtained by time-of-flight absorption imaging, $U = \int d^3\tilde{\mathbf{p}}\; n(\tilde{\mathbf{p}})\; \tilde{p}^2/2m$, where $\tilde{\mathbf{p}} = \mathbf{r}/t_f$ is the atomic momentum after the time of flight.

Time-of-flight imaging gives access to not only the static properties (density, atom number, condensed fraction, ...), but also the dynamic behavior (excitations, superfluidity, ...), the thermodynamic quantities (temperature, critical point, heat capacity, ...), and coherent features (phase, correlations, ...). The condition is, however, that the feature under study is made visible in the density distribution by an appropriate manipulation. For example, vortices can be generated by stirring the BEC with laser beams and detected through their characteristic doughnut-shaped density distribution.

15.3 Long-range forces in atom optics

For the purpose of the subsequent discussion, we classify the way atoms interact with each other into two categories. The first type of interaction is interpreted in terms of collisions, where the electromagnetic forces mutually exerted by the atoms may be understood as being mediated by the exchange of virtual photons. An example of a very long-range force of this category is the Coulomb repulsion of charged atoms, $V_c(R) \propto R^{-1}$. Clouds of up to 10 000 ions can be stored in electrodynamic traps (Bollinger *et al.* 2000; Drewsen *et al.* 2008), whose confinement counteracts the strong interionic Coulomb repulsion. At high temperatures, the ions are in a liquid-like phase. However, the ions can be laser-cooled, and below a certain critical temperature a phase transition to a solid structure, the so-called Wigner crystal, is experimentally observed. Long-range effects are also observed in clouds of neutral but polar atoms, where dipole–dipole interactions come into play, $V_{dd}(R) \propto R^{-3}$, whose range is much longer than that of van der Waals interactions.

The second type of atomic interaction is mediated by the exchange of real photons. This interaction is based on the following principle. An atom emitting a photon receives a photonic recoil. Any other atom which reabsorbs this photon again receives this recoil and is pushed away from the first atom. Since Rayleigh scattering is isotropic, the probability for an atom to receive a photon emitted by a particular atom decreases with the distance between them like R^{-2}. However, if dimensional constraints, for example those which arise when the high cooperativity of optical cavities is made use of, force scattering to occur only into a restricted solid angle, situations can arise where the force is *totally independent of the distance between the atoms.*

15.3.1 Dipole–dipole interactions

Dipole–dipole interactions are not properly long-ranged, but their range is well beyond contact interactions. The dipole–dipole interaction, $V_{dd}(R) \propto R^{-3}(1 - 3\cos 2\theta)$, is anisotropic. At low temperatures, this anisotropy determines global properties of the gas. The geometric shape of the cloud can, for instance, influence the total interaction energy and even the stability of the gas. Heteronuclear molecules can have huge electric dipole moments in deeply bound states, for example 4.2 debye for LiRb. Ultracold fermionic KRb molecules have recently been produced and transferred to low-lying vibrational states (Ospelkaus *et al.* 2003), so that molecular polar gases may soon become available.

Many atomic species (the paramagnetic ones) have, in their ground state, a magnetic dipole moment. The experimental challenge is due to the fact that, despite their short-range, contact interactions overwhelm the very weak dipole–dipole interaction. At the time of writing, a single experiment has been performed with spin-polarized condensed ^{52}Cr (Koch *et al.* 2008). Although this isotope possesses a particularly large dipole moment of $6\mu_B$, it was essential in the experiment to control the contact interaction by means of a Feshbach resonance. As mentioned in Section 15.2, BECs with a negative scattering length are unstable and collapse beyond a certain critical atom number. The dipole–dipole interaction now adds a mean-field energy which, owing to the anisotropy of this interaction, can be positive or negative. For cigar-shaped BECs (with the dipoles aligned along the symmetry axis), this energy is negative, because the dipoles attract each other. For pancake-shaped BECs, this energy is positive. Consequently, collapse is observed for cigar-shaped BECs with a small but *positive* scattering length, while pancake-shaped BECs with a *negative* scattering length are stabilized. The experiment is an impressive demonstration of a feature peculiar to long-range interacting systems: the shape dependence of the system's energy and stability. Similar effects are being studied in solid-state systems in the context of spin ice (see the chapter by Bramwell in this book).

An alternative approach is to *induce dipole–dipole interactions* by laser light. Depending on the particular geometry of the incident laser beams, interparticle potentials proportional to $-R^{-3}$ or even gravitation-like potentials proportional to $-R^{-1}$ can be generated (Giovanazzi *et al.* 2001). These attractive potentials can lead to self-binding of a BEC even in the absence of external trapping potentials, if the kinetic pressure of the zero-point energy and the contact interactions balance the attractive forces and prevent the BEC from collapsing. This dynamics demonstrates impressively how the stimulation of global phenomena in BECs can simulate the dynamics of astrophysical matter aggregations, for example Bose stars.

When the range of an interaction exceeds the mean interparticle distance, long-range pair correlations play a dominant role. They influence the structure factor of the condensate and produce a peak in the Bogoliubov dispersion relation at momentum values corresponding to the typical scale of the density fluctuations. These resonances in the dispersion relation are called *rotons* (O'Dell *et al.* 2003). While they have been observed in the strongly correlated superfluid liquid HeII, gaseous BECs are too dilute with regard to the limited range of s-wave collisions. In the presence of dipole–dipole interactions, however, the appearance of rotons is expected as a clear indication of a dynamics beyond the Bogoliubov mean-field theory.

15.3.2 Photon-mediated long-range interactions

Instabilities in 3D geometries. Photon-scattering-mediated forces are observed in magneto-optical traps. An MOT consists of an optical molasses (see Section 15.2.1) supplemented by a magnetic quadrupole field, which introduces a spatially dependent Zeeman shift giving rise to a spatially dependent restoring force. This results in spatial confinement of up to $N = 10^{10}$ atoms at densities of up to 10^{11} cm^{-3}.

A characteristic feature of MOTs is that for large sizes or densities, photons are scattered and reabsorbed by other atoms many times before they find their way out of the cloud. The bulk effect of this is an outward-directed radiation pressure force, which tends to blow up the cloud's size, until the Zeeman shift at its edge compensates the detuning of the cooling laser. When the threshold is crossed, the friction changes sign: the motion of atoms at the MOT edge is then accelerated instead of being damped, and these atoms are expelled outwards, decreasing the density of the cloud. Once the atomic density has decreased enough, multiple-scattering effects vanish and the MOT resumes its usual, low-density behavior. This in turn increases the penetration depth of the cooling laser beams and thus the restoring force pushing the atoms back to the center of the trap. Consequently, the density increases again, and the whole cycle starts over. Beyond a certain critical atom number, the signal suddenly starts to oscillate (Labeyrie *et al.* 2008). The dynamics is similar to what happens in pulsating stars, where the instability concerns an outer layer of the stellar envelope, whose opacity varies and determines the impact of radiation pressure. The instability results from a self-synchronization of the trajectories of an important fraction of the atoms.

Instabilities in 1D geometries. In some cases, long-range forces mediated by photon scattering produce anisotropic instabilities, which automatically reduce the dimensionality of the system by self-organization. This can happen, for example, when the shape of the atomic cloud is anisotropic (Inouye *et al.* 1999) or when the light fields interacting with the atoms are confined in an optical ring cavity (Kruse *et al.* 2003) (see also the chapter by Robb in this book).

Let us consider an atomic cloud irradiated from one direction by a pump laser. As long as the cloud is homogeneous, no light is backscattered, because the backscattered photons have random phases and interfere destructively. If, however, a tiny density fluctuation creates a small temporary inhomogeneity, light is scattered back. Interfering with the pump light, the backscattered probe gives rise to a weak modulation of the light intensity, i.e. a standing-wave fraction forming a periodic dipole potential. The atoms, sensing the dipole forces, are pulled towards the potential valleys, therefore arranging themselves into a periodic pattern, which dramatically increases the backscattering efficiency. This in turn amplifies the tendency of the atoms to self-organize, and so on in a exponential gain mechanism.

In fact, light scattering from a local density fluctuation does not occur exclusively in the backward direction, but in all directions. However, those directions are favored in which the gain is largest, for example because the cloud extends further in these direction (as in the case of superradiant Rayleigh scattering (Inouye *et al.* 1999)) or because cooperativity is enhanced by means of optical cavities (as in the case of collective atomic-recoil lasing (Kruse *et al.* 2003)). Once a mode has won the competition, the periodic atomic pattern suffices to enforce Bragg scattering into only this mode. The exponential gain is thus due to positive feedback between the directional bundling of Bragg-scattered light due to long-range order in the atomic cloud and the self-organized dynamic formation of an atomic lattice.

15.4 Collective atomic-recoil lasing

A paradigmatic example of self-organization in atom optics is CARL, which will be discussed in the remainder of this paper. The heart of the CARL experiments conducted at the University of Tübingen is an optical ring cavity, which consists of three supermirrors (see Fig. 15.3). Its round-trip length is 8 cm, its mode has a waist of 130 μm, and its finesse is 80 000. The cavity sits inside a vacuum vessel. A titanium–sapphire laser is phase-matched to one of the two cavity modes, generating a running wave inside the ring cavity. The intracavity power of about 1 W is obtained with only a few hundreds of microwatts of injected pump power. The light reflected from the cavity carries information about phase deviations between the laser and the cavity eigenfrequencies, which are utilized to stabilize the laser with respect to the cavity via a Pound–Drever–Hall-type servo control. The residual frequency fluctuations are much less than the 20 kHz linewidth of the cavity.

An experimental sequence starts by producing a gas of rubidium-85 atoms from a dispenser. A fraction of these atoms is then collected with a standard MOT before being transferred into the dipole potential of the ring cavity field. Typically, several million atoms are loaded into the dipole trap at temperatures of a few hundreds of microkelvin. The pump laser is tuned very far, i.e. 1 to 10 THz, to the red of the atomic D_1 resonance. A while after loading is completed and the atoms have found their thermal equilibrium, forming a homogeneous cloud, there comes a crucial step: the MOT laser beams are switched on again, but without the MOT's quadrupole magnetic field. In that way, an optical molasses is formed, in which the atoms move as if in a viscous fluid. The optical molasses thus induces a velocity-dependent friction force.

The experiment yields various signals. The field evolution is monitored via the

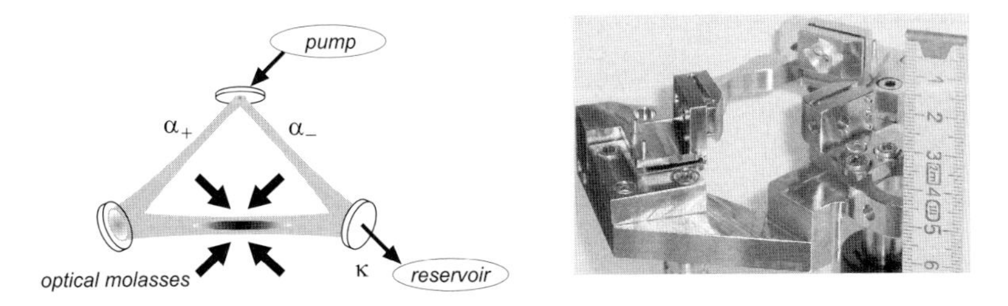

Fig. 15.3 Left: schematic drawing of laser-pumped ring cavity. The light modes $\alpha_\pm$ decay via transmission through the cavity mirrors into the reservoir of vacuum modes. An optical molasses exerts friction forces. Right: photograph of the cavity used in the experiments.

interference signal obtained by phase-matching the two counterpropagating ring cavity modes on a photodetector. Any phase shift of one beam with respect to the other translates into interference fringes in the photodetector signal. Spatial displacements of the atomic cloud are detected via time-of-flight absorption imaging. The imaging also gives access to the cloud's temperature.

15.4.1 CARL with thermal clouds

Figure 15.4(a) shows the time evolution of the beat signal. Initially, the signal only fluctuates slightly. However, as soon as the optical molasses is created by irradiation, strong oscillations with a fixed frequency appear, which persist for more than 100 ms. Absorption pictures reveal a net displacement of the atomic cloud even after a few milliseconds in the direction of the pump light.

The observations can be understood in terms of the self-organization process explained in Section 15.3.2. Backscattered light interferes with the pump light to form a standing-wave dipole potential. Since the scattering atoms are accelerated owing to photon recoil in the direction of the pump beam, the backscattered light is Doppler-shifted to the red of the pump light frequency, so that the standing wave propagates in the same direction. The time-varying interference resulting from the wave propagation causes the oscillations observed in the photodetector signal. The standing wave and the atoms drag and accelerate each other. The atoms behave as if they are surfing down a light wave which they have *created themselves*, and if no other forces are applied, the backscattered light frequency deviates more and more from the cavity resonance. This inhibits resonant enhancement of the probe beam and weakens the dipole potential. As a consequence, the accelerated atoms redistribute in phase space and the fluctuation disappears. However, if an optical molasses is now created, the atoms are slowed down and the Doppler shift of the probe beam is bounded. The propagation velocity corresponds to an equilibrium between the acceleration force exerted by the coherent backscattering and the velocity-dependent friction force exerted by the molasses (Cube *et al.* 2004). The atoms arrange themselves in a periodic lattice and synchronize their velocities.

Threshold behavior. This experiment demonstrates that, by a process mediated by collective atomic recoil, laser light can be generated. At the same time, the atoms

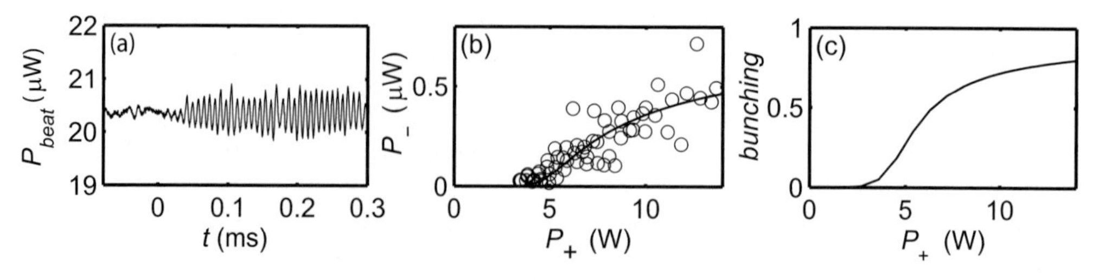

Fig. 15.4 (a) Sudden appearance of a pump–probe laser beat signal after irradiation of an optical molasses at time $t = 0$. (b) Measured and simulated threshold behavior of CARL with friction. (c) Simulation of the corresponding bunching parameter.

self-organize into a periodic lattice, while breaking translation symmetry. The role of the molasses in this process is twofold. On one hand, via dissipation of energy, it permits the system to reach a steady state. On the other hand, atomic momentum diffusion processes, which are intrinsically connected to optical molasses, limit the dissipative cooling at a specific equilibrium temperature. The interplay of dissipation and diffusion governs the thermodynamic phase transition, and should give rise to a threshold behavior of the CARL radiation. This is indeed what is observed in the experiment: a minimum pump laser power is needed to trigger the CARL and obtain probe laser emission, as can be seen from Fig. 15.4(b). The solid line shows a Fokker–Planck simulation (see below). These simulations also allow us to calculate an order parameter measuring the bunching of the atoms. The order parameter exhibited in Fig. 15.4(c) disappears below the threshold pump power, and approaches unity above threshold.

Obviously, the threshold should depend on some control parameters. Experiments have indeed verified that the threshold increases when the coupling strength is decreased, i.e. when the pump laser is detuned further away from the atomic resonance or when the atom number is decreased (Cube *et al.* 2004).

The CARL model. With N atoms in the cavity, the dynamics of the CARL can be described with $2N+1$ coupled degrees of freedom. The two counterpropagating cavity modes $\alpha_\pm$ are labeled by their electric field amplitude normalized to the amplitude generated by a single photon. We may, however, neglect depletion of the pump mode and set $\alpha_+ = \alpha_+^*$ and $\dot{\alpha}_+ = 0$. The modes can be considered as classical light fields because of the large cavity mode volume. The atomic excitation can be adiabatically eliminated in the limit of very large detunings of the pump laser from atomic resonances $\Delta_a \gg \Gamma$. The motional state of the atoms is described by their positions x_n scaled to the optical wavelength, $\theta_n \equiv 2kx_n$.

The CARL model consists of two equations of motion (Bonifacio and Salvo 1994),

$$\begin{aligned} \dot{\alpha}_- &= -\kappa\alpha_- - iNU_0 b\alpha_+ \,, \\ \ddot{\theta}_n &= -2i\omega_r U_0(\alpha_+\alpha_-^* e^{i\theta_n} - \alpha_+^*\alpha_- e^{-i\theta_n}) + \gamma_{\mathrm{fr}}\dot{\theta}_n + \gamma_{\mathrm{fr}}\xi_n(t) \,. \end{aligned} \tag{15.1}$$

The first equation, describing the evolution of the probe field α_-, is easily derived from a self-consistent treatment of the cavity modes: the change in the field amplitude after one round trip can be related to gain and losses. Losses occur via transmission of photons through the cavity mirrors at a damping rate κ. The probe mode is refilled by external pumping mediated by the presence of atoms at given positions, which shuffle photons from the pump into the probe mode. The shuffling is more efficient the larger the atomic bunching is, which is measured by the order parameter $b \equiv (1/N)\sum_m e^{i\theta_m}$ (also called the Debye–Waller factor). The presence of an atomic coordinate in the coupling term emphasizes the role of recoil in the scattering process. The field-to-field coupling strength (also called the single-photon light shift) $U_0 = g^2/\Delta_a$, which measures the cooperativity of the system, goes as the square of the atom–field coupling strength g, which has the meaning of a one-photon Rabi frequency.

The second equation governs the motional dynamics of the atoms, which can be calculated from the optical potential $\phi(X)$ knowing the field amplitudes. The dipole force $\ddot{\theta}_n = -\nabla_{X=x_n}\phi(X)$ has a simple interpretation in terms of the momentum recoil associated with photon-scattering processes between the cavity modes, for example when a photon is destroyed in the probe mode and recreated in the pump mode, or vice versa. $\omega_r \equiv 2\hbar k^2/m$ is the recoil shift. By exposing the atoms to laser cooling via an optical molasses, we exert a friction force with a friction coefficient γ_{fr}. The influence of the optical molasses is simply accounted for by an additional friction term in the equation of motion. But the molasses also introduces noise, giving rise to a random-walk motion of the atoms. This is accounted for by a Langevin force $\xi_n(t)$. The equations can be solved either by numerical simulation of the Langevin equation or by a Fokker–Planck treatment.

15.4.2 CARL and the Kuramoto model

Self-synchronization is generally driven by a feedback interaction between a macrosystem and an ensemble of microsystems. A macroscopic order parameter defines the boundary conditions for microscopic processes. On the other hand, these microscopic processes can, provided they act collectively, influence and shape the order parameter. This global feedback can give rise to instabilities and to long-range order. In the case of CARL, the order parameter is a bunching of atoms, the microscopic processes are light-scattering events, and the long-range order shows up as lasing and a simultaneous formation of an atomic lattice (see Fig. 15.5, left).

The simplest class of self-organizing systems, an ensemble of weakly and uniformly coupled limit cycle oscillators, is represented by the so-called Kuramoto model (Kuramoto 1984). Even when the eigenfrequencies of the oscillators are slightly different, a phase transition to a synchronized state takes place. However, if the coupling is weak, a critical number of oscillators is needed to perform the phase transition in a limited amount of time (see the chapter by Pikovsky in this book).

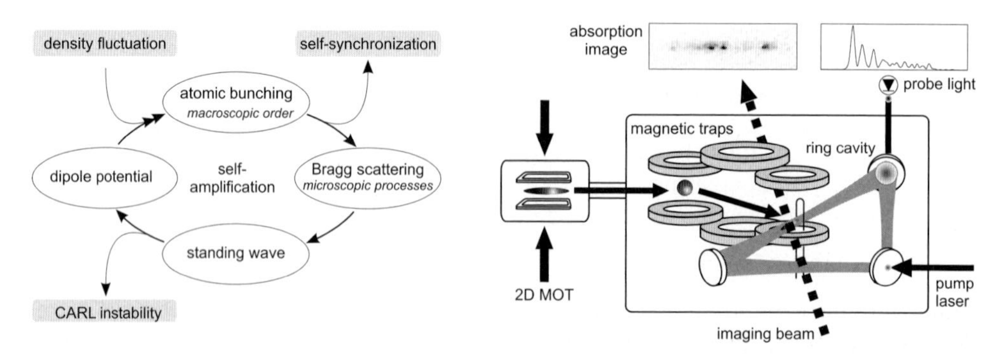

Fig. 15.5 Left: the feedback interaction of a macroscopic order parameter (atomic bunching) and an ensemble of microsystems (light-scattering processes) gives rise to the self-amplification at the origin of the CARL instability. Right: experimental setup for CARL on ultracold atoms, showing the 2D MOT, the current-carrying coils for the generation of magnetic trapping potentials, the ring cavity with its pump laser, the probe laser detector, and the absorption imaging system.

The correspondence between CARL and the Kuramoto model is drawn from the interpretation of the atomic positions, modulo the period of the standing light wave, as the phases of oscillators. The bunching of atoms then corresponds to a synchronization of oscillators. To see the analogy, we simplify the second CARL equation by neglecting the inertial term $\ddot{\theta}_n \equiv 0$, which is certainly a good assumption once the system has reached its steady state:

$$\dot{\theta}_n = i\omega_r U_0 \alpha_+ \gamma_{\text{fr}}^{-1}(\alpha_-^* e^{i\theta_n} - \alpha_- e^{-i\theta_n}) - \xi_n(t)\,. \tag{15.2}$$

Furthermore, in the steady state the amplitude of the probe field is constant, and only its phase propagates at constant velocity. This allows us to introduce a characteristic frequency ω_{ca} defined by $\dot{\alpha}_- = i\omega_{ca}\alpha_-$. This frequency, which corresponds to the frequency shift between the pump and probe beams and to the Doppler shift of the moving atoms, is (in the steady state) experimentally found to be much larger than the cavity decay rate, i.e. $\kappa \ll \omega_{ca}$, so that the first CARL equation becomes

$$\alpha_- \simeq \frac{-NU_0\alpha_+ b}{\omega_{ca}}\,. \tag{15.3}$$

Inserting this into the equation for the atomic phases (15.2) and introducing the Kuramoto coupling constant $K \equiv 2\omega_r N U_0^2 \alpha_+^2/\gamma_{\text{fr}}\omega_{ca}$, we recover the well-known Kuramoto equation (Cube *et al.* 2004),

$$\dot{\theta}_n = KN^{-1}\sum_m \sin(\theta_n - \theta_m) - \xi_n(t)\,. \tag{15.4}$$

While it is quite interesting that some features of CARL, in particular the self-synchronization behavior, can be mapped to the very simple Kuramoto model, great care should be taken when this is done, because the assumed simplifications do not apply below threshold, so that the CARL synchronization process is not quantitatively described. Furthermore, differently from the Kuramoto model, for CARL the collective frequency ω_{ca} is self-determined, i.e. it depends on parameters such as the coupling strength and the number of coupled oscillators. A thorough treatment of the CARL–Kuramoto correspondence can be found in Javaloyes *et al.* (2008).

15.4.3 CARL with condensates

Despite the fact that the CARL model deals with atoms, it describes a purely classical system. It would work with billiard balls, if a suitable coupling force were at hand. Hence, while the CARL system exhibits a dynamics which can give rise to very large-scale instabilities, the synchronizing particles are microscopic. Therefore, CARL ought to exhibit quantum features in some circumstances, in particular at ultralow temperatures, where a single photonic recoil noticeably accelerates an atom or when quantum statistics come into play. From this fact arise a variety of questions, which have just started to be addressed with a second generation of experiments carried out at Tübingen.

Experiment. Temperatures below the recoil limit are being reached with a completely new apparatus (Slama *et al.* 2007*a*,*b*), a schematic illustration of which is shown in Fig. 15.5 (right). A two-dimensional MOT provides a cold beam of ^{87}Rb atoms feeding a standard MOT sitting in an ultrahigh-vacuum chamber. The chamber contains a ring cavity and everything that is necessary to create a cold-atom cloud. From the MOT, the atoms are loaded, via intermediate quadrupolar magnetic fields, into a Ioffe–Pritchard-type microtrap located near the mode volume of a high-finesse ring cavity. Here, the atoms are evaporatively cooled to a few times 100 nK and overlapped with the cavity mode volume. Finally, a pump laser, which is red-detuned by about 1 THz from the atomic D_1 line, is suddenly irradiated into one of the two counterpropagating cavity modes. The reaction of the reverse cavity mode to this pumping is observed. Time-of-flight absorption images additionally map the atomic momentum distribution.

The main observation is the emission of superradiant bursts of light by the condensate into the reverse cavity mode. The time evolution of the emitted light is shown in Fig. 15.6(a) for a variety of temperatures of the atomic cloud. The height of the first peak, which appears to be very robust to perturbations, for example from spurious backscattering from imperfections on the mirror surfaces, was found to increase with pump laser power. Furthermore, the time delay before superradiance sets in was found to diminish, because seeding occurs faster with higher laser power.

We found that the purely classical CARL theory reproduces the observed superradiant light emission almost perfectly. In contrast, the classical model must fail to reproduce the discreteness of the momentum distribution observed in the absorption images taken after a time of flight and shown in Fig. 15.6(b) and (c). It is clear that the collective coupling splits the condensate into a distribution of higher-momentum states, all separated by $2\hbar k$. If the cloud is hotter than the recoil limit, which in our case corresponds to temperatures above the critical point, the momentum distribution is smeared out and the individual momentum states are indistinguishable, as seen in Fig. 15.6(d).

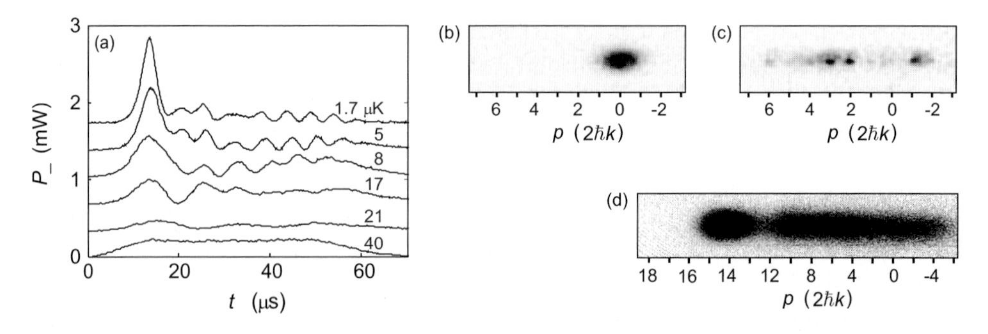

Fig. 15.6 (a) Measured time signal of the probe light power for different atomic temperatures. For clarity, the curves have been shifted by 0.35 mW from each other. (b) and (c) show the momentum distributions obtained via time-of-flight imaging before and after the CARL dynamics, respectively. (d) shows a momentum distribution obtained with 10 μK hot atoms.

CARL and superradiant Rayleigh scattering. The observed superradiant ringing is due to the same CARL instability as described in Section 15.4.1 for thermal clouds. However, in the absence of external friction forces no stationary state is reached, and the superradiant ringing vanishes after a finite coherence time. As long as the interaction is weak, the atoms scatter photons individually and are not influenced by the scattering of other atoms. This is even true for BECs, where all atoms are delocalized over the same region of space. If the atoms do not show long-range order, such individual scattering is manifested in a scattered light intensity which is proportional to the number of scatterers. In contrast, if the interaction is strong enough, the individual scattering events can be synchronized such that all atoms contribute in a cooperative way. Scattering then evolves as a global dynamics of the whole cloud, with the scattered intensity being enhanced by the phase-coherent emission of light fields and depending on the number of scatterers with a power law larger than one.

Another instability, which is closely related to CARL, has been found by Ketterle's group of (Inouye *et al.* 1999) and is termed *superradiant Rayleigh scattering.* In that experiment, a BEC was illuminated perpendicularly to its long axis by a short pulse of polarized laser light, which resulted in the spontaneous formation of momentum sidemodes, visible in time-of-flight images. This is due to the following gain process. A first scattering event produces a photon and a recoiling atom. The atom forms, together with the remaining part of the condensate, a standing matter wave, from which subsequent photons are Bragg-scattered in the same direction as the first photon. This self-amplifies the standing-wave contrast and the light intensity in the scattered optical mode.

SRyS and CARL share the same gain mechanism. The difference between the instabilities comes from the presence of an optical cavity in the latter system, whose main role is to enhance the coherence lifetime. Obviously, SRyS requires a stable, self-generated matter wave grating. Thermal motion blurs the grating and reduces the coherence time, simply because Rayleigh scattering is Doppler-broadened. Therefore, SRyS requires subrecoil temperatures, i.e. $k\sqrt{k_BT/m} < \omega_r$, which is the reason why it has until now been observed only in BECs.

For CARL, the presence of a cavity reconcentrates the spectrum of Rayleigh-scattered light by restricting the density of final states into which photons can be scattered to a narrow interval, $\kappa < \omega_r$, around the cavity modes. This spectral narrowing dramatically increases the lifetime of the system to a value limited only by the finesse of the cavity, and not by atomic motion. Consequently, CARL instabilities have been observed for temperatures as hot as 40 μK, as shown in Fig. 15.6(a), even though the contrast of the superradiant ringing fades out at higher temperatures. An interesting consequence of the observability of CARL with thermal atoms is the proof that superradiance is not conditional on quantum statistics.

CARL and SRyS are both collective phenomena, which is demonstrated by the fact that the intensity of the scattered light scales with the atom number as N^r, where $r > 1$. However, while for superradiance the intensity is proportional to N^2, for CARL, where the coherence is stored mainly in the optical modes, the intensity is proportional to $N^{4/3}$. These dependences have been verified by experiment (Slama *et al.* 2007*a*).

15.5 Conclusion

The advent of atom optics represents a unique opportunity to study features of and models developed for long-range interacting systems. The arguments and experiments presented in this chapter hopefully will convince the reader that the statistical mechanics of dilute gases is far from trivial, but shows a wealth of surprising phenomena and raises many questions.

The unprecedented purity of atom-optical systems and the perfect degree of control over their degrees of freedom has already led to the experimental observation of weak long-range effects, such as collapse induced by dipole–dipole interaction and spontaneous self-synchronization of atoms in light fields. The collective atomic-recoil laser, which I have discussed with special emphasis, is a paradigmatic example of synchronization in atom optics. It is governed by a collective behavior of atoms inside a high-finesse ring cavity and leads at the same time to the spontaneous arrangement of atoms into a periodic density grating and to the emission of superradiant bursts of light.

Atomic gases can be cooled to temperatures below the recoil limit, where the impact of light on the atomic motion can only be adequately described by quantum mechanics. Furthermore, at ultralow temperatures, the quantum statistical nature of the atoms, i.e. bosons or fermions, has an important impact on the dynamics of long-range interacting systems, as seen in the example of dipole–dipole interactions in Bose–Einstein condensates of chromium.

The experiments performed at Tübingen were the first to load BECs into optical cavities and to observe cavity-induced dynamics in BECs. However, in the collective dynamics of CARL, quantum statistics and quantum degeneracy do not play a role. The main motivation to extend the studies of CARL to the quantum regime is the enhancement of the coherence time of the collective dynamics, which may then allow the generation of entangled states between atoms and photons (Piovella *et al.* 2003). The prospects for studying the synchronization phenomenon itself are also very fascinating, in particular the impact of quantum noise on the synchronization process and the interplay of dynamical stabilization and quantum chaos (Zhirov and Shepelyansky 2006).

Obviously, cavities have a huge impact on light scattering. This feature can be used to introduce dissipative forces acting on atoms. Consequently, cavities have potential applications for new types of cooling, in particular of species that are difficult to cool by other means, for example molecules. Cavity dissipation may also be applied to damp collective excitations in superfluids in a controlled way, and it is even possible to merge independent BECs by mediation of a cavity (Jaksch *et al.* 2001).

References

Bagnato V. S. *et al.* (1987). *Phys. Rev. A*, **35**, 4354.
Bollinger J. J. *et al.* (2000). *Phys. Plasmas*, **7**, 7.
Bonifacio R. and Salvo L. D. (1994). *Nucl. Instrum. Methods*, **341**, 360.
Castin Y. and Dum R. (1996). *Phys. Rev. Lett.*, **77**, 5315.
Courteille P. W. *et al.* (1998). *Phys. Rev. Lett.*, **81**, 69.
Courteille P. W. *et al.* (2001). *Laser Phys.*, **11**, 659.

Cube C. V. *et al.* (2004). *Phys. Rev. Lett.*, **93**, 083601.
Dalfovo F. *et al.* (1999). *Rev. Mod. Phys.*, **71**, 463.
Drewsen M. *et al.* (2008). *AIP Conf. Proc.*, **970**, 295.
Ensher J. R. *et al.* (1996). *Phys. Rev. Lett.*, **77**, 4984.
Feshbach H. (1958). *Ann. Phys.*, **5**, 357.
Frisch O. R. (1933). *Z. Phys.*, **86**, 42.
Giovanazzi S. *et al.* (2001). *Phys. Rev. A*, **63**, 31603.
Grossmann S. and Holthaus M. (1995). *Phys. Lett. A*, **208**, 188.
Holland M. J. *et al.* (1997). *Phys. Rev. Lett.*, **78**, 3801.
Huang K. (1987). *Statistical Mechanics*, Wiley.
Inouye S. *et al.* (1999). *Science*, **285**, 571.
Jaksch D. *et al.* (2001). *Phys. Rev. Lett.*, **86**, 4733.
Javaloyes J. *et al.* (2008). *Phys. Rev. E*, **78**, 11108.
Koch T. *et al.* (2008). *Nature Phys.*, **4**, 218.
Kruse D. *et al.* (2003). *Phys. Rev. Lett.*, **91**, 183601.
Kuramoto Y. (1984). *Prog. Theor. Phys. Suppl.*, **79**, 223.
Labeyrie G. *et al.* (2008). *AIP Conf. Proc.*, **970**, 303.
Landau L. D. (1937). *Course of Theoretical Physics*, Vol. 5, *Statistical Physics*. Butterworth-Heinemann.
O'Dell D. *et al.* (2003). *Phys. Rev. Lett.*, **90**, 110402.
Ospelkaus S. *et al.* (2003). arXiv:0802.1093.
Parkins A. S. and Walls D. F. (1998). *Phys. Rep.*, **303**, 1.
Piovella N. *et al.* (2003). *Phys. Rev. A*, **67**, 013817.
Slama S. *et al.* (2007*a*). *Phys. Rev. Lett.*, **98**, 053603.
Slama S. *et al.* (2007*b*). *Phys. Rev. A*, **75**, 063620.
Stamper-Kurn D. M. and Ketterle W. (2000). *Proc. Les Houches Summer School*, Session LXXII.
Zhirov O. V. and Shepelyansky D. L. (2006). *Eur. J. Phys. D*, **38**, 375.

16
Collective instabilities in light–matter interactions

Gordon R. M. Robb

Scottish Universities Physics Alliance (SUPA),
Department of Physics, University of Strathclyde,
107 Rottenrow, Glasgow G4 0NG, United Kingdom

Fig. 16.1 A schematic diagram showing light amplification due to spatial self-organization of particles.

16.1 Introduction

The interaction between light and matter is a fundamental topic of classical and modern physics, covering a vast number of subdisciplines, for example spectroscopy, lasers, optical cooling, and particle acceleration. During an interaction involving light and an ensemble of particles, the particles can "communicate" via common scattered radiation fields. This indirect particle–particle interaction can produce a variety of *collective* interactions where the system emits or scatters coherently owing to a process of spatial self-organization or bunching, shown schematically in Fig. 16.1. The resulting behavior is very different from the single-particle regime, where the particles emit or scatter independently and incoherently.

From a general point of view, collective light–matter interactions are of considerable interest as examples of the collective or cooperative behavior in a many-body system that arise in a range of disciplines. In addition, collective radiation–particle interactions are of considerable practical interest owing to the evolution of the radiation fields and the particle dynamics during the interaction.

In this chapter, I discuss some details of some collective interactions between light and two different types of particles: electrons and atoms. I provide an outline derivation of some of the models used to describe collective interactions involving these media and some of the predictions of these models.

16.2 The free-electron laser (FEL)

16.2.1 FEL spontaneous emission

The high-gain free-electron laser (FEL) is an example of a collective interaction between light and relativistic free electrons. In an FEL amplifier, the beam of relativistic electrons passes through a magnetostatic, spatially periodic "wiggler" field to produce electromagnetic (EM) radiation, as shown schematically in Fig. 16.2. Free-electron lasers have a wide range of applications (spectroscopic, medical, etc.) principally because they are highly tunable, with the ability to produce coherent EM radiation across a broad range of the EM spectrum, from microwaves to hard X-rays. In addition to their wide tunability, they are also capable of producing very intense radiation, as unlike, for example, laser crystals the lasing medium cannot be damaged by high light intensities. An electron with charge $-e$ moving with a velocity $\vec{v}$ in a magnetic field $\vec{B}$ experiences a force $\vec{F}$ described by

$$\vec{F} = -e\vec{v} \times \vec{B},$$

which accelerates the electron and hence induces the electron to radiate. In a wiggler magnet, the magnetic field is spatially periodic with spatial period λ_w, which produces

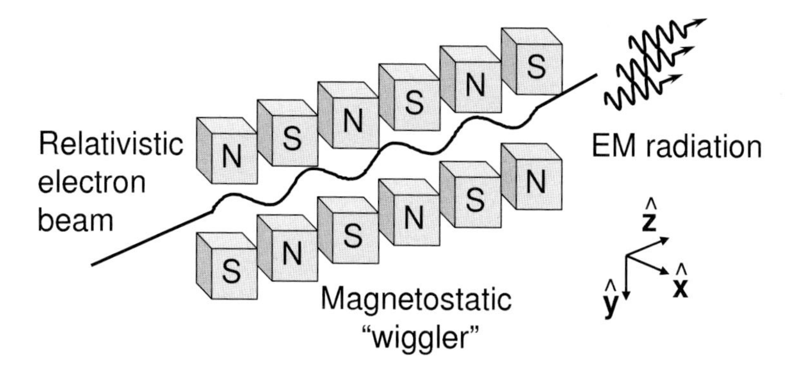

Fig. 16.2 A schematic diagram of a free-electron laser.

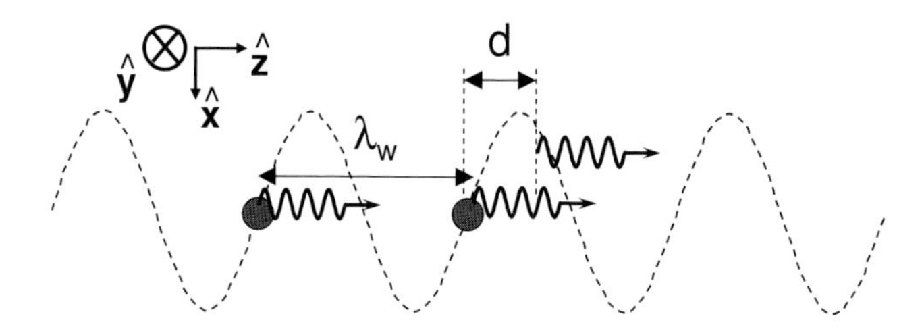

Fig. 16.3 Spontaneous emission in the FEL.

an oscillation with the same spatial period in the trajectory of the electron transverse to its propagation direction. Figure 16.3 shows the oscillatory trajectory of an electron passing through a wiggler magnet. Consider the electron radiating at two points in its trajectory, separated by one wiggler period λ_w. As the light travels faster than the electron, in the time taken for the electron to travel a distance λ_w along the axis of the wiggler, the light will have travelled a distance $\lambda_w c/v_\parallel$, where c is the speed of light in vacuo and $v_\parallel$ is the component of the electron velocity along the wiggler axis. The path difference d between the light emitted by the electron and the light emitted by the same electron one oscillation cycle previously is therefore

$$d = \lambda_w \left(\frac{c}{v_\parallel} - 1 \right).$$

It can be seen from this simple argument that there will be constructive interference if the path difference is equal to the radiation wavelength, i.e. $d = \lambda$, so the radiation wavelength at which spontaneous emission will be maximum is

$$\lambda = \lambda_w \left(\frac{c}{v_\parallel} - 1 \right) = \lambda_w \left(\frac{1 + a_w^2}{2\gamma^2} \right), \tag{16.1}$$

where $\gamma = \left(1 - v^2/c^2\right)^{1/2}$ is the relativistic factor and $a_w = eB_w/\lambda_w m_e c$ is the wiggler parameter, where m_e is the electron rest mass. The FEL resonant wavelength can therefore be tuned by varying one or more of the electron beam energy, the wiggler period, and the wiggler field strength. For long-wavelength FELs generating, for example, microwaves or millimeter waves, relatively low-energy electron beams of $< 1\,\mathrm{MeV}$ are commonly used, whereas for short-wavelength FELs generating, for example, X-rays, very high-energy beams of $> 1\,\mathrm{GeV}$ are necessary (LCLS CR 2002; TESLA TDR 2001).

16.2.2 FEL stimulated emission

Beam-wave energy exchange. Spontaneous FEL radiation is incoherent, as the electrons in the beam are randomly distributed, which leads to a random phase relation in the radiation emitted by them. In order to demonstrate coherent FEL emission, it is necessary to consider the influence of the radiation on the electrons. Consider the case of an electron moving in both a strong magnetostatic, helical wiggler and a circularly polarized electromagnetic wave. The combined electric and magnetic field which the electron experiences is therefore $\vec{E} = \vec{E}_r$ and $\vec{B} = \vec{B}_w + \vec{B}_r$, where the wiggler magnetic field $\vec{B}_w$ is defined as

$$\vec{B}_w = B_w \left(\cos(k_w z)\hat{x} + \sin(k_w z)\hat{y}\right),$$

where $k_w = 2\pi/\lambda_w$ and the radiation fields are defined as

$$\begin{aligned}\vec{E}_r &= -E_r \left(\sin\left(kz - \omega t\right)\hat{x} + \cos\left(kz - \omega t\right)\hat{y}\right),\\ \vec{B}_r &= \frac{E_r}{c}\left(\cos\left(kz - \omega t\right)\hat{x} - \sin\left(kz - \omega t\right)\hat{y}\right),\end{aligned}$$

where $k = 2\pi/\lambda$ is the radiation wavenumber, and $\omega = kc$ is the angular wave frequency. The electric field of the radiation can do work on the electrons and vice versa, so energy exchange between the electrons and the radiation can take place. The rate of change of electron beam energy will be equal to the work done by the electromagnetic radiation on the electrons, i.e.

$$mc^2 \frac{d\gamma}{dt} = -e\vec{E}\cdot\vec{v}. \tag{16.2}$$

Using the fact that canonical momentum is a conserved quantity and the electric field can be expressed in terms of the vector potential $\vec{A}$ as $\vec{E} = -\partial\vec{A}/\partial t$, it is possible to rewrite eqn (16.2) as

$$\frac{d\gamma}{dt} = \frac{e}{2mc^2}\frac{\partial}{\partial t}\left(\vec{A}\cdot\vec{A}\right). \tag{16.3}$$

As $\vec{A}$ consists of a contribution from the wiggler and the electromagnetic radiation, i.e. $\vec{A} = \vec{A}_w + \vec{A}_r$, then from eqn (16.3),

$$\frac{d\gamma}{dt} \propto \frac{\partial}{\partial t}\left(\vec{A}_w\cdot\vec{A}_r\right) \propto \sin\left(\left(k + k_w\right)z - \omega t\right). \tag{16.4}$$

It can be seen from eqn (16.4) that whether an electron gains or loses energy from the EM wave depends on the value of the phase variable

$$\theta = (k + k_w)\, z - \omega t\,.$$

A physical interpretation of this behavior is that the EM wave, with frequency and wavenumber (ω, k), and the magnetostatic wiggler "wave," with frequency and wavenumber $(0, k_w)$, interfere to produce a beat or ponderomotive wave with frequency and wavenumber $(\omega, k + k_w)$. The phase velocity of this ponderomotive wave is

$$v_{ph} = \frac{\omega}{k + k_w}\,,$$

and significant energy exchange between the electrons and the radiation will occur when $v_{\|} \approx v_{ph}$. This resonance condition can be shown (Bonifacio *et al.* 1990; Murphy and Pelligrini 1990) to be directly equivalent to the FEL spontaneous-emission relation in eqn (16.1), i.e.

$$\lambda \approx \lambda_w \frac{\left(1 + a_w^2\right)}{2\gamma^2}\,.$$

The high-gain regime and collective behavior. In the above, the evolution of the electromagnetic field was neglected, and therefore the model described was limited to cases of small beam–wave energy exchange, such that the amplitude and phase of the electromagnetic wave could be assumed approximately constant. In order to describe the beam–wave interaction self-consistently it is necessary to also describe the evolution of the electromagnetic field due to the oscillating electrons. A full derivation of this self-consistent model can be found in Bonifacio *et al.* (1990) and Murphy and Pelligrini (1990), of which only an outline is presented here. The starting point is Maxwell's wave equation

$$\left(\nabla^2 - \frac{1}{c^2}\frac{\partial^2}{\partial t^2}\right)\vec{E}_r = \mu_0 \frac{\partial \vec{J}}{\partial t}\,,$$

where the (transverse) current density $\vec{J}$ arises from the electron motion in the combined wiggler/EM wave and is described in terms of a collection of point charges so that

$$\vec{J} = -e\sum_{j=1}^{N} \vec{v}_j \delta(\vec{r} - \vec{r}_j(t))\,. \tag{16.5}$$

Expressing the electric field in terms of a field envelope which is slowly varying on the timescale and length scale of the EM wave period and wavelength, respectively, allows simplification of Maxwell's equation from second-order to first-order. Consistent with this, we also perform a spatial average over a radiation wavelength to ensure the source of the wave equation is a slowly varying continuous function. The end result is a set of coupled equations which describe the evolution of the electron dynamics (phase and energy) together with the EM field envelope:

$$\frac{d\theta_j}{d\bar{z}} = \bar{p}_j\,, \tag{16.6}$$

$$\frac{d\bar{p}_j}{d\bar{z}} = -\left(\bar{A}e^{i\theta_j} + \text{c.c.}\right), \tag{16.7}$$

$$\frac{d\bar{A}}{d\bar{z}} = \left\langle e^{-i\theta}\right\rangle, \tag{16.8}$$

where $\theta_j = (k + k_w)z - \omega t$ is the ponderomotive phase of electron j, $p_j = (\gamma_j - \gamma_R)/\rho_{\rm F}\gamma_R$, $|\bar{A}|^2 = \epsilon_0 |E_r|^2 / \rho_{\rm F} n_e \gamma_R m c^2$ is the scaled EM field intensity, n_e is the electron density, $\bar{z} = 4\pi\rho_{\rm F} z/\lambda_w$ is the scaled position in the wiggler, and the average in the angle brackets is given by $\langle \ldots \rangle \equiv (1/N)\sum_{j=1}^{N}(\ldots)_j$. Note that these equations have been written in a form involving no free parameters in terms of dimensionless quantities, which involve the dimensionless FEL parameter $\rho_{\rm F}$, defined as

$$\rho_F = \frac{1}{\gamma_R}\left(\frac{a_w \omega_p}{4ck_w}\right)^{2/3} \propto n_e^{1/3}, \tag{16.9}$$

where $\omega_p = \sqrt{n_e e^2/\epsilon_0 m}$ is the plasma frequency. In deriving eqns (16.6)–(16.8), the relative slippage of the EM radiation with respect to the electron beam, which arises from the fact that $v_{\|} < c$ and can give rise to superradiant FEL emission (Bonifacio *et al.* 1989), has been neglected.

Assuming an initial condition corresponding to equally distributed electron phases, a monoenergetic electron beam, and no electromagnetic wave, i.e.

$$\left\langle e^{-i\theta}\right\rangle = 0\,, \quad p_j = 0\,\forall j\,, \quad \bar{A} = 0\,,$$

then a linear stability analysis (Bonifacio *et al.* 1984, 1990; Murphy and Pelligrini 1990) shows that this initial condition is unstable to fluctuations in the EM wave amplitude and the electron phase, i.e. shot noise. A numerical solution of eqns (16.6)–(16.8) with initial conditions

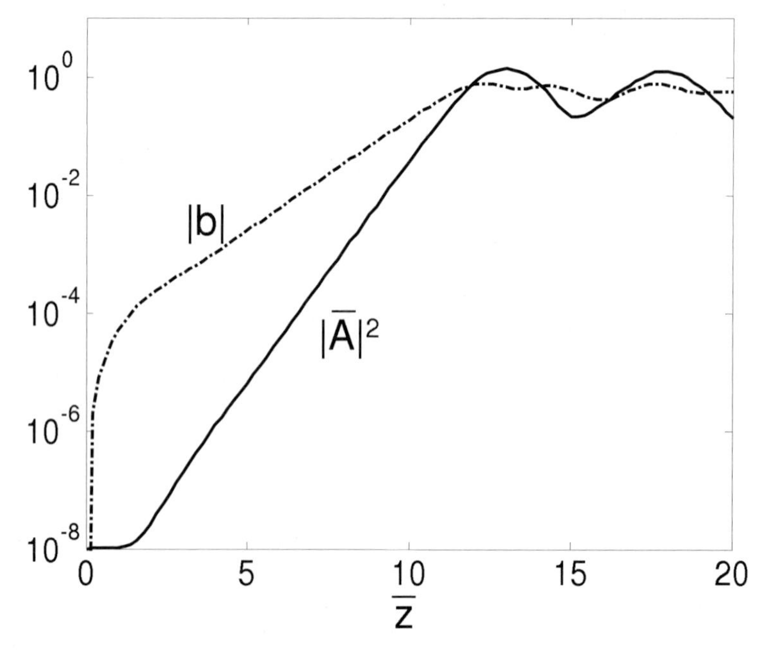

Fig. 16.4 Evolution of the scaled field intensity $|\bar{A}|^2$ and bunching factor $|b| = |\langle e^{-i\theta}\rangle|$ as a function of $\bar{z}$.

$$\langle e^{-i\theta}\rangle = 0\,, \quad p_j = 0\,\forall j\,, \quad |\bar{A}| = 10^{-4}$$

is shown in Fig. 16.4, showing exponential amplification of the EM wave intensity and of the bunching factor $|b| = |\langle e^{-i\theta}\rangle|$ before saturation when $|\bar{A}|^2 \approx 1.4$ and $|b| \approx 0.8$. This high-gain field amplification therefore occurs simultaneously with the development of strong electron bunching on the scale of the radiation wavelength. Note that because eqns (16.6)–(16.8) have no free parameters, this is the only solution for these initial conditions. Consequently, the fact that $|\bar{A}|^2$ always saturates at the same value tells us that the actual EM field intensity at saturation, I_{sat}, is proportional to $|E|^2 \propto \rho_{\mathrm{F}} n_e |\bar{A}^2| \propto N^{4/3}$. This scaling of the EM field intensity as N^{β}, where $\beta > 1$, is indicative of collective behavior and is a consequence of the fact that *all* electrons contribute to the EM field, which induces a coupling between them and allows them to emit collectively.

16.3 Collective atomic-recoil lasing (CARL)

16.3.1 Introduction

The concept of a collective interaction between light and collections of particles is common in beam and plasma physics involving moving charges, but a very different system which exhibits analogous behavior is that of a cold, neutral atomic gas. The field of cold-atom physics has undergone an explosion of interest over the last two decades, driven in part by the effort to realise a Bose–Einstein condensate (BEC). One consequence of this is that we now have a new and highly versatile experimental arena in which to investigate collective light–matter interactions. Usually the interaction between light and atoms is dominated by the internal atomic degrees of freedom, i.e. energy or momentum is exchanged primarily through the evolution of population differences and dipole moments, with the external or centre-of-mass degrees of freedom playing a secondary role, for example Doppler broadening. However, in cold atomic gases, it is possible for the atomic center-of-mass dynamics to take on the primary role in the interaction, and to be the main source of energy/momentum for exchange with electromagnetic waves. Several processes of this type have been considered by various authors and have given rise to a variety of terminology, including collective atomic-recoil lasing (CARL), superradiant Rayleigh scattering (SRS), and recoil-induced resonances (RIR). The similarities and differences between these interactions have been described in, for example, Berman (1999), Robb *et al.* (2005), and Piovella *et al.* (2001*a*). In the following sections, I consider only the CARL model.

16.3.2 Model

The experimental setup usually envisaged when investigating collective interactions between light and cold atoms is shown schematically in Fig. 16.5, with a cold atomic gas enclosed within an optical cavity, in which a strong pump beam and a weak probe/scattered beam with similar frequencies $\approx \omega$ are approximately counterpropagating. The presence of the optical cavity is not a necessary condition for the observation of collective light–matter interactions (Robb *et al.* 2005; Piovella *et al.* 2001*a*),

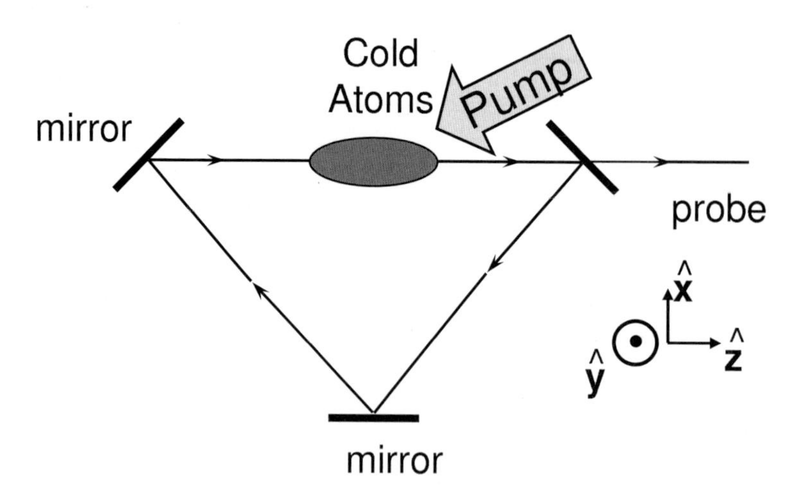

Fig. 16.5 Schematic of a CARL experimental setup.

but it is included here for convenience. The CARL model must self-consistently describe the internal atomic dynamics, the atomic center-of-mass dynamics, and the electromagnetic-wave/optical-field dynamics.

Atomic dynamics. A proper treatment of the internal atomic dynamics requires a quantum description of the relevant internal energy states, resulting in a set of coupled equations describing the evolution of the populations and coherences involved. In the case where the pump and probe have frequencies far-detuned from any atomic resonance, then the atom behaves quasi-classically as a linear dipole oscillator (Allen and Eberly 1987), in which the induced dipole moment is directly proportional to the electric field incident on the atom, i.e.

$$\vec{d} = \alpha \vec{E}\,, \tag{16.10}$$

where α is the polarizability of the atom. Closer to resonance, this approximation breaks down, and the situation is more complex, with both the internal and the center-of-mass dynamics playing significant roles (Robb *et al.* 2001; Bonifacio *et al.* 2000; Perrin *et al.* 2002).

The mechanical effect of light on atoms far from resonance is determined by the optical dipole force. The component of the optical dipole force acting along the cavity axis is given by

$$F_z = \vec{d} \cdot \frac{\partial \vec{E}}{\partial z}\,. \tag{16.11}$$

If, as in the configuration shown in Fig. 16.5, the pump and probe are counterpropagating, then we can express the electric fields involved as

$$\vec{E} = \vec{E}_{\text{pump}} + \vec{E}_{\text{probe}}\,, \tag{16.12}$$

$$\vec{E}_{\text{pump}} = \left(E_{\text{pump}}(t)e^{-i(kz+\omega t)} + \text{c.c.}\right)\hat{x}\,, \tag{16.13}$$

$$\vec{E}_{\text{probe}} = \left(E_{\text{probe}}(t)e^{i(kz-\omega t)} + \text{c.c.}\right)\hat{x}\,, \tag{16.14}$$

where we have assumed that the pump and probe are linearly polarized in the x-direction. Consequently, substituting for $\vec{d}$ and $\vec{E}$ in eqn (16.11) using eqn (16.10) and eqns (16.12)–(16.14), respectively, we obtain

$$F_z \propto \alpha \left(E^*_{\text{pump}}(t) E_{\text{probe}}(t) e^{2ikz} + \text{c.c.} \right), \tag{16.15}$$

i.e. the optical fields combine to produce a dynamic spatially periodic force with period $\lambda/2$.

Field dynamics. The final part of the model is that describing the evolution of the optical field. For simplicity it is assumed that the pump field is sufficiently strong that it is unaffected by interaction with the atoms, and can be considered as a constant parameter. The evolution of the probe field is described by Maxwell's wave equation

$$\left(\nabla^2 - \frac{1}{c^2} \frac{\partial^2}{\partial t^2} \right) \vec{E}_r = \mu_0 \frac{\partial^2 \vec{P}}{\partial t^2},$$

where the source term is a polarization, $\vec{P}$, which arises from the gas atoms, which can be treated as classical point dipoles so that

$$\vec{P} = \sum_{j=1}^{N} \vec{d}_j \delta(\vec{r} - \vec{r}_j(t)),$$

where $\vec{d}_j = \vec{d}(\vec{r} - \vec{r}_j)$ is defined by eqn (16.10). Assuming that the probe field is well described by a single cavity mode, as in eqn (16.14), and performing a slowly-varying-envelope approximation and spatial averaging similar to those performed for the FEL, then the equations describing the coupled atomic center-of-mass dynamics and probe field evolution can be written as

$$\frac{d\theta_j}{d\tau} = \bar{p}_j, \tag{16.16}$$

$$\frac{d\bar{p}_j}{d\tau} = -\left(\bar{A} e^{i\theta_j} + \text{c.c.} \right), \tag{16.17}$$

$$\frac{d\bar{A}}{d\tau} = \left\langle e^{-i\theta} \right\rangle - \bar{\kappa} \bar{A}, \tag{16.18}$$

where now $\theta_j = 2kz$ is the phase of atom j in the optical potential, $\bar{p}_j = mv_j/\hbar k \rho_C$ is the scaled momentum of atom j, $|\bar{A}|^2 = 2\epsilon_0 \left| E_{\text{probe}} \right|^2 / n_a \rho_C \hbar \omega$ is the scaled probe field intensity, n_a is the atomic number density in the cavity, $\tau = \omega_r \rho_C t$ is the scaled time, and $\omega_r = 2\hbar k^2/m$ is the recoil frequency. The parameter describing the CARL interaction, ρ_C, can be shown to be (Bonifacio and De Salvo Souza 1994; Bonifacio *et al.* 1994)

$$\rho_C = \left(\frac{\Omega_0}{2\Delta} \right)^{2/3} \left(\frac{\omega d^2 n_a}{2\hbar \epsilon_0 \omega_r^2} \right)^{1/3} \propto \left(\frac{P_0 n}{\Delta^2} \right)^{1/3}, \tag{16.19}$$

where Ω_0 is the Rabi frequency of the pump field with frequency $\omega = ck$, detuned from the atomic resonance by $\Delta = \omega - \omega_0$, d is the electric dipole moment, and P_0 is the pump power.

The parameter $\bar{\kappa} = \kappa_c/\omega_r\rho_C$ represents scaled cavity losses, where $\bar{\kappa}_c$ is the real cavity linewidth. Note that when $\bar{\kappa} \to 0$, eqns (16.16)–(16.18) become identical to eqns (16.6)–(16.8) used to describe the high-gain FEL, and so for $\tau \ll \bar{\kappa}^{-1}$ the probe field intensity will undergo the same evolution shown in Fig. 16.4, with exponential amplification of the probe field together with amplification of the bunching factor $|b|$. Because of their common behavior, CARL can be considered as an atomic analogue of the FEL. Optical-field amplification occurs at the expense of atomic momentum and the probe field intensity at saturation $\propto N^{4/3}$, indicating collective behavior. Recent experimental results using cold Rb gas trapped in a high-finesse optical cavity (Slama *et al.* 2007) (see also the chapter by Courteille in this book) are well described using this CARL model.

16.3.3 CARL as a model system for investigation of collective phenomena

CARL is of interest as a nonlinear optical phenomenon, but it is arguably of wider interest as an example of spontaneous ordering due to long-range coupling of atoms mediated by an optical "mean field." The versatility of CARL-like experiments involving optical interactions with cold atomic gases makes them a potential testing ground for the study of emergent phenomena in various mean-field or global-coupling models which are commonly used in, for example, plasma physics, condensed matter physics, mathematical biology, and neuroscience. In what follows I will describe cases where modification of the basic CARL setup described in the previous section can give rise to interesting collective phenomena which have direct relevance to effects in quite different systems.

"Viscous CARL" and collective synchronization. The "viscous CARL" setup is shown schematically in Fig. 16.6. The extra ingredient in these experiments with respect to those discussed previously is the presence of "optical molasses" beams, which act to cool the atomic gas. In the presence of these molasses beams, the force on each atom, given by eqn (16.17), must be modified to include a viscous or cooling force, $-\bar{\gamma}\bar{p}_j$, which damps the momentum of each atom, together with a stochastic scattering force with zero mean, $f(\tau)$, due to the random spontaneous emission of photons induced by the molasses beams:

$$\frac{d\bar{p}_j}{d\tau} = -\left(\bar{A}e^{i\theta_j} + \text{c.c.}\right) - \bar{\gamma}\bar{p}_j + f(\tau)\,. \tag{16.20}$$

In the limit of strong viscous damping, it is possible to move from a model of the interaction based on particles and described by eqns (16.16), (16.20), and (16.18) to a distribution function model involving a Fokker–Planck equation for the atomic center-of-mass dynamics coupled to the wave equation, i.e. (Robb *et al.* 2004)

$$\frac{\partial P}{\partial \tau} = \frac{\partial}{\partial \theta}\left[\left(\bar{A}e^{i\theta} + \text{c.c.}\right)P\right] + \bar{D}\frac{\partial^2 P}{\partial \theta^2}\,, \tag{16.21}$$

$$\frac{d\bar{A}}{d\tau} = \int_0^{2\pi} P(\theta,\tau)e^{-i\theta}\,d\theta - \bar{\kappa}\bar{A}\,, \tag{16.22}$$

where $\bar{D}$ is a diffusion coefficient proportional to the temperature T of the atomic gas.

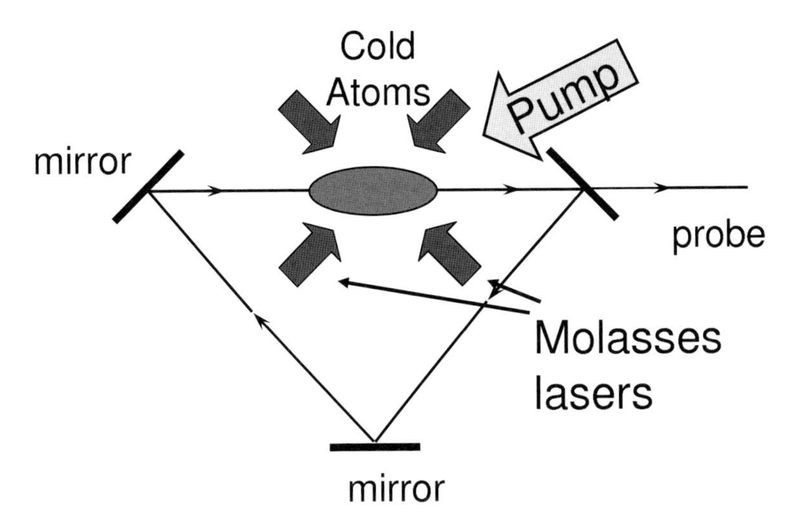

Fig. 16.6 Schematic of the "viscous CARL" experimental setup.

The distribution function $P(\theta, \tau)$ is normalized such that $\int_0^{2\pi} P(\theta, \tau)\, d\theta = 1$ and is periodic in θ with period 2π. Hence, eqns (16.21) and (16.22) can be written in terms of the spatial harmonics of $P(\theta, \tau)$ i.e. $P(\theta, \tau) = (1/2\pi) \sum_{n=-\infty}^{\infty} B_n(\tau) e^{in\theta}$, where $B_n(\tau) = \int_0^{2\pi} P(\theta, \tau) e^{-in\theta}\, d\theta$, so that

$$\frac{dB_n}{d\tau} = in\left(\bar{A}B_{n-1} + \bar{A}^* B_{n+1}\right) - n^2 \bar{D} B_n\,, \tag{16.23}$$

$$\frac{da}{d\tau} = B_1 - \bar{\kappa}\bar{A}\,. \tag{16.24}$$

Note that $B_{-n} = B_n^*$ and $B_0 = 1$. In particular, $b = |B_1| = |\langle e^{-i\theta} \rangle|$ is the bunching factor, describing the amplitude of the density grating.

It was shown in Robb *et al.* (2004) that the viscous CARL model exhibits a threshold behavior such that when κ and D satisfy

$$\bar{\kappa}\bar{D}\left(\bar{\kappa} + \bar{D}\right)^2 < 1\,,$$

the system described by eqns (16.23) and (16.24) exhibits an instability where the scattered field and the bunching factor are amplified exponentially to a steady, ordered state, as shown in Fig. 16.7. This behavior closely resembles that of the Kuramoto model (Kuramoto 1984; Javaloyes *et al.* 2004, 2008; von Cube *et al.* 2004), which has been used to describe collective synchronization between large populations of coupled oscillators as diverse as flashing fireflies and pacemaker cells in the sino-atrial node of the heart (Strogatz 2001).

CARL with a noisy pump field. The interplay between noise and nonlinearity has long been of interest to researchers from a broad range of disciplines, as it produces numerous fascinating and often counterintuitive phenomena. In many complex systems

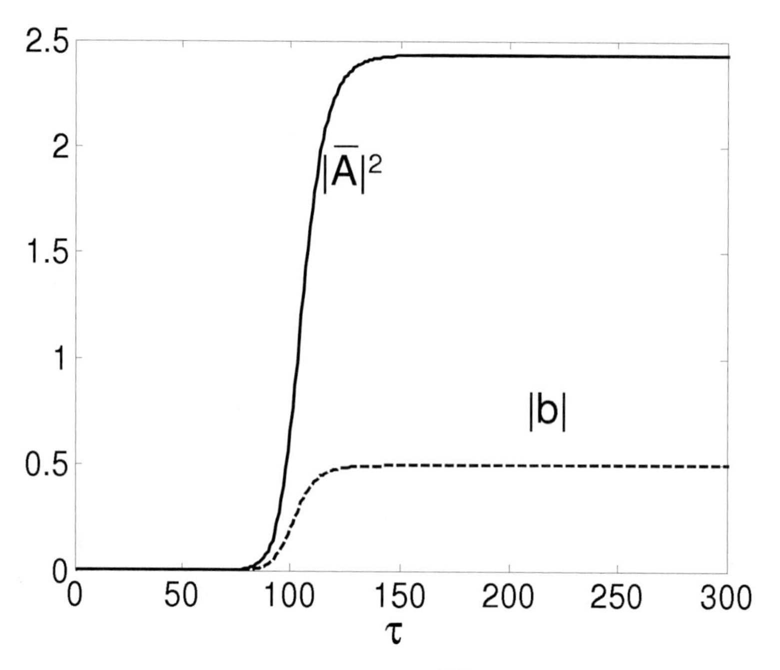

Fig. 16.7 Evolution of the scaled field intensity $|\bar{A}|^2$ and bunching factor $|b|$ as a function of τ when $\bar{\kappa} = 0.075$ and $\bar{D} = 1.5$.

in nature, the environment is intrinsically noisy. Here we study the effect of pump phase noise on the CARL interaction (Robb and Firth 2007).

The situation under consideration is that shown in Fig. 16.5. The difference from the cases considered in previous sections is that whereas the probe field is assumed coherent, the pump field is only partially coherent, its phase executing a random walk. The equations describing the interaction in this case are therefore the CARL equations, eqns (16.16)–(16.18), modified to allow for detuning between the pump and probe fields and to include the effect of a stochastic pump phase ϕ, i.e.

$$\frac{d\theta_j}{d\tau} = p_j \,, \tag{16.25}$$

$$\frac{dp_j}{d\tau} = -\left(\bar{A}e^{i(\theta_j - \phi)} + \text{c.c.}\right), \tag{16.26}$$

$$\frac{d\bar{A}}{d\tau} = \left\langle e^{-i(\theta - \phi)} \right\rangle + i\delta\bar{A}\,, \tag{16.27}$$

$$\frac{d\phi}{d\tau} = \xi(\tau)\,. \tag{16.28}$$

The partial coherence of the pump field is described using a phase diffusion model for the pump field phase ϕ, which is assumed to evolve according to eqn (16.28), where $\xi(\tau)$ is a Gaussian random variable with zero mean and variance Γ such that $\overline{\xi(\tau)} = 0$ and $\overline{\xi(\tau)\xi(\tau - T)} = 2\Gamma\delta(T)$. This corresponds to a pump field with a Lorentzian lineshape and a linewidth (phase diffusion rate) of $\Delta\omega_{\text{pump}} = \omega_r \rho_C \Gamma$.

The effect of pump phase diffusion on the nonlinear regime of the CARL interaction can be observed in Fig. 16.8(b), which shows the evolution of the scaled cavity mode intensity (averaged over 100 runs) for a partially coherent pump field with scaled phase diffusion rate (linewidth) $\Gamma = 5$, as calculated from the stochastic ordinary differential equations (16.25)–(16.27). It can be seen that the value of the cavity mode intensity at saturation $|\bar{A}_{\text{sat}}|^2 \to \delta$ when $\delta > 0$. In contrast to the case of coherent pumping there is no instability threshold in δ, with strong amplification of the probe field for $\delta > 2$. In fact, comparing these results with Fig. 16.8(a), it can be seen that for $\delta \geq 2$ with a partially coherent pump ($\Gamma = 5$), although the growth rate of the field is lower than in the case of the coherent pump, the cavity mode intensity exceeds that attained at saturation for the case of a coherent pump field.

Based on this simple argument, we therefore expect that saturation of the instability will occur when the scaled probe intensity $|A|^2 \approx \delta$. This implies that the real probe intensity I_{sat} is proportional to $n(\omega_2 - \omega_1)$, indicating that the atoms scatter effectively independently when a noisy pump is used. This simple mechanism for amplification is consistent with the results from numerical simulations shown in Fig. 16.8(b).

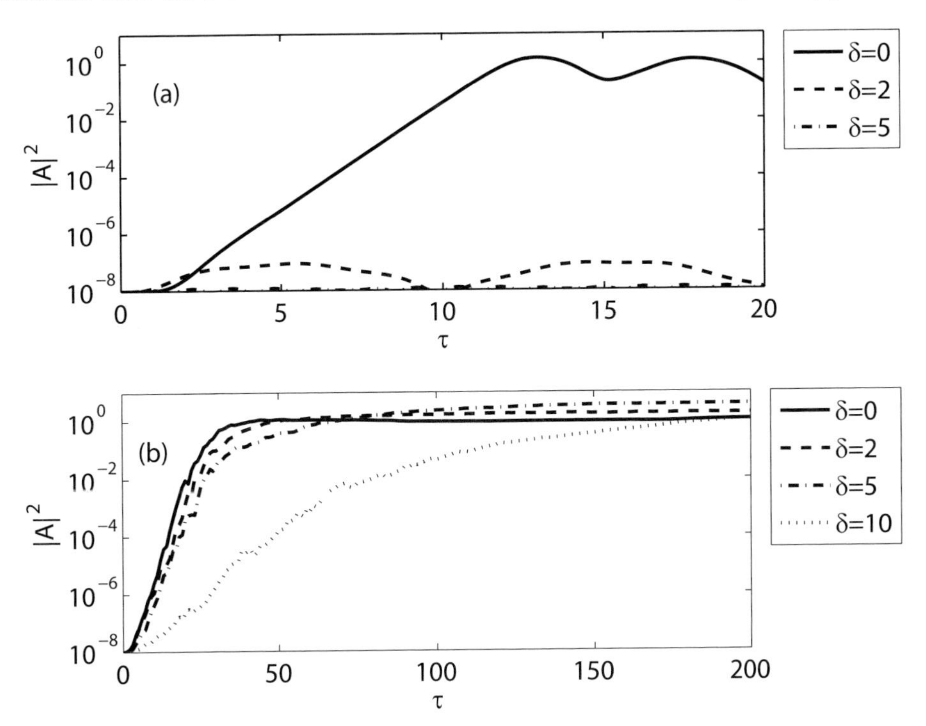

Fig. 16.8 Evolution of the scaled cavity mode intensity $|\bar{A}|^2$ (averaged over 100 runs) due to scattering of (a) a coherent pump ($\Gamma = 0$) and (b) a partially coherent pump ($\Gamma = 5$) by a cold gas for different values of the pump–probe detuning δ.

16.3.4 Quantum CARL

An exciting aspect of collective light–matter interactions involving cold atoms is that they offer the possibility of studying new quantum regimes and effects experimentally. It is now possible to cool atoms to temperatures less than the recoil temperature, $k_B T_R = (\hbar k)^2/2m$, associated with the recoil of a single photon. This condition is usually satisfied by Bose–Einstein condensates, and in what follows I will refer to the atomic gas as a (dilute) BEC. In order to describe a light–atom interaction such as CARL involving atomic gases at subrecoil temperatures, the atoms can no longer be treated as classical point particles, but must be treated as quantum wavepackets. I describe the procedure carried out in Piovella *et al.* (2001*b*) and Bonifacio *et al.* (2005*a*) to describe the collective interaction between light and an ultracold gas. The configuration we consider is as shown for classical CARL, i.e. Fig. 16.5. Using the atomic equations of motion, eqns (16.16) and (16.17) define a single-particle Hamiltonian which can be used to construct a Schrödinger equation for the single-particle wavefunction $\Psi(\theta, \tau)$ so that the equations describing the system are

$$i\frac{\partial \Psi(\theta,\tau)}{\partial \tau} = -\frac{1}{\rho_C}\frac{\partial^2 \Psi(\theta,\bar{t})}{\partial \theta^2} - \frac{i\rho_C}{2}\left[\bar{A}(\tau)e^{i\theta} - \text{c.c.}\right]\Psi(\theta,\tau)\,, \tag{16.29}$$

$$\frac{\partial \bar{A}}{\partial \tau} = \int_0^{2\pi} |\Psi(\theta,\tau)|^2 e^{-i\theta}\, d\theta + i\delta\bar{A}(\tau)\,, \tag{16.30}$$

where the classical average of eqn (16.18) has been replaced by a quantum mechanical ensemble average in eqn (16.30), and $\delta = (\omega_{\text{pump}} - \omega_{\text{probe}})/(\omega_r \rho_C)$ is the scaled detuning between the pump fields and the probe field. If the condensate is much longer than the generated radiation wavelength, then it can be assumed that the wavefunction $\Psi(\theta, \tau)$ is approximately periodic in θ with period 2π and so it can be expanded in a Fourier series:

$$\Psi(\theta,\tau) = \sum_{n=-\infty}^{\infty} c_n(\tau)e^{in\theta}\,. \tag{16.31}$$

By inserting eqn (16.31) into eqns (16.29) and (16.30), we obtain

$$\frac{dc_n}{d\tau} = -\frac{in^2}{\rho_C}c_n - \frac{\rho_C}{2}\left(\bar{A}c_{n-1} - \bar{A}^* c_{n+1}\right), \tag{16.32}$$

$$\frac{d\bar{A}}{d\tau} = \sum_{n=-\infty}^{\infty} c_n c_{n-1}^* + i\delta\bar{A}\,. \tag{16.33}$$

Equations (16.32) and (16.33) are the equations with which the remainder of this chapter is concerned. Note that from eqn (16.31) it follows that $|c_n|^2$ is the probability that a particle has a momentum $p = 2\hbar k n$, $b = \sum_{n=-\infty}^{\infty} c_n c_{n-1}^*$ is the bunching parameter, and $\langle p \rangle = \sum_{n=-\infty}^{\infty} n|c_n|^2$ is the average momentum of the atomic sample. Initially, we assume a quasi-equilibrium state with a very weak sum-frequency field, $|\bar{A}| \ll 1$, and all the particles at rest, i.e. in the state $n = 0$, with $|c_0|^2 = 1$ and $|c_m|^2 = 0$ for all $m \neq 0$. This is equivalent to assuming that the temperature of the system is zero initially.

Results from the quantum CARL model. It was shown in Piovella *et al.* (2001*b*) and Bonifacio *et al.* (2005*a*) from a linear analysis of eqns (16.32) and (16.33) that the equilibrium state with no generated field, $\bar{A} = 0$, and all atoms having zero momentum is unstable. The value of the pump–probe detuning δ at which maximum growth of the instability occurs was shown to be $1/\rho_{\mathrm{C}}$. In dimensional units, this corresponds to $\omega_{\mathrm{pump}} - \omega_{\mathrm{probe}} = \omega_r$ here, i.e. the probe frequency is downshifted from the pump frequency by the recoil frequency. Figure 16.9 shows the evolution of the generated field intensity as calculated numerically from eqns (16.32) and (16.33) when $\rho_{\mathrm{C}} = 10$, $\delta = 1/\rho_{\mathrm{C}}$, and $|\bar{A}| = 1 \times 10^{-4}$. On comparison with Fig. 16.9, it can be seen that the field intensity evolution is identical to that produced by the classical equations (16.16)–(16.18). The reason for the similarity between the evolution of the quantum and classical equations can be deduced from an inspection of the scaled variables. From the definition of $\bar{A}$ and ρ, it can be shown that

$$\rho_{\mathrm{C}} \left|\bar{A}\right|^2 = \frac{2\epsilon_0 |E_{\mathrm{probe}}|^2}{n_a \hbar \omega},$$

i.e. the value of $\rho_{\mathrm{C}}|\bar{A}|^2$ is the average number of probe photons per atom in the cavity. From Fig. 16.9, it can be seen that at saturation, $|\bar{A}|^2 \sim 1$. Consequently, for the example shown in Fig. 16.9 the number of generated probe photons is much larger than one, which implies a classical evolution of the instability. This is supported by an inspection of the atomic dynamics during the interaction. Figure 16.10 shows the evolution of the momentum distribution in the BEC during the interaction. It can be seen that many momentum states are occupied, with both $n > 0$ and $n < 0$ during the evolution of the collective instability. Note that the average momentum of the BEC

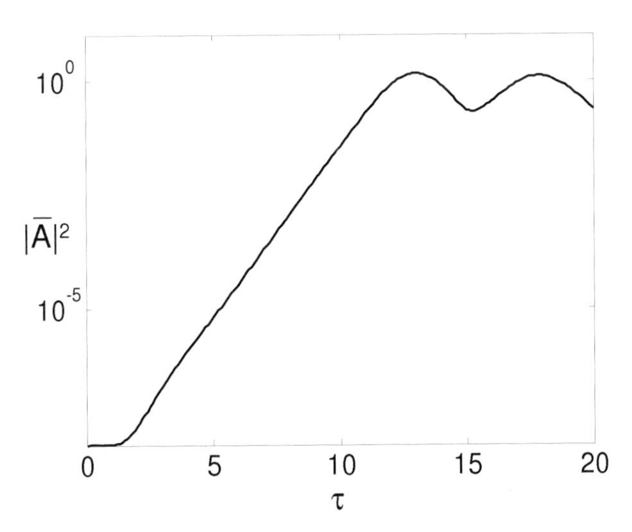

Fig. 16.9 Evolution of generated field intensity when $\rho_{\mathrm{C}} = 10$, $\delta = 1/\rho_{\mathrm{C}} = 0.1$, and $|\bar{A}(\tau = 0)| = 10^{-4}$.

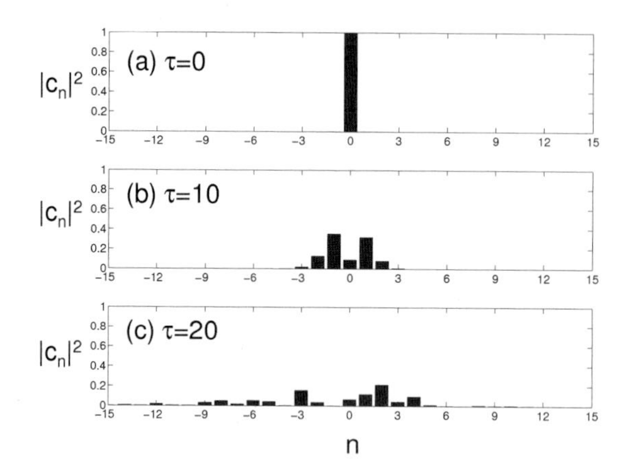

Fig. 16.10 Evolution of atomic momentum distribution when $\rho_C = 10$, $\delta = 1/\rho_C = 0.1$, and $|\bar{A}(\tau = 0)| = 10^{-4}$.

shifts to $n < 0$, as is necessary in order to maintain conservation of momentum, which can be written as

$$\langle p \rangle + \frac{\rho_C}{2} \left|\bar{A}\right|^2 = \text{constant}.$$

Let us now consider how the interaction is affected when $\rho_C < 1$. From the discussion above, we now expect to see quantum dynamics during the BEC–light interaction as on average less than one probe photon per atom will be generated. Figure 16.11 shows the evolution of the generated field intensity when $\rho_C = 0.1$ and $\delta = 1/\rho_C = 10$. It can be seen that the field evolution is now very different from the classical evolution of Fig. 16.9. Figure 16.12 shows the evolution of the BEC momentum distribution during the interaction. It can be seen that the interaction now involves only two momentum states: $n = 0$ and $n = -1$ (i.e. momenta $p = 0$ and $p = -2\hbar k$), which the atoms alternate between. It can be shown (Piovella *et al.* 2001*b*; Bonifacio *et al.* 2005*a*) that if all states other than these two are neglected, eqns (16.32) and (16.33) can be formally reduced to the Maxwell–Bloch equations describing the interaction between a two-level system and a quasi-resonant optical field. Consequently, the generated field intensity shown in Fig. 16.11 is a series of 2π hyperbolic-secant pulses.

It can be seen from the above that the quantum nature of the atom–light momentum exchange dramatically affects the nature of the interaction and the light emitted. Using the similarity of the CARL model to that of the FEL, this raises the interesting question of whether it may be possible to attain a similar quantum regime of FEL operation (QFEL), involving only two electron momentum states (Bonifacio *et al.* 2005*b*). Preliminary studies suggest that the radiation produced by such an FEL would have very attractive features for applications, particularly at X-ray wavelengths, for example greatly improved temporal coherence compared with conventional, classical FEL sources.

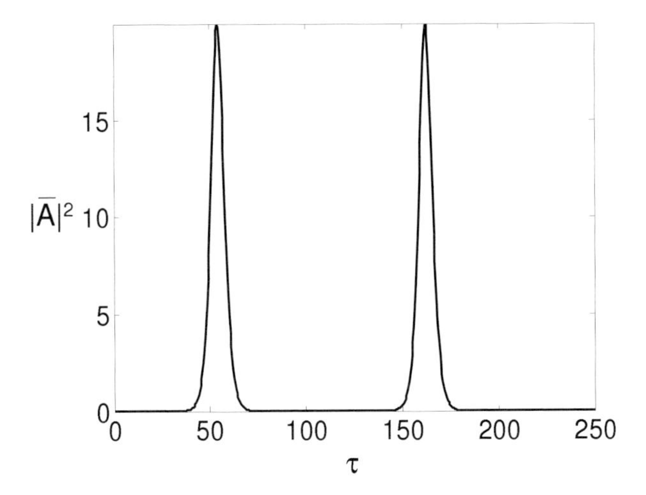

Fig. 16.11 Evolution of generated field intensity when $\rho_C = 0.1$, $\delta = 1/\rho_C = 10$, and $|\bar{A}(\tau = 0)| = 10^{-4}$.

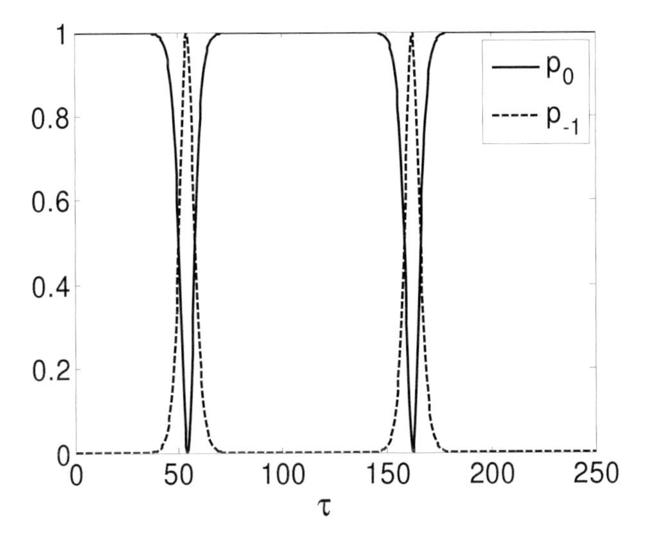

Fig. 16.12 Evolution of occupation probability of momentum states $n = 0$ (solid line) and $n = -1$ (dashed line) when $\rho_C = 0.1$, $\delta = 1/\rho_C = 10$, and $|\bar{A}(\tau = 0)| = 10^{-4}$.

16.4 Conclusion

In this chapter, I have discussed some details of some collective interactions between light and two different types of particles: electrons and atoms. The subject of collective light–matter interactions is a subject of interest in terms of both its optical applications, for example tunable, coherent light sources and new optical nonlinearities, and its cross-disciplinary relevance as a versatile testing ground for long-range interactions involving coupled many-body systems.

Although I have concentrated on interactions involving electrons and atoms here, there are several other systems where interactions similar to the kind described here have been considered, for example suspensions of nanoparticles (Wiggins *et al.* 2002; Robb and McNeil 2003). It is interesting to consider the extension of the concepts involved here to interactions involving other types of particle/medium and also other types of coupling.

Acknowledgments

I gratefully acknowledge the contribution of Rodolfo Bonifacio, Willie Firth, Brian McNeil, and Nicola Piovella to this work, and the Leverhulme Trust for support via research grant no. F/00273/I.

References

Allen L. and Eberly J. H. (1987). *Optical Resonance and Two-Level Atoms*, Dover, New York.

Berman P. R. (1999). *Phys. Rev. A*, **59**, 585.

Bonifacio R. and De Salvo Souza L. (1994). *Nucl. Instrum. Methods A*, **341**, 360.

Bonifacio R., Pellegrini C., and Narducci L. M. (1984). *Opt. Commun.*, **50**, 373.

Bonifacio R., McNeil B. W. J., and Pierini P. (1989). *Phys. Rev. A*, **40**, 4467.

Bonifacio R., Casagrande F., Oerchioni G., De Salvo Souza L., Pierini P., and Piovella N. (1990). *Riv. Nuovo Cimento*, **13**, 1.

Bonifacio R., De Salvo Souza L., Narducci L., and D'Angelo E. J. (1994). *Phys. Rev. A*, **50**, 1716.

Bonifacio R., McNeil B. W. J., Piovella N., and Robb G. R. M. (2000). *Phys. Rev. A*, **61**, 023807.

Bonifacio R., Cola M. M., Piovella N., and Robb G. R. M. (2005*a*). *Europhys. Lett.*, **69**, 55.

Bonifacio R., Piovella N., and Robb G. R. M. (2005*b*). *Nucl. Instrum. Methods A*, **543**, 645.

Javaloyes J., Perrin M., Lippi G. L., and Politi A. (2004). *Phys. Rev. A*, **70**, 023405.

Javaloyes J., Perrin M., and Politi A. (2008). *Phys. Rev. E*, **78**, 011108.

Kuramoto Y. (1984). *Prog. Theor. Phys. Suppl.*, **79**, 223.

LCLS CR (2002). SLAC Report No. SLAC-R-593; also available at http://lcls.slac.stanford.edu.

Murphy J. B. and Pelligrini C. (1990). *Laser Handbook* **6**, North-Holland, Amsterdam.

Perrin M., Lippi G.-L., and Politi A. (2002). *J. Mod. Opt.*, **49**, 419.

Piovella N., Bonifacio R., McNeil B. W. J., and Robb G. R. M. (2001*a*). *Opt. Commun.*, **187**, 165.

Piovella N., Gatelli M., and Bonifacio R. (2001*b*). *Opt. Commun.*, **194**, 167.

Robb G. R. M. and Firth W. J. (2007). *Phys. Rev. Lett.*, **99**, 253601.

Robb G. R. M. and McNeil B. W. J. (2003). *Phys. Rev. Lett.*, **90**, 123903.

Robb G. R. M., McNeil B. W. J., Bonifacio R., and Piovella N. (2001). *Opt. Commun.*, **194**, 151.

Robb G. R. M., Piovella N., Ferraro A., Bonifacio R., Courteille P. W., and Zimmermann C. (2004). *Phys. Rev. A*, **69**, 041403(R).

Robb G. R. M., Piovella N., and Bonifacio R. (2005). *J. Opt. B: Quantum Semiclass. Opt.*, **7**, 93.

Slama S., Bux S., Krenz G., Zimmermann C., and Courteille P. W. (2007). *Phys. Rev. Lett.*, **98**, 053603.

Strogatz S. H. (2001). *Nature*, **410**, 268.

TESLA TDR (2001). DESY Report No. DESY-2001-011; also available at http://xfel.desy.de.

von Cube C., Slama S., Kruse D., Zimmermann C., Courteille P. W., Robb G. R. M., Piovella N., and Bonifacio R. (2004). *Phys. Rev. Lett.*, **93**, 083601.

Wiggins S. M., Robb G. R. M., McNeil B. W. J., Jones D. R., Jaroszynski D. A., and Jamieson S. J. (2002). *J. Mod. Opt.*, **49**, 997.

Part VI

Dipolar interaction in condensed matter

17
Dipolar effects in condensed matter

Steven T. BRAMWELL

London Centre for Nanotechnology, University College London, 17–19 Gordon Street, London WC1H 0AH, United Kingdom

17.1 Introduction

This chapter is concerned with the two interactions

$$E_{ij}^{\text{electric}} = \frac{1}{4\pi\epsilon_0}\left[\frac{\mathbf{p}_i \cdot \mathbf{p}_j - 3(\mathbf{p}_i \cdot \hat{\mathbf{r}}_{ij})(\mathbf{p}_j \cdot \hat{\mathbf{r}}_{ij})}{|\mathbf{r}_{ij}|^3}\right], \tag{17.1}$$

$$E_{ij}^{\text{magnetic}} = \frac{\mu_0}{4\pi}\left[\frac{\mathbf{m}_i \cdot \mathbf{m}_j - 3(\mathbf{m}_i \cdot \hat{\mathbf{r}}_{ij})(\mathbf{m}_j \cdot \hat{\mathbf{r}}_{ij})}{|\mathbf{r}_{ij}|^3}\right], \tag{17.2}$$

which constitute, respectively, the electric and magnetic dipolar (or dipole–dipole) interactions. Here $\mathbf{p}_i$ is the electric dipole moment at position $\mathbf{r}_i$, $\mathbf{m}_i$ is the magnetic dipole moment at position $\mathbf{r}_i$, $\mathbf{r}_{ij} = \mathbf{r}_i - \mathbf{r}_j$, and the circumflex denotes a unit vector (note that we will always consider either an electric or a magnetic dipole: not both at the same time). We will mainly be concerned with the magnetic dipole interaction, but it will be useful to make occasional comparisons with the electric interaction, particularly when discussing the origin of the expression (17.2).

Feynman wrote that "there ought to be a good word, out of the Greek perhaps, to describe a law which reproduces the same law on a larger scale" (Feynman 1963). I think that a good word is still lacking, but a familiar, but nonetheless remarkable property of the dipolar interaction is that it does just what Feynman described: essentially the same law that governs the through-space magnetic interaction of two magnetic ions governs the interaction of two compass needles and governs the interaction of those compass needles with Planet Earth. This property is closely related to the fact that the dipolar law is an algebraic decay as $\sim 1/r^3$, which makes it both self-similar (a power law) and long-ranged. However, the top lines of eqns (17.1) and (17.2) give these dipolar interactions a complex angular dependence, which, as we shall see, leads to very subtle properties. When this is combined with the fact that dipolar interactions are *ubiquitous* in nature (as nearly all real materials exhibit one or the other at the atomic scale), one can see that the dipolar interaction is both important and potentially interesting as an example of a long-range interaction that is very real.

The aims of this chapter are to furnish you with some useful ideas about the dipole interaction, to explore the nature of its long-range character, and to illustrate some interesting physics. To fully comprehend the dipole interaction requires a mixture of electrodynamics, quantum mechanics, and statistical mechanics, as well an appreciation of experimental fact. The focus here is on ideas and basics, rather than on detailed methodology, so methods for summing dipole interactions (such as the Ewald method) are not treated in any detail: see Melko and Gingras (2004) and de'Bell *et al.* (2000) for recent reviews on such methods. Also, the experimental examples are drawn mainly from the author's own practical experience of dipolar effects in soft ferromagnets, low-dimensional magnets, and spin ice: some important dipolar systems, such as $LiHoF_4$ (Silevitch *et al.* 2007), are not discussed here. Likewise, we do not discuss some interesting aspects of dipolar physics, such as the dipolar glass (Ayton *et al.* 1995) and modulated phases (Garel and Doniach 1982).

The organization of the chapter is as follows. Section 17.2 introduces some classical electrodynamics and quantum mechanics of dipole moments: this should be familiar to many readers, but is reviewed in order to compare the electric and the magnetic case, a theme throughout this article. After this first section, the discussion centers on magnetic dipole–dipole interactions. Section 17.3 discusses simple model dipolar systems in one, two, and three dimensions, with a view to identifying the key characteristics of the dipole–dipole interaction. Section 17.4 then illustrates how dipolar effects are manifest in ferromagnets, through various experimental properties. Section 17.5, finally, discusses a dipolar system of topical interest that illustrates dipolar physics particularly well: so-called "spin ice." Some conclusions are drawn in Section 17.6.

17.2 Origin and properties of dipoles

17.2.1 Origin of the electric dipole–dipole interaction

To see why the electric dipole–dipole interaction takes the rather complex form of eqn (17.1), it is useful to review some elementary classical electrostatics. Thus, in this subsection we define the dipole moment, elucidate its interaction with an electric field and the field arising from it, and use these results to establish the interaction represented in eqn (17.1).

The electric dipole moment. The simplest dipole is a pair of equal and opposite charges q separated by a distance d that is small compared with the distance of observation. The dipole moment $\mathbf{p}$ is a vector that lies along the axis of the charge pair pointing towards the positive charge $(+ \leftarrow -)$, and has magnitude $|\mathbf{p}| = qd$. If we write $q = nq_0$ and $d = d_0/n$ then an *ideal dipole* is defined as qd in the limit of $n \to \infty$ (Böttcher 1952). For a more general static charge distribution $\rho(\mathbf{r})$ over a finite volume, the dipole moment is

$$\mathbf{p} = \int \mathbf{r}\rho(\mathbf{r})\, d^3r\,, \tag{17.3}$$

where the integral is over the whole volume. The charge pair dipole is a special case of this equation with $\rho(\mathbf{r}) = q\,[\delta(\mathbf{r} - \mathbf{r}_+) - \delta(\mathbf{r} - \mathbf{r}_-)]$: that is, $\mathbf{p} = q\mathbf{d}$, where $\mathbf{d} = \mathbf{r}_+ - \mathbf{r}_-$ is the vector separation of the positive and negative charges (see Fig. 17.1(a)). Note that, for an overall neutral system of charges, it follows from eqn (17.3) that the dipole moment is independent of the choice of origin (the proof is left as an exercise).

Energy of a dipole in a uniform field. When placed in a uniform electric field $\mathbf{E}$, the energy of the system of charges may be written

$$W = \int \rho(\mathbf{r})\Phi(\mathbf{r})\, d^3r\,, \tag{17.4}$$

where Φ is the electric potential from external sources (i.e. $\mathbf{E} = -\nabla\Phi$). Now we split the position vector $\mathbf{r}$ into two components, i.e. $\mathbf{r} \to \mathbf{r} + \mathbf{a}$, where the new $\mathbf{r}$ locates a fixed point within the charged volume, and $\mathbf{a}$ locates other points within that volume (see Fig. 17.1(b)). Next we expand the potential about the fixed point $\mathbf{r}$:

$$\Phi(\mathbf{r} + \mathbf{a}) = \Phi(\mathbf{r}) + \mathbf{a} \cdot \nabla\Phi + \dots\,, \tag{17.5}$$

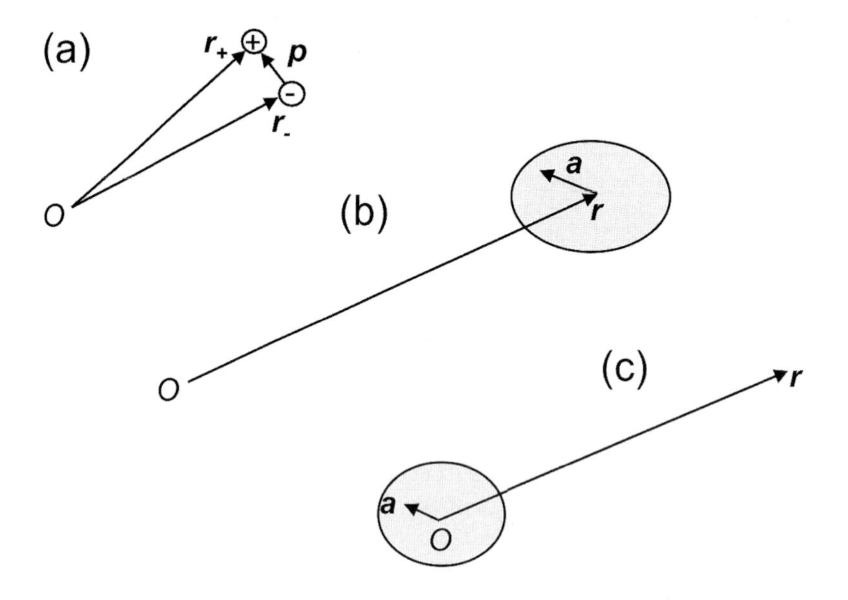

Fig. 17.1 Alternative coordinate systems referred to in this section. The shaded regions indicate volumes with finite charge density.

where the directional derivative is evaluated at point $\mathbf{r}$, and insert into the expression for W. As the field is uniform, the expansion of the potential may be stopped at the first derivative, so that

$$W = q\Phi + \mathbf{p} \cdot \nabla\Phi \,, \tag{17.6}$$

where $q = \int \rho(\mathbf{a})\, d^3a$ is the net charge. In the case where the system of charges is overall neutral, we may thus express the energy of the dipole as

$$W = -\mathbf{p} \cdot \mathbf{E} \,, \tag{17.7}$$

using the relation of field to potential.

Torque on a dipole in a uniform field. The torque on a dipole moment in a uniform electric field may be found by combining the Lorentz force

$$\mathbf{F} = q\mathbf{E} + q\mathbf{v} \times \mathbf{B} \tag{17.8}$$

for zero magnetic field (i.e. $\mathbf{F} = q\mathbf{E}$) with the equation for the torque on the system of charges,

$$\mathbf{T} = \int \mathbf{r} \times \mathbf{F}(\mathbf{r})\, d^3r \,, \tag{17.9}$$

which, using eqn (17.3), gives

$$\mathbf{T} = \mathbf{p} \times \mathbf{E} \,. \tag{17.10}$$

Potential due to a dipole. We again redefine our coordinate system so that we now shift the origin to a point within the system of charges, and use $\mathbf{a}$ to locate points within the charged volume and $\mathbf{r}$ to locate external points at a large distance away (see Fig. 17.1(c)). The potential at point $\mathbf{r}$ is then

$$\phi(\mathbf{r}) = \frac{1}{4\pi\epsilon_0}\int \frac{\rho(\mathbf{a})}{|\mathbf{r}-\mathbf{a}|}d^3a\,. \tag{17.11}$$

Expanding the reciprocal distance,

$$\frac{1}{|\mathbf{r}-\mathbf{a}|} = \frac{1}{r} - \mathbf{a}\cdot\nabla\frac{1}{r} + \ldots\,, \tag{17.12}$$

and substituting this into eqn (17.11), we find, using the definition of the dipole moment,

$$\phi(\mathbf{r}) = \frac{q}{4\pi\epsilon_0 r} - \frac{1}{4\pi\epsilon_0}\left(\mathbf{p}\cdot\nabla\frac{1}{r}\right) + \ldots\,, \tag{17.13}$$

which represents the first two terms in the *multipole expansion*. The first term is zero for an overall neutral system of charges, leaving the second term (if it is finite) as the dominant contribution to the potential. Now, $\nabla(1/r) = -\mathbf{r}/r^3$ (which may be proved by writing both $1/r$ and the gradient operator in Cartesian coordinates: this is left as an exercise). Substitution of this expression into the second term of eqn (17.13) leads to the result

$$\phi(\mathbf{r}) = \frac{1}{4\pi\epsilon_0}\frac{\mathbf{p}\cdot\mathbf{r}}{r^3}\,, \tag{17.14}$$

the potential due to a dipole at distance r.

Field due to a dipole. This may be found by differentiating the potential:

$$\mathbf{E}(\mathbf{r}) = -\nabla\phi(\mathbf{r}) = -\frac{1}{4\pi\epsilon_0}\left[\frac{1}{r^3}\nabla\mathbf{p}\cdot\mathbf{r} + \mathbf{p}\cdot\mathbf{r}\nabla\frac{1}{r^3}\right], \tag{17.15}$$

which, after the differentiation is performed (again left as an exercise), leads to the following expression for the field due to a dipole at distance r:

$$\mathbf{E}(\mathbf{r}) = \frac{1}{4\pi\epsilon_0}\left[\frac{-\mathbf{p}}{r^3} + \frac{3(\mathbf{p}\cdot\mathbf{r})\mathbf{r}}{r^5}\right]. \tag{17.16}$$

Dipole–dipole interaction energy. The energy of interaction of two dipoles may now be equated with the work done when one dipole is brought from an infinite distance into to the field of a second dipole placed at the origin:

$$W = -\mathbf{p}_1\cdot\mathbf{E}_2 = \frac{1}{4\pi\epsilon_0}\left[\frac{\mathbf{p}_1\cdot\mathbf{p}_2}{r_{12}^3} - \frac{3(\mathbf{p}_1\cdot\mathbf{r})(\mathbf{p}_2\cdot\mathbf{r})}{r_{12}^5}\right], \tag{17.17}$$

which is equivalent to eqn (17.1).

17.2.2 Origin of the magnetic dipole–dipole interaction

We wish to demonstrate why the magnetic dipole–dipole interaction in eqn (17.2) is of the same form as the electric, eqn (17.1), even though it is well known that magnetic dipoles are physically different from electric dipoles. The aim of this is to highlight the similarities and differences between electric and magnetic dipoles: so long as one is careful, it can be it can be very useful to think of magnetic dipoles as analogous to electric ones, as later examples will illustrate. Here we will derive the magnetic dipole–dipole interaction by a logical sequence analogous to what we used in the last section to derive the electric dipole–dipole interaction.

Magnetic dipole moment. In anticipation of a treatment of atoms and ions, we consider a static distribution of electrical currents confined in a volume of space, with no net flux in or out of the volume and no change of polarization with time. Now that we have thus excluded the free-current and polarization current density, the remaining current density can only circulate in the finite volume, and this circulation gives rise to a magnetic dipole moment (of course, classically such currents cannot persist, but we do not worry about this as quantum mechanics will remove that objection). In terms of the current density $\mathbf{J}$, the magnetic dipole moment may be defined as follows,

$$\mathbf{m} = \frac{1}{2}\int \mathbf{r} \times \mathbf{J}(\mathbf{r})\, d^3r\,, \tag{17.18}$$

which is independent of origin. This is reminiscent of the definition of the electric dipole moment $\mathbf{p}$, eqn (17.3), but note the factor of one-half in eqn (17.18), and the fact that $\mathbf{m}$ is an axial vector, whereas $\mathbf{p}$ is a polar vector.

Equation (17.18) is much more tricky to use than eqn (17.3). However, we can get the main results with the help of the two following properties:

$$\int \mathbf{r} \cdot \mathbf{J}(\mathbf{r})\, d^3r = 0\,, \tag{17.19}$$

which may be interpreted as the absence of radial current flow, and

$$\int (\mathbf{b} \cdot \mathbf{r})\mathbf{J}(\mathbf{r})\, d^3r = \mathbf{m} \times \mathbf{b}\,, \tag{17.20}$$

for an arbitrary vector $\mathbf{b}$. The latter has no quick and easy proof: see Jackson (1998) for its derivation. These two properties combine to make a third, which we will use below:

$$\int \mathbf{r} \times (\mathbf{J}(\mathbf{r}) \times \mathbf{b})\, d^3r = \int [\mathbf{J}(\mathbf{r})(\mathbf{r} \cdot \mathbf{b}) - \mathbf{b}(\mathbf{r} \cdot \mathbf{J})]\, d^3r = \mathbf{m} \times \mathbf{b}\,, \tag{17.21}$$

for a vector $\mathbf{b}$, where we have invoked a standard relationship for the vector triple product.

Torque on a magnetic dipole in a uniform magnetic field. In zero external electric field, the Lorentz force arising from the magnetic field $\mathbf{B}$ (eqn (17.8)) is

$$\mathbf{F} = q\mathbf{v} \times \mathbf{B} = \mathbf{J} \times \mathbf{B}\,, \tag{17.22}$$

so, using the property (17.21), the torque is

$$\mathbf{T} = \int \mathbf{r} \times (\mathbf{J}(\mathbf{r}) \times \mathbf{B})\, d^3r = \mathbf{m} \times \mathbf{B}\,, \tag{17.23}$$

in direct analogy with the torque on an electrical dipole, eqn (17.10).

Energy of a magnetic dipole in a uniform magnetic field. The torque may be interpreted as an energy gradient with respect to an angle θ that turns the magnetic dipole away from the field direction: $|\mathbf{T}| = mB\sin\theta$. Hence the work done is

$$W = -mB\cos\theta = -\mathbf{m}\cdot\mathbf{B}\,, \tag{17.24}$$

again in direct analogy with the electrical case, eqn (17.7). The rigorous proof of this equation is again a tricky business, best confined to the pages of Jackson (1998).

Vector potential due to a magnetic dipole. Adopting the Coulomb gauge, the vector potential may be written (in terms of the coordinate system of Fig. 17.1(c)) as

$$\mathbf{A}(\mathbf{r}) = \frac{\mu_0}{4\pi}\int \frac{\mathbf{J}(\mathbf{a})}{|\mathbf{r}-\mathbf{a}|}\, d^3a\,, \tag{17.25}$$

and we can again use the first two terms in the expansion of $1/|\mathbf{r}-\mathbf{a}|$ (eqn (17.12)) to find

$$\mathbf{A}(\mathbf{r}) = \frac{\mu_0}{4\pi}\left[\int \frac{\mathbf{J}(\mathbf{a})}{r}\, d^3a - \int \mathbf{J}(\mathbf{a})\left(\mathbf{a}\cdot\nabla\frac{1}{r}\right) d^3a\right]. \tag{17.26}$$

Hence, invoking eqn (17.20) and the absence of a net current,

$$\mathbf{A}(\mathbf{r}) = -\frac{\mu_0}{4\pi}\mathbf{m} \times \nabla\frac{1}{r}\,, \tag{17.27}$$

which, using the expression for $\nabla(1/r)$ (see above), becomes

$$\mathbf{A}(\mathbf{r}) = \frac{\mu_0}{4\pi}\frac{\mathbf{m}\times\mathbf{r}}{r^3}\,, \tag{17.28}$$

the vector potential due to a magnetic dipole at distance r.

Magnetic dipole field and dipole–dipole interaction. To derive an expression for the magnetic field arising from a magnetic dipole, we need to form $\mathbf{B} = \nabla \times \mathbf{A}$ in analogy to the electric case $\mathbf{E} = -\nabla\Phi$. Starting with eqn (17.27), a trick is to define $\nabla(1/r) \equiv \mathbf{b}$ and then to compare the expressions $\nabla \times (\mathbf{m} \times \mathbf{b})$ and $-\nabla(\mathbf{m}\cdot\mathbf{b})$. These triple products may be rearranged with use of the two formulae (1) $\nabla \times \mathbf{m} \times \mathbf{b} = (\mathbf{b}\cdot\nabla)\mathbf{m} - \mathbf{b}(\nabla\cdot\mathbf{m}) - (\mathbf{m}\cdot\nabla)\mathbf{b} + \mathbf{m}(\nabla\cdot\mathbf{b})$ and (2) $-\nabla(\mathbf{m}\cdot\mathbf{b}) = -\left[(\mathbf{b}\cdot\nabla)\mathbf{m} + (\mathbf{m}\cdot\nabla)\mathbf{b} + \mathbf{b}\times(\nabla\times\mathbf{m}) + \mathbf{m}\times(\nabla\times\mathbf{b})\right]$, which both equate to $-(\mathbf{m}\cdot\nabla)\mathbf{b}$ (proof of which is left as an exercise).

Using the expression for $\nabla(1/r)$, it is thus established that

$$\nabla \times \left(\frac{\mu_0}{4\pi} \frac{\mathbf{m} \times \mathbf{r}}{r^3} \right) = -\nabla \left(\frac{\mu_0}{4\pi} \frac{\mathbf{m} \cdot \mathbf{r}}{r^3} \right), \tag{17.29}$$

where the term in brackets on the right-hand side is analogous to eqn (17.14). Thus it straightforwardly follows that the the field due to a magnetic dipole and the magnetic dipole–dipole interaction have the same form as in the electric case (eqns (17.16) and (17.17)), with the substitutions $\epsilon_0 \to 1/\mu_0$ and $\mathbf{p} \to \mathbf{m}$. In this way, we finally establish eqn (17.2).

17.2.3 Magnetic versus electric dipoles

Similarities and differences. This chapter is mainly concerned with magnetic dipoles and their interactions, but in the preceding sections we have established that they are to a large extent equivalent to electric dipoles and their interactions. Recalling that we tacitly assumed that the distance of observation r was sufficiently large, the following are all equivalent: the dipolar field, the interaction between two dipoles (eqns (17.1) and (17.2)), the energy of the dipole in a uniform external field, and the torque and force (which is zero) on a dipole in a uniform external field. The equivalence breaks down when one considers movable dipoles, the force in a nonuniform field, or properties that depend on the dipolar field at $r = 0$. In the latter case a delta function needs to be added to the dipolar field, which has a different prefactor in the electric and magnetic cases (Griffiths 1992; Jackson 1998). As noted above, the symmetry of the magnetic dipole and the electric dipole are also different.

Experiments that distinguish types of dipoles. The fact that there are basically two types of dipoles—charge dipoles and Amperian current loop dipoles—has been appreciated for well over a century. However, the final proof that *microscopic* magnetic dipoles (those on electrons, protons, neutrons, etc.) are Amperian current loop dipoles did not emerge until the relatively recent period 1950–1980 (Vaidman 1990; Griffiths 1992). Well into that period, textbooks on magnetism still stressed the "magnetic charge" model, partly because this affords an easy way of visualizing and calculating magnetic properties and partly because there was no logical reason to reject it, as the question of the true nature of the microscopic magnetic dipole had yet to be resolved.

Bulk measurements on ferromagnets are incapable of rejecting the charge pair model for microscopic dipoles: indeed, they indicate many good reasons to adopt it (recall that Coulomb established the magnetic Coulomb Law as well as the electric law). The big breakthrough came with the invention of neutron scattering. As the neutron is a moving dipole that interacts with the nonuniform spatially varying field arising from the magnetic dipole moments in the sample, its scattering reflects the peculiarities of the current loop model. Thus the first polarized neutron-scattering experiments in 1951 rejected the charge pair model for the neutron once and for all (Hughes and Burgy 1951). Electron spin resonance experiments on the hyperfine coupling between proton and electron spins also reject the charge pair model of magnetic dipoles, as they probe the dipolar field at $r = 0$ (Griffiths 1992).

Effective magnetic charge in statistical mechanics. According to the previous discussion, as long as we steer clear of experiments that probe moving dipoles or the field at $r = 0$, the idea of magnetic dipoles as equivalent to electric dipoles is perfectly valid: for systems of dipoles that occupy definite positions in space (though may be free to reorient), the statistical mechanics should be equivalent in the two cases. Indeed, real materials can even behave as if they have unbound magnetic charges, as we will later illustrate. As mentioned, useful discussions of effective magnetic charges appear in most of the older textbook literature on magnetism, such as Morrish (1965).

Reality of dipoles: The multipole expansion. Although most experiments on magnetic or electric dipolar systems make the constituent dipoles look very real, the above derivations show that the dipole moment is in general just a convenient property by which to discuss a very complex reality: in particular, for the electric case it arises as the second term in the multipole expansion of the potential:

$$\Phi(\mathbf{r}) = \frac{1}{4\pi\epsilon_0}\left[\frac{q}{r} + \frac{\mathbf{p}\cdot\mathbf{r}}{r^3} + \frac{1}{2}\sum_{ij} Q_{ij}\frac{x_i x_j}{r^5} + \ldots\right], \tag{17.30}$$

where the Q's are quadrupole moments and the dots represent higher-order poles. In molecular systems, there is nothing special about the second term in this expansion: the quadrupole moments or higher-order poles may be more important. In magnetism there is also a multipole expansion (Jackson 1998), but the dipolar term tends to be very dominant. There are two reasons for this: one is the importance of the magnetic moment arising from electron spins, which is dipolar, and the second is the fact that atoms and ions are close to spherical, so their higher-order moments are small. Nevertheless the possibility of phenomena arising from higher-order magnetic poles is a real one: an example is the material NpO_2, which is thought to display ordering of magnetic octupoles (Di Matteo *et al.* 2007).

17.2.4 Magnetic dipoles in real systems

Pristine dipolar systems. Perhaps ironically, electric dipole systems tend to be less pure or pristine than magnetic ones. This is partly due to the competition of dipoles with higher-order poles and partly due to their interaction with the lattice, which can rarely be ignored. Magnetic dipoles, whether associated with nuclear or electronic spin, tend to be free from these complicating factors, but can nevertheless only be discussed with some caveats. In particular, nuclear-spin dipolar interactions are very weak, and electron-spin dipolar interactions generally compete with superexchange interactions, which in some materials can dominate. The best real examples of pure dipolar systems tend to be found among one particular class of magnetic material: the ionic salts of the lanthanide elements, where the electronic magnetic dipolar interaction is relatively strong and free from complicating factors. In this section we describe the basic theory of such systems, which also represents a qualitative introduction to a theory of ionic magnetism in general (White 1983).

Magnetomechanical parallelism. Consider a current in a magnetic ion,

$$\mathbf{j}(\mathbf{r}) = \sum_\alpha -e\mathbf{v}_\alpha \delta(\mathbf{r} - \mathbf{r}_\alpha) , \tag{17.31}$$

where α labels the electrons. The magnetic moment is

$$\vec{\mu} = \frac{1}{2} \int \mathbf{r} \times \mathbf{j}(\mathbf{r})\, d^3 r = \frac{-e}{2} \sum_\alpha \mathbf{r}_\alpha \times \mathbf{v}_\alpha , \tag{17.32}$$

but recall that the angular momentum is $\mathbf{l}_\alpha = \mathbf{r}_\alpha \times (m\mathbf{v}_\alpha)$, so

$$\vec{\mu} = \frac{-e}{2m} \sum_\alpha \mathbf{l}_\alpha = \frac{-e}{2m} \mathbf{L} , \tag{17.33}$$

where $\mathbf{L}$ is the total orbital angular momentum. In quantum mechanics, observables such as the angular momentum are replaced by operators, so we define the magnetic moment operator in terms of the angular momentum operator

$$\hat{\mu} = \frac{-e}{2m} \hat{L} , \tag{17.34}$$

which summarizes the concept of *magnetomechanical parallelism*: magnetic moment is proportional to angular momentum.

Now, we know that the angular momentum operator has the following two properties:

$$\langle \hat{L}^2 \rangle = \hbar^2 L(L+1) , \tag{17.35}$$

$$\langle \hat{L}_z \rangle = \hbar m_L , \tag{17.36}$$

where $m_L = L, L-1, \ldots, -L$ (here the angle brackets denote a quantum mechanical expectation value). This leads to

$$\langle \hat{\mu}^2 \rangle = \mu_B^2 L(L+1) , \tag{17.37}$$

$$\langle \hat{\mu}_z \rangle = -\mu_B m_L , \tag{17.38}$$

where $\mu_B = e\hbar/2m$ is the Bohr magneton. Thus the (square) magnitude and z-component of the magnetic moment are well defined in an atomic or ionic state $|L, m_L\rangle$.

It remains to include electron spin. It is found that the magnetic moment is

$$\hat{\mu} = \frac{-e}{2m} (\hat{L} + g\hat{S}) , \tag{17.39}$$

where $\hat{S}$ is the total spin angular momentum operator ($\hat{S} = \sum_\alpha \hat{s}_\alpha$), and $g = 2.002319\ldots$ is the electron g-factor. For a free atom or ion, we may form an overall angular momentum $\hat{J} = \hat{L} + \hat{S}$ and it is easily shown that $\hat{L} + 2\hat{S}$ has the same eigenvalues as the operator $g_J \hat{J}$, where g_J is the Landé g-factor (a function of J, L, S

varying between 1 and 2). Therefore the overall magnetic-moment operator may be expressed as

$$\hat{\mu} = -\frac{e}{2m} g_J \hat{J}\,, \tag{17.40}$$

and it has expectation values

$$\langle \hat{\mu}^2 \rangle = \mu_B^2 g_J^2 J(J+1)\,, \tag{17.41}$$

$$\langle \hat{\mu}_z \rangle = -\mu_B g_J m_J\,, \tag{17.42}$$

where, as usual, $m_J = J, J-1, \ldots, -J$. Thus the magnetic moment of a free atom or ion in a quantum state $|J, m_J\rangle$ is easily written down.

Effect of the crystal field. Lanthanide ions such as Ho^{3+} and Er^{3+} may be represented, to a first approximation, as free ions, as the valence f-electrons are relatively close to the core and do not partake in significant covalency: strong spin–orbit coupling means that J, m_J are well-defined quantum numbers. The surrounding negative ions in a crystal exert a nonuniform crystalline electric field that acts as a perturbation on the $|m_J\rangle$ manifold. The result is that the degeneracy of the $|m_J\rangle$ manifold in the free ion is lifted by the crystal field to give a ladder of states, each composed of mixtures of $|m_J\rangle$ states that satisfy the appropriate point symmetry of the crystal.

As a real example, we may consider the case of Ho^{3+} in the material holmium titanate, $Ho_2Ti_2O_7$ (a spin ice material: see later). The free ion has $L = 6, S = 2, J = 8$,

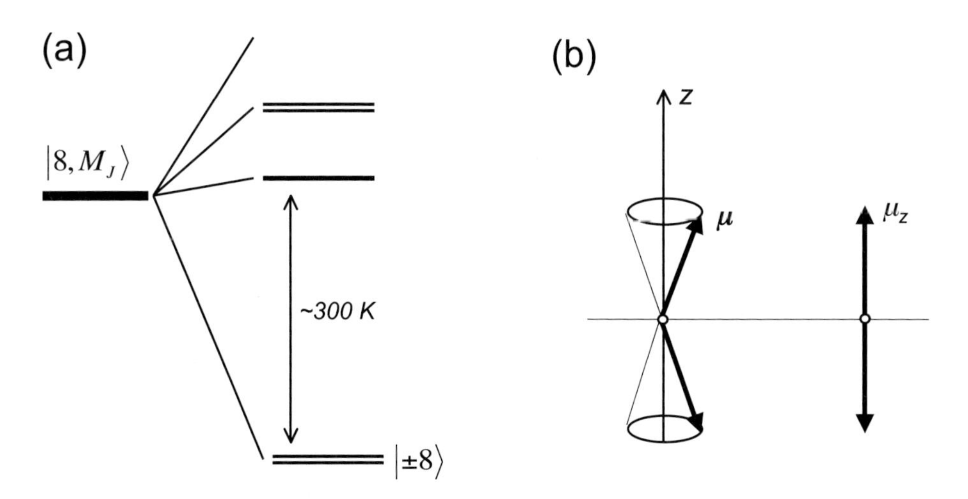

Fig. 17.2 Schematic illustration of the crystal field splitting of the $^5I_8$ free-ion state of the Ho^{3+} ion in $Ho_2Ti_2O_7$. (a) The ground state doublet is composed almost entirely of $|m_J = \pm 8\rangle$. (b) The corresponding magnetic moment μ has a very narrow cone of precession around its quantization axis, its z-component representing an almost perfect realization of an effective classical Ising moment.

so $m_J = 8, 7, 6, \ldots, -6, -7, -8$. The space symmetry of the crystal is cubic but the point symmetry at the Ho^{3+} is trigonal, and the crystal field has a very strong local axial component. The free-ion manifold splits such that the ground term is a doublet, composed almost entirely of $|m_J = \pm 8\rangle$, with the next excited state at about 300 K above the ground state (see Fig. 17.2(a)). Assuming that the ground term is exactly $\frac{1}{\sqrt{2}}(|8\rangle + |-8\rangle)$, the magnitude of the ionic magnetic moment is $10.65\mu_B$ and the z-component in the ground state is $\pm 10\mu_B$, where we have used $g_J = 1.25$. There are two points to notice. First the moment is very large: if a typical paramagnetic ion has $\mu \approx 1\mu_B$ then, from eqn (17.2), the dipole–dipole interaction between two holmium ions is 100 times stronger than that between typical magnetic ions. The dipole–dipole interaction in $Ho_2Ti_2O_7$ turns out to have a magnitude of about 1.5 K, which is significantly stronger than the exchange interaction between neighboring holmium ions. Second, the two states of the holmium ion have their magnetic moment almost parallel to the quantization axis (see Fig. 17.2(b)). It is the z-component of this moment that is observable in equilibrium statistical mechanics, so the moment is effectively Ising-like, and for many purposes may be taken to be a classical moment.

Note that in the case of d-block transition metals such as Cr, Cu, and Fe, the crystal field may be a perturbation of strength comparable to the spin–orbit coupling, which makes the crystal field analysis more difficult. Orbital angular momentum is largely quenched, moments are smaller, and anisotropies weaker. The dipole–dipole interaction is relatively weak, but this does not mean it can always be neglected: see the discussion in Section 17.4 on ferromagnetic materials.

Model magnets. The case of holmium titanate described above illustrates the general point that several lanthanide ions are associated with strong dipole–dipole interactions, strong anisotropies, and weak exchange. Their salts may therefore be used to represent (often classical) spin models of the dipole–dipole interaction (the "spin" generally has a significant orbital component). However, the typically strong anisotropies indicate that the most basic models are the Ising and XY models, rather than the Heisenberg model. It is important to emphasize that the spin symmetry is determined by the point symmetry of the crystal, not the space symmetry: cubic lattices, say, may have Ising-like or XY-like spins ($Er_2Ti_2O_7$ is an example of an XY system (Champion *et al.* 2003)).

17.3 Model systems

In this section we consider the basic properties of simple spin models with the dipole–dipole interaction of eqn (17.2).

17.3.1 Intrinsic properties of the interaction

Dipolar bond energy. Consider a single bond between two dipole moments $\vec{\mu}_{1,2}$. The classical interaction, eqn (17.2), may be written

$$E = E_1 + E_2\,, \tag{17.43}$$

where, in units of $\mu_0\mu^2/4\pi r_{12}^3$,

$$E_1 = \hat{\mu}_1 \cdot \hat{\mu}_2 \tag{17.44}$$

and

$$E_2 = -3\,(\hat{\mu}_1 \cdot \hat{r}_{12})\,(\hat{\mu}_2 \cdot \hat{r}_{12}) \tag{17.45}$$

(the circumflexes denote unit vectors). Consider the two states shown in Fig. 17.3(a) and (b), in which the "spins" are parallel and perpendicular, respectively, to the bond. When parallel, we have $E_1 = +1, E_2 = -3$, and $E = -2$; when perpendicular, $E_1 = -1$, $E_2 = 0$, and $E = -1$. Thus a "ferromagnetic" arrangement parallel to the bond minimizes the overall energy as well as E_2, but an "antiferromagnetic" arrangement (in *any* orientation) minimizes E_1.

Intrinsic frustration. The energy of a dipolar lattice system is just a sum over such bond energies, and hence a sum over E_1 and E_2. However, with a few exceptions, it is generally impossible to simultaneously minimize all pairwise interactions on any particular lattice: the dipole–dipole interaction may be said to be *intrinsically frustrated.* This is a most important observation that largely dominates dipolar physics. Furthermore, different lattices geometries exert a differing degree of frustration on E_1 and E_2.

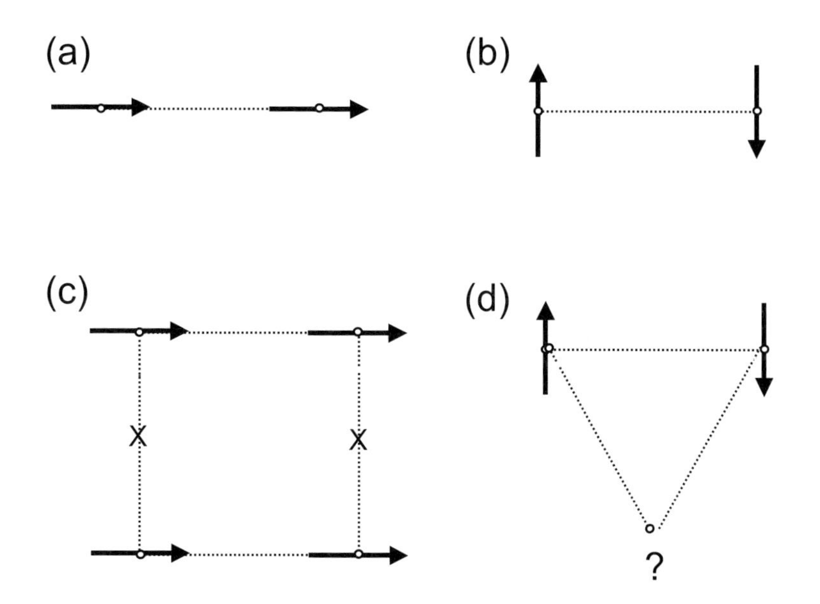

Fig. 17.3 Intrinsic frustration of the dipole–dipole interaction $E = E_1 + E_2$. For a pair of dipoles, E_2 is minimized with "ferromagnetic" alignment parallel to the bond (a), while E_1 is minimized by "antiferromagnetic" alignment (b). A square plaquette (c) frustrates the quasi-ferromagnetic interaction E_2 (as the bonds marked by crosses are not satisfied), while a triangular plaquette (d) frustrates the quasi-antiferromagnetic interaction E_1 (as the dipole marked by "?" is unconstrained).

To see this, consider an equilateral triangle and a square of spins (Fig. 17.3(c), (d)). E_1 is more strongly frustrated by the triangle, but E_2 is more strongly frustrated by the square (the proof of this statement is left as an exercise). Thus we might expect triangular geometries to favor ferromagnetism and square geometries to favor antiferromagnetism. This generalization is more or less true at a local level but there is an important complication to the naive argument: the fact that the interaction is intrinsically long-ranged.

Long-ranged nature. The long-ranged nature of the dipole–dipole interaction is a consequence of the r^{-3} decay of the interaction energy. Thus, a rough approximation to the energy per spin is

$$E/\text{spin} \sim \int \frac{1}{r^3} d^d r \,, \tag{17.46}$$

which may be written

$$E/\text{spin} \sim \int \frac{r^{d-1}}{r^3} dr \,, \tag{17.47}$$

which diverges logarithmically in three dimensions, though it converges in one and two dimensions. In reality, the anisotropy of the dipole–dipole interaction reduces the divergence. In three dimensions, the interaction is said to be *conditionally convergent*: for a ferromagnetic system, the convergence depends on the shape of the sample, as shown below. This intrinsically long-ranged nature combines with the intrinsic frustration to make the dipole–dipole interaction one of great subtlety, as the following examples will illustrate.

17.3.2 One dimension: The dipolar chain

Ground state. The simplest extended dipolar system is a one-dimensional lattice of equally spaced dipoles. It is easy to see that the ground state is ferromagnetic and that *all* bonds between dipoles (no matter how far apart) are at their minimum possible energy. Thus neither the intrinsic frustration nor the long-ranged nature of the dipole–dipole interaction is manifest in the one-dimensional ground state. Symmetry is broken at zero temperature as the net moment of the sample chooses between one of two equivalent directions; however, at any finite temperature the interaction is not sufficiently long-ranged to stabilize long-range order. Qualitatively, the system resembles the one-dimensional Ising ferromagnet.

Magnetic field external to a truncated chain. It is interesting to calculate the magnetic field arising from an ordered dipolar chain at a point exterior to it (see Fig. 17.4(a)). Referring to the figure, the scalar potential at point P is

$$\phi_P = \frac{-\mu_0}{4\pi} \sum_i (\delta\vec{m})_i \cdot \nabla \frac{1}{r} \,, \tag{17.48}$$

where $\delta\vec{m}_i$ is one of the microscopic dipoles. Let $\delta\vec{m} = (m/a)\delta\vec{a}$, where a is the lattice spacing and $\delta\vec{a}$ is a lattice vector. Passing from the sum to an integral, we find

$$\phi_P = \frac{-\mu_0}{4\pi} \frac{m}{a} \int_A^B \left(\nabla \frac{1}{r}\right) \cdot d\vec{a} = \frac{-\mu_0}{4\pi} \frac{m}{a} \int_A^B d\left(\frac{1}{r}\right) \,, \tag{17.49}$$

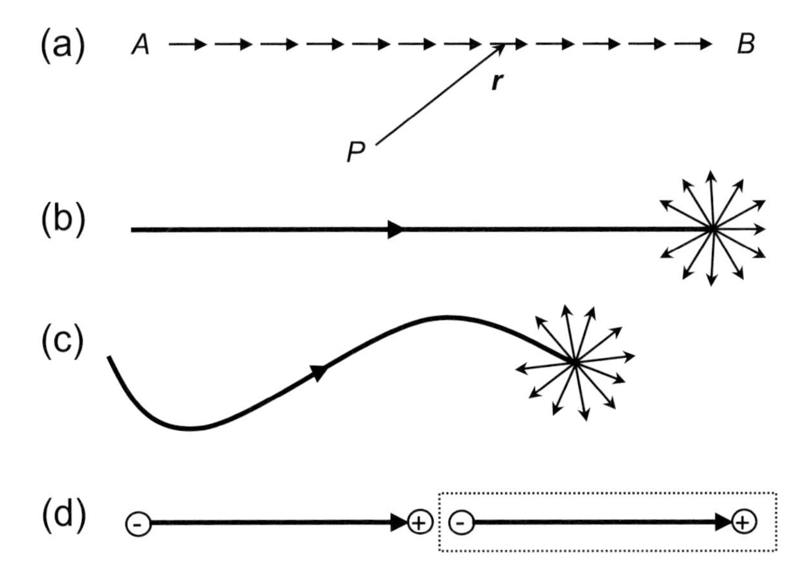

Fig. 17.4 Chain of dipoles. (a) Ground state and coordinate system for calculation of potential at point P. (b) Source of flux at the end of the chain. (c) Same result for a curved chain. (d) Gaussian surface that passes between two microscopic dipoles and contains no net charge.

where A and B denote the ends of the chain (see Fig. 17.4(a)). This integral may be evaluated to give

$$\phi = \frac{-\mu_0}{4\pi}\frac{m}{a}\left(\frac{1}{r_B} - \frac{1}{r_A}\right). \tag{17.50}$$

Now, if we consider a very long chain, and so let $r_A \to \infty$ and shift the origin to point B, we find

$$\phi'_P = \frac{\mu_0}{4\pi}\frac{m/a}{r_{PB}}, \tag{17.51}$$

which is exactly the potential for a monopole or "half dipole" at point B, the end of the chain!

Now, according to the arguments of Section 17.3, this result is equally true of a magnetic dipolar system and an electrical one: there is an effective monopole at the end of the chain (Fig. 17.4(b)). As the result depended on a line integral, it holds equally if the chain is "bendy" (see Fig. 17.4(c)). An argument similar to this is used in the construction of "Dirac strings" (Jackson 1998).

Compensating flux. One might pause to ask at this point, what has happened to Maxwell's law $\nabla \cdot \mathbf{B} = 0$, which forbids magnetic monopoles? In fact, there is no problem as a compensating flux runs along the dipolar chain itself. To see this, consider the Gaussian surface drawn in Fig. 17.4(d) and particularly the part of it that passes in between two dipoles. Clearly the "charge" inside is zero, so the total flux passing through the surface is zero. The monopole's worth of flux emanating from point B is therefore exactly balanced by flux passing along the chain or string itself.

In Dirac's argument, the compensating flux is considered unobservable. In a real dipolar chain, it is observable but only by a *microscopic* measurement. When viewed from any point exterior to the sample it is unobserved, so it looks very much like there is a magnetic monopole at the end of the sample (just like a hosepipe in the ground looks like a source of water, but in reality water is conserved). As already mentioned, the existence of (effective) magnetic charges is essentially what Coulomb established in his experiments on magnetic needles.

In three dimensions, the magnetic field is written $\mathbf{B} = \mu_0(\mathbf{H} + \mathbf{M})$. There are no monopoles in $\mathbf{B}$ but there may be effective monopoles in $\mathbf{H}$, with the necessary compensating flux provided by the magnetization $\mathbf{M}$. We will meet monopoles again in Section 17.5.

17.3.3 Two dimensions: The square lattice

In two dimensions, the intrinsic frustration of the dipole–dipole interaction is manifest, but the interaction is not yet truly long-ranged (see Section 17.3.1). This is amply demonstrated by the case of the square-lattice dipolar model.

Ground state. Consider a single square plaquette in this model. It is straightforward to show that the energy is minimized by the configurations shown in Fig. 17.5(a)–(c). In fact there is a continuous family of ground states parameterized by an angle ϕ: as illustrated in the figure, this corresponds to "winding" alternate sublattices of spins in opposite directions. This degeneracy survives to the whole square lattice at zero temperature, but, as shown by Prakash and Henley (1990), it is removed by thermal fluctuations at finite temperature $T = 0 + \delta$ to give a state of fourfold symmetry.

Critical properties. The square-lattice dipolar model remains ordered up to a critical temperature, but there is disagreement as to its precise critical properties, which must be considered an unsolved problem: see Taroni *et al.* (2008) for a summary. This example illustrates the fundamental complexity of dipolar problems: even the square-lattice model, which is very basic and highly relevant to experiment, remains poorly understood at the present time.

17.3.4 Three dimensions: The simple cubic lattice

In three dimensions, the intrinsic frustration and the conditionally convergent nature of the dipole–dipole interaction are both important. Hence the problem is rather subtle. We approach it by first considering the energy of a ferromagnetic configuration, without implying that this is necessarily the ground state.

Ferromagnetic configuration on a simple cubic lattice. This is illustrated in Fig. 17.5(d). The energy may be written, in an obvious notation, as

$$E = \frac{\mu_0 \mu^2}{8\pi} \sum_{ij} \frac{1}{r_{ij}^3} \left(1 - \frac{3z_{ij}^2}{r_{ij}^2} \right), \tag{17.52}$$

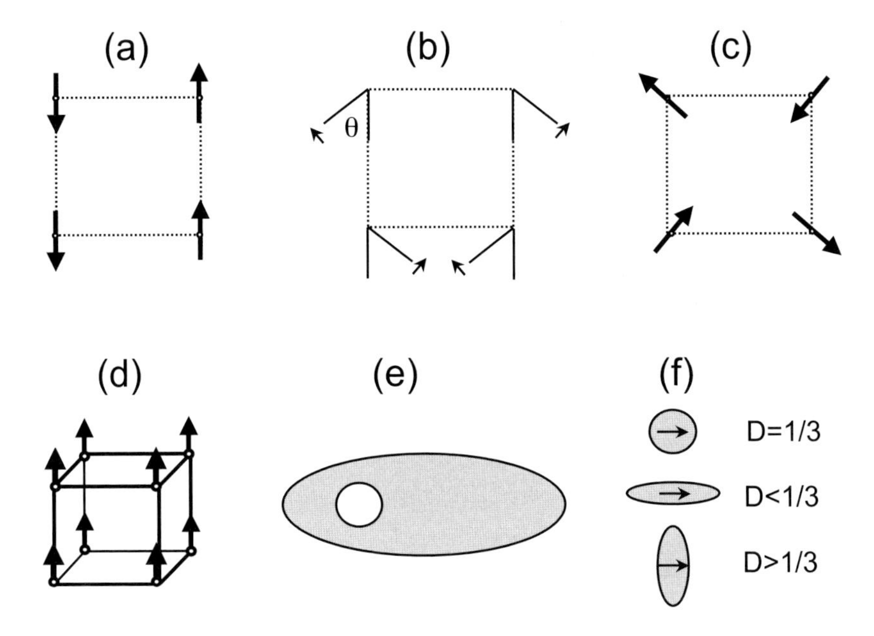

Fig. 17.5 Top row: dipoles on a square plaquette. (a) and (c) are equivalent ground states, part of a family parameterized by rotating the spins through an angle $\theta = 3\pi/4$ in the sense shown in (b). Bottom row: analysis of the dipolar energy of a ferromagnetic configuration on the simple cubic lattice (d). Spherical Lorentz cavity cut in an ellipsoidal sample (e), and demagnetizing factors for differently shaped ellipsoidal samples (f).

where the 8 in the denominator of the prefactor incorporates a factor of 2 that corrects for counting every bond twice. The expression may be further manipulated into the form

$$E = \frac{\mu_0 \mu^2}{8\pi} \sum_{ij} \left(\frac{x_{ij}^2 + y_{ij}^2 - 2z_{ij}^2}{r_{ij}^5} \right). \tag{17.53}$$

In general, this expression is difficult to evaluate but it becomes trivial for a spherical sample. In that case we have, by symmetry,

$$\sum_{ij} \frac{x_{ij}^2}{r_{ij}^5} = \sum_{ij} \frac{y_{ij}^2}{r_{ij}^5} = \sum_{ij} \frac{z_{ij}^2}{r_{ij}^5}, \tag{17.54}$$

and hence the energy E is equal to 0.

The fact that the energy is zero is not important: the key thing is that it does not depend on the size of the sample. Thus the interaction converges. However, one can already see from this simple argument that the energy is likely to depend on the sample shape, as we discuss below. Note also that the above argument could apply to the dipolar contribution to the energy of a spherical sample of a ferromagnet with moments aligned by exchange interactions.

Nonspherical sample. To treat a nonspherical sample, we start by writing the dipolar interaction as follows (White 1983):

$$E_{\text{dipolar}} = \frac{1}{2}\frac{\mu_0}{4\pi}\sum_{\mathbf{r}_i}\sum_{\mathbf{r}_j}\frac{\vec{\mu}_i\cdot\vec{\mu}_j}{|\mathbf{r}_i-\mathbf{r}_j|^3} - \frac{3(\vec{\mu}_i\cdot(\mathbf{r}_i-\mathbf{r}_j))(\vec{\mu}_j\cdot(\mathbf{r}_i-\mathbf{r}_j))}{|\mathbf{r}_i-\mathbf{r}_j|^5}, \tag{17.55}$$

where the factor of one-half again accounts for double counting. As an approximation, the sums may be decoupled:

$$\sum_{\mathbf{r}_i}\sum_{\mathbf{r}_j} \to \sum_{\mathbf{r}_i}\sum_{\mathbf{r}_k=\mathbf{r}_i-\mathbf{r}_j}, \tag{17.56}$$

whence the energy is

$$E_{\text{dipolar}} = \frac{\mu_0}{8\pi}\sum_{\mathbf{r}_i}\vec{\mu}_i\cdot\sum_{\mathbf{r}_k}\frac{\vec{\mu}_k}{r_k^3} - \frac{3\,(\vec{\mu}_k\cdot\mathbf{r}_k)\,\mathbf{r}_k}{r_k^5}. \tag{17.57}$$

Now, introducing the magnetization $\mathbf{M} = \sum\vec{\mu}/V$ and recalling the expression for the dipolar field $\mathbf{B}_k$, we find that the dipolar energy per unit volume is

$$\frac{E}{V} = -\frac{1}{2}\mathbf{M}\cdot\sum_k \mathbf{B}_k. \tag{17.58}$$

The dipolar field in this expression may be taken as the field experienced by a typical moment deep within the sample. We now consider an ellipsoidal sample, and construct a spherical "Lorentz cavity" around this typical spin (see Fig. 17.5(e)). The sum over k is now approximated as follows:

$$\sum_k \to \sum_{\text{in-sphere}} + \frac{N}{V}\int_{\text{sample}} - \frac{N}{V}\int_{\text{sphere}}, \tag{17.59}$$

where the discrete sum has already been shown to have the value zero (see above). To evaluate the integrals, the field is written

$$\mathbf{B} = -\frac{\mu_0}{4\pi}\nabla\left(\frac{\vec{\mu}\cdot\mathbf{r}}{r^3}\right) = -\frac{\mu_0}{4\pi}\nabla\left(\frac{\mu z}{r^3}\right), \tag{17.60}$$

which allows us to use the divergence theorem as follows:

$$\int\nabla\left(\frac{\mu z}{r^3}\right)d^3r = \int_{\text{surface}}\left(\frac{\mu z}{r^3}\right)d\mathbf{s} = \vec{\mu}\,(4\pi D_{\text{shape}}). \tag{17.61}$$

Here D is the demagnetizing factor, discussed below. Collecting together the various terms, we find

$$\frac{E}{V} = \frac{1}{2}\mu_0 M^2\left[D_{\text{surface}} - D_{\text{sphere}}\right], \tag{17.62}$$

so the energy is shape-dependent.

Ground states of the simple cubic lattice. The ground state is determined by the minimization of the sum of three factors: the discrete local dipolar sum, the Lorentz energy, and the demagnetizing energy. It is easy to see that in a *single* unit cell the interaction is minimized by the antiferromagnetic configuration pictured in Fig. 17.6(a). Here the extra energy of the transverse bonds outweighs the contribution of the Lorentz energy, which favors ferromagnetism, and also naturally minimizes the demagnetizing term as $M = 0$. Note that this agrees with the general idea put forward in Section 17.3.1 that square geometries favor antiferromagnetism.

Having gone to a great effort to calculate the energy of a ferromagnetic configuration on the simple cubic lattice, we might find it disappointing that this is not the ground state. However, as shown below, in certain lattice types the dipolar ground state is ferromagnetic, or nearly so, and we can use the analysis of this section to gain an insight into such problems. It is to these more general lattice types that we now turn.

17.3.5 Three dimensions: General lattices

The above results, and the various concepts arising, may be immediately generalized to lattices other than the simple cubic lattice, with the sum in eqn (17.59) taking values other than zero. Here we summarize the main results.

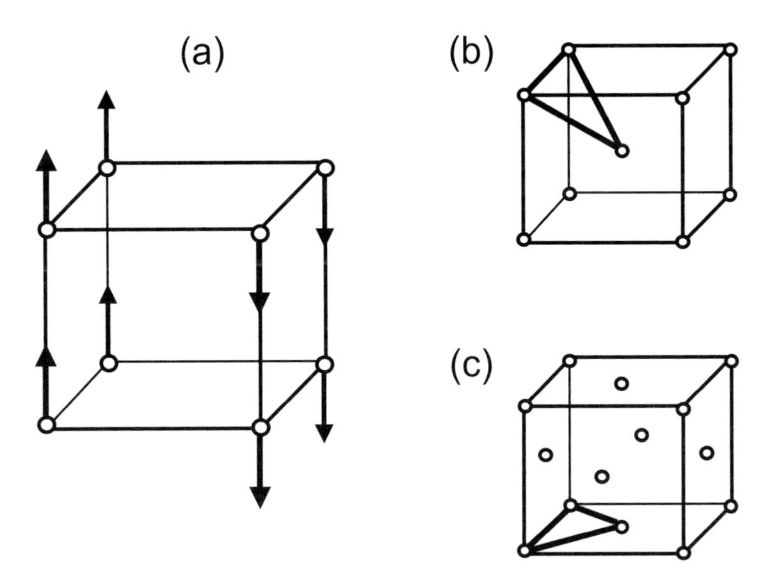

Fig. 17.6 (a) Ground state of Ising dipoles on the simple cubic lattice. (b) Body-centered lattice and (c) face-centered cubic lattice, indicating the local triangular geometry (thick black line) that frustrates the quasi-antiferromagnetic part of the dipole–dipole interaction.

Demagnetizing factor and dipolar energy. The demagnetizing factor D_{shape} can be evaluated using the integral of eqn (17.61). For a sphere, as shown for example in Kittel (1976), it takes the value of 1/3. For an prolate ellipsoid with the magnetization along the unique axis, it takes a value less than 1/3, tending to zero for a needle shape; for an oblate ellipsoid under the same condition, it takes a value greater than 1/3, tending to unity for a flat plate (see Fig. 17.5(f)).

Demagnetizing field and Lorentz cavity field. The competing terms in eqn (17.62) are associated with macroscopic fields $\mathbf{B} = 4\pi D\mathbf{M}$. The first term gives rise to the demagnetizing field and the second to the Lorentz cavity field. One might ask, does the dipolar energy depend on the spherical shape assumed for the Lorentz cavity? The answer is no, because $\left[\sum \cdots - (N/V) \int_{\text{cavity}}\right]$ is found to depend only on the crystal structure, not on the cavity shape nor the sample shape.

Dipolar energy of a ferromagnetic configuration. The dipolar energy may be written in the following form:

$$\frac{E}{V} = \frac{E_0}{V} + \frac{1}{2}\mu_0 M^2 D_{\text{surface}}\,, \tag{17.63}$$

where E_0, which comprises the summation and Lorentz contributions, is the convergent part of the dipolar sum. This equation approximates the dipolar energy of a ferromagnetic configuration on any lattice.

Interpretation of the dipolar energy. Equation (17.63) is essentially correct for any system with a bulk ferromagnetic component (there are a few caveats, such as that the sample should be approximately ellipsoidal). This is of course true for all material substances in an applied field (particularly paramagnetic substances), so the dipolar energy is potentially manifest in any bulk magnetic measurement.

The demagnetizing energy term $\frac{1}{2}\mu_0 M^2 D$ is essentially a mean-field frustrated term. Thus if we write $\mathbf{M} \propto \sum_i \vec{S}_i$ then a term in M^2 means effectively that every spin $\vec{S}$ interacts equally with any other spin, and the fact that this term is positive means that the ground state should minimize the magnetization. These factors are typical in frustrated simplex units such as triangles and tetrahedra, but here it applies to the whole lattice. There are a huge number of ways of achieving zero magnetization and each is sufficient to minimize this component of the dipolar energy.

The Lorentz energy term $-\frac{1}{6}\mu_0 M^2$, on the other hand, is negative and encourages the formation of a ferromagnetic configuration: however, this must be balanced against the discrete summation within the Lorentz cavity (see below) and, as mentioned above, is not strictly separable from it.

Validity of thermodynamics. At first sight the M^2 term in the dipolar energy might invalidate thermodynamics, where a thermodynamic variable such as the magnetization should not depend on sample shape. However, it has been shown that if derivatives of the free energy are formulated in terms of either the internal field $H_{\text{internal}} = H_{\text{external}} - DM$ or the local field $H_{\text{local}} = H_{\text{external}} + H_{\text{dipolar}}$ then these are shape-independent (Levy 1968). For practical magnetism, this is a very important fact as the conditional convergence of the dipolar interaction is factored out of the problem by applying a demagnetizing-field correction. Thus, dividing the internal field by the magnetization, we have for the small-field susceptibility

$$\chi^{-1} = \chi_{\text{observed}}^{-1} - D\,, \tag{17.64}$$

where $\chi = H_{\text{internal}}/M$ and $\chi_{\text{observed}} = H_{\text{external}}/M$. Defined in this way, the susceptibility χ is shape-independent.

Ground states of simple three-dimensional lattices. The ground state of the dipolar interaction on any particular lattice may be considered as arising from the competition of the discrete local dipolar sum, the Lorentz energy, and the demagnetizing energy. The latter is minimized in needle-shaped specimens, so we may just discuss the former two factors.

We have already seen that the simple cubic lattice, being based on square geometry, favors antiferromagnetism. However, for the body-centered cubic (bcc) and face-centered cubic (fcc) lattices, the basic motif of the lattice is now a triangle, rather than a square (see Fig. 17.6(b), (c)), and one would expect the antiferromagnetic ground state to be destabilized by the frustration of E_1 (eqn (17.43)). However, the energies of competing ground states are finely balanced. The bcc lattice has an antiferromagnetic ground state and the fcc lattice favors ferromagnetism, with the Lorentz term just tipping the balance. The fact that the fcc dipolar system is a ferromagnet has also been proved experimentally (Roser and Coruccini 1990).

Generally speaking, the calculation of dipolar ground states is a tricky business: the great work of Luttinger and Tsiza (1946) is the first you should consult if you are interested in the subject. The reader is also referred to a number of other works, both ancient and modern (Sauer 1939; de'Bell *et al.* 2000; Melko and Gingras 2004).

17.4 Dipolar effects in ferromagnets

In the last section we showed that the long-ranged nature of the dipole–dipole interaction is most important for ferromagnetic configurations. Thus it is interesting to consider in detail the practical consequences of the magnetic dipolar interaction in ferromagnets. Here the spins are strongly aligned by the (usually short-ranged) exchange forces $E_{\text{exchange}} = -J\mathbf{S}_i \cdot \mathbf{S}_j$. However, no matter how strong the exchange, the dipolar interaction always dominates over it on sufficiently long length scales! Thus it is wrong to call the dipolar interaction a "weak" interaction: it may be relatively weak as a *pairwise* interaction, but summed over the whole sample it generally dominates over the supposedly "strong" exchange interaction.

17.4.1 Ways of minimizing $(\mu_0/2)DM^2$

Shape anisotropy. The magnetization of a ferromagnet tends to orient itself in the direction that makes the demagnetizing factor as small as possible. For example, in a needle-shaped specimen it will tend to lie parallel to the long axis of the needle (Fig. 17.7(a)). This effect may be opposed or enhanced by magnetocrystalline anisotropy (see Section 17.2).

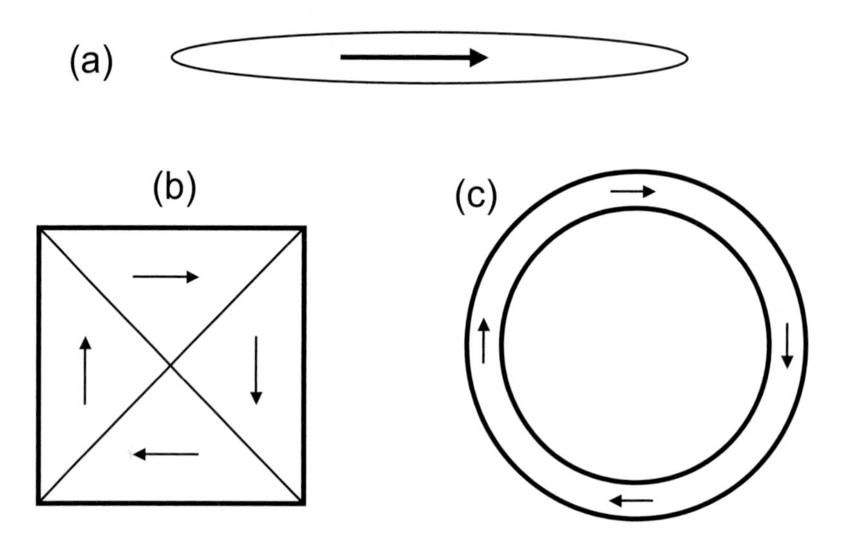

Fig. 17.7 Three ways that the dipolar energy may be minimized in a ferromagnetic sample: (a) orientation of the magnetic moment parallel to the long axis ("shape anisotropy"), (b) domain formation, (c) curling in a toroidal sample.

Domain formation. This is probably the best-known consequence of the dipolar interaction. If a magnet splits into two domains (to make $M = 0$) then the cost in exchange energy per spin scales as the number of bonds in an area $A = L^2$ divided by the total number of bonds, $\sim L^3$; that is, as $O(1/L)$ per spin. The relative benefit in reducing the dipolar energy, on the other hand, is $O(1)$ per spin, so the dipolar interaction dominates for sufficiently large sample size. In zero applied magnetic field, many magnets split into closed domains (Fig. 17.7(b)).

Curling. A third way of minimizing the dipolar energy may be called "curling." For example, in a toroidal sample the dipolar energy is minimized by the magnetization "curling" round the toroid (see Fig. 17.7(c)). This is discussed further below.

17.4.2 Griffiths' Theorem

Griffiths (1968) proved a rather general fact: that the free energy of a magnetic system is shape-independent in the thermodynamic limit. Because the energy of a ferromagnet is shape-dependent, Griffiths' Theorem has the consequence that the equilibrium state is not ferromagnetic: rather, it has $M = 0$ at some long length scale.

Griffiths' Theorem does not mean that dipolar ferromagnets do not exist. Instead, it means that *three-dimensional ferromagnetism is a mesoscopic phenomenon.* For a large enough length scale, some other state (antiferromagnetic, domain, or curling) must become the true ground state.

Proof of the theorem. To prove his theorem Griffiths replaced magnetic dipoles (spins) with charged spheres and used this construction to show that (1) the free energy per spin has a lower bound that is independent of sample shape, and (2) the free energy per spin decreases with increasing sample size. These two results imply that the free energy converges to a shape-independent value in the limit of large system size.

17.4.3 Experimental consequences of the dipolar interaction

Hard and soft ferromagnets. Roughly speaking, a hard ferromagnet is one with appreciable magnetocrystalline anisotropy (see Section 17.2), whereas a soft ferromagnet is one without. Domains are a property of hard ferromagnets, where the magnetocystalline anisotropy defines the sharpness of the domain walls. The realignment and elimination of domains in an applied magnetic field gives rise to the well-known hysteresis phenomenon, as well as effects such as "Barkhausen noise." In a soft ferromagnet the equilibrium state is generally unknown, but is probably akin to the curling mode of a toroidal sample (see above).

Phase transition in soft ferromagnets. At high temperature, in a soft ferromagnet, the magnetization in a weak applied field is uniform. At low temperature, it is in some kind of curling mode. To see the consequences of this, we follow Arrott (1968) in considering a toroidal sample with the field applied parallel to the axis of the toroid. Because the curling mode and the uniformly magnetized mode have different symmetries, there is a phase transition that separates them. It is straightforward to show (Wojtowicz and Rayl 1968) that the magnetization behaves as shown in Fig. 17.8(a) and that the phase diagram is as shown in Fig. 17.8(b).

These pictures may be compared with the bulk magnetization curves and phase diagram of a hypothetical ferromagnet with no dipolar interaction (Fig. 17.8(c), (d)). It is amusing to note that these are often presented as fact in textbooks, but the real facts are displayed in Figs. 17.8(a) and (b). Despite the universality of this behavior, surprisingly little is known about it. The author notes in passing that it is a reminiscent of a so-called Kasteleyn transition (Jaubert *et al.* 2008).

Soft ferromagnet: Macroscopic versus mesoscopic magnetization. Despite the peculiar behavior of the bulk magnetization of a ferromagnet, numerous neutron-scattering experiments of great accuracy yield the typical behavior of the idealized ferromagnet (Fig. 17.8(c), (d)) with the expected critical exponents in zero field. The reason is that neutron scattering effectively probes the (square of the) *mesoscopic* (but long-ranged) magnetization, which is generally uniform, while bulk methods probe the *macroscopic* magnetization, which is not. Consider the (necessarily positive) quantity $\sqrt{M^2}$. Its behavior as a function of field for a soft ferrromagnet is shown schematically in Fig. 17.9(a), where we distinguish between a mesoscopic and a macroscopic measurement. Thus the dipolar effects are only observable at very long length scales, of the order of the sample size.

Soft ferromagnet: Demagnetizing-field effects on the magnon spectrum. The dipolar interaction, being anisotropic and long-ranged, affects the magnon spectrum at small

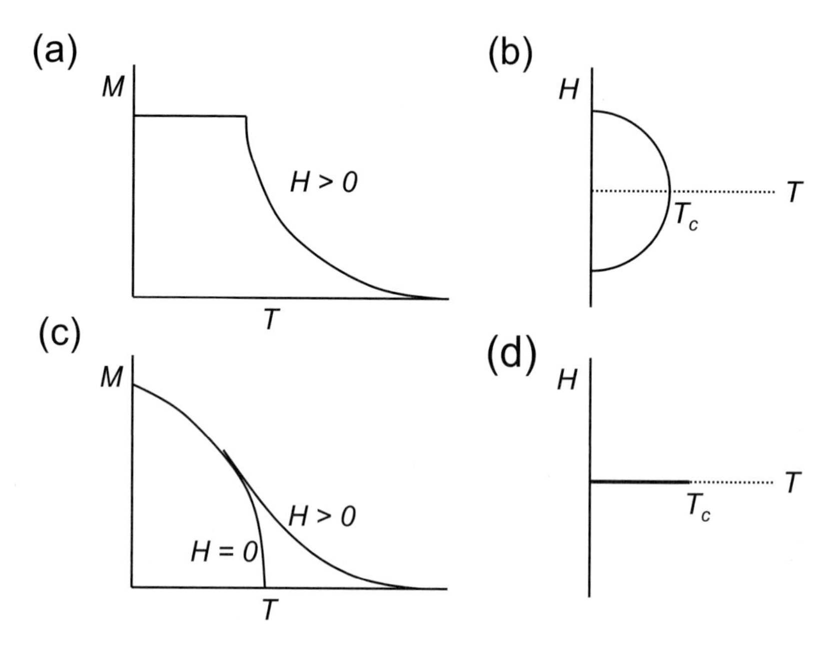

Fig. 17.8 (a) Schematic magnetization-versus-temperature curve for a real ferromagnet in the absence of anisotropy. (b) The corresponding field–temperature phase diagram. (c) Magnetization-versus-temperature curve for an idealized ferromagnet in the absence of both anisotropy and the dipole–dipole interaction. (d) The corresponding phase diagram.

wavevectors. As shown in Fig. 17.9(b), the spectrum develops a gap and a spread, with magnons propagating in different directions having different "dispersion" curves (Herring and Kittel 1951).

Magnetostatic modes. The true longest-wavelength excitations of a (soft) ferromagnet are not magnons but so-called magnetostatic modes, or Walker modes (Walker 1958). These are the solutions of the equations $\nabla \cdot (\mathbf{M}+\mathbf{H}) = 0$, $\nabla \times \mathbf{H} = 0$, and $d\mathbf{M}/dt = \gamma \mathbf{M} \times \mathbf{H}$, the latter being the Landau–Lifshitz equation (Gilbert 2004). Solving these equations gives a spectrum of nonuniform spin oscillations, quantized by the boundary conditions of the sample. They may be observed in microwave absorption experiments (see Fig. 17.9(c)).

Critical properties. The dipolar interaction alters the critical exponents of a ferromagnet. For example, the magnetization-versus-temperature exponent β for a three-dimensional ferromagnet is typically about $1/3$. This value is pushed up closer to 0.5 by

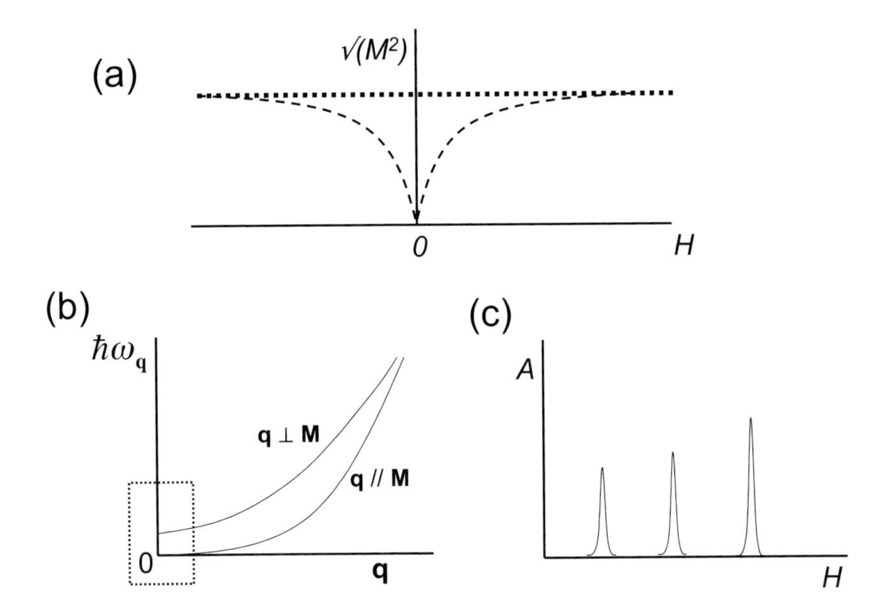

Fig. 17.9 (a) Some dipolar effects in ferromagnets (schematic). (a) The scalar magnetization measured by a mesoscopic probe (dotted line) is different from that measured by a bulk probe (dashed line). (b) The magnon spectrum develops a spread and a gap at small wavevectors; in the region of the dashed line, magnons give way to magnetostatic modes as the basic excitation. (c) Magnetostatic-mode spectrum: microwave absorption versus field.

the long-range dipole interaction. For further details, the reader is referred to Aharoni and Fisher (1973) and Als-Nielsen *et al.* (1974).

17.5 Spin ice

In the previous sections, we have emphasized two properties intrinsic to the many-body classical dipole–dipole interaction: first, the intrinsic frustration of the interaction and second, its long-ranged nature. We saw that both of these are most important in three dimensions, which is, of course, the natural dimensionality for bulk magnetic materials. In the last section we considered the effect of dipolar interactions on the properties of ferromagnets. In materials that are dominated by the dipole–dipole interaction, both ferromagnetism (in the sense discussed above) and antiferromagnetism are possible (see for example Silevitch *et al.* 2007; Champion *et al.* 2003; Stewart *et al.* 2004). However there is a third possibility—spin ice behavior—which is the subject of this section. Nowhere else in nature are the intrinsic characteristics and remarkable consequences of the dipole–dipole interaction better illustrated than in the case of spin ice (Bramwell and Gingras 2001; Bramwell *et al.* 2004).

17.5.1 Spin ice materials

The canonical spin ice materials are the rare earth oxides $Ho_2Ti_2O_7$ and $Dy_2Ti_2O_7$ (holmium and dysprosium titanate, respectively). These materials are ionic insulators and the only paramagnetic species is the rare earth ion, Ho^{3+} or Dy^{3+}. They exhibit an almost perfect spin ice state at temperatures below ~ 1 K.

Crystal structure and crystalline electric field. The magnetic ions occupy a pyrochlore lattice, a three-dimensional network of corner-linked tetrahedra, illustrated in Fig. 17.10(a). This lattice has face-centered cubic symmetry ($Fd\bar{3}m$), meaning that it is generated by combining a face-centered cubic Bravais lattice with a tetrahedral basis. However, the point symmetry at each site is trigonal ($\bar{3}m$), with the local unique axis chosen from the set of equivalent symmetry directions $\langle 111 \rangle$ (see Fig. 17.10(b)). The crystalline electric field is therefore highly anisotropic and has a strong axial component, the recipe for creating an effective Ising spin with large moment. The specific case of Ho^{3+} was described in Section 17.2. The moment is $10\mu_B$, the energy scale of the Ising behavior is about 300 K, and the dipolar coupling is about 1.5 K.

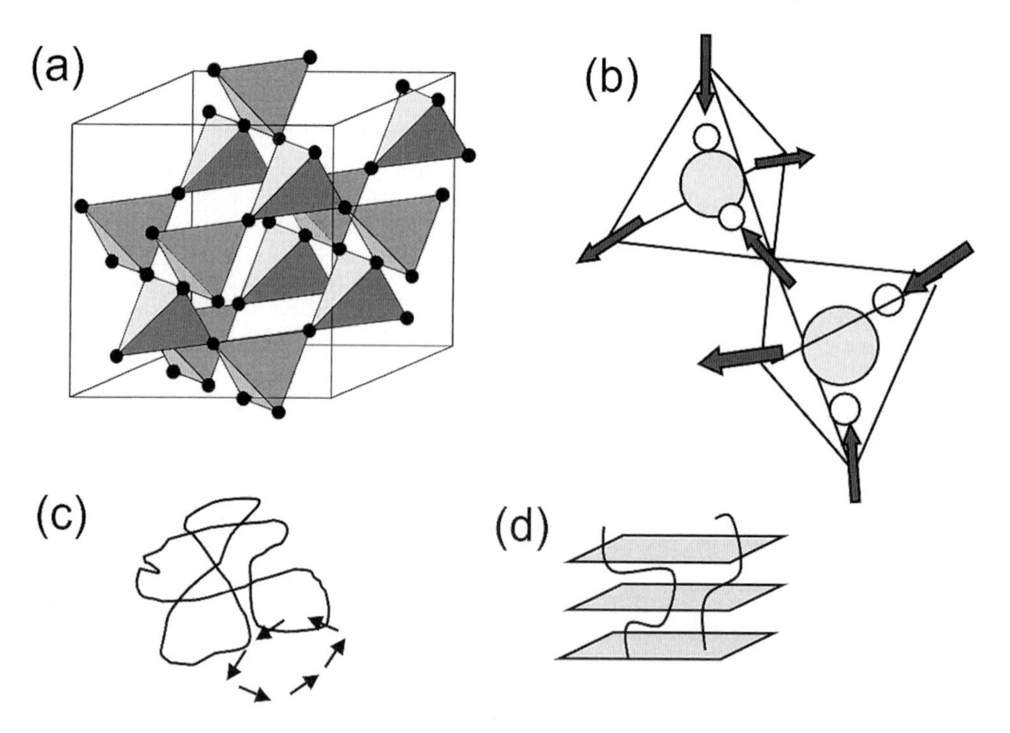

Fig. 17.10 (a) The pyrochlore lattice occupied by the magnetic ions in spin ice materials. (b) Fragment of either the water ice or the spin ice structure, showing oxygens (large circles), hydrogens (small circles), and hydrogen displacement vectors or spins (arrows). (c) The spin ice state may be pictured as composed of closed loops of spins of different sizes, which may be represented as a flux. (d) The number of lines of flux is conserved from plane to plane.

Thus, on the scale of the dipole–dipole coupling, the spins are essentially perfect Ising spins and, for equilibrium properties, may be treated as classical dipoles. There is a small correction to the dipolar coupling caused by a weak exchange interaction: this is unimportant and will not be discussed further here.

Ground state of a single tetrahedron. The important consequence of having a point symmetry that differs from the space symmetry is that the Ising spins have different local axes, each a member of the $\langle 111 \rangle$ set that points from the center of the tetrahedron to each vertex. Thus each spin points "in" or "out" of the tetrahedron. It is easy to show, using eqn (17.2), that the near-neighbor interaction is ferromagnetic. Thus a near-neighbor bond is satisfied by an in–out pair. However, it is impossible to simultaneously satisfy all six bonds on the tetrahedron, and the tetrahedron ground state has two spins in and two out (proof of this is left as an exercise). Thus the ground state is highly frustrated, with six (i.e. $4!/(2!2!)$) degenerate states, each having four "good" and two "bad" bonds (Fig. 17.10(b)).

17.5.2 The spin ice state

Water ice. Before constructing the ground state of the spin lattice, we shall briefly describe the problem of the ground state of water ice, H_2O, a classical problem in physical chemistry. In the early days of X-ray diffraction, Bragg deduced that the oxygen structure of (hexagonal) ice is a tetrahedral framework, but the diffraction techniques of the day could not locate hydrogens. In the late 1920s Giauque and colleagues measured the entropy of ice and discovered that it has a zero-point entropy: that is, the entropy tends to a finite value as the temperature approaches absolute zero (Giauque and Stout 1936). This contradicts stronger statements of the "third law" of thermodynamics, yet it is a very real fact (the experiments of Giauque are some of the most beautiful and accurate thermodynamic measurements ever performed). The explanation was famously given by Pauling (1935). He argued (following Bernal and Fowler) that the hydrogens must lie on the oxygen–oxygen lines of contact with two close and two far from each oxygen (Fig. 17.10(b)). This "ice rule" ensures that the the structure consists of water molecules H_2O, hydrogen-bonded together (a strong expectation from chemical-bonding arguments). However, as Pauling showed, the ice rules underconstrain the system so that there is a degree of choice as to which two hydrogens are near and which two are far, or, equivalently, a degree of choice in the water molecule orientation. Over the whole structure, this leads to a huge number of hydrogen configurations that are equally consistent with the ice rules. Pauling estimated this number to be $\Omega \sim (3/2)^N$, where N is the number of water molecules, which corresponds to a molar entropy $S = R\ln(3/2)$. Thus ice has a macroscopic ground state degeneracy. It is important to realize that the ice-type disorder is not simply random, but instead, is strongly correlated.

From water ice to spin ice. The ice-rules disorder of water ice may be described in terms of proton displacement vectors situated on the midpoints of the lines connecting oxygen centers (Fig. 17.10(b)). In this representation the ice rules correspond to two vectors pointing in and two pointing out, exactly the same rule that controls the magnetic-moment ordering in the spin ice materials. The lattice of the midpoints is, furthermore, a lattice of corner-linked tetrahedra: in cubic water ice it is just the pyrochlore lattice (cubic ice may be taken to have essentially the same properties as hexagonal ice). Thus, as far as the near-neighbor dipole–dipole interaction is concerned, there is an exact mapping of the problem of spin ordering in holmium and dysprosium titanate onto Pauling's ice model: hence the name "spin ice." One would expect equivalent properties, including the same correlations and the same zero-point entropy. Experiments on spin ice proceeded in reverse order to the original experiments on water ice: the initial discovery was by neutron scattering and final verification came from direct measurement of the entropy (Bramwell and Gingras 2001).

17.5.3 Spin ice as a vertex model

The near-neighbor spin ice model is a three-dimensional "16-vertex" model. The vertices in question are the 16 spin configurations of a single tetrahedron. These configurations are thus: the six two-in-two-out configurations, eight three-in-one-out or three-out-one-in configurations, and, finally, the two configurations that have all-in or all-out.

Vertex models of this type are important in statistical mechanics as they are among the rare set of models that invite exact solution. The results of such solutions are a set of phases that depend on the statistical weights of the vertices. One of the beauties of real spin ice is that the statistical weights can be tuned by changing the temperature or by applying magnetic fields along different crystallographic axes. In this way, a wide variety of different magnetic phases and phase transitions are realized (see below).

A second typical property of vertex models is that they admit many equivalent descriptions. Thus, by conceptually linking spins, one obtains a flux: with an additional rule about how flux should enter or leave a tetrahedron, an equivalent flux description is obtained (Fig. 17.10(c), (d)). Such flux is conserved from layer to layer in the pyrochlore structure: a kind of quantization. Thus it comes as no surprise that ice-type models can generally be mapped onto quantum models in one fewer dimension (Baxter 1982).

In terms of a flux description, spin ice has a kind of topological order: at equilibrium and at sufficiently low temperature it consists entirely of closed loops (Fig. 17.10(c)): the alternative open loops are not excluded energetically but they are excluded statistically. The loops may be broken by thermal excitations (see below).

Calculation of the entropy. We illustrate the calculation of the entropy of ice with a simplified ice model known as square ice, adapted to the spin ice case. This consists of a square array of corner-linked tetrahedra. Suppose we build up the array by adding successive tetrahedra (Fig. 17.11(a)). Before the addition of a particular tetrahedron, the number of possible configurations is Ω_0. The addition of a tetrahedron multiplies this number by 6 (i.e. the number of two-in-two-out states), but we also have to match two vertices, and the chance of doing that is roughly 1/2 for each vertex, so the new number of possible states is $\Omega_0 \times 6 \times (1/2)^2 = \Omega_0 \times (3/2)$. Building up the system in this way, we easily see that the total number of equivalent states is $\Omega = (3/2)^N$, where N is the number of squares, so the molar entropy is $R\ln(3/2)$. This argument is easily extended to the pyrochlore lattice, but for any real lattice it is only approximate. Correlations around closed loops in the lattice actually increase the degeneracy and the true entropy is slightly greater than $R\ln(3/2)$ (Nagle 1966; Lieb 1967).

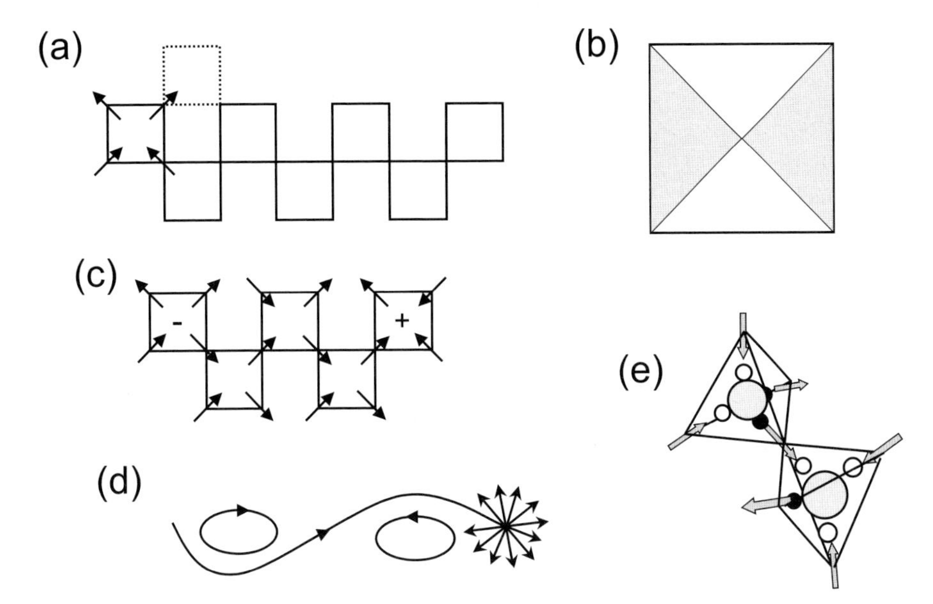

Fig. 17.11 (a) The square version of spin ice, with two spins in and two spins out per square: when an additional square is added (dotted square), it is constrained to match spins on the two vertices to which it connects. (b) Pinch point singularity at the Brillouin zone center. (c) Deconfined defects in square spin ice, labeled as $\pm$. (d) Monopole and associated Dirac string in spin ice, terminating a "broken spin loop." (e) Representation of spin ice in terms of charge pair dipoles: the lower tetrahedron contains a positive "monopole" defect.

Dipolar singularity in the scattering function. Although spin ice is disordered at all temperatures, the spin–spin correlation function adopts a form that is very different from that of a conventional paramagnet. Thus in the latter case the correlation function at long distance is like a screened Coulomb interaction $\langle \mathbf{S}_0 \cdot \mathbf{S}_r \rangle \sim (1/r)e^{-\kappa r}$, which Fourier transforms to give the famous Ornstein–Zernike scattering function $S(Q) = \chi_0 \kappa^2/(Q^2 + \kappa^2)$ (here κ is the inverse correlation length). In spin ice, in contrast, the long-distance spin–spin correlation function takes the form of a dipolar interaction (eqn (17.2)), so that both the spin–spin *interactions* and the spin–spin *correlations* are of dipolar form. The corresponding scattering function has a so-called pinch point singularity at the Brillouin zone center (Fig. 17.11(b)). Such singularities have recently been observed in neutron-scattering experiments (Fennell *et al.* 2007).

Youngblood and Axe (1981) described a simple calculation for the case of square ice that is easily adapted to illustrate the origin of the pinch point in a two-dimensional version of spin ice. In the following we recall that $S^{xx}(\mathbf{q}) = T\chi^{xx}_{\mathbf{q}}$, where $\chi^{xx}_{\mathbf{q}}$ is a component of the wavevector-dependent susceptibility tensor. Thus, the magnetization is Fourier analyzed as follows:

$$\mathbf{M}(\mathbf{r}) = \sum_{\mathbf{q}} \mathbf{M}_{\mathbf{q}} e^{i\mathbf{q}\cdot\mathbf{r}} . \tag{17.65}$$

Now we assume that the Helmholtz free energy is that of a paramagnet,

$$F = \frac{\chi_0^{-1}}{2} \sum_{\mathbf{q}} M_{\mathbf{q}}^2, \tag{17.66}$$

so that $\chi_{\mathbf{q}}^{xx} = \left[\partial^2 F / \partial (M_{\mathbf{q}}^x)^2\right]^{-1} = \chi_0$ (utilizing a thermodynamic relation). The ice rule, two spins in, two out, is like a zero-divergence constraint for the magnetization field, and may be introduced as $\nabla \cdot \mathbf{M}(\mathbf{r}) = \mathbf{0}$. This Fourier transforms as $\mathbf{q} \cdot M_{\mathbf{q}} = 0$, which may be written $q_x M_{\mathbf{q}}^x + q_y M_{\mathbf{q}}^y = 0$. Eliminating M^y and again forming the double differential, we easily find

$$\chi^{xx}(\mathbf{q}) = \chi_0 \left(\frac{q_y^2}{q_x^2 + q_y^2} \right), \tag{17.67}$$

which is singular at the origin and has the form depicted in Fig. 17.11, the so-called pinch point.

Field-induced phases. As described above, the statistical weights of the "vertices" of spin ice may be altered by changing the temperature T or magnetic field $\mathbf{H}$. This results in a wide variety of phases and phase transitions. We briefly describe the cases of three of the more interesting field directions (Bramwell *et al.* 2004).

The application of a field along [001] chooses an identical state on each equivalent tetrahedron and thus magnetizes the sample. However, at sufficiently low temperature, the magnetization proceeds via a phase transition, initially thought to be first-order, but now described as a "Kasteleyn" transition (Jaubert *et al.* 2008).

The application of a field along [110] selects two equivalent states per tetrahedron: at low temperature, the ice rules apply and the combination of these constraints causes the system to split into two sets of chains of spins, running parallel and perpendicular to the field. While the parallel chains are ordered, the perpendicular chains are largely ordered within the chain, but disordered between chains. Each perpendicular chain behaves much like the one-dimensional Ising dipolar ferromagnet described in Section 17.4. There is, however, no extensive entropy in this chain state as the degeneracy grows as N^α with $\alpha < 1$, meaning that $S \to 0$.

The application of a weak field along [111] pins one spin per tetrahedron but leaves three equivalent configurations of the other three. This latter set of spins occupy a two-dimensional "kagome" lattice in isolated sheets and there is an effective dimensional reduction. The remaining degeneracy is sufficient to give a zero-point entropy (less than the Pauling entropy) that has been calculated exactly and also measured experimentally (Udagawa *et al.* 2002; Aoki *et al.* 2004).The destruction of this state, known as the kagome ice state, by a weak transverse field proceeds via another Kasteleyn transition (Isakov *et al.* 2004), which has again been seen experimentally (Fennell *et al.* 2007). The application of a strong field along [111] eventually flips one spin per tetrahedron to give a three-in-one-out ground state or equivalent. The result is an ordered canted ferromagnetic state that breaks the ice rules (Bramwell *et al.* 2004). In the model there is no anomaly accompanying this ice-rules-breaking transition, but at sufficiently low temperature the real material shows a sharp first-order phase change with a giant spike in the entropy (Aoki *et al.* 2004). We return to discuss the nature of this transition below.

17.5.4 Influence of the long-ranged dipolar interaction

The previous sections have emphasized the similarities between real spin ice and the spin ice model, which is justified by truncating the dipole–dipole interaction at the nearest-neighbor term. However, the dipole interaction is long-ranged, so while it is obvious that the spin ice model is basically correct, it is not obvious a priori why the long-range part of the dipole interaction can be largely ignored. In fact, there is at least one property for which it cannot even be qualitatively ignored, namely the first-order nature of the ice-rules-breaking phase transition that terminates the kagome ice phase.

The dipolar spin ice model. The question of the fate of the long-range part of the interaction was answered by Gingras and coworkers, who studied a quasi-classical Hamiltonian that includes the whole dipolar interaction as well as local exchange terms (Melko and Gingras 2004). Using Ewald summation techniques, they were able to show that the long-range part of the interaction has a self-screening property: when summed over a sufficiently large number of spins, the long-range part behaves nearly, but not quite, like an effective near-neighbor coupling. The "not quite" gives a phase transition at very low temperature to an ordered state. This is not observed experimentally and has no relevance to the known physics of spin ice, beyond suggesting that the spin ice state may be metastable. The spin ice state is thus a (metastable) property of the many-body dipole–dipole interaction. The most accurate Hamiltonian to date for $Dy_2Ti_2O_7$ has been given by Yavors'kii *et al.* (2008), where the possible phase transition temperature is estimated as less than 60 mK.

Projective equivalence and emergent gauge. The near equivalence between the dipolar interaction (in the spin ice context) and the ice rules has been derived analytically and termed "projective equivalence" (Isakov *et al.* 2005). Recalling the arguments of Section 17.5.3, the dipole interaction is seen to impose on the magnetization field a divergence constraint and free energy that is equivalent to that of a magnetic field **B**. As this is a many-body effect that is not obviously coded into the dipole interaction, it may be termed an emergent gauge structure.

Magnetic monopoles. When the two-in-two-out ice rule is satisfied on every tetrahedron of the pyrochlore structure, as emphasized above, the coarse-grained magnetization field is solenoidal, or divergence-free: $\nabla \cdot \mathbf{M} = 0$. The ice rules may be broken by thermally activated spin flips, in which case neighboring tetrahedra have three-in-one-out and three-out-one-in, respectively. The zero-divergence condition is broken on *both* tetrahedra to create an effective pair of charges. This is most easily seen in the ice representation, where the charges correspond to real hydroxide and hydronium ions:

$$2H_2O = H_3O^+ + OH^- . \tag{17.68}$$

In ice these charges are deconfined (ice is a very good electrical conductor!), and the same thing can happen in principle in spin ice, as illustrated in Fig. 17.11(e). In the figure, further spin flips have moved the negative charge and positive charge apart.

The remarkable thing, as shown by Castelnovo *et al.* (2008), is that the dipole interaction ensures that the charges really *do* behave like magnetic charges, formed by "chopping a dipole in half"! The "length" of the dipoles is just the distance from the tetrahedron vertex to its center. Thus if the distance between tetrahedron centers is a, the charge is $Q = \pm 2\mu/a$. These charges interact via the law

$$V(\mathbf{r}_{ij}) = \frac{\mu_0}{4\pi}\frac{Q_i Q_j}{r_{ij}} \tag{17.69}$$

(for $r_{ij} \neq 0$), a perfect magnetic Coulomb Law!

Dumbbell model. It is of interest to sketch out some details of Castelnovo *et al.*'s argument. Like Griffiths (Section 17.4.2), they find it useful to represent magnetic dipoles in a charge picture: in this case as "dumbbells" with a charge $\pm q$ on either end. The length of these dumbbells is a and their dipole moment is qa. In addition to the Coulomb interaction for $r_{ij} \neq 0$,

$$V(\mathbf{r}_{ij}) = \frac{\mu_0}{4\pi}\frac{q_i q_j}{r_{ij}}\,, \tag{17.70}$$

for $r_{ij} = 0$ these authors introduce a "self-energy"

$$V(\mathbf{r}_{ij}) = V_0 q_i q_j\,, \tag{17.71}$$

where V_0 is tuned to make the near-neighbor spin–spin interaction correct (this also accounts for near-neighbor exchange, but this is not crucial to the argument). At long distance the fact that the dipoles are of finite length is unimportant, and the new interaction tends to the true dipole interaction. Thus the interaction is correct at near-neighbor and long-range, but is only an approximation at intermediate range. It may be resummed by introducing the charge on a tetrahedron (labeled α), $Q_\alpha = \sum q_\alpha$. Note that this introduces a factor of 2, because Q can take only the values $0, \pm 2q, \pm 4q$: this factor manifests itself by reducing the effective dipole length to $a/2$. The resummed Hamiltonian is

$$H_{\text{dipolar}} \approx \sum_\alpha \frac{1}{2}V_0 Q_\alpha^2 + \frac{\mu_0}{4\pi}\frac{Q_\alpha Q_\beta}{r_{\alpha\beta}}\,, \tag{17.72}$$

which has ground state $Q_\alpha = 0$ for all sites (i.e. the ice rules) and excited states $Q_{\alpha,\beta} \neq 0$: the monopole defects.

Properties of spin ice monopoles. When a thermally activated spin flip occurs, the two effective charges may unbind, as shown in Figs. 17.11(c) and (d), in the now familiar square representation of spin ice. In a flux picture one can immediately see that these create an effective Dirac string, reminiscent of that in the one-dimensional problem (Section 17.3.2), the main difference being that these poles live in three dimensions and are not confined to the surface of the sample. The chemical potential for the monopoles may be tuned by applying a field along the [111] direction, and Castelnovo *et al.* finally explained the first-order nature of the ice-rules-breaking phase transition seen at finite field along that direction (see above) as a condensation of positive and negative monopoles to form an effective ionic crystal. In recent work, a more direct signature of monopole existence has been identified by Jaubert and Holdsworth (2009), who showed that magnetic relaxation measurements on spin ice materials can be interpreted in terms of monopole diffusion.

What is special about spin ice monopoles? We have seen that breaking of the ice rules creates effective magnetic charges. In a coarse-grained description, the condition $\nabla \cdot \mathbf{M} = 0$ is broken. Thus, since $\mathbf{B} = \mu_0(\mathbf{H} + \mathbf{M})$, sources in $\mathbf{M}$ correspond to sinks in $\mathbf{H}$ and vice versa. Now, it can be shown that the magnetic potential or field exterior to any magnet can be written as a sum over effective magnetic charges (Morrish 1965). In most magnetic materials these charges mainly (though not exclusively) inhabit the surface. In spin ice, in zero magnetic field, they exclusively inhabit the bulk as there are no open loops (i.e. no magnetization). More importantly, the effective charges of ferromagnetism are really macroscopic objects that have no real meaning on the microscopic scale. The spin ice monopoles, on the other hand, exist on the scale of the lattice constant a, but of course have no meaning on smaller scales. There is no breaking of Maxwell's law as the $\mathbf{B}$-field remains divergence free.

Conduction in spin ice. Perhaps disappointingly, it seems very unlikely that spin ice can sustain a direct current of magnetic charges. The passage of, say, a positive a charge flips the spins in its wake and no other positive charge can pass that way (this can be appreciated by studying Fig. 17.11(c)). The same argument applies to proton conduction in water ice itself but there is a difference. In water ice there is another type of defect as well as an ionic defect, called a Bjerrum defect: the creation of these can "reset" the water molecules, allowing the passage of a direct current of protons (Eisenberg and Kauzmann 1969). This means that water ice is an excellent conductor. Indeed, it is rather like a proton semiconductor and can even be doped to make a kind of (rather pointless) p–n junction (Eigen and De Maeyer 1958)! In spin ice, there is no analogue of the Bjerrum defect and, of course, a charge imbalance cannot be created. Thus, on the one hand, spin ice is a much better realization of Pauling's ice model than is water ice, but on the other hand it is robbed of the interesting properties of direct current and doping.

17.5.5 Spin ice: Conclusions

In this section we have illustrated some of the physics of spin ice in order to illustrate the properties of the dipole interaction. Intrinsic frustration and the long-range nature are seen to play a role and the relationship with electrical charges and dipoles is very clearly highlighted. Excepting the slightly messy properties of Bjerrum defects and charge imbalance, spin ice and water ice are almost perfectly equivalent in their equilibrium statistical mechanics, with the substitutions $p \to \mu$ (here p is an electric dipole moment) and $\epsilon_0 \to 1/\mu_0$. This is a remarkable fact, but two caveats should be noted. First, the true physics of water ice is not necessarily dipolar (but this is largely compensated for by the projective equivalence). Second, Maxwell's Law does not become Gauss's Law!

17.6 Overall conclusions

In conclusion, I hope that I have managed to convince the reader that there is some interesting physics in the magnetic dipolar interaction. This is seen to stem from the combination of its long-range nature with the frustration that is intrinsic to its complicated angular dependence. To fully understand the dipole–dipole interaction one needs a combination of statistical mechanics, quantum theory, microscopic and macroscopic electrodynamics, field theory and numerical methodology. It is rather a subtle interaction, always present in real magnets, and capable of producing remarkable effects. Perhaps most importantly, it is very real, being important for any ferromagnet, and being ideally realized in certain ionic salts of rare earth elements. Systems such as these thus offer rich experimental possibilities for those interested in the study of long-range interacting systems.

References

Aharoni A. and Fisher M. E. (1973). *Phys. Rev. B*, **8**, 3323–3341.

Als-Nielsen J., Holmes L. M., and Guggenheim H. J. (1974). *Phys. Rev. Lett.*, **32**, 610–613.

Aoki H., Sakakibara T., Matsuhira K., and Hiroi Z. (2004). *J. Phys. Soc. Japan*, **73**, 2851–2856.

Arrott A. (1968). *Phys. Rev. Lett.*, **20**, 1029–1031.

Ayton G., Gingras M. J. P., and Patey G. N. (1995). *Phys. Rev. Lett.*, **75**, 2360–2363.

Baxter R. J. (1982). *Exactly Solved Models in Statistical Mechanics*, Academic Press.

Böttcher C. J. F. (1952). *Theory of Electric Polarisation*, Elsevier.

Bramwell S. T. and Gingras M. J. P. (2001). *Science*, **294**, 1495–1501.

Bramwell S. T., Holdsworth P. C. W., and Gingras M. J. P. (2004). Spin ice, in *Frustrated Spin Systems*, Ed. Diep H. T., pp. 367–456, World Scientific.

Castelnovo C., Moessner R., and Sondhi S. L. (2008). *Nature* **451**, 42–45.

Champion J. D. M., Harris M. J., Holdsworth P. C. W., Wills A. S., Balakrishnan G., Bramwell S. T., Čižmár E., Fennell T., Gardner J. S., Lago J., McMorrow D. F., Orendáč M., Orendáčová A., Paul D. M^{c}K., Smith R. I., Telling M. T. F., and Wildes A. (2003). *Phys. Rev. B*, **68**, 020401.

de'Bell K., MacIsaac A. B., and Whitehead, J. P. (2000). *Rev. Mod. Phys.*, **72** 225–257.

Di Matteo S., Magnani N., Wastin F., Rebizant J., and Cacciuffo R. (2007). *J. Alloys Compd.*, **444–445**, 278–280.

Eigen M. and De Maeyer L. (1958). *Proc. R. Soc. London, Ser. A*, **247**, 505–533.

Eisenberg D. and Kauzmann W. (1969). *The Structure and Properties of Water*, Oxford University Press.

Fennell T., Bramwell S. T., McMorrow D. F., Manuel P., and Wildes A. R. (2007). *Nature Phys.*, **3**, 566–572.

Feynman R. P. (1963). *The Feynman Lectures on Physics*, Addison-Wesley.

Garel T. and Doniach S. (1982). *Phys. Rev. B*, **26**, 325–329.

Giauque W. F. and Stout J. W. (1936). *J. Am. Chem. Soc.*, **58**, 1144–1150.

Gilbert T. L. (2004). *IEEE Trans. Magn.*, **40**, 3443–3449.

Griffiths R. B. (1968). *Phys. Rev.*, **176**, 655–659.

Griffiths D. J. (1992). *Am. J. Phys.*, **60**, 979–987.

Herring C. and Kittel C. (1951). *Phys. Rev.*, **81**, 869–880.

Hughes D. J. and Burgy M. J. (1951). *Phys. Rev.*, **81**, 498–503.

Isakov S. V., Raman K. S., Moessner R., and Sondhi S. L. (2004). *Phys. Rev. B*, **70**, 104418.

Isakov S. V., Moessner R., and Sondhi S. L. (2005). *Phys. Rev. Lett.*, **95**, 217201.

Jackson, J. D. (1998). *Classical Electrodynamics*, Wiley.

Jaubert L. D. C. and Holdsworth P. C. W. (2009). *Nature Phys.*, **5**, 258–261.

Jaubert L. D. C., Chalker J. T., Holdsworth P. C. W., and Moessner R. (2008). *Phys. Rev. Lett.*, **100**, 067207.

Kittel C. (1976). *Introduction to Solid State Physics*, 5th edn, Wiley.

Levy P. M. (1968). *Phys. Rev.*, **170**, 595–602.

Lieb E. H. (1967). *Phys. Rev.*, **162**, 162–172.

Luttinger J. M. and Tisza L. (1946). *Phys. Rev.*, **70**, 954–964.

Melko R. G. and Gingras M. J. P. (2004) *J. Phys.: Condens. Matter*, **16**, R1277–1319.

Morrish A. H. (1965). *The Physical Principles of Magnetism*, Wiley.

Nagle J. F. (1966). *J. Math. Phys.*, **7**, 1484–1491.

Pauling L. (1935). *J. Am. Chem. Soc.*, **57**, 2680–2684.

Prakash. S. and Henley C. L. (1990). *Phys. Rev. B* **42**, 6574–6589.

Roser M. R. and Coruccini L. R. (1990). *Phys. Rev. Lett.*, **65**, 1064–1067.

Sauer J. A. (1939). Phys. Rev, **57**, 142–146.

Silevitch D. M., Bitko D., Brooke J., Ghosh S., Aeppli G., and Rosenbaum T. F. (2007). *Nature*, **448**, 567–70.

Stewart J. R., Ehlers G., Wills A. S., Bramwell S. T., and Gardner J. S. (2004). *J. Phys.: Condens. Matter*, **16**, L321–L326.

Taroni A., Bramwell S. T., and Holdsworth P. C. W. (2008). *J. Phys.: Condens. Matter*, **20**, 275233.

Udagawa M., Ogata M., and Hiroi Z. (2002). *J. Phys. Soc. Japan*, **71**, 2365–2368.

Vaidman L. (1990). *Am. J. Phys.*, **58**, 978–983.

Walker L. R. (1958). *J. Appl. Phys.*, **29**, 318–323.

White R. M. (1983). *Quantum Theory of Magnetism*, 2nd edn, Springer-Verlag.

Wojtowicz P. J. and Rayl M. (1968). *Phys. Rev. Lett.*, **20**, 1489–1491.

Yavors'kii T., Fennell T., Gingras M. J. P., and Bramwell S. T. (2008). *Phys. Rev. Lett.*, **101**, 037204.

Youngblood R. W. and Axe J. D. (1981). *Phys. Rev. B*, **23**, 232–238.

18
Magnetic dipolar interactions, isolated systems and microcanonical ensemble

Bernard BARBARA

Laboratoire Louis Néel 25 Avenue des Martyrs, F-38042 Grenoble Cedex 9, France